NUCLEAR SYSTEMS I

NUCLEAR SYSTEMS I
Thermal Hydraulic Fundamentals

Neil E. Todreas
Mujid S. Kazimi

Massachusetts Institute of Technology

HEMISPHERE PUBLISHING CORPORATION
A Member of the Taylor & Francis Group
New York   Washington   Philadelphia   London

NUCLEAR SYSTEMS I Thermal Hydraulic Fundamentals

1 2 3 4 5 6 7 8 9 0   BRBR   8 9 8 7 6 5 4 3 2 1 0 9

This book was set in Times Roman by Waldman Graphics Inc. The production supervisor was Bermedica Production.
Cover design by Reneé E. Winfield.
Printer and binder was Braun-Brumfield, Inc.

Library of Congress Cataloging-in-Publication Data

Todreas, Neil E.
    Nuclear Systems / Neil E. Todreas, Mujid S. Kazimi.
      p.   cm.
    Includes bibliographical references.
    Contents: 1. Thermal hydraulic fundamentals.
    ISBN 0-89116-935-0 (v. 1)
    1. Nuclear reactors—Fluid dynamics. 2. Heat—Transmission.
3. Nuclear power plants. I. Kazimi, Mujid S. II. Title.
TK9202.T59 1989
621.48'3—dc20

89-20023
CIP

CONTENTS

PREFACE

This book can serve as a textbook for two to three courses at the advanced under-graduate and the graduate student level. It is also suitable as a basis for continuing education of engineers in the nuclear power industry, who wish to expand their knowledge of the principles of thermal analysis of nuclear systems. The book, in fact, was an outgrowth of the course notes used for teaching several classes at MIT over a period of nearly fifteen years.

The book is meant to cover more than thermal hydraulic design and analysis of the core of a nuclear reactor. Thus, in several parts and examples, other components of the nuclear power plant, such as the pressurizer, the containment and the entire primary coolant system are addressed. In this respect the book reflects the importance of such considerations in thermal engineering of a modern nuclear power plant. The traditional concentration on the fuel element in earlier textbooks was appropriate when the fuel performance had a higher share of the cost of electricity than in modern plants. The cost and performance of nuclear power plants has proven to be more influenced by the steam supply system and the containment building than previously anticipated.

The desirability of providing in one book the basic concepts as well as the complex formulations for advanced applications has resulted in a more comprehensive textbook than those previously authored in the field. The basic ideas of both fluid flow and heat transfer as applicable to nuclear reactors are discussed in Volume I. No assumption is made about the degree to which the reader is already familiar with the subject. Therefore, various reactor types, energy source distribution and fundamental laws of conservation of mass, momentum and energy are presented in early chapters. Engineering methods for analysis of flow hydraulics and heat transfer in single-phase as well as two-phase coolants are presented in later chapters. In Volume II, applications of these fundamental ideas to the multi-channel flow conditions in the reactor are described as well as specific design considerations such as natural convection and core thermal reliability. They are presented in a way that renders it possible to use the analytical development in simple exercises and as the bases for numerical computations similar to those commonly practiced in the industry.

A consistent nomenclature is used throughout the text and a table of the nomenclature is included in the appendix. Each chapter includes problems identified as to their topic and the section from which they are drawn. While the SI unit system is principally used, British Engineering Units are given in brackets for those results commonly still reported in the United States in this system.

ACKNOWLEDGMENTS

Much material in this book originated from lectures developed at MIT by Professor Manson Benedict with Professor Thomas Pigford for a subject in Nuclear Reactor Engineering and by Professors Warren Rohsenow and Peter Griffith for a subject in Boiling Heat Transfer and Two-Phase Flow. We have had many years of pleasant association with these men as their students and colleagues and owe a great deal of gratitude to them for introducing us to the subject material. The development of the book has benefited from the discussion and comments provided by many of our colleagues and students. In particular Professor George Yadigaroglu participated in the early stage of this work in defining the scope and depth of topics to be covered.

We are at a loss to remember all the other people who influenced us. However, we want to particularly thank those who were kind enough to review nearly completed chapters, while stressing that they are not to be blamed for any weaknesses that may still remain. These reviewers include John Bartzis, Manson Benedict, Greg Branan, Dae Cho, Michael Corradini, Hugo DaSilva, Michael Driscoll, Don Dube, Tolis Efthimiadis, Gang Fu, Elias Gyftopoulos, Steve Herring, John Kelly, Min Lee, Alan Levin, Joy Maneke, Mahmoud Massoud, John Meyer, Hee Cheon No, Klaus Rehme, Tae Sun Ro, Donald Rowe, Gilberto Russo, Robert Sawdye, Andre Schor, Nathan Siu, Kune Y. Suh and Robert Witt. Finally, we want to express our appreciation to all students at MIT who proof-tested the material at its various stages of development and provided us with numerous suggestions and corrections which have made their way into the final text.

Most of the figures in this book were prepared by a number of students using a microcomputer under the able direction of Alex Sich. Many others have participated in the typing of the manuscript. We offer our warmest thanks to Gail Jacobson, Paula Cornelio and Elizabeth Parmelee for overseeing preparation of major portions of the final text.

A generous grant from the Bernard M. Gordon (1948) Engineering Curriculum Development Fund at MIT was provided for the text preparation and for this we are most grateful.

Mujid S. Kazimi
Neil E. Todreas

To our families for their support in this endeavor
Carol, Tim and Ian
Nazik, Yasmeen, Marwan and Omar

THERMAL HYDRAULIC CHARACTERISTICS
OF POWER REACTORS

I INTRODUCTION

The energy source of a power reactor originates from the fission process within the fuel elements. Energy deposited in the fuel is transferred to the coolant by conduction, convection, and radiation.

This chapter presents the basic thermal hydraulic characteristics of power reactors. Study of these characteristics enables the student to appreciate the applications of the specialized techniques presented in the remainder of the text. The thermal hydraulic characteristics are presented as part of the description of the power cycle, core design, and fuel assembly design of these reactor types. Water-, gas-, and liquid-metal-cooled reactor types, identified in Table 1-1, include the principal nuclear power reactor designs currently employed in the world. Table 1-1 and others in Chapters 1 and 2 provide further detailed information useful for application to specific illustrative and homework examples.

II POWER CYCLES

In these plants a primary coolant is circulated through the reactor core to extract energy for ultimate conversion by a turbine process to electricity. Depending on the reactor design, the turbine may be driven directly by the primary coolant or by a secondary coolant that has received energy from the primary coolant. The number of

Table 1-1 Basic features of major power reactor types

Reactor type	Neutron spectrum	Moderator	Coolant	Fuel Chemical form	Approximate fissile content (all ^{235}U except LMFBR)
Water-cooled	Thermal				
PWR		H_2O	H_2O	UO_2	≈ 3% Enrichment
BWR		H_2O	H_2O	UO_2	≈ 3% Enrichment
PHWR (CANDU)		D_2O	D_2O	UO_2	Natural
SGHWR		D_2O	H_2O	UO_2	≈ 3% Enrichment
Gas-cooled	Thermal	Graphite			
Magnox			CO_2	U metal	Natural
AGR			CO_2	UO_2	≈ 3% Enrichment
HTGR			Helium	UC, ThO_2	≈ 7–20% Enrichment[a]
Liquid-metal-cooled	Fast	None	Sodium		
LMR				U/Pu metal; UO_2/PuO_2	≈ 15–20% Pu
LMFBR				UO_2/PuO_2	≈ 15–20% Pu

[a]Older operating plants have enrichments of more than 90%.

coolant systems in a plant equals the sum of one primary and one or more secondary systems. For the boiling water reactor (BWR) and the high-temperature gas reactor (HTGR) systems, which produce steam and hot helium by passage of primary coolant through the core, direct use of these primary coolants in the turbine is possible, leading to a single-coolant system. The BWR single-coolant system, based on the Rankine cycle (Fig. 1-1) is in common use. The Fort St. Vrain HTGR plant used a secondary water system in a Rankine cycle because the technology did not exist to produce a large, high-temperature, helium-driven turbine. Although the HTGR direct turbine system has not been built, it would use the Brayton cycle, as illustrated in Figure 1-2.

The pressurized water reactor (PWR) and the pressurized heavy water reactor (PHWR) are two-coolant systems. This design is necessary to maintain the primary

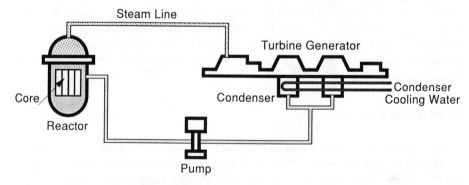

Figure 1-1 Direct, single-coolant Rankine cycle. (*Adapted courtesy of U.S. Department of Energy.*)

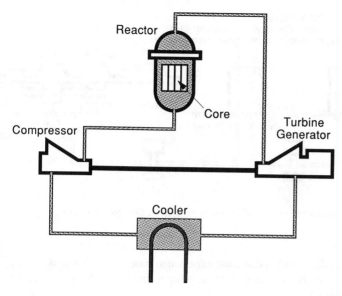

Figure 1-2 Direct, single-coolant Brayton cycle. *(Adapted courtesy of U.S. Department of Energy.)*

coolant conditions at a nominal subcooled liquid state. The turbine is driven by steam in the secondary system. Figure 1-3 illustrates the PWR two-coolant steam cycle.

The liquid metal fast breeder reactor (LMFBR) system employs three coolant systems: a primary sodium coolant system, an intermediate sodium coolant system, and a steam-water, turbine-condenser coolant system (Fig. 1-4). Three coolant systems are specified to isolate the radioactive primary sodium coolant from the steam–water circulating through the turbine, condenser, and associated conventional plant components. The liquid metal reactor (LMR) concept being developed in the United States

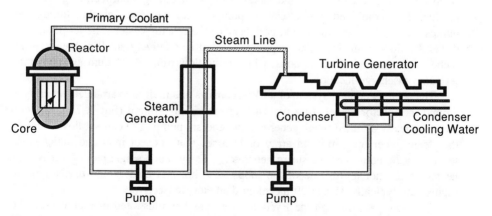

Figure 1-3 Two-coolant system steam cycle. *(Adapted courtesy of U.S. Department of Energy.)*

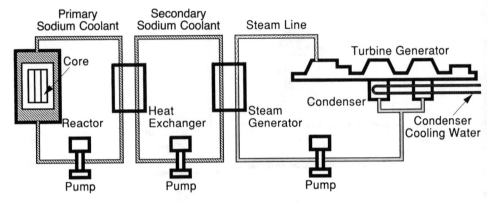

Figure 1-4 Three-coolant system steam cycle. *(Adapted courtesy of U.S. Department of Energy.)*

draws on LMFBR technology and operational experience base, but it is not designed as a breeder. Liquid-metal-cooled reactor characteristics and examples presented in this text are for the LMFBR.

The significant characteristics of the thermodynamic cycles used in these reactor types are summarized in Table 1-2. Thermodynamic analyses for typical Rankine and Brayton cycles are presented in Chapter 6.

III PRIMARY COOLANT SYSTEMS

The BWR single-loop primary coolant system is illustrated in Figure 1-5, and Figure 1-6 highlights the flow paths within the reactor vessel. The steam–water mixture first enters steam separators after exiting the core. After subsequent passage through steam dryers located in the upper portion of the reactor vessel, the steam flows directly to the turbine. The water, which is separated from the steam, flows downward in the periphery of the reactor vessel and mixes with the incoming main feed flow from the turbine. This combined flow stream is pumped into the lower plenum through jet pumps mounted around the inside periphery of the reactor vessel. The jet pumps are driven by flow from recirculation pumps located in relatively small-diameter ($\approx$ 20 inches) external recirculation loops, which draw flow from the plenum just above the jet pump discharge location.

The primary coolant system of a PWR consists of a multiloop arrangement arrayed around the reactor vessel. In a typical four-loop configuration (Fig. 1-7), each loop has a vertically oriented steam generator and coolant pump. The coolant flows through the steam generator within an array of U tubes that connect inlet and outlet plena located in the bottom of the steam generator. The system's single pressurizer is connected to the hot leg of one of the loops. The hot and cold (pump discharge) leg pipings are typically 31 and 29 inches in diameter, respectively.

The flow path through the PWR reactor vessel is illustrated in Figure 1-8. The inlet nozzles communicate with an annulus formed between the inside of the reactor

Table 1-2 Typical characteristics of the thermodynamic cycle for six reference power reactor types

Characteristic	BWR	PWR(W)	PHWR	HTGR	AGR	LMFBR
Reference design						
Manufacturer	General Electric	Westinghouse	Atomic Energy of Canada, Ltd.	General Atomic	National Nuclear Corp.	Novatome
System (reactor station)	BWR/6	(Sequoyah)	CANDU-600	(Fulton)[a]	HEYSHAM 2	(Superphenix)
Steam-cycle						
No. coolant systems	1	2	2	2	2	3
Primary coolant	H_2O	H_2O	D_2O	He	CO_2	Liq. Na
Secondary coolant	—	H_2O	H_2O	H_2O	H_2O	Liq. Na/H_2O
Energy conversion						
Gross thermal power, MW(th)	3579	3411	2180	3000	1550	3000
Net electrical power, MW(e)	1178	1148	638	1160	618	1200
Efficiency (%)	32.9	33.5	29.3	38.7	40.0	40.0
Heat transport system						
No. primary loops and pumps	2	4	2	6	8	4
No. intermediate loops	—	—	—	—	—	8
No. steam generators	—	4	4	6	4	8
Steam generator type	—	U tube	U tube	Helical coil	Helical coil	Helical coil
Thermal hydraulics						
Primary coolant						
Pressure (MPa)	7.17	15.5	10.0	4.90	4.30	~0.1
Inlet temp. (°C)	278	286	267	318	334	395
Ave. outlet temp. (°C)	288	324	310	741	635	545
Core flow rate (Mg/s)	13.1	17.4	7.6	1.42	3.91	16.4
Volume (L) or mass (kg)	—	3.06×10^5	1.20×10^5	(9550 kg)	5.3×10^6	(3.20×10^6 kg)
Secondary coolant						Na/H_2O
Pressure (MPa)	—	5.7	4.7	17.2	16.0	~0.1/17.7
Inlet temp. (°C)	—	224	187	188	156.0	345/235
Outlet temp. (°C)	—	273	260	513	541.0	525/487

Source: Knief [4], except AGR-HEYSHAM 2 data are from Alderson [1] and the PWR (W)-Sequoyah data from Coffey [3].
[a]Designed but not built.

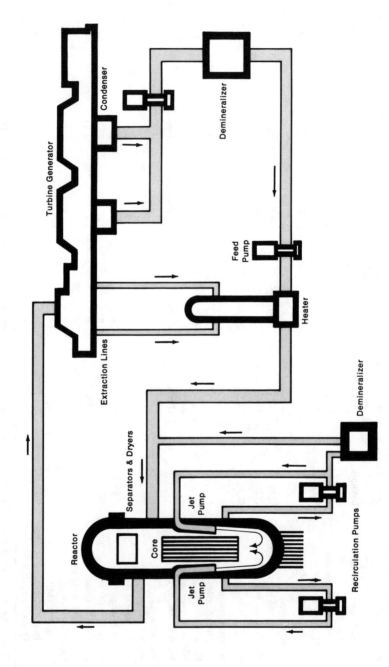

Figure 1-5 BWR single-loop primary coolant system. (*Courtesy of General Electric Company.*)

6

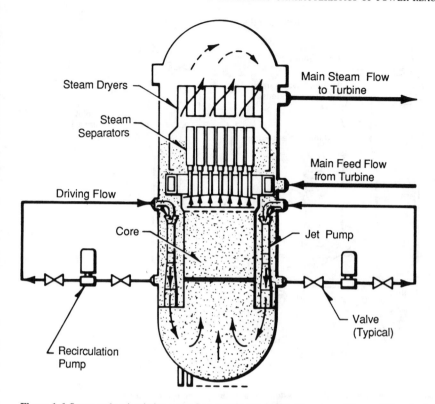

Figure 1-6 Steam and recirculation water flow paths in the BWR. *(Courtesy of General Electric Company.)*

vessel and the outside of the core support barrel. Coolant entering this annulus flows downward into the inlet plenum formed by the lower head of the reactor vessel. Here it turns upward and flows through the core into the upper plenum which communicates with the reactor vessel outlet nozzles.

The HTGR primary system is composed of several loops, each housed within a large cylinder of prestressed concrete. A compact HTGR arrangement as embodied in the modular high-temperature gas-cooled reactor (MHTGR) is illustrated in Figure 1-9. In this 588 MWe MHTGR arrangement [2] the flow is directed downward through the core by a circulator mounted above the steam generator in the cold leg. The reactor vessel and steam generator are connected by a short, horizontal cross duct, which channels two oppositely directed coolant streams. The coolant from the core exit plenum is directed laterally through the 47 inch diameter interior of the cross duct into the inlet of the steam generator. Coolant from the steam generator and circulator is directed laterally through the outer annulus (equivalent pipe diameter of approximately 46 inches) of the cross duct into the core inlet plenum.

LMFBR primary systems have been of the loop and pool types. The pool type configuration of the Superphenix reactor is shown in Figure 1-10, and its characteristics are detailed in Table 1-2. The coolant flow path is upward through the reactor

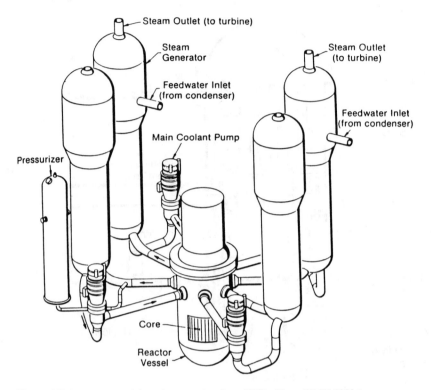

Figure 1-7 Arrangement of the primary system for a PWR. *(From WASH-1250.)*

core into the upper sodium pool of the main vessel. Coolant from this pool flows downward by gravity through the intermediate heat exchanger and discharges into a low-pressure toroidal plenum located on the periphery of the lower portion of the main vessel. Vertically oriented primary pumps draw the coolant from this low-pressure plenum and discharge it into the core inlet plenum.

IV REACTOR CORES

All these reactor cores except that of the HTGR are composed of assemblies of cylindrical fuel rods surrounded by coolant which flows along the rod length. The U.S. HTGR core consists of graphite moderator blocks that function as fuel assemblies. Within these blocks a hexagonal array of cylindrical columns is filled alternately with fuel and flowing helium coolant. The assemblies and blocks are described in detail in section V.

There are two design features that establish the principal thermal hydraulic characteristics of reactor cores: the orientation and the degree of hydraulic isolation of an

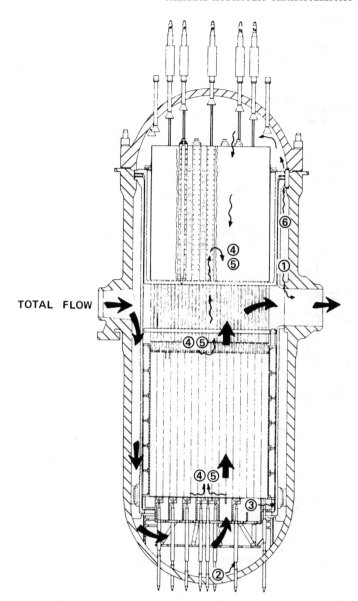

TOTAL FLOW

Figure 1-8 Flow path through a PWR reactor vessel. *Thick arrows,* main flow; *thin arrows,* bypass flow. *1,* through outlet nozzle clearance; *2,* through instrumented center guide tubes; *3,* through shroud-barrel annulus; *4,* through center guide tubes; *5,* through outer guide tubes; *6,* through alignment key-ways. *(Courtesy of Combustion Engineering.)*

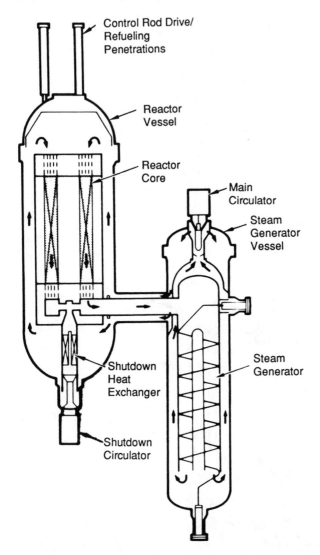

Figure 1-9 Modular HTGR primary coolant flow path. *(Courtesy of US-DOE.)*

assembly from its neighbors. It is simplest to adopt a reference case and describe the exceptions. Let us take as the reference case a vertical array of assemblies that communicate only at inlet and exit plena. This reference case describes the BWR, LMFBR, and the advanced gas reactor (AGR) systems. The HTGR is nominally configured in this manner also, although leakage between the graphite blocks which are stacked to create the proper core length, creates a substantial degree of communication between coolant passages within the core. The PHWR core consists of horizontal pressure tubes penetrating a low-pressure tank filled with heavy-water moderator. The fuel

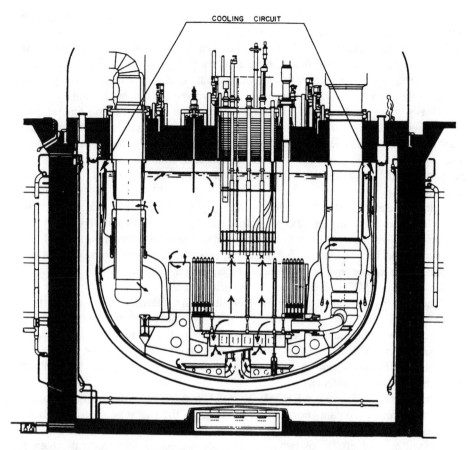

Figure 1-10 Primary system sodium flow path in the Superphenix reactor *(Courtesy of EDF.)*

assemblies housed within the pressure tubes are cooled by high-pressure heavy water, which is directed to and from the tubes by an array of inlet and outlet headers. Both the PHWR and the AGR are designed for on-line refueling.

The PWR assemblies are vertical but are not isolated hydraulically by enclosing ducts over the core length. In effect, fuel rods are grouped into assemblies only for handling and other structural purposes.

V FUEL ASSEMBLIES

The principal characteristics of power reactor fuel bundles are the array (geometric layout and rod spacing) and the method of fuel pin separation and support along their span. The light-water reactors (BWR and PWR), PHWR, AGR, and LMFBR all use fuel rods. The HTGR has graphite moderator blocks in which adjacent penetrating holes filled with fuel and flowing helium coolant exist.

Light-water reactors (LWRs), where the coolant also serves as the moderator, have small fuel-to-water volume ratios (commonly called the *metal-to-water ratio*) and consequently rather large fuel rod centerline-to-centerline spacing (commonly called the *rod pitch, P*). This moderate packing fraction permits use of a simple square array and requires a rod support scheme of moderate frontal area to yield low pressure drops. A variety of grid support schemes have evolved for this application.

Heavy-water reactors and advanced gas reactors are designed for on-line refueling and consequently consist of fuel assemblies stacked within circular pressure tubes. This circular boundary leads to an assembly design with an irregular array of rods. The on-line refueling approach has led to short fuel bundles in which the rods are supported at the assembly ends and at a center brace rather than by LWR-type grid spacers.

Liquid-metal-cooled fast-breeder reactors require no moderator and achieve high power densities by compact hexagonal fuel rod packing. With this tight rod-to-rod spacing, a lower pressure drop is obtained using spiral wire wrapping around each rod than could be obtained with a grid-type spacer. This wire wrap serves a dual function: as a spacer and as a promotor of coolant mixing within the fuel bundle. However, some LMFBR assemblies do use grid spacers.

The principal characteristics of the fuel for the six reference power reactor types are summarized in Table 1-3. The HTGR does not consist of an array of fuel rods within a coolant continuum. Rather, the HTGR blocks that contain fuel, coolant, and moderator can be thought of as inverted fuel assemblies. In these blocks, the fuel–moderator combination is the continuum that is penetrated by isolated, cylindrically shaped, coolant channels.

The LWRs (PWR and BWR), PHWR, AGR, and LMFBR utilize an array of fuel rods surrounded by coolant. For each of these arrays the useful geometric characteristics are given in Table 1-3 and typical subchannels identified. These subchannels are defined as coolant regions between fuel rods and hence are "coolant-centered" subchannels. Alternately, a "rod-centered" subchannel has been defined as that coolant region surrounding a fuel rod. This alternate definition is infrequently used.

A LWR Fuel Bundles: Square Arrays

A typical PWR fuel assembly for the LWR, including its grid-type spacer, is shown in Figure 1-11, along with the spring clips of the spacer, which contact and support the fuel rods. It is stressed that this figure represents only one of a variety of spacer designs now in use. The principal geometric parameters of the rods, their spacing, and the grid are defined in Figure 1-12, and the three types of subchannel commonly utilized are identified. Table 1-4 summarizes the number of subchannels of various-sized square arrays. Modern boiling water reactors of U.S. design utilize assemblies of 64 rods, whereas pressurized water reactor assemblies are typically composed of 225 to 289 rods. The formulas for subchannel and bundle dimensions, based on a PWR-type ductless assembly, are presented in Appendix J.

Table 1-3 Typical characteristics of the fuel for six reference power reactor types

Characteristic	BWR	PWR(W)	PHWR	HTGR	AGR	LMFBR[a]
Reference design						
Manufacturer	General Electric	Westinghouse	Atomic Energy of Canada, Ltd.	General Atomic	National Nuclear Corp.	Novatome
System (reactor station)	BWR/6	(Sequoyah)	CANDU-600	(Fulton)	HEYSHAM 2	(Superphenix)
Moderator	H_2O	H_2O	D_2O	Graphite	Graphite	—
Neutron energy	Thermal	Thermal	Thermal	Thermal	Thermal	Fast
Fuel production	Converter	Converter	Converter	Converter	Converter	Breeder
Fuel[b]						
Particles						
Geometry	Cylindrical pellet	Cylindrical pellet	Cylindrical pellet	Coated microspheres	Cylindrical pellet	Cylindrical pellet
Dimensions (mm)	$10.4D \times 10.4H$	$8.2D \times 13.5H$	$12.2D \times 16.4H$	400–800 μm D	$14.51D \times 14.51H$	$7.0\ D$
Chemical form	UO_2	UO_2	UO_2	UC/ThO_2	UO_2	PuO_2/UO_2
Fissile (wt% 1st core ave.)	$1.7\ ^{235}U$	$2.6\ ^{235}U$	$0.711\ ^{235}U$	$93\ ^{235}U$	$2.2\ ^{235}U$	15–18 ^{239}Pu
Fertile	^{238}U	^{238}U	^{238}U	Th	^{238}U	Depleted U
Pins						
Geometry	Pellet stack in clad tube	Pellet stack in clad tube	Pellet stack in clad tube	Cylindrical fuel stack	Pellet stack in clad tube	Pellet stack in clad tube
Dimensions (mm)	$12.27D \times 4.1\ mH$	$9.5D \times 4\ mH$	$13.1D \times 490L$	$15.7D \times 62L$	$14.89D \times 987H$	$8.65D \times 2.7\ mH$(C) $15.8D \times 1.95\ mH$(BR)
Clad material	Zircaloy-2	Zircaloy-4	Zircaloy-4	Graphite	Stainless steel	Stainless steel
Clad thickness (mm)	0.813	0.57	0.42	—	0.38	0.7
Assembly						
Geometry[c]	8×8 square rod array	17×17 square rod array	Concentric circles	Hexagonal graphite block	Concentric circles	Hexagonal rod array
Rod pitch (mm)	16.2	12.6	14.6		25.7	9.7 (C)/17.0 (BR)
No. rod locations	64	289	37	132 (SA)/76 (CA)[d]	37	271 (C)/91 (BR)
No. fuel rods	62	264	37	132 (SA)/76 (CA)[d]	36	271 (C)/91 (BR)
Outer dimensions (mm)	139	214	$102D \times 495L$	$360F \times 793H$	190.4 (inner)	$173F$
Channel	Yes	No	No	No	Yes	Yes
Total weight (kg)	273	—	—		342	—

Source: Knief [4] except AGR-HEYSHAM 2 data are from Alderson [1], and LMFBR pin and pellet diameters are from Vendryes [5].

[a]LMFBR-core (C), radial blanket (BR), axial blanket (BA).

[b]Fuel dimensions: diameter (D), height (H), length (L), (across the) flats (F), (width of) square (S).

[c]LWRs have utilized a range of number of rods.

[d]HTGR-standard assembly (SA), control assembly (CA).

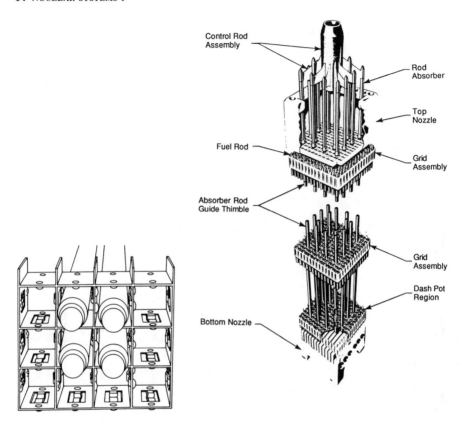

Portion of Spring Clip Grid
Assembly

Figure 1-11 Typical spacer grid for a light-water reactor fuel assembly. *(Courtesy of Westinghouse Electric Corporation.)*

Table 1-4 Subchannels for square arrays

Rows of rods	N_p Total no. of rods	N_1 No. of interior subchannels	N_2 No. of edge subchannels	N_3 No. of corner subchannels
1	1	0	0	4
2	4	1	4	4
3	9	4	8	4
4	16	9	12	4
5	25	16	16	4
6	36	25	20	4
7	49	36	24	4
8	64	49	28	4
N_{rows}	N_{rows}^2	$(N_{rows} - 1)^2$	$4(N_{rows} - 1)$	4

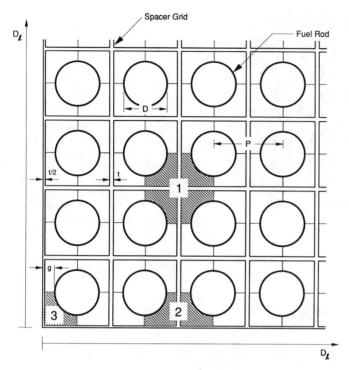

Figure 1-12 Typical fuel array for a light-water reactor. Subchannel designation: *1*, interior; *2*, edge; *3*, corner.

B PHWR and AGR Fuel Bundles: Mixed Arrays

The geometry and subchannel types for the PHWR and AGR fuel bundles are shown in Figure 1-13. Because these arrays are arranged in a circular sleeve, the geometric characteristics are specific to the number of rods in the bundle. Therefore the exact number of rods in the PHWR and the AGR bundle is shown.

C LMFBR Fuel Bundles: Hexagonal Arrays

A typical hexagonal array for a sodium-cooled reactor assembly with the rods wire-wrapped is shown in Figure 1-14. As with the light-water reactor, different numbers of rods are used to form bundles for various applications. A typical fuel assembly has about 271 rods. However, arrays of 7 to 331 rods have been designed for irradiation and out-of-pile simulation experiments of fuel, blanket, and absorber materials. The axial distance over which the wire wrap completes a helix of 360 degrees is called the lead length or axial pitch. Therefore axially averaged dimensions are based on averaging the wires over one lead length. The number of subchannels of various-sized hexagonal arrays are summarized in Table 1-5 and the dimensions for unit subchannels and the overall array in Appendix J.

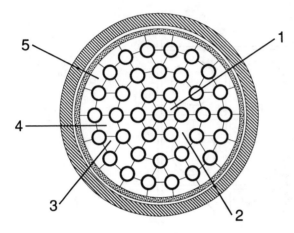

Figure 1-13 Fuel array of AGRs and PHWRs. Subchannel types: 1, interior first row (triangular); 2, interior second row (irregular); 3, interior third row (triangular); 4, interior third row (rhombus); 5, edge outer row. *Note:* Center pin is fueled in PHWR and unfueled in AGR.

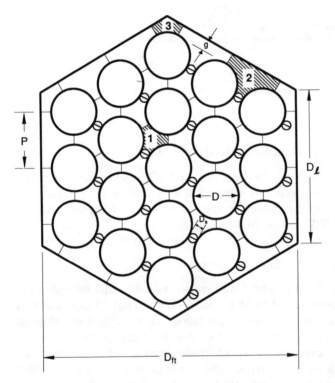

Figure 1-14 Typical fuel array for a liquid-metal-cooled fast breeder reactor. Subchannel designation: *1*, interior; *2*, edge; *3*, corner. $N = 19$ in this example. *Note:* The sectional view of the wire should strictly be elliptical.

Table 1-5 Subchannels for hexagonal arrays

Rings of rods	N_p Total no. of rods	N_{pS} No. of rods along a side	N_1 No. of interior subchannels	N_2 No. of edge subchannels	N_3 No. of corner subchannels
1	7	2	6	6	6
2	19	3	24	12	6
3	37	4	54	18	6
4	61	5	96	24	6
5	91	6	150	30	6
6	127	7	216	36	6
7	169	8	294	42	6
8	217	9	384	48	6
9	271	10	486	54	6
N_{rings}	$\sum\limits_{n=1}^{N_{rings}} 6n$	$N_{rings} + 1$	$6N_{rings}^2$	$6N_{rings}$	6

Additional useful relations between N_p, N_{pS}, and N_1 are as follows.

$$N_{pS} = \left[1 + \sqrt{1 + \frac{4}{3}(N_p - 1)} \right] \Big/ 2$$

$$N_p = 3N_{pS}(N_{pS} - 1) + 1$$

$$N_1 = 6(N_{pS} - 1)^2$$

$$N_2 = 6(N_{pS} - 1)$$

REFERENCES

1. Alderson, M. A. H. G. UKAEA. Personal communications, 6 October 1983 and 6 December 1983.
2. Breher, W., Neyland, A., and Shenoy, A. *Modular High-Temperature Gas-Cooled Reactor (MHTGR) Status.* GA Technologies, GA-A18878, May 1987.
3. Coffey, J. A. TVA. Personal communication, 24 January 1984.
4. Knief, R. A. *Nuclear Energy Technology: Theory and Practice of Commercial Nuclear Power.* New York: McGraw-Hill, 1981, pp. 12–14, 566–570.
5. Vendryes, G. A. Superphenix: a full-scale breeder reactor. *Sci. Am.* 236:26–35, 1977.

TWO

THERMAL DESIGN PRINCIPLES

I INTRODUCTION

The general principles of reactor thermal design are introduced in this chapter, with the focus on the parameters, design limits, and figures of merit by which the thermal design process is characterized. We do not attempt to present a procedure for thermal design because nuclear, thermal, and structural aspects are related in a complicated, interactive sequence. Specific design analysis techniques are detailed in Volume II.

II OVERALL PLANT CHARACTERISTICS INFLUENCED BY THERMAL HYDRAULIC CONSIDERATIONS

Thermal hydraulic considerations are important when selecting overall plant characteristics. Primary system temperature and pressure are key characteristics related to both the coolant selection and plant thermal performance. This thermal performance is dictated by the bounds of the maximum allowable primary coolant outlet temperature and the minimum achievable condenser coolant inlet temperature. Because this atmospheric heat sink temperature is relatively fixed, improved thermodynamic performance requires increased reactor coolant outlet temperatures. Figure 2-1 illustrates the relation among reactor plant temperatures for a typical PWR.

Bounds on the achievable primary outlet temperature depend on the coolant type. For liquid metals, in contrast to water, the saturated vapor pressure for a given tem-

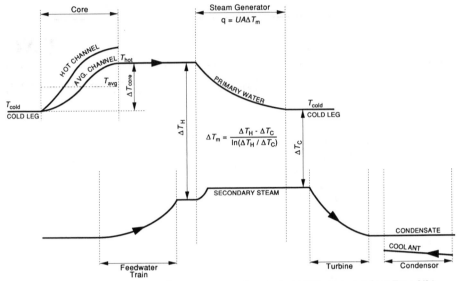

Figure 2-1 Relations among reactor plant temperatures for a typical PWR. (Adapted from Tong [4].)

perature is low, i.e., less than atmospheric pressure at outlet temperatures of interest of 500° to 550°C. Thus the outlet temperature for LMFBRs is not limited by the boiling point of the sodium coolant but, rather, by the creep lifetime characteristics of the stainless steel primary system material. For water-cooled reactors, on the other hand, high primary outlet temperatures require correspondingly high system pressures (7 to 15 MPa), which increases the stored energy in the primary coolant and requires increased structural piping and component wall thicknesses. Single-phase gas coolants offer the potential for high outlet temperatures without such inherently coupled high pressures. For these reactors the system pressure is dictated by the desired core heat transfer capabilities, as gas properties that enter these heat transfer correlations are strongly dependent on pressure. The resulting pressures are moderate, i.e., 4 to 5 MPa, whereas achievable outlet temperatures are high, i.e., 635° to 750°C. The numerical value of the plant thermal efficiency depends on the maximum temperature in the secondary or power-generation system. This temperature is lower than the reactor coolant outlet temperature owing to the temperature difference needed to transfer heat between the primary and secondary systems in the steam generator or intermediate heat exchanger. In the boiling water reactor a direct cycle is employed. The reactor outlet temperature is therefore identical (neglecting losses) to the inlet temperature to the turbine. This outlet temperature is also limited to the saturation condition, as BWRs do not operate under superheat conditions. In a typical BWR, however, the average outlet enthalpy achieved corresponds to an average quality of 15%. The PWR and BWR reactors achieve approximately equal thermal efficiencies, as the turbine steam conditions are comparable even though the primary system pressure and temperature conditions significantly differ (Table 2-1). Note that because of detailed differences in thermodynamic cycles the example PWR plant achieves a slightly higher

Table 2-1 Light-water reactor thermal conditions

Condition	PWR (Sequoyah)	BWR/6
Primary coolant outlet temperature	324°C	288°C
Primary coolant system pressure	15.5 MPa	7.17 MPa
Turbine steam saturation conditions		
Pressure	5.7 MPa	7.17 MPa
Temperature	272.3°C	287.5°C
Plant thermal efficiency	33.5%	32.9%

Source: Table 1-2.

plant thermal efficiency than does the BWR plant even though its steam temperature is lower.

Other plant characteristics are strongly coupled with thermal hydraulic considerations. Some notable examples are as follows.

1. *Primary coolant temperature*
   a. Corrosion behavior, though strongly dependent on water chemistry control, is also temperature-dependent.
   b. The reactor vessel resistance to brittle fracture degrades with accumulated neutron fluence. Vessel behavior under low-temperature, high-pressure transients from operating conditions is carefully evaluated to ensure that the vessel has retained the required material toughness over its lifetime.
2. *Primary system inventory*
   a. The time response during accidents and less severe transients strongly depends on coolant inventories. The reactor vessel inventory above the core is important to behavior in primary system rupture accidents. For a PWR, in particular, the pressurizer and steam generator inventories dictate transient response for a large class of situations.
   b. During steady-state operation the inlet plenum serves as a mixing chamber to homogenize coolant flow into the reactor. The upper plenum serves a similar function in multiple-loop plants with regard to the intermediate heat exchanger/steam generator while at the same time protecting reactor vessel nozzles from thermal shock in transients.
3. *System arrangement*
   a. The arrangement of reactor core and intermediate heat exchanger/steam generator thermal centers is crucial to the plant's capability to remove heat by natural circulation.
   b. Orientation of pump shafts and heat exchanger tubes coupled with support designs and impingement velocities is important relative to prevention of troublesome vibration problems.

The system arrangement issues are sometimes little appreciated if the design of a reactor type has been established for some years. For example, note that the dominant characteristics of the primary coolant system configuration for a typical U.S. designed

PWR (see Fig. 1-7) are cold leg pump placement, vertical orientation of pumps, vertical orientation of the steam generators, and elevation of the steam generator thermal center of gravity above that of the core. The pumps are located in the cold leg to take advantage of the increased subcooling, which increases the margin to onset of pump cavitation. Vertical orientation of the pumps provides for convenient accommodation of loop thermal expansion in a more compact layout and facilitates their serviceability, and the vertical steam generator orientation eases the difficulty in support design as well as eliminates a possible flow stratification problem. The relative vertical arrangements of thermal centers of steam generators above the core is established to provide the capability for natural circulation heat removal in the primary loop. These characteristics lead to creation of a loop seal of varying size in each arrangement between the steam generator exit and the pump inlet. This seal is of significance when considering system behavior under natural circulation, loss of coolant circumstances, or both.

III ENERGY PRODUCTION AND TRANSFER PARAMETERS

Energy production in a nuclear reactor core can be expressed by a variety of terms, reflecting the multidisciplinary nature of the design process. The terms discussed here are the following.

Volumetric energy (or heat) generation rate: $q'''(\vec{r})$
Surface heat flux: $\vec{q}''(S)$
Linear heat-generation rate or power rating: $q'(z)$
Rate of energy generation per pin: $\dot{q}$
Core power: $\dot{Q}$

The two additional parameters below are also commonly used. Because they are figures of merit of core thermal performance, they are discussed later (see section VI) after design limits and design margin considerations are presented.

Core power density: $\dot{Q}/V \equiv Q'''$
Core specific power: $\dot{Q}/$mass of heavy atoms

The relations among these terms must be well understood to ensure communication between the reactor physicist, the thermal designer, and the metallurgist or ceramicist. The reactor physicist deals with fission reaction rates, which lead to the volumetric energy generation rate: $q'''(\vec{r})$. The triple-prime notation represents the fact that it is an energy-generation rate per unit volume of the fuel material. Normally in this text a dot above the symbol is added for a rate, but for simplicity it is deleted here and in other energy quantities with primes. Hence the energy-generation rate is expressed per unit length cubed, as $q'''(\vec{r})$.

The thermal designer must calculate the fuel element surface heat flux, $\vec{q}''(S)$, which is related to $q'''(\vec{r})$ as

$$\iint_S \vec{q}''(S) \cdot \vec{n} \, dS = \iiint_V q'''(\vec{r}) dV \tag{2-1}$$

where S is the surface area that bounds the volume (V) within which the heat generation occurs.

Both the thermal and metallurgical designers express some fuel performance characteristics in terms of a linear power rating $q'(z)$ where

$$\int_L q'(z) dz = \iiint_V q'''(\vec{r}) dV \tag{2-2}$$

where L is the length of the volume (V) bounded by the surface (S) within which the heat generation occurs.

If the volume V is taken as the entire heat-generating volume of a fuel pin, the quantity $(\dot{q})$, unprimed, is the heat-generation rate in a pin, i.e.,

$$\dot{q} = \iiint_V q'''(\vec{r}) dV \tag{2-3}$$

Finally, the core power $(\dot{Q})$ is obtained by summing the heat generated per pin over all the pins in the core (N) assuming that no heat generation occurs in the nonfueled regions.

$$\dot{Q} = \sum_{n=1}^{N} \dot{q}_n \tag{2-4}$$

Actually, as discussed in Chapter 3, approximately 8% of the reactor power is generated directly in the moderator and structural materials.

These general relations take many specific forms, depending on the size of the region over which an average is desired and the specific shape (plate, cylindrical, spherical) of the fuel element. For example, considering a core with N cylindrical fuel pins, each having an active fuel length (L), core average values of the thermal parameters that can be obtained from Eqs. 2-1 to 2-4 are related as

$$\dot{Q} = N \langle \dot{q} \rangle = NL \langle q' \rangle = NL\pi D_{co} \langle q''_{co} \rangle = NL\pi R_{fo}^2 \langle q''' \rangle \tag{2-5}$$

where D_{co} = outside clad diameter; R_{fo} = fuel pellet radius; $\langle \rangle$ = a core volume averaged value. A more detailed examination of the above relations and the various factors affecting the heat-generation distribution in the reactor core are discussed in Chapter 3.

IV THERMAL DESIGN LIMITS

The principal design limits for the power reactors discussed in Chapter 1 are detailed here. All reactors except the HTGR employ a metallic clad to hermetically seal the cylindrical fuel. The HTGR uses graphite and silicon carbide barriers around the fuel particles to reduce the rate of diffusion of fission products out of the fuel.

A Fuel Pins with Metallic Cladding

For hermetically sealed fuel pins, thermal design limits are imposed to maintain the integrity of the clad. In theory, these limits should all be expressed in terms of structural design parameters, e.g., strain and fatigue limits for both steady-state and transient operation. However, the complete specification of limits in these terms is presently impractical because of the complex behavior of materials in radiation and thermal environments characteristic of power reactors. For this reason, design limits in power reactors have been imposed directly on certain temperatures and heat fluxes, although the long-term trend should be to transform these limits into more specific structural design terms.

The design limits for reactors that employ cylindrical, metallic clad, oxide fuel pellets are summarized in Table 2-2, which highlights the distinction between conditions that would cause damage (loss of clad integrity) and those that would exceed design limits. Also both PWRs and BWRs have hydrodynamic stability limits. Generally, these limits are not restrictive in the design of these reactors currently, so that such limits are not noted in Table 2-2. The inherent characteristics of light-water reactors (LWRs) limit clad temperatures to a narrow band above the coolant saturation temperature and thus preclude the necessity for a steady-state limit on clad midwall temperature. However, a significant limit on clad average temperature does exist in transient situations, specifically in the loss of coolant accident (LOCA). For this accident a number of design criteria are being imposed, key among which is maintenance of the Zircaloy clad below 2200°F to prevent extensive metal–water reaction from occurring.

Table 2-2 Typical thermal design limits

Characteristic	PWR	BWR	LMFBR
Damage limit	1% Clad strain *or* MDNBR ≤ 1.0	1% Clad strain *or* MCPR ≤ 1.0	0.7% Clad strain
Design limits			
Fuel centerline temperature			
Steady state	—	—	—
Transient	No incipient melt	No incipient melt	No incipient melt
Clad average temperature			
Steady state	—	—	1200–1300°F
Transient	<2200°F (LOCA)	<2200°F (LOCA)	1450°F (788°C) for anticipated transients 1600°F (871°C) for unlikely events
Surface heat flux			
Steady state	—	MCPR ≥ 1.2[a]	—
Transient	MDNBR ≥ 1.3 at 112% power	—	—

LOCA = loss of coolant accident; MDNBR = minimum departure from nucleate boiling ratio; MCPR = minimum critical power ratio.

[a]Corresponding value of minimum critical heat flux ratio is approximately 1.9.

The particular design limit that is governing reactor design varies with the reactor type and the continually evolving state of design methods. For example, for LWRs, fuel centerline temperature is typically maintained well within its design limit due to restrictions imposed by the critical heat flux limit. Furthermore, with the application of improved LOCA analysis methods during the mid-1980s, the LOCA-imposed limit on clad temperature has not been the dominant limit. Finally, for LWRs, the occurrence of excessive mechanical interaction between pellet and clad (pellet–clad interaction) has led to operational restrictions on allowable rates of change of reactor power and the extensive development of fuel and clad materials to alleviate these restrictions on reactor load following ability.

The critical heat flux (CHF) phenomenon results from a relatively sudden reduction of the heat transfer capability of the two-phase coolant. The resulting thermal design limit is expressed in terms of the departure from nucleate boiling condition for PWRs and the critical power condition for BWRs. For fuel rods, where the volumetric energy-generation rate $q'''(r, z, t)$ is the independent parameter, reduction in surface heat transfer capability for nominally fixed bulk coolant temperature (T_b) and heat flux causes the clad temperature to rise, i.e.,

$$T_{co} - T_b = \frac{q''}{h} = \frac{q''' R_{fo}^2}{h D_{co}} \tag{2-6}$$

where h = heat transfer coefficient. Physically, this reduction occurs because of a change in the liquid–vapor flow patterns at the heated surface. At low void fractions typical of PWR operating conditions, the heated surface, which is normally cooled by nucleate boiling, becomes vapor-blanketed, resulting in a clad surface temperature excursion by departure from nucleate boiling (DNB). At high void fractions typical of BWR operating conditions, the heated surface, which is normally cooled by a liquid film, overheats owing to film dryout (DRYOUT). The dryout phenomenon depends significantly on channel thermal hydraulic conditions upstream of the dryout location rather than on the local conditions at the dryout location. Because DNB is a local condition and DRYOUT depends on channel history, the correlations and graphical representations for DNB are in terms of heat flux ratios, whereas for DRYOUT they are in terms of power ratios.

These two mechanisms for the generic critical heat flux phenomenon are shown in Figure 2-2. Correlations have been established for both these conditions in terms of different operating parameters (see Chapter 12). Because these parameters change over the fuel length, different margins exist between the actual operating heat flux and the limiting heat flux for occurrence of DNB or DRYOUT. These differences are illustrated in Figure 2-3 for a typical DNB case. The ratio between the predicted correlation heat flux and the actual operating heat flux is called the departure from nucleate boiling ratio (DNBR). This ratio changes over the fuel length and reaches a minimum value, as shown in Figure 2-3, somewhere downstream of the peak operating heat flux location. An alternative representation in terms of bundle average conditions, which depend on total power input, exists for BWR dryout conditions. This representation is expressed as the critical power ratio (CPR) and is presented in Chapter 12.

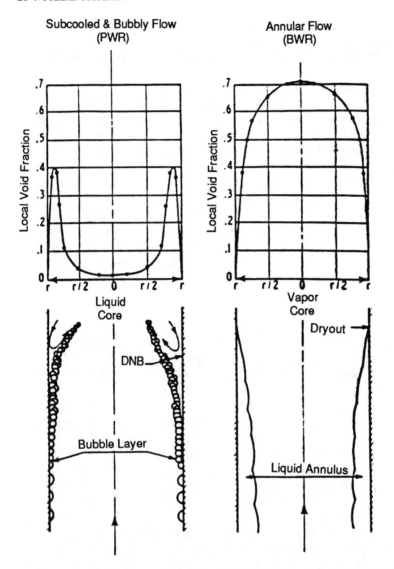

Figure 2-2 Critical heat flux mechanisms for PWR and BWR operating conditions. (From Tong [4].)

Critical condition limits are established for this minimum value of the appropriate ratio, i.e., MDNBR or MCPR (Table 2-2). Furthermore, prior to Brown's Ferry Unit 1, the BWR limit was applied to operational transient conditions, as is the present PWR practice. Subsequently it was applied to the 100% power condition. These limits for a BWR at 100% power and a PWR at 112% power allow for consistent overpower margin, as can be demonstrated for any specific case of prescribed axial heat flux distribution and coolant channel conditions.

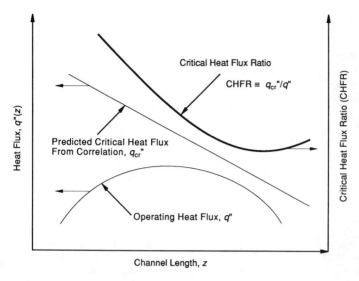

Figure 2-3 Critical heat flux ratio definition.

For the LMFBR, present practice is to require a level of subcooling such that the sodium temperature does not exceed its boiling temperature for transient conditions. Additionally, considerable effort is being applied to ensure that coolant voiding in accident situations can be satisfactorily accommodated. Hence the LMFBR limits are now placed on fuel and clad temperatures. At present, no incipient fuel melting is allowed. However, work is being directed at developing clad strain criteria. These criteria will reflect the fact that the stainless steel clad is operating in the creep regimen and that some degree of center fuel melting can be accommodated without clad failure.

B Graphite-Coated Fuel Pellets

The HTGR fuel is in the form of coated particles deposited in holes symmetrically drilled in graphite matrix blocks to provide passages for helium coolant. Two types of coated particle are used (Fig. 2-4). The BISO type has a fuel kernel surrounded by a low-density pyrolitic carbon buffer region, which is itself surrounded by a high-density, high-strength pyrolitic carbon layer. The TRISO type sandwiches a layer of silicon carbide between the two high-density pyrolitic carbon layers of the BISO type. In both fuel types the inner layer and the fuel kernel are designed to accommodate expansion of the particle and to trap gaseous fission products. The buffer layer acts to attenuate the fission fragment recoils. The laminations described also help to prevent crack propagation. Silicon carbide is used to supply dimensional stability and low diffusion rates. It has a greater thermal expansion rate than the surrounding pyrolitic carbon coating and thus is normally in compression. In both kinds of particles the total coating thickness is about 150μm. The TRISO coating is used for uranium fuel particles that are enriched to 93% ^{235}U. The BISO coating is used for thorium particles, which comprise the fertile material.

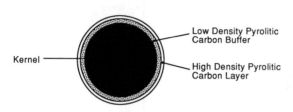

BISO PARTICLE

Kernel

Low Density Pyrolitic Carbon Buffer

High Density Pyrolitic Carbon Layer

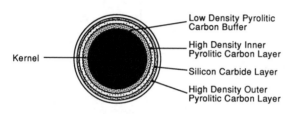

TRISO PARTICLE

Kernel

Low Density Pyrolitic Carbon Buffer

High Density Inner Pyrolitic Carbon Layer

Silicon Carbide Layer

High Density Outer Pyrolitic Carbon Layer

Figure 2-4 High-temperature gas reactor fuel particles. Typical dimensions: kernel diameters, 100 to 300 μm; total coating thickness, 50 to 190 μm.

Because fission gas release occurs in this system by diffusion through the coatings and directly from rupture of the coatings, limits are imposed to control these release rates such that steady-state fission product levels within the primary circuit do not exceed specified levels. These activity levels are established to ensure that radiation doses resulting from accidental release of the primary circuit inventory to the atmosphere are within regulations. These limits are on fuel particle center temperature.

100% power $\simeq$ 1300°C (2372°F)
Peak transient $\simeq$ 1600°C (2822°F) for short term

The full power limit minimizes steady-state diffusion, whereas the transient limit, based on in-pile tests, is imposed to minimize cracking of the protective coatings.

V THERMAL DESIGN MARGIN

A striking characteristic of thermal conditions existing in any core design is the large differences among conditions in different spatial regions of the core. The impact of the variation in thermal conditions for a typical core is shown in Figure 2-5. Starting with a core average condition such as linear power rating ($\langle q' \rangle$, defined by Eq. 2-5), nuclear power peaking factors, overpower factors, and engineering uncertainty factors are sequentially applied, leading to the limiting $\langle q' \rangle$ value. Each condition appearing in Figure 2-5 is clearly defined by a combination of the terms average, peak, nominal,

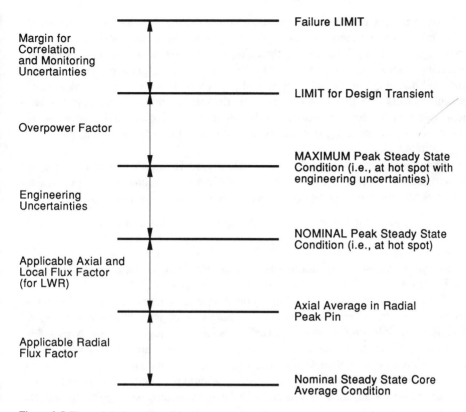

Figure 2-5 Thermal design nomenclature.

and maximum. No consistent set of definitions exists for these terms. However, a consensus of usage, which we adopt, is the following.

Average: Usually a core or pin axially averaged value. The distinction must be made or inferred from the application.

Peak: Sometimes referred to as hot spot; refers exclusively to a physical location at which the extreme value occurs; i.e., peak pin is that pin located where the radial power profile is greatest; peak linear power rating in the peak pin is that axial location in the peak pin of greatest linear power rating.

Nominal: Value of a parameter calculated using variables at their design-specified values.

Maximum: Value of a parameter calculated with allowance provided for deviation of the variables from their design-specified values.

Margin is provided between the transient operating LIMIT and the failure LIMIT (Fig. 2-5). This margin provides for uncertainties in the instrumentation used to monitor the operating condition and uncertainties in the correlations for transient limits.

It is apparent in Figure 2-5 that core power can be enhanced by flattening the shape of the power-generation rate, optimizing the power-to-flow ratio in every radial zone of the core, or both. Power flattening is achieved by a combination of steps involving reflector regions, enrichment zoning, shuffling of fuel assemblies with burnup, and burnable poison and shim control placement. In water reactors, local power peaking effects due to water regions comprise an additional factor that has been addressed by considerable attention to the detailed layout of the fuel/coolant lattice. However, power flattening may not be desirable if neutron leakage from the core is to be minimized.

Optimizing the power-to-flow ratio in the core is important for achieving the desired reactor vessel outlet temperature at maximum net reactor power, i.e., reactor power generation minus pumping power, where pumping power is expressed as:

$$\text{Pumping power} = \text{force through distance per unit time} \qquad (2\text{-}7)$$

$$\text{Pumping power} = (\Delta p)A_f V \qquad (2\text{-}8)$$

where Δp = pressure drop through the circuit (F/L^2); A_f = cross-sectional area of the coolant passage (L^2); and V = average coolant velocity (L/T).

Flow control involves establishing coolant flow rates across the core at the levels necessary to achieve equal coolant exit conditions in all assemblies and minimizing the amount of inlet coolant that bypasses the heated core regions. Some bypass flow is required, however, to maintain certain regions (e.g., the inner reactor vessel wall) at design conditions. Variations in energy generation within assemblies with burnup do make it difficult to maintain the power-to-flow ratio near unity. The assembly flow rates are normally adjusted by orificing. Orificing is accomplished by restricting the coolant flow in the assembly, usually at the inlet. Because the orifice devices are not typically designed to be changed during operation, deviations in the power-to-flow ratio from optimum over a full cycle are inevitable. Alternatively, large spatial variations in flow rate can also be accomplished using multiple inlet plena. This design approach, however, is complicated.

VI FIGURES OF MERIT FOR CORE THERMAL PERFORMANCE

The design performance of a power reactor can be characterized by two figures of merit: the power density (Q''') and the specific power. Table 2-3 tabulates the power density for the various power reactor concepts. The specific power can be calculated, as is shown from the other reactor parameters given.

Power density is the measure of the energy generated relative to the core volume. Because the size of the reactor vessel and hence the capital cost are nominally related to the core size, the power density is an indicator of the capital cost of a concept. For propulsion reactors, where weight and hence size are at a premium, power density is a relevant figure of merit.

Table 2-3 Typical core thermal performance characteristics for six reference power reactor types

Characteristic	BWR	PWR(W)	PHWR	HTGR	AGR	LMFBR[a]
Core						
Axis	Vertical	Vertical	Horizontal	Vertical	Vertical	Vertical
No. of assemblies						
Axial	1	1	12	8	8	1
Radial	748	193	380	493	332	364 (C) / 233 (BR)
Assembly pitch (mm)	152	215	286	361	460	179
Active fuel height (m)	3.81	3.66	5.94	6.30	8.296	1.0 (C) / 1.6 (C + BA)
Equivalent diameter (m)	4.70	3.37	6.29	8.41	9.458	3.66
Total fuel weight (ton)	156 UO_2	101 UO_2	98.4 UO_2	1.72 U / 37.5 Th	113.5 UO_2	32 MO_2
Reactor vessel						
Inside dimensions (m)	$6.05D \times 21.6H$	$4.83D \times 13.4H$	$7.6D \times 4L$	$11.3D \times 14.4H$	$20.25D \times 21.87H$	$21D \times 19.5H$
Wall thickness (mm)	152	224	28.6	4.72 m min	5.8 m	25
Material[b]	SS-clad carbon steel	SS-clad carbon steel	Stainless steel	Prestressed concrete	Concrete helical prestressed	Stainless steel
Other features			Pressure tubes	Steel liners	Steel lined	Pool type
Power density core average (kW/L)	54.1	105	12	8.4	2.66	280
Linear heat rate						
Core average (kW/m)	19.0	17.8	25.7	7.87	17.0	29
Core maximum (kW/m)	44.0	42.7	44.1	23.0	29.8	45
Performance						
Equilibrium burnup (MWD/T)	27,500	27,500	7500	95,000	18,000	100,000
Average assembly residence (full-power days)			470	1170	1320	
Refueling						
Sequence	¼ per yr	⅓ per yr	Continuous on-line	¼ per yr	Continuous on-line	Variable
Outage time (days)	30	30		14–20		32

Source: Knief [3], except AGR data are from Alderson [1] and Debenham [2].
[a] LMFBR: core (C), radial blanket (BR), axial blanket (BA).
[b] SS = stainless steel.

31

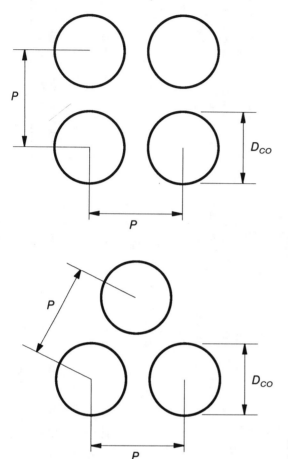

Figure 2-6 Square and triangular rod arrays.

The power density can be varied by changing the fuel pin arrangement in the core. For an infinite square array, shown in Figure 2-6, the power density (Q''') is related to the array pitch (P) as:

$$(Q''')_{\text{square array}} = \frac{4(1/4\ \pi R_{\text{fo}}^2)q'''\ dz}{P^2\ dz} = \frac{q'}{P^2} \tag{2-9}$$

whereas for an infinite triangular array, the comparable result is:

$$(Q''')_{\text{triangular array}} = \frac{3(1/6\ \pi R_{\text{fo}}^2)q'''\ dz}{\dfrac{P}{2}\left(\dfrac{\sqrt{3}}{2}P\right)dz} = \frac{q'}{\dfrac{\sqrt{3}}{2}P^2} \tag{2-10}$$

Comparing Eq. 2-9 and 2-10, we observe that the power density of a triangular array is 15.5% greater than that of a square array for a given pitch. For this reason, reactor concepts such as the LMFBR adopt triangular arrays, which are more complicated mechanically than square arrays. For light-water reactors, on the other hand, the simpler square array is more desirable, as the necessary neutron moderation can be provided by the looser-packed square array.

Specific power is the measure of the energy generated per unit mass of fuel material. It is usually expressed as watts per gram of heavy atoms. This parameter has direct implications on the fuel cycle cost and core inventory requirements. For the fuel pellet shown in Figure 2-7, the specific power, (watts per grams of heavy atoms), is:

$$\text{Specific power} = \frac{\dot{Q}}{\text{mass of heavy atoms}} = \frac{q'}{\pi R_{fo}^2 \, \rho_{\text{pellet}} f} = \frac{q'}{\pi (R_{fo} + \delta_g)^2 \rho_{\text{smeared}} f} \tag{2-11}$$

where

$$\rho_{\text{smeared}} = \frac{\pi R_{fo}^2 \, \rho_{\text{pellet}}}{\pi (R_{fo} + \delta_g)^2} \tag{2-12}$$

and

$$f = \text{mass fraction of heavy atoms in the fuel} = \frac{\text{grams of fuel heavy atoms}}{\text{grams of fuel}} \tag{2-13}$$

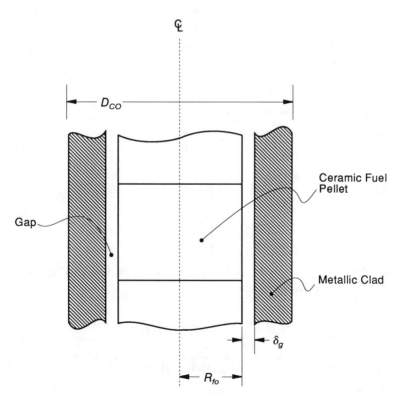

Figure 2-7 Typical power reactor fuel.

This definition of the *mass fraction* (f) is based on the following definition

1. Heavy atoms include all the U, Pu, or Th isotopes and are therefore composed of fissionable atoms (M_{ff}) and nonfissionable atoms (M_{nf}), where M is molecular weight.
2. Fuel is the entire fuel-bearing material, i.e., UO_2 but not the clad.

Thus for oxide fuel:

$$f = \frac{N_{ff}M_{ff} + N_{nf}M_{nf}}{N_{ff}M_{ff} + N_{nf}M_{nf} + N_{O_2}M_{O_2}} \tag{2-14}$$

where N = atomic density.

The enrichment (r) is the mass ratio of fissionable atoms to total heavy atoms, i.e.,

$$r = \frac{N_{ff}M_{ff}}{N_{ff}M_{ff} + N_{nf}M_{nf}} \tag{2-15}$$

and for later convenience:

$$1 - r = \frac{N_{nf}M_{nf}}{N_{ff}M_{ff} + N_{nf}M_{nf}} \tag{2-16}$$

It is useful to express f in terms of molecular weight and the enrichment. It follows from the observation that, for UO_2,

$$N_{O_2} = N_{ff} + N_{nf}$$

and manipulation of Eqs. 2-15 and 2-16 to yield:

$$N_{ff} = \left[r \frac{M_{nf}}{M_{ff}} \right] \frac{N_{nf}}{1 - r}; \quad N_{nf} = \left[(1 - r) \frac{M_{ff}}{M_{nf}} \right] \frac{N_{ff}}{r} \tag{2-17a,b}$$

$$N_{ff} + N_{nf} = \left[r \frac{M_{nf}}{M_{ff}} + (1 - r) \right] \frac{N_{nf}}{1 - r} \text{ and equally}$$

$$\left[(1 - r) \frac{M_{ff}}{M_{nf}} + r \right] \frac{N_{ff}}{r} \tag{2-18a,b}$$

Division of each term of Eq. 2-14 by N_{O_2} and substitution of Eqs. 2-17a,b and 2-18a,b yields the desired result, i.e.,

$$f_{UO_2} = \frac{\dfrac{r}{r + (1 - r)(M_{ff}/M_{nf})} M_{ff} + \dfrac{(1 - r)}{r(M_{nf}/M_{ff}) + (1 - r)} M_{nf}}{\dfrac{r}{r + (1 - r)(M_{ff}/M_{nf})} M_{ff} + \dfrac{(1 - r)}{r(M_{nf}/M_{ff}) + (1 - r)} M_{nf} + M_{O_2}} \tag{2-19}$$

which for the case where $M_{ff} \simeq M_{nf}$ simplifies to:

$$f_{UO_2} = \frac{rM_{ff} + (1 - r)M_{nf}}{rM_{ff} + (1 - r)M_{nf} + M_{O_2}} \tag{2-20}$$

The specific power in Eq. 2-11 has been expressed in terms of both the pellet and the smeared densities. The *smeared density* includes the void that is present as the gap between the fuel pellet and the clad inside diameter. The smeared density is the cold or hot density, depending on whether the gap (δ_g) is taken at the cold or hot condition. Smeared density is an important parameter associated with accommodation of fuel swelling with burnup.

Example 2-1 Power density and specific power for a PWR

PROBLEM Confirm the power density listed in Table 2-3 for the PWR case. Calculate the specific power of the PWR of Tables 2-3 and 1-3.

SOLUTION The power density is given by Eq. 2-9 as:

$$Q'''_{square\ array} = \frac{q'}{P^2}$$

From Table 2-3: $\langle q' \rangle_{core\ average} = 17.8\ \frac{kW}{m}$

Take $q' = \langle q' \rangle$, which yields a result for Q''' different from that in Table 2-3, as noted below.

From Table 1-3, $P = 12.6$ mm.

$$Q'''_{PWR} = \frac{17.8}{(12.6 \times 10^{-3})^2} = 0.112 \times 10^6\ \frac{kW}{m^3} = 112\ \frac{kW}{L}\ or\ \frac{MW}{m^3}$$

Observe that Table 2-3 lists the average power density as:

$$Q'''_{PWR} = 105\ \frac{kW}{L}$$

The difference arises because our calculation is based on the core as an infinite square array, whereas in practice a finite number of pins form each assembly so that edge effects within assemblies must be considered. The specific power is given by Eq. 2-11 as:

$$\frac{\dot{Q}}{Mass\ heavy\ atoms} = \frac{q'}{\pi R_{fo}^2\ \rho_{pellet}\ f}$$

Evaluating f using Eq. 2-20 for the enrichment of 2.6% listed in Table 1-3 yields:

$$f = \frac{0.026(235.0439) + 0.974(238.0508)}{0.026(235.0439) + 0.974(238.0508) + 2(15.9944)} = 0.8815$$

For a pellet density of 95% of the UO_2 theoretical density of 10.97 g/cm³ and a fuel pellet diameter of 8.2 mm, the specific power is:

$$\frac{\dot{Q}}{\text{Mass heavy atoms}} = \frac{1.78 \times 10^4 \; \dfrac{\text{W}}{\text{m}}}{\pi \left(\dfrac{8.2 \times 10^{-3}}{2}\right)^2 \left(\dfrac{0.95 \, (10.97)}{10^{-6}}\right) 0.8815 \; m^2 \; \dfrac{\text{g}}{\text{m}^3}}$$

$$= 36.70 \; \frac{\text{W}}{\text{g fuel}}$$

Alternately, Table 2-3 lists the total core loading of fuel material as 101×10^3 kg of UO_2. In this case

$$\frac{\dot{Q}}{\text{Mass heavy atoms}} = \frac{\text{core power}}{\text{fuel loading}} = \frac{\dot{Q}}{fM_{fm}}$$

If $\dot{Q}$ is evaluated from the PWR dimensions as in Tables 2-3 and 1-3,

$$\dot{Q} = q'LN = .0178 \; \frac{\text{MW}}{\text{m}} \; [3.66 \text{ m}][193(264)] = 3319 \text{ MWt}$$

Note: A thermal power of 3411 MWt is given in Table 1-2.

Then:

$$\frac{\dot{Q}}{\text{Mass heavy atoms}} = \frac{3319 \text{ MWt}}{0.8815(101 \times 10^3 \text{ kg})} = 37.28 \; \frac{\text{W}}{\text{g fuel}}$$

Unlike the case of power density, specific power is closely estimated, as the fuel mass, not the core volume, is utilized.

REFERENCES

1. Alderson, M. A. H. G. Personal communication, 6 October 1983.
2. Debenham, A. A. Personal communication, 5 August 1988.
3. Knief, R. A. *Nuclear Energy Technology: Theory and Practice of Commercial Nuclear Power.* New York: Hemisphere, 1981, pp. 566–570.
4. Tong, L. S. Heat transfer in water-cooled nuclear reactors. *Nucl. Eng. Design* 6:301–324, 1967.

PROBLEMS

Problem 2-1 Relations among fuel element thermal parameters in various power reactors (section III)

Compute the core average values of the volumetric energy-generation rate in the fuel (q''') and outside surface heat flux (q'') for the reactor types of Table 2-3. Use the core average linear power levels in Table 2-3 and the geometric parameters in Table 1-3.

Answer (for BWR): $q''' = 224$ MW/m^3
$q'' = 492.9$ kW/m^2

Problem 2-2 Minimum critical heat flux ratio in a PWR for a flow coastdown transient (section IV)

Describe how you would determine the minimum critical heat flux ratio versus time for a flow coastdown transient by drawing the relevant channel operating curves and the CHF limit curves for several time values. Draw your sketches in relative proportion and be sure to state all assumptions.

Problem 2-3 Minimum critical power ratio in a BWR (section IV)

Calculate the minimum critical power ratio for a typical 1000 MWe BWR operating at 100% power using the data in Tables 1-2, 1-3, and 2-3. Assume that:

1. The axial linear power shape can be expressed as

$$q'(z) = q'_{max} \exp(-\alpha z/L) \sin \frac{\pi z}{L}$$

where $\alpha = 1.96$.
2. The critical bundle power is 9319kW.

Answer: MCPR = 3.43

Problem 2-4 Pumping power for a PWR reactor coolant system (section V)

Calculate the pumping power under steady-state operating conditions for a typical PWR reactor coolant system. Assume the following operating conditions:

Core power = 3817 MWt
ΔT_{core} = 31°C
Reactor coolant system pressure drop = 778 kPa (113 psi)

Answer: Pumping power = 23.3 MW

Problem 2-5 Relations among thermal design conditions in a PWR (section V)

Compute the margin as defined in Figure 2-5 for a typical PWR having a core average linear power rate of 17.8 kW/m. Assume that the failure limit is established by centerline melting of the fuel at 70 kW/m. Use the following multiplication factors:

Radial flux = 1.55
Axial and local flux factor = 1.70
Engineering uncertainty factor = 1.05
Overpower factor = 1.15

Answer: Margin = 1.24

THREE

REACTOR HEAT GENERATION

I INTRODUCTION

Determination of the heat-generation distribution throughout the nuclear reactor is achieved via a neutronic analysis of the reactor. Accurate knowledge of the heat source is a prerequisite for analysis of the temperature field, which in turn is required for definition of the nuclear and physical properties of the fuel, coolant, and structural materials. Therefore coupling the neutronic and thermal analyses of a nuclear core is required for accurate prediction of its steady state as well as its transient conditions. For simplicity, the neutronic and thermal analyses during thermal design evaluations may not be coupled, in which case the level and distribution of the heat-generation rate are assumed fixed, and thermal analysis is carried out to predict the temperature field in the reactor core.

It should be noted that the operational power of the core is limited by thermal, not nuclear, considerations. That is, in practice, the allowable core power is limited by the rate at which heat can be transported from fuel to coolant without reaching, either at steady state or during specified transient conditions, excessively high temperatures, which would cause degradation of the fuel, structures, or both. The design limits are discussed in Chapter 2.

II ENERGY RELEASE AND DEPOSITION

A Forms of Released Energy

The energy released in a reactor is produced by exothermic nuclear reactions in which part of the nuclear mass is transformed to energy. Most of the energy is released when nuclei of heavy atoms split as they absorb neutrons. The splitting of these nuclei is called *fission*. A small fraction of the reactor energy comes from nonfission neutron capture in the fuel, moderator, coolant, and structural materials.

The fission energy, which is roughly 200 MeV (or 3.2×10^{-11} J) per fission, appears as kinetic energy of the fission fragments, kinetic energy of the newborn neutrons, and energy of emitted γ-rays. Many of the fission fragments are radioactive, undergoing β-decay accompanied by neutrons. The β-emission makes certain isotopes unstable, with resultant delayed neutron and γ-ray emission.

Immediately upon capture, the neutron-binding energy, which ranges from 2.2 MeV in hydrogen to 6 to 8 MeV in heavy materials, is released in the form of γ-rays. Many capture products are unstable and undergo decay by emitting β-particles, neutrons, and γ-rays. An approximate accounting of the forms of the released energy is given in Figure 3-1, and the energy distribution among these forms is outlined in Table 3-1.

The neutrons produced in the fission process have a relatively high kinetic energy and are therefore called *fast neutrons*. Most of these prompt neutrons have energies between 1 and 2 MeV, although some may have energies up to 10 MeV. The potential for a neutron to cause fission is improved if its energy is reduced to levels comparable to the surroundings by a slowing process, called *neutron moderation*. The slow neutrons, referred to as *thermal neutrons*, have energies in the range of 0.01 to 0.10 eV. The best moderating materials are those of low atomic masses. Hence moderators such as carbon, hydrogen, and deuterium have been used in power reactors that rely mostly on fission from slow (thermal) neutrons.

Table 3-1 Approximate distribution of energy release and deposition in thermal reactors

Type	Process	Percent of total released energy	Principal position of energy deposition
Fission			
I: instantaneous energy	Kinetic energy of fission fragments	80.5	Fuel material
	Kinetic energy of newly born fast neutrons	2.5	Moderator
	γ Energy released at time of fission	2.5	Fuel and structures
II: delayed energy	Kinetic energy of delayed neutrons	0.02	Moderator
	β-Decay energy of fission products	3.0	Fuel materials
	Neutrinos associated with β decay	5.0	Nonrecoverable
	γ-Decay energy of fission products	3.0	Fuel and structures
Neutron Capture			
III: instantaneous and delayed energy	Nonfission reactions due to excess neutrons plus β- and γ-decay energy of (n, γ) products	3.5	Fuel and structures
Total		100	

Source: Adapted from El-Wakil [2].

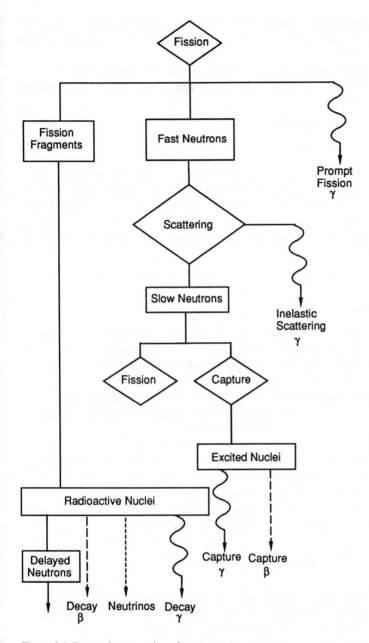

Figure 3-1 Forms of energy release in a reactor.

Splitting a fissile atom produces two smaller atoms and two or more neutrons. Uranium 235 is the only fissile material naturally present in extractable amounts, at roughly 0.7% of all uranium. Other fissile materials are generated by neutron capture in so-called fertile atoms. Thus fissile plutonium 239 and plutonium 241 are produced by neutron absorption in atoms of uranium 238 and plutonium 240, respectively. The only other practical fissile material is uranium 233, which is produced by neutron capture in thorium 232.

The energy release upon fission (E_f) is slightly dependent on the different fissile materials. Lamarsh [5] suggested the following relation for energy release due to fission by a thermal neutron:

$$E_f\ (^{233}U) = 0.98\ E_f\ (^{235}U)$$
$$E_f\ (^{239}Pu) = 1.04\ E_f\ (^{235}U)$$

In a typical LWR only about one-half of the neutrons are absorbed in fissile isotopes; the other one-half are captured by the fertile isotopes, control, and structural materials. The LWR fresh fuel is composed of uranium-based fuel, i.e., UO_2, slightly enriched in uranium 235. The plutonium produced in the LWR core during operation also participates in the energy-release process and may contribute up to 50% of the fission energy release.

B Energy Deposition

The fission products moving through the core material lose their energy through interaction with the surrounding matter. The energy loss rate depends mainly on the penetrating ability of the fission products. The neutrino energy is unrecoverable, as it does not interact with the surrounding materials. The fission fragments that carry most of the fission energy have a short range (<0.25 mm). The γ-rays' energy is released in structural as well as fuel materials. A considerable amount of the kinetic energy of neutrons is released in the moderator and the structure.

Let us consider the heat-generation rate per unit volume at any position, $q'''(\vec{r})$. It should be recognized that $q'''(\vec{r})$ is due to products of reaction events at all neighboring positions, as these reaction products pass through position $\vec{r}$. Let the volumetric heat-generation rate at position $\vec{r}$ due to reaction products of type i and energy E be $q'''_i(\vec{r},E)$. To obtain the total volumetric heat-generation rate at position $\vec{r}$ we must sum over all particle (and photon) types and energy spectrum:

$$q'''(\vec{r}) = \sum_i \int_o^\infty q'''_i(\vec{r},E)dE \tag{3-1}$$

Thus to calculate exactly the heat generation at a particular point of the reactor is difficult. However, the heat generation can be well approximated in the various parts of the reactor by established reactor physics analysis methods.

Information on the typical distribution of energy deposition in fuel and nonfuel materials in an LWR is given in Table 3-1. The amount of energy produced within the various parts of the reactor depends on reactor materials and geometry and hence on the reactor type. Roughly speaking, 87% of the total energy released per fission in an LWR is recovered in the fuel, 3% in the moderator, and 5% in the structure; 5% is unrecoverable (neutrino energy).

III HEAT-GENERATION PARAMETERS

A Heat Generation and Neutron Flux in Thermal Reactors

The heat-generation rate in the fuel, $q'''(\vec{r})$, is typically computed by assuming that the energy released by a fission reaction is recovered at the position of the fission event, except for the fraction carried away by neutrinos and the fraction deposited in nonfuel materials.* In other words, the spatial distribution of energy deposition from the fission fragments, γ's and β's, is assumed to follow the spatial distribution of the fission reaction rate $RR_f(\vec{r})$. Therefore the heat-generation rate $q'''(\vec{r},E)$ also becomes proportional to $RR_f(\vec{r})$.

We define the energy per fission reaction of isotope j, which is deposited in the fuel, as χ_f^j. For a typical thermal reactor, χ_f^{25} is about 190 MeV per fission.

Evaluation of the heat source distribution requires the knowledge of fission reaction rates, RR_f^j, which when summed over all atom types yield the total fission reaction rate, $RR_f(\vec{r})$:

$$RR_f(\vec{r}) = \sum_j RR_f^j(\vec{r}) \qquad (3\text{-}2)$$

Let $\sigma_a^j(E)$ be the microscopic absorption cross section of isotope j, which is the equivalent projected area of an atom for an absorption reaction. The cross section $\sigma_a^j(E)$ is proportional to the probability that one atom of type j absorbs an incident neutron of energy E. Because the absorption of a neutron by an atom can lead to either a fission process or a nonfission (or capture) process, the absorption cross section is the sum of the fission cross section σ_f and the capture cross section σ_c. The units of the microscopic cross section are normally given by square centimeters or barns (1 barn = 10^{-24} cm^2). Typical values of the absorption and fission cross sections of thermal neutrons are given in Table 3-2.

The macroscopic fission cross section is the sum of all microscopic cross sections for a fission reaction due to all atoms of type j within a unit volume interacting with an incident neutron per unit time. The macroscopic cross section is defined by:

$$\Sigma_f^j(\vec{r},E) \equiv N^j(\vec{r})\sigma_f^j(E) \qquad (3\text{-}3)$$

where N^j is the atomic density of isotope j. The atomic density (N^j) can be obtained from the mass density (ρ_j) using the relation:

$$N^j = A_v\,\rho_j/M_j \qquad (3\text{-}4)$$

where A_v = Avogadro's number for molecules in 1 gram mole (0.60225) $\times$ 10^{24} molecules/gram mole); M_j = molecular mass of the isotope j.

The fission rate of isotope j at position $\vec{r}$ due to the neutron flux $\phi(\vec{r},E)$ within the interval of neutron energy of E to E + dE is obtained from:

---

*In addition to the loss of the neutrino energy and the energy deposited outside the fuel, the magnitude of the energy released per reaction, which appears as heat in the fuel, should theoretically be reduced by conversion of a small part to potential energy of various metallurgical defects. It is, however, a negligible effect.

Table 3-2 Thermal (0.0253 eV) neutron cross sections

Material	Cross section (barns)	
	Fission σ_f	Absorption σ_a
Uranium 233	531	579
Uranium 235	582	681
Uranium 238	—	2.70[a]
Uranium, natural	4.2	7.6
Plutonium 239	743	1012
Boron	—	759
Cadmium	—	2450
Carbon	—	0.0034
Deuterium	—	0.0005
Helium	—	<0.007
Hydrogen	—	0.33
Iron	—	2.55
Oxygen	—	0.00027
Sodium	—	0.53
Zirconium	—	0.19

[a]The effective absorption cross section of ^{238}U in a typical LWR is substantially higher owing to the larger cross section at epithermal energies.

$$RR_f^j(\vec{r},E)dE = \Sigma_f^j(\vec{r},E)\,\phi(\vec{r},E)dE$$

The fission reaction rate for isotope j due to neutrons of all energies is:

$$RR_f^j(\vec{r}) = \int_0^\infty RR_f^j(\vec{r},E)\,dE \qquad (3\text{-}5)$$

The heat-generation rate per unit volume at $\vec{r}$ due to isotope j is:

$$q_j'''(\vec{r}) = \int_0^\infty \chi_f^j\,RR_f^j(\vec{r},E)dE \qquad (3\text{-}6)$$

Summing over all isotopes yields the total volumetric heat-generation rate:

$$q'''(\vec{r}) = \sum_j \int_0^\infty \chi_f^j\,RR_f^j(\vec{r},E)dE$$

or

$$q'''(\vec{r}) = \sum_j \int_0^\infty \chi_f^j\,\Sigma_f^j(\vec{r},E)\,\phi(\vec{r},E)dE \qquad (3\text{-}7)$$

In practice, the energy range is subdivided into a few intervals or groups. A multi-energy group model is then used to calculate the neutron fluxes; thus:

$$q'''(\vec{r}) = \sum_j \sum_{k=1}^{K} \chi_f^j\,\Sigma_{fk}^j(\vec{r})\phi_k(\vec{r}) \qquad (3\text{-}8)$$

where K = number of energy groups; Σ_{fk} and ϕ_k = equivalent macroscopic fission cross section and neutron flux, respectively, for the energy group k.

If we use a one energy group approximation, which gives good results for homogeneous thermal reactors at locations far from the reactor core boundaries, we have:

$$q'''(\vec{r}) = \sum_j \chi_f^j \Sigma_{fl}^j(\vec{r}) \, \phi_l(\vec{r}) \tag{3-9}$$

If we also assume uniform fuel material composition, then:

$$q'''(\vec{r}) = \sum_j \chi_f^j \Sigma_{fl}^j \phi_l(\vec{r}) \tag{3-10}$$

where Σ_{fl}^j is independent of position $\vec{r}$.

Assuming that $\chi_f^j = \chi_f$ for all fissile material, the volumetric heat-generation rate may be given by:

$$q'''(\vec{r}) = \chi_f \, \Sigma_{fl} \phi_l(\vec{r}) \tag{3-11}$$

where

$$\Sigma_{fl} = \sum_j \Sigma_{fl}^j \tag{3-12}$$

It should be noted that the thermal neutron flux in a typical LWR is only 15% of the total neutron flux. However, the fission cross section of ^{235}U and ^{239}Pu is so large at thermal energies that thermal fissions constitute 85 to 90% of all fissions.

Example 3-1 Determination of the neutron flux at a given power in a thermal reactor

PROBLEM A large PWR designed to produce heat at a rate of 3083 MW has 193 fuel assemblies each loaded with 517.4 kg of UO_2. If the average isotopic content of the fuel is 2.78 weight percent ^{235}U, what is the average thermal neutron flux in the reactor?

Assume uniform fuel composition, and χ_f = 190 MeV/fission (3.04×10^{-11} J/fission). Also assume that 95% of the reactor energy, i.e., of the recoverable fission energy is from the heat generated in the fuel. The effective thermal fission cross section of ^{235}U (σ_f^{25}) for this reactor is 350 barns. Note that this effective value of σ_f^{25} is smaller than that appearing in Table 3-2, as it is the average value of $\sigma_f^{25}(E)$ over the energy spectrum of the neutron flux, including the epithermal neutrons.

SOLUTION Use Eq. 3-11 to find $\phi_l(\vec{r})$.

$$\phi_l(\vec{r}) = \frac{q'''(\vec{r})}{\chi_f \, \Sigma_f} \tag{3-13a}$$

but because $\Sigma_f = \Sigma_f^{25}$ in this reactor, Σ_f^{25} is obtained from Eq. 3-3, yielding:

$$\phi_1(\vec{r}) = \frac{q'''(\vec{r})}{\chi_f \sigma_f^{25} N^{25}} \qquad (3\text{-}13b)$$

Consequently, the core-average value of the neutron flux in a reactor with uniform density of ^{235}U is obtained from the average heat-generation rate by:

$$\langle \phi_1 \rangle = \frac{\langle q''' \rangle}{\chi_f \sigma_f^{25} N^{25}} \qquad (3\text{-}13c)$$

Multiplying both the numerator and denominator by the UO_2 volume (V_{UO_2}) we get:

$$\langle \phi_1 \rangle = \frac{\langle q''' \rangle V_{UO_2}}{\chi_f \sigma_f^{25} N^{25} V_{UO_2}} = \frac{\dot{Q}}{\chi_f \sigma_f^{25} N^{25} V_{UO_2}} \qquad (3\text{-}13d)$$

To use the above equation only N^{25} needs to be calculated. All other values are given. The value of N^{25} can be obtained from the uranium atomic density if the ^{235}U atomic fraction (a) is known:

$$N^{25} = aN^U \qquad (3\text{-}14)$$

The uranium atomic density is equal to the molecular density of UO_2, as each molecule contains one uranium atom ($N^U = N^{UO_2}$). Thus from Eqs. 3-4 and 3-14:

$$N^{25} = a\frac{A_v \rho_{UO_2}}{M_{UO_2}} \qquad (3\text{-}15a)$$

Multiplying each side of Eq. 3-15a by V_{UO_2} we get:

$$N^{25}V_{UO_2} = aN^{UO_2} V_{UO_2} = \frac{a(A_v)m_{UO_2}}{M_{UO_2}} \qquad (3\text{-}15b)$$

Now

$$m_{UO_2} = 193 \text{ (assemblies)} \times 517.4 \left[\frac{\text{kg } UO_2}{\text{assembly}}\right] \times \frac{1000 \text{ g}}{\text{kg}} = 9.9858 \times 10^7 \text{ g}$$

The molecular mass of UO_2 is calculated from:

^{235}U M_{25} = 235.0439 [g/mole]
^{238}U M_{28} = 238.0508
Oxygen M_O = 15.9994

$$M_{UO_2} = M_U + 2M_O \qquad (3\text{-}16a)$$
$$M_U = a M_{25} + (1 - a)M_{28} \qquad (3\text{-}16b)$$
$$\therefore M_{UO_2} = a M_{25} + (1 - a)M_{28} + 2M_O \qquad (3\text{-}16c)$$

To obtain the value of a, we use the known ^{235}U weight fraction, or enrichment (r):

$$r = \frac{aM_{25}}{M_U} = \frac{aM_{25}}{aM_{25} + (1 - a)M_{28}} \qquad (3\text{-}17)$$

Equation 3-17 can be solved for a

$$a = \frac{r}{r + \frac{M_{25}}{M_{28}}(1 - r)} = \frac{0.0278}{0.0278 + \frac{235.0439}{238.0508}(0.9722)} = 0.028146 \quad (3\text{-}18)$$

From Eq. 3-16c:

$$M_{UO_2} = 0.028146(235.0439) + (0.971854)(238.0508) + 2(15.9994)$$
$$= 237.9657 + 2(15.9994) = 269.9645$$

Then from Eq. 3-15b:

$$N^{25}V_{UO_2} = \frac{0.028146\left(0.60225 \times 10^{24} \frac{\text{atoms}}{\text{g} \cdot \text{mole}}\right)(9.9858 \times 10^7 \text{ g})}{269.9645 \text{ g/mole}}$$
$$= 6.27 \times 10^{27} \text{ atoms U}^{25}$$

Now using Eq. 3-13d, the value for the average flux is calculated:

$$\langle \phi_1 \rangle = \frac{0.95 \, (3083 \text{ MW})\left(10^6 \frac{\text{W}}{\text{MW}}\right)}{\left[3.04 \times 10^{-11} \frac{\text{J}}{\text{fission}}\right]\left[350 \times 10^{-24} \frac{\text{cm}^2}{\text{atom-neutron}}\right][6.270 \times 10^{27} \text{ atom}]}$$

Answer: $\langle \phi_1 \rangle = 4.38 \times 10^{13}$ neutron/cm^2 · s

B Relation Between Heat Flux, Volumetric Heat Generation, and Core Power

1 Single pin parameters. Three thermal parameters, introduced in Chapter 2, are related to the volumetric heat generation in the fuel: (1) the fuel pin power or rate of heat generation ($\dot{q}$); (2) the heat flux ($\vec{q}''$), normal to any heat transfer surface of interest that encloses the fuel (e.g., the heat flux may be defined at the inner and outer surfaces of the cladding or the surface of the fuel itself); (3) the power rating per unit length (linear heat-generation rate) of the pin (q').

At steady state the three quantities are related by Eqs. 2-1, 2-2, and 2-3, which can be combined and applied to the nth fuel pin to yield:

$$\dot{q}_n = \iiint\limits_{V_{fn}} q''' \, (\vec{r})dV = \iint\limits_{S_n} \vec{q}'' \cdot \vec{n} \, dS = \int\limits_L q' \, dz \quad (3\text{-}19)$$

where V_{fn} = volume of the energy-generating region of a fuel element; $\vec{n}$ = outward unity vector normal to the surface S_n surrounding V_{fn}; L = length of the active fuel element.

It is also useful to define the mean heat flux through the surface of our interest:

$$\{q''\}_n = \frac{1}{S_n} \iint_{S_n} \vec{q}''(\vec{r}) \cdot \vec{n} \, dS = \frac{\dot{q}}{S_n} \tag{3-20}$$

where $\{ \} = $ a surface averaged quantity.

The mean linear power rating of the fuel element is obtained from:

$$q'_n = \frac{1}{L} \int_L q' dz = \frac{\dot{q}_n}{L} \tag{3-21}$$

Let us apply the relations of Eq. 3-19 to a practical case. For a cylindrical fuel rod of pellet radius R_{fo}, outer clad radius R_{co}, and length L, the total rod power is related to the volumetric heat-generation rate by:

$$\dot{q}_n = \int_{-L/2}^{L/2} \int_0^{R_{fo}} \int_0^{2\pi} q'''(r,\theta,z) r \, d\theta \, dr \, dz \tag{3-22a}$$

The pin power can be related to the heat flux at the cladding outer surface (q''_{co}) by:

$$\dot{q}_n = \int_{-L/2}^{L/2} \int_0^{2\pi} q''_{co}(\theta,z) R_{co} \, d\theta \, dz \tag{3-22b}$$

Here we have neglected axial heat transfer through the ends of the rod and heat generation in the cladding and gap. Finally, the rod power can be related to the linear power by:

$$\dot{q}_n = \int_{-L/2}^{L/2} q'(z) dz = q'_n L \tag{3-22c}$$

The mean heat flux through the outer surface of the clad is according to Eq. 3-20:

$$\{q''_{co}\}_n = \frac{1}{2\pi R_{co}L} \int_{-L/2}^{L/2} \int_0^{2\pi} q''_{co}(\theta,z) R_{co} \, d\theta \, dz = \frac{\dot{q}_n}{2\pi R_{co}L} \tag{3-23}$$

It should be recognized that the linear power at any axial position is equal to the heat flux integrated over the perimeter:

$$q'(z) = \int_0^{2\pi} q''_{co}(\theta,z) R_{co} \, d\theta \tag{3-24a}$$

The linear power can also be related to the volumetric heat generation by:

$$q'(z) = \int_0^{R_{fo}} \int_0^{2\pi} q'''(r,\theta,z) \, r \, d\theta \, dr \tag{3-24b}$$

Then, for any fuel rod:

$$\dot{q}_n = L q'_n = L 2\pi R_{co}\{q''_{co}\}_n = L \pi R_{fo}^2 \langle q''' \rangle_n \tag{3-25}$$

where the average volumetric generation rate in the pin is:

$$\langle q''' \rangle_n = \frac{\dot{q}_n}{\pi R_{fo}^2 L} = \frac{\dot{q}_n}{V_{fn}} \tag{3-26}$$

2 Core power and fuel pin parameters. Consider a core consisting of N fuel pins. The overall power generation in the core is then:

$$\dot{Q} = \sum_{n=1}^{N} \dot{q}_n + \dot{Q}_{nonfuel} \tag{3-27}$$

Defining γ as the fraction of power generated in the fuel:

$$\dot{Q} = \frac{1}{\gamma} \sum_{n=1}^{N} \dot{q}_n \tag{3-28}$$

From Eqs. 3-25 and 3-28, for N fuel pins of identical dimensions:

$$\dot{Q} = \frac{1}{\gamma} \sum_{n=1}^{N} \dot{q}_n = \frac{1}{\gamma} \sum_{n=1}^{N} L\,q_n' = \frac{1}{\gamma} \sum_{n=1}^{N} L\,2\pi R_{co}\{q_{co}''\}_n = \frac{1}{\gamma} \sum_{n=1}^{N} L\pi R_{fo}^2 \langle q''' \rangle_n \tag{3-29}$$

We can define core-wide thermal parameters for an average pin as:

$$\frac{\dot{Q}}{N} = \frac{1}{\gamma} \langle \dot{q} \rangle = \frac{L}{\gamma} \langle q' \rangle = \frac{L}{\gamma} 2\pi R_{co} \langle q_{co}'' \rangle = \frac{L}{\gamma} \pi R_{fo}^2 \langle q''' \rangle \tag{3-30}$$

When all the energy release is assumed to occur in the fuel, Eq. 3-30 becomes identical to Eq. 2-5.

The core-average volumetric heat generation rate in the fuel is given by:

$$\langle q''' \rangle = \frac{\gamma \dot{Q}}{V_{fuel}} = \frac{\gamma \dot{Q}}{N V_{fn}} \tag{3-31}$$

The core-wide average fuel volumetric heat-generation rate $\langle q''' \rangle$ should not be confused with the core power density Q''', defined in Chapter 2 as:

$$Q''' = \dot{Q}/V_{core} \tag{3-32}$$

which takes into account the volume of all the core constituents: fuel, moderator, and structures.

Example 3-2 Heat transfer parameters in various power reactors

PROBLEM For the set of reactor parameters given below, calculate for each reactor type:

1. Equivalent core diameter and core length
2. Average core power density Q''' (MW/m³)
3. Core-wide average linear heat-generation rate of a fuel rod, $\langle q' \rangle$ (kW/m)

4. Core-wide average heat flux at the interface between the rod and the coolant $\langle q''_{co} \rangle$ (MW/m^2)

Quantity	PWR	BWR	PHWR[a] (CANDU)	LMFBR[b]	HTGR[c]
Core power level (MWt)	3800	3579	2140	780	3000
% of power deposited in fuel rods	96	96	95	98	100
Fuel assemblies/core	241	732	12 × 380 = 4560	198	8 × 493 = 3944
Assembly lateral spacing (mm)	207 (square pitch)	152 (square pitch)	280 (square pitch)	144 (across hexagonal flats)	361 (across hexagonal flats)
Fuel rods/assembly	236	62	37	217	72
Fuel rod length (mm)	3810	3760	480	914	787
Fuel rod diameter (mm)	9.7	12.5	13.1	5.8	21.8

[a]CANDU, 12 fuel assemblies are stacked end to end at 380 locations. CANDU fuel rods are oriented horizontally.
[b]Blanket assemblies are excluded from the LMFBR calculation.
[c]HTGR, eight fuel assemblies are stacked end to end at 493 locations. Fuel rod dimensions given actually refer to coolant holes.

SOLUTION Only the *PWR* case is considered in detail here; the results for the other reactors are summarized.

1. Equivalent core diameter and length calculation

$$\text{Fuel assembly area} = (0.207 \text{ m})^2 = 0.043 \text{ m}^2$$
$$\text{Core area} = (0.043 \text{ m}^2)(241 \text{ fuel assemblies}) = 10.36 \text{ m}^2$$
$$\text{Equivalent circular diameter: } \frac{\pi D^2}{4} = 10.36 \text{ m}^2$$
$$\therefore D = 3.64 \text{ m}$$
$$\text{Core length } (L) = 3.81 \text{ m}$$
$$\text{Total core volume} = 3.81 \frac{\pi (3.64)^2}{4} = 39.65 \text{ m}^3$$

2. Average power density in the core from Eq. 3-32:

$$Q''' = \frac{\dot{Q}}{\pi R^2 L} = \frac{\dot{Q}}{V_{core}} = \frac{3800 \text{ MW}}{39.65 \text{ m}^3} = 95.85 \text{ MW/m}^3$$

3. Average linear heat generation rate in a fuel rod can be obtained from Eq. 3-30 as:

$$\langle q' \rangle = \frac{\gamma \dot{Q}}{NL} = \frac{0.96 (3800 \text{ MW})}{(236 \text{ rods/assembly})(241 \text{ assemblies})(3.81 \text{ m/rod})} = 16.8 \text{ kW/m}$$

4. Average heat flux at the interface between a rod and the coolant: From Eq. 3-30 we can obtain $\langle q''_{co} \rangle$ as:

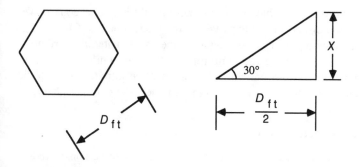

Figure 3-2 Hexagonal assembly.

$$\langle q''_{co} \rangle = \frac{\gamma \dot{Q}}{NL2\pi R_{co}} = \frac{\gamma \dot{Q}}{NL\pi D_{co}} = \frac{\langle q' \rangle}{\pi D_{co}}$$

$$= \frac{(16.8 \text{ kW/m})(10^{-3} \text{ MW/kW})}{\pi(0.0097 \text{ m})} = 0.552 \text{ MW/m}^2$$

Calculations for the other reactor types are left as an exercise for the reader. The solutions are given below.

Quantity	BWR	CANDU	LMFBR	HTGR
Equivalent core diameter (m)	4.64	6.16	2.13	8.42
Core length (m)	3.76	5.76	0.914	6.3
Core power density (MW/m³)	56.3	12.5	239.5	8.55
Average linear heat generation of a fuel rod (kW/m)	20.1	25.1	19.9	13.4
Average heat flux at the interface of fuel rod and coolant (MW/m²)	0.512	0.61	1.09	0.20

Note: Area for hexagon-shaped assembly (Fig. 3-2) is given by:

$$\text{Area hexagon} = \frac{12}{2} (D_{ft}/2)(D_{ft}/2)(\tan 30°) = \frac{\sqrt{3}}{2} D_{ft}^2$$

where D_{ft} = distance across hexagonal flats.

IV POWER PROFILES IN REACTOR CORES

We shall consider simple cases of reactor cores to form an appreciation of the overall power distribution in various geometries. The simplest core is one in which the fuel is homogeneously mixed with the moderator and uniformly distributed within the core volume. Consideration of such a core is provided here as a means to establish the general tendency of neutron flux behavior. In a neutronically heterogeneous power reactor the fuel material is dispersed in lumps within the moderator (see Chapter 1).

In practice, different strategies may be sought for the fissile material distribution in the core. For example, to burn the fuel uniformly in the core, uniform heat generation is desired. Therefore various enrichment zones may be introduced, with the highest enrichment located at the low neutron flux region near the core periphery.

A Homogeneous Unreflected Core

In the case of a homogeneous unreflected core, the whole core can be considered as one fuel element. Using the one energy group scheme, it is clear from Eq. 3-11 that $q'''(\vec{r})$ is proportional to $\phi_1(\vec{r})$.

Solving the appropriate one-group neutron diffusion equation, simple analytical expressions for the neutron flux, and hence the volumetric heat-generation rate, have been obtained for simple geometries. The general distribution is given by*:

$$q'''(\vec{r}) = q'''_{max} F(\vec{r}) \tag{3-34}$$

where q'''_{max} is the heat generation at the center of the homogeneous core. Expressions for $F(\vec{r})$ are given in Table 3-3. Thus for a cylindrical core:

$$q'''(r,z) = q'''_{max} J_o\left(2.4048 \frac{r}{R_e}\right) \cos\left(\frac{\pi z}{L_e}\right) \tag{3-35}$$

where r and z are measured from the center of the core.

The shape of q''' as a function of r is shown in Figure 3-3. It is seen that the neutron flux becomes zero at a small distance δR from the actual core boundary. The

Table 3-3 Distribution of heat generation in a homogeneous unreflected core

Geometry	Coordinate	$q'''(\vec{r})/q'''_{max}$ or $F(\vec{r})$	$q'''_{max}/\langle q'''\rangle$ (ignoring extrapolation lengths)
Infinite slab	x	$\cos\dfrac{\pi x}{L_e}$	$\dfrac{\pi}{2}$
Rectangular parallelepiped	x,y,z	$\cos\left(\dfrac{\pi x}{L_{xe}}\right)\cos\left(\dfrac{\pi y}{L_{ye}}\right)\cos\left(\dfrac{\pi z}{L_{ze}}\right)$	$\dfrac{\pi^3}{8}$
Sphere	r	$\dfrac{\sin\left(\dfrac{\pi r}{R_e}\right)}{\pi r/R_e}$	$\dfrac{\pi^2}{3}$
Finite cylinder	r,z	$J_o\left(2.405\dfrac{r}{R_e}\right)\cos\left(\dfrac{\pi z}{L_e}\right)$	$2.32\left(\dfrac{\pi}{2}\right)$

$L_e = L + 2\delta L$; $R_e = R + \delta R$; L_e, R_e = extrapolated dimensions; L, R = fuel physical dimensions. Source: Rust [8].

*Note that in the homogeneous reactor $\langle q'''\rangle$ and Q''' are identical.

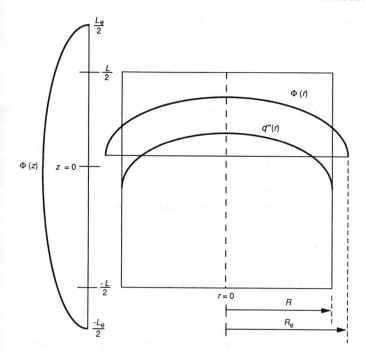

Figure 3-3 Neutron flux and heat-generation rate profiles in a homogeneous cylindrical reactor.

distances δR and δL are called the *extrapolation lengths* and are usually small relative to L and R, respectively.

The overall core heat generation rate is given by:

$$\dot{Q} = q_{max}''' \iiint\limits_{V_{core}} F(\vec{r})dV \tag{3-36}$$

In real reactors, the higher burnup of fuel at locations of high neutron fluxes leads to flattening of radial and axial power profiles.

B Homogeneous Core with Reflector

For a homogeneous core with a reflector it is also possible to use a one-group scheme inside the reactor. For the region near the boundary between the core and the reflector, a two-group approximation is usually required. An analytical expression for q''' in this case is more difficult. The radial shape of the thermal neutron flux is shown in Figure 3-4.

C Heterogeneous Core

In the case of a heterogeneous thermal reactor, heat is produced mainly in the fuel elements, and the thermal neutron flux is generated in the moderator.

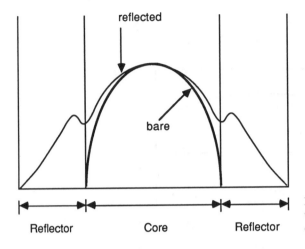

Figure 3-4 Effect of neutron reflector on the thermal neutron flux radial distribution.

In most power reactors, there are large numbers of rods. Thus with little error we can approximate the profile of the heat-generation rate in a fresh core with uniform enrichment by the previous expressions for a homogeneous core, provided that now $q'''(r)$ is understood to represent the heat-generation rate in a fuel rod that is a distance r from the center of the reactor core. However, in practice, a fresh core may have zones of variable enrichment, which negates the possibility of using the homogeneous core expressions. With fuel burnup, various amounts of plutonium and fission products are introduced. A typical power distribution in a PWR mid-burnup core is shown in Figure 3-5. It is clear that no analytic expression can easily describe the spatial distribution.

D Effect of Control Rods

The control rods depress the neutron flux radially and axially. Thus the radial power profile is also depressed near the rods (Fig. 3-6). In many reactors, some control material is uniformly mixed with the fuel in several fuel pins so that it is burned up as the fuel burns, thereby readily compensating for loss of fuel fissile content. In addition, soluble poison is routinely used for burnup cycle control in PWRs.

Example 3-3 Local pin power for a given core power

PROBLEM For a heavy-water-moderated reactor with uniform distribution of enriched UO_2 fuel in a cylindrical reactor core, calculate the power generated in a fuel rod located half-way between the centerline and the outer boundary. The important parameters for the core are as follows.

1. Core diameter (R) = 8 ft (2.44 m)
2. Core height (L) = 20 ft (6.10 m)
3. Fuel pellet outside diameter = 0.6 in. (1.524 cm)
4. Maximum thermal-neutron flux = 10^{13} neutrons/cm^2 · s

Center-line

	A	B	C	D	E	F	G	H
A (Center-line)	7FG 27841 35997 0.72	9AH 10506 20929 0.94	8GB 22820 31824 0.81	9FG 8021 18967 1.03	9FF 12007 22721 1.03	8AH 20927 29456 0.82	88G 23452 32601 0.93	10AH 0 10505 1.12
B	9HA 10506 20928 0.94	9BG 12827 23454 0.97	9CH 9884 20535 0.98	8DH 19247 28621 0.86	9DH 8086 19249 1.08	8EG 21118 30125 0.88	10BG 0 12826 1.36	10BH 0 10847 1.16
C	8GB 22820 31820 0.81	9HC 9884 20530 0.98	9GB 12827 22827 0.91	8FG 18968 27840 0.81	9EG 10514 21117 1.02	8CH 20532 29348 0.86	8DG 23409 32506 0.92	10CH 0 9883 1.03
D	9FG 8021 18956 1.02	8HD 19236 28594 0.86	8GF 19382 28183 0.80	8FF 22722 31183 0.79	9BH 10847 21660 1.07	9DG 12829 23410 1.07	10DG 0 12830 1.37*	10DH 0 8086 0.83
E	9FF 12007 22714 1.03	9HD 8086 19238 1.08	9GE 10514 21098 1.02	9HB 10847 21656 1.07	9GF 8021 19382 1.15	8BH 21656 31071 0.95	10EG 0 10516 1.10	
F	8HA 20926 29454 0.82	8GE 21099 30105 0.88	8HC 20527 29340 0.86	9GD 12829 23408 1.07	8HB 21651 31067 0.95	10FF 0 12010 1.25	10FG 0 8022 0.82	
G	88G 23452 32601 0.93	10GB 0 12825 1.36	8GD 23407 32504 0.92	10GD 0 12829 1.37*	10GE 0 10516 1.10	10GF 0 8022 0.82		
H	10HA 0 10505 1.12	10HB 0 10847 1.16	10HC 0 9883 1.03	10HD 0 8086 0.83				

*Maximum Relative Power

Fuel Lots 7, 8, 9, 10   Initially 3.20 w/o U-235

Cycle Average Burnup = 10,081 MWd/MT
Cycle Thermal Energy = 896.8 GWd

Key

1AA	Assembly Number
0	BOC Burnup, MWd/MT
17302	EOC Burnup, MWd/MT
1.04	BOC Relative Power (Assembly/Average)

Figure 3-5 Typical PWR assembly power and burnup distribution, assuming fresh fuel is introduced at the outer core locations. *(From Benedict et al. [1].)*

Assume that the extrapolated dimensions can be approximated by the physical dimensions and that enrichment and average moderator temperature are such that:

$$q'''(\text{Btu/hr ft}^3) = 6.99 \times 10^{-7} \, \phi \, (\text{neutrons/cm}^2 \cdot \text{s}) \text{ at every position}$$
$$q'''(\text{W/m}^3) = 7.27 \times 10^{-6} \, \phi \, (\text{neutrons/cm}^2 \cdot \text{s})$$

SOLUTION This reactor can be approximated as a homogeneous unreflected core in the form of a finite cylinder. From Table 3-3:

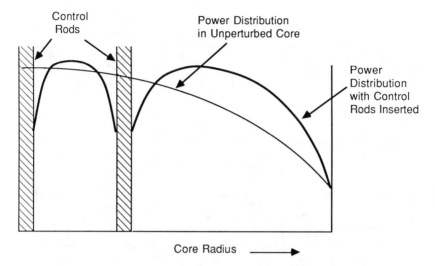

Figure 3-6 Radial power profile in a cylindrical reactor with inserted control rods.

$$q'''(r,z) = q'''_{max} J_o \left(2.4048 \frac{r}{R_e} \right) \cos \left(\frac{\pi z}{L_e} \right) \tag{3-37}$$

The value of q'''_{max} is readily established as

$$q'''_{max} = 6.99 \times 10^{-7} \phi_{max} = 6.99 \times 10^6 \, \text{Btu/hr ft}^3 \, (72.68 \, \text{MW/m}^3)$$

for $r/R_e = 0.5$.

$$q'''(r,z) = q'''_{max} J_o(1.202) \cos \left(\frac{\pi z}{L_e} \right)$$

For the heat generated from a single fuel rod, Eq. 3-22a yields:

$$\dot{q}_n = \int_{-L/2}^{L/2} q'''_n (r,z) A \, dz$$

where $A = \frac{\pi D_{fo}^2}{4}$. Taking Le as L

$$\dot{q}_n = \int_{-L/2}^{L/2} q'''_{max} A J_o(1.202) \cos \left(\frac{\pi z}{L} \right) dz$$

$$= q'''_{max} A J_o(1.202) \frac{L}{\pi} \sin \frac{\pi z}{L} \Bigg|_{-L/2}^{L/2}$$

$$= q'''_{max} A J_o(1.202) \frac{2L}{\pi}$$

The value of $J_o(1.202) = 0.6719$ [10]. Hence

$$\dot{q}_n = (6.99 \times 10^6) \frac{\pi \left(\frac{0.6}{12}\right)^2}{4} (0.6719) \frac{2}{\pi} \quad (20)$$

Answer: $\dot{q}_n = 1.17 \times 10^5$ Btu/hr (34.5 kW)

V HEAT GENERATION WITHIN A FUEL PIN

A Fuel Pins of Thermal Reactors

Consider a cylindrical fuel pin inside the reactor. The heat-generation rate at a particular point of the rod, $q'''(r,\theta,z)$ depends on the position of the rod in the reactor and the concentration of the various fissionable materials at this point.

In thermal reactors with uniform fuel enrichment, the profile of the heat-generation rate follows approximately the thrermal neutron flux. For single-phase cooled reactors, in many cases the axial profile of q''' can be approximated by a cosine function, i.e.,

$$q'''(z) \sim \cos\left(\frac{\pi z}{L_e}\right) \quad (3\text{-}38)$$

As mentioned before, for fresh fuel this formula gives adequate results for illustrative purposes.

Radially within the fuel element, q''' is expected to be reduced at the center because of thermal neutron flux depression. This depression is often neglected for small-diameter low-absorbing fuel rods but should not be ignored for thick, highly absorbing rods.

B Fuel Pins of Fast Reactors

The dependence of q''' on z is similar to that of thermal reactors. However, the shape of q''' as a function of r within a rod is different because in fast reactors energetic neutrons, even fission neutrons, contribute to the fission reaction directly without being slowed. The radial profile of q''' tends to be flatter than in thermal and epithermal systems. Fortunately, in the real cases the diameter of the fuel rod is relatively small and the mean free path of the neutrons relatively large. Therefore the assumption that q''' is independent of r and θ is a good approximation.

VI HEAT GENERATION WITHIN THE MODERATOR

The energy deposition in the moderator mainly comes from (1) neutron slowing by scattering due to collisions with the nuclei of the moderator material; and (2) γ-ray absorption. The dominant mechanism of heat production is neutron slowing due to elastic scattering. Neutrons also lose energy through inelastic collisions as a result of excitation of the target nuclei. However, for light nuclei, moderation by inelastic scattering is less important than by elastic scattering. With heavy nuclei, such as those involved in structures, the inelastic scattering is the principal mechanism for neutron moderation.

With elastic scattering the energy lost from the neutrons appears as kinetic energy of the struck nucleus. With inelastic scattering the energy lost by the neutrons appears as γ-rays.

Let $\Sigma_{s,e\ell}(\vec{r},E)$ be the macroscopic elastic scattering cross section of neutrons of energy E at position $\vec{r}$, and let $\phi(\vec{r},E)$ be the neutron flux at this position and energy. Hence the elastic scattering reaction rate $RR_{s,e\ell}(\vec{r},E)dE$ within the energy interval E to $E + dE$ is given by:

$$RR_{s,e\ell}(\vec{r},E)dE = \Sigma_{s,e\ell}(\vec{r},E)\ \phi(\vec{r},E)dE \tag{3-39}$$

If $\Delta E(E)$ is the mean energy loss per collision at neutron energy E, the heat generation due to $\phi(r,\vec{E})$ is:

$$q'''_{e\ell}(\vec{r},E)dE = \Delta E(E)\Sigma_{s,e\ell}(\vec{r},E)\ \phi(\vec{r},E)dE \tag{3-40}$$

Define the mean logarithmic energy decrement per collision (ξ) by

$$\xi \equiv \overline{\ell n\ \frac{E}{E'}} \tag{3-41}$$

where E' = neutron energy after one collision (i.e., $\Delta E = E - E'$). By expanding the logarithm in terms of ΔE, we get:

$$\ell n\ \frac{E}{E'} \simeq \frac{\Delta E}{E} + \frac{1}{2}\left(\frac{\Delta E}{E}\right)^2 + \frac{1}{3}\left(\frac{\Delta E}{E}\right)^3 + \ldots\ldots \tag{3-42}$$

Assuming $\dfrac{\Delta E}{E}$ is sufficiently small, we can approximate:

$$\overline{\ell n\ \frac{E}{E'}} \simeq \frac{\Delta E}{E} \tag{3-43}$$

Then Eq. 3-40 becomes:

$$q'''_{e\ell}(\vec{r},E)dE = \xi E\ \Sigma_{s,e\ell}(\vec{r},E)\ \phi(\vec{r},E)dE \tag{3-44}$$

Usually the moderator can be considered a homogeneous material, so that Σ_s is independent of $\vec{r}$ within the moderator. Then:

$$q'''_{e\ell}(\vec{r},E)dE = \xi E\ \Sigma_{s,e\ell}(E)\ \phi(\vec{r},E)dE \tag{3-45}$$

The heat-generation rate from neutron elastic scatterings at all energies is:

$$q'''_{e\ell}(\vec{r}) = \int_{E_c}^{\infty} \xi E\ \Sigma_{s,e\ell}(E)\ \phi(\vec{r},E)dE \tag{3-46}$$

where E_c = an energy level under which the energy loss by neutrons is negligible (e.g., $E_c = 0.1$ eV).

In order to assess the conditions for the validity of Eq. 3-43, consider isotropic elastic scattering in the center of mass system. Let j be a particular isotope. Then [3]:

$$\overline{\Delta E} = \frac{E - \alpha_j E}{2} \tag{3-47}$$

where

$$\alpha_j = \left(\frac{A^j - 1}{A^j + 1}\right)^2 \tag{3-48}$$

and A^j is the atomic mass number. Then

$$\frac{\overline{\Delta E}}{E} = \frac{1 - \alpha_j}{2} \tag{3-49}$$

We see that for all possible values of A^j, $0 \le \alpha_j < 1$, and $\dfrac{\overline{\Delta E}}{E}$ ranges from 0 to 0.5. Thus because elastic scattering is close to being isotropic, $\dfrac{\overline{\Delta E}}{E}$ is small; therefore it can be concluded that Eq. 3-43 is a good approximation to Eq. 3-42.

VII HEAT GENERATION IN THE STRUCTURE

The main sources of heat generation in the structure are (1) γ-ray absorption; (2) elastic scattering of neutrons; and (3) inelastic scattering of neutrons.

A γ-Ray Absorption

The photon "population" at the particular point $\vec{r}$ of the structure is mainly due to: (1) γ-rays born somewhere in the fuel, which arrive without scattering at position $\vec{r}$ within the structure; (2) γ-rays born within the structural materials, which arrive unscattered at $\vec{r}$; and (3) scattered photons (Compton effect).

Consider the quantity $N_\gamma(\vec{r},E)dE$, which is the photon density at a particular position r within the structure having energy between E and $E + dE$. The energy flux is defined as:

$$I_\gamma(\vec{r},E)dE \equiv EN_\gamma(\vec{r},E)dE \tag{3-50}$$

The absorption rate is described by application of the linear energy absorption coefficient $\mu_a(E)$ as follows:

$$q_\gamma''' \, (\vec{r},E)dE = \mu_a \, (E) \, I_\gamma(\vec{r},E)dE \tag{3-51}$$

where $q_\gamma'''(\vec{r},E)$ = the absorbed energy density per unit time from the γ-ray energy flux within the interval E to $E + dE$; $\mu_a(E)$ = a function of the material, as can be found in Table 3-4. The total heat-generation rate then is:

Table 3-4 Linear γ-ray attenuation and absorption coefficients

γ-Ray energy (MeV)	Coefficient (m^{-1})			
	Water	Iron	Lead	Concrete
0.5				
μ	9.66	65.1	164	20.4
μ_a	3.30	23.1	92.4	7.0
1.0				
μ	7.06	46.8	77.6	14.9
μ_a	3.11	20.5	37.5	6.5
1.5				
μ	5.74	38.1	58.1	12.1
μ_a	2.85	19.0	28.5	6.0
2.0				
μ	4.93	33.3	51.8	10.5
μ_a	2.64	18.2	27.3	5.6
3.0				
μ	3.96	28.4	47.7	8.53
μ_a	2.33	17.6	28.4	5.08
5.0				
μ	3.01	24.6	48.3	6.74
μ_a	1.98	17.8	32.8	4.56
10.0				
μ	2.19	23.1	55.4	5.38
μ_a	1.65	19.7	41.9	4.16

Source: Templin [11]

$$q_\gamma''' (\vec{r}) = \int_0^{E_\infty} q_\gamma'''(\vec{r},E)dE = \int_0^{E_\infty} \mu_a(E)I_\gamma(\vec{r},E)dE \qquad (3\text{-}52)$$

where E_∞ is selected at a sufficiently high value.

$I_\gamma(\vec{r},E)$ can be found by solving the appropriate transport equation. However, practical calculation of γ-ray attenuation is often greatly simplified by the use of so-called buildup factors and the uncollided γ-ray flux. With this procedure the simplified transport equation is first solved neglecting the scattering process to yield the uncollided flux $I_\gamma^*(\vec{r},E)$. For example, for a plane geometry the energy of the uncollided γ-ray flux is obtained from:

$$I_\gamma^* = I_\gamma^\circ \, e^{-\mu(E)x} \qquad (3\text{-}53)$$

where $\mu(E)$ = linear attenuation coefficient for photons at energy E due to absorption and scattering (Table 3-4); I_γ° = unattenuated γ-ray flux. If $I_\gamma (\vec{r},E')$ is the real energy flux at a point $\vec{r}$ resulting from I_γ°, the buildup factor (B) is defined as:

$$B(\vec{r},\mu,E) = \frac{\int_0^{E_\infty} \mu_a(E')I_\gamma(\vec{r},E')dE'}{\mu_a(E)I_\gamma^*(\vec{r},E)} \qquad (3\text{-}54)$$

Note that by definition $B(\vec{r},E)$ is greater than unity and depends primarily on the boundary conditions, and the energy level of the uncollided photons (through the linear attenuation coefficient and the scattered photon source distribution). The values for the cases of interest are tabulated elsewhere [4, 6]. Utilizing this definition of B, Eq. 3-51 becomes:

$$q_\gamma''' \, (\vec{r},E) \, dE \; = \; B\mu_a(E) \, I_\gamma^*(\vec{r},E)dE \qquad (3\text{-}55)$$

The following are mathematical expressions for simplified cases for the evaluation of Eq. 3-55 when $I_\gamma^*(\vec{r},E)$ can be analytically evaluated.

1. Point isotropic source emitting S photons of energy E_o per second. In this case:

$$q_\gamma''' \, (r) \; = \; S \, B \, \mu_a(E_o) \, E_o \, \frac{e^{-\mu r}}{4\pi r^2} \qquad (3\text{-}56)$$

  where r is the distance from the point source.
2. Infinite plane source emitting S photons with energy E_o per unit time per unit surface in the positive direction of the x-axis.

$$q_\gamma''' \, (x) \; = \; S \, B \, \mu_a(E_o) \, E_o e^{-\mu x} \qquad (3\text{-}57)$$

3. Plane isotropic source emitting S photons of energy E_o per unit source area, per second in all directions.

$$q_\gamma''' \, (x) \; = \; S \, B\mu_a(E_o) \, \frac{E_o}{2} \int_{\mu x}^{\infty} \frac{e^{-t}}{t} \, dt \qquad (3\text{-}58)$$

B Neutron Slowing

In the structure, the neutrons slow by (1) elastic scattering and (2) inelastic scattering. For elastic scattering we can use the approximation of Eq. 3-46:

$$q_{e\ell}''' \, (\vec{r}) \; = \; \int_{E_c}^{\infty} \xi \, E\Sigma_{s,e\ell} \, (E) \, \phi(\vec{r},E)dE$$

For inelastic scattering, the approach is more complicated owing to the generation of γ-rays. Taking into consideration that (1) inelastic scattering heating is not large compared with γ-heating, and (2) the γ-rays due to inelastic scattering are of moderate energies and therefore are absorbed in relatively short distances [3], the heat can be assumed to be released at the point of the inelastic scattering event. Then $q_{i\ell}''' \, (\vec{r})$ is given by an expression similar to that of Eq. 3-46:

$$q_{i\ell}''' \, (\vec{r}) \; = \; \int_{E_c}^{\infty} Ef(E)\Sigma_{s,i\ell}(E)\phi(\vec{r},E)dE \qquad (3\text{-}59)$$

where $f(E) =$ the fraction of the neutron energy E lost in the collision. The parameter f is a function of E and the material composition.

Finally, the heat generation within the structure is:

$$q'''(\vec{r}) \; = \; q_\gamma''' \, (\vec{r}) + q_{e\ell}''' \, (\vec{r}) + q_{i\ell}''' \, (\vec{r}) \qquad (3\text{-}60)$$

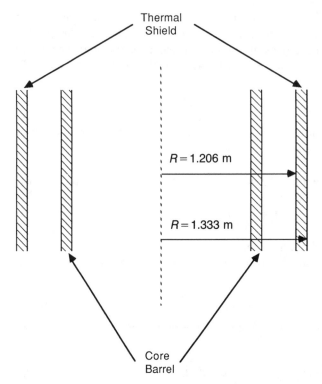

Figure 3-7 Thermal shield.

Example 3-4 Power deposition in a thermal shield

PROBLEM In a PWR the core is surrounded by a thermal shield (Fig. 3-7) to protect the pressure vessel from γ-ray heating and neutron-induced radiation damage. For an iron thermal shield with the radiation values given below, calculate the volumetric heat generation rate in the shield at its outermost position.

Assume that the core of the reactor is equivalent to an infinite plane source and that the shield can be treated as a slab owing to the small thickness-to-radius ratio.

Note that at energies well above the inelastic scattering threshold the total cross section is due to scattering. For steel, the inelastic scattering rate may be assumed to be equal to the elastic scattering rate:

$$\Sigma_{s,e\ell} \approx \Sigma_{s,i\ell} \tag{3-61}$$

The total neutron cross section can be approximated by the sum of both scatterings:

$$\Sigma_T \approx \Sigma_{s,e\ell} + \Sigma_{s,i\ell} \tag{3-62}$$

Use the following values for the radiation flux parameters.

* γ Radiation	* Neutron Radiation
$E_{\mathrm{o}} = 2$ MeV	$\phi_{\mathrm{fast}} = 10^{14}$ neutrons/cm^2 · s
$S = 10^{14}$ γ/cm^2 s	Effective neutron energy $= 0.6$ MeV
$B_{\mathrm{Fe}} = 4.212$	Neutron total cross section $= \sigma_{\mathrm{T}}$ $(0.1 < E < 15) = 3$ barns
	$f(2$ MeV$) = 0.1$

SOLUTION Equation 3-57 may be used to find $q_{\gamma}'''(x)$.

$$q_{\gamma}'''(x) = SB\mu_{\mathrm{a}}(E_{\mathrm{o}})E_{\mathrm{o}}e^{-\mu x}$$

where x is measured from the inner wall:

$$x = 1.333 \text{ m} - 1.206 \text{ m} = 0.127 \text{ m} = 12.7 \text{ cm}$$

From Table 3-4, μ_{a} for iron at $E_{\mathrm{o}} = 2.0$ MeV is 0.182 cm^{-1} and μ for iron at $E_{\mathrm{o}} = 2.0$ MeV is 0.333 cm^{-1}.

$$\therefore q_{\gamma}'''(x) = (10^{14})(4.212)(0.182)(2)e^{-0.333(12.7)} = 2.23 \times 10^{12} \text{ MeV/cm}^3 \cdot \text{s}$$

To solve for the elastic and inelastic scattering of the neutrons, Eqs. 3-46 and 3-59 must be solved:

$$q_{e\ell}'''(\vec{r}) = \int_{E_c}^{\infty} \xi \, E\Sigma_{\mathrm{s},e\ell}(E) \, \phi(r,E)dE \qquad (3\text{-}46)$$

$$q_{i\ell}'''(\vec{r}) = \int_{E_c}^{\infty} Ef(E)\Sigma_{\mathrm{s},i\ell}(E)\phi(\vec{r},E)dE \qquad (3\text{-}59)$$

Using a one-group approximation for the integral and defining ϕ_{fast} as the fast neutron flux, the above equations reduce to:

$$q_{e\ell}'''(r) = \overline{\xi}\,\overline{E}\,\Sigma_{\mathrm{s},e\ell}\phi_{\mathrm{fast}} \qquad (3\text{-}63)$$

$$q_{i\ell}'''(r) = f(\overline{E})\overline{E}\,\Sigma_{\mathrm{s},i\ell}\phi_{\mathrm{fast}} \qquad (3\text{-}64)$$

where $\overline{E}$ is the effective energy of the flux.

Now, from Eqs. 3-42 and 3-49:

$$\xi = \overline{\ln \frac{E}{E'}} = \frac{\overline{\Delta E}}{E} = \frac{1}{2}(1 - \alpha)$$

where $\alpha = \left(\dfrac{A - 1}{A + 1}\right)^2$ (Eq. 3-48).

For iron, $A = 55.85$ so that $\alpha = 0.9309$ and:

$$\xi = \frac{1}{2}\left(1 - \left(\frac{54.85}{56.85}\right)^2\right) = 0.0346$$

The total removal cross section can be obtained from:

$$\Sigma_T = \frac{\rho_{Fe} A_v}{M_{Fe}} \sigma_T = \frac{7.87 \text{ g/cm}^3 (0.6022 \times 10^{24} \text{ atom/mole})}{55.85 \text{ g/mole}} (3 \times 10^{-24} \text{ cm}^2)$$

$$= 0.254 \text{ cm}^{-1}$$

Therefore, with the assumption that $\Sigma_{s,el} = \Sigma_{s,i\ell}$, each would equal one-half of Σ_T:

$$\Sigma_{s,i\ell} = \Sigma_{s,e\ell} \simeq 0.5 \, \Sigma_T = 0.127 \text{ cm}^{-1}$$

Now the heat-generation rate due to neutron scattering from this monoenergetic neutron flux can be calculated:

$$
\begin{aligned}
q'''_{e\ell}(\vec{r}) &= \bar{\xi} \bar{E} \, \Sigma_{s,e\ell} \, \phi_{fast} \\
&= (0.0346)(0.6 \text{ MeV})(0.127 \text{ 1/cm})(10^{14} \text{ neutron/cm}^2 \cdot \text{s}) \quad (3\text{-}63) \\
&= 0.26 \times 10^{12} \text{ MeV/cm}^3 \cdot \text{s}
\end{aligned}
$$

$$
\begin{aligned}
q'''_{i\ell}(\vec{r}) &= f(\bar{E}) \bar{E} \, \Sigma_{s,i\ell} \, \phi_{fast} \\
&= (0.1)(0.6 \text{ MeV})(0.127 \text{ 1/cm})(10^{14} \text{ neutron/cm}^2 \cdot \text{s}) \quad (3\text{-}64) \\
&= 0.76 \times 10^{12} \text{ MeV/cm}^3 \cdot \text{s}
\end{aligned}
$$

$$
\begin{aligned}
\therefore q'''(\vec{r}) &= q'''_\gamma(\vec{r}) + q'''_{e\ell}(\vec{r}) + q'''_{i\ell}(\vec{r}) \\
&= 2.23 \times 10^{12} + 0.26 \times 10^{12} + 0.76 \times 10^{12} \\
&= 3.25 \times 10^{12} \text{ MeV/cm}^3 \cdot \text{s} = 0.52 \text{ W/cm}^3
\end{aligned}
$$

Note that the heat deposition due to γ-rays is the principal source of heat generation in the structure. In fact, because of the high buildup factor (B) for iron, the calculated heat generation due to γ-rays may exceed the incident flux of photons. A more refined transport calculation in which the photon energy and scattering properties are accounted for in detail is needed if exact prediction of the heat-generation level is desired.

VIII SHUTDOWN HEAT GENERATION

It is important to evaluate the heat generated in a reactor after shutdown for determining cooling requirements under normal conditions and accident consequences following abnormal events. Reactor shutdown heat generation is the sum of heat produced from the following: (1) fissions from delayed neutron or photoneutron emissions; and (2) decay of fission products, fertile materials, and other activation products from neutron capture. These two sources initially contribute equal amounts to the shutdown heat generation. However, within minutes from shutdown fissions from delayed neutron emission are reduced to a negligible amount.

A Fission Heat After Shutdown

The heat generated from fissions by delayed neutrons is obtained by solving the neutron kinetic equations after a large negative insertion of reactivity. Assuming a single group of delayed neutrons, the time-dependent neutron flux can be given by [5]:

$$\phi(t) = \phi_o \left[\frac{\beta}{\beta - \rho} e^{-\gamma_1 t} - \frac{\rho}{\beta - \rho} e^{-\frac{(\beta - \rho)t}{\ell}} \right] \qquad (3\text{-}65)$$

where ϕ_o = steady-state neutron flux prior to shutdown; β = total delayed neutron fraction; ρ = step reactivity change; ℓ = prompt neutron lifetime; γ_1 = decay constant for longest-lived delayed neutron precursor; t = time after initiation of the transient.

Substituting typical values for a ^{235}U-fueled, water-moderated reactor of γ_1 = 0.0124 s^{-1}, β = 0.006, and ℓ = 10^{-4} s into Eq. 3-65 for a reactivity insertion of ρ = -0.09, the fractional power, which is proportional to the flux, is given by:

$$\frac{\dot{Q}}{\dot{Q}_o} = 0.0625\, e^{-0.0124t} + 0.9375\, e^{-960t} \qquad (3\text{-}66)$$

where t is in seconds.

The second term in Eq. 3-66 becomes negligible in less than 0.01 second. Consequently, the reactor power decreases exponentially over a period of approximately 80 seconds, which is about the half-life of the longest-lived delayed neutron precursor.

B Heat from Fission Product Decay

The major source of shutdown heat generation is fission product decay. Simple, empirical formulas for the rate of energy release due to β and γ emissions from decaying fission products are given by [3]:

$$\begin{aligned} \beta \text{ energy release rate} &= 1.40\, t'^{-1.2} \text{ MeV/fission} \cdot \text{s} \\ \gamma \text{ energy release rate} &= 1.26\, t'^{-1.2} \text{ MeV/fission} \cdot \text{s} \end{aligned} \qquad (3\text{-}67)$$

where t' = time after the occurrence of fission in seconds.

The equations above are accurate within a factor of 2 for 10 s $< t' <$ 100 days. Integrating the above equations over the reactor operation time yields the rate of decay energy released from fission products after a reactor has shut down.

Assuming 200 MeV are released for each fission, 3.1×10^{10} fissions per second would be needed to produce 1 watt of operating power. Thus a fission rate of $3.1 \times 10^{10}\, q_o'''$ fissions/cm$^3 \cdot$ s is needed to produce q_o''' W/cm^3. The decay heat at a time τ seconds after reactor startup due to fissions occurring during the time interval between τ' and $\tau' + d\tau'$ is given by the following (see Figure 3-8 for time relations).

$$dP_\beta = 1.40(\tau - \tau')^{-1.2} (3.1 \times 10^{10}) q_o''' d\tau' \text{ MeV/cm}^3 \cdot \text{s} \qquad (3\text{-}68a)$$

$$dP_\gamma = 1.26(\tau - \tau')^{-1.2} (3.1 \times 10^{10}) q_o''' d\tau' \text{ MeV/cm}^3 \cdot \text{s} \qquad (3\text{-}68b)$$

For a reactor operating at a constant power level over the period τ_s, we integrate Eqs. 3-68a and 3-68b to get the decay heat from all fissions:

$$P_\beta = 2.18 \times 10^{11} q_o''' [(\tau - \tau_s)^{-0.2} - \tau^{-0.2}] \text{ MeV/cm}^3 \cdot \text{s} \qquad (3\text{-}69a)$$

$$P_\gamma = 1.95 \times 10^{11} q_o''' [(\tau - \tau_s)^{-0.2} - \tau^{-0.2}] \text{ MeV/cm}^3 \cdot \text{s} \qquad (3\text{-}69b)$$

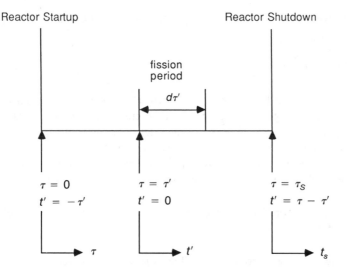

Figure 3-8 Time intervals. τ = time after reactor startup; $t' = \tau - \tau'$ = time after fission; $t_s = \tau - \tau_s$ = time after shutdown.

The decay heat level may be expressed as a fraction of the constant operating power level (P_o)—which is associated with the steady-state volumetric heat-generation rate (q_o''')—by multiplying P_γ and P_β in Eqs. 3-69a and 3-69b by 1.602×10^{-13} to convert the units to W/cm^3—and rearranging to obtain:

$$\frac{P_\beta}{P_o} = 0.035 \, [(\tau - \tau_s)^{-0.2} - \tau^{-0.2}] \tag{3-70a}$$

$$\frac{P_\gamma}{P_o} = 0.031 \, [(\tau - \tau_s)^{-0.2} - \tau^{-0.2}] \tag{3-70b}$$

The total fission power decay heat rate (P) is then given by:

$$\frac{P}{P_o} = 0.066[(\tau - \tau_s)^{-0.2} - \tau^{-0.2}] \tag{3-70c}$$

This equation may also be written as

$$\frac{P}{P_o} = 0.066[t_s^{-0.2} - (t_s + \tau_s)^{-0.2}]$$

Although all the energy from the β-particles is deposited in the fuel material, depending on the reactor configuration, only a fraction of the γ energy is deposited in the fuel material. The rest is deposited within the structural materials of the core and the surrounding supporting structures.

The decay heat rate predicted by Eq. 3-70c is plotted in Figure 3-9 as a function of time after reactor shutdown for various times of reactor operation. Note that for

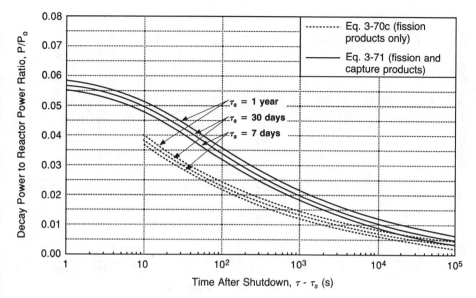

Figure 3-9 Decay heat rate from empirical relations as a function of shutdown time.

operating times (τ_s) of a few days or longer, the initial decay power is independent of reactor operating time. However, the reactor operating time is relatively important for determining the long-term decay heat.

For comparison, another equation that was later experimentally obtained from a 1-in. diameter uranium rod is also plotted in Figure 3-9. The resulting equation is given by [3]:

$$\frac{P}{P_o} = 0.1[(\tau - \tau_s + 10)^{-0.2} - (\tau + 10)^{-0.2} + 0.87(\tau + 2 \times 10^7)^{-0.2}$$
$$- 0.87(\tau - \tau_s + 2 \times 10^7)^{-0.2}]$$

(3-71)

This equation, as may be observed in Figure 3-9, predict higher decay powers due to, among other factors, the inclusion of the decay heat of the actinides ^{239}U and ^{289}Np along with decay of ^{235}U fission products. The effect of neutron capture in fission products is to increase the decay heat on the order of a few percent, depending on the level of burnup and the operating time.

C ANS Standard Decay Power

In 1961 the data from several experiments were combined to provide a more accurate method for predicting fission product decay heat power [9]. The results (Figure 3-10) were adopted in 1971 by the American Nuclear Society (ANS) as the basis for a draft standard (ANS-5.1/N18.6) for reactor shutdown cooling requirements. The curve, which spans a time range from 1 second to 10^9 seconds, refers to reactors initially

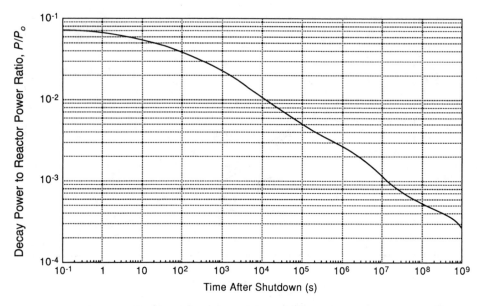

Figure 3-10 Fission-product decay heat power as a function of time after shutdown. *(From ANS-5.1/ N18.6.)*

fueled with uranium and operated at a constant power (P_o) for an infinite* period before being instantaneously shut down. The value of P/P_o for reactors operated for a finite period (τ_s) may be obtained from Figure 3-10 by subtracting the value of P/P_o at the time $\tau_s + t_s$ from the value of P/P_o at the time t_s, where t_s is the cooling time after shutdown.

Some important observations may be derived from Figure 3-10. Consider the values of P/P_o for reactor operating times of 20 days, 200 days, and infinity and for various cooling times. These values are given in Table 3-5. For all three reactor operating times, the fractional decay power at 1 hour is approximately 1.3%. For cooling times less than 1 day the ratio P/P_o is independent of τ_s for $\tau_s > 20$ days. This situation is due to the rapid decay of fission products of short half-lives, which reach their saturation values during short periods of operation. The decay heat at longer cooling times is due to fission products with longer half-lives. Because the amount of these products present at shutdown is dependent on the reactor operation time (τ_s) the decay heat after cooling times longer than 1 day is also dependent on the operating time. The uncertainty associated with the 1971 proposed ANS standard was given as

$$t_s < 10^3 \text{ seconds} \qquad +20\%, \ -40\%$$
$$10^3 < t_s < 10^7 \text{ seconds} \qquad +10\%, \ -20\%$$
$$t_s > 10^7 \text{ seconds} \qquad +25\%, \ -50\%$$

*An infinite period is considered to occur when all fission products have reached saturation levels.

Table 3-5 Decay heat power after shutdown

	Fraction of thermal operating power (P/P_o) at various times after shutdown (cooling period)		
Operating time	1 Hour	1 Day	100 Days
20 Days	0.013	2.5×10^{-3}	3×10^{-4}
200 Days	0.013	4.1×10^{-3}	5×10^{-4}
Infinite	0.013	5.1×10^{-3}	12×10^{-4}

Investigation of the inaccuracies in the 1971 ANS proposed standard due to the assumptions that (1) decay heats from different fission products are equal and (2) neutron capture effects are negligible led to a new standard that was developed in 1979 and reaffirmed in 1985. The revised ANS standard [8], which consists of equations based on summation calculations, explicitly accounts for decay heat from ^{235}U, ^{238}U, and ^{239}Pu fission products. Neutron capture in fission products is included through a correction factor multiplier. The new standard is also capable of accounting for changes in fissile nuclides with fuel life. Accuracy within the first 10^4 seconds after shutdown was emphasized in the new standard's development for accident consequence evaluation.

The old ANS standard is compared with the data on decay heat associated with one fission from ^{238}U, ^{235}U, and ^{239}Pu atoms in Figure 3-11. As shown, the ^{238}U data are higher for the first 20 seconds, whereas the ^{239}Pu and ^{235}U data are lower. For decay times less than 10^3 seconds, the new standard predicts decay powers lower than those of the old standard, whereas for decay times of more than 10^4 seconds the revised standard may give higher decay powers. This increase is due to the inclusion of neutron capture effects.

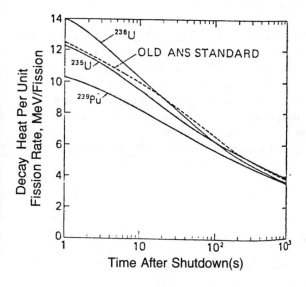

Figure 3-11 Comparison of the ANS 1971 revised standard with the ANS 1978 standard for irradiation time of 10^{13} seconds [8]. Decay power can be obtained by multiplying the above shown values by the fission rate during operation.

Table 3-6 Comparison of the 1971 and 1979 ANS decay heat standards
A. Reactor operating history

Interval	Duration (days)	Power	Fractional power during interval			
			^{235}U	^{239}Pu	^{238}U	^{241}Pu
1	300	Full	0.8	0.13	0.06	0.01
2	60	Zero				
3	300	Full	0.6	0.29	0.07	0.04
4	60	Zero				
5	300	Full	0.40	0.42	0.08	0.10

B. Decay heat power based on above history

Time after shutdown (s)	1979 Revised standard[a]				Decay power ratio (Revised/old)	
	Capture multiplier	Uncertainty 1σ (%)	$\dfrac{P}{P_o}$		Nominal[b]	Upper bound[c]
1	1.005	5.1	0.0577		0.925	0.860
10	1.005	3.5	0.0448		0.890	0.808
10^2	1.005	3.1	0.0295		0.909	0.802
10^3	1.006	3.1	0.0178		1.02	0.879

Source: Schrock [8].
[a]Based on $E_f = 200$ MeV/fission.
[b]1979 Revised standard nominal from column 4. Old standard nominal from standard curve.
[c]New standard: $(1 + 2\sigma) \times$ nominal; old standard: $1.2 \times$ nominal.

A more meaningful comparison can be made by choosing a specific reactor application. Such a comparison is presented in Table 3-6, where the new standard upper bound was chosen as $1 + 2\sigma$ times the nominal results, where σ is the standard deviation as prescribed by the ANS standard. This comparison shows that for typical end-of-life core composition, the lower-decay heat from ^{239}Pu plays an important role in reducing the decay heat power predicted by the 1979 standard and that the capture effect is small for a short time after shutdown. Calculated uncertainties in the 1979 standard are much lower than in the older standard, giving upper bound decay power results that are lower by 12 to 19%. A similar comparison made for "younger" fuel would show somewhat smaller differences because the change in the ^{239}Pu role is the major effect. Because of the sharp increase in the capture effect for $t_s > 10^4$ seconds, the revised standard may give higher decay power than did the old standard for long time after shutdown.

REFERENCES

1. Benedict, M., Pigford, T. H., and Levi, H. W. *Nuclear Chemical Engineering*. New York: McGraw-Hill, 1981.
2. El-Wakil, M. M. *Nuclear Heat Transport*. Scranton, PA: International Textbook Company, 1971.

3. Glasstone, S., and Sesonske, A. *Nuclear Reactor Engineering*. New York: Van Nostrand Reinhold, 1967 (2nd ed.), 1981 (3rd ed.).
4. Goldstein, H. *Fundamental Aspects of Reactor Shielding*. Reading, MA: Addison Wesley, 1959.
5. Lamarsh, J. *Nuclear Reactor Theory*. Reading, MA: Addison Wesley, 1966.
6. Leipunskii, O., Noroshicov, B. V., and Sakharov, V. N. *The Propagation of Gamma Quanta in Matter*. Oxford: Pergamon Press, 1965.
7. Rust, J. H. *Nuclear Power Plant Engineering*. Buchanan, GA: Haralson Publishing, 1979.
8. Schrock, V. E. A revised ANS standard for decay heat from fission products. *Nucl. Technol.* 46:323, 1979; and ANSI/ANS-5.1-1979: *Decay Heat Power in Light Water Reactors*. Hinsdale, IL: American Nuclear Society, 1979.
9. Shure, K. *Fission Product Decay Energy*. USAEC report WAPD-BT-24, 1961, Pittsburgh, PA, pp. 1–17, 1961.
10. Spiegel, U. S. *Mathematical Handbook,* Outline Series. New York: McGraw-Hill: 1968, p. 111.
11. Templin, L. T. (ed.). *Reactor Physics Constants* (2nd ed.). Argonne, IL: ANL-5800. 1963.

PROBLEMS

Problem 3-1 Thermal design parameters for a cylindrical fuel pin (section III)
Consider the PWR reactor of Example 3-1.

1. Evaluate the average thermal neutron flux if the enrichment of the fuel is 3.25%.
2. Evaluate the average power density of the fuel in MW/m^3. Assume the fuel density is 90% of theoretical density.
3. Calculate the average linear power of the fuel, assuming there are 207 fuel rods per assembly. Assume the fuel rod length to be 3810 mm.
4. Calculate the average heat flux at the cladding outer radius, when the cladding diameter is 10 mm.

Answers:

1. $\langle\phi\rangle = 3.75 \times 10^{13}$ neutrons/cm$^2 \cdot$ s
2. $\langle q'''\rangle = 289.6$ MW/m^3
3. $\langle q'\rangle = 19.24$ kW/m
4. $\langle q''\rangle = 612.7$ kW/m^2

Problem 3-2 Power profile in a homogeneous reactor (section IV)
Consider an ideal core with the following characteristics: The ^{235}U enrichment is uniform throughout the core, and the flux distribution is characteristic of an unreflected, uniformly fueled cylindrical reactor, with extrapolation distances δz and δR of 10 cm. How closely do these assumptions allow prediction of the following characteristics of a PWR?

1. Ratio of peak to average power density and heat flux?
2. Maximum heat flux?
3. Maximum linear heat generation rate of the fuel rod?
4. Peak-to-average enthalpy rise ratio, assuming equal coolant mass flow rates in every fuel assembly?
5. Temperature of water leaving the central fuel assembly?

Calculate the heat flux on the basis of the area formed by the cladding outside diameter and the active fuel length. Use as input only the following values.

Total power = 3411 MWt (Table 1-2)
Equivalent core diameter = 3.37 m (Table 2-3)
Active length = 3.66 m (Table 2-3)

Fraction of energy released in fuel = 0.974
Total number of rods = 50,352 (Tables 2-2, 2-3)
Rod outside diameter = 9.5 mm (Table 1-3)
Total flow rate = 17.4×10^3 kg/s (Table 1-2)
Inlet temperature = 286°C (Table 1-2)
Core average pressure = 15.5 MPa (Table 1-2)

Answers:

1. $\phi_o/\overline{\phi} = 3.11$
2. $q''_{max} = 1.88$ MW/m^2
3. $q'_{max} = 56.1$ kW/m
4. $(\Delta h)_{max}/\Delta\overline{h} = 2.08$
5. $(T_{out})_{max} = 344.9$°C

Problem 3-3 Power generation in a thermal shield (section VII)
Consider the heat-generation rate in the PWR core thermal shield discussed in Example 3-4.

1. Calculate the total power generation in the thermal shield if it is 4.0 m high.
2. How would this total power change if the thickness of the shield is increased from 12.7 cm to 15 cm?

Assume uniform axial power profile.

Answers:

1. $\dot{Q} = 23.85$ MW
2. $\dot{Q} = 24.37$ MW

Problem 3-4 Decay heat energy (section VIII)
Using Eq. 3-70c, evaluate the energy generated in a 3000 MWth LWR after the reactor shuts down. The reactor operated for 1 year at the equivalent of 75% of total power.

1. Consider the following time periods after shutdown:
   a. 1 Hour
   b. 1 Day
   c. 1 Month
2. How would your answers be different if you had used Eq. 3-71 (i.e., would higher or lower values be calculated)?

Answers:

1a. 0.113 TJ: 1TJ = 10^{12} J
1b. 1.24 TJ
1c. 13.1 TJ
2. Higher

Problem 3-5: Effect of continuous refueling on decay heat (section VIII)
Using Eq. 3-70c, estimate the decay heat rate in a 3000 MWth reactor in which 3.2% ^{235}U-enriched UO_2 assemblies are being fed into the core. The burned-up fuel stays in the core for 3 years before being replaced. Consider two cases:

1. The core is replaced in two batches every 18 months.
2. The fuel replacement is so frequent that refueling can be considered a continuous process. (*Note:* The PHWR reactors and some of the water-cooled graphite-moderated reactors in the Soviet Union are effectively continuously refueled.)

Compare the two situations at 1 minute, 1 hour, 1 day, 1 month, and 1 year.
Answers:

	Case 1	Case 2
1 minute:	$P = 81.9$ MW	$P = 81.0$ MW
1 hour:	$P = 33.2$ MW	$P = 32.2$ MW
1 day:	$P = 15.0$ MW	$P = 14.1$ MW
1 month:	$P = 4.97$ MW	$P = 4.26$ MW
1 year:	$P = 1.28$ MW	$P = 0.963$ MW

FOUR

TRANSPORT EQUATIONS FOR SINGLE-PHASE FLOW

I INTRODUCTION

Thermal analyses of power conversion systems involve the solution of transport equations of mass, momentum, and energy in forms that are appropriate for the system conditions. Engineering analysis often starts with suitably tailored transport equations. These equations are achieved by simplifying the general equations, depending on the necessary level of resolution of spatial distributions, the nature of the fluids involved (e.g., compressibility), and the numerical accuracy required for the analysis. The general forms of the transport equations for single-phase flow and many of the simplifying assumptions are presented in this chapter.

The basic assumption made here is that the medium can be considered a continuum. That is, the smallest volume of concern contains enough molecules to allow each point of the medium to be described on the basis of average properties of the molecules. Hence unique values for temperature, velocity, density, and pressure (collectively referred to as field variables) can be assumed to exist at each point in the medium of consideration. Differential equations of conservation of mass, momentum, and energy in a continuum can then be developed to describe the average molecular values of the field variables. The equations of a continuum do not apply when the mean free path of molecules is of a magnitude comparable to the dimension of the volume of interest. Under these conditions few molecules would exist in the system, and averaging loses its meaning when only a few molecules are involved. In that case statistical distribution of the molecular motion should be used for the description of the macroscopic behavior of the medium. Because a cube of gas whose side is 1 μm contains 2.5×10^7 molecules at normal temperature and pressure, the continuum condition is readily met in most practical systems. An example of a situation where the continuum equations should not be applied is that of rarefield gases, as in fusion vacuum equipment or for space vehicles flying at the edge of the atmosphere. (The mean free path of air at atmospheric standard conditions is 6×10^{-8} m, whereas at 100 miles' elevation it is 50 m.)

A Equation Forms

Two approaches are used to develop the transport equations: an integral approach and a differential approach. The integral approach can be further subdivided into two categories: the lumped parameter approach and the distributed parameter integral approach.

The *integral approach* addresses the behavior of a system of a specific region or mass. With the *lumped parameter integral approach,* the medium is assumed to occupy one or more compartments, and within each compartment the spatial distributions of the field variables and transport parameters of the material are ignored when setting up the integral equations. Conversely, with the *distributed parameter integral approach* the spatial dependence of the variables within the medium is taken into consideration when obtaining the equations.

The *differential approach* is naturally a distributed parameter approach but with balance equations for each point and not for an entire region. Integrating the differential equations over a volume yields integral distributed parameter equations for the volume. However, integral distributed parameter equations for a volume may also be formulated on the basis of conveniently assumed spatial behavior of the parameters in the volume. In that case the information about the point-by-point values of the parameter would not be as accurate as predicted by a differential approach.

The integral equations can be developed for two types of system: a control mass or a control volume (Table 4-1). With the *control mass approach,* the boundary of the system is the boundary of the mass, and no mass is allowed to cross this boundary (a closed system formulation). With the *control volume approach* mass is allowed to cross the boundaries of the system (an open system formulation), as illustrated in Figure 4-1. In general, the boundary surface surrounding a control mass or a control volume may be deformable, although in practice rigid boundaries are encountered for most engineering applications.

Table 4-1 Classification of transport equations

Integral	Differential
Lumped parameter	
Control mass	
Control volume	
Distributed parameter	
Control mass	Lagrangian equations
Control volume	Eulerian equations

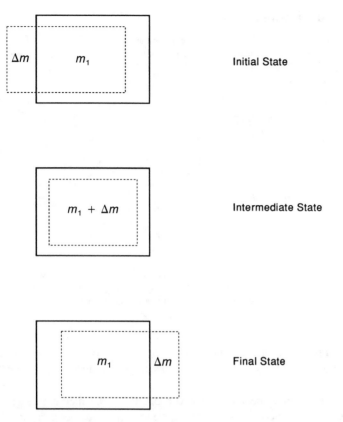

Figure 4-1 Control mass and control volume boundaries. *Solid line* indicates the control volume. *Dotted line* indicates the control mass. Note that at the intermediate stage the control mass boundary is identical to the control volume boundary.

The system of differential equations used to describe the control mass is called Lagrangian. With the *Lagrangian system* the coordinates move at the flow velocity (as if attached to a particular mass), and thus the spatial coordinates are not independent of time. Eulerian equations are used to describe transport equations as applied to a control volume. The *Eulerian system* of equations can be derived for a coordinate frame moving at any velocity. When the coordinate frame's origin is stationary in space at a particular position, the resulting equations are the most often used form of the Eulerian equations.

Applications of the integral forms of these equations are found in Chapters 6 and 7, and the differential forms are in Chapters 8, 9, 10, and 13.

B Intensive and Extensive Properties

System properties whose values are obtained by the summation of their values in the components of the system are called *extensive properties*. These properties depend on

Table 4-2 Extensive and intensive properties[a]

Property	Total value	Specific value per unit mass	Specific value per unit volume
Mass	m	1	ρ
Volume	V	$v \equiv \dfrac{V}{m} \equiv 1/\rho$	1
Momentum	$m\vec{v}$	$\vec{v}$	$\rho\vec{v}$
Kinetic energy	$\frac{1}{2} m\, v^2$	$\frac{1}{2} v^2$	$\frac{1}{2} \rho v^2$
Potential energy	mgz	gz	ρgz
Internal energy	$U = mu$	u	ρu
Stagnation internal energy	$U^\circ = mu^\circ$	$u^\circ \equiv u + \frac{1}{2} v^2$	ρu°
Enthalpy	$H = mh$	$h = u + pv$	ρh
Stagnation enthalpy	$H^\circ = mh^\circ$	$h^\circ = h + \frac{1}{2} v^2$	ρh°
Total energy	$E = me$	$e \equiv u^\circ + gz$	ρe
Entropy	$S = ms$	s	ρs
Temperature	T	T	T
Pressure	p	p	p
Velocity	$\vec{v}$	$\vec{v}$	$\vec{v}$

[a]Properties that are independent of the size of the system are called intensive properties. Intensive properties include $\vec{v}$, p, and T, as well as the specific values of the thermodynamic properties.

the extent of a system. Properties that are independent of the size of the system are called *intensive properties*. Table 4-2 lists some of the properties appearing in the transport equations.

For a control mass (m), let C be an extensive property of the medium (e.g., volume, momentum, or internal energy). Also, denote the specific value per unit mass of the extensive property as c. The lumped parameter approach is based on the assumption that the medium is of uniform properties, and thus $C = mc$. For a distributed parameter approach, the mass density (ρ) and c are not necessarily uniform, and $C = \iiint\limits_{V} \rho c\, dV$.

II MATHEMATICAL RELATIONS

It is useful to remind the reader of some of the basic mathematics encountered in this chapter.

A Time and Spatial Derivative

Consider a property c as a function of time and space. The time rate of change of the property c as seen by an observer at a fixed position in space is denoted by the partial time derivative $\partial c/\partial t$.

The time rate of change of a property c as seen by an observer moving at a velocity $\vec{v}_o$ within a medium at rest is given by the total time derivative dc/dt. In Cartesian coordinates the total derivative is related to the partial derivative by

$$\frac{dc}{dt} = \frac{\partial c}{\partial t} + \frac{\partial x}{\partial t}\frac{\partial c}{\partial x} + \frac{\partial y}{\partial t}\frac{\partial c}{\partial y} + \frac{\partial z}{\partial t}\frac{\partial c}{\partial z} = \frac{\partial c}{\partial t} + v_{ox}\frac{\partial c}{\partial x} + v_{oy}\frac{\partial c}{\partial y} + v_{oz}\frac{\partial c}{\partial z} \quad (4\text{-}1)$$

where v_{ox}, v_{oy}, and v_{oz} are the components of $\vec{v}_o$ in the x, y, and z directions. The general relation between the total and the partial derivatives for a property c is given by:

$$\frac{dc}{dt} = \frac{\partial c}{\partial t} + \vec{v}_o \cdot \nabla c \quad (4\text{-}2)$$

If the frame of reference for the coordinates is moving at the flow velocity $\vec{v}$, the rate of change of c as observed at the origin (the observer) is given by the substantial derivative Dc/Dt. A better name for this derivative is *material derivative*, as suggested by Whitaker [7]. It is given by:

$$\frac{Dc}{Dt} = \frac{\partial c}{\partial t} + \vec{v} \cdot \nabla c \quad (4\text{-}3)$$

The right hand side of Eq. 4-3 describes the rate of change of the variable in Eulerian coordinates, and the left hand side describes the time rate of change in Lagrangian coordinates.

In a steady-state flow system, the rate of change of c at a fixed position $\left(\dfrac{\partial c}{\partial t}\right)$ is zero. Hence at the fixed position an observer does not see any change in c with time. However, the fluid experiences a change in its property c as it moves because of the nonzero values of $\vec{v} \cdot \nabla c$. Thus in the Lagrangian system $\dfrac{Dc}{Dt}$ is nonzero, and the observer moving with the flow does see a change in c with time.

Note that:

$$\frac{Dc}{Dt} = \frac{dc}{dt} + (\vec{v} - \vec{v}_o) \cdot \nabla c \quad (4\text{-}4)$$

Thus the total derivative becomes identical to the substantial derivative when $\vec{v}_o = \vec{v}$.

For a vector, $\vec{c}$, the substantial derivative is obtained by applying the operator to each component so that when they are summed we get:

$$\frac{D\vec{c}}{Dt} = \frac{Dc_i}{Dt}\vec{i} + \frac{Dc_j}{Dt}\vec{j} + \frac{Dc_k}{Dt}\vec{k} \quad (4\text{-}5)$$

where $\vec{i}$, $\vec{j}$, and $\vec{k}$ = the unit vectors in the Cartesian directions x, y, and z.

Example 4-1: Various time derivatives

PROBLEM Air bubbles are being injected at a steady rate into a water channel at various positions such that the bubble population along the channel is given by:

$$N_b = N_{bo} \left[1 + \left(\frac{z}{L} \right)^2 \right]$$

where L = channel length along the axis z; N_{bo} = bubble density at the channel inlet ($z = 0$).

What is the observed rate of change of the bubble density by

1. A stationary observer at $z = 0$?
2. An observer moving in the channel with a constant speed v_0?

SOLUTION The time rate of change of the bubble density as observed by a stationary observer at $z = 0$ is:

$$\frac{\partial N_b}{\partial t} = 0$$

However, for the observer moving along the channel with a velocity v_0 m/s, the rate of change of the bubble density is:

$$\frac{dN_b}{dt} = \frac{\partial N_b}{\partial t} + v_0 \frac{\partial N_b}{\partial z} = v_0 N_{bo} \left(\frac{2z}{L^2} \right)$$

B Gauss's Divergence Theorem

If V is a closed region in space, totally bounded by the surface S, the volume integral of the divergence of a vector $\vec{c}$ is equal to the total flux of the vector at the surface S:

$$\iiint_V (\nabla \cdot \vec{c}) dV = \oiint_S \vec{c} \cdot \vec{n} \, dS \tag{4-6}$$

where $\vec{n}$ = unit vector directed normally outward from S.

This theorem has two close corollaries for scalars and tensors:

$$\iiint_V \nabla c \, dV = \oiint_S c \, \vec{n} \, dS \tag{4-7}$$

and

$$\iiint_V \nabla \cdot \overline{\overline{c}} \, dV = \oiint_S (\overline{\overline{c}} \cdot \vec{n}) dS \tag{4-8}$$

The last equation is also applicable to a dyadic product of two vectors $\vec{c}_1 \vec{c}_2$.

C Leibnitz's Rules

The general Leibnitz rule for differentiation of an integral of a function (f) is given by [3]:

$$\frac{d}{d\lambda} \int_{a(\lambda)}^{b(\lambda)} f(x,\lambda)dx = \int_{a(\lambda)}^{b(\lambda)} \frac{\partial f(x,\lambda)}{\partial \lambda} dx + f(b,\lambda) \frac{db}{d\lambda} - f(a,\lambda) \frac{da}{d\lambda} \qquad (4\text{-}9)$$

It should be noted that for this rule to be applicable the functions f, a, and b must be continuously differentiable with respect to λ. Also the function f and $\frac{\partial f}{\partial \lambda}$ should be continuous with respect to x between $a(\lambda)$ and $b(\lambda)$.

This rule can be used to yield a general relation between the time differential of a volume integral and the surface velocity of the volume as follows. Let x be the Cartesian coordinate within a plate of a surface area A and extending from $x = a(t)$ to $x = b(t)$, as seen in Figure 4-2. According to Eq. 4-9, the time derivative of an integral of the function $f(x,t)$ over x is given by:

$$\frac{d}{dt} \int_{a(t)}^{b(t)} f(x,t)A \, dx = \int_{a(t)}^{b(t)} \frac{\partial f(x,t)}{\partial t} A dx + f(b,t) A \frac{db}{dt} - f(a,t) A \frac{da}{dt} \qquad (4\text{-}10)$$

The extension of Eq. 4-10 to three-dimensional integration of a function $f(\vec{r},t)$ over a volume V surrounded by the surface S yields [5]:

$$\frac{d}{dt} \iiint_V f(\vec{r},t)dV = \iiint_V \frac{\partial f(\vec{r},t)}{\partial t} dV + \oiint_S f(\vec{r},t) \, \vec{v}_s \cdot \vec{n} \, dS \qquad (4\text{-}11)$$

where $\vec{v}_s$ = local and instantaneous velocity of the surface S.

The velocity $\vec{v}_s$ may be a function of the spatial coordinate (if the surface is deforming) and time (if the volume is accelerating or decelerating). For a deformable volume, whether the center of the volume is stationary or moving, $\vec{v}_s$ is not equal to zero.

Equation 4-11 is the *general transport theorem* [6]. A special case of interest is when the volume under consideration is a material volume V_m (i.e., a volume encom-

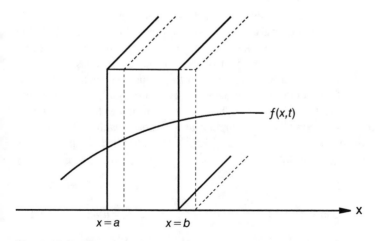

Figure 4-2 Plate moving in the x direction.

passing a certain mass) within the surface S_m. In that case the surface velocity represents the fluid velocity, i.e., $\vec{v}_s = \vec{v}$, and the left hand side represents the substantial derivative of a volume integral (refer to Eq. 4-4). In this case Eq. 4-11 becomes:

$$\frac{D}{Dt} \iiint_{V_m} f(\vec{r},t)dV = \iiint_{V_m} \frac{\partial f(\vec{r},t)}{\partial t} dV + \oiint_{S_m} f(\vec{r},t)\vec{v} \cdot \vec{n} \, dS \qquad (4\text{-}12)$$

This equation is the *Reynolds transport theorem*. Note that the integration in Eq. 4-12 is performed over a material volume V_m bounded by the surface S_m. The Reynolds transport theory is useful for transforming the derivatives of integrals from material-based coordinates (Lagrangian left hand side of Eq. 4-12) into spatial coordinates (Eulerian right hand side of Eq. 4-12).

Another special case of Eq. 4-11 is that of a fixed volume (i.e., a stationary and nondeformable volume) where $\vec{v}_s = 0$, in which case Eq. 4-11 reduces to:

$$\frac{d}{dt} \iiint_V f(\vec{r},t)dV = \iiint_V \frac{\partial f(\vec{r},t)}{\partial t} dV = \frac{\partial}{\partial t} \iiint_V f(\vec{r},t)dV \qquad (4\text{-}13)$$

The last equality is obtained because the volume in this case is time-independent.

The total rate of change of the integral of the function $f(\vec{r},t)$ can be related to the material derivative at a particular instant when the volume boundaries of V and V_m are the same, so that by subtracting Eq. 4-11 from Eq. 4-12:

$$\frac{D}{Dt} \iiint_V f(\vec{r},t) \, dV = \frac{d}{dt} \iiint_V f(\vec{r},t)dV + \oiint_S f(\vec{r},t)(\vec{v} - \vec{v}_s) \cdot \vec{n}dS \qquad (4\text{-}14)$$

where $\vec{v} - \vec{v}_s$ = relative velocity of the material with respect to the surface of the control volume:

$$\vec{v}_r = \vec{v} - \vec{v}_s \qquad (4\text{-}15)$$

The first term on the right hand side of Eq. 4-14 is the "time rate of change" term, and the second term is the "flux" term, both referring to the function $f(\vec{r},t)$ within the volume V. The time rate of change term may be nonzero for two distinct conditions: if the integrand is time-dependent or the volume is not constant. A deformable control surface is not a sufficient condition to produce a change in the total volume, as expansion of one region may be balanced by contraction in another. Note also that only the component of $\vec{v}_r$ normal to the surface S contributes to the flux term.

The general transport theorem can be applied to a vector as well as to a scalar. The extension to vectors is easily obtained if Eq. 4-11 is applied to three scalar functions (f_x, f_y, f_z), then multiplied by $\vec{i}$, $\vec{j}$, and $\vec{k}$, respectively, and summed to obtain:

$$\frac{d}{dt} \iiint_V \vec{f}(\vec{r},t)dV = \iiint_V \frac{\partial \vec{f}(\vec{r},t)}{\partial t} dV + \oiint_S \vec{f}(\vec{r},t)(\vec{v}_s \cdot \vec{n})dS \qquad (4\text{-}16)$$

The reader should consult the basic fluid mechanics textbooks for further discussion and detailed derivation of the transport theorems presented here [e.g., 2, 4, 6].

Example 4-2: Difference in total and substantial derivatives

PROBLEM Consider loss of coolant from a pressure vessel. The process can be described by observing the vessel as a control volume or the coolant as a control mass. Assume the coolant leaves the vessel at an opening of area A_1 with velocity v_1. Evaluate the rate of change of the mass remaining in the vessel.

SOLUTION The solid line in Figure 4-3 defines the control volume around the vessel, and the dashed line surrounds the coolant mass (i.e., material volume).

The observed rate of change of mass in the material volume is $\dfrac{Dm}{Dt} = 0$. We can

obtain $\dfrac{dm}{dt}$ from $\dfrac{Dm}{Dt}$ by considering Eq. 4-14 and the fact that the total mass is the

integral of the coolant density (ρ) over the volume of interest to get:

$$\frac{dm}{dt} = \frac{Dm}{Dt} - \iint \rho \vec{v}_r \cdot \vec{n} \, dS$$

However $\vec{v}_r = \vec{v}_1 - 0$ over the opening A_1 and zero elsewhere around the surface of the vessel. Therefore if the coolant density at the opening is ρ_1, we get:

$$\frac{dm}{dt} = 0 - \rho_1 \vec{v}_1 \cdot \vec{A}_1$$

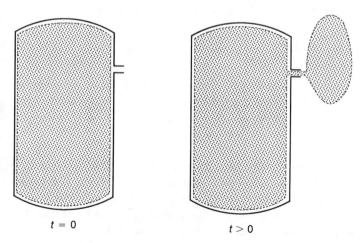

$t = 0$

$t > 0$

Figure 4-3 Boundaries of the vessel control volume (*solid line*) and the coolant control mass (*dashed line*) after a loss of coolant accident.

III LUMPED PARAMETER INTEGRAL APPROACH

A Control Mass Formulation

1 Mass. By definition, the control mass system is formed such that no mass crosses the boundaries. Hence the conservation of mass implies that the mass (m) in the system volume (V_m), is constant, or:

$$\frac{Dm}{Dt} = \frac{D}{Dt} \iiint\limits_{V_m} \rho dV = 0 \tag{4-17a}$$

Because we are treating mass and energy as distinct quantities, this statement ignores any relativistic effects such as those involved in nuclear reactions, where mass is transformed into energy. From Eq. 4-14, taking $f(\vec{r},t)$ equal to $\rho(\vec{r},t)$, we can recast Eq. 4-17a into

$$\frac{d}{dt} \iiint\limits_{V} \rho dV + \iint\limits_{S} \rho \vec{v}_r \cdot \vec{n} dS = 0 \tag{4-17b}$$

For a control mass, the material expands at the rate of change of the volume it occupies, so that $\vec{v}_r = 0$ and:

$$\frac{Dm}{Dt} = \frac{d}{dt} \iiint\limits_{V_m} \rho dV = \left(\frac{dm}{dt}\right)_{c.m.} = 0 \tag{4-17c}$$

where c.m. = control mass.

2 Momentum. The momentum balance (Newton's second law of motion) applied to the control mass equates the rate of change of momentum to the net externally applied force. Thus when all parts of the mass (m) have the same velocity ($\vec{v}$):

$$\frac{Dm\vec{v}}{Dt} = \left(\frac{dm\vec{v}}{dt}\right)_{c.m.} = \sum_k \vec{F}_k = \sum_k m\vec{f}_k \tag{4-18}$$

The forces $\vec{F}_k$ may be body forces arising from gravitational, electrical, or magnetic effects, or surface forces such as those exerted by pressure. The forces are positive if acting in the positive direction of the coordinates. In each case $\vec{f}_k$ is the force per unit mass.

3 Energy. The *energy equation*, the first law of thermodynamics, states that the rate of change of the stored energy in a control mass should equal the difference between the rate of energy addition (as heat or as work) to the control mass and the rate of energy extraction (as heat or as work) from the control mass. For systems that do not involve nuclear or chemical reactions, stored energy is considered to consist of the thermodynamic internal energy and the kinetic energy. Whereas stored energy is a state property of the control mass, heat and work are the forms of energy transfer to or from the mass. By convention, heat *added to* a system is considered positive, but work *provided by* the system to the surroundings (i.e., extracted from the system) is considered positive. Thus:

$$\frac{DU^\circ}{Dt} = \left(\frac{dU^\circ}{dt}\right)_{c.m.} = \left(\frac{dQ}{dt}\right)_{c.m.} - \left(\frac{dW}{dt}\right)_{c.m.} \tag{4-19a}$$

The change in the stored energy associated with internally distributed chemical and electromagnetic interactions are not treated in this book. Stored energy changes due to nuclear interactions are treated as part of the internal energy in Chapter 6 only. However, as the nuclear internal energy changes lead to volumetric heating of the fuel, it is more convenient to account for this energy source by an explicit heat source term on the right hand side of Eq. 4-19a:

$$\left(\frac{dU^\circ}{dt}\right)_{c.m.} = \left(\frac{dQ}{dt}\right)_{c.m.} + \left(\frac{dQ}{dt}\right)_{gen} - \left(\frac{dW}{dt}\right)_{c.m.} \tag{4-19b}$$

where, for the lumped parameter approach, the internal energy is assumed uniform in the control mass and is given by:

$$U^\circ = mu^\circ = m\left(u + \frac{1}{2}v^2\right)$$

We may subdivide the work term into shaft work, expansion work by surface forces (which can be normal forces such as those exerted by pressure or tangential forces, e.g., shear forces), and work associated with the body forces (e.g., gravity, electrical, or magnetic forces). Thus if the body forces consist of a single force $(\vec{f})$ per unit mass, the work term can be decomposed into:

$$\left(\frac{dW}{dt}\right)_{c.m.} = \left(\frac{dW}{dt}\right)_{shaft} + \left(\frac{dW}{dt}\right)_{normal} + \left(\frac{dW}{dt}\right)_{shear} + \vec{v} \cdot m\vec{f} \tag{4-20}$$

The energy effects arising from a body force may be included in Eqs. 4-19a and 4-19b as part of the stored energy term instead of the work term only when the force is associated with a time-independent field, e.g., gravity. To illustrate, consider $\vec{f}$ the force per unit mass due to spatial change in a force field ψ per unit mass such that:

$$\vec{f} = -\nabla\psi \tag{4-21}$$

Using Eq. 4-3 we get:

$$\vec{v} \cdot m\vec{f} = -m\vec{v} \cdot \nabla\psi = -m\frac{D\psi}{dt} + m\frac{\partial\psi}{\partial t} \tag{4-22a}$$

When the field is time-independent, Eq. 4-22a reduces to:

$$\vec{v} \cdot m\vec{f} = -m\frac{D\psi}{Dt} = -\frac{D[m\psi]}{Dt} \left(\text{because } \frac{Dm}{Dt} = 0\right) \tag{4-22b}$$

Thus for a control mass:

$$\vec{v} \cdot m\vec{f} = -m\left(\frac{d\psi}{dt}\right)_{c.m.} = -\left(\frac{d[m\psi]}{dt}\right)_{c.m.} \tag{4-22c}$$

For gravity:

$$\psi = gz \text{ and } \vec{f} = \vec{g} \tag{4-23}$$

If the only body force is gravity, Eqs. 4-22c and 4-23 can be used to write the energy Eq. 4-19b in the form:

$$\left(\frac{d}{dt}[mu^\circ]\right)_{c.m.} = \left(\frac{dQ}{dt}\right)_{c.m.} + \left(\frac{dQ}{dt}\right)_{gen} - \left(\frac{dW}{dt}\right)_{shaft}$$
$$- \left(\frac{dW}{dt}\right)_{normal} - \left(\frac{dW}{dt}\right)_{shear} - \left(\frac{d(mgz)}{dt}\right)_{c.m.}$$

which can be rearranged to give:

$$\left(\frac{d}{dt}[m(u^\circ + gz)]\right)_{c.m.} = \left(\frac{dQ}{dt}\right)_{c.m.} + \left(\frac{dQ}{dt}\right)_{gen} - \left(\frac{dW}{dt}\right)_{shaft}$$
$$- \left(\frac{dW}{dt}\right)_{normal} - \left(\frac{dW}{dt}\right)_{shear} \tag{4-24a}$$

or equivalently in terms of the total energy (E):

$$\left(\frac{d}{dt}[E]\right)_{c.m.} = \left(\frac{dQ}{dt}\right)_{c.m.} + \left(\frac{dQ}{dt}\right)_{gen} - \left(\frac{dW}{dt}\right)_{shaft}$$
$$- \left(\frac{dW}{dt}\right)_{normal} - \left(\frac{dW}{dt}\right)_{shear} \tag{4-24b}$$

These forms of the energy equation hold for reversible and irreversible processes. For a reversible process, when the only normal force on the surface is due to pressure (p), $\left(\frac{dW}{dt}\right)_{normal}$ can be replaced by $p\left(\frac{dV}{dt}\right)_{c.m.}$, the work involved in changing the volume of this mass. In this case Eq. 4-24b takes the form:

$$\left(\frac{d}{dt}[E]\right)_{c.m.} = \left(\frac{dQ}{dt}\right)_{c.m.} + \left(\frac{dQ}{dt}\right)_{gen} - \left(\frac{dW}{dt}\right)_{shaft} - p\left(\frac{dV}{dt}\right)_{c.m.} - \left(\frac{dW}{dt}\right)_{shear} \tag{4-24c}$$

4 Entropy. The *entropy equation*, an expression of the second law of thermodynamics, states that the net change of the entropy of a control mass interacting with its surroundings and the entropy change of the surroundings of the system should be equal to or greater than zero. Because there is no mass flow across the control mass boundary, the entropy exchanges with the surroundings are associated with heat interaction with the surroundings. The entropy equation then is:

$$\frac{DS}{Dt} = \left(\frac{dS}{dt}\right)_{c.m.} \geq \frac{\left(\dfrac{dQ}{dt}\right)_{c.m.}}{T_s} \tag{4-25a}$$

where T_s = temperature at the location where the energy (Q) is supplied as heat.

The equality holds if the control mass undergoes a reversible process, whereas the inequality holds for an irreversible process. For a reversible process, the control mass temperature (T) would be equal to heat supply temperature (T_s). For the inequality case, the excess entropy is that produced within the control mass by irreversible action. Thus we can write an entropy balance equation as:

$$\left(\frac{dS}{dt}\right)_{c.m.} = \dot{S}_{gen} + \frac{\left(\dfrac{dQ}{dt}\right)_{c.m.}}{T_s} \tag{4-25b}$$

where $\dot{S}_{gen}$ = rate of entropy generation due to irreversibilities.

For a reversible adiabatic process, Eq. 4-25b indicates that the entropy of the control mass is unchanged, i.e., that the process is isentropic:

$$\left(\frac{dS}{dt}\right)_{c.m.} = 0 \quad \text{reversible adiabatic process} \tag{4-26}$$

It is also useful to introduce the definition of the availability function (A). The availability function is a characteristic of the state of the system and the environment which acts as a reservoir at constant pressure (p_o) and constant temperature (T_o). The availability function of the control mass is defined as:

$$A \equiv E + p_o V - T_o S \tag{4-27}$$

The usefulness of this function derives from the fact that the change in the availability function from state to state yields the maximum useful work ($W_{u,max}$), obtainable from the specified change in state, i.e.,

$$\left(\frac{dW}{dt}\right)_{u,max} \equiv -\frac{dA}{dt} = -\frac{d}{dt}(E + p_o V - T_o S) \tag{4-28}$$

For a change from state 1 to 2

$$W_{u,max\ 1\to2} \equiv A_1 - A_2 = E_1 - E_2 + p_o(V_1 - V_2) - T_o(S_1 - S_2) \tag{4-29}$$

B Control Volume Formulation

1 Mass. The conservation of mass in the control volume approach requires that the net mass flow rate into the volume equals the rate of change of the mass in the volume.

As already illustrated in Example 4-2, Eq. 4-14 can be specified for the mass balance, i.e., $f(\vec{r},t) = \rho$, in a lumped parameter system to get:

$$\frac{Dm}{Dt} = \left(\frac{dm}{dt}\right)_{c.v.} - \sum_{i=1}^{I} \dot{m}_i \tag{4-30a}$$

Because $\dfrac{Dm}{Dt} = 0$:

$$\left(\frac{dm}{dt}\right)_{c.v.} = \sum_{i=1}^{I} \dot{m}_i \tag{4-30b}$$

where c.v. = control volume; I = number of the gates i through which flow is possible.

Note that $\dot{m}_i$ is defined positive for mass flow into the volume and is defined negative for mass flow out of the volume as follows:

$$\dot{m}_i = - \iint_{S_i} \rho \vec{v}_r \cdot \vec{n} dS \tag{4-31a}$$

For uniform properties at S_i:

$$\dot{m}_i = -\rho(\vec{v} - \vec{v}_s)_i \cdot \vec{S}_i \tag{4-31b}$$

2 Momentum. The momentum law, applied to the control volume, accounts for the rate of momentum change because of both the accumulation of momentum due to net influx and the external forces acting on the control volume. By specifying Eq. 4-14 to the linear momentum [$f(r,t) = \rho\vec{v}$] of a lumped parameter system and using Eq. 4-31, we get:

$$\left(\frac{dm\vec{v}}{dt}\right)_{c.m.} = \frac{Dm\vec{v}}{Dt} = \left(\frac{dm\vec{v}}{dt}\right)_{c.v.} - \sum_{i=1}^{I} \dot{m}_i\vec{v}_i \tag{4-32}$$

Equations 4-18 and 4-32 can be combined to give:

$$\left(\frac{dm\vec{v}}{dt}\right)_{c.v.} = \sum_{i=1}^{I} \dot{m}_i\vec{v}_i + \sum_{k} m \vec{f}_k \tag{4-33}$$

3 Energy. The energy equation (first law of thermodynamics) applied to a control volume takes into consideration the rate of change of energy in the control volume due to net influx and any sources or sinks within the volume.

By specifying Eq. 4-14 to the stagnation energy, i.e., $f(\vec{r},t) = \rho u^\circ$, and applying Eq. 4-31a, we get the total change in the stagnation energy in the volume due to the flow in and out of the volume:

$$\frac{DU^\circ}{Dt} = \left(\frac{dU^\circ}{dt}\right)_{c.v.} - \sum_{i=1}^{I} \dot{m}_i u_i^\circ \tag{4-34}$$

Hence the first law of thermodynamics applied to this system (the system here is the control volume) is:

$$\left(\frac{dU^\circ}{dt}\right)_{c.v.} = \sum_{i=1}^{I} \dot{m}_i \, u_i^\circ + \left(\frac{DQ}{Dt}\right) - \left(\frac{DW}{Dt}\right) \tag{4-35}$$

The heat addition term does not have a component that depends on the flow into the volume. This fact is emphasized here by dropping the subscript c.v. for this term:

$$\frac{DQ}{Dt} = \left(\frac{dQ}{dt}\right)_{c.v.} = \frac{dQ}{dt} \tag{4-36}$$

The work term in Eq. 4-35 should account for the work done by the mass in the control volume as well as the work associated with mass flow into and out of the volume. Thus:

$$\frac{DW}{Dt} = \left(\frac{dW}{dt}\right)_{c.v.} - \sum_{i=1}^{I} \dot{m}_i (pv)_i \tag{4-37}$$

The shaft, normal, and shear components of the work must be evaluated independently (Fig. 4-4). For a time-independent force field, as in the case of gravity, the work associated with the body force can be added to the stored internal energy. Now, similar to Eq. 4-24b, we can write an equation for the total energy change in the volume as:

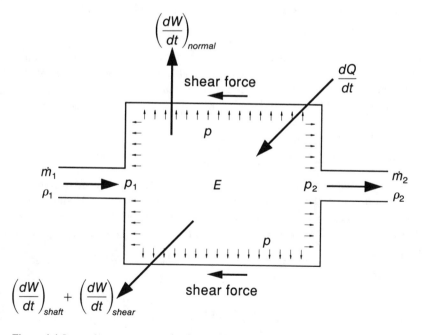

Figure 4-4 Lumped parameter control volume with a moving boundary.

$$\left(\frac{dE}{dt}\right)_{c.v.} = \sum_{i=1}^{I} \dot{m}_i \left(u_i^\circ + \frac{p_i}{\rho_i} + gz_i\right) + \frac{dQ}{dt} + \left(\frac{dQ}{dt}\right)_{gen}$$
$$- \left(\frac{dW}{dt}\right)_{shaft}^{c.v.} - \left(\frac{dW}{dt}\right)_{normal} - \left(\frac{dW}{dt}\right)_{shear} \tag{4-38}$$

When the only surface normal force is that exerted by pressure and for a reversible process, $\left(\frac{dW}{dt}\right)_{normal}^{c.v.}$ can be replaced by $p\left(\frac{dV}{dt}\right)_{c.v.}$.

For a stationary control volume, and given that $u_i^\circ + \frac{p_i}{\rho_i} = h_i^\circ$, we get

$$\left(\frac{\partial E}{\partial t}\right)_{c.v.} = \sum_{i=1}^{I} \dot{m}_i(h_i^\circ + gz_i) + \frac{dQ}{dt} + \left(\frac{dQ}{dt}\right)_{gen} - \left(\frac{dW}{dt}\right)_{shaft}$$
$$- \left(\frac{\partial W}{\partial t}\right)_{normal}^{c.v.} - \left(\frac{dW}{dt}\right)_{shear} \tag{4-39}$$

Example 4-3 Determining the pumping power

PROBLEM Obtain an expression for the power requirements of a PWR pump in terms of the flow rate.

SOLUTION The PWR pump is modeled here as an adiabatic, steady-state pump with no internal heat sources. The control volume is taken as the pump internal volume and is thus fixed, i.e., stationary and nondeformable. Pump inlet and outlet are at the same elevation, and the difference in kinetic energy between inlet and outlet streams is negligible.

With these assumptions, Eq. 4-39 becomes:

$$0 = \dot{m}(h_{in} - h_{out}) - \dot{W}_{shaft} - \dot{W}_{shear}$$

In order to evaluate the pumping work (i.e., W_{shaft} and W_{shear}), we need to compute the $h_{in} - h_{out}$ term. It is done by determining the thermodynamic behavior of the fluid undergoing the process. Pumping is assumed to be isothermal and the fluid to be incompressible. Because $h(T,p) = u(T,p) + p/\rho(T,p)$, $h(T,p) = u(T) + p/\rho(T)$ when $\frac{\partial \rho}{\partial p} = 0$. Thus $h_{in} - h_{out} = u_{in}(T) - u_{out}(T) - (p/\rho)_{in} - (p/\rho)_{out} = (p_{out} - p_{in})/\rho$. Hence:

$$\text{Pump power} = -\dot{W}_{shaft} - \dot{W}_{shear} = \dot{m}(h_{out} - h_{in}) = \frac{\dot{m}}{\rho}(p_{out} - p_{in})$$

4 Entropy. The entropy equation (second law of thermodynamics) applied to a control volume is obtained from examination of the condition already introduced for a control mass (Eq. 4-25b):

$$\left(\frac{dS}{dt}\right)_{c.m.} = \dot{S}_{gen} + \frac{\left(\frac{dQ}{dt}\right)_{c.m.}}{T_s}$$

Now by Eq. 4-14, when $f(\vec{r},t)$ is taken as ρs, we get:

$$\left(\frac{dS}{dt}\right)_{c.m.} = \left(\frac{dS}{dt}\right)_{c.v.} - \sum_{i=1}^{I} \dot{m}_i s_i \tag{4-40a}$$

For a stationary control volume:

$$\left(\frac{dS}{dt}\right)_{c.m.} = \left(\frac{\partial S}{\partial t}\right)_{c.v.} - \sum_{i=1}^{I} \dot{m}_i s_i \tag{4-40b}$$

Hence applying Eqs. 4-36 and 4-40a, Eq. 4-25b can be recast in the form

$$\left(\frac{\partial S}{\partial t}\right)_{c.v.} = \sum_{i=1}^{I} \dot{m}_i s_i + \dot{S}_{gen} + \frac{\frac{dQ}{dt}}{T_s} \tag{4-41}$$

It is also desirable to define the maximum useful work and the lost work (or irreversibility) for a control volume with respect to the environment acting as a reservoir at pressure p_o and temperature T_o. (Useful work may be defined with respect to any reservoir. Most often the reservoir is chosen to be the environment.) The actual work is given by:

$$\left(\frac{dW}{dt}\right)_{actual} \equiv \left(\frac{dW}{dt}\right)_{shaft} + \left(\frac{dW}{dt}\right)_{normal} \tag{4-42}$$

Note that a certain amount of the work is required to displace the environment. Therefore the useful part of the actual work is given by:

$$\left(\frac{dW}{dt}\right)_{u,actual} = \left(\frac{dW}{dt}\right)_{actual} - p_o \frac{dV}{dt} \tag{4-43}$$

Realizing that for a stationary volume $\frac{dV}{dt} = \frac{\partial V}{\partial t}$, the useful actual work can be obtained by substituting from Eqs. 4-39 and 4-42 into 4-43 as:

$$\left(\frac{dW}{dt}\right)_{u,actual} = \sum_{i=1}^{I} \dot{m}_i(h_i^\circ + gz_i) + \frac{dQ}{dt} + \left(\frac{dQ}{dt}\right)_{gen} \\ - \left(\frac{\partial(E + p_o V)}{\partial t}\right)_{c.v.} - \left(\frac{dW}{dt}\right)_{shear} \tag{4-44}$$

Now multiplying Eq. 4-41 by T_o and subtracting from Eq. 4-44, we obtain, after rearrangement:

$$\left(\frac{dW}{dt}\right)_{u,actual} = -\left[\frac{\partial(E + p_oV - T_oS)}{\partial t}\right]_{c.v.} + \sum_{i=1}^{I} \dot{m}_i(h° - T_os + gz)_i$$

$$+ \left(1 - \frac{T_o}{T_s}\right)\frac{dQ}{dt} + \left(\frac{dQ}{dt}\right)_{gen} - \left(\frac{dW}{dt}\right)_{shear} - T_o\dot{S}_{gen} \qquad (4\text{-}45)$$

The maximum useful work $\left(\dfrac{dW}{dt}\right)_{u,max}$ is given by Eq. 4-45 when $\dot{S}_{gen}$ is zero:

$$\left(\frac{dW}{dt}\right)_{u,max} = -\left[\frac{\partial(E + p_oV - T_oS)}{\partial t}\right]_{c.v.} + \sum_{i=1}^{I} \dot{m}_i (h° - T_os + gz)_i$$

$$+ \left(1 - \frac{T_o}{T_s}\right)\frac{dQ}{dt} + \left(\frac{dQ}{dt}\right)_{gen} - \left(\frac{dW}{dt}\right)_{shear} \qquad (4\text{-}46)$$

The irreversibility (lost work) is then:

$$i \equiv \left(\frac{dW}{dt}\right)_{u,max} - \left(\frac{dW}{dt}\right)_{u,actual} \equiv T_o\dot{S}_{gen} \qquad (4\text{-}47)$$

Evaluating $\left(\dfrac{dW}{dt}\right)_{u,max}$ from Eq. 4-46 and $\left(\dfrac{dW}{dt}\right)_{u,actual}$ from Eq. 4-44, the irreversibility becomes:

$$i = T_o\left(\frac{\partial S}{\partial t}\right)_{c.v.} - T_o\sum_{i=1}^{I} \dot{m}_i \, s_i - \frac{T_o}{T_s}\frac{dQ}{dt} \qquad (4\text{-}48)$$

where $\partial S/\partial t$ involves only the temporal variations in the entropy (zero at steady state).

Example 4-4 Setting up the energy equation for a PWR pressurizer

PROBLEM Consider the pressurizer of a PWR plant that consists of a large steel container partially filled with water (liquid); the rest of the volume is filled with steam (vapor) (Fig. 4-5). The pressurizer is connected at the bottom to the hot leg of the plant primary system and draws spray water from the cold leg for injection at the top. A heater is installed in the lower part to vaporize water, if needed, for pressure regulation.

Consider the energy equations for a control volume and answer the following questions:

1. If the pressurizer is the control volume, can we apply Eq. 4-39?
   Answer: Yes, because the pressurizer is a stationary volume.
2. How many "gates" does the pressurizer have for flow into it (or out of it).
   Answer: Two: the nozzle for the spray and the connection to the primary.
3. What are the contributing factors to the heat input term $\dfrac{dQ}{dt}$?

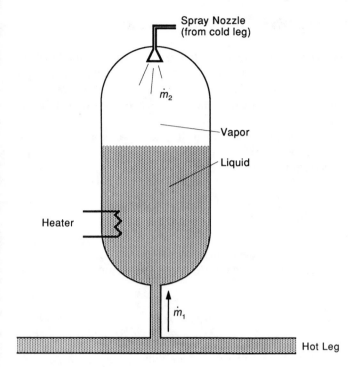

Figure 4-5 PWR pressurizer.

Answer: Any heat provided by the electric heater; also the heat extracted from or conducted into the vessel walls.

4. Does the value of $\left(\dfrac{\partial E^\circ}{\partial t}\right)$ equal the value of $\dfrac{\partial U^\circ}{\partial t}$ in this case?

Answer: Not necessarily; because $E^\circ = U^\circ + mgz$, both m and z (the effective height of the center of mass) may be changing.

5. Should the work terms $\left(\dfrac{dW}{dt}\right)_{\text{shaft}}$, $p\left(\dfrac{\partial V}{\partial t}\right)_{\text{c.v.}}$, and $\left(\dfrac{dW}{dt}\right)_{\text{shear}}$ be included in this pressurizer analysis?

Answer:

$\left(\dfrac{dW}{dt}\right)_{\text{shaft}} = 0$, as no motor exists.

$p\left(\dfrac{\partial V}{\partial t}\right)_{\text{c.v.}} = 0$; the volume considered is nondeformable and stationary.

$\left(\dfrac{dW}{dt}\right)_{\text{shear}} < 0$, because some work has to be consumed to cause motion. This term is small for slow-moving media.

IV DISTRIBUTED PARAMETER INTEGRAL APPROACH

With the distributed parameter integral approach the properties of the fluid are not assumed uniform withn the control volume. As a general law, the rate of change of an extensive property (C) is manifested in changes in the local distribution of the mass, the property, or both and is governed by the net effects of the external factors contributing to changes in C.

Recall from Eq. 4-12 that the rate of change of the extensive property C can be written as the sum of the volume integral of the local rate of change of the property and the net efflux from the surface:

$$\frac{DC}{Dt} = \iiint_{V_m} \frac{\partial}{\partial t} (\rho c)\, dV + \oiint_{S_m} (\rho c)\vec{v} \cdot \vec{n} dS \qquad (4\text{-}49)$$

where c = specific value of the property per unit mass; ρ = mass density; $\vec{v}$ = mass velocity at the boundary S_m, which surrounds the material volume V_m.

The rate of change of the quantity C due to external volumetric effects as well as surface effects can be expressed as:

$$\frac{DC}{Dt} = \iiint_{V_m} \rho\phi\, dV + \oiint_{S_m} \vec{J} \cdot \vec{n}\, dS \qquad (4\text{-}50)$$

where ϕ = rate of introduction of c per unit mass within the volume V_m (a "body" factor); $\vec{J} \cdot \vec{n}$ = rate of loss of C per unit area of S_m due to surface effects (a "surface" factor).

Combining Eqs. 4-49 and 4-50, we get the general form for the integral transport equation as:

$$\iiint_{V_m} \frac{\partial}{\partial t} (\rho c)dV + \oiint_{S_m} (\rho c)\vec{v} \cdot \vec{n}\, dS = \iiint_{V_m} \rho\phi\, dV + \oiint_{S_m} \vec{J} \cdot \vec{n} dS \qquad (4\text{-}51)$$

It is now assumed that this general integral balance is applicable to any arbitrary volume. Therefore Eq. 4-51 can be written as:

$$\iiint_{V} \frac{\partial(\rho c)}{\partial t}\, dV + \oiint_{S} (\rho c)\vec{v} \cdot \vec{n} dS = \iiint_{V} \rho\phi dV + \oiint_{S} \vec{J} \cdot \vec{n} dS \qquad (4\text{-}52)$$

Substituting for the first term on the left hand side of Eq. 4-52 from Eq. 4-11 we obtain:

$$\frac{d}{dt} \iiint_{V} (\rho c)\, dV + \oiint_{S} (\rho c)(\vec{v} - \vec{v}_s) \cdot \vec{n} dS = \iiint_{V} \rho\phi V + \oiint_{S} \vec{J} \cdot \vec{n} dS \qquad (4\text{-}53)$$

When the body force is due only to gravity, the general Eqs. 4-52 and 4-53 can be specified for mass, momentum, and energy by specifying c, $\vec{J}$, and ϕ as follows:

For the mass equation:

$$c = 1 \qquad \vec{J} = 0 \qquad \phi = 0 \qquad (4\text{-}54)$$

For the momentum equation:

$$c = \vec{v} \qquad \overline{\overline{J}} = \overline{\overline{\tau}} - p\overline{\overline{I}} \qquad \phi = \vec{g} \qquad (4\text{-}55)$$

For the energy equation:

$$c = u^\circ = u + \frac{v^2}{2} \qquad \vec{J} = -\vec{q}'' + (\overline{\overline{\tau}} - p\overline{\overline{I}}) \cdot \vec{v} \qquad \phi = \frac{q'''}{\rho} + \vec{g} \cdot \vec{v} \quad (4\text{-}56)$$

where $\overline{\overline{\tau}}$ = a stress tensor; $\overline{\overline{I}}$ = a unity tensor; $\vec{g}$ = gravitational acceleration; $\vec{q}''$ = surface heat flux; q''' = volumetric heat-generation rate.

Substituting from Eq. 4-54 into Eq. 4-53, we get the integral mass balance equation:

$$\frac{d}{dt}\iiint\limits_V \rho dV + \oiint\limits_S \rho(\vec{v} - \vec{v}_s) \cdot \vec{n} dS = 0 \qquad (4\text{-}57)$$

Note that:

$$\iiint\limits_V \rho dV = (m)_{\text{c.v.}} \qquad (4\text{-}58)$$

and from Eq. 4-31a:

$$\iint\limits_{S_i} \rho(\vec{v} - \vec{v}_s) \cdot \vec{n} dS_i = -\dot{m}_i \qquad (4\text{-}59)$$

where i identifies a localized opening at the boundary such that $(\vec{v} - \vec{v}_s)_i \neq 0$. Thus Eq. 4-57 can be cast in a rearranged form of Eq. 4-30b:

$$\left(\frac{d}{dt} m\right)_{\text{c.v.}} - \sum_i \dot{m}_i = 0$$

For momentum balance, we substitute from Eq. 4-55 into Eq. 4-53 to get:

$$\frac{d}{dt}\iiint\limits_V \rho\vec{v} dV + \oiint\limits_S \rho\vec{v}(\vec{v} - \vec{v}_s) \cdot \vec{n} dS = \oiint\limits_S (\overline{\overline{\tau}} - p\overline{\overline{I}}) \cdot \vec{n} dS + \iiint\limits_V \rho\vec{g} dV \tag{4-60}$$

If $\vec{v}$ is uniform within V, we can recast Eq. 4-60 in a form similar to Eq. 4-33 as:

$$\left(\frac{d}{dt} m\vec{v}\right)_{\text{c.v.}} - \sum_i \dot{m}_i \vec{v}_i = \sum_j m\vec{f}_j + m\vec{g} = \sum_k m\vec{f}_k = \sum_k \vec{F}_k \quad (4\text{-}61)$$

where $\vec{f}_j$ is the force at the surface portion j and is obtained from:

$$\vec{f}_j = \frac{1}{m}\iint\limits_{S_j} (\overline{\overline{\tau}} - p\overline{\overline{I}}) \cdot \vec{n} dS \qquad (4\text{-}62)$$

For energy balance, the parameters c, ϕ, and $\vec{J}$ from Eq. 4-56 are substituted in Eq. 4-53 to get:

$$\frac{d}{dt} \iiint_V \rho u^\circ dV + \oiint_S \rho u^\circ (\vec{v} - \vec{v}_s) \cdot \vec{n} dS =$$

$$\oiint_S [-\vec{q}'' + (\overline{\overline{\tau}} - p\overline{\overline{I}}) \cdot \vec{v}] \cdot \vec{n} dS + \iiint_V (q''' + \rho \vec{g} \cdot \vec{v}) dV \qquad (4\text{-}63)$$

If u° is uniform within V, then Eq. 4-63 can be recast as:

$$\left(\frac{d}{dt} m u^\circ\right)_{c.v.} - \sum_i \dot{m}_i u^\circ_i = \frac{dQ}{dt} - \frac{dW}{dt} \qquad (4\text{-}64)$$

where

$$\frac{dQ}{dt} = \iiint_V q''' dV - \oiint_S \vec{q}'' \cdot \vec{n} dS \qquad (4\text{-}65)$$

and

$$\frac{dW}{dt} = + \oiint_S [(p\overline{\overline{I}} - \overline{\overline{\tau}}) \cdot \vec{v}] \cdot \vec{n} dS - \iiint_V \rho \vec{g} \cdot \vec{v} dV \qquad (4\text{-}66)$$

If the pressure-related energy term is moved to the left hand side of Eq. 4-63, we get:

$$\frac{d}{dt} \iiint_V \rho u^\circ dV + \oiint_S \rho \left(u^\circ + \frac{p}{\rho}\right)(\vec{v} - \vec{v}_s) \cdot \vec{n} dS + \oiint_S p\vec{v}_s \cdot \vec{n} dS =$$

$$\frac{dQ}{dt} + \oiint_S \overline{\overline{\tau}} \cdot \vec{v} \cdot \vec{n} dS + \iiint_V \rho \vec{g} \cdot \vec{v} dV \qquad (4\text{-}67)$$

The various forms of the integral transport equations are summarized in Table 4-3, 4-4, and 4-5.

V DIFFERENTIAL CONSERVATION EQUATIONS

It is possible to derive the differential transport equations using the integral transport equation (Eq. 4-51) as follows. Apply Gauss's theorem to the material volume (V_m) so that the last term on the right hand side of Eq. 4-51 is transformed to a volume integral:

$$\oiint_{S_m} \vec{J} \cdot \vec{n} dS = \iiint_{V_m} \nabla \cdot \vec{J} dV \qquad (4\text{-}68)$$

as well as the last term on the left hand side of Eq. 4-51:

$$\oiint_{S_m} \rho c\vec{v} \cdot \vec{n} dS = \iiint_{V_m} \nabla \cdot (\rho c\vec{v}) dV \qquad (4\text{-}69)$$

Substituting from Eqs. 4-68 and 4-69 into Eq. 4-51, we get:

$$\iiint_{V_m} \left[\frac{\partial(\rho c)}{\partial t} + \nabla \cdot (\rho c\vec{v})\right] dV = \iiint_{V_m} [\nabla \cdot \vec{J} + \rho \phi] dV \qquad (4\text{-}70)$$

Table 4-3 Various forms of integral transport equations of mass

Deformable control volume

$$\frac{d}{dt} \iiint_V \rho dV + \oiint_S \rho \vec{v}_r \cdot \vec{n}\, dS = 0$$

$$\iiint_V \frac{\partial \rho}{\partial t} dV + \oiint_S \rho \vec{v}_s \cdot \vec{n}\, dS + \oiint_S \rho \vec{v}_r \cdot \vec{n}\, dS = 0$$

Nondeformable control volume ($\vec{v}_s = 0, \vec{v}_r = \vec{v}$)

$$\frac{d}{dt} \iiint_V \rho dV + \oiint_S \rho \vec{v} \cdot \vec{n}\, dS = 0$$

or

$$\iiint_V \frac{\partial \rho}{\partial t} dV + \oiint_S \rho \vec{v} \cdot \vec{n}\, dS = 0$$

Steady flow

$$\oiint_S \rho \vec{v} \cdot \vec{n}\, dS = 0$$

Steady uniform flow (single inlet and outlet systems)

$$\rho_1 \vec{v}_1 \cdot \vec{A}_1 = \rho_2 \vec{v}_2 \cdot \vec{A}_2 = \dot{m}$$

$\vec{v}_r$ = velocity of fluid relative to control volume surface = $\vec{v} - \vec{v}_s$
$\vec{v}_s$ = velocity of control surface boundary.
$\vec{v}$ = velocity of fluid in a fixed coordinate system.

For Eq. 4-70 to be true for any arbitrary volume (V_m) and if ρ, c, $\vec{v}$, and ϕ are continuous functions of time and space, the integrands of both sides of Eq. 4-70 must be equal. Hence for a continuum, the local instantaneous transport equation can be written as:

$$\frac{\partial}{\partial t}[\rho c] + \nabla \cdot [\rho c \vec{v}] = \nabla \cdot \vec{J} + \rho \phi \qquad (4\text{-}71)$$

where c, $\vec{J}$, and ϕ can be specified from Eqs. 4-54, 4-55, and 4-56.

The differential formulations can also be derived by considering an infinitesimal elementary control volume. The shape of this volume should be selected according to the system of coordinates. In the derivation here, only Cartesian coordinates are considered. In Cartesian coordinates the control volume is an elementary cube (Fig. 4-6). Also, the control volume is small enough that the variables can be considered uniform over the control surfaces that bound it.

A Conservation of Mass

Denote v_x, v_y, and v_z the three components of the velocity vector $\vec{v}$. The mass equation can be written as:

Table 4-4 Various forms of integral transport equations of momentum

Deformable control volume

$$\frac{d}{dt} \iiint_V \rho \vec{v}\, dV + \oiint_S \rho \vec{v}(\vec{v}_r \cdot n)\, dS = \Sigma \vec{F}$$

or

$$\iiint_V \frac{\partial}{\partial t}(\rho\vec{v})\, dV + \oiint_S \rho\vec{v}(\vec{v}_s \cdot \vec{n})\, dS + \oiint_S \rho\vec{v}(\vec{v}_r \cdot \vec{n})\, dS = \Sigma\vec{F}$$

Nondeformable control volume ($\vec{v}_s = 0,\ \vec{v}_r = \vec{v}$)

$$\frac{d}{dt} \iiint_V \rho\vec{v}\, dV + \oiint_S \rho\vec{v}(\vec{v} \cdot \vec{n})\, dS = \Sigma\vec{F}$$

or

$$\iiint_V \frac{\partial}{\partial t}(\rho\vec{v})\, dV + \oiint_S \rho\vec{v}(\vec{v} \cdot \vec{n})\, dS = \Sigma\vec{F}$$

Steady flow

$$\oiint_S \rho\vec{v}(\vec{v} \cdot \vec{n})\, dS = \Sigma\vec{F}$$

Steady uniform flow (single inlet and outlet systems)

$$\rho_2(\vec{v}_2 \cdot \vec{A}_2)v_2 - \rho_1(\vec{v}_1 \cdot \vec{A}_1)\vec{v}_1 = \dot{m}(\vec{v}_2 - \vec{v}_1) = \Sigma\vec{F}$$

$\vec{v}_r$, $\vec{v}_s$, and $\vec{v}$ are defined in the footnotes to Table 4-3.

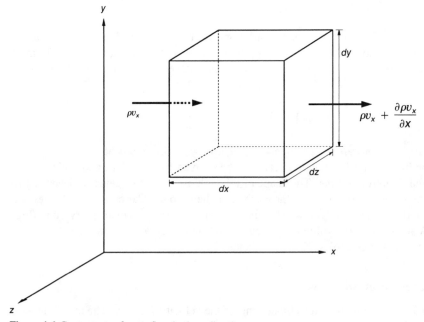

Figure 4-6 Components of mass flow in the *x* direction.

Table 4-5 Various forms of integral transport equations of energy

Deformable control volume

$$\frac{dQ}{dt} - \frac{dW^+}{dt} = \frac{d}{dt}\iiint_V \rho u^\circ dV + \oiint_S \rho\left[u^\circ + \frac{p}{\rho}\right]\vec{v}_r \cdot \vec{n}\, dS + \oiint_S p\, \vec{v}_s \cdot \vec{n}\, dS$$

$$\frac{dQ}{dt} - \frac{dW^+}{dt} = \iiint_V \frac{\partial}{\partial t}\rho u^\circ\, dV + \oiint_S \rho\left[u^\circ + \frac{p}{\rho}\right]\vec{v}_r \cdot \vec{n}\, dS + \oiint_S \rho\left[u^\circ + \frac{p}{\rho}\right]\vec{v}_s \cdot \vec{n}\, dS$$

Nondeformable control volume $(\vec{v}_s = 0,\ \vec{v}_r = \vec{v})$

$$\frac{dQ}{dt} - \frac{dW^+}{dt} = \frac{d}{dt}\iiint_V \rho u^\circ\, dV + \oiint_S \rho\left[u^\circ + \frac{p}{\rho}\right]\vec{v} \cdot \vec{n}\, dS$$

$$\frac{dQ}{dt} - \frac{dW^+}{dt} = \iiint_V \frac{\partial}{\partial t}\rho u^\circ\, dV + \oiint_S \rho\left[u^\circ + \frac{p}{\rho}\right]\vec{v} \cdot \vec{n}\, dS$$

Steady flow

$$\frac{dQ}{dt} - \frac{dW^+}{dt} = \oiint_S \rho\left[u^\circ + \frac{p}{\rho}\right]\vec{v} \cdot \vec{n}\, dS$$

Steady uniform flow (single inlet and outlet systems)[a]

$$\frac{dQ}{dt} - \left(\frac{dW}{dt}\right)_{shaft} - \left(\frac{dW}{dt}\right)_{shear} = \left[\left(\frac{v_2^2}{2} + \frac{p_2}{\rho_2} + gz_2 + u_2\right) - \left(\frac{v_1^2}{2} + \frac{p_1}{\rho_1} + gz_1 + u_1\right)\right]\dot{m}$$

$\vec{v}_r, \vec{v}_s, \vec{v}$ are defined in the footnotes to Table 4-3.

$$\frac{dW^+}{dt} = \left(\frac{dW}{dt}\right)_{shaft} + \left(\frac{dW}{dt}\right)_{shear} + \left(\frac{dW}{dt}\right)_{body\ force}.$$

[a]Gravity is the only body force.

[Rate of change of mass in control volume] =
   [mass flow rate into control volume] − [mass flow rate out of control volume]

$$\frac{\partial \rho}{\partial t}(dxdydz) = \rho v_x(dydz) - \left[\rho v_x + \frac{\partial}{\partial x}(\rho v_x)\, dx\right](dydz) +$$

$$\rho v_y(dxdz) - \left[\rho v_y + \frac{\partial}{\partial y}(\rho v_y)dy\right](dxdz) +$$

$$\rho v_z(dxdy) - \left[\rho v_z + \frac{\partial}{\partial z}(\rho v_z)dz\right](dxdy)$$

or, after simplification:

$$\frac{\partial \rho}{\partial t} = -\frac{\partial}{\partial x}(\rho v_x) - \frac{\partial}{\partial y}(\rho v_y) - \frac{\partial}{\partial z}(\rho v_z) \qquad (4\text{-}72)$$

which when rearranged and written in vector-algebra notation becomes:

$$\frac{\partial \rho}{\partial t} + \nabla \cdot (\rho\vec{v}) = 0 \qquad (4\text{-}73)$$

This expression is the Eulerian form of the mass conservation equation.

Expanding the second term of Eq. 4-73 as:

$$\nabla \cdot (\rho \vec{v}) = \rho(\nabla \cdot \vec{v}) + \vec{v} \cdot \nabla \rho$$

and substituting for $\nabla \cdot (\rho \vec{v})$ in Eq. 4-73, we get:

$$\frac{D\rho}{Dt} + \rho(\nabla \cdot \vec{v}) = 0 \tag{4-74}$$

where

$$\frac{D\rho}{Dt} = \frac{\partial \rho}{\partial t} + \vec{v} \cdot \nabla \rho$$

Equation 4-74 is the Lagrangian form of the mass conservation equation.

When the density does not change appreciably in the domain of interest, and when small density changes do not affect appreciably the behavior of the system, we can assume that the density is constant.

For such an incompressible fluid the continuity equation (4-74) simplifies to:

$$\nabla \cdot \vec{v} = 0; \quad \rho = \text{constant} \tag{4-75}$$

B Conservation of Momentum

The momentum equation expresses mathematically the fact that the rate of change of momentum in the control volume equals the momentum flow rate into the control volume minus the momentum flow rate out of the control volume plus the net external force on the control volume. Both body forces and surface forces are to be included (Figs. 4-7 and 4-8).

The forces that must be accounted for include, in addition to gravitational, electrical, or magnetic forces, three surface forces on each face: one normal and two tangential. The normal forces are caused by the pressure and internal frictional effects that act to elongate the fluid element. The shear (tangential) forces are due to internal friction, which attempts to rotate the fluid element. Thus for the momentum in the x direction:

$$
\begin{aligned}
\frac{\partial \rho v_x}{\partial t} (dxdydz) = {} & \rho v_x v_x \, dydz - \left(\rho v_x v_x + \frac{\partial \rho v_x v_x}{\partial x} dx \right) dydz \\
& + \rho v_y v_x \, dxdz - \left(\rho v_y v_x + \frac{\partial \rho v_y v_x}{\partial y} dy \right) dxdz \\
& + \rho v_z v_x \, dydx - \left(\rho v_z v_x + \frac{\partial \rho v_z v_x}{\partial z} dz \right) dydx \\
& + \left(\sigma_x + \frac{\partial \sigma_x}{\partial x} dx \right) dydz - \sigma_x \, dydz \\
& + \left(\tau_{yx} + \frac{\partial \tau_{yx}}{\partial y} dy \right) dxdz - \tau_{yx} \, dxdz
\end{aligned}
$$

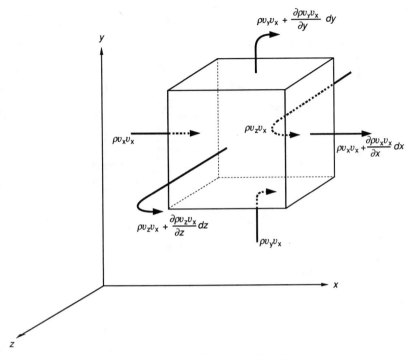

Figure 4-7 Components of the momentum efflux in the x direction.

$$+ \left(\tau_{zx} + \frac{\partial \tau_{zx}}{\partial z} \, dz \right) dydx \; - \; \tau_{zx} \, dydx$$

$$+ \; \rho f_x \, dxdydz$$

which upon simplification and rearrangement leads to:

$$\frac{\partial}{\partial t} (\rho v_x) + \frac{\partial}{\partial x} (\rho v_x v_x) + \frac{\partial}{\partial y} (\rho v_x v_y) + \frac{\partial}{\partial z} (\rho v_x v_z) = \frac{\partial \sigma_x}{\partial x} + \frac{\partial \tau_{yx}}{\partial y} + \frac{\partial \tau_{zx}}{\partial z} + \rho f_x$$

$$(4\text{-}76)$$

The left hand side of Eq. 4-76 represents the rate of momentum change for any fluid packet as it flows with the stream in the x direction. The three-dimensional change in the momentum can be written as:

$$\text{Rate of momentum change} = \frac{\partial}{\partial t} \rho \vec{v} + \nabla \cdot \rho \vec{v} \vec{v} \qquad (4\text{-}77)$$

where $\vec{v}\vec{v}$ is a dyadic product of all the velocity components and is given in Cartesian coordinates by:

$$\vec{v}\vec{v} = \begin{pmatrix} v_x v_x & v_x v_y & v_x v_z \\ v_y v_x & v_y v_y & v_y v_z \\ v_z v_x & v_z v_y & v_z v_z \end{pmatrix} \qquad (4\text{-}78)$$

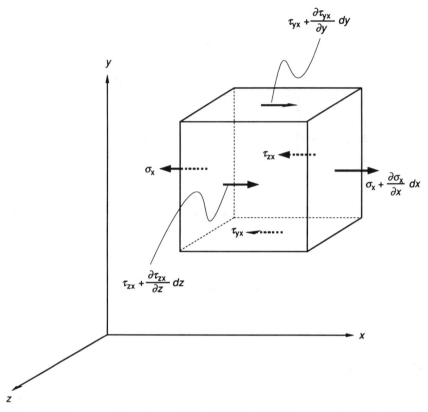

Figure 4-8 Components of stress tensor in the x direction. (The shear stress is positive when pointing in a positive direction on a positive surface or in a negative direction on a negative surface.)

The normal stress σ_x can be expanded into a pressure component and an internal friction component so that:

$$\begin{aligned}
\sigma_x &= -p + \tau_{xx} \\
\sigma_y &= -p + \tau_{yy} \\
\sigma_z &= -p + \tau_{zz}
\end{aligned} \tag{4-79}$$

Substituting Eq. 4-79 into Eq. 4-76 and utilizing Eq. 4-77, it is possible to write the three-dimensional equation of the momentum in vector form as:

$$\frac{\partial}{\partial t} \rho \vec{v} + \nabla \cdot \rho \vec{v}\vec{v} = -\nabla p + \nabla \cdot \overline{\overline{\tau}} + \rho \vec{f} = -\nabla \cdot (p \overline{\overline{I}} - \overline{\overline{\tau}}) + \rho \vec{f} \tag{4-80}$$

where the tensor $\overline{\overline{\tau}}$ is given by

$$\overline{\overline{\tau}} = \begin{pmatrix} \tau_{xx} & \tau_{xy} & \tau_{xz} \\ \tau_{yx} & \tau_{yy} & \tau_{yz} \\ \tau_{zx} & \tau_{zy} & \tau_{zz} \end{pmatrix} \tag{4-81}$$

By taking moments about the axis at the position (x,y,z), it is found that for equilibrium conditions:

$$\tau_{xy} = \tau_{yx}; \quad \tau_{xz} = \tau_{zx}; \quad \tau_{yz} = \tau_{zy}$$

That is, the tensor $\overline{\overline{\tau}}$ is symmetric.

By expanding the left hand side of Eq. 4-80 and utilizing the continuity Eq. 4-73, we get:

$$\frac{\partial}{\partial t} \rho\vec{v} + \nabla \cdot \rho\vec{v}\vec{v} = \rho \frac{\partial}{\partial t}\vec{v} + \vec{v}\frac{\partial \rho}{\partial t} + \rho\vec{v} \cdot \nabla\vec{v} + \vec{v}\nabla \cdot \rho\vec{v} \tag{4-82}$$

$$= \rho \frac{\partial \vec{v}}{\partial t} + \rho\vec{v} \cdot \nabla\vec{v} = \rho \frac{D\vec{v}}{Dt}$$

Equation 4-80 can therefore be written as:

$$\rho \frac{D\vec{v}}{Dt} = -\nabla p + \nabla \cdot \overline{\overline{\tau}} + \rho\vec{f} \tag{4-83}$$

This expression is the Lagrangian form of the momentum equation. It is easily seen that it is a restatement of Newton's law of motion, whereby the left hand side is the mass times the acceleration, and the right hand side is the sum of the forces acting on that mass. To solve the momentum equation and obtain the velocity field $\vec{v}(\vec{r},t)$, the density ρ and forces ∇p and $\vec{f}$ must be specified, and the internal stress force $\nabla \cdot \overline{\overline{\tau}}$ must be stated in terms of the velocity field, i.e., the velocity or velocity gradients and fluid properties. In other words, a constitutive relation for $\overline{\overline{\tau}}$ is needed for mathematic closure. In practice, the mass Eq. 4-74 and a state relation of pressure and density are solved together with the momentum equation, thus providing the means to specify $\rho(\vec{r},t)$ and $\nabla p(\vec{r},t)$. This subject is discussed in Chapter 9.

Because all fluids are isotropic with respect to stress–strain behavior, the stresses in the three directions must be related to material properties that are independent of coordinates. Thus if the fluids are assumed to follow the Newtonian laws of viscosity, the friction terms are given by [1]:

$$\tau_{ii} = 2\mu \left(\frac{\partial v_i}{\partial x_i}\right) - \left(\frac{2}{3}\mu - \mu'\right)(\nabla \cdot \vec{v}) \tag{4-84}$$

$$\tau_{ij} = \tau_{ji} = \mu \left(\frac{\partial v_i}{\partial x_j} + \frac{\partial v_j}{\partial x_i}\right) \tag{4-85}$$

where μ = ordinary "dynamic viscosity"; μ' = "bulk viscosity."

For dense gases and fluids, μ' is negligible. It is identically zero for low density monoatomic gases. It is only important for acoustic or shock-wave problems in gases where it produces viscous decay of the high-frequency waves. Thus in general we can take $\mu' = 0$ so that Eq. 4-84 becomes:

$$\tau_{ii} = 2\mu \frac{\partial v_i}{\partial x_i} - \frac{2}{3}\mu\nabla \cdot \vec{v} \tag{4-86}$$

When Eqs. 4-79, 4-85, and 4-86 are substituted into Eq. 4-76, we get:

$$\frac{\partial}{\partial t}(\rho v_x) + \frac{\partial}{\partial x}(\rho v_x v_x) + \frac{\partial}{\partial y}(\rho v_x v_y) + \frac{\partial}{\partial z}(\rho v_x v_z) =$$

$$- \frac{\partial p}{\partial x} + \frac{\partial}{\partial x}\left[2\mu \frac{\partial v_x}{\partial x} - \frac{2}{3}\mu\,(\nabla \cdot \vec{v})\right] \qquad (4\text{-}87)$$

$$+ \frac{\partial}{\partial y}\left[\mu\left(\frac{\partial v_x}{\partial y} + \frac{\partial v_y}{\partial x}\right)\right] + \frac{\partial}{\partial z}\left[\mu\left(\frac{\partial v_x}{\partial z} + \frac{\partial v_z}{\partial x}\right)\right] + \rho f_x$$

This expression is the *Navier-Stokes equation* for the momentum balance in the x direction. The momentum equation in the y and z directions can be obtained by permutation of x, y, and z in Eq. 4-87. The resultant equations can be written in vector form as:

$$\frac{\partial}{\partial t}(\rho\vec{v}) + \nabla \cdot \rho\vec{v}\vec{v} = -\nabla p - \nabla \mathrm{x}[\mu\nabla \mathrm{x}\vec{v}] + \nabla\left[\frac{4}{3}\mu\nabla \cdot \vec{v}\right] + \rho\vec{f} \qquad (4\text{-}88)$$

The momentum equations, the continuity equation, the equations specifying the pressure-density and viscosity-density relation, and the initial and boundary conditions are sufficient to define the velocity, density, and pressure distributions in the medium.

For an incompressible fluid ($\nabla \cdot \vec{v} = 0$) with a constant viscosity, the Navier-Stokes equation in the x direction simplifies to:

$$\rho\frac{\partial v_x}{\partial t} + \rho\nabla \cdot v_x\vec{v} = -\frac{\partial p}{\partial x} + \mu\nabla^2 v_x + \rho f_x; \quad \rho = \text{constant}$$
$$\mu = \text{constant} \qquad (4\text{-}89)$$

where

$$\nabla^2 v_x = \frac{\partial^2 v_x}{\partial x^2} + \frac{\partial^2 v_x}{\partial y^2} + \frac{\partial^2 v_x}{\partial z^2}$$

Note that:

$$\nabla \cdot \rho\vec{v}\vec{v} = \rho\vec{v} \cdot \nabla\vec{v} + \vec{v}(\nabla \cdot \rho\vec{v})$$

so that for a constant density fluid:

$$\nabla \cdot \rho\vec{v}\vec{v} = \rho\vec{v} \cdot \nabla\vec{v}$$

However,

$$\rho\nabla \cdot v_x\vec{v} = \rho\vec{v} \cdot \nabla v_x + \rho v_x(\nabla \cdot \vec{v})$$

so that for a constant-density fluid:

$$\rho\nabla \cdot v_x\vec{v} = \rho\vec{v} \cdot \nabla v_x$$

In vector form, the momentum balance equation for a fluid with constant viscosity and density is then given by simplification of Eq. 4-88:

$$\rho\frac{\partial\vec{v}}{\partial t} + \rho\vec{v} \cdot \nabla\vec{v} = -\nabla p + \mu\nabla^2\vec{v} + \rho\vec{f}; \quad \rho = \text{constant}$$
$$\mu = \text{constant} \qquad (4\text{-}90a)$$

or

$$\rho \frac{D\vec{v}}{Dt} = -\nabla p + \mu \nabla^2 \vec{v} + \rho \vec{f}; \; \rho = \text{constant} \tag{4-90b}$$
$$\mu = \text{constant}$$

For flow that can be considered inviscid (i.e., of negligible viscosity effects), the second term on the right hand side in Eqs. 4-90a and 4-90b disappears. Hence Eq. 4-90b becomes:

$$\rho \frac{D\vec{v}}{Dt} = -\nabla p + \rho \vec{f}; \quad \text{inviscid flow} \tag{4-91}$$

Equation 4-91 is the equation derived by Euler in 1755 for inviscid flow.

Note that if the only body force present is the gravity,

$$\vec{f} = \vec{g} \tag{4-92}$$

Example 4-5 Shear stress distribution in one-dimensional, Newtonian fluid flow

PROBLEM Consider uniaxial flow of an incompressible Newtonian fluid in a duct of rectangular cross section (Fig. 4-9). The flow is in the axial direction and is given by:

$$\vec{v} = v_x \vec{i} = v_{max} \left[1 - \left(\frac{y}{L_y} \right)^2 \right] \left[1 - \left(\frac{z}{L_z} \right)^2 \right] \vec{i}$$

Evaluate the maximum shear stress in the fluid.

SOLUTION From Eq. 4-85 the shear stress components are given by

$$\tau_{xy} = \mu \left(\frac{\partial v_x}{\partial y} + 0 \right) = -\mu \, v_{max} \left[1 - \left(\frac{z}{L_z} \right)^2 \right] \left[\frac{2y}{L_y^2} \right]$$

$$\tau_{xz} = \mu \left(\frac{\partial v_x}{\partial z} + 0 \right) = -\mu \, v_{max} \left[1 - \left(\frac{y}{L_y} \right)^2 \right] \left[\frac{2z}{L_z^2} \right]$$

$$\tau_{yz} = 0$$

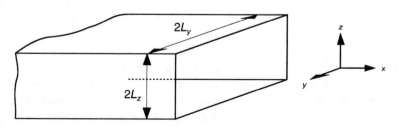

Figure 4-9 Flow channel for example 4-5.

From Eq. 4-84:

$$\tau_{xx} = 0 \quad \text{because} \quad \frac{\partial v_x}{\partial x} = 0 \quad \text{and} \quad \nabla \cdot \vec{v} = 0$$

∴ the maximum values of the shear components are:

$$(\tau_{xy})_{max} = \pm \frac{2\mu\, v_{max}}{L_y} \quad \text{at } y = \mp L_y \quad \text{and} \quad z = 0$$

$$(\tau_{xz})_{max} = \pm \frac{2\mu\, v_{max}}{L_z} \quad \text{at } y = 0 \quad \text{and} \quad z = \mp L_z$$

C Conservation of Energy

1 Stagnation internal energy equation. The energy equation expresses the fact that the rate of change of total internal energy in an infinitesimal volume must be equal to the rate at which internal energy is brought into the volume by the mass inflow, minus that removed by mass outflow, plus the heat transported diffusively or generated, minus the work performed by the medium in the volume and the work needed to put the flow through the volume. Again the stagnation energy (u°) refers to the internal energy and the kinetic energy (the potential energy is included in the work term).

Now for the volume $dxdydz$:

$$\left(\frac{\partial}{\partial t}\, \rho u^{\circ}\right) dxdydz = -\left(\frac{\partial \rho v_x u^{\circ}}{\partial x}\, dx\right) dydz - \left(\frac{\partial \rho v_y u^{\circ}}{\partial y}\, dy\right) dxdz$$

$$-\left(\frac{\partial \rho v_z u^{\circ}}{\partial z}\, dz\right) dxdy - \left(\frac{\partial q_x''}{\partial x}\, dx\right) dydz - \left(\frac{\partial q_y''}{\partial y}\, dy\right) dxdz - \left(\frac{\partial q_z''}{\partial z}\, dz\right) dxdy$$

$$+\, (q''')\, dxdydz + \left(\frac{\partial}{\partial x}\, (\sigma_x v_x + \tau_{xy} v_y + \tau_{xz} v_z) dx\right) dydz$$

$$+\left(\frac{\partial}{\partial y}\, (\sigma_y v_y + \tau_{yx} v_x + \tau_{yz} v_z) dy\right) dxdz + \left(\frac{\partial}{\partial z}\, (\sigma_z v_z + \tau_{zx} v_x + \tau_{zy} v_y) dz\right) dxdy$$

$$+\, (v_x \rho f_x + v_y \rho f_y + v_z \rho f_z)\, dxdydz \tag{4-93}$$

Dividing both sides of Eq. 4-93 by $dxdydz$ and using the vector notation, we get:

$$\frac{\partial}{\partial t}\, \rho u^{\circ} = -\nabla \cdot \rho u^{\circ} \vec{v} - \nabla \cdot \vec{q}'' + q''' - \nabla \cdot p\vec{v} + \nabla \cdot (\overline{\overline{\tau}} \cdot \vec{v}) + \vec{v} \cdot \rho \vec{f} \tag{4-94}$$

The first term on the left hand side is the local rate of change of the stagnation internal energy. The first term on the right represents the net change in the internal energy per unit time due to convection; the second term is the net heat transport rate by conduction and radiation (if present); the third term is the internal heat-generation rate. The fourth, fifth, and sixth terms are the work done on the fluid by the pressure, viscous forces, and body forces, respectively per unit time.

Grouping the first term on the right side and that on the left side of Eq. 4-94 and expanding both terms we get:

$$\frac{\partial}{\partial t} \rho u^\circ + \nabla \cdot \rho u^\circ \vec{v} = \rho \frac{\partial u^\circ}{\partial t} + u^\circ \frac{\partial \rho}{\partial t} + \rho \vec{v} \cdot \nabla u^\circ + u^\circ \nabla \cdot \rho \vec{v}$$

$$= \rho \frac{\partial u^\circ}{\partial t} + \rho \vec{v} \cdot \nabla u^\circ = \rho \frac{Du^\circ}{Dt} \tag{4-95}$$

Now, Eq. 4-94 may be rewritten as:

$$\rho \frac{Du^\circ}{Dt} = - \nabla \cdot \vec{q}'' + q''' - \nabla \cdot p\vec{v} + \nabla \cdot (\overline{\overline{\tau}} \cdot \vec{v}) + \vec{v} \cdot \rho \vec{f} \tag{4-96}$$

Example 4-6 Energy equation for one-dimensional newtonian fluid flow

PROBLEM For the velocity profile of Example 4-5, determine the rate of the volumetric work production in the coolant and compare it to the power density in the PWR core at steady state.

SOLUTION Because:

$$\vec{v} = v_x \vec{i} = v_{max} \left[1 - \left(\frac{y}{L_y} \right)^2 \right] \left[1 - \left(\frac{z}{L_z} \right)^2 \right] \vec{i}$$

the values of the three work terms in Eq. 4-96 can be determined as follows:

$$*\nabla \cdot p\vec{v} = \frac{\partial}{\partial x} p v_x = v_x \frac{\partial p}{\partial x}$$

$$*\nabla \cdot (\overline{\overline{\tau}} \cdot \vec{v}) = \nabla \cdot \begin{pmatrix} 0 & \tau_{xy} & \tau_{xz} \\ \tau_{xy} & 0 & 0 \\ \tau_{xz} & 0 & 0 \end{pmatrix} \cdot \begin{pmatrix} v_x \\ 0 \\ 0 \end{pmatrix} = \frac{\partial}{\partial x}(0) + \frac{\partial}{\partial y}(\tau_{xy} v_x) + \frac{\partial}{\partial z}(\tau_{xz} v_x)$$

$$*\vec{v} \cdot \rho \vec{f} = v_x \rho g_x$$

At steady state, Eq. 4-96 specified to our system leads to:

$$\rho v_x \frac{\partial u^\circ}{\partial x} = - \nabla \cdot \vec{q}'' + q''' - v_x \frac{\partial p}{\partial x} + \frac{\partial}{\partial y}(\tau_{xy} v_x) + \frac{\partial}{\partial z}(\tau_{xz} v_x) + v_x \rho g_x$$

The shear stresses τ_{xy} and τ_{xz} have been evaluated in example 4-5. Applying them here and differentiating, we get:

$$\rho v_x \frac{\partial u^\circ}{\partial x} = - \nabla \cdot \vec{q}'' + q''' - v_x \frac{\partial p}{\partial x}$$

$$- \mu \, v_{max}^2 \left[1 - \left(\frac{z}{L_z} \right)^2 \right]^2 \frac{\partial}{\partial y} \left[\frac{2y}{L_y^2} \left\{ 1 - \left(\frac{y}{L_y} \right)^2 \right\} \right]$$

$$- \mu \, v_{max}^2 \left[1 - \left(\frac{y}{L_y} \right)^2 \right]^2 \frac{\partial}{\partial z} \left[\frac{2z}{L_z^2} \left\{ 1 - \left(\frac{z}{L_z} \right)^2 \right\} \right]$$

$$+ v_x \rho g_x = - \nabla \cdot \vec{q}'' + q''' - v_x \frac{\partial p}{\partial x}$$

$$- \mu v_{max}^2 \left\{ \left[1 - \left(\frac{z}{L_z} \right)^2 \right]^2 \left[\frac{2}{L_y^2} - \frac{6y^2}{L_y^4} \right] \right.$$

$$+ \left. \left[1 - \left(\frac{y}{L_y} \right)^2 \right]^2 \left[\frac{2}{L_z^2} - \frac{6z^2}{L_z^4} \right] \right\} + v_x \rho g_x \qquad (4\text{-}96)$$

Now consider two points within the channel:

1. $y = L_y$ and $z > 0$
2. $y = z = 0$

For $y = L_y$ and $z > 0$, $v_x = 0$ and Eq. 4-96 becomes:

$$0 = - \nabla \cdot \vec{q}'' + q''' - 0 - \mu v_{max}^2 \left\{ \left[1 - \left(\frac{z}{L_z} \right)^2 \right] \left[\frac{2}{L_y^2} - \frac{6}{L_y^2} \right] - 0 \right\} + 0$$

For $y = z = 0$, $v_x = v_{max}$, and Eq. 4-96 becomes:

$$\rho v_{max} \frac{\partial u^\circ}{\partial x} = - \nabla \cdot q'' + q''' - v_{max} \frac{\partial p}{\partial x} - \mu v_{max}^2 \left\{ \frac{2}{L_y^2} + \frac{2}{L_z^2} \right\} + v_{max} \rho g_x$$

Hence there is internal heat generation and consequent molecular heat diffusion due to the shear work term as well as to the other work terms. However, in energy conversion systems this work term is small. For example if $v_{max} = 10$ m/s, $\mu = 100 \ \mu Pa \cdot s$ (water at 300°C and 15 MPa), and $L_y = L_z = 0.01$ m (representative of subchannels in PWRs), the largest value of the work term is:

$$\frac{4 \ \mu v_{max}^2}{L_y^2} = \frac{4 \times 100 \times 10^{-6} \times (10)^2}{(0.01)^2} \frac{(Pa \cdot s)(m^2/s^2)}{(m^2)} = 400 \ Pa/s = 400 \ W/m^3$$

This amount is small relative to the power density in a PWR of about 100 MW/m³ (see Example 3-2). It should be mentioned that the numbers given here are for illustration only. In a real PWR at the given velocity, the shape of the velocity profile in a subchannel does not correspond to the parabolic shape.

2 Stagnation enthalpy equation. The explicit term $\nabla \cdot p\vec{v}$ can be eliminated if the energy equation is cast in terms of the rate of change of enthalpy instead of internal energy. If we expand this term as:

$$\nabla \cdot p\vec{v} = \nabla \cdot \left(\frac{p}{\rho} \right) \rho\vec{v} = \left(\frac{p}{\rho} \right) \nabla \cdot \rho\vec{v} + \rho\vec{v} \cdot \nabla \left(\frac{p}{\rho} \right)$$

and apply the continuity equation, Eq. 4-73:

$$\frac{\partial \rho}{\partial t} + \nabla \cdot \rho\vec{v} = 0$$

we get:

$$\nabla \cdot p\vec{v} = -\left(\frac{p}{\rho}\right)\frac{\partial \rho}{\partial t} + \rho\vec{v} \cdot \nabla \left(\frac{p}{\rho}\right) \qquad (4\text{-}97)$$

Considering that:

$$\frac{\partial \left(\frac{p}{\rho}\rho\right)}{\partial t} = \left(\frac{p}{\rho}\right)\frac{\partial \rho}{\partial t} + \rho\frac{\partial \left(\frac{p}{\rho}\right)}{\partial t} \qquad (4\text{-}98)$$

we get:

$$\left(\frac{p}{\rho}\right)\frac{\partial \rho}{\partial t} = \frac{\partial p}{\partial t} - \rho\frac{\partial}{\partial t}\left(\frac{p}{\rho}\right) \qquad (4\text{-}99)$$

Substituting from Eq. 4-99 into Eq. 4-97, we get:

$$\nabla \cdot p\vec{v} = -\frac{\partial p}{\partial t} + \rho\frac{\partial}{\partial t}\left(\frac{p}{\rho}\right) + \rho\vec{v} \cdot \nabla \left(\frac{p}{\rho}\right) = -\frac{\partial p}{\partial t} + \rho\frac{D}{Dt}\left(\frac{p}{\rho}\right) \qquad (4\text{-}100)$$

Substituting for $\nabla \cdot p\vec{v}$ in Eq. 4-96 and rearranging, we get:

$$\rho\frac{Dh^\circ}{Dt} = -\nabla \cdot \vec{q}'' + q''' + \frac{\partial p}{\partial t} + \nabla \cdot (\overline{\overline{\tau}} \cdot \vec{v}) + \vec{v} \cdot \rho\vec{f} \qquad (4\text{-}101)$$

where $h^\circ \equiv u^\circ + \dfrac{p}{\rho}$.

3 Kinetic energy equation. It is possible to relate the change in kinetic energy to the work done on the fluid by multiplying both sides of Eq. 4-83 by $\vec{v}$ to obtain the mechanical energy equation:

$$\rho\vec{v} \cdot \frac{D\vec{v}}{Dt} = -\vec{v} \cdot \nabla p + \vec{v} \cdot (\nabla \cdot \overline{\overline{\tau}}) + \vec{v} \cdot \rho\vec{f}$$

or

$$\rho\frac{D}{Dt}\left(\frac{1}{2}v^2\right) = -\vec{v} \cdot \nabla p + \vec{v} \cdot (\nabla \cdot \overline{\overline{\tau}}) + \vec{v} \cdot \rho\vec{f} \qquad (4\text{-}102)$$

However, for a symmetrical stress tensor:

$$\overline{\overline{\tau}} : \nabla\vec{v} \equiv (\overline{\overline{\tau}} \cdot \nabla) \cdot \vec{v} = \nabla \cdot (\overline{\overline{\tau}} \cdot \vec{v}) - \vec{v} \cdot (\nabla \cdot \overline{\overline{\tau}}) \qquad (4\text{-}103)$$

Therefore Eq. 4-102 may be written as:

$$\rho\frac{D}{Dt}\left(\frac{1}{2}v^2\right) = -\vec{v} \cdot \nabla p + \nabla \cdot (\overline{\overline{\tau}} \cdot \vec{v}) - (\overline{\overline{\tau}} : \nabla\vec{v}) + \vec{v} \cdot \rho\vec{f}$$

which can be rearranged as:

$$\rho \frac{D}{Dt}\left(\frac{1}{2}v^2\right) = \nabla \cdot [(\overline{\overline{\tau}} - p\overline{\overline{I}}) \cdot \vec{v}] + p\nabla \cdot \vec{v} + \vec{v} \cdot \rho\vec{f} - (\overline{\overline{\tau}} : \nabla\vec{v}) \quad (4\text{-}104)$$

where $\overline{\overline{I}}$ is the unity tensor.

4 Thermodynamic energy equations. By subtracting Eq. 4-104 from Eq. 4-101, we get the energy equation in terms of the enthalpy (h)—not the stagnation enthalpy $(h°)$:

$$\rho \frac{Dh}{Dt} = -\nabla \cdot \vec{q}'' + q''' + \frac{Dp}{Dt} + (\overline{\overline{\tau}}:\nabla\vec{v}) \quad (4\text{-}105)$$

By a manipulation similar to that of Eq. 4-95, it is seen that Eq. 4-105 may also be written in the form:

$$\rho \frac{Dh}{Dt} = \frac{\partial}{\partial t}(\rho h) + \nabla \cdot (\rho h\vec{v}) = -\nabla \cdot \vec{q}'' + q''' + \frac{Dp}{Dt} + (\overline{\overline{\tau}}:\nabla\vec{v}) \quad (4\text{-}106)$$

If we define:

$$\phi \equiv (\overline{\overline{\tau}}:\nabla\vec{v}) = \text{dissipation function} \quad (4\text{-}107)$$

then Eq. 4-106 may be written in a form often encountered in thermal analysis textbooks:

$$\rho \frac{Dh}{Dt} = \frac{\partial}{\partial t}(\rho h) + \nabla \cdot (\rho h\vec{v}) = -\nabla \cdot \vec{q}'' + q''' + \frac{Dp}{Dt} + \phi \quad (4\text{-}108)$$

By applying the definition of the enthalpy:

$$h \equiv u + \frac{p}{\rho}$$

and rearranging Eq. 4-108, we get a similar energy equation for the internal energy u:

$$\frac{\partial}{\partial t}(\rho u) + \nabla \cdot (\rho u\vec{v}) = -\nabla \cdot \vec{q}'' + q''' - p\nabla \cdot \vec{v} + \phi \quad (4\text{-}109)$$

or

$$\rho \frac{Du}{Dt} = -\nabla \cdot \vec{q}'' + q''' - p\nabla \cdot \vec{v} + \phi \quad (4\text{-}110)$$

5 Special forms. To solve Eq. 4-96, 4-101, 4-106, or 4-109, the viscosity term must be explicitly written in terms of the velocity field and fluid properties. In most cases the rate of heat addition due to viscous effects can be neglected compared to the other terms, and the energy equation 4-105 may be given by:

$$\rho \frac{Dh}{Dt} = -\nabla \cdot \vec{q}'' + q''' + \frac{Dp}{Dt}; \text{ inviscid flow} \quad (4\text{-}111)$$

Furthermore, for incompressible as well as inviscid flow, the energy equation can be simplified, as ϕ and $\nabla \cdot \vec{v}$ equal 0. Thus Eq. 4-110 becomes:

$$\rho \frac{Du}{Dt} = q''' - \nabla \cdot \vec{q}''; \text{ incompressible and inviscid flow} \qquad (4\text{-}112)$$

Note also that the heat flux $\vec{q}''$ may be due to both conduction and radiation, so that in general:

$$\vec{q}'' = \vec{q}''_c + \vec{q}''_r \qquad (4\text{-}113)$$

where $\vec{q}''_c$ and $\vec{q}''_r$ denote the conduction and radiation heat fluxes, respectively. Within dense materials in most engineering analysis $\vec{q}''_r$ can be neglected, although the transmissivity of radiation within a given medium is dependent on the wavelength of the radiation.

Fourier's law of conduction gives the molecular heat flux as linearly proportional to the gradient of temperature.

$$\vec{q}''_c = - k\nabla T \qquad (4\text{-}114)$$

In heat transfer problems it is often convenient to cast Eq. 4-108 in terms of temperature rather than enthalpy. For a pure substance the enthalpy is a function of only two thermodynamic properties, e.g., temperature and pressure:

$$dh = \left. \frac{\partial h}{\partial T} \right|_p dT + \left. \frac{\partial h}{\partial p} \right|_T dp = c_p dT + \left. \frac{\partial h}{\partial p} \right|_T dp \qquad (4\text{-}115)$$

Also, for a pure substance the first law of thermodynamics as applied to a unit mass can be written as:

$$dh = Tds + dp/\rho$$

so that:

$$\left. \frac{\partial h}{\partial p} \right|_T = T \left. \frac{\partial s}{\partial p} \right|_T + \frac{1}{\rho} \qquad (4\text{-}116)$$

Defining β as the volumetric thermal expansion coefficient, we have:

$$\beta \equiv - \frac{1}{\rho} \left. \frac{\partial \rho}{\partial T} \right|_p \qquad (4\text{-}117)$$

From Maxwell's thermodynamic relations we know that:

$$\left. \frac{\partial s}{\partial p} \right|_T = - \left. \frac{\partial (1/\rho)}{\partial T} \right|_p = \frac{1}{\rho^2} \left. \frac{\partial \rho}{\partial T} \right|_p = - \frac{\beta}{\rho} \qquad (4\text{-}118)$$

Substituting from Eq. 4-118 into Eq. 4-116, we get:

$$\left. \frac{\partial h}{\partial p} \right|_T = - \frac{\beta T}{\rho} + \frac{1}{\rho} \qquad (4\text{-}119)$$

Now we substitute Eq. 4-119 into Eq. 4-115 to get:

$$dh = c_p dT + (1 - \beta T) \frac{dp}{\rho} \tag{4-120a}$$

Note that for an ideal gas, Eq. 4-120a reduces to:

$$dh = c_p dT \tag{4-120b}$$

Applying Eq. 4-120a to a unit mass in the Lagrangian formulation we get:

$$\rho \frac{Dh}{Dt} = \rho c_p \frac{DT}{Dt} + (1 - \beta T) \frac{Dp}{Dt} \tag{4-121}$$

If we introduce Eqs. 4-113, 4-114, and 4-121 into Eq. 4-108, the energy equation takes a form often used for solving heat transfer problems:

$$\rho c_p \frac{DT}{Dt} = -\nabla \cdot \vec{q}'' + q''' + \beta T \frac{Dp}{Dt} + \phi \tag{4-122a}$$

or

$$\rho c_p \frac{DT}{Dt} = \nabla \cdot k\nabla T - \nabla \cdot \vec{q}_r'' + q''' + \beta T \frac{Dp}{Dt} + \phi \tag{4-122b}$$

The energy term due to thermal expansion of fluids is often small compared to the other terms. An exception is near the front of a shock wave, where $\frac{Dp}{Dt}$ can be of a high magnitude.

For stagnant fluids or solids, neglecting compressibility and thermal expansion (ρ is constant) and the dissipation function, the energy equation is given by:

$$\rho c_p \frac{\partial T}{\partial t} = \nabla \cdot k\nabla T + q''' \tag{4-123}$$

Note also that, for incompressible materials, $c_p = c_v$. This is the reason behind neglecting the difference between c_p and c_v when solving heat transfer problems for liquids and solids but not for gases.

D Summary of equations

Useful differential forms of the relevant conservation equations are compiled in Tables 4-6 to 4-12.

VI TURBULENT FLOW

In turbulent flow, it can be assumed that the moving medium properties exhibit fluctuations in time, which are the result of random movement of eddies (or pockets) of fluid. The local fluid motion can still be described instantaneously by the equations

Table 4-6 Differential fluid transport equations in vector form

Generalized form

Unsteady term + convection term = diffusion term + source term

$$\frac{\partial}{\partial t}[\rho c] + \nabla \cdot \rho c \vec{v} \qquad = \nabla \cdot \vec{J} \qquad + \rho \phi \qquad \text{(Eq. 4-71)}$$

Note: $\dfrac{\partial}{\partial t}[\rho c] + \nabla \cdot \rho c \vec{v} = \rho \dfrac{Dc}{Dt}$

Equation	c	J	ϕ^a	Eq.
Continuity	1	0	0	4-73
Linear momentum	$\vec{v}$	$\overline{\overline{\tau}} - p\overline{\overline{I}}$	$\vec{g}$	4-80
Stagnation internal energy	u°	$(\overline{\overline{\tau}} - p\overline{\overline{I}}) \cdot \vec{v} - \vec{q}''$	$\dfrac{q'''}{\rho} + \vec{v} \cdot \vec{g}$	4-96
Energy				
Internal energy	u	$-\vec{q}''$	$\dfrac{1}{\rho}(q''' - p\nabla \cdot \vec{v} + \phi)$	4-109
Enthalpy	h	$-\vec{q}''$	$\dfrac{1}{\rho}\left(q''' + \dfrac{Dp}{Dt} + \phi\right)$	4-108
Kinetic energy	$\dfrac{1}{2}v^2$	$(\overline{\overline{\tau}} - p\overline{\overline{I}}) \cdot \vec{v}$	$\dfrac{1}{\rho}(\vec{v} \cdot \rho\vec{g} - \phi + p\nabla \cdot \vec{v})$	4-104

[a] Assuming gravity is the only body force.

Table 4-7 Differential equation of continuity

Vector form

$$\frac{\partial \rho}{\partial t} + \nabla \cdot (\rho \vec{v}) = 0 \qquad \text{(Eq. 4-73)}$$

or

$$\frac{D\rho}{Dt} + \rho(\nabla \cdot \vec{v}) = 0 \qquad \text{(Eq. 4-74)}$$

Cartesian

$$\frac{\partial \rho}{\partial t} + \frac{\partial}{\partial x}(\rho v_x) + \frac{\partial}{\partial y}(\rho v_y) + \frac{\partial}{\partial z}(\rho v_z) = 0$$

Cylindrical

$$\frac{\partial \rho}{\partial t} + \frac{1}{r}\frac{\partial}{\partial r}(\rho r v_r) + \frac{1}{r}\frac{\partial}{\partial \theta}(\rho v_\theta) + \frac{\partial}{\partial z}(\rho v_z) = 0$$

Spherical

$$\frac{\partial \rho}{\partial t} + \frac{1}{r^2}\frac{\partial}{\partial r}(\rho r^2 v_r) + \frac{1}{r \sin \theta}\frac{\partial}{\partial \theta}(\rho v_\theta \sin \theta) + \frac{1}{r \sin \theta}\frac{\partial}{\partial \phi}(\rho v_\phi) = 0$$

Table 4-8 Differential equation of continuity for incompressible materials

Vector form

$$\nabla \cdot \vec{v} = 0 \qquad \text{(Eq. 4-75)}$$

Cartesian

$$\frac{\partial}{\partial x}(v_x) + \frac{\partial}{\partial y}(v_y) + \frac{\partial}{\partial z}(v_z) = 0$$

Cylindrical

$$\frac{1}{r}\frac{\partial}{\partial r}(rv_r) + \frac{1}{r}\frac{\partial(v_\theta)}{\partial \theta} + \frac{\partial(v_z)}{\partial z} = 0$$

Spherical

$$\frac{1}{r^2}\frac{\partial}{\partial r}(r^2 v_r) + \frac{1}{r \sin\theta}\frac{\partial}{\partial \theta}(v_\theta \sin\theta) + \frac{1}{r \sin\theta}\frac{\partial}{\partial \phi}(v_\phi) = 0$$

Table 4-9 Differential equation of motion[a]

Vector form

$$\rho \frac{D\vec{v}}{Dt} = -\nabla p + \nabla \cdot \overline{\overline{\tau}} + \rho\vec{f} \qquad \text{(Eq. 4-83)}$$

or

$$\rho \frac{D\vec{v}}{Dt} = -\nabla p - \nabla \times [\mu \nabla \times \vec{v}] + \nabla \left[\frac{4}{3}\mu \nabla \cdot \vec{v}\right] + \rho\vec{f} \qquad \text{(Eq. 4-88)}$$

Note:

$$\rho \frac{D\vec{v}}{Dt} = \frac{\partial}{\partial t}\rho\vec{v} + \nabla \cdot \rho\vec{v}\vec{v} \qquad \text{(Eq. 4-82)}$$

Cartesian

$$\frac{D}{Dt} = \frac{\partial}{\partial t} + v_x\frac{\partial}{\partial x} + v_y\frac{\partial}{\partial y} + v_z\frac{\partial}{\partial z}$$

$$\rho \frac{Dv_x}{Dt} = -\frac{\partial p}{\partial x} + \frac{\partial}{\partial x}\left[2\mu\frac{\partial v_x}{\partial x} - \frac{2}{3}\mu\nabla\cdot\vec{v}\right] + \frac{\partial}{\partial y}\left[\mu\left(\frac{\partial v_x}{\partial y} + \frac{\partial v_y}{\partial x}\right)\right]$$
$$+ \frac{\partial}{\partial z}\left[\mu\left(\frac{\partial v_x}{\partial z} + \frac{\partial v_z}{\partial x}\right)\right] + \rho f_x$$

$$\rho \frac{Dv_y}{Dt} = -\frac{\partial p}{\partial y} + \frac{\partial}{\partial y}\left[2\mu\frac{\partial v_y}{\partial y} - \frac{2}{3}\mu\nabla\cdot\vec{v}\right] + \frac{\partial}{\partial z}\left[\mu\left(\frac{\partial v_y}{\partial z} + \frac{\partial v_z}{\partial y}\right)\right]$$
$$+ \frac{\partial}{\partial x}\left[\mu\left(\frac{\partial v_x}{\partial y} + \frac{\partial v_y}{\partial x}\right)\right] + \rho f_y$$

$$\rho \frac{Dv_z}{Dt} = -\frac{\partial p}{\partial y} + \frac{\partial}{\partial z}\left[2\mu\frac{\partial v_z}{\partial z} - \frac{2}{3}\mu\nabla\cdot\vec{v}\right] + \frac{\partial}{\partial x}\left[\mu\left(\frac{\partial v_z}{\partial x} + \frac{\partial v_x}{\partial z}\right)\right]$$
$$+ \frac{\partial}{\partial y}\left[\mu\left(\frac{\partial v_y}{\partial z} + \frac{\partial v_z}{\partial y}\right)\right] + \rho f_z$$

Table 4-9 (*Continued*)

Cylindrical

$$\frac{D}{Dt} = \frac{\partial}{\partial t} + v_r \frac{\partial}{\partial r} + \frac{v_\theta}{r} \frac{\partial}{\partial \theta} + v_z \frac{\partial}{\partial z}$$

$$\rho \left[\frac{Dv_r}{Dt} - \frac{v_\theta^2}{r} \right] = -\frac{\partial p}{\partial r} + \frac{\partial}{\partial r} \left[2\mu \frac{\partial v_r}{\partial r} - \frac{2}{3} \mu \nabla \cdot \vec{v} \right] + \frac{1}{r} \frac{\partial}{\partial \theta} \left[\mu \left(\frac{1}{r} \frac{\partial v_r}{\partial \theta} + \frac{\partial v_\theta}{\partial r} - \frac{v_\theta}{r} \right) \right]$$

$$+ \frac{\partial}{\partial z} \left[\mu \left(\frac{\partial v_r}{\partial z} + \frac{\partial v_z}{\partial r} \right) \right] + \frac{2\mu}{r} \left(\frac{\partial v_r}{\partial r} - \frac{1}{r} \frac{\partial v_\theta}{\partial \theta} - \frac{v_r}{r} \right) + \rho f_r$$

$$\rho \left[\frac{Dv_\theta}{Dt} + \frac{v_r v_\theta}{r} \right] = -\frac{1}{r} \frac{\partial p}{\partial \theta} + \frac{1}{r} \frac{\partial}{\partial \theta} \left[\frac{2\mu}{r} \frac{\partial v_\theta}{\partial \theta} - \frac{2}{3} \mu \nabla \cdot \vec{v} \right] + \frac{\partial}{\partial z} \left[\mu \left(\frac{1}{r} \frac{\partial v_z}{\partial \theta} + \frac{\partial v_\theta}{\partial z} \right) \right]$$

$$+ \frac{\partial}{\partial r} \left[\mu \left(\frac{1}{r} \frac{\partial v_r}{\partial \theta} + \frac{\partial v_\theta}{\partial r} - \frac{v_\theta}{r} \right) \right] + \frac{2\mu}{r} \left[\frac{1}{r} \frac{\partial v_r}{\partial \theta} + \frac{\partial v_\theta}{\partial r} - \frac{v_\theta}{r} \right] + \rho f_\theta$$

$$\rho \frac{Dv_z}{Dt} = -\frac{\partial p}{\partial z} + \frac{\partial}{\partial z} \left[2\mu \frac{\partial v_z}{\partial z} - \frac{2}{3} \mu \nabla \cdot \vec{v} \right] + \frac{1}{r} \frac{\partial}{\partial r} \left[\mu r \left(\frac{\partial v_r}{\partial z} + \frac{\partial v_z}{\partial r} \right) \right]$$

$$+ \frac{1}{r} \frac{\partial}{\partial \theta} \left[\mu \left(\frac{1}{r} \frac{\partial v_z}{\partial \theta} + \frac{\partial v_\theta}{\partial z} \right) \right] + \rho f_z$$

Spherical

$$\frac{D}{Dt} = \frac{\partial}{\partial t} + v_r \frac{\partial}{\partial r} + \frac{v_\theta}{r} \frac{\partial}{\partial \theta} + \frac{v_\theta}{r \sin \theta} \frac{\partial}{\partial \phi}$$

$$\rho \left[\frac{Dv_r}{Dt} - \frac{v_\theta^2 + v_\phi^2}{r} \right] = -\frac{\partial p}{\partial r} + \frac{\partial}{\partial r} \left[2\mu \frac{\partial v_r}{\partial r} - \frac{2}{3} \mu \nabla \cdot \vec{v} \right]$$

$$+ \frac{1}{r} \frac{\partial}{\partial \theta} \left[\mu \left\{ r \frac{\partial}{\partial r} \left(\frac{v_\theta}{r} \right) + \frac{1}{r} \frac{\partial v_r}{\partial \theta} \right\} \right]$$

$$+ \frac{1}{r \sin \theta} \frac{\partial}{\partial \phi} \left[\mu \left\{ \frac{1}{r \sin \theta} \frac{\partial v_r}{\partial \phi} + r \frac{\partial}{\partial r} \left(\frac{v_\phi}{r} \right) \right\} \right]$$

$$+ \frac{\mu}{r} \left[4 \frac{\partial v_r}{\partial r} - \frac{2}{r} \frac{\partial v_\theta}{\partial \theta} - \frac{4 v_r}{r} - \frac{2}{r \sin \theta} \frac{\partial v_\phi}{\partial \phi} - \frac{2 v_\theta \cot \theta}{r} \right.$$

$$\left. + r \cot \theta \frac{\partial}{\partial r} \left(\frac{v_\theta}{r} \right) + \frac{\cot \theta}{r} \frac{\partial v_r}{\partial \theta} \right] + \rho f_r$$

$$\rho \left[\frac{Dv_\theta}{Dt} + \frac{v_r v_\theta}{r} - \frac{v_\phi^2 \cot \theta}{r} \right] = -\frac{1}{r} \frac{\partial p}{\partial \theta} + \frac{1}{r} \frac{\partial}{\partial \theta} \left[\frac{2\mu}{r} \left(\frac{\partial v_\theta}{\partial \theta} + v_r \right) - \frac{2}{3} \mu \nabla \cdot \vec{v} \right]$$

$$+ \frac{1}{r \sin \theta} \frac{\partial}{\partial \phi} \left[\mu \left\{ \frac{\sin \theta}{r} \frac{\partial}{\partial \theta} \left(\frac{v_\phi}{\sin \theta} \right) + \frac{1}{r \sin \theta} \frac{\partial v_\theta}{\partial \phi} \right\} \right]$$

$$+ \frac{\partial}{\partial r} \left[\mu \left\{ r \frac{\partial}{\partial r} \left(\frac{v_\theta}{r} \right) + \frac{1}{r} \frac{\partial v_r}{\partial \theta} \right\} \right]$$

$$+ \frac{\mu}{r} \left[2 \left(\frac{1}{r} \frac{\partial v_\theta}{\partial \theta} - \frac{1}{r \sin \theta} \frac{\partial v_\phi}{\partial \phi} - \frac{v_\theta \cot \theta}{r} \right) \cdot \cot \theta \right.$$

$$\left. + 3 \left\{ r \frac{\partial}{\partial r} \left(\frac{v_\theta}{r} \right) + \frac{1}{r} \frac{\partial v_r}{\partial \theta} \right\} \right] + \rho f_\theta$$

Table 4-9 (*Continued*)

$$\rho\left[\frac{Dv_\phi}{Dt} + \frac{v_\phi v_r}{r} + \frac{v_\theta v_\phi \cot\theta}{r}\right] = -\frac{1}{r\sin\theta}\frac{\partial p}{\partial\phi}$$

$$+ \frac{1}{r\sin\theta}\frac{\partial}{\partial\phi}\left[\frac{2\mu}{r}\left(\frac{1}{\sin\theta}\frac{\partial v_\phi}{\partial\phi} + v_r + v_\theta\cot\theta\right) - \frac{2}{3}\mu\nabla\cdot\vec{v}\right]$$

$$+ \frac{\partial}{\partial r}\left[\mu\left\{\frac{1}{r\sin\theta}\frac{\partial v_r}{\partial\phi} + r\frac{\partial}{\partial r}\left(\frac{v_\phi}{r}\right)\right\}\right]$$

$$+ \frac{1}{r}\frac{\partial}{\partial\theta}\left[\mu\left\{\frac{\sin\theta}{r}\frac{\partial}{\partial\theta}\left(\frac{v_\phi}{\sin\theta}\right) + \frac{1}{r\sin\theta}\frac{\partial v_\theta}{\partial\phi}\right\}\right]$$

$$+ \frac{\mu}{r}\left[3\left\{\frac{1}{r\sin\theta}\frac{\partial v_r}{\partial\phi} + r\frac{\partial}{\partial r}\left(\frac{v_\phi}{r}\right)\right\}\right.$$

$$\left. + 2\cot\theta\left\{\frac{\sin\theta}{r}\frac{\partial}{\partial\theta}\left(\frac{v_\phi}{\sin\theta}\right) + \frac{1}{r\sin\theta}\frac{\partial v_\theta}{\partial\phi}\right\}\right] + \rho f_\phi$$

[a] $\mu' = 0$ in Eq. 4-84.

Table 4-10 Differential equation of motion for incompressible fluids[a]

Vector form

$$\rho\frac{D\vec{v}}{Dt} = -\nabla p + \mu\nabla^2\vec{v} + \rho\vec{f} \qquad\qquad \text{(Eq. 4-90b)}$$

Cartesian

$$\rho\left(\frac{\partial v_x}{\partial t} + v_x\frac{\partial v_x}{\partial x} + v_y\frac{\partial v_x}{\partial y} + v_z\frac{\partial v_x}{\partial z}\right) = -\frac{\partial p}{\partial x} + \mu\left(\frac{\partial^2 v_x}{\partial x^2} + \frac{\partial^2 v_x}{\partial y^2} + \frac{\partial^2 v_x}{\partial z^2}\right) + \rho f_x$$

$$\rho\left(\frac{\partial v_y}{\partial t} + v_x\frac{\partial v_y}{\partial x} + v_y\frac{\partial v_y}{\partial y} + v_z\frac{\partial v_y}{\partial z}\right) = -\frac{\partial p}{\partial y} + \mu\left(\frac{\partial^2 v_y}{\partial x^2} + \frac{\partial^2 v_y}{\partial y^2} + \frac{\partial^2 v_y}{\partial z^2}\right) + \rho f_y$$

$$\rho\left(\frac{\partial v_z}{\partial t} + v_x\frac{\partial v_z}{\partial x} + v_y\frac{\partial v_z}{\partial y} + v_z\frac{\partial v_z}{\partial z}\right) = -\frac{\partial p}{\partial z} + \mu\left(\frac{\partial^2 v_z}{\partial x^2} + \frac{\partial^2 v_z}{\partial y^2} + \frac{\partial^2 v_z}{\partial z^2}\right) + \rho f_z$$

Cylindrical

$$\rho\left[\frac{\partial v_r}{\partial t} + v_r\frac{\partial v_r}{\partial r} + \frac{v_\theta}{r}\frac{\partial v_r}{\partial\theta} + v_z\frac{\partial v_r}{\partial z} - \frac{v_\theta^2}{r}\right] =$$

$$-\frac{\partial p}{\partial r} + \mu\left[\frac{\partial^2 v_r}{\partial r^2} + \frac{1}{r}\frac{\partial v_r}{\partial r} + \frac{1}{r^2}\frac{\partial^2 v_r}{\partial\theta^2} + \frac{\partial^2 v_r}{\partial z^2} - \frac{v_r}{r^2} + \frac{2}{r^2}\frac{\partial v_\theta}{\partial\theta}\right] + \rho f_r$$

$$\rho\left[\frac{\partial v_\theta}{\partial t} + v_r\frac{\partial v_\theta}{\partial r} + \frac{v_\theta}{r}\frac{\partial v_\theta}{\partial\theta} + v_z\frac{\partial v_\theta}{\partial z} + \frac{v_r v_\theta}{r}\right] =$$

$$-\frac{1}{r}\frac{\partial p}{\partial\theta} + \mu\left[\frac{\partial^2 v_\theta}{\partial r^2} + \frac{1}{r}\frac{\partial v_\theta}{\partial r} + \frac{1}{r^2}\frac{\partial^2 v_\theta}{\partial\theta^2} + \frac{\partial^2 v_\theta}{\partial z^2} + \frac{2}{r^2}\frac{\partial v_r}{\partial\theta} - \frac{v_\theta}{r^2}\right] + \rho f_\theta$$

$$\rho\left[\frac{\partial v_z}{\partial t} + v_r\frac{\partial v_z}{\partial r} + \frac{v_\theta}{r}\frac{\partial v_z}{\partial\theta} + v_z\frac{\partial v_z}{\partial z}\right] = -\frac{\partial p}{\partial z} + \mu\left[\frac{\partial^2 v_z}{\partial r^2} + \frac{1}{r}\frac{\partial v_z}{\partial r} + \frac{1}{r^2}\frac{\partial v_z}{\partial\theta^2} + \frac{\partial^2 v_z}{\partial z^2}\right] + \rho f_z$$

Table 4-10 (*Continued*)

Spherical

$$\rho\left[\frac{\partial v_r}{\partial t} + v_r\frac{\partial v_r}{\partial r} + \frac{v_\theta}{r}\frac{\partial v_r}{\partial \theta} + \frac{v_\phi}{r\sin\theta}\frac{\partial v_r}{\partial \phi} - \frac{v_\theta^2 + v_\phi^2}{r}\right] = -\frac{\partial p}{\partial r} + \mu\left[\frac{1}{r^2}\frac{\partial}{\partial r}\left(r^2\frac{\partial v_r}{\partial r}\right)\right.$$

$$\left.+ \frac{1}{r^2\sin\theta}\left(\sin\theta\frac{\partial v_r}{\partial\theta}\right) + \frac{1}{r^2\sin^2\theta}\frac{\partial^2 v_r}{\partial\phi^2} - \frac{2v_r}{r^2} - \frac{2}{r^2}\frac{\partial v_\theta}{\partial\theta} - \frac{2v_\theta\cot\theta}{r^2} - \frac{2}{r^2\sin\theta}\frac{\partial v_\phi}{\partial\phi}\right] + \rho f_r$$

$$\rho\left[\frac{\partial v_\theta}{\partial t} + v_r\frac{\partial v_\theta}{\partial r} + \frac{v_\theta}{r}\frac{\partial v_\theta}{\partial \theta} + \frac{v_\phi}{r\sin\theta}\frac{\partial v_\theta}{\partial \phi} + \frac{v_r v_\theta}{r} - \frac{v_\phi^2\cot\theta}{r}\right] = -\frac{1}{r}\frac{\partial p}{\partial \theta} + \mu\left[\frac{1}{r^2}\frac{\partial}{\partial r}\left(r^2\frac{\partial v_\theta}{\partial r}\right)\right.$$

$$\left.+ \frac{1}{r^2\sin\theta}\frac{\partial}{\partial\theta}\left(\sin\theta\frac{\partial v_\theta}{\partial\theta}\right) + \frac{1}{r^2\sin^2\theta}\frac{\partial^2 v_\theta}{\partial\phi^2} + \frac{2}{r^2}\frac{\partial v_r}{\partial\theta} - \frac{v_\theta}{r^2\sin^2\theta} - \frac{2\cos\theta}{r^2\sin^2\theta}\frac{\partial v_\phi}{\partial\phi}\right] + \rho f_\theta$$

$$\rho\left[\frac{\partial v_\phi}{\partial t} + v_r\frac{\partial v_\phi}{\partial r} + \frac{v_\theta}{r}\frac{\partial v_\phi}{\partial \theta} + \frac{v_\phi}{r\sin\theta}\frac{\partial v_\phi}{\partial \phi} + \frac{v_\phi v_r}{r} + \frac{v_\theta v_\phi\cot\theta}{r}\right] = -\frac{1}{r\sin\theta}\frac{\partial p}{\partial \phi}$$

$$+ \mu\left[\frac{1}{r^2}\frac{\partial}{\partial r}\left(r^2\frac{\partial v_\phi}{\partial r}\right) + \frac{1}{r^2\sin\theta}\frac{\partial}{\partial\theta}\left(\sin\theta\frac{\partial v_\phi}{\partial\theta}\right) + \frac{1}{r^2\sin^2\theta}\frac{\partial^2 v_\phi}{\partial\phi^2} - \frac{v_\phi}{r^2\sin^2\theta} + \frac{2}{r^2\sin^2\theta}\frac{\partial v_r}{\partial\theta}\right.$$

$$\left.+ \frac{2\cos\theta}{r^2\sin^2\theta}\frac{\partial v_\theta}{\partial\phi}\right] + \rho f_\phi$$

[a] $\mu' = 0$; μ = constant; ρ = constant.

Table 4-11 Differential equations of energy[a]

Vector form

$$\rho\frac{Dh}{Dt} = -\nabla\cdot\vec{q}'' + q''' + \frac{Dp}{Dt} + \phi \qquad\qquad \text{(Eq. 4–108)}$$

or

$$\rho\frac{Du}{Dt} = -\nabla\cdot\vec{q}'' + q''' - p\nabla\cdot\vec{v} + \phi \qquad\qquad \text{(Eq. 4–110)}$$

$$\rho c_p\frac{DT}{Dt} = \nabla\cdot k\nabla T - \nabla\cdot\vec{q}_r'' + q''' + \beta T\frac{Dp}{Dt} + \phi \qquad\qquad \text{(Eq. 4–122b)}$$

$$\phi = \bar{\bar{\tau}}:\nabla\vec{v} = \nabla\cdot(\bar{\bar{\tau}}\cdot\vec{v}) - \vec{v}\cdot(\nabla\cdot\bar{\bar{\tau}}) \qquad\qquad \text{(Eq. 4–103)}$$

Cartesian

$$\rho\frac{Dh}{Dt} = \frac{\partial}{\partial x}\left(k\frac{\partial T}{\partial x}\right) + \frac{\partial}{\partial y}\left(k\frac{\partial T}{\partial y}\right) + \frac{\partial}{\partial z}\left(k\frac{\partial T}{\partial z}\right) - \nabla\cdot\vec{q}_r'' + q''' + \frac{Dp}{Dt} + \phi$$

Cylindrical

$$\rho\frac{Dh}{Dt} = \frac{1}{r}\frac{\partial}{\partial r}\left(rk\frac{\partial T}{\partial r}\right) + \frac{1}{r^2}\frac{\partial}{\partial \theta}\left(k\frac{\partial T}{\partial \theta}\right) + \frac{\partial}{\partial z}\left(k\frac{\partial T}{\partial z}\right) - \nabla\cdot\vec{q}_r'' + q''' + \frac{Dp}{Dt} + \phi$$

Spherical

$$\rho\frac{Dh}{Dt} = \frac{1}{r^2}\frac{\partial}{\partial r}\left(r^2 k\frac{\partial T}{\partial r}\right) + \frac{1}{r^2\sin\theta}\frac{\partial}{\partial \theta}\left(k\sin\theta\frac{\partial T}{\partial \theta}\right) + \frac{1}{r^2\sin\theta}\frac{\partial}{\partial \phi}\left(k\frac{\partial T}{\partial \phi}\right) - \nabla\cdot\vec{q}_r'' + q''' + \frac{Dp}{Dt} + \phi$$

Table 4-12 Differential equations of energy for incompressible materials[a]

Vector form

$$\rho \frac{Dh}{Dt} = - \nabla \cdot \vec{q}'' + q''' + \phi$$

$$\rho \frac{Du}{Dt} = - \nabla \cdot \vec{q}'' + q''' + \phi$$

$$\rho\, c_{\mathrm{p}} \frac{DT}{Dt} = - \nabla \cdot \vec{q}'' + q''' + \phi$$

$$\phi = \overline{\overline{\tau}} : \vec{\nabla v} = \nabla \cdot (\overline{\overline{\tau}} \cdot \vec{v}) - \vec{v} \cdot (\nabla \cdot \overline{\overline{\tau}})$$

Special case for k = constant

$$\rho \frac{Dh}{Dt} = k\nabla^2 T - \nabla \cdot \vec{q}''_r + q''' + \phi$$

Cartesian

$$\rho \frac{Dh}{Dt} = k \left(\frac{\partial^2 T}{\partial x^2} + \frac{\partial^2 T}{\partial y^2} + \frac{\partial^2 T}{\partial z^2} \right) - \nabla \cdot \vec{q}''_r + q''' + \phi$$

$$\phi = 2\mu \left[\left(\frac{\partial v_x}{\partial x} \right)^2 + \left(\frac{\partial v_y}{\partial y} \right)^2 + \left(\frac{\partial v_z}{\partial z} \right)^2 + \frac{1}{2} \left(\frac{\partial v_x}{\partial y} + \frac{\partial v_y}{\partial x} \right)^2 + \frac{1}{2} \left(\frac{\partial v_x}{\partial z} + \frac{\partial v_z}{\partial x} \right)^2 \right.$$
$$\left. + \frac{1}{2} \left(\frac{\partial v_y}{\partial z} + \frac{\partial v_z}{\partial y} \right)^2 \right]$$

Cylindrical

$$\rho \frac{Dh}{Dt} = k \left(\frac{\partial^2 T}{\partial r^2} + \frac{1}{r} \frac{\partial T}{\partial r} + \frac{1}{r^2} \frac{\partial^2 T}{\partial \theta^2} + \frac{\partial^2 T}{\partial z^2} \right) - \nabla \cdot \vec{q}''_r + q''' + \phi$$

$$\phi = \mu \left[2 \left\{ \left(\frac{\partial v_r}{\partial r} \right)^2 + \left(\frac{1}{r} \frac{\partial v_\theta}{\partial \theta} + \frac{v_r}{r} \right)^2 + \left(\frac{\partial v_z}{\partial z} \right)^2 \right\} + \left(\frac{\partial v_z}{\partial \theta} + \frac{\partial v_\theta}{\partial z} \right)^2 + \left(\frac{\partial v_r}{\partial z} + \frac{\partial v_z}{\partial r} \right)^2 \right.$$
$$\left. + \left(\frac{1}{r} \frac{\partial v_r}{\partial \theta} + \frac{\partial v_\theta}{\partial r} - \frac{v_\theta}{r} \right)^2 \right]$$

Spherical

$$\rho \frac{Dh}{Dt} = k \left[\frac{1}{r^2} \frac{\partial}{\partial r} \left(r^2 \frac{\partial T}{\partial r} \right) + \frac{1}{r^2 \sin \theta} \frac{\partial}{\partial \theta} \left(\sin \theta \frac{\partial T}{\partial \theta} \right) + \frac{1}{r^2 \sin^2 \theta} \frac{\partial^2 T}{\partial \phi^2} \right] - \nabla \cdot \vec{q}''_r + q''' + \phi$$

$$\phi = \mu \left[2 \left\{ \left(\frac{\partial v_r}{\partial r} \right)^2 + \left(\frac{1}{r} \frac{\partial v_\theta}{\partial \theta} + \frac{v_r}{r} \right)^2 + \left(\frac{1}{r \sin \theta} \frac{\partial v_\phi}{\partial \phi} + \frac{v_r}{r} + \frac{v_\theta \cot \theta}{r} \right)^2 \right\} \right.$$
$$+ \left\{ \frac{1}{r \sin \phi} \frac{\partial v_\theta}{\partial \phi} + \frac{\sin \theta}{r} \frac{\partial}{\partial \theta} \left(\frac{v_\phi}{\sin \theta} \right) \right\}^2 + \left\{ \frac{1}{r \sin \theta} \frac{\partial v_r}{\partial \phi} + r \frac{\partial}{\partial r} \left(\frac{v_\phi}{r} \right) \right\}^2$$
$$\left. + \left\{ r \frac{\partial}{\partial r} \left(\frac{v_\theta}{r} \right) + \frac{1}{r} \frac{\partial v_r}{\partial \theta} \right\}^2 \right]$$

[a] $\mu' = 0$; ρ = constant.

of section V; but in fact, the practical characteristics of the moving fluid require time-averaging manipulation of these equations. It is also possible to space-average the local equations to obtain volume-average or area-average properties. Such space averaging is particularly useful for the two-phase flow condition, described in Chapter

5. Furthermore, applications to reactor core analysis by the porous media approach (see Volume II) draw on such space averaging.

For any property c, a time-averaged value can be obtained by:

$$\bar{c} \equiv \frac{1}{\Delta t} \int_{t-\Delta t/2}^{t+\Delta t/2} c \, dt \tag{4-124}$$

Thus the instantaneous value of the property may be written as:

$$c \equiv \bar{c} + c' \tag{4-125}$$

The period Δt is chosen to be sufficiently large to smooth out c but sufficiently small with respect to the transient time constants of the flow system under consideration.

Let us average 4-71 over the period Δt to get:

$$\overline{\frac{\partial[\rho c]}{\partial t}} + \overline{\nabla \cdot [\rho c \vec{v}]} = \overline{\nabla \cdot \vec{J}} + \overline{\rho \phi} \tag{4-126}$$

Because Δt is sufficiently small:

$$\overline{\frac{\partial[\rho c]}{\partial t}} = \frac{\partial \overline{[\rho c]}}{\partial t} \tag{4-127a}$$

For a stationary system of coordinates, time and space are independent variables, so that from Eq. 4-11, because $v_s = 0$, we get:

$$\overline{\nabla \cdot [\rho c \vec{v}]} = \nabla \cdot \overline{[\rho c \vec{v}]} \tag{4-127b}$$

and

$$\overline{\nabla \cdot \vec{J}} = \nabla \cdot \overline{\vec{J}} \tag{4-127c}$$

Applying Eqs. 4-127a, 4-127b, and 4-127c to Eq. 4-126 we get:

$$\frac{\partial \overline{[\rho c]}}{\partial t} + \nabla \cdot \overline{[\rho c \vec{v}]} = \nabla \cdot \overline{\vec{J}} + \overline{\rho \phi} \tag{4-128}$$

Assume that the density, velocity, and stagnation internal energy can be represented as a sum of a time-averaged value and a perturbation such that:

$$\rho \equiv \bar{\rho} + \rho' \tag{4-129a}$$
$$\vec{v} \equiv \vec{\bar{v}} + + \vec{v}' \tag{4-129b}$$
$$u^\circ \equiv \overline{u^\circ} + u^{\circ\prime} \tag{4-129c}$$

It is clear that:

$$\overline{c'} = 0 \quad \text{and} \quad \overline{\bar{c} \, c'} = 0 \tag{4-130}$$

Therefore Eq. 4-128 can be expanded to:

$$\frac{\partial[\bar{\rho}\,\bar{c}]}{\partial t} + \frac{\partial[\overline{\rho'c'}]}{\partial t} + \nabla \cdot [\bar{\rho}\,\bar{c}\,\vec{\bar{v}}] +$$

$$\nabla \cdot [\overline{\rho'c'}\,\vec{\bar{v}} + \overline{\rho'\vec{v}'}\,\bar{c} + \bar{\rho}\,\overline{c'\vec{v}'} + \overline{\rho'c'\vec{v}'}] = \nabla \cdot \overline{\vec{J}} + \bar{\rho}\,\bar{\phi} + \overline{\rho'\phi'} \tag{4-131}$$

Equation 4-131 can be rearranged to collect the fluctuation terms on the right hand side:

$$\frac{\partial[\bar{\rho}\,\bar{c}]}{\partial t} + \nabla \cdot [\bar{\rho}\,\bar{c}\,\vec{\bar{v}}] = \nabla \cdot \vec{\bar{J}} + \bar{\rho}\,\bar{\phi} + \left\{ \overline{\rho'\phi'} - \frac{\partial[\overline{\rho'c'}]}{\partial t} - \nabla \cdot \vec{\bar{J}}^{t} \right\} \quad (4\text{-}132)$$

where

$$\vec{\bar{J}}^{t} = \overline{\rho'c'}\,\vec{\bar{v}} + \overline{\rho'\vec{v}'}\,\bar{c} + \bar{\rho}\,\overline{c'\vec{v}'} + \overline{\rho'c'\vec{v}'} \quad (4\text{-}133)$$

If it is assumed that the density fluctuations are small or that the fluid is incompressible, $\rho' = 0$ and the turbulent transport Eq. 4-132 reduces to:

$$\frac{\partial[\rho\bar{c}]}{\partial t} + \nabla \cdot [\rho\bar{c}\,\vec{\bar{v}}] = \nabla \cdot \vec{\bar{J}} + \rho\bar{\phi} - \nabla \cdot [\overline{\rho c'\vec{v}'}]; \rho' = 0 \quad (4\text{-}134)$$

The mass equation can be obtained by specifying c, $\vec{J}$, and ϕ, from Table 4-6, in Eq. 4-134 to get the turbulent flow equations of an incompressible fluid.
The *mass equation* becomes:

$$\frac{\partial[\rho]}{\partial t} + \nabla \cdot [\rho\vec{\bar{v}}] = 0 \quad (4\text{-}135)$$

The *momentum equation* becomes:

$$\frac{\partial[\rho\vec{\bar{v}}]}{\partial t} + \nabla \cdot [\rho\vec{\bar{v}}\,\vec{\bar{v}}] = \nabla \cdot [\overline{\overline{\bar{\tau}}} - \bar{p}\,\overline{\overline{I}}] + \rho\vec{g} - \nabla \cdot [\overline{\rho\vec{v}'\vec{v}'}] \quad (4\text{-}136)$$

The *stagnation energy equation* becomes:

$$\frac{\partial[\rho\,\bar{u}^{\circ}]}{\partial t} + \nabla \cdot [\rho\bar{u}^{\circ}\vec{\bar{v}}] = \nabla \cdot [-\vec{\bar{q}}'' + (\overline{\overline{\bar{\tau}}} - p\,\overline{\overline{I}}) \cdot \vec{\bar{v}}]$$
$$+ \bar{q}''' + \rho\vec{g} \cdot \vec{\bar{v}} - \nabla \cdot [\overline{\rho u^{\circ'}\vec{v}'}] \quad (4\text{-}137)$$

The solution of Eqs. 4-136 and 4-137 requires identification of constitutive relations between the terms, including the fluctuating variables and the behavior of the average variables. That is, it is required that $[\overline{\rho\vec{v}'\vec{v}'}]$, $(\overline{\overline{\bar{\tau}}} - p\,\overline{\overline{I}}) \cdot \vec{\bar{v}}$, and $[\overline{\rho u^{\circ'}\vec{v}'}]$ be specified in terms of ρ, $\vec{\bar{v}}$, and $\bar{u}^{\circ}$ for the above set of equations to be solvable. The constitutive equations are discussed in Chapters 9 and 10.

REFERENCES

1. Bird, R. B., Stewart, W. E., and Lightfoot, E. N. *Transport Phenomena*. New York: Wiley, 1960.
2. Currie, I. G. *Fundamental Mechanics of Fluids*. New York: McGraw-Hill, 1974.
3. Hildebrand, F. B. *Advanced Calculus for Applications*. Englewood Cliffs, NJ: Prentice Hall, 1962.
4. Potter, M. C., and Foss, J. R. *Fluid Mechanics*. New York: Wiley, 1975.
5. Slattery, J. C., and Gaggioli, R. A. The macroscopic angular momentum balance. *Chem. Eng. Sci.* 17:873–895, 1962.

6. Whitaker, S. *Introduction to Fluid Mechanics*. Huntington, NY: Krieger, 1981.
7. Whitaker, S. Laws of continuum physics for single phase, single component systems. In G. Hetsroni (ed.), *Handbook of Multiphase Systems*. Washington, D.C.: Hemisphere, 1982.

PROBLEMS

Problem 4-1 Various time derivatives (section II)
    Air bubbles are being injected at a steady rate into the bottom of a vertical water channel of height L. The bubble rise velocity with respect to the water is v_b. The water itself is flowing vertically at a velocity v_ℓ. What is the velocity of the bubbles at the channel exit as observed by the following:

1. A stationary observer
2. An observer who moves upward with velocity v_ℓ
3. An observer who moves upward with velocity $2v_\ell$

   Answers:

1. $v_\ell + v_b$
2. v_b
3. $v_b - v_\ell$

Problem 4-2 Conservation of energy in a control volume (section III)
    A mass of 9 kg of gas with an internal energy of 1908 kJ is at rest in a rigid cylinder. A mass of 1.0 kg of the same gas with an internal energy of 95.4 kJ and a velocity of 30 m/s flows into the cylinder. In the absence of heat transfer to the surroundings and with negligible change in the center of gravity, find the internal energy of the 10 kg of gas finally at rest in the cylinder. The absolute pressure of the flowing gas crossing the control surface is 0.7 MPa, and the specific volume of the gas is 0.00125 m³/kg.
    Answer: U_f = 2004.7 kJ

Problem 4-3 Process-dependent heat addition to a control mass (section III)
    Two tanks, A and B (Fig. 4-10), each with a capacity of 20 ft³ (0.566 m³), are perfectly insulated from the surroundings. A diatomic perfect gas is initially confined in tank A at a pressure of 10 atm abs and a temperature of 70°F (21.2°C). Valve C is initially closed, and tank B is completely evacuated.

1. If valve C is opened and the gas is allowed to reach the same temperature in both tanks, what are the final pressure and temperature?
2. If valve C is opened just until the pressure in the two tanks is equalized and is then closed, what are the final temperature and pressure in tanks A and B? Assume no transfer of heat between tank A and tank B.

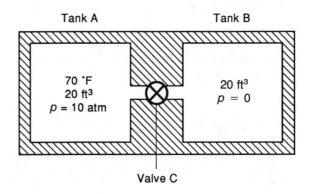

Figure 4-10 Initial state of perfectly insulated tanks for problem 4-3.

Note that for a diatomic gas, $c_v = (5/2)$ R.

Answers:

1. $p = 5$ atm; $T = 530°R$
2. $p = 5$ atm; $T_A = 434.8°R$, $T_B = 678.3°R$

Problem 4-4 Control mass energy balance (section III)

A perfectly insulated vessel contains 20 lb (9.1 kg) of water at an initial temperature of 500°R (277.8°K). An electric immersion heater with a mass of 1 lb (0.454 kg) is also at an initial temperature of 500°R (277.8°K). The water is slowly heated by passing an electric current through the heater until both water and heater attain a final temperature of 600°R (333.3°K). The specific heat of the water is 1 Btu/lb °R (4.2 kJ/kg K) and that of the heater is 0.12 Btu/lb °R (0.504 kJ/kg°K). Disregard volume changes and assume that the temperatures of water and heater are the same at all stages of the process.

Calculate $\int \dfrac{dQ}{T}$ and ΔS for:

1. The water as a system
2. The heater as a system
3. The water and heater as the system

Answers:

1. $\Delta S = \int \dfrac{dQ}{T} = 3.646$ Btu/°R
2. $\Delta S = 0.0219$ Btu/°R

   $\int \dfrac{dQ}{T} = -3.646$ Btu/°R
3. $\Delta S = 3.668$ Btu/°R

   $\int \dfrac{dQ}{T} = 0$

Problem 4-5 Qualifying a claim against the first and second laws of thermodynamics (section III)

An engineer claims to have invented a new compressor that can be used with a small gas-cooled reactor. The CO_2 used for cooling the reactor enters the compressor at 200 psi (1.378 MPa) and 120°F (48.9°C) and leaves the compressor at 300 psi (2.067 MPa) and 20°F ($-6.7°C$). The compressor requires no input power but operates simply by transferring heat from the gas to a low-temperature reservoir surrounding the compressor. The inventor claims that the compressor can handle 2 lb (0.908 kg) of CO_2 per second if the temperature of the reservoir is $-140°F$ ($-95.6°C$) and the rate of heat transfer is 60 Btu/s (63.6 kJ/s). Assuming that the CO_2 enters and leaves the device at very low velocities and that no significant elevation changes are involved:

1. Determine if the compressor violates the first or second law of thermodynamics.
2. Draw a schematic of the process on a T-s diagram.
3. Determine the change in the availability function (A) of the fluid flowing through the compressor.
4. Using the data from items 1 through 3, determine if the compressor is theoretically possible.

Answers:

1. The compressor does not violate either the first or the second law.
3. $\Delta A = 7.28$ Btu/lb

Problem 4-6 Determining rocket acceleration from an energy balance (section III)

A rocket ship is traveling in a straight line through outer space, beyond the range of all gravitational forces. At a certain instant, its velocity is V, its mass is M, the rate of consumption of propellant is P, the rate of energy liberation by the chemical reaction is $\dot{Q}_R$. From the first law of thermodynamics alone (i.e., without using a force balance on the rocket), derive a general expression for the rate of change of velocity with time as a function of the above parameters and the discharged gas velocity V_d and enthalpy h_d.

Answer:

$$\frac{dV}{dt} = \frac{1}{MV}\left[\dot{Q}_R - M\frac{du}{dt} + \left(u + \frac{V^2}{2} - h_d - \frac{V_d^2}{2}\right)P\right]$$

Problem 4-7 Momentum balance for a control volume (section III)

A jet of water is directed at a vane (Fig. 4-11) that could be a blade in a turbine. The water leaves the nozzle with a speed of 15 m/s and mass flow of 250 kg/s; it enters the vane tangent to its surface (in the x direction). At the point the water leaves the vane, the angle to the x direction of 120°. Compute the resultant force on the vane if:

1. The vane is held constant.
2. The vane moves with a velocity of 5 m/s in the x direction.

Answers:

1. $F_x = 1875$ N, $F_y = 3247.6$ N
2. $F_x = 833.3$ N, $F_y = 1443.4$ N

Problem 4-8 Internal conservation equations for an extensive property (section IV)

Consider a coolant in laminar flow in a circular tube. The one-dimensional velocity is given by

$$\vec{v} = v_{max}\left[1 - \left(\frac{r}{R}\right)^2\right]\vec{i}_z$$

where $v_{max} = 2.0$ m/s; $R =$ radius of the tube $= 0.05$ m; $\vec{i}_z =$ a unit vector in the axial direction. Assume the fluid density is uniform within the tube ($\rho_o = 300$ kg/m³).

1. What is the coolant flow rate in the tube?
2. What is the coolant average velocity (V) in the tube?
3. What is the true kinetic head of this flow? Does the kinetic head equal $\frac{1}{2}\rho_o V^2$?

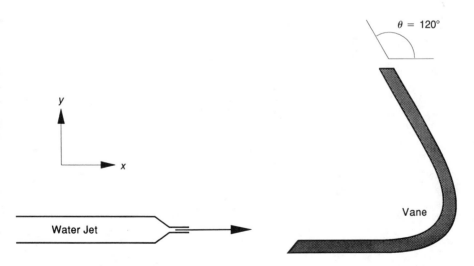

Figure 4-11 Jet of water directed at a vane.

Answers:

1. $Q = 7.854 \times 10^{-3} \, \text{m}^3/\text{s}$
2. $V = 1 \, \text{m/s}$
3. Kinetic head $= 1.33 \, \dfrac{\rho_o V^2}{2}$

Problem 4-9 Differential transport equations (section V)

Can the following sets of velocities belong to possible incompressible flow cases?

1. $v_x = x + y + z^2$
   $v_y = x - y + z$
   $v_z = 2xy + y^2$
2. $v_x = xyzt$
   $v_y = -xyzt^2$
   $v_z = \dfrac{z^2}{2}(xt^2 - yt)$

Answers:

1. Yes
2. Yes

TRANSPORT EQUATIONS FOR TWO-PHASE FLOW

I INTRODUCTION

In principle, transport of mass, momentum, and energy can be formulated by taking balances over a control volume or by integrating the point equations over the desired region. It has been found that integrating (averaging) the point continuum equations over the desired volume is a more accurate and insightful, although more lengthy, approach to the formulation. Because of its simplicity, the balance approach has been useful in well known one-dimensional applications, e.g., the separate flow model of Martinelli and Nelson [4]. However, the averaging approach, which was developed later, can better account for the interfacial conditions at the gas–liquid boundaries.

The formulation of appropriate models for two-phase flow in nuclear, chemical, and mechanical systems has been an area of substantial interest since the mid-1960s. However, the proliferation of two-phase modeling efforts has made it difficult for newcomers to identify the basic assumptions and the limitations of the abundant models. This chapter provides a framework for defining the relations between the more established approach of taking balances to derive control volume equations that the reader may find in Wallis [7] or Collier [1] and the more recent approaches of averaging the local and instantaneous balance equations in time and/or space as described by Ishii [3] or Delhaye et al. [2].

A Macroscopic Versus Microscopic Information

The fundamental difficulty of two-phase flow description arises from the multiplicity of internal configurations that must be taken into consideration. A sensor at a localized position in the flow channel may feel the presence of one phase continuously, as with annular flow, or the two phases intermittently, as with bubbly flow. On the other hand, the viewers of an x-ray snapshot of a cross section of a two-phase flow channel can observe only the space-averaged behavior of the mixture of two phases and cannot identify the local behavior of each point in the field. It is this macroscopic or averaged two-phase flow behavior that is of interest in practice. However, the interaction between the two phases or between the fluids and the structures depends to a great extent on the microscopic behavior. Therefore the basic question is how to predict the practically needed (and measurable) macroscopic behavior on the basis of whatever microscopic behavior may exist—and to the degree of accuracy required for the application.

With a single-phase flow, bridging between the microscopic phenomena and macroscopic description of the flow is easier. The observer of a point moving within the single-phase volume of consideration sees one continuous material. With two-phase flow, the local observer encounters interphase surfaces that lead to jump conditions between the two phases. The observer of a large volume may be able to consider the two phases as a single fluid but only after the properties of the two phases are properly averaged and the jump conditions are accounted for. These jump conditions describe mass, momentum, or energy exchange between the two phases.

B Multicomponent Versus Multiphase Systems

It is also beneficial to note that a two-phase system can be classified as a one-component or a multicomponent system, where "component" refers to the chemical species of the substance. Thus steam–water is a one-component, two-phase system, whereas nitrogen–water is a two-component, two-phase system. In the latter case mass transfer between the two phases can often be ignored, but momentum and energy transfer must be accounted for. It is possible that a mixture of various chemical components constitutes one of the phases, e.g., air and steam mixture in a reactor containment following a loss of coolant accident (LOCA). A multicomponent phase is often treated as a homogeneous mixture. Obviously, a multicomponent phase may be treated as a multifluid problem at the expense of greater complexity.

Because chemical reactions between the phases are the exception in nuclear engineering applications, they are not considered here. Thus the two-phase flow equations formulated here are still a special case of a more general formulation that may consider chemically reactive two phases.

C Mixture Versus Multifluid Models

A wide variety of possibilities exist for choosing the two-phase flow model. They range from describing the two-phase flow as a pseudo single-phase fluid (mixture) to a multifluid flow (e.g., liquid film, vapor, and droplets). Generally, as the two-phase flow model becomes more complex, more constitutive equations are required to represent the interactions between the fluids.

The homogeneous equilibrium model (HEM) is the simplest of the mixture models. It assumes that there is no relative velocity between the two phases (i.e., homogeneous flow) and that the vapor and liquid are in thermodynamic equilibrium. In this case the mass, momentum, and energy balance equations of the mixture are sufficient to describe the flow. The HEM assumptions are clearly limiting but may be adequate for certain flow conditions. Extensions of this model to include relative velocity, or slip, between the two phases and thermal nonequilibrium (e.g., subcooled boiling) effects are possible using externally supplied, usually empirical, constitutive relations. The HEM model has the advantage of being useful in analytical studies, as demonstrated in several chapters in this book.

Mixture models other than HEM add some complexity to the two-phase flow description. By allowing the vapor and liquid phases to have different velocities but constraining them to thermal equilibrium, these methods allow for more accurate velocity predictions. Alternatively, by allowing one phase to depart from thermal equilibrium, enthalpy prediction can be improved. In this case four or even five transport equations may be needed as well as a number of externally supplied relations to specify the interaction between the two phases. A well known example for such mixture models is the thermal equilibrium drift flux model, which allows the vapor velocity to be different from that of the liquid by providing an algebraic relation for the velocity differential between the two phases (three-equation drift flux model). By considering two mass equations but using only mixture momentum and energy equations, one of the phases may depart from thermal equilibrium (i.e., a four-equation drift flux model).

In the two-fluid model, three conservation equations are written for both the vapor and the liquid phases. Hence the model is often called the six-equation model. This model allows a more general description of the two-phase flow. However, it also requires a larger number of constitutive equations. The most important relations are those that represent the transfer of mass (Γ), transfer of energy (Q_s), and transfer of momentum (F_s) across liquid–vapor interfaces. The advantage of using this model is that the two phases are not restricted to prescribed temperature or velocity conditions. Extensions of the two-fluid model to multifluid models—in which vapor bubbles, a continuous liquid, a continuous vapor, and liquid droplets are described by separate sets of conservation equations—are also possible but have not been as widely applied as the simpler two-fluid model.

The combination of a mixture transport equation and one separate phase equation can also be used to replace two separate phase equations. Table 5-1 summarizes a variety of possible models to describe the two phase-flow problem. Note that there are six conservation equations and imposed restrictions for all models except the three-fluid model. Note also that models requiring a larger number of constitutive relations, though not reducing the number of imposed restrictions, are generally not worth exploiting. Thus the four-equation model B is less useful than the four-equation model A. For the five-equation models, model C is less useful than model B.

The choice among the alternative models discussed above depends on the nature of the problem to be solved. In LWR applications, the three-equation HEM model may be adequate for predicting the pressure drop in a flow channel under high-pressure steady-state conditions. Calculating the void distribution requires a specified relative velocity or a four-equation mixture model because of the tendency of the vapor to

Table 5-1 Two-phase flow models: $p_\ell = p_v = p_s$

Two-phase flow model	Conservation equations				Imposed restrictions			Constitutive laws					
								Wall		Interphase			
	Mass	Energy	Mom.	Total No.	Phase enthalpy or temperatures	Phase velocities	Total No.	Mom. F_w	Energy Q_w	Mass Γ	Energy Q_s	Mom. F_s	Total No.
General 3-equation models	1	1	1	3	T_v and T_ℓ specified	Specified	3	1	1	0	0	0	2
Homogeneous equilibrium	1	1	1	3	T_v and T_ℓ equilibrium	Equal	3	1	1	0	0	0	2
Equilibrium drift flux	1	1	1	3	T_v and T_ℓ equilibrium	Specified drift flux	3	1	1	0	0	0	2
4-Equation models													
A	2	1	1	4	T_v or T_ℓ	Slip relation	2	1	1	1	0	0	3
B	1	2	1	4	T_v or T_ℓ equilibrium	Slip relation	2	1	2	1[a]	1	0	5
C	1	1	2	4	T_v and T_ℓ equilibrium	None	2	2	1	1[a]	0	1	5
5-Equation models													
A	2	2	1	5	None	Slip relation	1	1	2	1	1	0	5
B	2	1	2	5	T_v or T_ℓ equilibrium	None	1	2	1	1	0	1	5
C	1	2	2	5	T_v or T_ℓ equilibrium	None	1	2	2	1[a]	1	1	7
Two-fluid	2	2	2	6	None	None	0	2	2	1	1	1	7
Three-fluid: continuous liquid, vapor, and liquid drops	3	3	3	9	None	None	0	3	3	2	2	2	12

[a] Γ is needed whenever Q_s or F_s is needed.

move faster than the liquid. Fast transients with sudden pressure changes are best handled by the six-equation, two-fluid model because of the expected large departure from equilibrium conditions.

II AVERAGING OPERATORS FOR TWO-PHASE FLOW

Given the nature of two-phase flow, the field parameters of interest are usually averaged quantities in either space or time and often in both. Thus it is important to introduce certain averaging mathematical operators here as well as define the more basic parameters that appear in the two-phase flow transport equations.

A Phase Density Function

If a phase (k) is present at a point $(\vec{r})$, a phase density function (α_k) can be defined as follows:

$$\left.\begin{array}{l} \alpha_k(\vec{r},t) \equiv 1 \text{ if point } \vec{r} \text{ is occupied by phase } k \\ \alpha_k(\vec{r},t) \equiv 0 \text{ if point } \vec{r} \text{ is not occupied by phase } k \end{array}\right\} \tag{5-1}$$

B Volume-Averaging Operators

Any volume (V) can be viewed as divided into two domains (V_k) (where $k = v$ or ℓ), each containing one of the two phases. We can then define two instantaneous volume-averaging operators acting on any parameter (c); one over the entire volume:

$$\langle c \rangle \equiv \frac{1}{V} \iiint_V c\,dV \tag{5-2}$$

and the other over the volume occupied by the phase k:

$$\langle c \rangle_k \equiv \frac{1}{V_k} \iiint_{V_k} c\,dV = \frac{1}{V_k} \iiint_V c\alpha_k\,dV \tag{5-3}$$

C Area Averaging Operators

The description of the two-phase flow conditions at the boundary of a control volume requires the space- and time-averaged values of a property on the surface area surrounding the control volume. Thus we introduce here the definition of such an area average for a parameter c:

$$\{c\} \equiv \frac{1}{A} \iint_A c\,dA \tag{5-4}$$

It should also be clear that the average of parameter c over the area occupied only by the phase k may be obtained from:

$$\{c\}_k \equiv \frac{1}{A_k} \iint_{A_k} c \, dA \equiv \frac{1}{A_k} \iint_A c\alpha_k dA \qquad (5\text{-}5)$$

D Local Time-Averaging Operators

Because the two phases may intermittently pass through a point $\vec{r}$, a time-average process can be defined as follows:

$$\tilde{c} \equiv \frac{1}{\Delta t^*} \int_{t-\Delta t^*/2}^{t+\Delta t^*/2} c \, dt \qquad (5\text{-}6)$$

where Δt^* is chosen large enough to provide a meaningful statistical count but short enough that the flow conditions are not substantially altered during the observation of a transient flow condition. Defining a suitable Δt^* can be a practical problem in fast transients.

It is also possible to define a phase time-averaging operator by averaging only over that portion of time within Δt^* when a single phase occupies the position $\vec{r}$ so that:

$$\tilde{c}^k \equiv \int_{t-\Delta t^*/2}^{t+\Delta t^*/2} c\alpha_k dt \div \int_{t-\Delta t^*/2}^{t+\Delta t^*/2} \alpha_k \, dt \qquad (5\text{-}7)$$

Note that Δt^* is long compared to a short time period Δt that may be chosen for averaging the turbulent effects (i.e., high-frequency fluctuations) within each single phase as was done in Chapter 4, Section VI. To filter out the high-frequency single-phase fluctuation, an average value of c has been defined by:

$$\bar{c} \equiv \frac{1}{\Delta t} \int_{t-\Delta t/2}^{t+\Delta t/2} c \, dt \qquad (4\text{-}124)$$

E Commutativity of Space- and Time-Averaging Operations

Vernier and Delhaye [6] argued that the volumetric average of $\tilde{c}$ must be identical to the time average of $\langle c \rangle$, so that:

$$\langle \tilde{c} \rangle = \overline{\langle c \rangle} \qquad (5\text{-}8)$$

Sha et al. [5] suggested that Eq. 5-8 is strictly valid only for one-dimensional flow with uniform velocity. Their restriction has not yet been widely evaluated. Here, we adopt the view that commutativity of the time and space operators is valid for the Eulerian coordinate system where the time and space act as independent variables.

Similarly, for averaging the turbulent equations of a single phase:

$$\langle \bar{c} \rangle = \overline{\langle c \rangle} \qquad (5\text{-}9)$$

III VOLUME-AVERAGED PROPERTIES

A Void Fraction

1 Instantaneous space-averaged void fraction. The fraction of the control volume
(V) that is occupied by the phase k at a given time can be obtained from:

$$\langle \alpha_k \rangle = \frac{1}{V} \iiint_V \alpha_k \, dV = \frac{V_k}{V} = \frac{V_k}{V_k + V_{k'}} \tag{5-10}$$

The volume fraction of the gaseous phase is normally called the *void fraction* and is
referred to as simply $\langle \alpha \rangle$, that is:

$$\langle \alpha \rangle \equiv \langle \alpha_v \rangle \tag{5-11a}$$

Similarly, $\langle 1 - \alpha \rangle$ or $\langle \alpha_\ell \rangle$ is used to refer to the liquid volume fraction.

$$\langle 1 - \alpha \rangle \equiv \langle \alpha_\ell \rangle \tag{5-11b}$$

A spatial void fraction may be defined for an area or a line, as well as for a
volume.

2 Local Time-Averaged Void Fraction. Because any point is instantaneously occupied only by one phase, the phase fraction at point $\vec{r}$ in a flow channel is present
only in a time-averaged sense and is defined by:

$$\tilde{\alpha}_k = \frac{1}{\Delta t^*} \int_{t - \Delta t^*/2}^{t + \Delta t^*/2} \alpha_k \, dt \tag{5-12}$$

where $\tilde{\alpha}_v$ = a local time-averaged void fraction.

3. Space- and time-averaged void fraction. The two-phase flow is sufficiently turbulent to make the value of $\langle \alpha \rangle$ a fluctuating value, even at steady-state flow conditions. It is useful to introduce the time-averaged value as the basis for describing the
measurable quantities such as pressure gradient and mass fluxes. Thus the time- and
volume-averaged void fraction is defined as:

$$\overline{\langle \alpha \rangle} = \frac{1}{\Delta t^*} \int_{t - \Delta t^*/2}^{t + \Delta t^*/2} \langle \alpha \rangle \, dt \tag{5-13}$$

From Eq. 5-8, the time average of $\langle \alpha_k \rangle$ defined by Eq. 5-10 is equal to the volume
average of $\tilde{\alpha}_k$ defined by Eq. 5-12. Therefore:

$$\overline{\langle \alpha \rangle} = \overline{\langle \alpha_v \rangle} = \langle \tilde{\alpha}_v \rangle \tag{5-14a}$$

and

$$\overline{\langle 1 - \alpha \rangle} = \overline{\langle \alpha_\ell \rangle} = \langle \tilde{\alpha}_\ell \rangle \tag{5-14b}$$

B Volumetric Phase Averaging

1 Instantaneous volumetric phase averaging. It is common to assume that the spatial variation of the phase properties can be ignored within a control volume, so that average properties within the phase can be used to derive the balance equations, as done later in Sections V and VI. Therefore it is desirable to derive an expression for the average value of a property, i.e., $\langle c_k \rangle_k$ where the phasic property c_k has been averaged over only that volume occupied by phase k (V_k). Note that the parameter c of Eq. 5-2 is now considered as a property of phase k. We express $\langle c_k \rangle_k$ as:

$$\langle c_k \rangle_k \equiv \frac{1}{V_k} \iiint_{V_k} c_k dV \equiv \frac{1}{V_k} \iiint_{V} \alpha_k c_k dV \qquad (5\text{-}15)$$

However, the average of a phase property c_k over the entire volume is given by:

$$\langle c_k \rangle \equiv \frac{1}{V} \iiint_{V} \alpha_k c_k dV \equiv \frac{1}{V} \iiint_{V_k} c_k dV \qquad (5\text{-}16)$$

From Eqs. 5-15, 5-16 and 5-10 it is seen that:

$$\langle c_k \rangle = \langle c_k \rangle_k \frac{V_k}{V} = \langle c_k \rangle_k \langle \alpha_k \rangle \qquad (5\text{-}17)$$

Thus for any phasic property (c_k) the instantaneous phasic-averaged value $\langle c_k \rangle_k$ is related to the instantaneous volume-averaged value $\langle c_k \rangle$ by the volumetric phase fraction $\langle \alpha_k \rangle$. For example, if the property c_k is the liquid phase or the vapor phase density (ρ_ℓ or ρ_v), Eq. 5-17 yields:

$$\langle \rho_\ell \rangle = \langle \rho_\ell \rangle_\ell (1 - \alpha) \qquad (5\text{-}18a)$$

$$\langle \rho_v \rangle = \langle \rho_v \rangle_v \langle \alpha \rangle \qquad (5\text{-}18b)$$

Note that $\langle \rho_\ell \rangle_\ell$ and $\langle \rho_v \rangle_v$ are merely the thermodynamic state values found in handbooks, whereas $\langle \rho_\ell \rangle$ and $\langle \rho_v \rangle$ depend also on the values of $\langle \alpha \rangle$ and $\langle 1 - \alpha \rangle$ in the volume of interest.

2 Time averaging of volume-averaged quantities. The time average of the product of two variables does not equal the product of the time average of the two variables. Assuming that the instantaneous values of $\langle c_k \rangle_k$ and $\langle \alpha_k \rangle$ can be given as the sum of a time-average component and a fluctuation component, we get, using Eq. 5-17:

$$\overline{\langle c_k \rangle} = \overline{\langle c_k \rangle_k \langle \alpha_k \rangle} = \overline{[\overline{\langle c_k \rangle_k} + \langle c_k \rangle_k'][\overline{\langle \alpha_k \rangle} + \langle \alpha_k \rangle']}$$

Because $\overline{\overline{\langle c \rangle}} = \overline{\langle c \rangle}$:

$$\overline{\langle c_k \rangle} = \overline{\langle c_k \rangle_k} \; \overline{\langle \alpha_k \rangle} + \psi' \qquad (5\text{-}19)$$

where:

$$\psi' = \overline{\overline{\langle c_k \rangle_k} \langle \alpha_k \rangle'} + \overline{\langle c_k \rangle_k' \overline{\langle \alpha_k \rangle}} + \overline{\langle c_k \rangle_k' \langle \alpha_k \rangle'} \qquad (5\text{-}20)$$

But $\langle c \rangle' = 0$, so:

$$\psi' = \overline{\langle c_k \rangle'_k \langle \alpha_k \rangle'} \tag{5-21}$$

Clearly the term ψ' is the result of the time fluctuations in $\langle c_k \rangle$ and $\langle \alpha_k \rangle$ and may become important in high flows or rapidly varying flow conditions. Therefore experimentally obtained values of $\overline{\langle c_k \rangle}$ should not always be readily assumed to equal $\overline{\langle c_k \rangle_k} \, \overline{\langle \alpha_k \rangle}$ unless fluctuations in $\langle c_k \rangle_k$ or $\langle \alpha_k \rangle$ can be ignored.

C Static Quality

An important variable in many two-phase situations is the mass fraction of vapor in a fixed volume. This fraction, called the *static quality*, is typically written without the brackets for volume averages as: x_{st}. It is given at any instant by:

$$x_{st} = \frac{m_v}{m_v + m_\ell} \tag{5-22}$$

$$x_{st} = \frac{\langle \rho_v \rangle V}{(\langle \rho_v \rangle + \langle \rho_\ell \rangle) V} \tag{5-23}$$

Note that x_{st} is a volume-average property by definition; therefore $\langle x_{st} \rangle \equiv x_{st}$.

Using Eq. 5-17, we can recast Eq. 5-23 as:

$$x_{st} = \frac{\langle \rho_v \rangle_v \langle \alpha \rangle}{\langle \rho_v \rangle_v \langle \alpha \rangle + \langle \rho_\ell \rangle_\ell \langle 1 - \alpha \rangle} \tag{5-24}$$

D Mixture Density

The two-phase mixture density in a volume can be given by:

$$\langle \rho \rangle = \frac{m_v + m_\ell}{V}$$

and by using Eqs. 5-16 and 5-17 we get:

$$\langle \rho \rangle = \langle \rho_v \rangle_v \langle \alpha \rangle + \langle \rho_\ell \rangle_\ell \langle 1 - \alpha \rangle \tag{5-25}$$

Consequently, the average phasic density may be given by (using Eqs. 5-24 and 5-25):

$$\langle \rho_v \rangle_v = \frac{x_{st} \langle \rho \rangle}{\langle \alpha \rangle} \tag{5-26a}$$

or

$$\langle \rho_\ell \rangle_\ell = \frac{(1 - x_{st}) \langle \rho \rangle}{\langle 1 - \alpha \rangle} \tag{5-26b}$$

IV AREA-AVERAGED PROPERTIES

A Area-Averaged Phase Fraction

The instantaneous fraction of an area occupied by phase k is given by:

$$\{\alpha_k\} \equiv \frac{A_k}{A} = \frac{1}{A} \iint_A \alpha_k dA \tag{5-27}$$

Because the area not occupied by phase k must be occupied by the other phase k', the fraction of area occupied by phase k can also be given as:

$$\{\alpha_k\} \equiv \frac{A_k}{A} = \frac{A_k}{A_k + A_{k'}} \tag{5-28}$$

If space- and time-averaging commutativity are applied, the time- and area-averaged phase fraction can be given by:

$$\overline{\{\alpha_k\}} \equiv \{\tilde{\alpha}_k\} \tag{5-29}$$

Similar to Eq. 5-17, it is possible to relate the phase property average to the total area property average by:

$$\{c_k\} = \{c_k\}_k \{\alpha_k\} \tag{5-30}$$

Similar to Eq. 5-19, the time- and area-averaged property $\overline{\{c_k\}}$ may be obtained from the following manipulations:

$$\overline{\{c_k\}} = \overline{\{c_k\}_k \{\alpha_k\}} = \overline{\overline{\{c_k\}}_k \overline{\{\alpha_k\}}} + \overline{\overline{\{c_k\}}_k \{\alpha_k\}'} + \overline{\{c_k\}'_k \overline{\{\alpha_k\}}} + \overline{\{c_k\}'_k \{\alpha_k\}'} \tag{5-31a}$$

However,

$$\overline{\{\alpha_k\}'} = \overline{\{c\}'_k} = 0 \text{ and } \overline{\overline{\{c_k\}}} = \overline{\{c_k\}}$$

Therefore the last equation becomes:

$$\overline{\{c_k\}} = \overline{\{c_k\}_k}\overline{\{\alpha_k\}} + \overline{\{c_k\}'_k\{\alpha_k\}'} \tag{5-31b}$$

So that when turbulent fluctuations in either of $\{\alpha_k\}$ or $\{c_k\}$ are ignored:

$$\overline{\{c_k\}_k} = \frac{\overline{\{c_k\}}}{\overline{\{\alpha_k\}}}; \text{ turbulent fluctuations ignored.} \tag{5-32}$$

Example 5-1 Time average of area-averaged void fraction

PROBLEM Consider a series of cylindrical bubbles, each of length ℓ_2 and diameter d_b, moving at a velocity V_b in a cylindrical pipe. The pipe diameter is D, and its length is L.

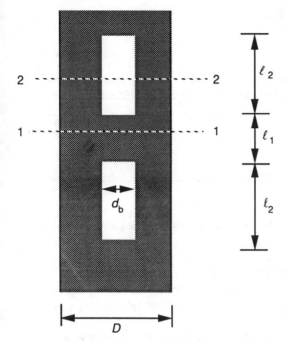

Figure 5-1 Bubble and pipe characteristics for Example 5-1.

If the bubbles are separated by a distance $\ell_1 = 0.5\ell_2$, and the bubble diameter $d_b = 0.4D$, evaluate the area-averaged void fraction at cross sections 1 and 2 at the instant illustrated in Figure 5-1. Also, evaluate the time- and area-averaged void fraction at positions 1 and 2.

SOLUTION The instantaneous area-averaged void fractions at the positions indicated in Figure 5-1 are obtained by applying Eq. 5-28:

$$\{\alpha\}_1 = 0$$

$$\{\alpha\}_2 = \frac{\pi d_b^2/4}{\pi D^2/4} = (0.4)^2 = 0.16$$

The time- and area-averaged values are equal: $\overline{\{\alpha\}}_1 = \overline{\{\alpha\}}_2 = \overline{\{\alpha\}}$ and are obtained by applying Eq. 5-6:

$$\overline{\{\alpha\}} = \frac{1}{\Delta t^*} \int_{t-\Delta t^*/2}^{t+\Delta t^*/2} \{\alpha\} \, dt = \frac{1}{\Delta t^*} \int_0^{\Delta t^*} \{\alpha\} \, dt$$

$$\overline{\{\alpha\}} = \frac{(0)\dfrac{\ell_1}{V_b} + (0.16)\dfrac{\ell_2}{V_b}}{(\ell_1 + \ell_2)/V_b}$$

$$\overline{\{\alpha\}} = 0.16\left(\frac{\ell_2}{\ell_1 + \ell_2}\right) = 0.16\left(\frac{1}{0.5 + 1}\right) = 0.1067$$

Note that Δt^* is chosen such that:

$$\Delta t^* > \frac{\ell_1 + \ell_2}{V_b}$$

in order to properly average the two-phase properties. Under transient conditions, it is important to select an averaging time less than the characteristic bubble residence time in the tube so that:

$$\Delta t^* < \frac{L}{V_b}$$

For steady-state conditions this condition is immaterial, as the behavior of the two-phase system is periodic.

Example 5-2 Equivalence of time–area and area–time sequence of operators

PROBLEM Consider vertical flow of a gas–liquid mixture between two parallel plates with two types of gas bubble with plate geometry as shown in Figure 5-2a. The size and spacing of the bubbles are indicated on the figure. The bubble velocities are related by $V_A = 2V_B$. Six probes measure the phase–time relations: r_1 to r_6. The measured signals appear in Figure 5-2b.

Determine the area average of the time-averaged local void fractions, at r_1 to r_6, i.e., $\{\tilde{\alpha}_v\}$. Also, determine the time average of the instantaneous area-averaged void fraction at the plane of the probes, i.e., $\overline{\{\alpha_v\}}$. Are they equal for the stated conditions?

SOLUTION The local time-averaged void fraction can be written from Eq. 5-12 as:

$$\tilde{\alpha}_k = \frac{1}{\Delta t^*} \int_0^{\Delta t^*} \alpha_k dt; \quad \alpha_k = 1 \text{ if phase } k \text{ is present}$$

where $\Delta t^* \gg \tau$, the time it takes one bubble to pass by the probe. Therefore

$\tilde{\alpha}_{v1} = 0$
$\tilde{\alpha}_{v2} = 0.5$
$\tilde{\alpha}_{v3} = 0.5$
$\tilde{\alpha}_{v4} = 0$
$\tilde{\alpha}_{v5} = 0.5$
$\tilde{\alpha}_{v6} = 0$

At the plane of the sensors, the area average of the local time-averaged void fraction is:

$$\{\tilde{\alpha}_v\} = \frac{1}{6\ell} \int_0^{6\ell} \tilde{\alpha}_v \, d\ell = \frac{1}{6\ell} \sum_j \tilde{\alpha}_{vj} \, \ell_j = [0.5(2\ell + \ell)] \frac{1}{6\ell} = 0.25$$

Next consider the average of the instantaneous void fraction at the plane of the sensors for the time shown ($t = 0$):

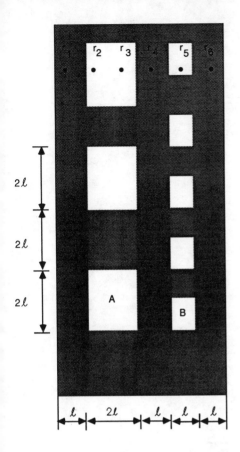

(a) Bubble Configuration in Channel

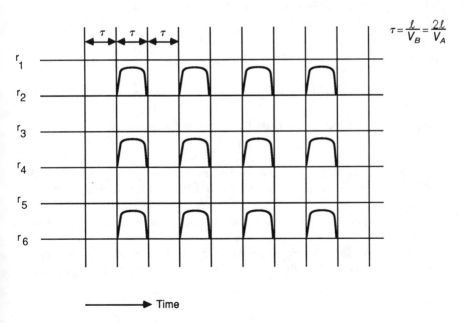

$$\tau = \frac{\ell}{V_B} = \frac{2\ell}{V_A}$$

Time

(b) Vapor Sensor Output

$$\{\alpha_v\} = \frac{\sum_j \alpha_v \ell_j}{6\ell} = 0.5$$

If we observe the time behavior of $\{\alpha_v\}$ we get:

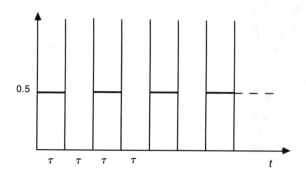

which implies that:

$$\overline{\{\alpha_v\}} = \frac{1}{\Delta t^*} \int_0^{\Delta t^*} \{\alpha_v\} \, dt = \frac{0.5\tau + 0\tau}{2\tau} = 0.25$$

Therefore:

$$\{\tilde{\alpha}_v\} = \overline{\{\alpha_v\}}$$

which is what Vernier and Delhaye [6] proved mathematically.

B Flow Quality

The instantaneous mass flow rate of phase k through an area $\vec{A}_j$ can be obtained from:

$$\dot{m}_{kj} = \iint\limits_{A_j} \alpha_k \rho_k \vec{v}_k \cdot \vec{n} \, dA_j = \{\alpha_k \rho_k \vec{v}_k\}_j \cdot \vec{A}_j \qquad (5\text{-}33a)$$

Using Eq. 5-30, the instantaneous mass flow rate can be written as:

$$\dot{m}_{kj} = \{\rho_k \vec{v}_k\}_{kj} \{\alpha_k\}_j \cdot \vec{A}_j \qquad (5\text{-}33b)$$

The time-averaged flow rate of phase k is then obtained by integrating Eq. 5-33b over the time Δt^*. Performing this integration we get, according to Eq. 5-31b:

$$\overline{\dot{m}_{kj}} = \overline{\{\rho_k \vec{v}_k\}_{kj}} \, \overline{\{\alpha_k\}_j} \cdot \vec{A}_j; \text{ turbulent fluctuation in } \{\alpha_k\} \text{ or } \{\rho_k \vec{v}_k\}_k \text{ is ignored.} \quad (5\text{-}34)$$

The vapor mass flow fraction of the total flow is called the *flow quality* (x). To define the flow quality in a one-dimensional flow, e.g., in the z direction, we write:

$$x_z \equiv \dot{m}_{vz}/(\dot{m}_{vz} + \dot{m}_{\ell z}) \qquad (5\text{-}35)$$

Let:

$$\dot{m} = \dot{m}_{vz} + \dot{m}_{\ell z} \tag{5-36}$$

If the total flow rate is assumed nonfluctuating, i.e.,

$$\overline{\dot{m}_z} = \dot{m}_z$$

for the time-averaged value of the quality we get:

$$\tilde{x}_z = \tilde{\dot{m}}_{vz}/(\tilde{\dot{m}}_{vz} + \tilde{\dot{m}}_{\ell z}) = \tilde{\dot{m}}_{vz}/\dot{m}_z \tag{5-37}$$

The flow quality is often used in the analysis of predominantly one-dimensional two-phase flow. Many correlations that are extensively used in two-phase flow and heat transfer are given in terms of the flow quality. The flow quality becomes particularly useful when thermodynamic equilibrium between the two phases is assumed. In that case the flow quality can be obtained from the energy balances of the flow, although additional information or correlations are needed to calculate the void fraction. With two-dimensional flow the quality at a given plane has two components and is not readily defined by a simple energy balance. Therefore for two- or three-dimensional flow, the flow quality becomes less useful because of the scarcity of literature on the effect of the vectorial nature of the quality on the flow conditions.

The concept of flow quality breaks down when the net flow rate is zero or when there is countercurrent flow of two phases (extensions to these cases are possible but not very useful).

C Mass Fluxes

The superficial mass fluxes or mass velocities of the phases are defined as the phase flow rates per unit cross-sectional area, e.g.,

$$G_{\ell z} \equiv \frac{\dot{m}_{\ell z}}{A_z} = \frac{\dot{m}_z(1 - x_z)}{A_z} = G_{mz}(1 - x_z) \tag{5-38a}$$

and

$$G_{vz} = \frac{\dot{m}_{vz}}{A_z} = \frac{\dot{m}_z x_z}{A_z} = G_{mz} x_z \tag{5-38b}$$

The superficial mass velocity can be expressed also in terms of the actual phasic velocity by specifying $\dot{m}_{vz}$ from Eq. 5-33b:

$$G_{kz} \equiv \frac{\{\rho_k \vec{v}_k\}_{kz} \{\alpha_k\}_z \cdot \vec{A}_z}{A_z} = \{\rho_k v_{kz}\}_{kz} \{\alpha_k\}_z \tag{5-39}$$

The mass flux (G_m) is defined as the average mass flow rate per unit flow area. Thus in the z direction:

$$G_{mz} \equiv \frac{\dot{m}_z}{A_z} = \frac{\dot{m}_{vz} + \dot{m}_{\ell z}}{A_z} = G_{vz} + G_{\ell z} \qquad (5\text{-}40a)$$

Utilizing Eq. 5-39, the mass flux of the mixture can also be given by:

$$G_{mz} = \{\rho_v v_{vz}\}_{vz} \{\alpha_v\}_z + \{\rho_\ell v_{\ell z}\}_{\ell z} \{\alpha_\ell\}_z \qquad (5\text{-}40b)$$

which, using Eq. 5-30, can be written as:

$$G_{mz} = \{\rho_v \alpha \, v_{vz}\}_z + \{\rho_\ell (1 - \alpha) v_{\ell z}\}_z \qquad (5\text{-}40c)$$

Time-averaging G_{mz}, using Eq. 5-40b and ignoring time fluctuations in $[\alpha_k]$, we obtain:

$$\tilde{G}_{mz} = \frac{\tilde{\dot{m}}_z}{A_z} = \overline{\{\rho_v \, v_{vz}\}_{vz}\{\alpha\}}_z + \overline{\{\rho_\ell v_{\ell z}\}_{\ell z}\{1 - \alpha\}}_z \qquad (5\text{-}41)$$

D Volumetric Fluxes and Flow Rates

The instantaneous volumetric flow rate of phase k may be related to the volumetric flux (or superficial velocity) over the area A_j by:

$$Q_{kj} = \{\vec{j}_k\}_j \cdot \vec{A}_j \qquad (5\text{-}42)$$

Note that for one-dimensional flow in the z direction, when the spatial variation in ρ_k can be neglected over the area A_z, we get:

$$\{j_v\}_z = \frac{Q_{vz}}{A_z} = \frac{\dot{m}_{vz}}{\rho_v A_z} = \frac{G_{vz}}{\rho_v} = \frac{G_{mz} x_z}{\rho_v} \qquad (5\text{-}43a)$$

Similarly:

$$\{j_\ell\}_z = \frac{Q_{\ell z}}{A} = \frac{\dot{m}_{\ell z}}{\rho_\ell A_z} = \frac{G_{\ell z}}{\rho_\ell} = \frac{G_{mz}(1 - x_z)}{\rho_\ell} \qquad (5\text{-}43b)$$

and

$$\{j\}_z = \frac{Q_{\ell z} + Q_{vz}}{A_z} = \frac{Q_z}{A_z} = G_m \left[\frac{1 - x_z}{\rho_\ell} + \frac{x_z}{\rho_v} \right] \qquad (5\text{-}44)$$

Because the volumetric flow rates Q_ℓ and Q_v are simply related to the mass flow rates and the phasic densities from Eqs. 5-43a and 5-43b, the volumetric fluxes $\{j_k\}$ can be considered known if the mass flow rates of the two phases—or, alternatively, the total mass flow rate and the quality—are known.

The local volumetric flux (or superficial velocity) of phase k is obtained from:

$$\vec{j}_k = \alpha_k \vec{v}_k \qquad (5\text{-}45)$$

Space- and time-averaged values of the volumetric fluxes can be obtained starting with either operation, as space and time commutation can be assumed:

$$\overline{\{\vec{j}_k\}} = \{\overline{\vec{j}_k}\} = \overline{\{\alpha_k \, \vec{v}_k\}} \qquad (5\text{-}46)$$

Similar to Eq. 5-31b, by ignoring time fluctuations in $\{\alpha_k\}$ or $\{\vec{v}_k\}$ Eq. 5-46 can be shown to yield:

$$\overrightarrow{\{\vec{j}_k\}} = \{\alpha_k\} \overrightarrow{\{\vec{v}_k\}}_k \tag{5-47}$$

E Velocity (Slip) Ratio

The time- and space-averaged, or macroscopic, velocity ratio, often referred to as *slip ratio* (S), is defined as:

$$S_i \equiv \frac{\{\bar{v}_{vi}\}_v}{\{\bar{v}_{\ell i}\}_\ell} \tag{5-48}$$

S_i is a directional value, as it depends on the velocity in the direction i.

For spatially uniform liquid and vapor densities and when time fluctuations in $\{\alpha_k\}$ or $\{v_{ki}\}$ are ignored, Eqs. 5-43a, 5-43b, and 5-47 can be used to obtain:

$$S_i = \frac{\{\bar{j}_v\}_i}{\{\bar{j}_\ell\}_i} \frac{\{1 - \alpha\}_i}{\{\alpha\}_i} = \frac{x_i}{1 - x_i} \frac{\rho_\ell}{\rho_v} \frac{\{1 - \alpha\}_i}{\{\alpha\}_i} \tag{5-49}$$

This equation is a commonly used relation between x_i, $\{\alpha_v\}_i$, and S_i. If any two are known, the third can be determined. Experimental void fraction data for one-dimensional flows have been used in the past to calculate the velocity ratio and then correlate S_i in terms of the relevant parameters. However, $\vec{S}$ cannot be easily related to meaningful parameters in three-dimensional flows. For one-dimensional flows, several models can be used to estimate S_i, as discussed in Chapter 11.

F Mixture Density Over an Area

Analogous to Eq. 5-25, we define the mixture average density of an area by:

$$\rho_m \equiv \{\rho\} = \{\alpha\}\{\rho_v\}_v + \{1 - \alpha\}\{\rho_\ell\}_\ell \tag{5-50a}$$

Note that this density can be called the *static density* of the mixture because it is not affected by the flow velocity. Using Eq. 5-30, we can write ρ_m as:

$$\rho_m = \{\alpha\rho_v\} + \{(1 - \alpha)\rho_\ell\} \tag{5-50b}$$

G Volumetric Flow Ratio

The volumetric flow ratio of vapor to the total mixture is given by:

$$\{\beta_z\} = \frac{Q_{vz}}{Q_{vz} + Q_{\ell z}} = \frac{\{j_v\}_z}{\{j\}_z} \tag{5-51}$$

H Flow Thermodynamic Quality

The flow (mixing-cup) enthalpy is defined for flow in the direction "i" as:

$$(h_m^+)_i \equiv x_i h_v + (1 - x_i) h_\ell \tag{5-52}$$

Table 5-2 One-dimensional relations between two phase parameters

	$\dot{m}$	G	Q	$\{\alpha\}$
$\dot{m}$	$\dot{m} = \dot{m}_k + \dot{m}_{k'}$			
G	$G_k = \dfrac{\dot{m}_k}{A}$	$G_m = G_k + G_{k'}$		
Q	$Q_k = \dfrac{\dot{m}_k}{\rho_k}$	$Q_k = \dfrac{G_k A}{\rho_k}$	$Q = Q_k + Q_{k'}$	
$\{\alpha\}$	$\{\alpha_k\} = \left(1 + \dfrac{\dot{m}_{k'}\{\rho v\}_k}{\dot{m}_k\{\rho v\}_{k'}}\right)^{-1}$	$\{\alpha_k\} = \left(1 + \dfrac{G_{k'}\{\rho v\}_k}{G_k\{\rho v\}_{k'}}\right)^{-1}$	$\{\alpha_k\} = \left(1 + \dfrac{Q_{k'}\{v\}_k}{Q_k\{v\}_{k'}}\right)^{-1}$	—
$\{\beta\}$	$\{\beta_k\} = \dfrac{\left(\dfrac{\dot{m}}{\rho}\right)_k}{\left(\dfrac{\dot{m}}{\rho}\right)_k + \left(\dfrac{\dot{m}}{\rho}\right)_{k'}}$	$\{\beta_k\} = \dfrac{\left(\dfrac{G}{\rho}\right)_k}{\left(\dfrac{G}{\rho}\right)_k + \left(\dfrac{G}{\rho}\right)_{k'}}$	$\{\beta_k\} = \dfrac{Q_k}{Q}$	$\{\beta_k\} = \left(1 + \dfrac{\{\alpha v\}_k}{\{\alpha v\}_k}\right)$
$\{v\}$	$\{v_k\} = \dfrac{\dot{m}_k}{\{\rho A\}_k}$	$\{v_k\} = \dfrac{G_m x_k}{\{\rho a\}_k}$	$\{v_k\} = \dfrac{Q_k}{A_k}$	$\dfrac{\{v_k\}}{\{v_{k'}\}} = \dfrac{\{\alpha\rho\}_{k'}\, x_k}{\{\alpha\rho\}_k\, x_k}$
$\{j\}$	$\{j_k\} = \dfrac{\dot{m}_k}{\rho_k A}$	$\{j_k\} = \dfrac{G_k}{\rho_k}$	$\{j_k\} = \dfrac{Q_k}{A}$	$\{j_k\} = \{(\alpha v)_k\}$
x_{st}	$\{x_{stk}\} = \left(1 + \dfrac{\dot{m}_{k'}\{v_k\}}{\dot{m}_k\{v_{k'}\}}\right)^{-1}$	$\{x_{stk}\} = \left(1 + \dfrac{G_{k'}\{v_k\}}{G_k\{v_{k'}\}}\right)^{-1}$	$\{x_{stk}\} = \dfrac{Q_k}{\{v_k\}A}\dfrac{\rho_k}{\rho_m}$	$\{x_{stk}\} = \dfrac{\{(\alpha\rho)_k\}}{\rho_m}$
x	$x_k = \dfrac{\dot{m}_k}{\dot{m}}$	$x_k = \dfrac{G_k}{G_m}$	$x_k = \left[1 + \dfrac{(Q\rho)_{k'}}{(Q\rho)_k}\right]^{-1}$	$x_k = \left(1 + \dfrac{\{(\alpha\rho v)_{k'}\}}{\{(\alpha\rho v)_k\}}\right)^{-1}$

All entries above the diagonal can be obtained by inverting corresponding existing entries in this table.

The flow thermodynamic (or equilibrium) quality is given by relating the flow enthalpy to the saturation liquid and vapor enthalpies as:

$$(x_e)_i = \frac{(h_m^+)_i - h_f}{h_g - h_f} \tag{5-53}$$

where h_f and h_g = saturation specific enthalpies of the liquid and vapor, respectively. Thus the flow quality can be used to define the equilibrium quality:

$$(x_e)_i = \frac{x_i h_v + (1 - x_i)h_\ell - h_f}{h_g - h_f} \tag{5-54}$$

Under thermal equilibrium conditions:

$$h_\ell = h_f \text{ and } h_v = h_g; \text{ therefore } (x_e)_i = (x)_i$$

H Summary of Useful Relations for One-Dimensional Flow

Table 5-2 presents a summary of useful relations between parameters in one-dimensional flow. It may be helpful to summarize the relations between the void fraction

$\{\beta\}$	$\{v\}$	$\{j\}$	x_{st}	x
—				
$\dfrac{\{v_k\}}{\{v_{k'}\}} = \dfrac{\{\beta_k\}\,\{\alpha_{k'}\}}{\{\beta_{k'}\}\,\{\alpha_k\}}$	—			
$\{j_k\} = \{\beta_k\}\{j\}$	$\{j_k\} = \{(v\alpha)_k\}$	$\{j\} = \{j_k\} + \{j_{k'}\}$		
$\{x_{stk}\} = \dfrac{\{\beta_k\}\{j\}\,\rho_k}{\{v_k\}\,\rho_m}$	see $\{x_{stk}\}$ vs. Q, β, or j	$\{x_{stk}\} = \dfrac{\{j_k\}\,\rho_k}{\{v_k\}\,\rho_m}$	—	
$x_k = \left(1 + \dfrac{\{(\beta\rho)_{k'}\}}{\{(\beta\rho)_k\}}\right)^{-1}$	see x_k vs. α or x_{stk}	$x_k = \left(1 + \dfrac{\{(j\rho)_{k'}\}}{\{(j\rho)_k\}}\right)^{-1}$	$x_k = \left[1 + \dfrac{(x_{st}v)_{k'}}{(x_{st}v)_k}\right]^{-1}$	—

$\{\alpha\}$ and the static quality (x_{st}) at an area (*static parameters*), with the volumetric flow vapor fraction (β) and the flow quality (x) (i.e., the dynamic parameters). Assuming uniform phase densities across an area, Eq. 5-49 leads to:

$$\{\alpha\} = \frac{1}{1 + \dfrac{1 - x}{x}\dfrac{\rho_v}{\rho_\ell}\,S} \tag{5-55}$$

and

$$\frac{x}{1 - x} = \frac{\{\alpha\}\,\rho_v}{(1 - \{\alpha\})\rho_\ell}\,S \tag{5-56}$$

However, the area-averaged static quality (x_{st}) can be defined (see Eq. 5-26a) for space-independent density ρ_v and ρ_ℓ as:

$$x_{st} \equiv \frac{\{\alpha\}\rho_v}{\rho_m} \tag{5-57}$$

which yields:

$$\frac{x_{st}}{1 - x_{st}} = \frac{\{\alpha\}\rho_v}{(1 - \{\alpha\})\rho_\ell} \tag{5-58}$$

Comparing Eqs. 5-56 and 5-58, it is obvious that if $S = 1$, $x = x_{st}$. Consider also the volumetric flow fraction given by

$$\{\beta\} = \frac{\{j_v\}}{\{j\}} = \frac{1}{1 + \dfrac{\{j_\ell\}}{\{j_v\}}} \tag{5-59}$$

Because for space-independent ρ_v and ρ_ℓ:

$$\{j_\ell\} = \frac{(1 - x)G_m}{\rho_\ell} \text{ and } \{j_v\} = \frac{xG_m}{\rho_v}$$

we get:

$$\{\beta\} = \frac{1}{1 + \dfrac{1 - x}{x}\dfrac{\rho_v}{\rho_\ell}} \tag{5-60}$$

By comparing Eqs. 5-55 and 5-60, we get:

$$\{\alpha\} = \{\beta\} \text{ if } S = 1$$

The assumption of $S = 1$ is referred to as the *homogeneous flow assumption*. Generally $S \neq 1$, as is discussed in Chapter 11.

A summary of the relations of x, x_{st}, $\{\alpha\}$, and $\{\beta\}$ appear in Figure 5-3.

V MIXTURE EQUATIONS FOR ONE-DIMENSIONAL FLOW

The one-dimensional case is presented first as it is often encountered when describing energy equipment, and because it is easier to establish than the multidimensional cases. A formal derivation of the one-dimensional equations as a special case of the three-dimensional equations is presented in section VII, below.

A Mass Continuity Equation

Consider the one-dimensional flow through a plane of an area A_z at a position z along a channel as illustrated in Figure 5-4. The mass balance equation can be written as:

$$\frac{\partial}{\partial t} \iint_{A_z} \rho \, dA_z + \frac{\partial}{\partial z} \iint_{A_z} \rho v_z \, dA_z = 0 \tag{5-61}$$

or

$$\frac{\partial}{\partial t} \{\rho_v \alpha + \rho_\ell(1 - \alpha)\} A_z + \frac{\partial}{\partial z} \{\rho_v \alpha v_{vz} + \rho_\ell(1 - \alpha) v_{\ell z}\} A_z = 0 \tag{5-62}$$

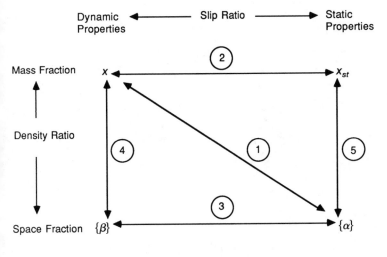

$$\frac{1-\{\alpha\}}{\{\alpha\}} = \frac{1-x}{x}\frac{\rho_v}{\rho_\ell} S \qquad (1)$$

$$\frac{1-x_{st}}{x_{st}} = \frac{1-x}{x} S \qquad (2)$$

$$\frac{1-\{\alpha\}}{\{\alpha\}} = \frac{1-\{\beta\}}{\{\beta\}} S \qquad (3)$$

$$\frac{1-\{\beta\}}{\{\beta\}} = \frac{1-x}{x}\frac{\rho_v}{\rho_\ell} \qquad (4)$$

$$\frac{1-\{\alpha\}}{\{\alpha\}} = \frac{1-x_{st}}{x_{st}}\frac{\rho_v}{\rho_\ell} \qquad (5)$$

$S = 1$ for Homogeneous flow

Figure 5-3 Useful relations for one-dimensional flow and spatially uniform phase densities.

where v_{vz} and $v_{\ell z}$ = vapor and liquid velocities in the axial direction. Eq. 5-62 can also be written as:

$$\frac{\partial}{\partial t}(\rho_m A_z) + \frac{\partial}{\partial z}(G_m A_z) = 0 \qquad (5\text{-}63)$$

where ρ_m and G_m = average mixture density and mass flux over the area A_z, defined in Eqs. 5-50b and 5-40c, respectively.

B Momentum Equation

Consider next the momentum equation, which can be obtained from equating the rate of momentum change to the net forces (shown in Figure 5-4 in the axial direction:

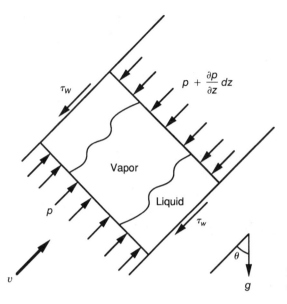

Figure 5-4 One-dimensional flow in a channel.

$$\frac{\partial}{\partial t} \iint\limits_{A_z} \rho v_z dA_z + \frac{\partial}{\partial z} \iint\limits_{A_z} \rho v_z^2 dA_z = -\iint\limits_{A_z} \frac{\partial p}{\partial z} dA_z - \int\limits_{P_z} \tau_w dP_z - \iint\limits_{A_z} \rho g \cos \theta dA$$

(5-64)

where P_z = perimeter of the channel at the location z, and θ = angle between the axis of the channel and the upward vertical direction. Thus:

$$\frac{\partial}{\partial t} \{\rho_v \alpha v_{vz} + \rho_\ell (1 - \alpha) v_{vz}\} A_z + \frac{\partial}{\partial z} \{\rho_v \alpha v_{vz}^2 + \rho_\ell (1 - \alpha) v_{\ell z}^2\} A_z =$$

$$- \left(\frac{\partial \{p\} A_z}{\partial z}\right) - \int\limits_{P_z} \tau_w \, dP_z - \{\rho_v \alpha + \rho_\ell (1 - \alpha)\} g \cos \theta A_z$$

(5-65)

If it is assumed that the pressure p is uniform within the area A_z, we can write Eq. 5-65 as:

$$\frac{\partial}{\partial t} (G_m A_z) + \frac{\partial}{\partial z} \left(\frac{G_m^2}{\rho_m^+} A_z\right) = -\frac{\partial (pA_z)}{\partial z} - \int\limits_{P_z} \tau_w dP_z - \rho_m g \cos \theta A_z \quad (5\text{-}66)$$

where ρ_m is defined in Eqs. 5-50a and 5-50b and ρ_m^+ = dynamic (or mixing cup) density given by:

$$\frac{1}{\rho_m^+} \equiv \frac{1}{G_m^2} \{\rho_v \alpha v_{vz}^2 + \rho_\ell (1 - \alpha) v_{\ell z}^2\} \quad (5\text{-}67)$$

Note that because the velocity in a general case has three components, the dynamic density is, in effect, a directional quantity.

C Energy Equation

Finally, for the energy balance we get:

$$\frac{\partial}{\partial t} \iint\limits_{A_z} \rho u^\circ \, dA_z + \frac{\partial}{\partial z} \iint\limits_{A_z} \rho h^\circ \, v_z dA_z = q' + \iint\limits_{A_z} q''' \, dA_z \tag{5-68}$$

where q' = linear heat addition rate from the walls, q''' = volumetric heat generation rate in the coolant, and the axial heat conduction and the work terms are ignored. Note that in this formulation there is no expansion work term, as the control volume is fixed and the surface forces due to pressure and wall shear p and τ_w are acting on a stationary surface. In reality, the internal shear leads to conversion of some of the mechanical energy into an internal heat source, which is often negligibly small. The wall shear stress τ_w indirectly affects u° by influencing the pressure in the system. Finally, the work due to the gravity force is considered negligible. A much more elaborate derivation is given in section VII.

Equation 5-68 may be written as:

$$\frac{\partial}{\partial t} \iint\limits_{A_z} \rho h^\circ \, dA_z + \frac{\partial}{\partial z} \iint\limits_{A_z} \rho h^\circ \, v_z dA_z = \frac{\partial}{\partial t} \iint\limits_{A_z} \rho(pv) \, dA_z + q' + \iint\limits_{A_z} q''' dA_z \tag{5-69}$$

or

$$\frac{\partial}{\partial t} (\{\rho_v \alpha h_v^\circ + \rho_\ell (1 - \alpha) h_\ell^\circ\} A_z) + \frac{\partial}{\partial z} (\{\rho_v \alpha h_v^\circ v_{vz} + \rho_\ell (1 - \alpha) h_\ell^\circ v_{\ell z}\} A_z) =$$
$$\left(\frac{\partial p}{\partial t}\right) A_z + q' + \iint\limits_{A_z} q''' \, dA_z \tag{5-70}$$

where p was assumed uniform over A_z. Ignoring the kinetic energy of both phases, Eq. 5-70 may be written as:

$$\frac{\partial}{\partial t} (\rho_m h_m A_z) + \frac{\partial}{\partial z} (G_m h_m^+ A_z) = \left(\frac{\partial p}{\partial t}\right) A_z + q' + \iint\limits_{A_z} q''' \, dA_z \tag{5-71}$$

where

$$h_m = \frac{1}{\rho_m} \{\rho_v \alpha h_v + \rho_\ell (1 - \alpha) h_\ell\} \tag{5-72}$$

is the static mixture enthalpy averaged over the area A, and:

$$h_m^+ = \frac{1}{G_m} (\{\rho_v \alpha \, h_v v_{vz} + \rho_\ell (1 - \alpha) h_\ell v_{\ell z}\}) \tag{5-73}$$

is the dynamic (or mixing cup) average enthalpy of the flowing mixture (similar to Eq. 5-52), which is a directional quantity.

The balance equations presented above do not impose restrictions on the velocities or the temperatures of the two phases. However, various simplifications can be imposed if deemed accurate for the case of interest. For example, if v_{vz} and $v_{\ell z}$ were taken to be equal:

$$\rho_m^+ = \rho_m \text{ for equal velocities} \tag{5-74}$$

and

$$h_m^+ = h_m \text{ for equal velocities} \tag{5-75}$$

Furthermore, when saturated conditions are assumed to apply, Eqs. 5-63, 5-66, and 5-71 in ρ_m, G_m, h_m, and p can be solved when: (1) q', q''', and τ_w are specified in relation to the other variables, and (2) the state equation for the fluid $\rho_m(p,h_m)$ is used. Note that these equations can be extended to apply to the single-phase liquid, saturated mixture, and single-phase vapor flow sections of the channel.

Various approaches to the specification of τ_w are discussed in Chapter 11.

VI CONTROL–VOLUME INTEGRAL TRANSPORT EQUATIONS

A two-fluid (separate flow) approach is used to derive the instantaneous transport equations for liquid and vapor. Considering each of the phases to be a continuum material, we can write the mass, momentum, and energy equations by applying the relations derived for deformable volumes in section IV of Chapter 4.

We shall consider two volumes V_v and V_ℓ (Fig. 5-5), which are joined to obtain the two-fluid equations in the total volume, i.e.,

$$V = V_\ell + V_v \tag{5-76}$$

A Mass Balance

1 Mass balance for volume V_k. The rate of change of the mass in the control volume V_k, depicted in Figure 5-5, should be equal to the net convection of phase k mass into the volume and due to the change of phase. Applying Eq. 4-57 to the control volume of the phase k we obtain:

$$\frac{d}{dt} \iiint_{V_k} \rho_k \, dV + \iint_{S_k} \rho_k(\vec{v}_k - \vec{v}_s) \cdot \vec{n} \, dS = 0$$

Let us divide the surface S_k into a fixed surface A_{kj} at the control volume boundary (across which convection may occur) and the deformable surface A_{ks} at the liquid–vapor interface (across which change of phase occurs). Thus we get:

$$\frac{d}{dt} \iiint_{V_k} \rho_k \, dV + \iint_{A_{kj}} \rho_k(\vec{v}_k - \vec{v}_s) \cdot \vec{n} \, dS + \iint_{A_{ks}} \rho_k(\vec{v}_k - \vec{v}_s) \cdot \vec{n} \, dS = 0 \tag{5-77}$$

$$\underbrace{\begin{array}{c}\text{Rate of mass}\\\text{increase in}\\\text{the control}\\\text{volume}\end{array}}_{} + \underbrace{\begin{array}{c}\text{Rate of mass}\\\text{loss by}\\\text{convection}\end{array}}_{} + \underbrace{\begin{array}{c}\text{Rate of mass}\\\text{loss by change}\\\text{of phase}\end{array}}_{} = 0$$

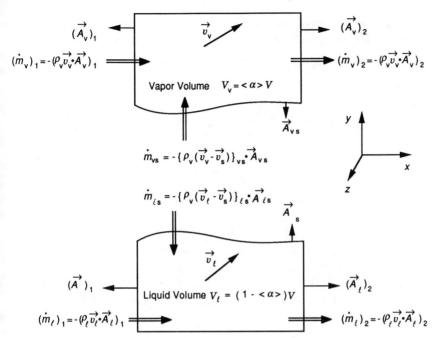

Figure 5-5 Mass balance components in a two-phase volume. The geometric control volume is composed of the phasic subvolumes separated by the wavy lines. *Single arrows* represent vectors, which are positive in the direction of the coordinate axes. *Double arrows* are scalars, which are positive when adding to the control volume content.

Note that $\vec{v}_s = 0$ at the A_{kj} surface, as the total control volume is taken as stationary and nondeformable. Thus using Eqs. 5-15 and 5-5, Eq. 5-77 may be written in the form:

$$\frac{d}{dt}\left[\langle \rho_k \rangle_k V_k\right] + \sum_j \{\rho_k \vec{v}_k\}_{kj} \cdot \vec{A}_{kj} + \{\rho_k(\vec{v}_k - \vec{v}_s)\}_{ks} \cdot \vec{A}_{ks} = 0 \qquad (5\text{-}78)$$

where $\vec{v}_k$ in the last term = phase k velocity at the vapor–liquid interface, and $\vec{v}_s$ = velocity of the interface itself. The vector designating an area $\vec{A}_k$ is normal to the area pointing outward of the subvolume containing the phase k.

Denoting the rate of mass addition due to phase change by $\dot{m}_{ks}$, it is seen that

$$\dot{m}_{ks} = -\{\rho_k(\vec{v}_k - \vec{v}_s)\}_{ks} \cdot \vec{A}_{ks} \qquad (5\text{-}79)$$

If $(\vec{v}_k - \vec{v}_s)_{ks}$ is positive, its dot product with $\vec{A}_{ks}$ takes the sign of $\vec{A}_{ks}$. The area $\vec{A}_{ks}$ is considered positive if it points in a positive coordinate direction. On a positive surface, the right-hand side of Eq. 5-79 renders $\dot{m}_{ks}$ negative, which implies that mass is being lost through the surface.

2 Mass balance in the entire volume V. Let us now write the mass equation of phase k in terms of the total volume and surface area, recalling that:

$$V_k = \langle \alpha_k \rangle V \text{ and } \vec{A}_{kj} = \{\alpha_k\}_j \vec{A}_j$$

We can rewrite Eq. 5-78 as:

$$\frac{d}{dt} [\langle \rho_k \rangle_k \langle \alpha_k \rangle V] + \sum_j \{\rho_k \vec{v}_k\}_{kj} \cdot \{\alpha_k\}_j \vec{A}_j = \dot{m}_{ks} \tag{5-80}$$

Let m_k be the mass of phase k in the control volume V:

$$m_k = \langle \rho_k \rangle_k \langle \alpha_k \rangle V \tag{5-81}$$

and $\dot{m}_{kj}$ be the mass flow rate of phase k through the area A_j:

$$\dot{m}_{kj} = - \{\rho_k \vec{v}_k\}_{kj} \cdot \vec{A}_{kj} = - \{\rho_k \vec{v}_k\}_{kj} \cdot \{\alpha_k\}_j \vec{A}_j \tag{5-82}$$

Note that in Figure 5-5 the vector $\vec{A}_{k1}$ is pointing in a negative direction, whereas the vector $\vec{A}_{k2}$ is in a positive direction. Hence $\vec{v}_k \cdot \vec{A}_{k1}$ is negative and $\vec{v}_k \cdot \vec{A}_{k2}$ is positive. The use of the negative sign on the right-hand side of Eq. 5-82 makes $\dot{m}_{kj}$ positive for inflow and negative for outflow. We can recast Eq. 5-80 into the form:

$$\frac{d}{dt} m_k - \sum_j \dot{m}_{kj} = \dot{m}_{ks} \tag{5-83}$$

For a mixture we get:

$$\frac{d}{dt} (m_v + m_\ell) - \sum_j (\dot{m}_{\ell j} + \dot{m}_{vj}) = \dot{m}_{vs} + \dot{m}_{\ell s} \tag{5-84}$$

It should also be remembered that because the frame of reference of the control volume is stationary, by using Eq. 4-13, Eq. 5-83 reduces to

$$\frac{\partial}{\partial t} m_k - \sum_j \dot{m}_{kj} = \dot{m}_{ks} \tag{5-85a}$$

or

$$\frac{\partial}{\partial t} m_k = \sum_j \dot{m}_{kj} + \dot{m}_{ks} \tag{5-85b}$$

Recognizing that x is the flow quality, Eq. 5-85b may be written for a stationary control volume as:

For vapor:
$$\frac{\partial}{\partial t} m_v = \sum_j (x\dot{m})_j + \dot{m}_{vs} \tag{5-86a}$$

For liquid:
$$\frac{\partial}{\partial t} m_\ell = \sum_j [(1 - x)\dot{m}]_j + \dot{m}_{\ell s} \tag{5-86}$$

3 Interfacial jump condition. Because there is no mass source or sink at the interface, all the vapor added by change of phase should appear as a loss to the liquid and vice versa. This describes the interphase "jump condition." If $\dot{m}_{\ell v}$ is the rate of change of phase to liquid into vapor:

$$\dot{m}_{vs} = -\dot{m}_{\ell s} \equiv \dot{m}_{\ell v} \tag{5-87}$$

It is also useful to define the rate of phase addition by a change in state per unit volume (i.e., vaporization or condensation) so that:

$$\Gamma_k = \frac{\dot{m}_{ks}}{V} \tag{5-88}$$

From Eq. 5-87, it is clear that the jump condition can also be written as:

$$\Gamma_v = -\Gamma_\ell = \Gamma \tag{5-89}$$

4 Simplified form of the mixture equation. The mass balance for the two-phase mixture is obtained by combining Eqs. 5-86a, 5-86b, and 5-87 to obtain an equation for the rate of change of the mixture density $\langle \rho \rangle$:

$$\frac{\partial}{\partial t}[m_v + m_\ell] = \sum_j \dot{m}_j \tag{5-90}$$

or

$$\frac{\partial}{\partial t}[\langle \rho \rangle V] = \sum_j \dot{m}_j \tag{5-91}$$

where

$$\langle \rho \rangle = \langle \rho_v \rangle_v \langle \alpha \rangle + \langle \rho_\ell \rangle_\ell \langle 1 - \alpha \rangle \tag{5-25}$$

When the phase densities are assumed uniform in the volume, we get:

$$\langle \rho \rangle = \alpha \rho_v + (1 - \alpha)\rho_\ell \tag{5-92}$$

B Momentum Balance

1 Momentum balance for volume V_k. The momentum balance for each phase is an application of Newton's law of motion to the deformable volumes of the vapor and liquid phases. The forces acting on each volume are illustrated in Figure 5-6.

Applying Eq. 4-60 to the control volume V_k we obtain:

$$\frac{d}{dt}\iiint_{V_k} \rho_k \vec{v} \, dV + \oiint_{S_k} \rho_k \vec{v}_k (\vec{v}_k - \vec{v}_s) \cdot \vec{n} \, dS =$$

$$\oiint_{S_k}(\bar{\bar{\tau}}_k - p_k \bar{\bar{I}}) \cdot \vec{n} \, dS + \iiint_{V_k} \rho_k \vec{g} \, dV \tag{5-93}$$

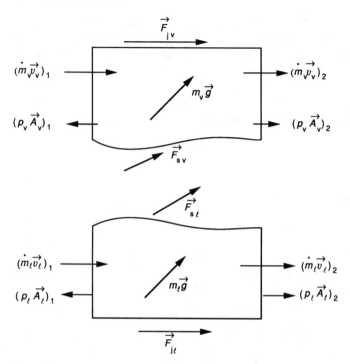

Figure 5-6 Control volume momentum balance components. The arrows represent vectors which are positive in the direction of the coordinate axes.

Again, consider the surface S_k to be divided into a number of fixed bounding areas (A_{kj}) and a deformable area (A_{ks}). Applying the volume- and area-averaging operations, and rearranging the above equation, we get:

$$\frac{d}{dt}[\langle \rho_k \vec{v}_k \rangle_k V_k] = -\sum_j \{\rho_k \vec{v}_k \vec{v}_k\}_{kj} \cdot \vec{A}_{kj} - \{\rho_k \vec{v}_k (\vec{v}_k - \vec{v}_s)\}_{ks} \cdot \vec{A}_{ks}$$
$$+ \sum_j \vec{F}_{jk} - \sum_j \{p_k\}_{kj} \vec{A}_{kj} + \vec{F}_{sk} + \langle \rho_k \rangle_k \vec{g} V_k \tag{5-94}$$

where

$$\vec{F}_{jk} = \iint\limits_{A_{kj}} \overline{\overline{\tau}}_k \cdot \vec{n} dS \tag{5-95}$$

represents the shear forces at all surfaces surrounding the phase except the vapor–liquid interface, and:

$$\vec{F}_{sk} = \iint\limits_{A_{ks}} (\overline{\overline{\tau}}_k - p_k \overline{\overline{I}}) \cdot \vec{n} dS \tag{5-96}$$

represents the shear and normal forces at the vapor–liquid interface. Equation 5-94 represents the following balance:

| Rate of change of momentum within control volume | = Rate of momentum addition by convection | + Rate of momentum addition by change of phase | + Net shear forces at the fixed boundaries A_{kj} | + Net pressure forces at the fixed boundaries A_{kj} | + Net shear and pressure forces at the vapor–liquid interface | + Gravity force |

For the stationary control volume V_k, when the velocities $\vec{v}_k$ are assumed position-independent within the volume and at each area A_{kj} and A_{sk} of the surrounding surface, we can reduce Eq. 5-94 to the lumped parameter form:

$$\frac{\partial}{\partial t}[(m_k \vec{v}_k)] = \sum_j (\dot{m}_k \vec{v}_k)_j + \dot{m}_{ks}\, \vec{v}_{ks} + \sum_j \vec{F}_{jk} - \sum_j (p_k \vec{A}_k)_{kj} + \vec{F}_{sk} + m_k \vec{g}$$

(5-97)

where $\vec{v}_{ks}$ = velocity of the vaporized mass at the phase k side of the interface S.

2 Momentum balance in the entire volume V. When the two volumes V_v and V_ℓ are added together to form the volume V, we can recast Eq. 5-94 for the phase k in the form:

$$\frac{d}{dt}[\langle \rho_k \vec{v}_k \rangle_k \langle \alpha_k \rangle V] = -\sum_j \{\rho_k \vec{v}_k \vec{v}_k\}_{kj} \cdot \{\alpha_k\}_j \vec{A}_j - \{\rho_k \vec{v}_k (\vec{v}_k - \vec{v}_s)\}_{ks} \cdot \vec{A}_{ks}$$
$$+ \sum_j \vec{F}_{jk} - \sum_j \{p_k\}_{kj}\{\alpha_k\}_j \vec{A}_j + \vec{F}_{sk} + \langle \rho_k \rangle_k \langle \alpha_k \rangle \vec{g}\, V$$

(5-98)

where:

$$\sum_{k=1}^{2} V_k = V$$

$$\sum_{k=1}^{2} \vec{A}_{kj} = \vec{A}_j$$

$$\sum_{k=1}^{2} \vec{A}_{ks} = 0 \quad \text{(because the interface areas are equal but have directionally opposite normal vectors)}$$

Using Eqs. 5-79 and 5-88:

$$-\{\rho_k \vec{v}_k (\vec{v}_k - \vec{v}_s)\}_{ks} \cdot \vec{A}_{ks} = \dot{m}_{ks} \vec{v}_{ks} = \Gamma_k \vec{v}_{ks} V$$

(5-99)

then by Eq. 5-82 we get:

$$-\{\rho_k \vec{v}_k \vec{v}_k\}_{kj} \cdot \{\alpha_k\}_j \vec{A}_j = (\dot{m}_k \vec{v}_k)_j$$

(5-100)

Finally, we can simplify the situation by assuming uniform ρ_k and $\vec{v}_k$ within each volume, so that the averaging operations are not needed. Eq. 5-98 can then be written for a fixed control volume as:

$$\frac{\partial}{\partial t}(\alpha_k \rho_k \vec{v}_k)V = \sum_j (\dot{m}_k \vec{v}_k)_j + \Gamma_k \vec{v}_{ks}V + \sum_j \vec{F}_{jk}$$
$$- \sum_j (\alpha_k p_k \vec{A})_j + \vec{F}_{sk} + \alpha_k \rho_k \vec{g}V \qquad (5\text{-}101)$$

The mixture equation is obtained by adding the vapor and liquid momentum equations (Eq. 5-101):

$$\frac{\partial}{\partial t}[\alpha \rho_v \vec{v}_v + (1 - \alpha)\rho_\ell \vec{v}_\ell]V = \sum_j [\dot{m}_v \vec{v}_v + \dot{m}_\ell \vec{v}_\ell]_j$$
$$+ \Gamma(\vec{v}_{vs} - \vec{v}_{\ell s})V + \sum_j (\vec{F}_{jv} + \vec{F}_{j\ell})$$
$$- \sum_j [\alpha p_v + (1 - \alpha)p_\ell]_j \vec{A}_j + \vec{F}_{sv}$$
$$+ \vec{F}_{s\ell} + [\alpha \rho_v + (1 - \alpha)\rho_\ell]\vec{g}V \qquad (5\text{-}102)$$

3 Interfacial jump condition. If the surface tension force (and hence the surface deformation energy) is negligible, no net momentum is accumulated at the interface. Thus the net momentum change plus the interface forces can be expected to vanish:

$$\Gamma(\vec{v}_{vs} - \vec{v}_{\ell s})V + \vec{F}_{sv} + \vec{F}_{s\ell} = 0 \qquad (5\text{-}103)$$

Several assumptions have been used in the literature with regard to $\vec{v}_{vs}$ and $\vec{v}_{\ell s}$. The most common approaches are as follows. Assume either:

$$\vec{v}_{vs} = \vec{v}_{\ell s}; \text{ hence } \vec{F}_{sv} = -\vec{F}_{s\ell} \qquad (5\text{-}104a)$$

or

$$\vec{v}_{vs} = \vec{v}_v \text{ and } \vec{v}_{\ell s} = \vec{v}_\ell; \text{ hence } \vec{F}_{sv} = -\vec{F}_{s\ell} - \Gamma(\vec{v}_v - \vec{v}_\ell)V \qquad (5\text{-}104b)$$

In equation 5-104a, it is often assumed that

$$\vec{v}_{vs} = \vec{v}_{\ell s} = \eta \vec{v}_v + (1 - \eta)\vec{v}_\ell \qquad (5\text{-}104c)$$

where $\eta = 0$ if $\Gamma < 0$; or $\eta = 1$ if $\Gamma > 0$.

The last formulation avoids penetration of the interface in a direction not compatible with Γ.

4 Common assumptions. It is common to assume that if the flow area A_j is horizontal the pressure would be uniform across the area and hence:

$$(p_v)_j = (p_\ell)_j = p_j \qquad (5\text{-}105)$$

Although this assumption is often extended to all areas irrespective of their inclination, it may be a poor approximation for a vertical flow area with stratified flow where significant pressure variations due to gravity may exist.

The shear forces at the boundary areas A_j are usually assumed to constitute friction terms with the wall* that can be considered additive:

$$\sum_j (\vec{F}_{jv} + \vec{F}_{j\ell}) = \vec{F}_w \qquad (5\text{-}106)$$

5 Simplified forms of the mixture equation. Imposing the conditions of Eqs. 5-103, 5-105, and 5-106 on Eq. 5-102, we get a simplified mixture momentum balance equation:

$$\frac{\partial}{\partial t} [\alpha \rho_v \vec{v}_v + (1 - \alpha) \rho_\ell \vec{v}_\ell] V = \sum_j [x \vec{v}_v + (1 - x) \vec{v}_\ell]_j \dot{m}_j$$
$$+ \vec{F}_w - \sum_j p_j \vec{A}_j + \rho_m \vec{g} V \qquad (5\text{-}107)$$

where x_j = flow quality in the direction perpendicular to $\vec{A}_j$.

Because the mass flux $\vec{G}_m$ can be defined for spatially uniform properties as:

$$\vec{G}_m = \alpha \rho_v \vec{v}_v + (1 - \alpha) \rho_\ell \vec{v}_\ell \qquad (5\text{-}108)$$

and

$$\dot{m}_j = (\vec{G}_m \cdot \vec{A})_j \qquad (5\text{-}109)$$

It is possible to write Eq. 5-107 in terms of the mass flux $\vec{G}_m$ as:

$$\frac{\partial}{\partial t} (\vec{G}_m V) = \sum_j [x \vec{v}_v + (1 - x) \vec{v}_\ell]_j (\vec{G}_m \cdot \vec{A})_j + \vec{F}_w - \sum_j p_j \vec{A}_j + \rho_m \vec{g} V \quad (5\text{-}110)$$

Example 5-3 Modeling of a BWR suppression pool transient

PROBLEM Consider the simplified diagram of a BWR suppression pool shown in Figure 5-7. The function of the suppression pool is to act as a means of condensing the steam emerging from the primary system during a LOCA. In a postulated accident, a sudden rupture of a BWR recirculation pipe causes a jet of somewhat high-pressure steam–water to enter into the suppression pool from a connecting pipe. This jet is preceded by the air initially in the pipe, which is noncondensible. Because of the rapid nature of this transient, there is some concern that the water slug impacted by the jet may exert a large pressure spike on the bottom of the suppression pool.

Describe how you would use Eq. 5-94 to obtain an approximate answer for the value of the pressure spike. Consider a control volume drawn around the water slug

---

*It is assumed that the shear forces due to velocity gradients in the fluid at open portions of the surface area are much smaller than those at fluid–solid surfaces.

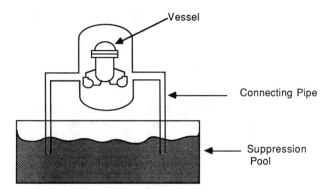

Figure 5-7 Simplified diagram of the suppression pool in a BWR.

between the exit of the connecting pipe and the bottom of the suppression pool, as shown in Figure 5-8, and make the following assumptions:

1. The nature of this transient is sufficiently fast that a water slug mass does not have time to deform from a roughly cylindrical configuration assumed to have the same diameter as the connecting pipe.
2. Within the time interval of interest, the momentum of the water within the control volume remains fairly constant.

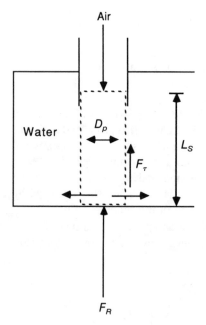

Figure 5-8 Control volume for analysis of BWR suppression pool pressure spike.

3. All the water in the slug is carried to the bottom of the pool, where it is deflected to move parallel to the bottom of the pool.
4. The distance between the connecting pipe outlet and the bottom of the suppression pool is on the order of meters.

SOLUTION Let us consider the terms in Eq. 5-94 for the water slug momentum in the vertical direction.

1. By assumption 2, $\dfrac{d}{dt} [\langle \rho_k \vec{v}_k \rangle V_k] = 0$.

2. $-\sum_j \{\rho_k \vec{v}_k \vec{v}_k\}_{kj} \cdot \vec{A}_{kj} = \rho_w V_w^2 A_p = \dot{m} \vec{V}_w$, where the subscript w = water, and $\dot{m}_w$ = downward flow rate of the water. Note here that:

a. We consider the flow at the pipe outlet because the flow at this part of the control volume surface is in the vertical direction.
b. The sign convention makes the flow positive because it is at an inlet to the control volume.

3. $- \{\rho_k \vec{v}_k (\vec{v}_k - \vec{v}_s)\}_{ks} \cdot \vec{A}_{ks} = 0$ because this term represents momentum transfer by change of phase, and there is no phase change between water and air. If the air were replaced by steam, this term may become important because of the expected condensation.

4. $\sum_j \vec{F}_{jk} \simeq 0$ = the force exerted on the control volume by shear at the ends of the control volume

5. $\sum_j \{p_k\}_{kj} \vec{A}_{kj} = \vec{F}_R - p_{air} \pi \dfrac{D_p^2}{4} \vec{n}_z$, where $\vec{F}_R$ = force sought at the bottom, and $\vec{n}_z$ = vertical unity vector.

6. $\vec{F}_{sk} = -\vec{F}_\tau = -\tau_s(\pi D_p L_s) \vec{n}_z$ is the shear force at the interface of the water slug and the pool.

7. $\langle \rho_k \rangle_k \vec{g} V_k \simeq 0$, as the column is relatively short by assumption 4.

Equation 5-94 then becomes

$$0 = \dot{m} \vec{V}_w - 0 + 0 - \vec{F}_R + p_{air} \frac{\pi D_p^2}{4} \vec{n}z - \vec{F}_\tau + 0$$

$$\vec{F}_R = \dot{m} \vec{V}_w + p_{air}\pi \frac{D_p^2}{4} \vec{n}_z - \tau_s(\pi D_p L_s)\vec{n}_z$$

C Energy Balance

1 Energy balance for the volume V_k. The energy balance for each phase is an application of the first law of thermodynamics to the phase subvolume. Equation 4-63 can be applied to the vapor volume shown in Figure 5-9 to obtain:

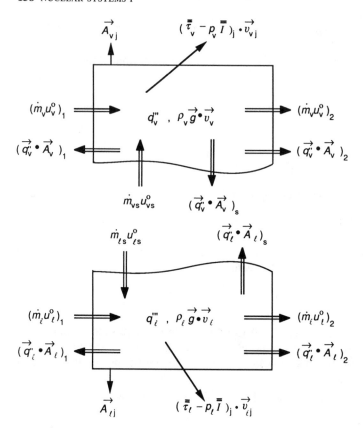

Figure 5-9 Forces and heat fluxes at the vapor–liquid interface. *Single arrows* represent vectors, which are positive in the positive direction of the coordinate axis. *Double arrows* represent scalars, which are positive when adding to the control volume content.

$$\frac{d}{dt} \iiint_{V_k} \rho_k u_k^\circ \, dV + \iint_{S_k} \rho_k u_k^\circ (\vec{v}_k - \vec{v}_s) \cdot \vec{n} dS =$$

$$\oiint_{S_k} \left[-\vec{q}_k' + (\overline{\overline{\tau}}_k - p_k \overline{\overline{I}}) \cdot \vec{v}_k \right] \cdot \vec{n} dS + \iiint_{V_k} \rho_k \vec{g} \cdot \vec{v}_k dV + \iiint_{V_k} q_k''' \, dV$$

$$(5\text{-}111)$$

A heat addition term due to both internal (to the fluid k) and surface heat addition can be defined as:

$$\left(\frac{dQ}{dt}\right)_k = \iiint_{V_k} q_k''' \, dV - \oiint_{S_k} \vec{q}_k'' \cdot \vec{n} \, ds \qquad (5\text{-}112)$$

where $\vec{q}_k'' \cdot \vec{n}$ = outward heat flux from the surface.

A work term due to shear forces can be defined as:

$$\left(\frac{dW}{dt}\right)_{k,shear} = -\oint_{S_k} (\bar{\bar{\tau}}_k \cdot \vec{v}_k) \cdot \vec{n} \, dS \tag{5-113}$$

A work term due to the gravity (body) force can be defined as:

$$\left(\frac{dW}{dt}\right)_{k,gravity} = -\iiint_{V_k} (\rho_k \vec{g} \cdot \vec{v}_k \, dV) \tag{5-114}$$

Recalling that the surface S can be divided into a number of fixed bounding surfaces A_{kj} and the deformable interface A_{ks}, Eq. 5-111 can be recast using Eqs. 5-112, 5-113, and 5-114 in the form:

$$\begin{aligned}
\frac{d}{dt} [\langle \rho_k u_k^\circ \rangle_k V_k] = &- \sum_j \{\rho_k u_k^\circ (\vec{v}_k - \vec{v}_s) \cdot \vec{n}\}_{kj} A_{kj} \\
&- \{\rho_k u_k^\circ (\vec{v}_k - \vec{v}_s) \cdot \vec{n}\}_{ks} A_{ks} \\
&- \sum_j \{p_k \vec{v}_k \cdot \vec{n}\}_{kj} A_{kj} - \{p_k \vec{v}_k \cdot \vec{n}\}_{ks} A_{ks} \\
&+ \left(\frac{dQ}{dt}\right)_k - \left(\frac{dW}{dt}\right)_{k,shear} - \left(\frac{dW}{dt}\right)_{k,gravity}
\end{aligned} \tag{5-115}$$

Note that

$$\{p_k \vec{v}_k \cdot \vec{n}\}_{ks \text{ or } kj} = \{p_k \vec{v}_s \cdot \vec{n}_s\}_{ks \text{ or } kj} + \{p_k(\vec{v}_k - \vec{v}_s) \cdot n_j\}_{ks \text{ or } kj} \tag{5-116}$$

The energy balance for V_k (Eq. 5-114) can then be written in terms of the total enthalpy:

$$\begin{aligned}
\frac{d}{dt} [\langle \rho_k u_k^\circ \rangle_k V_k] = &- \sum_j \{\rho_k h_k^\circ (\vec{v}_k - \vec{v}_s) \cdot \vec{n}\}_{kj} A_{kj} - \sum_j \{p_k \vec{v}_s \cdot \vec{n}\}_{kj} A_{kj} \\
&- \{\rho_k h_k^\circ (\vec{v}_k - \vec{v}_s) \cdot \vec{n}\}_{ks} A_{ks} - \{p_k \vec{v}_s \cdot \vec{n}\}_{ks} A_{ks} \\
&+ \left(\frac{dQ}{dt}\right)_k - \left(\frac{dW}{dt}\right)_{k,shear} - \left(\frac{dW}{dt}\right)_{k,gravity}
\end{aligned} \tag{5-117}$$

For a fixed control volume $\{\vec{v}_s\}_{kj} = 0$, then we have

$$\begin{aligned}
\frac{\partial}{\partial t} [\langle \rho_k u_k^\circ \rangle_k V_k] = &- \sum_j \{\rho_k h_k^\circ (\vec{v}_k) \cdot \vec{n}\}_{kj} A_{kj} \\
&- \{\rho_k h_k^\circ (\vec{v}_k - \vec{v}_s) \cdot \vec{n}\}_{ks} A_{ks} \\
&- \{p_k \vec{v}_s \cdot \vec{n}\}_{ks} A_{ks} \\
&+ \left(\frac{dQ}{dt}\right)_k - \left(\frac{dW}{dt}\right)_{k,shear} - \left(\frac{dW}{dt}\right)_{k,gravity}
\end{aligned} \tag{5-118}$$

For uniform properties within the volume and at each gate area we can write the lumped parameter form of the energy equation (Eq. 5-118) as:

$$\frac{\partial}{\partial t}(m_k u_k^\circ) = \sum_j (\dot{m}_k h_k^\circ)_j + (\dot{m}_k h_k^\circ)_s + \left(\frac{\partial Q}{\partial t}\right)_k$$
$$- \left(\frac{\partial W}{\partial t}\right)_{k,c.v.} - \left(\frac{dW}{dt}\right)_{k,gravity} \tag{5-119}$$

where:

$$\left(\frac{\partial W}{\partial t}\right)_{k,c.v.} = \left(\frac{\partial W}{\partial t}\right)_{k,shear} + \{p_k \vec{v}_s \cdot \vec{n}\}_{ks} A_{ks} \tag{5-120}$$

Note that $\left(\frac{\partial W}{\partial t}\right)_{k,c.v.}$ can be written in terms of the work required to change the vapor volume given the various forces surrounding the vapor volume. The control volume surface and the phase interface area are moving at velocity $\vec{v}_j$ and $\vec{v}_s$, respectively. Thus:

$$\left(\frac{\partial W}{\partial t}\right)_{k,c.v.} = \sum_j \left(\frac{\partial W}{\partial t}\right)_{k,c.v.,j} + \left(\frac{\partial W}{\partial t}\right)_{k,c.v.,s} \tag{5-121}$$

where for uniform properties over each area:

$$\left(\frac{\partial W}{\partial t}\right)_{k,c.v.,j} = -\sum_j (\overline{\overline{\tau}}_k \cdot \vec{v}_k)_j \cdot \vec{n}_{kj} A_{kj} \tag{5-122}$$

and

$$\left(\frac{\partial W}{\partial t}\right)_{k,c.v.,s} = -(\overline{\overline{\tau}}_k \cdot \vec{v}_k)_s \cdot \vec{n}_{ks} A_s + (p_k \vec{v}_s) \cdot \vec{n}_{ks} A_s \tag{5-123}$$

Also, the heat addition term can be divided into three parts:

$$\left(\frac{\partial Q}{\partial t}\right)_k = \dot{Q}_k - \sum_j (\vec{q}_k'' \cdot \vec{n}_k)_j A_{kj} - (\vec{q}_k'' \cdot \vec{n}_k)_s A_s \tag{5-124}$$

where $\vec{q}_{kj}''$ and $\vec{q}_{ks}''$ = heat transfer rates at the boundary surfaces j and s surrounding the phasic volume, respectively, in the positive coordinate direction, and $\dot{Q}_k$ = rate of volumetric heat generation in the fluid k.

Let us now substitute from Eqs. 5-122 and 5-124 into 5-119 to obtain, after rearranging the terms:

$$\frac{\partial}{\partial t}(m_k u_k^\circ) = \sum_j (\dot{m}_k h_k^\circ)_j + \dot{Q}_k - \sum_j (\vec{q}_k'' \cdot \vec{n}_k)_j A_{kj}$$
$$- \sum_j (\overline{\overline{\tau}}_k \cdot \vec{v}_k)_j \cdot \vec{n}_{kj} A_{kj} - \left(\frac{dW}{dt}\right)_{k,gravity} + (\dot{m}_k h_k^\circ)_s \tag{5-125}$$
$$- (\vec{q}_k'' + \overline{\overline{\tau}}_k \cdot \vec{v}_k - p_k \vec{v}_s) \cdot \vec{n}_{ks} A_{ks}$$

Note that the last two terms represent the energy exchange rate between the interior of a phase and the vapor–liquid interface due to change of phase, heat transport, and work of surface forces.

2 Energy equations for the total volume V. The two-phase mixture energy equation can be obtained by adding the vapor and liquid equations of 5-125:

$$
\begin{aligned}
\frac{\partial}{\partial t}(m_v u_v^{\circ} + m_\ell u_\ell^{\circ}) = & \sum_j [\dot{m}_\ell h_\ell^{\circ} + \dot{m}_v h_v^{\circ}]_j \\
& + \dot{Q}_\ell + \dot{Q}_v - \sum_j (\vec{q}_\ell'' \alpha_\ell + \vec{q}_v'' \alpha_v)_j \cdot \vec{n}_j A_j \\
& - \sum_j [(\overline{\overline{\tau}}_\ell \cdot \vec{v}_\ell)\alpha_\ell + (\overline{\overline{\tau}}_v \cdot \vec{v}_v)\alpha_v]_j \cdot \vec{n}_j A_j \\
& - \left(\frac{dW}{dt}\right)_{v,\text{gravity}} - \left(\frac{dW}{dt}\right)_{\ell,\text{gravity}} + (\dot{m}_\ell h_\ell^{\circ})_s \\
& + (\dot{m}_v h_v^{\circ})_s - \{[\vec{q}_v'' + \overline{\overline{\tau}}_v'' \cdot \vec{v}_v - p_v \vec{v}_s]\} \cdot \vec{n}_{vs} A_{vs} \\
& + \{[\vec{q}_\ell'' + \overline{\overline{\tau}}_\ell \cdot \vec{v}_\ell - p_\ell \vec{v}_s]\} \cdot \vec{n}_{\ell s} A_{\ell s}
\end{aligned}
\tag{5-126}
$$

where it has been noted that:

$$
(\vec{q}_k'' \cdot \vec{n}_k)_j A_{kj} = \vec{q}_k'' \alpha_k \cdot \vec{n}_j A_j
$$

and

$$
\overline{\overline{\tau}}_k \cdot \vec{v}_k \cdot \vec{n}_{kj} A_{kj} = \overline{\overline{\tau}}_k \cdot \vec{v}_k \alpha_k \cdot \vec{n}_j A_j
$$

3 Jump condition. It should be clear that no energy sources or sinks should exist at the interface (when the surface tension is ignored), i.e., at the interface all energy exchange should add up to zero. Applying the conditions:

$$
\dot{m}_{vs} = -\dot{m}_{\ell s} = \dot{m}_{\ell v}, \quad \vec{n}_{vs} A_{vs} = -\vec{n}_{\ell s} A_{\ell s}, \text{ and } A_{vs} = A_{\ell s} \equiv A_s
$$

we get:

$$
\dot{m}_{\ell v}(h_v^{\circ} - h_\ell^{\circ}) - [(\vec{q}_v'' - \vec{q}_\ell'')_s + (\overline{\overline{\tau}}_v \cdot \vec{v}_v - \overline{\overline{\tau}}_\ell \cdot \vec{v}_\ell)_s - (p_v - p_\ell)_s \vec{v}_s] \cdot \vec{n}_{vs} A_s = 0
\tag{5-127}
$$

Several simplifications can be made. The most common assumption used is that the term related to the interfacial forces can be ignored, which reduces Eq. 5-127 to:

$$
\dot{m}_{\ell v} = (\vec{q}_v'' - \vec{q}_\ell'')_s \cdot \vec{n}_{vs} A_s / (h_v^{\circ} - h_\ell^{\circ})_s
\tag{5-128}
$$

The terms $(\vec{q}_v'')_s$ and $(\vec{q}_\ell'')_s$ have to be externally supplied as constitutive equations.

Substituting from Eq. 5-127 into Eq. 5-126, the mixture energy equation can be written as:

$$
\begin{aligned}
\frac{\partial}{\partial t}(mu_m^{\circ}) = & \sum_j [\{(1 - x)h_\ell^{\circ} + xh_v^{\circ}\}\dot{m}]_j + \dot{Q} - \sum_j \vec{q}_j'' \cdot \vec{n}_j A_j \\
& - \sum_j \overline{\overline{\tau}}_{\text{eff}} \cdot \vec{j} \cdot \vec{A}_j - \left(\frac{dW}{dt}\right)_{\text{gravity}}
\end{aligned}
\tag{5-129}
$$

where
$$\dot{Q} = \dot{Q}_\ell + \dot{Q}_v$$
$$\vec{q}''_j = (\vec{q}''_\ell \alpha_\ell + \vec{q}''_v \alpha_v)_j$$
$$\bar{\bar{\tau}}_{\text{eff}} \cdot \vec{j} \cdot \vec{A}_j = [\bar{\bar{\tau}}_\ell \cdot \vec{v}_\ell \alpha_\ell + \bar{\bar{\tau}}_v \cdot \vec{v}_v \alpha_v]_j \cdot \vec{n}_j A_j$$

VII ONE-DIMENSIONAL SPACE-AVERAGED TRANSPORT EQUATIONS

In this section the differential form of the space-averaged equations is derived from the integral transport equations. No time averaging is invoked.

A Mass Equations

From Eq. 5-80, when V is considered equal to a small volume $A_z \Delta z$, we get:

$$\frac{d}{dt} [\langle \rho_k \rangle_k \langle \alpha_k \rangle A_z \Delta z] + \{\rho_k \vec{v}_k\}_{kz^+} \cdot \{\alpha\}_{z^+} \vec{A}_{z^+} - \{\rho_k \vec{v}_k\}_{kz^-} \cdot \{\alpha_k\}_z \vec{A}_{z^-} = \dot{m}_{ks} = \Gamma_k A_z \Delta z$$

(5-130)

In the limit of infinitesimal Δz and for a fixed frame of coordinates, we get:

$$\frac{\partial}{\partial t} (\{\rho_k \alpha_k\} A_z) + \frac{\partial}{\partial z} (\{\rho_k v_{kz}\}_k \{\alpha_k\} A_z) = \Gamma_k A_z$$

(5-131)

Because Δ_z has been taken infinitesimally small, the volume average quantities are equivalent to the area-averaged ones over the area A_z. Note that Γ_k is then also the average over A_z. Now, with the application of Eq. 5-30, Eq. 5-131 can be written as:

$$\frac{\partial}{\partial t} (\{\rho_k \alpha_k\} A_z) + \frac{\partial}{\partial z} (\{\rho_k v_{kz} \alpha_k\} A_z) = \Gamma_k A_z$$

(5-132)

By adding the phasic equations (Eq. 5-132) and applying the condition $\Gamma_v + \Gamma_\ell = 0$ we get the mixture equation:

$$\frac{\partial}{\partial t} (\rho_m A_z) + \frac{\partial}{\partial z} (G_m A_z) = 0$$

(5-63)

where
$$\rho_m = \{\rho_v \alpha\} + \{\rho_\ell (1 - \alpha)\}$$

(5-50b)

and
$$G_m = \{\rho_v v_{vz}\}_v \{\alpha\} + \{\rho_\ell v_{\ell z}\}_\ell \{(1 - \alpha)\}$$

(5-40b)

For a constant area channel, Eq. 5-132 leads to the following equations for vapor and liquid:

Vapor
$$\frac{\partial}{\partial t} \{\rho_v \alpha\} + \frac{\partial}{\partial z} \{\rho_v v_{vz} \alpha\} = \Gamma_v$$

(5-133a)

Liquid
$$\frac{\partial}{\partial t} \{\rho_\ell (1 - \alpha)\} + \frac{\partial}{\partial z} \{\rho_\ell v_{\ell z} (1 - \alpha)\} = \Gamma_\ell$$

(5-133b)

B Momentum Equation

From Eq. 5-98, when V is considered equal to $A_z\Delta z$ and the limit of infinitesimal volume is applied we get for a fixed frame of reference:

$$\frac{\partial}{\partial t}\left(\{\rho_k v_{kz}\}_k \{\alpha_k\} A_z\right) + \frac{\partial}{\partial z}\left(\{\rho_k v_{kz}^2\}_k \{\alpha_k\} A_z\right) = \iint_{A_z} \Gamma_k \, (\vec{v}_{ks})_{av} \cdot d\vec{A}_z$$

$$+ \iint_{A_z} \vec{F}_{wk}'' \cdot d\vec{A}_z - \frac{\partial}{\partial z}\left(\{p_k\}_z \{\alpha_k\}\right) A_z + \iint_{A_z} \vec{F}_{sk}'' \cdot d\vec{A}_z + \{\rho_k \alpha_k\} \vec{g} \, A_z \tag{5-134}$$

where:

$$\Gamma_k(\vec{v}_{ks})_{av} = -\frac{1}{V}\{\rho_k \vec{v}_k(\vec{v}_k - \vec{v}_s)\}_{ks} \cdot \vec{A}_{ks} \tag{5-135}$$

$$\vec{F}_{wk}'' = \frac{1}{V}\sum_j \vec{F}_{kj} \tag{5-136}$$

$$\vec{F}_{sk}''' = \frac{1}{V}\vec{F}_{sk} \tag{5-137}$$

If we again apply Eq. 5-30 we get:

$$\frac{\partial}{\partial t}\left(\{\rho_k v_{kz}\alpha_k\} A_z\right) + \frac{\partial}{\partial z}\left(\{\rho_k v_{kz}^2\alpha_k\} A_z\right) = \{\Gamma_k \vec{v}_{ks} \cdot \vec{n}_z\} A_z$$

$$+ \{\vec{F}_{wk}'' \cdot \vec{n}_z\} A_z - \frac{\partial}{\partial z}\left(\{p_k\alpha_k\}_z \, A_z\right) + \{\vec{F}_{sk}''' \cdot \vec{n}_z\} \cdot A_z + \{\rho_k\alpha_k\}\vec{g} \cdot \vec{n}_z A_z \tag{5-138}$$

By adding Eq. 5-138 as applied to each phase and applying the jump condition at the interface:

$$\Gamma_v \vec{v}_{vs} \cdot \vec{n}_z + \vec{F}_{sv}''' \cdot \vec{n}_z + \Gamma_\ell \vec{v}_{\ell s} \cdot \vec{n}_z + \vec{F}_{s\ell}''' \cdot \vec{n}_z = 0 \tag{5-139}$$

we get:

$$\frac{\partial}{\partial t}(G_m A_z) + \frac{\partial}{\partial z}\left(\frac{G_m^2 A_z}{\rho_m^+}\right) = -F_{wz}'' A_z - \frac{\partial}{\partial z}(\{p\}A_z) - \rho_m g A_z \cos\theta \tag{5-140}$$

where

$$\frac{1}{\rho_m^+} \equiv \frac{1}{G_m^2}\left[\{\rho_v v_{vz}^2 \alpha\} + \{\rho_\ell v_{\ell z}^2(1-\alpha)\}\right] \tag{5-67}$$

$$F_{wz}'' \equiv -\left(\vec{F}_{wv}'' \cdot \vec{n}_z + \vec{F}_{w\ell}'' \cdot \vec{n}_z\right) \tag{5-141}$$

$$\{p\} \equiv \{p_v\alpha\} + \{p_\ell(1-\alpha)\} \tag{5-142}$$

and θ is the angle between the flow direction and the vertical direction.

The assumption of $p_v = p_\ell = p$ can usually be applied in well mixed one-dimensional two-phase flow. Special relations are developed to relate F_{wz}'' to the average flow parameters such as G_m and ρ_m. It should be noted that F_{wz}'' is the net force per unit volume due to shear forces at the walls. Thus it can be obtained from:

$$F_{wz}''' = \frac{1}{A_z} \int \tau_w \, dP_z \tag{5-143}$$

where the shear stress at the walls (τ_w) is a function of the flow conditions, as is discussed in Chapter 11.

C Energy Equations

Again considering $V = A_z \Delta z$ and the limit of infinitesimal Δz, we can get from Eq. 5-118 a one-dimensional energy equation. Let us recall that:

$$\left(\frac{dQ}{dt} \right)_k = \iiint_{V_k} q_k''' \, dV - \oiint_{S_k} \vec{q}_k'' \cdot \vec{n} \, dS \tag{5-112}$$

and

$$\left(\frac{dW}{dt} \right)_{k,shear} = - \oiint_{S_k} (\overline{\overline{\tau}}_k \cdot \vec{v}_k) \cdot \vec{n}_k \, dS \tag{5-113}$$

Note that

$$\frac{1}{V} (p_k \vec{v}_s \cdot \vec{n}_s)_{ks} A_{ks} = p_k \frac{\partial}{\partial t} (\alpha_k) \tag{5-144}$$

Then for a fixed frame of reference we get:

$$\frac{\partial}{\partial t} (\{\rho_k u_k^{\circ} \alpha_k\} A_z) + \frac{\partial}{\partial z} (\{\rho_k h_k^{\circ} v_{kz} \alpha_k\} A_z) = \{\Gamma_k h_{ks}^{\circ}\} A_z - \left\{ p_k \frac{\partial \alpha_k}{\partial t} \right\} A_z$$

$$+ \{q_k''' \alpha_k\} A_z - (q_{wk}'' \alpha_{wk} P_w) - (q_{sk}'' P_s) - \frac{\partial}{\partial z} (q_{kz}'' \alpha_k A_z) \tag{5-145}$$

$$+ \frac{\partial}{\partial z} (\{(\tau_{xz} v_x)_k + (\tau_{yz} v_y)_k + (\tau_{zz} v_z)_k\}_k \{\alpha_k\} A_z) - \{\rho_k g \, v_{kz} \, \alpha_k\} A_z$$

where q_{wk}'' = heat flux from phase k to the wall in the A_z plane, P_w = wall perimeter in the A_z plane, and P_s = interphase perimeter in the A_z plane.

For one-dimensional flow in a uniform area channel, A_z is constant. The axial heat conduction and the shear effect are small so that both can be neglected. In addition, p_k may be assumed constant in the channel, i.e., $p_v = p_\ell = p$. Thus Eq. 5-145 can be reduced to:

$$\frac{\partial}{\partial t} \{\rho_k u_k^{\circ} \alpha_k\} + \frac{\partial}{\partial z} \{\rho_k h_k^{\circ} v_{kz} \alpha_k\} = \Gamma_k h_{ks}^{\circ} - p \frac{\partial \alpha_k}{\partial t} + \{q_k''' \alpha_k\}$$

$$- q_{wk}'' \alpha_{wk} \frac{P_w}{A_z} - \{\rho_k g v_{kz} \alpha_k\} + \{Q_{sk}^*\} \tag{5-146}$$

where Q_{sk}^* is given by:

$$Q_{sk}^* = -\{\vec{q}_k''\}_s \cdot \vec{n}_{ks} A_{sk}/V = \{\vec{q}_k''\}_s \cdot \vec{n} \, P_S/A_z \qquad (5\text{-}147)$$

and where the jump condition is given by:

$$\sum_{k=1}^{2} (\Gamma_k h_{ks}^\circ + Q_{sk}^*) = 0 \qquad (5\text{-}148)$$

For the one-dimensional mixture equation we add the phasic equations to obtain

$$\frac{\partial}{\partial t}\left\{\rho_m\left[h_m + \frac{1}{2}(v^2)_m\right] - p\right\} + \frac{\partial}{\partial z}\left\{G_m\left[h_m^+ + \frac{1}{2}(v^2)_m^+\right]\right\} =$$

$$q_m''' - q_w'' \frac{P_w}{A_z} - gG_m \cos\theta \qquad (5\text{-}149)$$

where

$$\rho_m = \{\alpha_v \rho_v + \alpha_\ell \rho_\ell\} \qquad (5\text{-}50b)$$
$$G_m = \{\alpha_v \rho_v v_{vz} + \alpha_\ell \rho_\ell v_{\ell z}\} \qquad (5\text{-}40c)$$
$$h_m = \{\rho_v h_v \alpha_v + \rho_\ell h_\ell \alpha_\ell\}/\rho_m \qquad (5\text{-}150)$$
$$h_m^+ = \{\rho_v h_v v_{vz} \alpha_v + \rho_\ell h_\ell v_{\ell z} \alpha_\ell\}/G_m \qquad (5\text{-}151)$$
$$(v^2)_m = \{\alpha_v \rho_v v_v^2 + \alpha_\ell \rho_\ell v_\ell^2\}/\rho_m \qquad (5\text{-}152)$$
$$(v^2)_m^+ = \{\alpha_v \rho_v v_v^3 + \alpha_\ell \rho_\ell v_\ell^3\}/G_m \qquad (5\text{-}153)$$
$$q_m''' = q_v''' \alpha_v + q_\ell''' \alpha_\ell \qquad (5\text{-}154)$$

When v_{kz} is uniform within A_z, it is possible to obtain an equation for the kinetic energy of each phase by multiplying the one-dimensional momentum equation for the phase (Eq. 5-138) by v_{kz} to get:

$$v_{kz}\left(v_{kz}\frac{\partial \rho_k \alpha_k}{\partial t}A_z + \rho_k \alpha_k A_z \frac{\partial v_{kz}}{\partial t} + v_{kz}\frac{\partial \rho_k \alpha_k v_{kz}A_z}{\partial z} + \rho_k \alpha_k v_{kz}A_z \frac{\partial v_{kz}}{\partial z}\right) =$$

$$- v_{kz}F_{wkz}''' A_z - v_{kz}\frac{\partial p\alpha_k}{\partial z}A_z - \rho_k \alpha_k v_{kz}gA_z + \vec{v}_{kz}\cdot\left(\Gamma_k \vec{v}_{ks} + \vec{F}_{sk}'''\right)A_z \qquad (5\text{-}155)$$

The left-hand side (LHS) can be simplified using the mixture mass balance equation to:

$$\rho_k \alpha_k A_z \frac{\partial(v_{kz}^2/2)}{\partial t} + \rho_k \alpha_k v_{kz}A_z \frac{\partial(v_{kz}^2/2)}{\partial z} + v_{kz}^2 \Gamma_k A_z \qquad (5\text{-}156)$$

Hence by adding to the left-hand side the term:

$$\frac{v_{kz}^2}{2}\left(\frac{\partial}{\partial t}\rho_k \alpha_k + \frac{\partial}{\partial z}\rho_k \alpha_k v_k - \Gamma_{vk}\right)A_z$$

Table 5-3 One-dimensional transport equations for uniform density within each phasic region

Mass equations

Phase

$$\frac{\partial}{\partial t}\{\rho_k\alpha_k\}A_z + \frac{\partial}{\partial z}\{\rho_k v_{kz}\alpha_k\}A_z = \Gamma_k A_z$$

Jump condition

$$\sum_{k=1}^{2}\Gamma_k = 0$$

Mixture

$$\frac{\partial}{\partial t}(\rho_m A_z) + \frac{\partial}{\partial z}(G_m A_z) = 0$$

where

$$\rho_m = \{\rho_v\alpha\} + \{\rho_\ell(1-\alpha)\}$$
$$G_m = \{\rho_v v_{vz}\alpha\} + \{\rho_\ell v_{\ell z}(1-\alpha)\}$$

Momentum equations

Phase

$$\frac{\partial}{\partial t}\{\rho_k v_{kz}\alpha_k\}A_z + \frac{\partial}{\partial z}\{\rho_k v_{kz}^2\alpha_k\}A_z = \{\Gamma_k \vec{v}_{ks}\cdot\vec{n}_z\}A_z + \{\vec{F}_{wk}'''\cdot\vec{n}_z\}A_z$$
$$- \frac{\partial}{\partial z}\{p_k\alpha_k\}A_z + \{\vec{F}_{sk}'''\cdot\vec{n}_z\}A_z + \{\rho_k\alpha_k\}\vec{g}\cdot\vec{n}_z A_z$$

Jump condition

$$\sum_{k=1}^{2}(\Gamma_k \vec{v}_{ks}\cdot\vec{n}_z + \vec{F}_{sk}'''\cdot\vec{n}_z) = 0$$

Mixture

$$\frac{\partial}{\partial t}(G_m A_z) + \frac{\partial}{\partial z}\left(\frac{G_m^2 A_z}{\rho_m^+}\right) = -F_{wz}''A_z - \frac{\partial}{\partial z}\{p\}A_z - \rho_m g A_z \sin\theta$$

where:

$$\frac{G_m^2}{\rho_m^+} \equiv \{\rho_v v_{vz}^2\alpha\} + \{\rho_\ell v_{\ell z}^2(1-\alpha)\}$$
$$\{p\} \equiv \rho_v\alpha + \rho_\ell(1-\alpha)$$
$$F_{wz}'' = -(\vec{F}_{wv} + \vec{F}_{w\ell})\cdot\vec{n}_z$$

Energy equation

Phase

$$\frac{\partial}{\partial t}\{\rho_k u_k^\circ\alpha_k\}A_z + \frac{\partial}{\partial z}\{\rho_k h_k^\circ v_{kz}\alpha_k\}A_z = \Gamma_k h_{ks}^\circ A_z - p\frac{\partial\alpha_k}{\partial z}A_z$$
$$+ \{q_k''\alpha_k\}A_z - q_{wk}''\alpha_{wk}P_w - \{\rho_k g v_{kz}\alpha_k\}A_z + \{Q_{sk}^*\}A_z$$

Jump

$$\sum_{k=1}^{2}(\Gamma_k h_{ks}^\circ + Q_{ks}^*) = 0$$

Mixture

$$\frac{\partial}{\partial t}(\rho_m h_m - p)A_z + \frac{\partial}{\partial z}(G_m h_m^+)A_z = q_w''A_z - q_w''P_w + \frac{G_m}{\rho_m}\left(F_{wz}'' + \frac{\partial p}{\partial z}\right)$$

where

$$h_m = \{\rho_v h_v\alpha_v + \rho_\ell h_\ell\alpha_\ell\}/\rho_m$$
$$h_m^+ = \{\rho_v h_v v_{vz}\alpha_v + \rho_\ell h_\ell v_{\ell z}\alpha_\ell\}/G_m$$

which equals zero (from the mass balance), we get:

$$\frac{\partial}{\partial t}[(\rho_k\alpha_k v_{kz}^2/2)A_z] + \frac{\partial}{\partial z}[\rho_k\alpha_k v_{kz}(v_{kz}^2/2)A_z] + \frac{v_{kz}^2}{2}\Gamma_k A_z =$$
$$- v_{kz}F_{wkz}'''A_z - v_{kz}\frac{\partial p\alpha_k}{\partial z}A_z - \rho_k\alpha_k v_{kz}gA_z + \vec{v}_{kz}\cdot(\Gamma_k\vec{v}_{ks} + \vec{F}_{sk}''')A_z$$
(5-157)

If we add the equations of kinetic energy for vapor and liquid we get:

$$\frac{\partial}{\partial t}\rho_m(v_m^2/2)A_z + \frac{\partial}{\partial z}G_m(v_m^2/2)A_z = - v_{vz}F_{wvz}'''A_z + v_{\ell z}F_{w\ell z}'''A_z$$

$$- \left(v_{vz}\frac{\partial p\alpha_v}{\partial z} + v_{\ell z}\frac{\partial p\alpha_\ell}{\partial z}\right)A_z - (\rho_v\alpha_v v_{vz}g + \rho_\ell\alpha_\ell v_{\ell z}g)A_z$$
(5-158)

$$+ \vec{v}_{vz}\cdot[\Gamma_v(\vec{v}_{vs} - \vec{v}_{vz}) + \vec{F}_{sv}''']A_z + \vec{v}_{\ell z}\cdot[\Gamma_\ell(\vec{v}_{\ell s} - \vec{v}_{\ell z}) + \vec{F}_{s\ell}''']A_z$$

Let us simplify by assuming that the last two terms can be dropped owing to cancellation of each other. Then, by substracting Eq. 5-158 from Eq. 5-149, we obtain an energy equation for the mixture enthalpy:

$$\frac{\partial}{\partial t}(\rho_m h_m - p)A_z + \frac{\partial}{\partial z}(G_m h_m^+ A_z) = q_m''' A_z - q_w'' P_w$$

$$+ v_{vz}\left(F_{wvz}''' + \frac{\partial p\alpha_v}{\partial z}\right)A_z \quad (5\text{-}159)$$

$$+ v_{\ell z}\left(F_{w\ell z}''' + \frac{\partial p\alpha_\ell}{\partial z}\right)A_z$$

Equation 5-159 is an approximate one, as a term pertaining to the kinetic energy exchange at the liquid–vapor interface was neglected. However, it is interesting to note that the gravity term has dropped from the equation. The last two terms are the heat addition terms, included because of wall friction and static pressure changes. When these terms are small, and for models where the vapor and liquid velocities are not easily identifiable, it is possible to approximate the energy equation as:

$$\frac{\partial}{\partial t}[(\rho_m h_m - p)A_z] + \frac{\partial}{\partial z}(G_m h_m^+ A_z) = q_m''' A_z - q_w'' P_w$$
$$+ \frac{G_m}{\rho_m}\left(F_{wz}''' + \frac{\partial p}{\partial z}\right)A_z$$
(5-160)

Table 5-3 summarizes the one-dimensional relations for a single-phase situation as well as for the two-phase mixture.

REFERENCES

1. Collier, J. G. *Convective Boiling and Condensation* (2nd ed.). New York: McGraw-Hill, 1980.
2. Delhaye, J. M., Giot, M., and Reithmuller, M. L. *Thermodynamics of Two-Phase Systems for Industrial Design and Nuclear Engineering.* New York: McGraw-Hill, 1981.

3. Ishii, M. *Thermo-fluid Dynamic Theory of Two-Phase Flow*. Eyrolles, Paris: Scientific and Medical Publications of France, 1975.

4. Martinelli, R. C., and Nelson, D. B. Prediction of pressure drop during forced circulation boiling of water. *Trans. ASME*. 49:695–702, 1948.

5. Sha, W. T., Chao, B. T., and Soo, S. L. *Time Averaging of Local Volume-Averaged Conservation Equations of Multiphase Flow*. ANL-83-49, July 1983.

6. Vernier, Ph., and Delhay, J. M. General two-phase flow equations applied to the thermohydraulics of boiling water nuclear reactors. *Energie Primaire* 4:5–46, 1968.

7. Wallis, G. B. *One Dimensional Two-Phase Flow*. New York: McGraw-Hill, 1969.

PROBLEMS

Problem 5-1 Area averaged parameters (section IV)

In a BWR assembly it is estimated that the exit quality is 0.15 and the mass flow rate is 17.5 kg/s. If the pressure is 7.2 MPa, and the slip ratio can be given as $S = 1.5$, determine $\{\alpha\}$, $\{\beta\}$, $\{j_v\}$, G_v and G_ℓ. The flow area of the assembly is 1.2×10^{-2} m^2.

Answers:

$\{\alpha\} = 0.6968$
$\{\beta\} = 0.7751$
$\{j_v\} = 5.80$ m/s
$G_v = 218.75$ kg/m^2 s
$G_\ell = 1239.6$ kg/m^2 s

Problem 5-2 Momentum balance for a two-phase jet load (section V)

Calculate the force on a wall subjected to a two-phase jet (Fig. 5-10) that has the following parameters:

Mass flux at exit $(G_m) = 10.75 \times 10^3$ kg/m^2 s
Exit diameter $(D) = 0.3$ m
Upstream pressure $(p) = 7.2$ MPa
Pressure at throat $(p_o) = 3.96$ MPa
Exit quality $(x_o) = 0.68$
Slip ratio $(S_o) = 1.5$

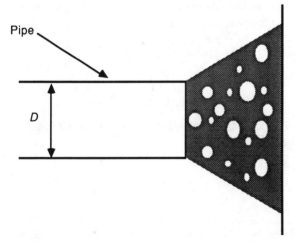

Pipe

D

Figure 5-10 Two-phase jet impacting a wall.

Water

Steam

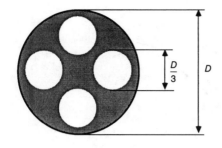

Figure 5-11 Plan view of a high-pressure tube.

Answer: Force = 0.416 MN

Problem 5-3 Estimating phase velocity differential (section IV)
A vertical tube is operating at high-pressure conditions as follows (Fig. 5-11):

Operating Conditions
Pressure (p) = 7.4 MPa
Mass flux (G) = 2000 kg/m² s
Exit quality (x_e) = 0.0693
Geometry
D = 10.0 mm
L = 3.66 m
Saturated water properties at 7.4 MPa
h_f = 1331.33 kJ/kg
h_g = 2759.60 kJ/kg
h_{fg} = 1448.27 kJ/kg
v_f = 0.001381 m³/kg
v_g = 0.02390 m³/kg
v_{fg} = 0.02252 m³/kg

Assuming that thermal equilibrium between steam and water has been attained at the tube exit, find:

1. The tube exit cross-sectional averaged true and superficial vapor velocities; i.e., find $\{v_v\}_v$ and $\{j_v\}$.
2. The difference between the tube exit cross-sectional averaged vapor and liquid velocities; i.e., find $\{v_v\}_v - \{v_\ell\}_\ell$ at the exit.
3. The difference between the tube exit cross-sectional averaged vapor and liquid superficial velocities; i.e., find $\{j_v\} - \{j_\ell\}$.
   Answers:

1. $\{v_v\}_v$ = 7.45 m/s, $\{j_v\}$ = 3.25 m/s
2. 4.89 m/s
3. 1.636 m/s

Problem 5-4 Torque on vessel due to jet from hot leg break (section V)
Calculate the torque on a pressure vessel when a break develops in the hot leg as shown in Figure 5-12. The conditions of the two-phase emerging jet are:

G_{cr} = 10.75 × 10³ kg/m² s
x_{cr} = 0.68
L = 5m
Angle θ = 70°
S = 1.5
p_o = 3.96 MPa

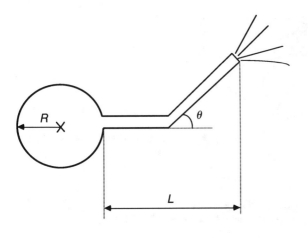

Figure 5-12 Two-phase jet at a pipe break.

$p = 7.2$ MPa

Flow area at the rupture ≈ 0.08 m^2

Answer: Torque $= 2.8$ MN.m

Problem 5-5 Interfacial terms in the momentum equation (section VI)

In a two-fluid model the momentum equation of the vapor in one-dimensional flow may be written as:

$$\frac{\partial}{\partial t}(\alpha \rho_v V_v A) = -\frac{\partial}{\partial z}(\alpha G_v V_v A) + \Gamma A V_{vs} - \frac{\partial p}{\partial z}A - F_{wv} - F_{sv} - \alpha \rho_v g\, A \cos \theta$$

where Γ = rate of mass exchange between vapor and liquid; F_{wv} = rate of momentum loss at the wall due to friction; A = flow area; V_{vs} = vapor velocity at the liquid–vapor interface; and F_{sv} = rate of momentum exchange between vapor and liquid.

Given the following:

Steady-state flow conditions in a tube
Tube diameter is D
Uniform axial heat flux q''
Annular flow conditions prevail

1. Write appropriate mass balance and energy balance equations for vapor to complete the model. State any assumptions you made.
2. Provide an expression for Γ. Justify your answer.

Answer: 2. $\Gamma = \dfrac{4q''}{Dh_{fg}}$

THERMODYNAMICS OF NUCLEAR ENERGY CONVERSION SYSTEMS: NONFLOW AND STEADY FLOW: FIRST AND SECOND LAW APPLICATIONS

I INTRODUCTION

The working forms of the first and second laws for the control mass and control volume approaches are summarized in Table 6-1. There are many applications of these laws to the analysis of nuclear systems. It is of prime importance that the reader not only develop the proficiency to apply these laws to new situations but also recognize which approach (control mass or control volume) is more convenient for the formulation of the solution of specific problems.

The elementary application of these laws avoids the time-dependent prediction of parameters. Usually processes are either modeled as nontransient by specifying the initial and end states or as steady-state processes. This choice is not dictated by inherent limitations in the first and second laws but, rather, by the complexity involved when describing the heat and work rate terms, $\dot{Q}$ and $\dot{W}$, which appear in these laws. For example, the analytic description of $\dot{Q}$ requires definition of the heat transfer rates which, in many processes, are complex and available in empirical form only.

Nonflow and steady-flow processes are discussed here. Variable-flow processes are discussed in Chapter 7. The examples analyzed in this chapter as nonflow processes are essentially transient processes. However, as for any engineering analysis, it is important to consider the question posed and then to model the process in the simplest form, consistent with the required objective and the information available. Therefore

Table 6-1 Summary of the working forms of the first and second laws of thermodynamics

Parameter	First law	Second law
Control mass	$\dot{U}^\circ_{c.m.} = \dot{Q}_{c.m.} - \dot{W}_{c.m.}$ (Eq. 4-19a)	$\dot{S}_{c.m.} = \dot{S}_{gen} + \dfrac{\dot{Q}_{c.m.}}{T_s}$ (Eq. 4-25b)

For a process involving finite changes between states 1 and 2, Eq. 4-19a becomes

$$U_2 - U_1 = Q_{1 \to 2} - W_{1 \to 2} \qquad \text{(Eq. 6-1)}$$

if kinetic energy differences are negligible

Convention: W_{out} and Q_{in} are positive

where T_s is the temperature at which heat is supplied

Control Mass

$Q_{1 \to 2}$

$U_2 - U_1$

$W_{1 \to 2}$

Control volume (stationary)

$$\dot{E}_{c.v.} = \sum_{i=1}^{I} \dot{m}_i (h_i^\circ + gz_i) + \dot{Q} + \dot{Q}_{gen} - \dot{W}_{shaft} - \dot{W}_{normal} - \dot{W}_{shear}$$
$$\text{(Eq. 4-39)}$$

$$\dot{S}_{c.v.} = \sum_{i=1}^{I} \dot{m}_i s_i + \dot{S}_{gen} + \frac{\dot{Q}}{T_s}$$
$$\text{(Eq. 4-41)}$$

Neglecting shear work and differences in kinetic and potential energy, and treating $\dot{Q}_{gen}$ as part of $\dot{U}$, Eq. 4-39 becomes

$$\dot{U}_{c.v.} = \sum_{i=1}^{I} \dot{m}_i h_i + \dot{Q} - \dot{W}_{shaft} - \dot{W}_{normal} \qquad \text{(Eq. 6-2)}$$

Convention: W_{out} and Q_{in} are positive

$\dot{m}_i$

$\dot{W}_s$

Control Volume Surface

Control Volume

Component

$\dot{U}_{cv}$

$\dot{Q}$

$\dot{m}_o$

Table 6-2 Examples covered in Chapters 6 and 7

Process	Control mass	Control volume
Nonflow	Ch. 6, Sect. II: expansion work from a fuel–coolant interaction process	—
Steady flow	—	Ch. 6, Sect. IV: reactor system thermodynamic efficiency and irreversibility analysis
Nonsteady flow	Ch. 7, Sect. II: reactor containment pressurization from loss of coolant accident	Ch. 7, Sect. II: reactor containment pressurization from loss of coolant accident
	—	Ch. 7, Sects. III and IV: pressurizer response to load change

when illustrating the control mass and control volume approaches, the examples are modeled in different ways to obviate or include the time-dependent description of the relevant rate processes. In doing so, the differences in the results achieved when complex processes are modeled in these fundamentally different ways are illustrated. Table 6-2 summarizes the approaches and examples covered here and in Chapter 7.

II NONFLOW PROCESS

The essential step in the nonflow approach is to carefully define the control mass and the sign convention for the interactions. A sketch of this control mass, complete with the relevant energy flows is useful. Also desirable is a state diagram using suitable thermodynamic properties of the initial and final equilibrium states of the process the control mass undergoes. Finally, if the control mass undergoes a known reversible process, it is useful to represent the process path on the state diagram.

To determine the energy flows, it is necessary to specify the time base over which the analysis is applicable. Usually a fixed time period is not specified; rather, it is stated or implicitly assumed that the analysis applies over a period sufficient for a desired transition between fixed thermodynamic states for the control mass. In that way we do not explicitly involve the rate processes that would require detailed heat transfer and fluid mechanics information. The price for this simplification is that we do not learn anything of the transient aspects of the process, only the relation between the end states and, if the process path is defined, the work and heat interactions of the control mass.

The first law (Table 6-1) for a control mass undergoing a process involving finite changes between state "1" and state "2" is

$$U_2 - U_1 = Q_{1 \to 2} - W_{1 \to 2} \tag{6-1}$$

where positive W = work done by the control mass, and positive Q = heat transferred into the control mass.

As an illustration of application of this law to a control mass, let us examine the thermal interaction between a hot liquid and a more volatile cold liquid. This phenomenon, a molten fuel–coolant interaction, is of interest when evaluating the integrity of the containment under hypothetical accident conditions. It can be postulated to occur in both light-water- and sodium-cooled reactors only in unlikely situations. In both reactors, the hot liquid would consist of the molten fuel, cladding, and structural materials from the partially melted core; and the volatile, cold liquid would be the coolant, water, or sodium. We are interested in evaluating the effect of this interaction on the containment by calculating the expansion work resulting from the mixing of these hot and cold liquids. The relevant properties for these materials are presented in Table 6-3.

To estimate this expansion work, assume that the process occurs in two steps. In the first step, an equilibrium temperature is found for the fuel and coolant under the condition of no expansion by the more volatile coolant or the fuel. This constant-volume condition is reasonable if the thermal equilibration time is small (≤ 1 ms), which, however, is not the case if film boiling occurs when the fuel mixes with the coolant. In the second step, work occurs when the coolant and fuel are assumed to expand reversibly to a prescribed state. This expansion may take place: (1) without heat transfer between the fuel and the coolant; or (2) with heat transfer between the fuel and the coolant so that thermal equilibrium is maintained during the expansion.

With these somewhat artificial prescriptions for the interaction, we have implicitly prescribed time periods for the two steps by defining the intermediate state (i.e., that corresponding to the equilibrium temperature at the initial volume) and the final state. In this manner the truly transient fuel–coolant interaction process has been idealized as a nonflow process. In fact, these time periods are not real because the processes occurring during the two steps are not physically distinct. Furthermore, there are heat transfer and expansion work rates that characterize this process. However, for convenience, our idealization of this process between end states allows us to establish conservative bounds on the expansion work. The analysis of this process that follows generally adopts the approach of Hicks and Menzies [1].

A Fuel–Coolant Thermal Interaction

1 Step I: Coolant and fuel equilibration at constant volume. We may define either one control mass, consisting of the combined mass of the fuel and the coolant, or two control masses, one consisting of the mass of the fuel and the other consisting of the mass of the coolant. We choose the second representation because it is easier to define the T–s diagrams for a one-component (two-phase) fluid (i.e., the coolant or fuel alone) then it is to define the T–s diagram for the two-component coolant–fuel mixture.

The control masses do not provide work to the environment during the constant-volume thermal equilibration process. To analyze the interaction conservatively, we assume that all the heat transferred from the fuel during the process is transferred to the coolant. Finally, because the thermal equilibration process is so fast, the decay heat generation rate is low enough to be neglected during the process. With these assumptions, the control mass and the process representation for the coolant alone may be sketched as in Figure 6-1.

Table 6-3 Properties for fuel–coolant interaction examples

Parameter	Symbol	Units	Sodium	Water	Fuel: UO$_2$ or mixed oxide
Mass (typically primary system inventory)	m_c or m_f	kg	3500	4000	40,000
Initial temperature	T_l	°K	600	400	3100
Initial density	ρ_l	kg/m^3	835	945	~8000
Saturation pressure	p_{sat}	MPa	$\ln p_{sat}\ (\text{MPa}) = 8.11 - \dfrac{12{,}016}{T(K)}$	$\ln p_{sat}\ (\text{MPa}) = 10.55 - \dfrac{4798}{T(K)}$	—
Gas constant	R	J/kg °K	361	462	31
Vapor specific heat ratio	γ	—	1.15	1.3	~1.06 to 1.07
Liquid specific heat at constant volume*	c_{vc} or c_{vf}	J/kg °K	1300	4184	560
Vapor specific heat at constant pressure	c_{pc} or c_{pf}	J/kg °K	$\dfrac{\gamma R}{\gamma - 1} = 2767$	2003	~500
Vapor specific heat at constant volume	c_{vc} or c_{vf}	J/kg °K	$\dfrac{R}{\gamma - 1} = 2410$	1540	~475
Latent heat of vaporization	h_{fg}	J/kg	2.9×10^6	1.9×10^6	~1.9×10^6
Coolant critical point properties					
Pressure	p_{crit}	MPa	40.0	22.1	—
Temperature	T_{crit}	°K	2733	647.3	—
Density	ρ_{crit}	kg/m^3	818	317	—
Internal energy	u_{crit}	J/kg	4.29×10^6	2.03×10^6	—

*For nearly incompressible liquids, the constant pressure and constant volume specific heats can be taken as equal.

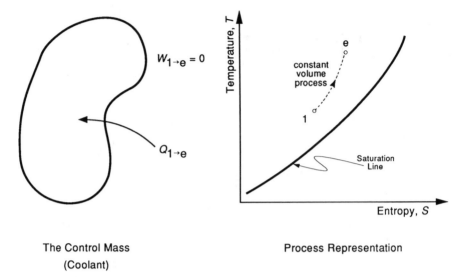

The Control Mass
(Coolant)

Process Representation

Figure 6-1 Coolant control mass behavior in the constant-volume thermal interaction with fuel: step I.

To evaluate Q, we identify the fuel as another control mass undergoing a constant-volume cooling process to the equilibrium temperature. Analogous sketches for this fuel control mass are shown in Figure 6-2.

Expressing the first law for each control mass, undergoing a change of state from 1 to e, neglecting potential and kinetic energy changes we obtain:

$$\text{Coolant } \Delta E_c = \Delta U_c = Q_{1 \to e} \qquad (6\text{-}2)$$
$$\text{Fuel } \Delta E_f = \Delta U_f = -Q_{1 \to e} \qquad (6\text{-}3)$$

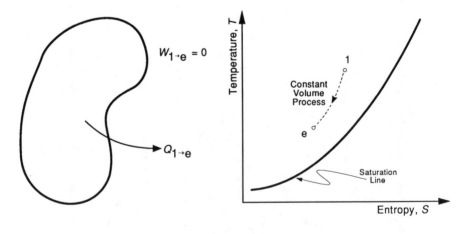

The Control Mass
(Fuel)

Process Representation

Figure 6-2 Fuel control mass behavior in the constant-volume thermal interaction with coolant: step I.

Using the state equation that expresses internal energy as a function of temperature, we obtain:

$$\Delta U_c = m_c c_{v_c} (T_e - T_{1_c})$$
$$\Delta U_f = m_f c_{v_f} (T_e - T_{1_f})$$

The interaction process of the combined coolant–fuel system is adiabatic. Hence:

$$m_c c_{v_c} (T_e - T_{1_c}) = m_f c_{v_f}(T_{1_f} - T_e)$$

or:

$$T_e = \frac{\left(\dfrac{m_f c_{v_f}}{m_c c_{v_c}}\right) T_{1_f} + T_{1_c}}{1 + \dfrac{m_f c_{v_f}}{m_c c_{v_c}}} \tag{6-4}$$

Because this equilibrium state is to the left of the saturated liquid line on a T–s plot for each component, the static quality of each fluid is by definition either 1.0 or 0 depending on whether T_e is above or below the critical temperature.

Example 6-1 Determination of the equilibrium temperature

PROBLEM Compute the equilibrium temperature achieved by constant volume mixing of (1) sodium and mixed oxide (UO_2-15 weight percent PuO_2) and (2) water and UO_2 for the parameters of Table 6-3.

SOLUTION For the sodium and mixed oxide Eq. 6-4 yields:

$$T_e = \frac{\left[\dfrac{40,000\,(560)}{3500\,(1300)}\right] 3100 + 600}{1 + \dfrac{40,000\,(560)}{3500\,(1300)}} = 2678°K$$

Because the critical temperature of sodium is 2733°K, $(x_{st})_e = 0$. The quality is unsubscripted with regard to coolant versus fuel because the fuel remains a subcooled liquid in the examples of the fuel–coolant interaction in this chapter. For simplicity the subscript st is deleted in the remainder of this section. For the water and UO_2 combination:

$$T_e = \frac{\left[\dfrac{40,000\,(560)}{400\,(4184)}\right] 3100 + 400}{1 + \dfrac{40,000\,(560)}{4000\,(4184)}} = 1945°K$$

and $x_e = 1.0$, as T_e is above the critical temperature of water, 647°K.

2 Step II: Coolant and fuel expanded as two independent systems, isentropically and adiabatically.

Consider the fuel and the coolant each undergoing an adiabatic, isentropic, and hence reversible expansion. Let us first estimate the work done by the coolant when expanding to 1 atmosphere. The control mass and the process representation for the coolant are shown in Figure 6-3.

Note that when expanding to the final state (state 2) some coolant remains in liquid form. Assume that this liquid occupies negligible volume and is incompressible.

From the first law for our control mass neglecting potential and kinetic energy changes we obtain:

$$W_{e \to 2} = -\Delta U_{e2} \equiv U_e - U_2 \tag{6-5}$$

In order to obtain an analytic solution for the W_{e2}, the internal energy is next expressed by an approximate equation of state, which allows for expansion into the two-phase region. Thus the change in internal energy is expressed as:

$$\Delta u \equiv c_v \Delta T + \Delta(x u_{fg}) \equiv c_v \Delta T + \Delta[x(h_{fg} - p v_{fg})] \tag{6-6}$$

where $u_{fg} = u_g - u_f$, $h_{fg} = h_g - h_f$, and $v_{fg} = v_g - v_f$; c_v and c_p without further subscript refer to liquid specific heats, whereas vapor specific heats are designated c_{v_v} and c_{p_v}. Considering the change in specific internal energy (Δu) as defined in Eq. 6-6, the work $W_{e \to 2}$ can be written as:

$$W_{e \to 2} = -m_c \Delta u = -m_c (u_2 - u_e) \tag{6-7}$$
$$W_{e \to 2} = m_c[c_v(T_e - T_2) + x_e(h_{fg} - p v_{fg})_e - x_2(h_{fg} - p v_{fg})_2]$$

If we neglect the liquid volume at state 2, this equation can be written as:

$$W_{e \to 2} = m_c\{c_v(T_e - T_2) + [(x h_{fg})_e - (x h_{fg})_2] - [(x p v_{fg})_e - (x p v_g)_2]\} \tag{6-9}$$

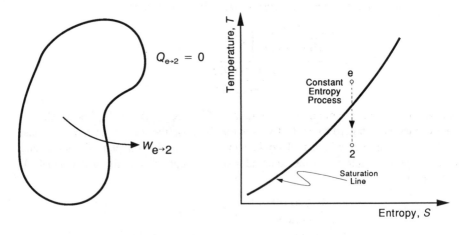

The Control Mass
(Coolant)

Process Representation

Figure 6-3 Coolant control mass behavior in the adiabatic, isentropic expansion: step II.

To evaluate $W_{e\to 2}$, we must first determine x_2, which is done as follows. For a pure substance the following relation exists between thermodynamic properties for two infinitesimally close equilibrium states:

$$Tds = dh - vdp \tag{6-10a}$$

which for an isentropic process can be rewritten as:

$$\left(\frac{dh}{T}\right)_s = v\left(\frac{dp}{T}\right)_s \tag{6-10b}$$

Analogous to the treatment of the internal energy by an approximate equation of state (Eq. 6-7), take h_{fg} and c_p independent of temperature over the range of interest and express the enthalpy change for this isentropic expansion as:

$$dh = c_p dT + h_{fg} dx \tag{6-11}$$

Neglecting the liquid specific volume and applying the perfect gas result:

$$v \equiv v_f + xv_{fg} \approx xv_g \approx x\frac{RT}{p} \tag{6-12}$$

In an isentropic process, the differential of $p(T,s)$ is

$$dp = \left(\frac{\partial p}{\partial T}\right)_s dT$$

If we assume state e is almost at the saturation line, so that the expansion is almost totally under the saturation dome, we can express $\left(\frac{\partial p}{\partial T}\right)_s$ utilizing the Clausius-Clapeyron relation, which links parameters along the saturation line to the enthalpy and volume of vaporization. Hence:

$$dp = \left(\frac{\partial p}{\partial T}\right)_s dT \simeq \left(\frac{\partial p}{\partial T}\right)_{sat} dT = \frac{h_{fg}}{Tv_{fg}} dT = \frac{ph_{fg}}{RT^2} dT \tag{6-13}$$

where the last step utilizes the assumptions of Eq. 6-12. Substituting Eqs. 6-11, 6-12, and 6-13 into Eq. 6-10 and a rearranging, we obtain:

$$h_{fg}\frac{dx}{T} + c_p\frac{dT}{T} - h_{fg}\, x\, \frac{dT}{T^2} = 0$$

which can be written as:

$$d\left(\frac{x}{T}\right) = -\frac{c_p}{h_{fg}}\frac{dT}{T}$$

where h_{fg} and c_p have been taken constant.

Integrating this result between the initial equilibrium temperature of the fuel and coolant and the final coolant temperature, we obtain the expression defining x_2 in terms of the fuel/coolant equilibrium temperature and the final coolant temperature.

$$x_2 = T_2 \left(\frac{x_e}{T_e} + \frac{c_p}{h_{fg}} \ln \frac{T_e}{T_2} \right) \qquad (6\text{-}14)$$

The work has been calculated for expansion of only the coolant. Next consider the work from the fuel as it reversibly changes volume. The fuel cooldown to a partial pressure which is negligible consistent with a total pressure of one atmosphere would terminate at a temperature different from that achieved in the coolant-only expansion, as Figure 6-4 illustrates. However the work associated with fuel cooldown is small because the fuel remains a subcooled liquid. Hence, it is not evaluated here.

Example 6-2 Determination of final quality and work done by sodium coolant

PROBLEM Evaluate the final quality (x_2) and the work done by the coolant for the sodium case taking the final state (state 2) as the saturation temperature (1154°K) at 1 atmosphere. Use the properties of Table 6-3.

SOLUTION Utilizing Eq. 6-14 and $x_e = 0$ from Example 6-1 yields:

$$x_2 = 1154 \left[0 + \frac{1300}{2.9(10^6)} \ln \frac{2678}{1154} \right] = 0.435$$

Returning to Eq. 6-9, the expansion work done by the coolant is:

$$W = 3500[1300(2678 - 1154) - 0.435(2.9)(10^6)$$
$$+ 0.435(1.013)(10^5)4.11] = 3153 \text{ MJ}$$

where:

$$v_{g_2} \approx \left(\frac{RT}{p} \right)_2 = \frac{361(1154)}{1.013 \times 10^5} = 4.11 \text{ m}^3/\text{kg}$$

and

$$x_e = 0.$$

Examination of the water–fuel case for the properties given in Table 6-3 using Eq. 6-14 shows that the final state is superheated. Because the equilibrium condition has been already shown to be supercritical, use of Eq. 6-7 is inappropriate and Eq. 6-9 for the expansion work must be rederived.

3 Step III: Coolant and fuel expanded as one system in thermal equilibrium, adiabatically and isentropically. Let us now calculate the work considering that the sodium and the fuel remains in thermal equilibrium as the mixture expands adiabatically and isentropically to the final pressure. This process furnishes the upper bound of the expansion work. In this expansion process the coolant–fuel system expands adiabatically and isentropically in thermal equilibrium. Consequently, heat is being removed from the fuel and added to the coolant so that the coolant entropy increases and the fuel entropy decreases an equal amount. Hence the coolant passes into the

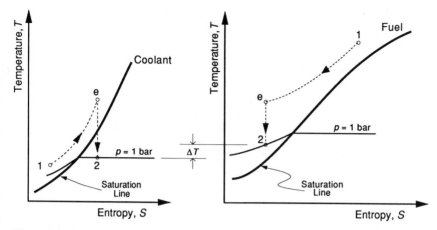

Figure 6-4 Fuel control mass and coolant control mass behavior in the adiabatic isentropic expansion: step II.

two-phase region and may achieve a superheated vapor state. On the other hand, the fuel remains a subcooled liquid throughout the expansion process because the equilibrium temperature is below the fuel boiling point even at 1 atmosphere. The control mass and process representations for the coolant and for the fuel as separate control masses are shown in Figure 6-5, which illustrates the entropy increase of the coolant and the equal entropy decrease of the fuel. For illustration, the final coolant state is shown as superheated vapor. The work ($W_{e \to 2}$) is the expansion work performed by the coolant–fuel system. Although the entropy of each component of the mixture changes, the entropy of the mixture is constant, as Figure 6-5 illustrates.

In the analysis we assume that the liquid phase of each component occupies negligible volume and is incompressible. Eq. 6-5 is the relevant first-law formulation but now is applied to the mixture. Considering the mixture specific internal energy (u_m), Eq. 6-5 is written as:

$$W_{e \to 2} = -m_m \Delta u_m = -(m_c \Delta u_c + m_f \Delta u_f) \tag{6-15}$$

To express the coolant internal energy change (Δu_c), the final state of the coolant (i.e., whether it is a two-phase mixture or a superheated vapor) must be known. Therefore we must evaluate the final coolant quality (x_2) to see if it is less than or greater than unity. (Note that there is no need to subscript the quality as x_{c_2}, as the fuel is always a subcooled liquid.) To do so, we repeat the procedure of Section 2, but now consider the mixture expansion as adiabatic and isentropic.

Expressing Eq. 6-10a for the mixture in this isentropic process:

$$T_m ds_m = dh_m - v_m \, dp_m = 0 \tag{6-16}$$

From the definition of mixture enthalpy

$$\Delta h_m = \frac{(m_c c_{p_c} + m_f c_{p_f}) \Delta T + m_c \Delta (x \, h_{fg_c})}{m_c + m_f} \tag{6-17}$$

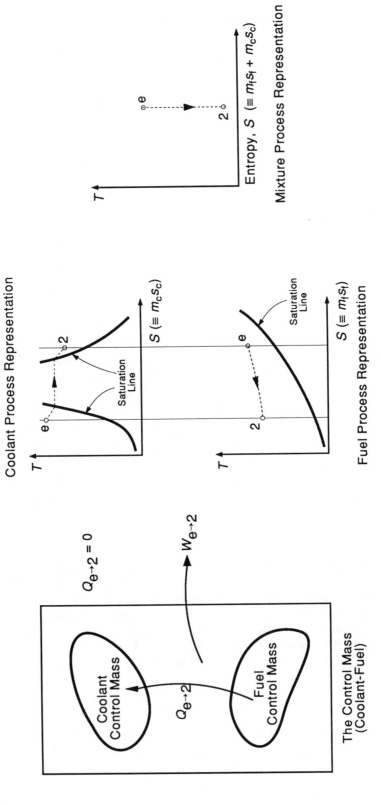

Figure 6-5 Coolant–fuel control mass and coolant–fuel interaction during the adiabatic, isentropic control mass expansion: step II.

Differentiating, assuming h_{fg} constant:

$$dh_m = \frac{(m_c c_{p_c} + m_f c_{p_f})dT + m_c h_{fg_c} dx}{m_c + m_f} \tag{6-18}$$

Neglecting the fuel volume and the liquid coolant volume:

$$v_m = \frac{x m_c v_{g_c}}{m_c + m_f} = \frac{x R_c T}{p}\left(\frac{m_c}{m_c + m_f}\right) \tag{6-19}$$

Utilizing the Clausius-Clapeyron relation in the same manner as previously, reexpress Eq. 6-13 for the mixture as follows noting that the fuel remains as a subcooled liquid:

$$dp_m = \left(\frac{\partial p}{\partial T}\right)_{sat} dT = \frac{p h_{fg_c}}{R_c} \frac{dT}{T^2} \tag{6-20}$$

Substituting the above three relations into the state principle (Eq. 6-16) and rearranging, we obtain:

$$(m_c c_{p_c} + m_f c_{p_f})\frac{dT}{T} + m_c h_{fg_c} d\left(\frac{x}{T}\right) = 0 \tag{6-21}$$

Integrating this result between the initial equilibrium temperature of the fuel and coolant and final mixture temperature, we obtain:

$$x_2 = T_2\left[\frac{x_e}{T_e} + \left(\frac{m_c c_{p_c} + m_f c_{p_f}}{m_c h_{fg_c}}\right) \ln \frac{T_e}{T_2}\right] \tag{6-22}$$

If evaluation of x_2 indicates that the final state is a superheated vapor, i.e., $x_2 > 1$, Δu_c is written as:

$$\Delta u_c = \Delta u_c (x_e = 0 \text{ to } x = 1) + \Delta u_c (x = 1 \text{ to the actual state 2})$$

which can be written as follows utilizing Eq. 6-7 and the numerical values of x_e and x:

$$\Delta u_c = [c_v(T - T_e) + h_{fg} - pv_{fg}]_c + c_{v_c}(T_2 - T) \tag{6-23}$$

where T = temperature corresponding to x equal to unity, and the subscript on c_v in the second term explicitly indicates that the coolant is in vapor form.

The change in fuel internal energy is

$$\Delta u_f = c_{v_f}(T - T_e) + c_{v_f}(T_2 - T) \tag{6-24}$$

where c_{v_f} is the specific heat at constant volume for liquid fuel.

To evaluate Δu_c and Δu_f, the final temperature (T_2) must be determined in the superheat region. It is done again using the state principle (Eq. 6-16) but now considering only that portion of the expansion process in which the coolant state is a superheated vapor. Therefore Eq. 6-18 is written as:

$$dh_m = \frac{(m_c c_{p_{vc}} + m_f c_{p_f})\, dT}{m_c + m_f} \tag{6-25}$$

where again $c_{p_{vc}}$ explicitly indicates that the coolant is in vapor form, and Eq. 6-19 is written for $x = 1$ as:

$$v_m = \frac{R_c T}{p} \left(\frac{m_c}{m_c + m_f} \right) \tag{6-26}$$

Substituting Eqs. 6-25 and 6-26 into Eq. 6-16, we obtain the following relation between mixture temperature and pressure:

$$\left(\frac{m_c c_{p_{vc}} + m_f c_{p_f}}{m_c R_c} \right) \frac{dT}{T} = \frac{dp}{p} \tag{6-27}$$

Integration of Eq. 6-27 to the unknown final state p_2, T_2 yields:

$$T_2 = T \left(\frac{p_2}{p} \right)^{1/n} \tag{6-28}$$

where $n =$ the term in brackets in Eq. 6-27, which can be written in terms of coolant vapor properties as:

$$n \equiv \frac{m_c \dfrac{\gamma R_c}{\gamma - 1} + m_f c_{p_f}}{m_c R_c} = \frac{m_c \, c_{p_{vc}} + m_f \, c_{p_f}}{m_c \, R_c} \tag{6-29}$$

Although $T (x = 1)$ can be determined by expressing Eq. 6-22 for $x = 1$ and this corresponding T, the corresponding p is the coolant partial pressure, not the mixture pressure. However, because the fuel is liquid, its partial pressure is negligible. Therefore we can evaluate T_2 from Eq. 6-28 with negligible error. Finally, returning to evaluation of the expansion work, it can be expressed by applying Eqs. 6-23 and 6-24 to 6-15, yielding:

$$W = W(\text{to } x = 1) + W(x = 1 \text{ to } T_2) = m_c[c_v(T_e - T) - h_{fg} + R_c T]_c \tag{6-30}$$
$$+ m_f c_{v_f}(T_e - T) + m_c c_{v_{vc}}(T - T_2) + m_f c_{v_f}(T - T_2)$$

where the coolant vapor is taken as perfect gas allowing (pv) to be replaced by $R_c T$.

Example 6-3 Determination of final quality, temperature, and expansion work for the sodium–mixed oxide combination

PROBLEM For the sodium–mixed oxide combination, evaluate the final quality and temperature (x_2 and T_2) and the expansion work using the parameters of Table 6-3.

SOLUTION First evaluate x_2 from Eq. 6-22, assuming initially that state 2 is a two-phase state at 1 atmosphere. For this assumption T_2 is the sodium saturation temperature corresponding to 1 atmosphere or 1154°K. The quality (x_2) is obtained from Eq. 6-22 as:

$$x_2 = 1154 \left[0 + \frac{40,000(560) + 3500(1300)}{(3500) \, 2.9(10^6)} \ln \frac{2678}{1154} \right] = 2.58$$

Because $x_2 > 1$, this final state is superheated, and our original assumption is false. Therefore it becomes necessary to find the state at which $x = 1$. Returning to Eq. 6-22 but defined for the state $x = 1$ and the corresponding temperature T, yielding:

$$1 = T\left[0 + \frac{40,000(560) + 3500(1300)}{(3500)2.9(10^6)}\ln\frac{2678}{T}\right]$$

Solving for T yields 2250°K, which has a corresponding sodium vapor pressure of 158 bars.

T_2 can now be evaluated from Eq. 6-28 utilizing n obtained from Eq. 6-29. These steps yield:

$$T_2 = 2250\left(\frac{1}{158}\right)^{(1/25.4)} = 1843°K$$

because:

$$n = \frac{3.5\,(2767) + 40(560)}{3.5(361)} = 25.4$$

Finally, evaluating the expansion work by Eq. 6-30 yields:

$$
\begin{aligned}
W &= 3500[1300(2678 - 2250) - 2.9 \times 10^6 + (361)(2250)] \\
&\quad + 40,000(560)(2678 - 2250) + 3500(2410)(2250 - 1843) \\
&\quad + 40,000(560)(2250 - 1843) \\
&= -5360 + 9587 + 3433 + 9117 \\
&= 16,777 \text{ MJ}
\end{aligned}
$$

Note that the work evaluated by Example 6-3 is greater than that from Example 6-2 because work is dependent on the process path and the initial and end states. Although both processes have the same initial fuel and coolant states and are reversible, the process paths as well as the final states of the fuel and coolant are different. Consequently, there is no reason to expect that expansion work integrals performed to different final states should be the same. For the adiabatic and thermal equilibrium cases, the work is evaluated by Eq. 6-9 and 6-30, respectively. A comparison of the numerical values of the terms of each equation is given in Table 6-4 to illustrate the origin of the differences in net work. In the thermal equilibrium case, the terms involving coolant properties and those involving fuel properties should not be interpreted as the work contributions due to the coolant and the fuel separately.

Table 6-4 Comparison between results of Examples 6-2 and 6-3

Parameter	Fuel and coolant expansion: two independent systems (Eq. 6-9)	Fuel-coolant expansion: one-system (Eq. 6-30)
Terms involving coolant properties	+ 3153 MJ	− 5360 + 3433 = − 1927 MJ
Terms involving fuel properties	0	+ 9587 + 9117 = + 18,704 MJ
Net work	+ 3153 MJ	+ 16,777 MJ

III THERMODYNAMIC ANALYSIS OF NUCLEAR POWER PLANTS

The analysis of nuclear power plants represents a prime application of the methods for thermodynamic analysis of steady-flow processes. The results of such analyses determine the relation between the mixed mean outlet coolant temperatures (or enthalpy for BWR) of the core through the primary and secondary systems to the generation of electricity at the turbine. The variety of possible reactor systems, with their associated coolants, leads to a corresponding multiplicity of primary and secondary system configurations. In addition, because gaseous reactor coolants can be used directly to drive electric turbines, the Brayton cycle can be considered for gas-cooled reactor systems, whereas systems that use steam-driven electric turbines employ a Rankine cycle. The various cycles employed with the principal reactor types have been described in Chapter 1. In this section the cycles used for the various reactor types and the methods of thermodynamic analysis of these cycles are described.

The Rankine and Brayton cycles are constant-pressure heat addition and rejection cycles for steady-flow operation. They differ regarding the phase changes the working fluid undergoes. In the Rankine cycle the working fluid is vaporized and condensed, whereas in the Brayton cycle the working fluid remains a single gaseous phase as it is heated and cooled. In central station nuclear Rankine cycles, water is employed as the working fluid, whereas helium is used in proposed Brayton cycles.

The PWR and BWR employ the water Rankine cycle. Because the PWR limits the reactor coolant to a nominal saturated mixed mean core outlet condition, the vapor that drives the turbine must be generated in a steam generator in a secondary system. A simplified pressurized water reactor two-coolant system is illustrated in Figure 6-6. A steam generator links these primary and secondary systems. Figure 6-7 illustrates the temperature distribution within such a recirculating PWR steam generator. The mixed mean core outlet condition (state 5) is dictated by the allowable core performance, particularly material corrosion limits.

As the temperature is raised, the primary loop pressure must also be raised to maintain state 5 at the nominally saturated condition. Establishment of the optimum

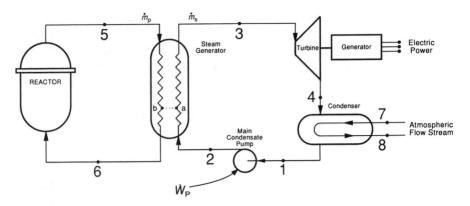

Figure 6-6 Simplified PWR plant.

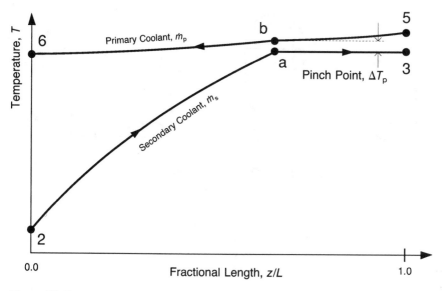

Figure 6-7 Temperature distribution within the steam generator of the simplified PWR plant of Example 6-4.

primary pressure in current PWRs at 2250 psia (15.5 MPa) has involved many considerations, among them the piping and reactor vessel wall dimensions and the pressure dependence of the critical heat flux limit. The secondary side temperature and hence pressure are related to these primary side conditions, as suggested in Figure 6-7. This relation is demonstrated in section IV, below.

The BWR employs a direct Rankine cycle. The reactor is itself the steam generator, so that the mixed mean core outlet is also nominally saturated. The outlet temperature and pressure conditions are thus established by both the allowable core and secondary loop design conditions. As Table 2-1 illustrates, the Rankine working fluid conditions (at turbine entrance) for both PWRs and BWRs are close, i.e., 5.7 MPa and 7.17 MPa, respectively.

The simple Brayton cycle is illustrated in Figure 6-8. There are four components, as in the secondary system of the PWR plant operating under the Rankine cycle. The pictured Brayton cycle compressor and heat exchanger perform functions analogous to those of the Rankine cycle condensate pump and condensor. In practice, the maximum temperature of the Brayton cycle (T_3) is set by turbine blade and gas-cooled reactor core material limits far higher than those for the Rankine cycle, which is set by liquid-cooled reactor core materials limits. The Brayton cycle in its many possible variations is presented in Section VI, below.

It is useful to assess the thermodynamic performance of components and cycles using nondimensional ratios called *efficiencies*. Three definitions for efficiency are considered:

Thermodynamic efficiency (or effectiveness) (ζ)
Isentropic efficiency (η_s)
Thermal efficiency (η_{th})

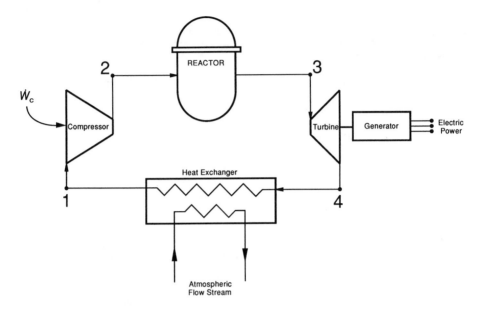

Figure 6-8 Simple Brayton cycle.

The *thermodynamic efficiency* or effectiveness is defined as:

$$\zeta \equiv \frac{\dot{W}_{u,actual}}{\dot{W}_{u,max}} \qquad (6.31)$$

where $\dot{W}_{u,max}$ is defined by Eq. 4-46, which for the useful case of a fixed, nondeformable control volume with zero shear work and negligible kinetic and potential energy differences between the inlet and outlet flow streams takes the form:

$$\dot{W}_{u,max} = \left[\frac{\partial(U - T_oS)}{\partial t}\right] + \sum_{i=1}^{I} \dot{m}_i(h - T_os)_i + \left(1 - \frac{T_o}{T_s}\right)\dot{Q} \qquad (6-32)$$

where T_s = temperature at which heat is supplied and $\left(\dfrac{dQ}{dt}\right)_{gen}$ is treated as part of $\dfrac{dU}{dt}$.

The *isentropic efficiency* is defined as:

$$\eta_s = \left(\frac{\dot{W}_{u,actual}}{\dot{W}_{u,max}}\right)_{\dot{Q}=0} \qquad (6-33)$$

Obviously, for an adiabatic control volume, $\zeta = \eta_s$. The quantity $\dot{W}_{u,max|\dot{Q}=0}$ is the useful, maximum work associated with a reversible adiabatic (and hence isentropic) process for the control volume. It can be expressed from Eq. 6-32 as:

$$\dot{W}_{u,max|\dot{Q}=0} = -\left[\frac{\partial U}{\partial t}\right]_{c.v.} + \sum_{i=1}^{I} \dot{m}_i h_{is} \qquad (6\text{-}34)$$

which for steady-state conditions becomes:

$$\dot{W}_{u,max|\dot{Q}=0} = \sum_{i=1}^{I} \dot{m}_i h_{is} \qquad (6\text{-}35)$$

The useful, actual work of an adiabatic nondeformable control volume with zero shear and negligible kinetic and potential energy differences between the inlet and outlet flow streams for steady-state conditions from Eq. 4-44 (treating $\dot{Q}_{gen}$ as part of $\dot{E}$) is:

$$\dot{W}_{u,act} = \sum_{i=1}^{I} \dot{m}_i h_i \qquad (6\text{-}36)$$

and the isentropic efficiency becomes:

$$\eta_s = \frac{\displaystyle\sum_{i=1}^{I} \dot{m}_i h_i}{\displaystyle\sum_{i=1}^{I} \dot{m}_i h_{is}} \qquad (6\text{-}37)$$

Finally, the *thermal efficiency* is defined as:

$$\eta_{th} = \frac{\dot{W}_{u,act}}{\dot{Q}_{in}} \qquad (6\text{-}38)$$

where $\dot{Q}_{in}$ = rate of heat addition to the control volume. For adiabatic systems the thermal efficiency is not a useful measure of system performance.

These efficiencies are now evaluated for nuclear plants. Typical plants employing the Rankine and Brayton cycles have been presented in Figures 6-6 and 6-8, respectively. Such plants have two interactions with their surroundings: a net work output as electricity and a flow cooling stream that is in mutual equilibrium with the atmosphere. Hence the availability of the inlet and the outlet streams are zero, i.e.,

$$\sum_{i=1}^{I} \dot{m}_i (h - T_o s)_i = 0$$

Note, however, that the energy and entropy of the stream change as the stream passes through the plant condenser or heat exchanger, but the stream remains at pressure p_o and temperature T_o.

In section IV, each component of typical plants is analyzed using a stationary, nondeformable control volume with zero shear work and negligible kinetic and potential energy differences between the inlet and the outlet flow streams. These results are then utilized for evaluating the entire nuclear plant modeled by one adiabatic control volume, as illustrated in Figure 6-9A. Two other control volume representations of the complete nuclear plant are illustrated in Figure 6-9B,C.

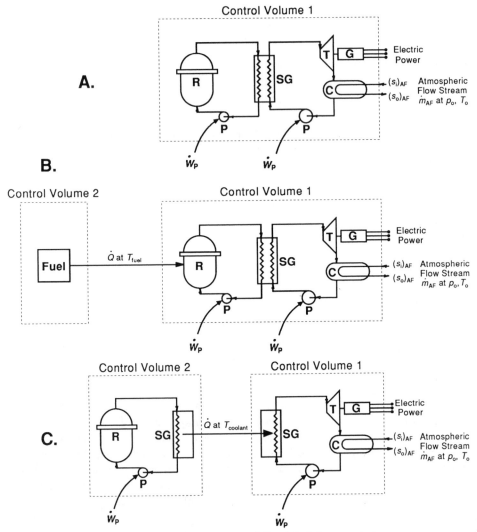

Figure 6-9 Alternative control volume representations of a batch-fueled reactor plant.

Let us now contrast the computed effectiveness and thermal efficiencies for the three nuclear plant representations of Figure 6-9. In section IV, the maximum useful work for control volume 1 of Figure 6-9A is shown equal to the fission rate, which equals the coolant enthalpy rise across the reactor. Hence:

$$\dot{W}_{u,max} = \dot{m}_p(h_{out} - h_{in})_{reactor}$$

(see Eq. 6-53). The effectiveness of the nuclear plant is then:

$$\zeta = \frac{\dot{W}_{u,actual}}{\dot{m}_p(h_{out} - h_{in})_{reactor}} \qquad (6\text{-}39)$$

In Figure 6-9B the same nuclear plant is modeled by two control volumes. Control volume 1 encompasses the reactor plant except for the fuel loading, which is segregated into the second control volume. The energy liberated by fission is transferred from the second control volume to the first control volume as heat ($\dot{Q}$) at the temperature of the fuel. Because $\dot{Q}$ equals the fission rate:

$$\dot{Q} = \dot{m}_p (h_{out} - h_{in})_{reactor} \tag{6-40}$$

and

$$\eta_{th|c.v.1 \atop \text{(Fig. 6-9B)}} = \frac{\dot{W}_{u,actual}}{\dot{Q}} = \zeta_{c.v.1 \atop \text{(Fig. 6-9A)}} \tag{6-41}$$

Finally, the same nuclear plant is alternatively modeled as shown in Figure 6-9C, with the reactor (including the fuel) and the primary side of the steam generator located within control volume 2. In this case, the same heat transfer rate ($\dot{Q}$) to control volume 1 occurs across the steam generator but now at the coolant temperature, so that:

$$\eta_{th|c.v.1 \atop \text{(Fig. 6-9C)}} = \eta_{th|c.v.1 \atop \text{(Fig. 9B)}} = \zeta_{c.v.1 \atop \text{(Fig. 6-9A)}} \tag{6-42}$$

Figure 6-9A is the only control volume configuration in which the entire nuclear plant is contained within one control volume. The effectiveness of the portions of the reactor plant contained within the three control volumes designated c.v.1 are not identical because the maximum, useful work associated with each of these control volumes is different. In section IV.B we show how transfer of the fission energy to the reactor coolant may be modeled as a series of three heat interactions that bring the fission energy from $T_{fission}$ to $T_{coolant}$. Consequently:

$$\dot{W}_{u,max|c.v.1 \atop \text{(Fig. 6-9A)}} > \dot{W}_{u,max|c.v.1 \atop \text{(Fig. 6-9B)}} > \dot{W}_{u,max|c.v.1 \atop \text{(Fig. 6-9C)}} \tag{6-43}$$

Hence the effectiveness of the portion of the plant within control volume 1 in each of the three configurations are related as:

$$\zeta_{|c.v.1 \atop \text{(Fig. 6-9C)}} > \zeta_{|c.v.1 \atop \text{(Fig. 6-9B)}} > \zeta_{c.v.1 \atop \text{(Fig. 6-9A)}} \tag{6-44}$$

In this text, we often refer to the entire nuclear plant, and therefore the representation in Figure 6-9A appears and the term thermodynamic efficiency or effectiveness is used. However, by virtue of the equalities in Eq. 6-42, it is also possible and useful to alternatively refer to the thermal efficiency of the cycle, which is also done. The previous examples illustrate the conditions under which the computation of various efficiencies lead to the same numerical values.

Finally, it is of interest to recall a case in which the effectiveness and the thermal efficiency do not coincide: For the case of reversible cycles operating between a heat source at T_{high} and a reservoir at T_{low}, the thermal efficiency equals the Carnot efficiency. On the other hand, the effectiveness equals unity, as $\dot{W}_{u,max} = \dot{W}_{u,actual}$ for these cycles.

IV THERMODYNAMIC ANALYSIS OF A SIMPLIFIED PWR SYSTEM

The steady-state thermodynamic analysis of power plant systems is accomplished by considering in turn the components of these systems and applying to each the control volume form of the first and second laws. Let us examine the simplified PWR two-coolant system shown in Figure 6-6. We wish to develop the procedures for evaluating the thermodynamic states and system flow rates so that the overall plant thermodynamic efficiency and sources of lost work or irreversibilities can be computed from a combined first and second law analysis. At the outset assume for simplicity that pressure changes occur only in the turbine and condensate pumps.

A First Law Analysis of a Simplified PWR System

Consider any one component with multiple inlet and outlet flow streams operating at steady state and surround it with a nondeformable, stationary control volume (Fig. 6-10). Applying the first law (Eq. 6-2) to this control volume we obtain:

$$\sum_{k=1}^{I} (\dot{m}h)_{in,k} - \sum_{k=1}^{I} (\dot{m}h)_{out,k} = \dot{W}_{shaft} - \dot{Q} \qquad (6\text{-}45)$$

where the notation of the summations has been generalized from that of Eq. 4-30b to specifically identify the multiple inlet and outlet flow streams, where for steady state:

$$\dot{m} = \sum_{k=1}^{I} \dot{m}_{in,k} = \sum_{k=1}^{I} \dot{m}_{out,k}$$

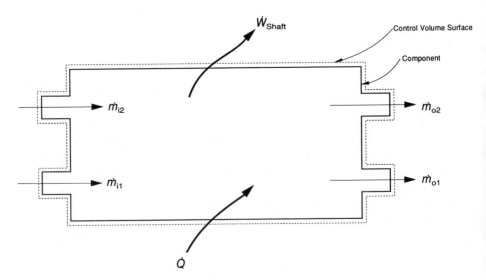

Figure 6-10 Generalized fixed-volume component with surrounding control volume.

Table 6-5 Forms of first law (Eq. 6-2) for PWR plant components

Component	Desired quantity	Assumptions	Resultant equation	Consequence of sign convention on the desired parameter
Turbine	$\dot{W}_{shaft}$	$\dot{Q}_2 = 0$ $\Delta(\frac{1}{2}v^2) = 0$	$\dot{W}_T^* = [\dot{m}(h_{in} - h_{out})]_T$ $\quad = [\dot{m}\eta_T(h_{in} - h_{out,s})]_T$	$\dot{W}_T$ Positive (work out)
Pump	$\dot{W}_{shaft}$	$\dot{Q}_2 = 0$ $\Delta(\frac{1}{2}v^2) = 0$	$\dot{W}_P = [\dot{m}(h_{in} - h_{out})]_P$ $\quad = \left[\dfrac{\dot{m}}{\eta_P}(h_{in} - h_{out,s}) \right]_P$	$\dot{W}_P$ Negative (work in)
Condensor steam generator, any feed water heater	h_o of one stream, e.g., h_{o1}	$\dot{Q} = 0$ $\dot{W}_{shaft} = 0$ $\Delta(\frac{1}{2}v^2) = 0$	$h_{out,1} = $ $\dfrac{\sum\limits_{k=1}^{I}(\dot{m}h)_{in,k} - \sum\limits_{k=2}^{I}(\dot{m}h)_{out,k}}{\dot{m}_{out,1}}$	$h_{out,1} - h_{in,1}$ positive (stream 1 is heated)

*Because the control volumes are nondeformable, the shaft is dropped on $\dot{W}$ and replaced by a subscript describing the component.

Equation 6-45 can be further specialized for the components of interest. In particular for the components of Figure 6-6, heat addition from an external source can be neglected. For all components except the turbine and the pump, no shaft work exists. Table 6-5 summarizes the results of specializing Eq. 6-45 to the components of Figure 6-6. The adiabatic turbine actual work is related to the ideal work by the component isentropic efficiency defined by Eq. 6-33. From Equation 6-37, which is a specialized form of Eq. 6-33, the turbine efficiency is:

$$\eta_T = \frac{h_{in} - h_{out}}{h_{in} - h_{out,s}} \tag{6-46}$$

where

$$h_{out,s} = f[p(h_{out}), s_{in}]. \tag{6-47}$$

Because work is supplied to the pump, the isentropic efficiency of the pump is analogously defined as:

$$\eta_P = \frac{\text{ideal work required}}{\text{actual work required}} = \frac{h_{in} - h_{out,s}}{h_{in} - h_{out}} \tag{6-48}$$

The primary and secondary flow rates are related by application of the first law to the component they both flow through, i.e., the steam generator. Considering only one primary and one secondary flow stream, the first law can be used to express the ratio of these two flow rates. The first law (Eq. 6-45) for the steam generator yields:

$$\frac{\dot{m}_p}{\dot{m}_s} = \frac{(h_{out} - h_{in})_s}{(h_{in} - h_{out})_p} \tag{6-49}$$

where subscripts p and s = primary and secondary flow streams, respectively.

It is often useful to apply Eq. 6-49 to several portions of the steam generator. In Figure 6-7 temperatures are plotted versus fractional length assuming constant pressure for a recirculating-type PWR steam generator producing saturated steam. The minimum primary to secondary temperature difference occurs at the axial location of the onset of bulk boiling in the secondary side. All temperatures are related to this minimum, or pinch point, temperature difference. The steam generator heat transfer area is inversely related to this minimum temperature difference for a given heat exchange capacity. Also the irreversibility of the steam generator is directly related to this temperature difference. Therefore the specification of the pinch point temperature difference is an important design choice based on tradeoff between cost and irreversibility.

The ratio of flow rates in terms of the pinch point temperature difference (ΔT_p) is, from the first law:

$$\frac{\dot{m}_p}{\dot{m}_s} = \frac{h_3 - h_a}{\bar{c}_p[T_5 - (T_a + \Delta T_p)]} = \frac{h_a - h_2}{\bar{c}_p[(T_a + \Delta T_p) - T_6]} \tag{6-50}$$

where $\bar{c}_p$ = average coolant specific heat over the temperature range of interest.

For the reactor plant, $\dot{W}_{u,actual}$ is the net rate of work generated, i.e.,

$$\dot{W}_{u,actual} = \dot{W}_T + \dot{W}_P = [\dot{m}_s(h_{in} - h_{out})]_T + [\dot{m}_s(h_{in} - h_{out})]_P \tag{6-51}$$

$$= [n_T\dot{m}_s(h_{in} - h_{out,s})]_T + \left[\frac{\dot{m}_s}{\eta_P}(h_{in} - h_{out,s})\right]_P \tag{6-52}$$

For a pump $h_{in} < h_{out,s}$, whereas for a turbine $h_{in} > h_{out,s}$. Hence $\dot{W}_P$ is negative, indicating, per our sign convention presented in Tables 6-1 and 6-5, that work is supplied to the pump. Analogously, $\dot{W}_T$ is positive, and the turbine delivers work. The familiar expression of the maximum useful work of the nuclear plant that is derived in Section IV.B is

$$\dot{W}_{u,max} = \dot{m}_p(h_{out} - h_{in})_R = \dot{m}_p(h_5 - h_6) \tag{6-53}$$

for Figure 6-6. It is desirable to reexpress $\dot{W}_{u,max}$ in terms of secondary side conditions. To do it, apply the first law (Eq. 6-45) to the PWR steam generator of Figure 6-6, yielding:

$$\dot{m}_p(h_{in} - h_{out})_{SGp} = \dot{m}_s(h_{out} - h_{in})_{SGs} \tag{6-54}$$

From Figure 6-6 observe that the primary pump work is neglected so that $(h_{out})_R$ is identically $(h_{in})_{SGp}$ and $(h_{in})_R$ is identically $(h_{out})_{SGp}$. Hence from Eqs. 6-53 and 6-54:

$$\dot{W}_{u,max} = \dot{m}_p (h_{out} - h_{in})_R = [\dot{m}_s (h_{out} - h_{in})]_{SGs} \tag{6-55}$$

or

$$\dot{W}_{u,max} = \dot{m}_p(h_5 - h_6) = \dot{m}_s(h_3 - h_2) \tag{6-56}$$

for Figure 6-6. Hence utilizing Eqs. 6-31, 6-51 and 6-55, the overall nuclear plant thermodynamic efficiency or effectiveness (ζ) is:

Table 6-6 PWR operating conditions for Example 6-4

State	Temperature °R (°K)	Pressure psia (kPa)	Condition
1	—	1 (6.89)	Saturated liquid
2	—	1124 (7750)	Subcooled liquid
3	—	1124 (7750)	Saturated vapor
4	—	1 (6.89)	Two-phase mixture
5	1078.2 (599)	2250 (15,500)	Subcooled liquid
6	1016.9 (565)	2250 (15,500)	Subcooled liquid
7			Subcooled liquid
8			Subcooled liquid
a	—	1124 (7750)	Saturated liquid
b	$T_a + 26$ ($T_a + 14.4$)	2250 (15,500)	Subcooled liquid

$$\zeta = \frac{[\dot{m}_s(h_{in} - h_{out})]_T + [\dot{m}_s(h_{in} - h_{out})]_P}{[\dot{m}_s(h_{out} - h_{in})]_{SGs}} \tag{6-57}$$

For the plant of Figure 6-6, with the entire secondary flow passing through a single turbine, Eq. 6-57 becomes:

$$\zeta = \frac{h_3 - h_4 + h_1 - h_2}{h_3 - h_2} \tag{6-58}$$

Example 6-4 Thermodynamic analysis of a simplified PWR plant

PROBLEM The PWR plant of Figure 6-6 operates under the conditions given in Table 6-6. Assume that the turbine and pump have isentropic efficiencies of 85%.

1. Draw the temperature-entropy $(T–s)$ diagram for this cycle.
2. Compute the ratio of the primary to secondary flow rates.
3. Compute the nuclear plant thermodynamic efficiency.
4. Compute the cycle thermal efficiency.

SOLUTION
1. The $T–s$ diagram is sketched in Figure 6-11. Note that states 4 and 2 reflect the fact that the turbine and the pump are not 100% efficient. Also, states on the primary side of the steam generator are shown at higher temperatures than the corresponding (same x/L position) states on the second side.

The T versus x/L diagram is the same as Figure 6-7.

If a primary system main coolant pump were to be included, the reactor system and its representation on a $T–s$ diagram would be as sketched in Figure 6-12, where the primary system pressure loss is taken as 0.48 MPa (70 psia).

2. From Eq. 6-50:

$$\frac{\dot{m}_p}{\dot{m}_s} = \frac{h_3 - h_a}{\bar{c}_p[T_5 - (T_a + \Delta T_p)]}$$

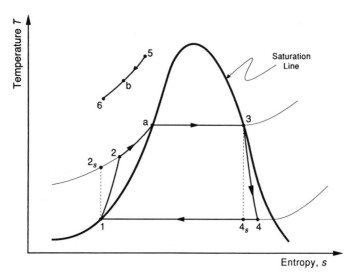

Figure 6-11 *T–s* diagram for PWR cycle, Example 6-4.

From steam tables:

$h_3 = h_g$(sat. at 1124 psia) $= 1187.29$ BTU/lbm (2.771 MJ/kg)
$h_a = h_f$(sat. at 1124 psia) $= 560.86$ BTU/lbm (1.309 MJ/kg)
$T_a = $ sat. liquid at 1124 psia $= 1018.8$ R (566.0°K)

Hence:

$$\frac{\dot{m}_p}{\dot{m}_s} = \frac{1187.29 - 560.86}{1.424\ [1078.2 - (1018.8 + 26)]} = 13.18$$

in SI:

$$\frac{\dot{m}_p}{\dot{m}_s} = \frac{2.77 \times 10^6 - 1.309 \times 10^6}{5941\ [599 - (566 + 14.4)]} = 13.18$$

3. The nuclear plant thermodynamic efficiency (ζ) for the plant within control volume 1 of Figure 6-9A, is given by Eq. 6-39. Draw a control volume around the PWR plant of interest in Figure 6-6. The result is an arrangement identical to that of Figure 6-9A. For this PWR plant, Eq. 6-39 reduces to:

$$\zeta = \frac{h_3 - h_4 + h_1 - h_2}{h_3 - h_2} \tag{6-58}$$

Proceed to obtain the required enthalpies.
From the steam tables:

$$h_1 = h_f \text{ (sat. at 1 psia)} = 69.74 \text{ BTU/lbm (0.163 MJ/kg)}$$

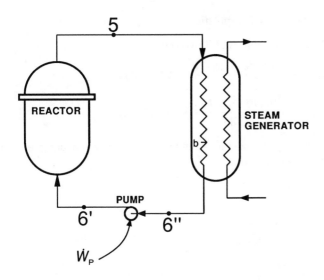

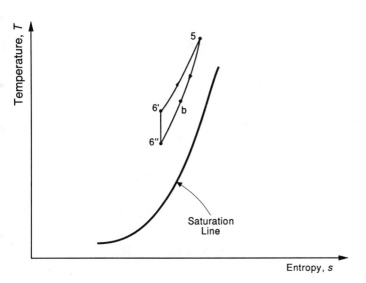

Figure 6-12 Primary system condition of Example 6-4.

From the definition of component isentropic efficiencies:

$$h_2 = h_1 + \frac{h_{2s} - h_1}{\eta_P}$$

$$h_4 = h_3 - \eta_T(h_3 - h_{4s})$$

State 2:
$$s_{2s} = s_1 = 0.13266 \text{ BTU/lbm°R (557 J/kg K)}$$
$$p_{2s} = p_2 = 1124 \text{ psia (7.75MP}_a\text{)}$$

From the subcooled liquid tables by interpolation:

$$h_{2s} = 73.09 \text{ BTU/lbm } (0.170 \text{ MJ/kg})$$

$$h_2 = 69.74 + \frac{73.09 - 69.74}{0.85} = 73.69 \text{ BTU/lbm}$$

$$= 0.163 + \frac{0.170 - 0.163}{0.85} = 0.171 \text{ MJ/kg}$$

Alternately, considering the liquid specific volume constant and the compression process isothermal, h_2 can be evaluated as:

$$h_2 = h_1 + v_1 \frac{(p_2 - p_1)}{\eta_P}$$

$$= 69.74 + \frac{0.016136(1123)}{0.85} \frac{144}{778} = 73.69 \text{ BTU/lbm}$$

$$= 0.163 + \frac{7.75 - 0.0069}{0.85(993)} = 0.171 \text{ MJ/kg}$$

State 4:

$$s_{4s} = s_3 = 1.3759 \text{ BTU/lbm } (5756 \text{ J/kg°K})$$

$$x_{4s} = \frac{s_{4s} - s_f}{s_{fg}} = \frac{1.3759 - 0.13266}{1.8453} = 0.674$$

$$h_{4s} = h_f + x_{4s}h_{fg} = 69.74 + 0.674(1036.0)$$
$$= 768.00 \text{ BTU/lbm } (1.79 \text{ MJ/kg})$$
$$h_4 = h_3 - \eta_T (h_3 - h_{4s}) = 1187.29 - 0.85(1187.29 - 768.00)$$
$$= 830.9 \text{ BTU/lbm } (1.94 \text{ MJ/kg})$$

Substituting into Eq. 6-58, the thermodynamic efficiency can be evaluated as:

$$\zeta = \frac{1187.29 - 830.9 + 69.74 - 73.69}{1187.29 - 73.69} \quad \text{(English units)} = 0.317$$

$$= \frac{2.771 - 1.94 + 0.163 - 0.171}{2.771 - 0.171} \quad \text{(SI units)} = 0.317$$

4. The plant thermal efficiency (η_{th}) is defined by Eq. 6-38 and may be determined for the portion of the plant within control volumes 1 of Figure 6-9B,C. In either case, from Eq. 6-40, $\dot{Q}$ in these figures equals the enthalpy rise of the coolant across the reactor. From Eq. 6-42, the thermal efficiencies of the portions of the plants within control volumes 1 of Figure 6-9B,C are equal to the thermodynamic efficiency of the plant within control volume 1 of Figure 6-9A:

$$\underset{\text{(Fig. 6-9C)}}{\eta_{th|c.v.1}} = \underset{\text{(Fig. 6-9B)}}{\eta_{th|c.v.1}} = \underset{\text{(Fig. 6-9A)}}{\zeta_{c.v.1}} \qquad (6\text{-}42)$$

The thermal efficiencies of these portions of the plant are commonly referred to as the *cycle thermal efficiency* (because they are numerically equal). This nomenclature

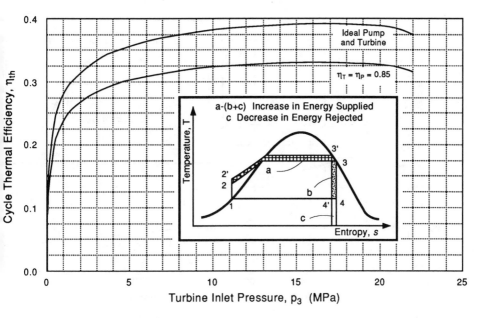

Figure 6-13 Thermal efficiency of Rankine cycle using saturated steam for varying turbine inlet pressure. Turbine inlet: saturated vapor. Exhaust pressure: 7kPa.

is used in the remainder of the chapter for the various reactor plants considered. Because the PWR plant of Figure 6-6 is just a specific case of the general plant illustrated in Figure 6-9, we can directly state that the thermal efficiency of the PWR cycle equals the therodynamic efficiency of the plant; i.e., $\eta_{th} = \zeta$ for the PWR plant.

Improvements in thermal efficiency for this simple saturated cycle are achievable by (1) increasing the pressure (and temperature) at which energy is supplied to the working fluid in the steam generator (i.e., path 2-3, the steam generator inlet and turbine inlet conditions) and (2) decreasing the pressure (and temperature) at which energy is rejected from the working fluid in the condenser (i.e., path 4-1, the turbine outlet and pump inlet conditions). For the cycle of Figure 6-6, assuming an ideal turbine and condenser, the insets in Figures 6-13 and 6-14 illustrate cycles in which each of these strategies is demonstrated.

The effect on cycle thermal efficiency of increasing the pressure at which energy is supplied as heat and of decreasing the pressure at which energy is rejected as heat is illustrated in Figures 6-13 and 6-14, respectively. The cross-hatched areas in the insets represent changes in energy supplied to and rejected from the working fluid.

B Combined First and Second Law or Availability Analysis of a Simplified PWR System

Consider again the stationary, nondeformable control volume of Figure 6-10 operating at steady state with inlet and outlet flow streams at prescribed states. Neglect shear work and differences between kinetic and potential energies of the inlet and outlet

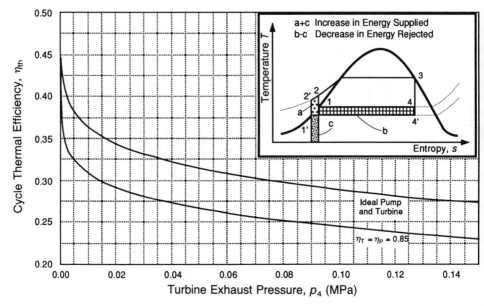

Figure 6-14 Thermal efficiency of Rankine cycle for a saturated turbine inlet state for varying turbine outlet pressure. Turbine inlet: 7.8 MPa saturated vapor.

streams. The maximum useful work obtainable from this control volume, which is in the form of shaft work because the control volume is nondeformable, can be expressed from Eq. 4-46 as follows:

$$\dot{W}_{u,max} = \sum_{i=1}^{I} \dot{m}_i(h - T_o s)_i + \left(1 - \frac{T_o}{T_s}\right)\dot{Q} \tag{6-59}$$

where T_s = temperature at which $\dot{Q}$ is supplied and $\left(\dfrac{dQ}{dt}\right)_{gen}$ is treated as part of $\dfrac{dE}{dt}$.

For the same control volume under the same assumptions, the *lost work* or the *irreversibility* is given by Eqs. 4-47 and 4-48 as:

$$\dot{W}_{lost} \equiv \dot{I} = -T_o \sum_{i=1}^{I} \dot{m}_i s_i - \frac{T_o}{T_s}\dot{Q} = T_o \dot{S}_{gen} \tag{6-60}$$

Let us now restrict the above results to adiabatic conditions, i.e., $\dot{Q} = 0$, and examine two cases: a control volume with shaft work and a control volume with no shaft work. Also recall that because the control volume is nondeformable,

$$\left(\frac{dW}{dt}\right)_{normal} = 0.$$

Case I: $\dot{W}_{shaft} \neq 0$

From the first law, Eq. 4-39,

$$\dot{W}_{shaft} = \sum_{i=1}^{I} \dot{m}_i h_i \tag{6-61a}$$

From Eq. 4-42,

$$\dot{W}_{actual} = \dot{W}_{shaft} \tag{6-62a}$$

Now the irreversibility is given by Eq. 6-60 as:

$$\dot{I} = - T_o \sum_{i=1}^{I} \dot{m}_i s_i = T_o \dot{S}_{gen} \tag{6-63a}$$

However, the irreversibility is also equal from Eq. 4-47 to:

$$\dot{I} = \dot{W}_{u,max} - \dot{W}_{actual} = \sum_{i=1}^{I} \dot{m}_i (h - T_o s)_i - \dot{W}_{shaft} \tag{6-63b}$$

where the maximum useful work has been given by Eq. 6-59 with $\dot{Q} = 0$ as:

$$\dot{W}_{u,max} = \sum_{i=1}^{I} \dot{m}_i (h - T_o s)_i \tag{6-63c}$$

Case II: $\dot{W}_{shaft} \equiv 0$
From the first law, Eq. 4-39,

$$\sum_{i=1}^{I} \dot{m}_i h_i = 0 \tag{6-61b}$$

From Eq. 4-42,

$$\dot{W}_{actual} = 0 \tag{6-62b}$$

Hence the maximum useful work and the irreversibility, expressed by Eqs. 6-59 and 6-60, respectively, are equal and given by:

$$\dot{W}_{u,max} = \dot{I} = - T_o \sum_{i=1}^{I} \dot{m}_i s_i = T_o \dot{S}_{gen} \tag{6-64}$$

Let us now apply these results to PWR system components.

1 Turbine and pump. The turbine and pump have finite shaft work (case I) and single inlet and exit streams. Hence the irreversibilities and the maximum useful work can be expressed from Eqs. 6-63a and 6-63c, respectively, as:

$$\dot{I} = \dot{m} T_o (s_{out} - s_{in}) \tag{6-65}$$

and

$$\dot{W}_{u,max} = \dot{m}[(h_{in} - h_{out}) - T_o(s_{in} - s_{out})] \tag{6-66}$$

2 Steam generator and condenser. The steam generator and condenser have zero shaft work (case II) and two inlet and exit streams. Hence the irreversibilities and the maximum useful work can be expressed from Eq. 6-64 as:

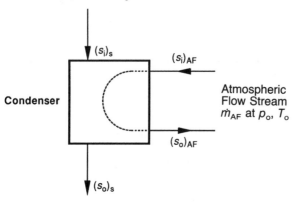

Secondary Coolant
Flow Stream, $\dot{m}_s$

$(s_i)_s$

$(s_i)_{AF}$

Condenser

Atmospheric
Flow Stream
$\dot{m}_{AF}$ at p_o, T_o

$(s_o)_{AF}$

$(s_o)_s$

Figure 6-15 Flow streams for a condenser.

$$\dot{I} = \dot{W}_{u,max} = T_o \left(\sum_{k=1}^{2} \dot{m}_{out,k} s_{out,k} - \sum_{k=1}^{2} \dot{m}_{in,k} s_{in,k} \right) \qquad (6\text{-}67)$$

It is useful to expand on this result for the condenser. Figure 6-15 illustrates a condenser cooled by an atmospheric flow stream ($\dot{m}_{AF}$) that is drawn from and returns to the environment, which acts as a reservoir at p_o, T_o. For the condenser conditions specified in Figure 6-15, the irreversibility of the condenser can be expressed from Eq. 6-67 as:

$$\dot{I} = \dot{m}_{AF} \, T_o \Delta s_{AF} + \dot{m}_s \, T_o \Delta s_s \qquad (6\text{-}68)$$

For this constant pressure atmospheric flow stream interaction, Eq. 6-10a can be written as:

$$T ds = dh \qquad (6\text{-}10b)$$

which can be expressed for the atmospheric coolant stream as:

$$\dot{m}_{AF} \, T_o \Delta s_{AF} = \dot{m}_{AF} \Delta h_{AF} \qquad (6\text{-}69)$$

From an energy balance on the condenser:

$$\dot{m}_{AF} \, \Delta h_{AF} = -\dot{m}_s \Delta h_s \qquad (6\text{-}70)$$

Substituting Eq. 6-70 into Eq. 6-69, we obtain the desired result:

$$\dot{m}_{AF} T_o \, \Delta s_{AF} = -\dot{m}_s \Delta h_s \qquad (6\text{-}71)$$

The irreversibility can be rewritten using Eq. 6-70 as:

$$\dot{I} = T_o \, \dot{m}_s \left[-\frac{\Delta h_s}{T_o} + \Delta s_s \right] \qquad (6\text{-}72a)$$

which, referring to Figure 6-6, becomes:

$$\dot{I} = T_o \dot{m}_s \left[- \left(\frac{h_1 - h_4}{T_o} \right) + s_1 - s_4 \right] \qquad (6\text{-}72\text{b})$$

The condenser irreversibility also can be obtained directly from an availability balance. The irreversibility of the condenser is simply the difference between the availability of the inlet and exit secondary coolant flow streams because there is no change of availability in the atmospheric flow stream. Note that there is, however, a change of energy and entropy in the atmospheric flow stream. This availability difference is evaluated from the definition of the availability function A given by Eq. 4-27 as:

$$\dot{I} = \dot{m}_{in}(h_{in} - T_o s_{in}) - \dot{m}_{out}(h_{out} - T_o s_{out})$$

For the state of the secondary coolant flow stream, this result is identical to Eq. 6-72b.

3 Reactor irreversibility. To proceed, we first compute the maximum work that could be done by the fissioning fuel. Upon fissioning and with respect to fission products, each fission fragment at steady state may be regarded as having a kinetic energy of about 100 MeV. If it were a perfect gas in a thermodynamic equilibrium state, the fragment would have a temperature T_a such that $\frac{3}{2} kT_a = 100$ MeV. Maximum work is done by the fragment in a reversible process that brings the fragment to a state of mutual equilibrium with the environment at $T_o = 298°K$ or $kT_o = 0.025$ eV. At this state, the energy of the fragment is $u_o = 0.0375$ eV. Thus the maximum work per unit mass of fission fragments is:

$$W_{u,max} = h_a - h_o - T_o(s_a - s_o)$$

which can be approximated as:

$$W_{u,max} \simeq u_a - u_o - cT_o \ln \frac{T_a}{T_o} = (u_a - u_o) \left[1 - \frac{kT_o \ln(T_a/T_o)}{k(T_a - T_o)} \right] \qquad (6\text{-}73)$$

For $T_a \gg T_o$, Eq. 6-73 becomes:

$$W_{u,max} \simeq u_a$$

The maximum work is thus approximately equal to the fission energy (u_a). In other words, all the fission energy may be considered to be available as work.

We compute the rate of lost work in several steps. The maximum useful power available as a result of fission equals the fission power ($\dot{u}_a$) and is delivered to the fuel rods considered at a constant fuel temperature (T_{fo}). Because this fuel temperature is far below T_a, for the purpose of this computation the fuel temperature and the clad surface temperature are taken to be T_{fo}. The transfer of power from the fission fragments to the fuel rods is highly irreversible, as $T_{fo} \ll T_a$. The maximum power available from the fuel rod at fixed surface temperature T_{fo} is:

$$(\dot{W}_{u,max})_{fuel} = (\dot{W}_{u,max})_{fission} \left(1 - \frac{T_o}{T_{fo}} \right) \qquad (6\text{-}74)$$

and therefore the irreversibility is:

$$\dot{I}_{\text{fuel}} = (\dot{W}_{\text{u,max}})_{\text{fission}} - (\dot{W}_{\text{u,max}})_{\text{fuel}} = (\dot{W}_{\text{u,max}})_{\text{fission}} \frac{T_o}{T_{\text{fo}}} \qquad (6\text{-}75)$$

From the fuel rods, all the fission power is transferred to the coolant. The maximum work available from the coolant is:

$$(\dot{W}_{\text{u,max}})_{\text{coolant}} = \dot{m}_p \left[(h_{\text{out}} - h_{\text{in}})_R - T_o(s_{\text{out}} - s_{\text{in}})_R \right] \qquad (6\text{-}76)$$

so here the irreversibility is:

$$\dot{I}_{\text{coolant}} = (\dot{W}_{\text{u,max}})_{\text{fuel}} - (\dot{W}_{\text{u,max}})_{\text{coolant}} = (\dot{W}_{\text{u,max}})_{\text{fission}} \left[\frac{T_o(s_{\text{out}} - s_{\text{in}})_R}{(h_{\text{out}} - h_{\text{in}})_R} - \frac{T_o}{T_{\text{fo}}} \right]$$
$$(6\text{-}77)$$

because $(\dot{W}_{\text{u, max}})_{\text{fission}}$ equals the fission power $\dot{U}_a$, which in turn equals $\dot{m}_p (h_{\text{out}} - h_{\text{in}})_R$, by application of the first law (Eq. 4-39) to the reactor control volume, which is adiabatic, is nondeformable, and has no shaft work—and where kinetic energy, potential energy, and shear work terms are neglected. Hence the irreversibility of the reactor is:

$$\dot{I}_R = \dot{I}_{\text{fission}} + \dot{I}_{\text{fuel}} + \dot{I}_{\text{coolant}} \approx 0 + (\dot{W}_{\text{u,max}})_{\text{fission}} \frac{T_o}{T_{\text{fo}}}$$
$$+ (\dot{W}_{\text{u,max}})_{\text{fission}} \left[\frac{T_o(s_{\text{out}} - s_{\text{in}})_R}{(h_{\text{out}} - h_{\text{in}})_R} - \frac{T_o}{T_{\text{fo}}} \right]$$

Again, because $(\dot{W}_{\text{u,max}})_{\text{fission}}$ equals $\dot{m}_p (h_{\text{out}} - h_{\text{in}})_R$, the above result becomes:

$$\dot{I}_R = \dot{m}_p T_o (s_{\text{out}} - s_{\text{in}})_R \qquad (6\text{-}78)$$

Eq. 6-78 is a particular form of the general control volume entropy generation equation as expressed by Eq. 6-60. Eq. 6-78 states that for a steady-state, adiabatic reactor control volume the change of entropy of the coolant stream equals the entropy generation within the reactor.

4 Plant irreversibility. Finally, consider the entire nuclear plant within one control volume as illustrated in Figure 6-9A. For this adiabatic, nondeformable, stationary control volume, the maximum useful work, given by Eq. 6-34, is:

$$\dot{W}_{\text{u,max}}_{\substack{\text{reactor} \\ \text{plant}}} = - \left[\frac{\partial(U - T_o S)}{\partial t} \right]_{\substack{\text{reactor} \\ \text{plant}}} + \dot{m}_{\text{AF}}[(h_{\text{in}} - h_{\text{out}}) - T_o(s_{\text{in}} - s_{\text{out}})]_{\text{AF}}$$
$$(6\text{-}79)$$

because $\dot{Q} = 0$ and differences in kinetic and potential energy as well as shear work are neglected. The first term of the right-hand side of Eq. 6-79 involves only the fission process within the reactor, because for all other components the time rate of change of $U - TS$ is zero. Hence:

$$-\left[\frac{\partial(U - T_0S)}{\partial t}\right]_{\substack{\text{reactor} \\ \text{plant}}} = -\left[\frac{\partial(U - T_0S)}{\partial t}\right]_{\text{reactor}} = -\left(\frac{\partial U}{\partial t}\right)_{\text{reactor}} \qquad (6\text{-}80a)$$

because for the reactor $\dot{I}_R = T_0\dot{S}_{\text{gen}}$ and $\dot{Q} = 0$; hence Eq. 6-79 becomes:

$$\left(\frac{\partial S}{\partial t}\right)_{\text{reactor}} = \dot{m}_p(s_{\text{in}} - s_{\text{out}})_R + \frac{\dot{I}_R}{T_0} = 0$$

utilizing Eq. 6-78 for $\dot{I}_R$.

Again, the first law (Eq. 4-39) for the reactor control volume, which is adiabatic, is nondeformable, and has no shaft work, becomes (treating $\dot{Q}_{\text{gen}}$ as part of $\dot{E}$):

$$\left(\frac{\partial U}{\partial t}\right)_{\text{reactor}} = \dot{m}_{\text{in}}h_{\text{in}} - \dot{m}_{\text{out}}h_{\text{out}} \qquad (6\text{-}80b)$$

Hence the change in reactor internal energy can be expressed in terms of steam generator stream enthalpy differences as:

$$-\left(\frac{\partial U}{\partial t}\right)_{\text{reactor}} = \dot{m}_p(h_{\text{in}} - h_{\text{out}})_{\text{SGp}} = \dot{m}_s(h_{\text{out}} - h_{\text{in}})_{\text{SGs}} \qquad (6\text{-}81)$$

Furthermore, from Eq. 6-69, the second term on the right-hand side of Eq. 6-79 is zero. Hence applying Eq. 6-81 to Eq. 6-79 yields:

$$[\dot{W}_{u,\text{max}}]_{\substack{\text{reactor} \\ \text{plant}}} = \dot{m}_p(h_{\text{in}} - h_{\text{out}})_{\text{SGp}} = \dot{m}_s(h_{\text{out}} - h_{\text{in}})_{\text{SGs}} \qquad (6\text{-}55)$$

which is the result earlier presented in Eqs. 6-53 through 6-55. The irreversibility from Eq. 4-48 is:

$$\dot{I}_{\substack{\text{reactor} \\ \text{plant}}} = -T_0 \sum_{i=1}^{I} \dot{m}_i s_i = T_0 \dot{m}_{\text{AF}}(s_{\text{out}} - s_{\text{in}})_{\text{AF}} \qquad (6\text{-}82)$$

which by Eqs. 6-69 and 6-70 yields:

$$\dot{I}_{\substack{\text{reactor} \\ \text{plant}}} = -\dot{m}_s \Delta h_s = \dot{m}_s(h_{\text{in}} - h_{\text{out}})_s \qquad (6\text{-}83)$$

Referring to Figure 6-6, Eq. 6-83 becomes:

$$\dot{I}_{\substack{\text{reactor} \\ \text{plant}}} = \dot{m}_s(h_4 - h_1) \qquad (6\text{-}84)$$

Example 6-5 Second law thermodynamic analysis of a simplified PWR cycle

PROBLEM Consider the PWR plant of Figure 6-6, which was analyzed in Example 6-4. The plant operating conditions from Example 6-4 are summarized in Table 6-7 together with additional state conditions that can be obtained from the results of Example 6-4. We wish to determine the magnitude of the irreversibility for each

Table 6-7 PWR operating conditions for Example 6-5

Point	Flow rate	Pressure (psia)	Temperature °F	Temperature °R	Enthalpy (BTU/lb)	Entropy (BTU/lb °R)
1	1.0	1	101.7	561.3	69.74	0.13266
2	1.0	1124	103.0	562.6	73.69	0.13410
3	1.0	1124	559.1	1018.7	1187.29	1.37590
4	1.0	1	101.7	561.3	830.70	1.48842
5	13.18	2250	618.5	1078.1	640.31	0.83220
6	13.18	2250	557.2	1016.8	555.80	0.75160

*Flow rate relative to 1 lb of steam through the turbine.

component. This analysis is a useful step for assessing the incentive for redesigning components to reduce the irreversibility.

SOLUTION First evaluate the irreversibility of each component. These irreversibilities sum to the irreversibility of the reactor plant, which may be independently determined by considering the entire plant within one control volume. T_o is the reservoir temperature taken at 539.6°R (298.4°K).

For the steam generator, utilizing Eq. 6-67 and dividing by $\dot{m}_s$:

$$I_{SG} = \dot{I}_{SG}/\dot{m}_s = T_o \frac{\dot{m}_p}{\dot{m}_s} (s_6 - s_5) + T_o(s_3 - s_2)$$

$$= 539.6(13.18)(0.7516 - 0.8322) + (539.6)(1.3759 - 0.1341) \quad (6\text{-}85)$$

$$= -573.22 + 670.08$$

$$= 96.85 \text{ BTU/lbm steam } (0.226 \text{ MJ/kg})$$

For the turbine, using Eq. 6-65 and dividing by $\dot{m}_s$:

$$I_T = \dot{I}_T/\dot{m}_s = T_o(s_4 - s_3)$$

$$= 539.6 (1.4884 - 1.3759) \quad (6\text{-}86)$$

$$= 60.71 \text{ BTU/lbm steam } (0.142 \text{ MJ/kg})$$

For the condenser, using Eq. 6-72b and dividing by $\dot{m}_s$:

$$I_{CD} = \frac{\dot{I}_{CD}}{\dot{m}_s} = T_o \left(\frac{h_4 - h_1}{T_o} + s_1 - s_4 \right)$$

$$= 539.6 \left(\frac{830.7 - 69.74}{539.6} + 0.13266 - 1.48842 \right) \quad (6\text{-}87)$$

$$= 29.39 \text{ BTU/lbm steam } (6.86 \times 10^{-2} \text{ MJ/kg})$$

For the pump, using Eq. 6-65 and dividing by $\dot{m}_s$:

$$I_P = \frac{\dot{I}_P}{\dot{m}_s} = T_o (s_2 - s_1)$$

$$= 539.6 (0.1341 - 0.13266) \quad (6\text{-}88)$$

$$= 0.78 \text{ BTU/lbm steam } (1.81 \times 10^{-2} \text{ MJ/kg})$$

For the reactor, using Eq. 6-78 and dividing by $\dot{m}_s$:

$$I_R = \frac{\dot{I}_R}{\dot{m}_s} = \frac{\dot{m}_p}{\dot{m}_s} T_o (s_5 - s_6)$$

$$= 13.18(539.6)(0.8322 - 0.7516) \tag{6-89}$$
$$= 573.22 \text{ BTU/lbm steam } (1.33 \text{ MJ/kg})$$

This reactor irreversibility can be broken into its two nonzero components, I_{fuel} and $I_{coolant}$, i.e., irreversibility associated with fission power transfer from the fission products to the fuel rod and from the fuel rod to the coolant. These irreversibility components are expressed using Eqs. 6-75 and 6-77 assuming a fuel surface temperature of 1760°R (977.6°K) as:

$$I_{fuel} = \frac{\dot{I}_{fuel}}{\dot{m}_s} = \frac{(\dot{W}_{u,max})_{fission}}{\dot{m}_s} \frac{T_o}{T_{fo}}$$

$$= \left(\frac{\dot{m}_p}{\dot{m}_s}\right)(h_{out} - h_{in})_R \frac{T_o}{T_{fo}}$$

$$= (13.18)(640.31 - 555.8)\frac{539.6}{1760}$$

$$= 34.14 \text{ BTU/lbm steam } (7.97 \times 10^{-2} \text{ MJ/kg})$$

$$I_{coolant} = \frac{\dot{I}_{coolant}}{\dot{m}_s}$$

$$= \frac{(\dot{W}_{u,max})_{fission}}{\dot{m}_s}\left[\left[\frac{T_o(s_{out} - s_{in})_R}{(h_o - h_{in})_R}\right] - \frac{T_o}{T_{fo}}\right]$$

$$= \left(\frac{\dot{m}_p}{\dot{m}_s}\right)T_o\left[(s_{out} - s_{in})_R - \frac{(h_{out} - h_{in})_R}{T_{fo}}\right]$$

$$= (13.18)\,539.6\left[0.8322 - 0.7516 - \left(\frac{640.31 - 555.80}{1760}\right)\right]$$

$$= 539.08 \text{ BTU/lbm steam } (1.25 \text{ MJ/kg})$$

The sum of the numerical values of the irreversibilities of all components is:

$I_R = 573.22$
$I_{SG} = 96.85$
$I_T = 60.71$
$I_{CD} = 29.39$
$I_P = 0.78$
Total $= 760.95$ BTU/lbm steam

which equals (considering round-off error) the irreversibility of the reactor plant evaluated in Eq. 6-84 as:

$$I_{RP} = \frac{\dot{I}_{RP}}{\dot{m}_s} = h_4 - h_1 = 830.70 - 69.74 = 760.96 \text{ BTU/lbm steam}$$

The generality of this result may be illustrated by summing the expressions for the irreversibility of each component and showing that the sum is identical to the irreversibility of the reactor plant. Proceeding in this way we obtain:

$$
\begin{aligned}
\dot{I}_{RP} &= \dot{W}_{u,max} - \dot{W}_{net} \\
&= \dot{m}_s \left[h_3 - h_2 - [h_3 - h_4 - (h_2 - h_1)] \right] \\
&= \dot{m}_s (h_4 - h_1)
\end{aligned}
\tag{6-84}
$$

$$
\dot{I}_R = T_o \dot{m}_p (s_5 - s_6) \tag{6-89}
$$

$$
\dot{I}_{SG} = T_o \dot{m}_p (s_6 - s_5) + T_o \dot{m}_s (s_3 - s_2) \tag{6-85}
$$

$$
\dot{I}_T = T_o \dot{m}_s (s_4 - s_3) \tag{6-86}
$$

$$
\dot{I}_P = T_o \dot{m}_s (s_2 - s_1) \tag{6-88}
$$

$$
\dot{I}_{CD} = T_o \dot{m}_s (s_1 - s_4) + \dot{m}_s (h_4 - h_1) \tag{6-87}
$$

Hence:

$$
\dot{I} = \dot{m}_s (h_4 - h_1) = \dot{I}_{RP} \tag{6-84}
$$

V MORE COMPLEX RANKINE CYCLES: SUPERHEAT, REHEAT, REGENERATION, AND MOISTURE SEPARATION

Improvements in overall cycle performance can be accomplished by superheat, reheat, and regeneration, as each approach leads to a higher average temperature at which heat is received by the working fluid. Furthermore, to minimize erosion of the turbine blades, the turbine expansion process can be broken into multiple stages with intermediate moisture separation. The superheat and reheat options also allow maintenance of higher exit steam quality, thereby decreasing the liquid fraction in the turbine.

The superheat, reheat, regeneration, and moisture separation options are illustrated in subsequent figures for ideal turbines. In actual expansion processes, the coolant entropy increases, and the exit steam is dryer, although the shaft work generated per unit mass is decreased. Although each process is discussed independently, in practice they are utilized in varying combinations. Superheat is accomplished by heating the working fluid into the superheated vapor region. A limited degree of superheat is being achieved in PWR systems employing once-through versus recirculating steam generators. Figure 6-16 illustrates the heat exchanger and turbine processes in a superheated power cycle.

Reheat is a process whereby the working fluid is returned to a heat exchanger after partial expansion in the turbine. As shown in Figure 6-17, the fluid can be reheated to the maximum temperature and expanded. Actual reheat design conditions may differ with application.

Regeneration is a process by which the colder portion of the working fluid is heated by the hotter portion of the working fluid. Because this heat exchange is internal to the cycle, for single-phase fluids (e.g., Brayton cycle) the average temperature of external heat rejection is reduced and the average temperature of external heat addition

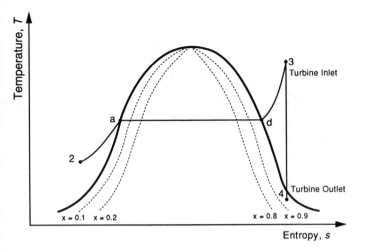

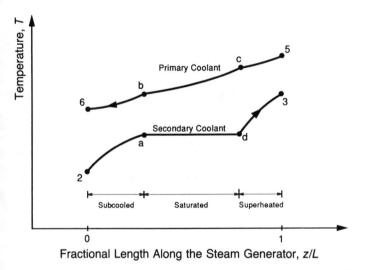

Figure 6-16 Heat exchanger and turbine processes in a superheated power cycle.

is increased. For a two-phase fluid (e.g., Rankine cycle) only the average temperature of external heat addition is changed, i.e., increased. In either case the cycle thermal efficiency increases. For the limiting case of this heat exchange occurring at infinitesimal temperature differences—an ideal regeneration treatment—it can be shown that the cycle thermal efficiency equals the Carnot efficiency of a cycle operating between the same two temperature limits. In practice for a Rankine cycle, this regenerative process is accomplished by extracting steam from various turbine stages and directing it to a series of heaters where the condensate or feedwater is preheated. These feedwater heaters may be open (OFWH), in which the streams are directly

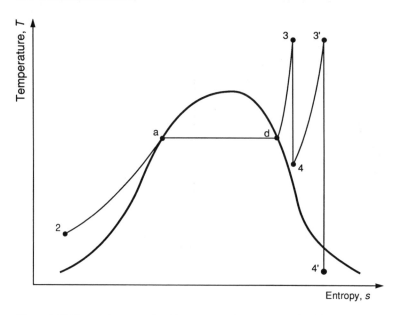

Figure 6-17 Heat exchanger and turbine processes in a power cycle with superheat and reheat.

mixed, or closed (CFWH) in which the heat transfer occurs through tube walls. Figure 6-18 illustrates a system and associated T–s diagram for the case of two open feedwater heaters. In this case a progressively decreasing flow rate passes through later stages of the turbine. Hence the total turbine work is expressed as:

$$\dot{W}_T = \dot{m}_s(h_3 - h_{3'}) + (\dot{m}_s - \dot{m}_{3'})(h_{3'} - h_{3''}) + \dot{m}_4(h_{3''} - h_4) \qquad (6\text{-}90)$$

The exit enthalpy from each feedwater heater can be expressed by specializing the relations of Table 6-5, i.e., for heater No. 1.

For OFWH No. 1:

$$h_z = \frac{\dot{m}_{3''}h_{3''} + \dot{m}_4 h_{1'}}{\dot{m}_{3''} + \dot{m}_4} = \frac{\dot{m}_{3''}h_{3''} + \dot{m}_4 h_{1'}}{\dot{m}_s - \dot{m}_{3'}} \qquad (6\text{-}91)$$

For OFWH No. 2:

$$h_y = \frac{\dot{m}_{3'}h_{3'} + (\dot{m}_s - \dot{m}_{3'})h_{z'}}{\dot{m}_s} \qquad (6\text{-}92)$$

Figures 6-19 and 6-20 illustrate a system and the associated T–s diagram for the case of two closed feedwater heaters. Steam traps at the bottom of each heater allow the passage of saturated liquid only. The condensate from the first heater is returned to the main condensate line by a drip pump. For the last heater (No. 2) the condensate drip is flashed back to the adjacent upstream heater (No. 1). Observe on the T–s diagram that certain temperature differentials must be maintained. In this case, exit enthalpies of each CFWH and the mixing tee can be written:

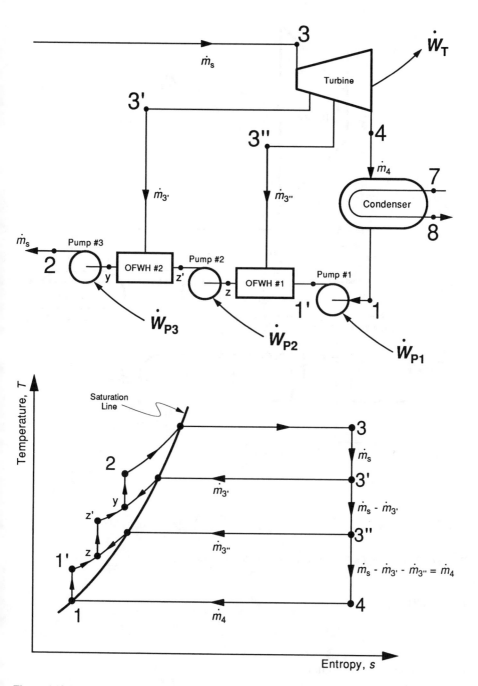

Figure 6-18 Power cycle with open feedwater heaters.

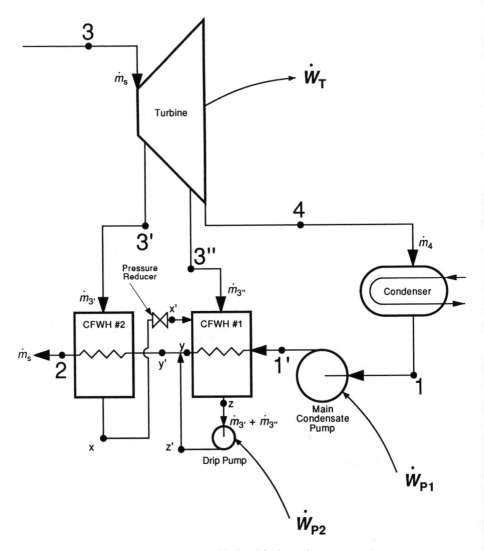

Figure 6-19 Portion of Rankine power cycle with closed feedwater heaters.

For CFWH No. 1:

$$h_y = \frac{\dot{m}_{3''}h_{3''} + \dot{m}_4 h_{1'} + \dot{m}_{3'}h_{x'} - (\dot{m}_{3'} + \dot{m}_{3''})h_z}{\dot{m}_4}$$

(6-93)

For the mixing tee, at position y':

$$h_{y'} = \frac{\dot{m}_4 h_y + (\dot{m}_{3'} + \dot{m}_{3''})h_{z'}}{\dot{m}_s}$$

(6-94)

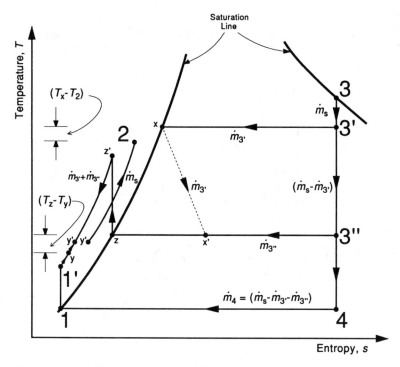

Figure 6-20 Portion of Rankine power cycle with closed feedwater heaters. *Note:* Process y' to 2 offset from constant pressure line 1'yy'z'2 for clarity.

For CFWH No. 2:

$$h_2 = \frac{\dot{m}_{3'}(h_{3'} - h_x) + \dot{m}_s h_{y'}}{\dot{m}_s} \tag{6-95}$$

From these feedwater heater illustrations it can be observed that the OFWH approach requires a condensate pump for each heater, whereas the CFWH approach requires only one condensate pump plus a smaller drip pump. However, higher heat transfer rates are achievable with the OFWH. Additionally, the OFWH permits deaeration of the condensate. For most applications closed heaters are favored, but for purposes of feedwater deaeration at least one open heater is provided.

The final process considered is moisture separation. The steam from the high-pressure turbine is passed through a moisture separator, and the separated liquid is diverted to a feedwater heater while the vapor passes to a low-pressure turbine. The portion of the power cycle with a moisture separator and the associated T–s diagram are shown in Figure 6-21. The moisture separator is considered ideal, producing two streams of fluid, a saturated liquid flow rate of magnitude $(1 - x_{3'})\dot{m}_s$ and enthalpy

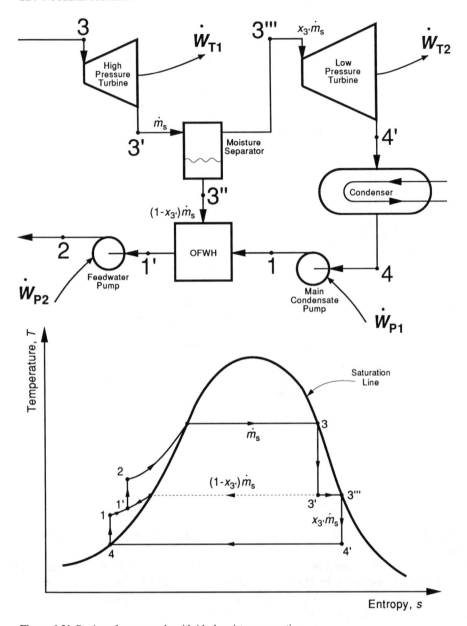

Figure 6-21 Portion of power cycle with ideal moisture separation.

$h_{3''} = h_f$ (at $p_{3'}$), and a saturated vapor flow rate of magnitude $x_3 \dot{m}_s$ and enthalpy $h_{3'''} = h_g$ (at $p_{3'}$). In this case the feedwater heater exit enthalpy is given as

$$h_{1'} = \frac{h_{3''}(1 - x_{3'})\dot{m}_s + h_1 x_{3'} \dot{m}_s}{\dot{m}_s} \qquad (6\text{-}96)$$

Example 6-6 Thermodynamic analysis of a PWR cycle with moisture separation and one stage of feedwater heating

PROBLEM This example demonstrates the advantage in cycle performance gained by adding moisture separation and open feedwater heating to the cycle of Example 6-4. The cycle, as illustrated in Figure 6-22, operates under the conditions given in Table 6-8. All states are identical to those in Example 6-4 except state 2, as states 9 through 13 have been added. The flow rate at state 11, which is diverted to the open feedwater heater, is sufficient to preheat the condensate stream at state 13 to the conditions of state 10. Assume that all turbines and pumps have isentropic efficiencies of 85%.

1. Draw the temperature–entropy (T–s) diagram for this cycle, as well as the temperature–fractional length (T versus z/L) diagram for the steam generator.
2. Compute the ratio of primary to secondary flow rates.
3. Compute the cycle thermal efficiency.

SOLUTION

1. The T–s diagram is sketched in Figure 6-23. The T versus z/L diagram is the same as that in Figure 6-7.

2. $\dfrac{\dot{m}_p}{\dot{m}_s} = \dfrac{h_3 - h_a}{\bar{c}_p[T_5 - (T_a + \Delta T_p)]} = 13.18$, as states 3, 5, a, and b were specified the same as in Example 6-4.

3. $\eta_{th} = \dfrac{\dot{m}_s(h_3 - h_9) + (1 - f)\dot{m}_s(h_{11} - h_4)}{\dot{m}_s(h_3 - h_2)}$, neglecting pump work.

Table 6-8 PWR operating conditions for Example 6-6

State	Temperature °F (°K)	Pressure psia (MPa)	Condition
1		1 (6.89 × 10⁻³)	Saturated liquid
2		1124 (7.75)	Subcooled liquid
3		1124 (7.75)	Saturated vapor
4		1 (6.89 × 10⁻³)	Two-phase mixture
5	618.5 (599)	2250 (15.5)	Subcooled liquid
6	557.2 (565)	2250 (15.5)	Subcooled liquid
7			Subcooled liquid
8			Subcooled liquid
a		1124 (7.75)	Saturated liquid
b	$T_a + 26$ (14.4)	2250 (15.5)	Subcooled liquid
9		50 (0.345)	Two-phase mixture
10		50 (0.345)	Saturated liquid
11		50 (0.345)	Saturated vapor
12		50 (0.345)	Saturated liquid
13		50 (0.345)	Subcooled liquid

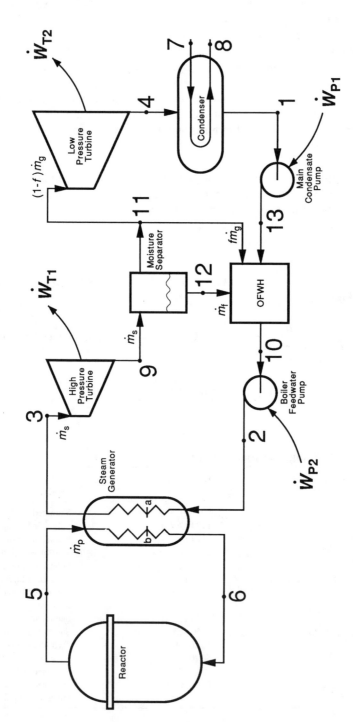

Figure 6-22 PWR cycle analyzed in Example 6-6.

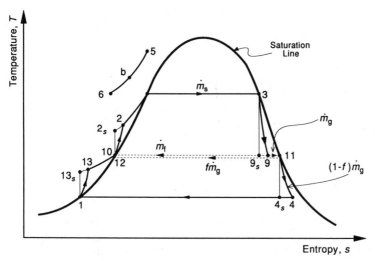

Figure 6-23 PWR cycle with moisture separation and one stage of feedwater heating.

From the problem statement, the following enthalpies can be directly determined:

$h_3 = 1187.29$ BTU/lbm or 2.77 MJ/kg (Example 6-4)
$h_{11} = 1174.40$ BTU/lbm (2.74 MJ/kg)
$h_{12} = h_{10} = 250.24$ BTU/lbm (0.584 MJ/kg)

The following parameters must be calculated: h_9, $\dot{m}_g$, h_4, f, and h_2.

$$h_9 = h_3 - \eta_T(h_3 - h_{9s})$$

where

$$h_{9s} = h_f + x_{9s}h_{fg} = h_f + \left(\frac{s_{9s} - s_f}{s_{fg}}\right)h_{fg}$$

Hence:

$$h_{9s} = 250.24 + \left(\frac{1.3759 - 0.41129}{1.2476}\right)924.2$$
$$= 964.81 \text{ BTU/lbm (2.25 MJ/kg)}$$
$$h_9 = 1187.29 - 0.85(1187.29 - 964.81)$$
$$= 998.18 \text{ BTU/lbm (2.330 MJ/kg)}$$
$$\dot{m}_g = x_9\dot{m}_s = \frac{h_9 - h_f}{h_{fg}}\dot{m}_s = \frac{998.18 - 250.24}{924.2}\dot{m}_s = 0.81\ \dot{m}_s \text{ (English units)}$$
$$= \frac{2.331 - 0.584}{2.156}\dot{m}_s = 0.81\ \dot{m}_s \text{ (SI units)}$$
$$\dot{m}_f = \left(1 - \frac{\dot{m}_g}{\dot{m}_s}\right)\dot{m}_s = 0.19\ \dot{m}_s$$

$$h_4 = h_{11} - \eta_T(h_{11} - h_{4s}) = 1174.4 - 0.85(1174.4 - 926.61)$$
$$= 963.77 \text{ BTU/lbm } (2.25 \text{ MJ/kg})$$

where:

$$h_{4s} = h_f + x_{4s}h_{fg} = h_f + \left(\frac{s_{4s} - s_f}{s_{fg}}\right)h_{fg}$$

$$= 69.74 + \left(\frac{1.6589 - 0.13266}{1.8453}\right)1036.0$$

$$= 926.61 \text{ BTU/lbm } (2.16 \text{ MJ/kg})$$

$$h_{13} = h_1 + \frac{v_1(p_{13} - p_1)}{\eta_p}$$

$$= 69.91 \text{ BTU/lbm } (0.163 \text{ MJ/kg}) \text{ (same as state 2, Example 6-4)}$$

Applying the first law to the feedwater heater:

$$\dot{m}_f h_{12} + f\dot{m}_g h_{11} + (1 - f)\dot{m}_g h_{13} = \dot{m}_s h_{10}$$

$$f = \frac{\dot{m}_s h_{10} - \dot{m}_f h_{12} - \dot{m}_g h_{13}}{\dot{m}_g(h_{11} - h_{13})} = \frac{h_{10} - h_{13}}{h_{11} - h_{13}} = \frac{250.24 - 69.91}{1174.4 - 69.91} = 0.16$$

$$h_2 = h_{10} + \frac{v_{10}(p_2 - p_{10})}{\eta_p}$$

$$= 250.24 + \frac{0.017269(1124 - 50)}{0.85}\left(\frac{144 \text{ in}^2/\text{ft}^2}{778 \text{ lb}_f - \text{ft/BTU}}\right)$$

$$= 254.28 \text{ BTU/lbm } (0.594 \text{ MJ/kg})$$

Hence:

$$\eta_{th} = \left[\frac{(1187.29 - 998.18) + (1 - 0.16)0.81(1174.4 - 963.77)}{1187.29 - 254.28}\right] 100 \text{ English units}$$

$$= \left[\frac{2.77 - 2.330 + (1 - 0.16)\, 0.81\, (2.74 - 2.25)}{2.79 - 0.594}\right] 100 \text{ SI units}$$

$$= 35.6\%$$

VI SIMPLE BRAYTON CYCLE

Reactor systems that employ gas coolants offer the potential for operating as direct Brayton cycles by passing the heated gas directly into a turbine. This Brayton cycle is ideal for single-phase, steady-flow cycles with heat exchange and therefore is the basic cycle for modern gas turbine plants as well as proposed nuclear gas-cooled reactor plants. As with the Rankine cycle, the remaining processes and components are, sequentially, a heat exchanger for rejecting heat, a compressor, and a heat source,

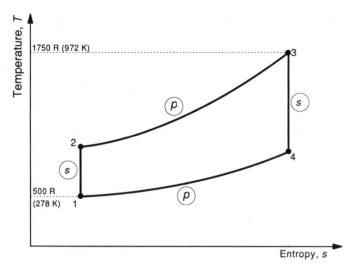

Figure 6-24 Temperature–entropy (T–s) plot of the simple Brayton cycle. $\textcircled{p}$ Constant pressure process. $\textcircled{s}$ Constant entropy process.

which in our case is the reactor. The ideal cycle is composed of two reversible constant-pressure heat-exchange processes and two reversible, adiabatic work processes. However, because the working fluid is single phase, no portions of the heat-exchange processes are carried out at constant temperature as in the Rankine cycle. Additionally, because the volumetric flow rate is higher, the compressor work, or "backwork," is a larger fraction of the turbine work than is the pump work in a Rankine cycle. This large backwork has important ramifications for the Brayton cycle.

The T–s plot of the simple Brayton cycle of Figure 6-8 is shown in Figure 6-24. The pressure or compression ratio of the cycle is defined as:

$$r_p = \frac{p_2}{p_1} \equiv \frac{p_3}{p_4} \qquad (6\text{-}97)$$

For isentropic processes with a perfect gas, the thermodynamic states are related in the following manner:

$$Tv^{\gamma-1} = \text{constant} \qquad (6\text{-}98)$$

$$Tp^{\frac{1-\gamma}{\gamma}} = \text{constant} \qquad (6\text{-}99)$$

where $\gamma \equiv c_p/c_v$. For a perfect gas, because enthalpy is a function of temperature only and the specific heats are constant:

$$\Delta h = c_p \Delta T \qquad (6\text{-}100)$$

Applying this result for Δh (Eq. 6-100) and the isentropic relations for a perfect gas (Eqs. 6-98 and 6-99) to the specialized forms of the first law of Table 6-5, the Brayton cycle can be analyzed. The following equations are in terms of the pressure ratio (r_p) and the lowest and highest cycle temperatures (T_1 and T_3) both of which are generally known quantities.

The turbine and compressor works are given as follows. For a perfect gas:

$$\dot{W}_T = \dot{m}c_p(T_3 - T_4) = \dot{m}c_pT_3\left(1 - \frac{T_4}{T_3}\right) \tag{6-101a}$$

Furthermore, for an isentropic process, application of Eq. 6-99 to Eq. 6-101a yields:

$$\dot{W}_T = \dot{m}c_pT_3\left[1 - \frac{1}{(r_p)^{\frac{\gamma-1}{\gamma}}}\right] \tag{6-101b}$$

Analogously for the compressor:

$$\dot{W}_{CP} = \dot{m}c_p(T_2 - T_1) = \dot{m}c_pT_1\left[\frac{T_2}{T_1} - 1\right] = \dot{m}c_pT_1[(r_p)^{\frac{\gamma-1}{\gamma}} - 1] \tag{6-102}$$

The heat input from the reactor and the heat rejected by the heat exchanger are:

$$\dot{Q}_R = \dot{m}c_p(T_3 - T_2) = \dot{m}c_pT_1\left[\frac{T_3}{T_1} - (r_p)^{\frac{\gamma-1}{\gamma}}\right] \tag{6-103}$$

$$\dot{Q}_{HX} = \dot{m}c_p(T_4 - T_1) = \dot{m}c_pT_3\left[\frac{1}{(r_p)^{\frac{\gamma-1}{\gamma}}} - \frac{T_1}{T_3}\right] \tag{6-104}$$

As in Eq. 6-56, the maximum useful work of the reactor plant is equal to the product of the system flow rate and the coolant enthalpy rise across the core. Hence:

$$\dot{W}_{u,\,max} \equiv \dot{Q}_R = \dot{m}c_pT_1\left[\frac{T_3}{T_1} - (r_p)^{\frac{\gamma-1}{\gamma}}\right] \tag{6-105}$$

The Brayton nuclear plant thermodynamic efficiency is then:

$$\zeta = \frac{\dot{W}_T - \dot{W}_{CP}}{\dot{W}_{u,max}} = \frac{T_3\left[1 - \dfrac{1}{(r_p)^{\frac{\gamma-1}{\gamma}}}\right] - T_1(r_p)^{\frac{\gamma-1}{\gamma}}\left[1 - \dfrac{1}{(r_p)^{\frac{\gamma-1}{\gamma}}}\right]}{T_1\left[\dfrac{T_3}{T_1} - (r_p)^{\frac{\gamma-1}{\gamma}}\right]}$$

$$= 1 - \frac{1}{(r_p)^{\frac{\gamma-1}{\gamma}}} \tag{6-106}$$

It can be further shown that the optimum pressure ratio for maximum net work is:

$$(r_p)_{optimum} = \left(\frac{T_3}{T_1}\right)^{\frac{\gamma}{2(\gamma-1)}} \tag{6-107}$$

The existence of an optimum compression ratio can also be seen on a T–s diagram by comparing the enclosed areas for cycles operating between fixed temperature limits of T_1 and T_3.

Example 6-7 First law thermodynamic analysis of a simple Brayton cycle
PROBLEM Compute the cycle efficiency for the simple Brayton cycle of Figures 6-8 and 6-24 for the following conditions:

1. Helium as the working fluid taken as a perfect gas with

$c_p = 1.25$ BTU/lbm °R (5230 J/kg °K)
$\gamma = 1.658$
$\dot{m}$ in lbm/s (English units) or kg/s (SI units)

2. Pressure ratio of 4.0
3. Maximum and minimum temperatures of 1750°R (972°K) and 500°R (278°K), respectively

SOLUTION

$$\frac{\dot{W}_T}{\dot{m}} = c_p T_3 \left[1 - \frac{1}{(r_p)^{\frac{\gamma-1}{\gamma}}} \right] = 1.25\,(1750) \left[1 - \frac{1}{(4.0)^{0.397}} \right]$$

$$= 925.9 \text{ BTU/lbm } (2.150 \text{ MJ/kg})$$

$$\frac{\dot{W}_{CP}}{\dot{m}} = c_p T_1 \left[(r_p)^{\frac{\gamma-1}{\gamma}} - 1 \right] = 1.25\,(500)\,[(4.0)^{0.397} - 1]$$

$$= 458.67 \text{ BTU/lbm } (1.066 \text{ MJ/kg})$$

$$\frac{\dot{W}_{u,\,max}}{\dot{m}} = c_p T_1 \left[\frac{T_3}{T_1} - (r_p)^{\frac{\gamma-1}{\gamma}} \right] = 1.25\,(500) \left[\frac{1750}{500} - (4.0)^{0.397} \right]$$

$$= 1103.8 \text{ BTU/lbm } (2.560 \text{ MJ/kg})$$

$$\zeta = \frac{(\dot{W}_T - \dot{W}_{CP})/\dot{m}}{\dot{W}_{u,max}/\dot{m}} = \left(\frac{925.9 - 458.7}{1103.8} \right) 100 \text{ (English units)}$$

$$= \left(\frac{2.15 - 1.066}{2.56} \right) 100 \text{ (SI units)}$$

$$= 42.3\%$$

Analogous to the PWR cycle demonstrated in Example 6-5, the Brayton cycle thermal efficiency equals the Brayton nuclear plant thermodynamic efficiency, i.e.,

$$\eta_{th} = \zeta$$

For ideal Brayton cycles, as Eq. 6-106 demonstrates, the thermodynamic efficiency increases with the compression ratio. The thermal efficiency of a Carnot cycle

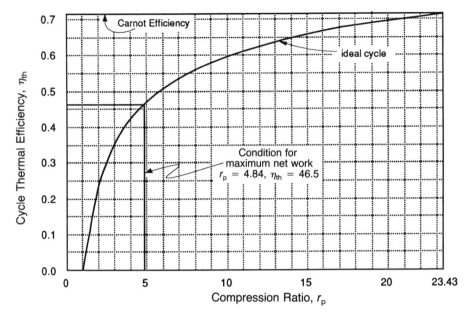

Figure 6-25 Thermal efficiency of an ideal Brayton cycle as a function of the compression ratio. $\gamma = 1.658$.

operating between the maximum and minimum temperatures specified is:

$$\eta_{\text{Carnot}} = \frac{T_{\text{max}} - T_{\text{min}}}{T_{\text{max}}} = \frac{972°\text{K} - 278°\text{K}}{972°\text{K}} = 0.714$$

Figure 6-25 illustrates the dependence of cycle efficiency on the compression ratio for the cycle of Example 6-7. The pressure ratio corresponding to the maximum net work is also identified. This optimum pressure ratio is calculated from Eq. 6-107 for a perfect gas as:

$$(r_p)_{\text{optimum}} = \left(\frac{T_3}{T_1}\right)^{\frac{\gamma}{2(\gamma - 1)}} = \left(\frac{972}{278}\right)^{\frac{1.658}{2(0.658)}} = 4.84$$

VII MORE COMPLEX BRAYTON CYCLES

In this section the various realistic considerations are included in the analysis and are illustrated through a number of examples.

Example 6-8 Brayton cycle with real components

PROBLEM Compute the thermal efficiency for the cycle depicted in Figure 6-26 if the isentropic efficiencies of the compressor and the turbine are each 90%. All other conditions of Example 6-7 apply.

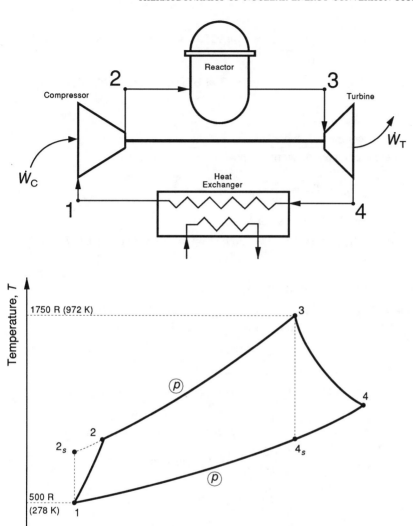

Figure 6-26 Brayton cycle with real components: Example 6-8.

SOLUTION

For $\dot{W}_T$:

$$\eta_T = \frac{\text{actual work out of turbine}}{\text{ideal turbine work}} = \frac{\dot{W}_T}{\dot{W}_{Ti}} = \frac{\dot{m}c_p(T_3 - T_4)}{\dot{m}c_p(T_3 - T_{4s})}$$

$$\therefore \dot{W}_T = \eta_T \dot{W}_{Ti} = \eta_T \dot{m}c_p(T_3 - T_{4s}) = \eta_T \dot{m}c_p T_3 \left(1 - \frac{T_{4s}}{T_3}\right)$$

$$= \eta_T \dot{m} c_p T_3 \left[1 - \frac{1}{(r_p)^{\frac{\gamma-1}{\gamma}}} \right] = \eta_T \dot{m} \, 925.9 = (0.9)(925.9) \, \dot{m}$$

$$= 833.3 \, \dot{m} \text{ BTU/s } (1.935 \, \dot{m} \text{ MJ/s or MW})$$

For $\dot{W}_{CP}$:

$$\eta_{CP} = \frac{\text{ideal compressor work}}{\text{actual compressor work}} = \frac{\dot{W}_{CPi}}{\dot{W}_{CP}} = \frac{\dot{m} c_p (T_{2s} - T_1)}{\dot{m} c_p (T_2 - T_1)}$$

$$\dot{W}_{CP} = \frac{\dot{m}}{\eta_{CP}} c_p (T_{2s} - T_1) = \frac{\dot{m}}{\eta_{CP}} c_p T_1 \left(\frac{T_{2s}}{T_1} - 1 \right) = \frac{\dot{m} \, 458.7}{0.9}$$

$$= 509.7 \, \dot{m} \text{ BTU/s } (1.184 \, \dot{m} \text{ MW})$$

$$\dot{W}_{NET} = \dot{W}_T - \dot{W}_{CP} = \dot{m} (833.3 - 509.7) = \dot{m} \, 323.6 \text{ BTU/s } (\dot{m} \, 0.752 \text{ MW})$$

$$\dot{Q}_R = \dot{m} c_p (T_3 - T_2)$$

To evaluate $\dot{Q}_R$ it is necessary to find T_2.

In Example 6-7 we found $\dot{W}_{CPi} = \dot{m} c_p (T_{2s} - T_1) = 458.7 \, \dot{m}$ BTU/s $= 1.066 \, \dot{m}$ MW, so that from the expressions above for $\dot{W}_{CP}$ and η_{CP}:

$$T_2 - T_1 = \frac{\dot{W}_{CP}}{\dot{m} c_p} = \frac{\dot{W}_{CPi}}{\dot{m} c_p \eta_{CP}} = \frac{458.7}{1.25(0.9)} = 407.7°R \, (226.5°K)$$

$$T_2 = 407.7 + T_1 = 407.7 + 500 = 907.7°R \, (504.3°K)$$

$$\dot{Q}_R = (1.25)(T_3 - T_2) \dot{m} = 1.25 \, (1750 - 907.7) \, \dot{m}$$
$$= 1052.9 \, \dot{m} \text{ BTU/s } (2.45 \, \dot{m} \text{ MW})$$

$$\eta_{th} = \frac{\dot{W}_{NET}}{\dot{Q}_R} = \left(\frac{323.6}{1052.9} \right) 100 \text{ (English units)}$$

$$= \left(\frac{0.752}{2.45} \right) 100 \text{ (SI units)}$$

$$= 30.7\%$$

Example 6-9 Brayton cycle considering duct pressure losses

PROBLEM Compute the cycle thermal efficiency considering pressure losses in the reactor and heat exchanger processes as well as 90% isentropic turbine and compressor efficiencies. The cycle is illustrated in Figure 6-27. The pressure losses are characterized by the parameter β where:

$$\beta \equiv \left(\frac{p_4 \, p_2}{p_1 \, p_3} \right)^{\frac{\gamma-1}{\gamma}} = 1.05$$

All other conditions of Example 6-7 apply.

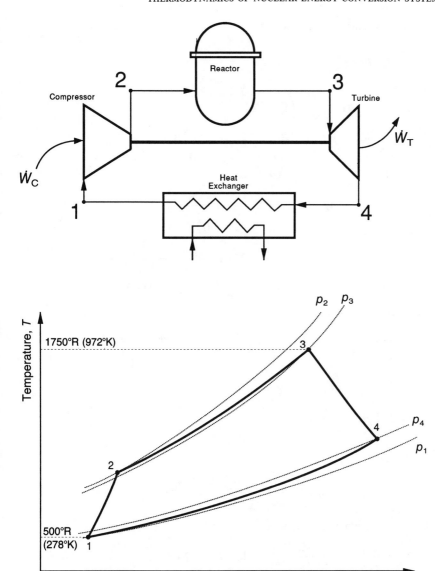

Figure 6-27 Brayton cycle considering duct pressure losses: Example 6-9.

$$\dot{W}_T = \eta_T \dot{m} c_p T_3 \left(1 - \frac{T_{4s}}{T_3} \right) = \eta_T \dot{m} c_p T_3 \left[1 - \frac{1}{\left(\dfrac{p_3}{p_4} \right)^{\frac{\gamma-1}{\gamma}}} \right]$$

SOLUTION Because β is defined as:

$$\left(\frac{p_4}{p_1} \cdot \frac{p_2}{p_3}\right)^{\frac{\gamma-1}{\gamma}}$$

$$\therefore \left(\frac{p_4}{p_3}\right)^{\frac{\gamma-1}{\gamma}} = \frac{\left(\dfrac{p_4}{p_1} \cdot \dfrac{p_2}{p_3}\right)^{\frac{\gamma-1}{\gamma}}}{\left(\dfrac{p_2}{p_1}\right)^{\frac{\gamma-1}{\gamma}}} = \frac{\beta}{(r_p)^{\frac{\gamma-1}{\gamma}}}$$

$$\therefore \dot{W}_T = \eta_T \dot{m} c_p T_3 \left[1 - \frac{\beta}{(r_p)^{\frac{\gamma-1}{\gamma}}}\right]$$

$$= 0.9 \, \dot{m} \, (1.25)(1750) \left[1 - \frac{1.05}{(4)^{0.397}}\right]$$

$$= 776.5 \, \dot{m} \text{ BTU/s } (1.803 \, \dot{m} \text{ MW})$$

Again as in Example 6-8:

$$\dot{W}_{CP} = \frac{\dot{m}c_p}{\eta_{CP}}(T_{2s} - T_1) = \dot{m}\frac{c_p T_1}{\eta_{CP}}\left(\frac{T_{2s}}{T_1} - 1\right) = \frac{\dot{m}c_p T_1}{\eta_{CP}}\left[\left(\frac{p_2}{p_1}\right)^{\frac{\gamma-1}{\gamma}} - 1\right]$$

$$= \frac{\dot{m}(1.25)(500)}{0.9}(1.7338 - 1.0)$$

$$= 509.7 \, \dot{m} \text{ BTU/s } (1.184 \, \dot{m} \text{ MW})$$

Now $\dot{Q}_R = \dot{m}c_p(T_3 - T_2)$
We know that

$$\eta_{CP} = \frac{\dot{m}c_p(T_{2i} - T_1)}{\dot{m}c_p(T_2 - T_1)} = \frac{\dot{W}_{CPi}}{\dot{W}_{CP}}$$

where $\dot{W}_{CPi}$ was calculated in Example 6-7.

$$\therefore T_2 - T_1 = \frac{\dot{W}_{CPi}}{c_p \eta_{CP}} = \frac{458.7}{(0.9)(1.25)} = 407.7°R \ (226.5°K)$$

$$\therefore T_2 = 500 + 407.7 = 907.7°R \ (504.3°K)$$

$$\therefore \dot{Q}_R = \dot{m}c_p \ (1750 - 907.7) = 1052.9 \, \dot{m} \text{ BTU/s } (2.45 \, \dot{m} \text{ MW})$$

$$\dot{W}_{NET} = \dot{W}_T - \dot{W}_{CP} = \dot{m} \ (776.5 - 509.6) = 266.9 \, \dot{m} \text{ BTU/s } (0.620 \, \dot{m} \text{ MW})$$

$$\eta_{th} = \frac{\dot{W}_{NET}}{\dot{Q}_R} = \left(\frac{266.9}{1052.9}\right) 100 \text{ (English units)}$$

$$= \left(\frac{0.620}{2.45}\right) 100 \text{ (SI units)}$$

$$= 25.3\%$$

Examples 6-8 and 6-9 demonstrate that consideration of real component efficiencies and pressure losses cause the cycle thermal efficiency to decrease dramatically. Employment of regeneration, if the pressure ratio allows, reverses this trend, as demonstrated next in Example 6-10, which considers both ideal and real turbines and compressors.

Example 6-10 Brayton cycle with regeneration for ideal and then real components

PROBLEM Compute the cycle thermal efficiency first for ideal turbines and compressors but with the addition of a regenerator of effectiveness 0.75. The cycle is illustrated in Figure 6-28. Regenerator effectiveness is defined as the actual preheat temperature change over the maximum possible temperature change, i.e.,

$$\xi = \frac{T_5 - T_2}{T_4 - T_2}$$

All other conditions of Example 6-7 apply.

SOLUTION

$$\dot{W}_{Cp} = \dot{m}c_p(T_2 - T_1) = \dot{m}c_pT\left[\left(\frac{p_2}{p_1}\right)^{\frac{\gamma-1}{\gamma}} - 1\right]$$

$$= \dot{m}\ 458.6\ \text{BTU/sec}\ (\dot{m}\ 1.066\ \text{MW})$$

(as in Example 6-7). Likewise:

$$\dot{W}_T = \dot{m}c_p(T_3 - T_4) = \dot{m}c_pT_3\left[1 - \frac{1}{(r_p)^{\frac{\gamma-1}{\gamma}}}\right]$$

$$= \dot{m}\ 925.9\ \text{BTU/s}\ (\dot{m}\ 2.150\ \text{MW})$$

$$\dot{Q}_R = \dot{m}c_p(T_3 - T_5)$$

$$\xi\ \substack{\text{(effectiveness} \\ \text{of regenerator)}} = \frac{T_5 - T_2}{T_4 - T_2} = 0.75$$

$$\therefore T_5 = (T_4 - T_2)(0.75) + T_2 = 0.75\ T_4 + 0.25\ T_2$$

Writing T_4 in terms of T_3 and r_p, T_2 in terms of T_1 and r_p, and noting that processes 1-2 and 3-4 are isentropic, obtain:

$$T_5 = (0.75)\left[\frac{T_3}{(r_p)^{\frac{\gamma-1}{\gamma}}}\right] + 0.25T_1(r_p)^{\frac{\gamma-1}{\gamma}}$$

$$= (0.75)(0.5767)(1750) + (0.25)(500)(1.7338)$$

$$= 973.7°\text{R}\ (540.9°\text{K})$$

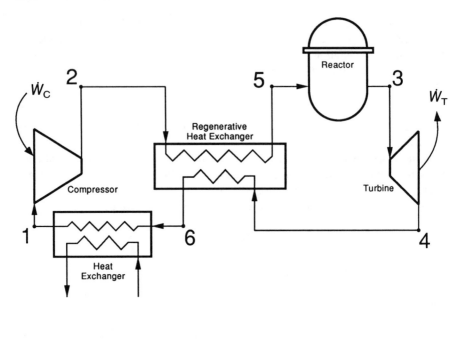

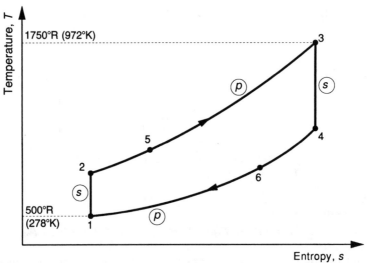

Figure 6-28 Brayton cycle with ideal components and regeneration: Example 6-10.

Hence:

$$\dot{Q}_R = \dot{m}c_p\,(1750 - 973.7) = 970.4\,\dot{m}\ \text{BTU/s} \ (2.254\,\dot{m}\ \text{MW})$$

$$\dot{W}_{NET} = \dot{W}_T - \dot{W}_{CP} = \dot{m}\,(925.9 - 458.7) = \dot{m}\,467.2\ \text{BTU/s} \ (1.084\,\dot{m}\ \text{MW})$$

$$\eta_{th} = \frac{\dot{W}_{NET}}{\dot{Q}_R}$$

$$= \left(\frac{467.2}{970.4}\right) 100 \text{ (English units)}$$

$$= \left(\frac{1.084}{2.254}\right) 100 \text{ (SI units)}$$

$$= 48.1\%$$

Repeating these calculations for real compressors and turbines with component isentropic efficiencies of 90%, the thermal efficiency is reduced to 36.4%. Table 6-9 lists the intermediate calculations under Example 6-10B.

It is interesting to compare these results with those of Examples 6-7 and 6-8, which are identical except for regeneration. One sees that the use of regeneration has increased the efficiency for the cases of ideal components (48.1% versus 42.3%) and real components (36.4% versus 30.7%). In each situation regeneration has caused the thermal efficiency of the cycle to increase by decreasing $\dot{Q}_R$ ($\dot{W}_{NET}$ being the same), a result that is possible only if T_4, the turbine exit, is hotter than T_2, the compressor exit.

This condition is not the case for a sufficiently high pressure ratio (e.g., $p_2 = 8 p_1$). In this case (Example 6-11 in Table 6-9), the cycle thermal efficiency is reduced to 38.4%.

A loss of thermal efficiency has resulted, from the introduction of a regenerator compared to the same cycle operated at a lower compression ratio, i.e., Example 6-10A (38.4% versus 48.1%). An additional comparison is between this case and the case of an ideal cycle without regenerator but operated at the same high compression ratio of 8. Again the efficiency of this case is less than this ideal cycle without regeneration (38.4% versus 56.2%—where the value of 56.2% can be obtained from Equation 6-106 or, equivalently, Figure 6-25). Physically both comparisons reflect the fact that for this case (Example 6-11) r_p is so high that T_2 is higher than T_4; i.e., the exhaust gases are cooler than those after compression. Heat would thus be transferred to the exhaust gases, thereby requiring $\dot{Q}_R$ to increase and η_{th} to decrease. Obviously, for this high compression ratio, inclusion of a regenerator in the cycle is not desirable. Major process variations possible for cycles with high compression ratios are the intermediate extraction and cooling of gases undergoing compression (to reduce compressor work) and the intermediate extraction and heating of gases undergoing expansion (to increase turbine work). These processes are called *intercooling* and *reheat*, respectively, and are analyzed next in Example 6-14. Intermediate cases of intercooling only and reheat only are included in Table 6-9 as Examples 6-12 and 6-13, respectively.

Example 6-14 Brayton cycle with reheat and intercooling

Note: Examples 6-11, 6-12, and 6-13 appear only in Table 6-9.

SOLUTION Calculate the thermal efficiency for the cycle employing both intercooling and reheat as characterized below. The cycle is illustrated in Figure 6-29. All other conditions of Example 6-7 apply.

Table 6-9 Results of Brayton cycle cases of Examples 6-7 through 6-14

Parameter	Ex. 6-7	Ex. 6-8	Ex. 6-9	Ex. 6-10A	Ex. 6-10B	Ex. 6-11	Ex. 6-12	Ex. 6-13	Ex. 6-14
$\beta = \left(\dfrac{p_2 p_4}{p_3 p_1}\right)^{\frac{\gamma-1}{\gamma}}$	1.0	1.0	1.05	1.0	1.0	1.0	1.0	1.0	1.0
Component isentropic efficiency (η_s)	1.0	0.9	0.9	1.0	0.9	1.0	1.0	1.0	1.0
Regenerator effectiveness (ξ)	—	—	—	0.75	0.75	0.75	—	—	—
Pressure ratio (r_p)	4	4	4	4	4	8	4	4	4
Intercooling	—	—	—	—	—	—	$\dfrac{p'_1}{p_1} = \dfrac{1}{2}\dfrac{p_2}{p_1}$ $T''_1 = T_1$	—	$\dfrac{p'_1}{p_1} = \dfrac{1}{2}\dfrac{p_2}{p_1}$ $T''_1 = T$
Reheat	—	—	—	—	—	—	—	$\dfrac{p'_3}{p_4} = \dfrac{1}{2}\dfrac{p_3}{p_4}$ $T''_3 = T_3$	$\dfrac{p'_3}{p_4} = \dfrac{1}{2}\dfrac{p_3}{p_4}$ $T''_3 = T_3$
Turbine work ($\dot{W}_T/\dot{m}$)									
BTU/lbm	925.9	833.3	776.5	925.9	833.3	1229.4	925.9	1052.5	1052.5
MJ/kg	2.150	1.935	1.803	2.150	1.935	2.855	2.150	2.444	2.444
Compressor work ($\dot{W}_c/\dot{m}$)									
BTU/lbm	458.7	509.7	509.7	458.7	509.7	801.9	395.96	458.7	395.96
MJ/kg	1.066	1.184	1.184	1.066	1.184	1.862	0.920	1.066	0.920
Net work ($\dot{W}_{NET}/\dot{m}$)									
BTU/lbm	467.2	323.6	266.9	467.2	323.6	427.5	529.9	593.8	656.5
MJ/kg	1.084	0.752	0.620	1.084	0.752	0.993	1.23	1.378	1.524
Heat in ($\dot{Q}_R/\dot{m}$)									
BTU/lbm	1103.0	833.3	1052.9	970.4	888.1	1112.2	1364.5	1630.1	1890.8
MJ/kg	2.560	1.935	2.45	2.254	2.062	2.583	3.169	3.786	4.391
Cycle thermal efficiency (η_{th}) (%)	42.3	30.7	25.3	48.1	36.4	38.4	38.8	36.4	34.7

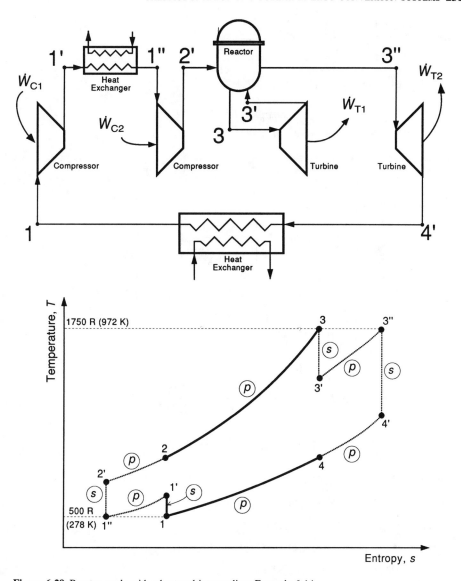

Figure 6-29 Brayton cycle with reheat and intercooling: Example 6-14.

$$\text{Intercooling:} \frac{p_1'}{p_1} = \frac{p_2}{p_1'} = r_p' \qquad T_1'' = T_1$$

$$\text{Reheat:} \frac{p_3'}{p_4} = \frac{p_3}{p_3'} = r_p' \qquad T_3'' = T_3$$

SOLUTION

$$\dot{W}_{CP} = \dot{m}c_p(T_1' - T_1) + \dot{m}c_p(T_2' - T_1'')$$

$$\dot{W}_{CP} = \dot{m}c_pT_1\left(\frac{T_1'}{T_1} - 1\right) + \dot{m}c_pT_1''\left(\frac{T_2'}{T_1''} - 1\right)$$

$$= \dot{m}c_pT_1[(r_p')^{\frac{\gamma-1}{\gamma}} - 1] + \dot{m}c_pT_1''[(r_p')^{\frac{\gamma-1}{\gamma}} - 1]$$

$$= 2\dot{m}c_pT_1[(r_p')^{\frac{\gamma-1}{\gamma}} - 1] = 2\dot{m}c_pT_1[(2)^{0.397} - 1]$$

$$= 395.96 \ \dot{m} \ \text{BTU/s} \ (0.920 \ \dot{m} \ \text{MW})$$

$$\dot{W}_T = \dot{m}c_p(T_3 - T_3') + \dot{m}c_p(T_3'' - T_4') = \dot{m}c_pT_3\left(1 - \frac{T_3'}{T_3}\right) + \dot{m}c_pT_3''\left(1 - \frac{T_4'}{T_3''}\right)$$

Again for the isentropic case:

$$\dot{W}_T = 2\dot{m}c_pT_3\left[1 - \frac{1}{(r_p)^{\frac{\gamma-1}{\gamma}}}\right]$$

$$= 2\dot{m} \ (1.25) \ 1750 \left[1 - \frac{1}{(2)^{0.397}}\right]$$

$$= 1052.5 \ \dot{m} \ \text{BTU/s} \ (2.444 \ \dot{m} \ \text{MW})$$

$$\frac{T_3'}{T_3} = \frac{1}{(r_p')^{\frac{\gamma-1}{\gamma}}}$$

$$\therefore T_3' = \frac{T_3}{(r_p')^{\frac{\gamma-1}{\gamma}}} = \frac{1750}{(2)^{0.397}} = \frac{1750}{1.317}$$

$$= 1329.0°R \ (738.3°K)$$

where $T_2' = T_1'' (r_p)^{\frac{\gamma-1}{\gamma}}$ and $T_1'' = T_1 = 500°R \ (278°K)$.

$$\therefore T_2' = (500°R) \ 2^{0.397} = 658.4°R \ (365.8°K)$$

$$\dot{Q}_R = \dot{m}c_p(T_3 - T_2) + \dot{m}c_p(T_3'' - T_3')$$

$$\dot{Q}_R = \dot{m}c_p \ [(1750 - 658.4) + (1750 - 1329.0)]$$

$$= 1890.8 \ \dot{m} \ \text{BTU/s} \ (4.391 \ \dot{m} \ \text{MW})$$

$$\dot{W}_{NET} = \dot{W}_T - \dot{W}_{CP} = \dot{m} \ (1052.5 - 395.96) = 656.5 \ \dot{m} \ \text{BTU/s} \ (1.524 \ \text{MW})$$

$$\eta_{th} = \frac{\dot{W}_{NET}}{\dot{Q}_R}$$

$$= \left(\frac{656.5}{1890.8}\right) 100 \ \text{(English units)}$$

$$= \left(\frac{1.524}{4.391}\right) 100 \ \text{(SI units)}$$

$$= 34.7\%$$

Compared with Example 6-7, both the net work and the maximum useful work have increased, but the thermal efficiency has decreased. An increase in the thermal efficiency would result if regeneration were employed. These changes are the combined result of the intercooling and reheating processes. They can best be understood by examining each separately with reference to the results of Examples 6-12 and 6-13, respectively.

In comparison with Example 6-7 (which is the same except for intercooling), we see that intercooling has decreased the compressor work and hence raised the $\dot{W}_{NET}$. However, the cycle efficiency has decreased to 38.8% from 42.3%, as the cycle with intercooling has a smaller pressure ratio than the basic cycle. However, by lowering inlet temperature to the reactor, i.e., T_2 to T'_2, a portion of this added energy can be supplied regeneratively, thereby offering the possibility to recoup some of the decrease in efficiency.

As we see in comparison with Example 6-7, reheat also decreases cycle thermal efficiency (36.4% versus 42.3%) because the cycle with reheat has a smaller pressure ratio than the basic cycle itself. However, the turbine exit temperature has been raised from T_4 to T'_4, offering the possibility of supplying some of the energy addition from states 2 to 3 regeneratively with its associated improvement in cycle thermal efficiency.

REFERENCE

1. Hicks, E. P., and Menzies, D. C. Theoretical studies on the fast reactor maximum accident." In: *Proceedings of the Conference on Safety, Fuels, and Core Design in Large Fast Power Reactors.* ANL-7120, 1965, pp. 654–670.

PROBLEMS

Problem 6-1 Work output of a fuel–water interaction (section II)
Compute the work done by a fuel–water interaction assuming that the 40,000 kg of mixed oxide fuel and 4000 kg of water expand independently and isentropically to 1 atmosphere. Assume that the initial fuel and water conditions are such that equilibrium mixture temperature (T_e) achieved is 1945°K. Other water conditions are as follows: $T_{initial} = 400°K$; $\rho_{initial} = 945$ kg/m³; $c_v = 4184$ J/kgK. *Caution:* Eq. 6-9 is inappropriate for these conditions, as the coolant at state e is supercritical.
Answer: 1.67×10^{10} J

Problem 6-2 Evaluation of alternate ideal Rankine cycles (section III)
Three alternative steam cycles illustrated in Figure 6-30 are proposed for a nuclear power station capable of producing either saturated steam or superheated steam at a temperature of 293°C. The condensing steam temperature is 33°C.

1. Assuming ideal machinery, calculate the cycle thermal efficiency and steam rate (kg steam/kWe-hr) for each cycle using the steam tables.
2. Compare the cycle thermal efficiencies calculated with the Carnot efficiency.
3. For each cycle, compare the amount of heat added per unit mass of working fluid in the legs 3→4 and 4→1.
4. Briefly compare advantages and disadvantages of each of these cycles. Which would you use?

Answers: $\eta_{th} = 38.2\%, 45.9\%, 36.8\%$
Steam rate = 3.60, 5.38, 3.54 kg steam/(kWe-hr)

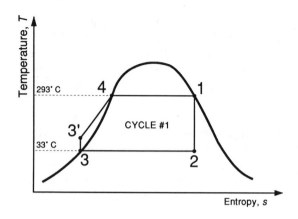

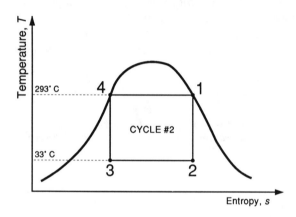

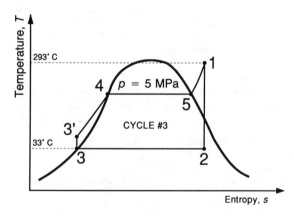

Figure 6-30 Alternate ideal Rankine cycles.

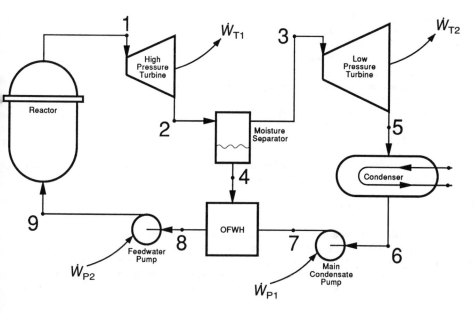

Figure 6-31 BWR plant.

Problem 6-3 Availability analysis of a simplified BWR (section IV)

A BWR system with a one-stage moisture separation is shown in Figure 6-31. The conditions in the table may be assumed.

Points	p(kPa)	Condition
1	6890	Saturated vapor
2	1380	
3	1380	Saturated vapor
4	1380	Saturated liquid
5	6.89	
6	6.89	Saturated liquid
7	1380	
8	1380	
9	6890	

Turbine isentropic efficiency = 90%
Pump isentropic efficiency = 85%
Environmental temperature = 30°C

1. Calculate the cycle thermal efficiency.
2. Recalculate the thermal efficiency of the cycle assuming that the pumps and turbines have isentropic efficiency of 100%.
3. Calculate the lost work due to the irreversibility of each component in the cycle and show numerically that the available work equals the sum of the lost work and the net work.

Answers: $\eta_{th} = 34.2\%$

$\qquad \eta_{th\ max} = 37.7\%$

$\qquad \Sigma\dot{I} = 1.65$ MJ/kg

$\qquad \dot{W}_{NET} = 0.86$ MJ/kg

Problem 6-4 Analysis of a steam turbine (section IV)

In the test of a steam turbine the following data were observed:

$h_1 = 3000$ kJ/kg; $p_1 = 10$ MPa; $V_1 = 150$ m/s

$h_2 = 2600$ kJ/kg; V_2 is negligible, $p_2 = 0.5$ MPa

$Z_2 = Z_1$ and $W_{1,2} = 384.45$ kJ/kg

1. Assume steady flow, and determine the heat transferred to the surroundings per kilogram of a steam.
2. What is the quality of the exit steam?

Answers: $Q = -26.8$ kJ/kg

$\qquad x = 92.9\%$

Problem 6-5 Advantages of moisture separation and feedwater heating (section V)

A simplified BWR system with moisture separation and an open feedwater heater is described in Problem 6-3. Compute the improvement in thermal efficiency that results from the inclusion of these two components in the power cycle. Do you think the thermal efficiency improvement from these components is sufficient to justify the capital investment required? Are there other reasons for having moisture separation?

Answer: η_{th} changes from 36.7% to 37.8%

Problem 6-6 Ideal Brayton cycle (section VI)

The Brayton cycle shown in Figure 6-32 operates using CO_2 as a working fluid with compresser and turbine isentropic efficiencies of 1.0. Calculate the thermal efficiency of this cycle when the working fluid is modeled as:

1. A perfect gas.
2. A real fluid (see below for extracted values from Keenan and Kay's gas tables).
3. A real fluid and the compressor and turbine both have isentropic efficiencies of 0.95.

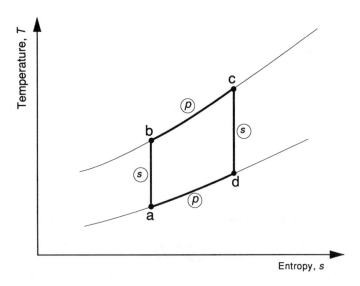

Figure 6-32 Ideal Brayton cycle. State a: $p = 20$ psia, $T = 90°F$. State c: $p = 100$ psia, $T = 1000°F$.

The parameters needed for a real fluid (from Keenan and Kay's gas tables) are shown in the following table.

Parameter	a	b	c	d
T	90		1000	
p	20	100	100	20
p_r	0.16108	0.8054	31.5	6.3
$\bar{h}$	4146.0	6239.8	14,204	10,077

where T = temperature,°F; p_r = relative pressure*; $\bar{h}$ = enthalpy per mole, BTU/lb-mole.

*The ratio of the pressures p_a and p_b corresponding to the temperatures T_a and T_b, respectively, for an isentropic process is equal to the ratio of the relative pressures p_{ra} and p_{rb} as tabulated for T_a and T_b, respectively. Thus:

$$\left(\frac{p_a}{p_b}\right)_{s=\text{constant}} = \frac{p_{ra}}{p_{rb}}$$

Answers: η_{th} = 31.0%, 25.5%, and 21.9%, respectively

Problem 6-7 Complex real Brayton cycle (section VII)

A gas-cooled reactor is designed to heat helium gas to a maximum temperature of 1000°F. The helium flows through a gas turbine, generating work to run the compressors and an electric generator, and then through a regenerative heat exchanger and two stages of compression with precooling to 100°F before entering each compressor. Each compressor and the turbine have an isentropic efficiency of 85%, and the pressure drop factor β is equal to 1.05. Each compression stage has a pressure ratio (r_p) of 1.27. The heat exchanger effectiveness (ξ) is 0.90.

Determine the cycle thermal efficiency. The Brayton cycle system is illustrated in Figure 6-33. The pressure drop factor β is defined as:

$$\beta \equiv \left(\frac{p_4}{p_6} \cdot \frac{p_7}{p_1} \cdot \frac{p_2}{p_3}\right)^{\frac{\gamma-1}{\gamma}} \equiv 1.05$$

Answer: η_{th} = 14.1%

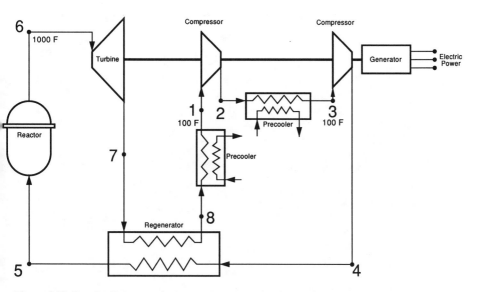

Figure 6-33 Complex Brayton cycle.

SEVEN

THERMODYNAMICS OF NUCLEAR ENERGY CONVERSION SYSTEMS: NONSTEADY FLOW FIRST LAW ANALYSIS

I INTRODUCTION

Clear examples of time-varying flow processes relevant to nuclear technology are the (1) pressurization of the containment due to postulated rupture of the primary or secondary coolant systems, (2) response of a PWR pressurizer to turbine load changes, and (3) BWR suppression pool heatup by addition of primary coolant. Unlike the steady-flow analysis, the variable-flow analyses can be performed with equal ease by either the control mass or the control volume approach. These approaches are demonstrated for the containment example. The pressurizer example is solved using the control volume approach. The suppression pool case is given as Problem 7-5.

II CONTAINMENT PRESSURIZATION PROCESS

The analysis of the rapid mixing of a noncondensible gas and a flashing liquid has application in reactor safety; e.g., for the light-water reactor, one postulated accident is the release of primary or secondary coolant within the containment. The magnitude of the peak pressure and the time to peak pressure are of interest for structural considerations of the containment.

The fluid released in the containment can be due to the rupture of either the primary or secondary coolant loops. In both cases the assumed pipe rupture begins the blowdown. The final state of the water–air mixture depends on several other factors: (1) the initial thermodynamic state and mass of water in the reactor and the

Table 7-1 Factors to consider during analysis of coolant system ruptures

Possible heat sinks	Possible heat sources	Possible fluid added from external sources
Primary system rupture		
Containment walls and other cool surfaces	Stored heat Decay heat	Emergency core cooling water
Active containment heat removal systems—air coolers, sprays, heat exchangers	Other energy sources in core (e.g., Zr-H$_2$O reactions, H$_2$ explosion)	Feedwater (BWR)
Steam generator secondary side	Steam generator secondary side	
Secondary system rupture		
Containment walls and other cool surfaces	Primary coolant through steam generator	Condensate makeup (PWR)
Active containment heat removal systems—air coolers, sprays, heat exchangers		

air in the containment; (2) the rate of release of fluid into the containment and the possible heat sources or sinks involved; (3) the likelihood of exothermic chemical reactions; and (4) the core decay heat. Table 7-1 lists the various external factors in the blowdown process that could affect the peak pressure. Figure 7-1 is a general pictorial representation of the factors involved in the containment response to the loss of primary coolant. We include only heat loss to structure and heat gain from the core in our discussion below. Furthermore, potential and kinetic energy effects are neglected.

A Analysis of Transient Conditions

We perform the analysis of transient conditions using the control mass and control volume approaches to illustrate that the two techniques are equally applicable to variable-flow processes.

1 Control mass approach. Let us define the thermodynamic system of interest as composed of three subsystems: containment air of mass (m_a), water vapor initially in the air of the containment (m_{wc_i}), and water initially in the primary (or secondary) system depending on rupture assumption (m_{wp}). At any given time, of the mass m_{wp}, the portion m_{wpd} has discharged into the containment and the portion m_{wpr} remains in the primary system. Hence the total primary system inventory $m_{wp} = m_{wpd} + m_{wpr}$.

For calculation purposes, it can be assumed that each mass in the containment exists at the total containment pressure (p_T) and therefore at a partial volume as seen in Figure 7-2, where the free containment volume (V_c) is given by:

$$V_c = V_a + V_{wc_i} + V_{wpd} \tag{7-1}$$

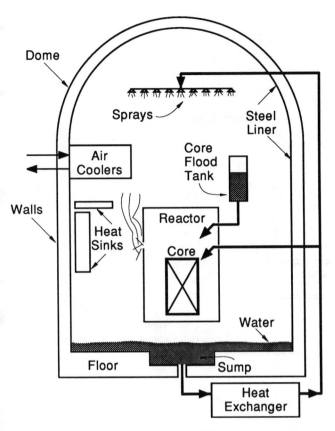

Figure 7-1 Containment features in a loss of primary coolant accident.

Application of the first law (Eq. 6-1) to each of these subsystems (m_a, m_{wc_1}, m_{wpd}, and m_{wpr}), neglecting potential and kinetic energy effects, leads to:

For m_a:

$$\frac{d(m_a u_a)}{dt} = \dot{Q}_{wc_1-a} + \dot{Q}_{wpd-a} - \dot{Q}_{a-st} - p_T \frac{dV_a}{dt} \qquad (7\text{-}2a)$$

For m_{wc_1}:

$$\frac{d(m_{wc_1} u_{wc_1})}{dt} = \dot{Q}_{wpd-wc_1} - \dot{Q}_{wc_1-a} - \dot{Q}_{wc_1-st} - p_T \frac{dV_{wc_1}}{dt} \qquad (7\text{-}2b)$$

For m_{wpd}:

$$\frac{d(m_{wpd} u_{wpd})}{dt} = \dot{Q}_{wpr-wpd} - \dot{Q}_{wpd-wc_1} - \dot{Q}_{wpd-a} - \dot{Q}_{wpd-st} - p_T \frac{dV_{wpd}}{dt} \qquad (7\text{-}2c)$$

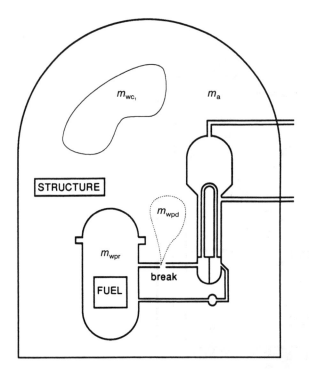

Figure 7-2 Control mass for transient analysis of containment conditions.

For m_{wpr}:

$$\frac{d(m_{wpr}u_{wpr})}{dt} = \dot{Q}_{n-wpr} - \dot{Q}_{wpr-wpd} - p_T\frac{dV_{wpr}}{dt} \qquad (7\text{-}2d)$$

where $\dot{Q}_{i-j}$ = rate of heat transfer from subsystem i to subsystem j. Heat transfer from mass wpr to the containment is assumed to occur only to mass wpd: p_T = total containment pressure; V_{wpd} = instantaneous partial volume of the portion of m_{wp} (i.e., m_{wpd}) discharged into the containment; V_{wpr} = volume of the portion of m_{wp} (i.e., m_{wpr}) remaining in the primary system. Subscripts a, wc_1, wpd, wpr, n, and st refer to air, initial containment water, coolant discharged from primary system, coolant remaining in primary system, core fuel, and structures, respectively.

These equations do not explicitly reflect the sign of the time derivative of the volume terms, i.e., dV_i/dt. This determination is obtained by applying the volume constraint for the total free volume available to the subsystem, i.e., the free containment and the primary system volumes:

$$\frac{d(V_a + V_{wc_1} + V_{wpd} + V_{wpr})}{dt} = 0 \qquad (7\text{-}3)$$

The time derivative of the volume of the coolant remaining in the primary system is zero, although the relative volumes of liquid and vapor of this subsystem do change with time.

$$\frac{d(V_{wpr})}{dt} = 0 \qquad (7\text{-}4)$$

On the other hand, the sign of the heat transfer rate terms do explicitly indicate whether heat flow is into or out of the control mass or system. When Eqs. 7-2a, 7-2b, 7-2c, and 7-2d are added together, and the volume constraint of Eq. 7-3 is applied, we obtain:

$$\frac{d}{dt}(m_a u_a + m_{wc_1} u_{wc_1} + m_{wpd} u_{wpd} + m_{wpr} u_{wpr}) = \dot{Q}_{n-wpr} - \sum_i \dot{Q}_{i-st} \qquad (7\text{-}5)$$

where the $\sum_i$ represents the three subsystems that comprise the containment atmosphere, i.e, a, wc_1, and wpd.

Upon integration of Eq. 7-5 from the time of break occurrence (1) to a later time (2) during the discharge process, we get:

$$U_2 - U_1 = Q_{n-wpr} - \sum_i Q_{i-st} \qquad (7\text{-}6)$$

where $U_2 = m_a u_{a_2} + (m_{wc_1} + m_{wpd_2})u_{wc_2} + m_{wpr_2}u_{wp_2}$
$U_1 = m_a u_{a_1} + m_{wc_1}u_{wc_1} + m_{wpr_1}u_{wp_1}$

because at the time of break occurrence m_{wpr_1}, which is identical to m_{wp}, exists at internal energy u_{wp_1}. If time 1 is taken as an arbitrary later time after the break, U_1 should be expressed as:

$$U_1 = m_a u_{a_1} + (m_{wc_1} + m_{wpd_1})\, u_{wc_1} + m_{wpr_1}u_{wp_1} \qquad (7\text{-}7)$$

Equation 7-6 is the desired result for transient analysis of containment conditions. Specification of $m_{wpd}(t)$ and $u_{wp}(t)$ are required. The discharged primary mass, $m_{wpd}(t)$, is obtained by integrating the break flow rate, $\dot{m}(t)$ over the interval 1 to 2, i.e.,

$$m_{wpd_2} = m_{wpd_1} + \int_1^2 \dot{m}(t)dt \qquad (7\text{-}8)$$

where the break rate is obtained from a critical flow analysis as presented in Chapter 11.

The primary system internal energy $u_{wp}(t)$ is obtained from the integrated discharge rate, the primary system volume, and an assumption on how to calculate the state of the coolant remaining in the primary system. For example, if the remaining coolant is assumed homogenized and undergoing a reversible, adiabatic expansion, the internal energy $u_{wp}(t)$ can be obtained from the initial known entropy s_{wp_1} and the calculable time-dependent specific volume $v_{wp}(t)$. Finally, Q_{n-wpr} and $\sum_i Q_{i-st}$ are obtained by transient thermal analyses.

2 Control volume approach. Consider now the control volume of Figure 7-3. Upon rupture, the system coolant flows into the control volume at the rate $\dot{m}(t)$. The control

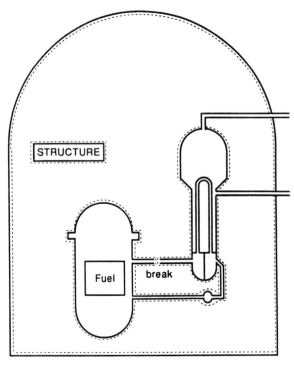

STRUCTURE

Fuel

break

Control Volume Boundary

Figure 7-3 Control volume for transient analysis of containment conditions.

volume shape remains constant with time, and there is no shaft work. Therefore the first law for a control volume (Eq. 6-2 of Table 6-1) is written as:

$$\dot{U}_{c.v.} = \dot{m}(t)h_p(t) + \dot{Q}_{wpr-c} - \dot{Q}_{c-st} \qquad (7-9)$$

where the second term on the right-hand side represents heat flow into the control volume at the system discharge plane, and the third term on the right-hand side is heat flow from the control volume into structures.

Integrating between times 1 and 2, Eq. 7-9 becomes:

$$U_2 - U_1 = + \int_1^2 h_p(t)\dot{m}(t)\, dt + Q_{wpr-c} - Q_{c-st} \qquad (7-10)$$

where now for the control volume:

$$U_2 = m_a u_{a_2} + (m_{wc_1} + m_{wpd_2})u_{wc_2}$$
$$U_1 = m_a u_{a_1} + m_{wc_1} u_{wc_1}$$

Eq. 7-10 becomes:

$$m_a u_{a_2} + (m_{wc_1} + m_{wpd_2}) u_{wc_2} = m_a u_{a_1} + m_{wc_1} u_{wc_1} \tag{7-11}$$

$$+ \int_1^2 h_p(t) \dot{m}(t) \, dt + Q_{wpr-c} - Q_{c-st}$$

Equation 7-6 should be identical to Eq. 7-11, as both the control mass and the control volume approaches should give identical results. To show this equivalence, we must show that the following terms are equal, i.e.,

$$m_{wpr_2} u_{wp_2} - m_{wpr_1} u_{wp_1} = - \int_1^2 h_p(t) \dot{m}(t) dt + Q_{n-wpr} - Q_{wpr-c} \tag{7-12}$$

as $\sum_i Q_{i-st}$ is equal to Q_{c-st} by definition.

This result follows from applying the first law to a new control volume, the primary coolant system volume. This volume remains constant with time.

For this control volume, the first law (Eq. 6-2) becomes

$$\dot{U}_{c.v.} = - h_p(t) \dot{m}(t) + \dot{Q}_{n-wpr} - \dot{Q}_{wpr-c} \tag{7-13}$$

where the signs on the right-hand side of this equation follow from the observations that the discharge flow is out of the control volume, and heat flows into this control volume from the core fuel and out this control volume into the containment. Integrating over the time interval 1 to 2 yields:

$$m_{wpr_2} u_{wp_2} - m_{wpr_1} u_{wp_1} = - \int_1^2 h_p(t) \dot{m}(t) dt + Q_{n-wpr} - Q_{wpr-c}$$

This result is identical to Eq. 7-12.

Containment condition histories can be evaluated by Eq. 7-11 in the following manner. First obtain the break flow rate, $\dot{m}(t)$, from a critical flow analysis and the heat transferred from the coolant remaining in the vessel to the containment and the heat transferred to containment structures Q_{wpr-c} and Q_{c-st}, respectively, by transient thermal analyses. Primary system enthalpy is obtained in a manner analogous to that described for primary system internal energy after Eq. 7-8.

B Analysis of Final Equilibrium Pressure Conditions

Determination of the transient pressure must be based on a variable-flow analysis such as that just described. If we wish to simplify the analysis, we can ask a simpler question and accept a more approximate answer. For example, consider only final conditions upon completion of the blowdown process and establishment of pressure equilibrium between the contents of the containment vessel and the primary system. We retain provision for heat transfer to the containment but only as the total heat transferred rather than the rate of heat transfer.

1 Control mass approach. As in section II.A, we define our thermodynamic system of interest as the mass of air and water vapor in the containment and all the water coolant in the primary or secondary system, as Figure 7-4 illustrates.

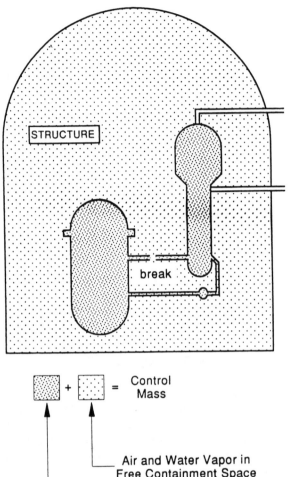

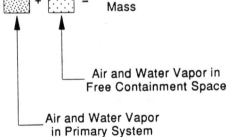

Air and Water Vapor in
Free Containment Space

Air and Water Vapor
in Primary System

Figure 7-4 Control mass for final
containment conditions.

Equation 7-6 is still applicable, but now state 2 is the state after completion of blowdown and achievement of pressure equilibrium. Hence u_{wc_2} and u_{wp_2} are identical, and at state 1 all the primary coolant is in the primary system so that Eq. 7-6 becomes:

$$U_2 - U_1 = Q_{n-wpr} - \sum_i Q_{i-st} \tag{7-14a}$$

where $U_2 = m_a u_{a_2} + (m_{wc_1} + m_{wp}) u_{wc_2}$
$U_1 = m_a u_{a_1} + m_{wc_1} u_{wc_1} + m_{wp} u_{wp_1}$

2 Control volume approach. For the control volume approach we simply draw the control volume around the containment and primary system, as illustrated in Figure 7-5. For this control volume there are no entering or exiting flow streams, no shaft

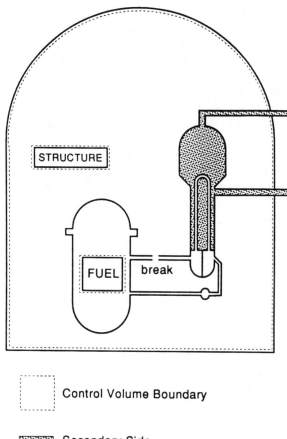

Control Volume Boundary

Figure 7-5 Control volume for final
containment conditions.

Secondary Side
(Note that the secondary system is
excluded from the Control Volume.)

work, and no expansion work. The first law for control volume (Eq. 6-2 of Table 6-1) reduces directly to Eq. 7-14a, where the heat loss to the structures is reexpressed as Q_{c-st}, i.e.,

$$U_2 - U_1 = Q_{n-wpr} - Q_{c-st} \qquad (7\text{-}14b)$$

3 Governing equations for determination of final conditions. Although Eq. 7-14b has been identified as the governing equation, by specifying the example more fully we can transform this result and introduce subsidiary relations in working forms suitable for numerical manipulation. Let us assume that the following initial and final conditions are known:

1. The initial state (designated "1") for the primary or secondary system fluid: For example, for a PWR the primary system water at operating conditions is subcooled at a known pressure (p_{w_1}) and temperature (T_{w_1}), with a known mass (m_w).

2. The initial mass (m_a) and thermodynamic state of the air. The amount of water vapor (relative humidity) initially in the air is known and can be included in the analysis, although its effect is usually small.

3. The final assumed containment condition, which depends on the reason for the analysis. If we are analyzing a preexisting plant, the containment volume is known and the peak pressure is sought. If, on the other hand, we are designing a new plant, we can specify the peak pressure as a design limit and seek the containment volume needed to limit the pressure to this peak value.

We can now elaborate on the governing equations. For simplicity, redesignate the water initially in the containment air as m_{wa}. Also note that the ruptured system can be either the primary or secondary system, so that for these general equations replace the subscript p with sys. Finally treat the internal energy of the air in terms of its specific heat-temperature product.

Reexpressing the energy balance of Eq. 7-14b in this nomenclature yields:

$$m_w (u_{w_2} - u_{w_1}) + m_a c_{va} (T_2 - T_{a_1}) = Q_{n-wsysr} - Q_{c-st} \qquad (7\text{-}15)$$

and

$$m_w u_{w_1} \equiv m_{wa} u_{wa_1} + m_{wsys} u_{wsys_1}$$

$$m_w u_{w_2} \equiv (m_{wa} + m_{wsys})\, u_{w_2}$$

where m_w = mass of water, which is composed of water vapor initially in the air and water or water and steam initially in the failed system, i.e., $m_{wa} + m_{wsys}$; m_a = mass of air in containment; $u = u(T,v)$ = internal energy per unit mass defined with respect to a reference internal energy $u_0(T_0,v_0)$ per unit mass; u_{w_1} = internal energy of the water initially in the containment air and the water initially in the failed system, i.e., u_{wa_1} and u_{wsys_1}; c_{va} = specific heat of air at constant volume; T_{a_1} = initial air temperature; T_2 = final temperature for the air–water mixture in the containment. The initial air temperature and initial water internal energy are known, whereas the final equilibrium state (T_2, u_2) is unknown.

Additional relations are needed that relate the properties of the fluids being mixed—water and air—to the total pressure and volume of the mixture. If the small volume occupied by liquid water is neglected, we can assume that the air occupies the same total volume (V_T) the liquid water plus water vapor occupy, i.e., V_T equals the free containment volume (V_c) plus the system, which is either the secondary system (V_s) or the primary system (V_p) depending on the problem definition. Further assume that the water vapor and liquid exist at the partial pressure of the saturated water vapor. Actually, whereas the water vapor and air are intermingled gases, each exerting its partial pressure, the liquid is agglomerated and at a pressure equal to the total pressure. Then from Dalton's law of partial pressures:

$$p_2 = p_{w_2} (T_2) + p_{a_2} \qquad (7\text{-}16)$$

where p_2 = final equilibrium pressure of the mixture; p_{w_2} = partial pressure of the saturated water vapor corresponding to T_2; p_{a_2} = partial pressure of air corresponding to T_2; and from the associated fact that each mixture component occupies the total volume:

$$V_T = m_{w_2} v_{w_2} (T_{2,sat}) \simeq m_a v_a(T_2, p_{a_2}) \tag{7-17}$$

Introducing the definition of the steam static quality (x_{st}) in the containment and treating air as a perfect gas, Eq. 7-17 becomes:

$$V_T = m_{w_2} [v_{f_2} + x_{st} v_{fg_2} (T_{2,sat})] \simeq \frac{m_a R_a T_2}{p_{a_2}} \tag{7-18}$$

Equations 7-15, 7-16, and 7-18 are used to find the final equilibrium state.

Establishment of the initial air pressure (p_{a_1}) in the containment should consider the water vapor present. This correction on p_{a_1} is minor but illustrates the use of Dalton's law of partial pressures. The initial conditions are characteristically stated in terms of a relative humidity (ϕ), the dry bulb temperature (T_{a_1}), and the total pressure (p_1). From the definition of relative humidity (ϕ), the saturated water vapor pressure for the given initial condition (p_{wa_1}) is given by:

$$p_{wa_1} = \phi p_{sat} (T_{a_1}) \tag{7-19}$$

Figure 7-6 illustrates this relation. Therefore by Dalton's law of partial pressures:

$$p_{a_1} = p_1 - p_{wa_1} \tag{7-20}$$

and by the perfect gas law:

$$m_a = \frac{p_{a_1} V_c}{R_a T_{a_1}} \tag{7-21}$$

Finally, we relate the final partial pressure of air to this initial air pressure by neglecting the difference in volume available to the air at the final versus the initial condition.

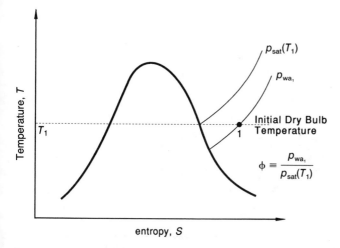

Figure 7-6 Initial water vapor pressure from relative humidity.

Hence again from the gas law:

$$p_{a_2} \simeq p_{a_1} \frac{T_2}{T_1} \qquad (7\text{-}22)$$

4 Individual cases. We can now treat two general cases for the final condition in the containment vessel.

    1. Saturated water mixture in equilibrium with the air. It is the expected result in the containment after a postulated large primary system pipe rupture. Because the heat addition from the core is relatively small, we neglect it here for simplicity. The process representation for this case is shown in Figure 7-7.

    2. Superheated steam in equilibrium with the air. This case requires that heat be added to the thermodynamic system. Such a situation could occur upon rupture of a PWR main steam line, as the intact primary system circulates through the steam generator and adds significant heat to the secondary coolant, which is blowing down into the containment. The process representation for this case is shown in Figure 7-8, where the water vapor path is illustrated as one of entropy increase.

    In both cases heat losses to active heat removal systems and to the structure within containment are neglected. Therefore the results obtained from this simplified analysis should be conservative, i.e., overprediction of peak pressure or containment volume. If heat transfer to structures is included, they need to be input in practice in a manner that takes into account their transient character.

    Example 7-1 Containment pressurization: saturated water in equilibrium with air resulting from a PWR primary system rupture

PROBLEM A *Find the peak pressure given the containment volume.*

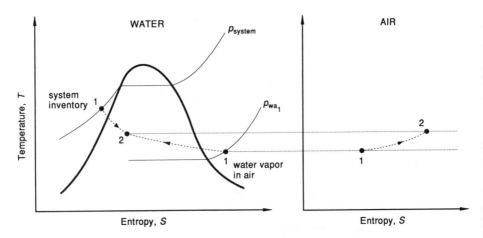

Figure 7-7 Process representation: saturated water mixture in equilibrium with air.

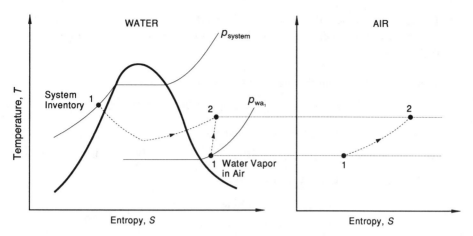

Figure 7-8 Process representation: superheated steam in equilibrium with air.

SOLUTION Equations 7-15 and 7-18 are the governing equations, and numerical values for containment conditions are drawn from Table 7-2. There are several ways in which the final pressure can be determined. One method is a trial and error solution using the steam tables.

The approach is to assume a final temperature (T_2) and from Eq. 7-18 calculate the static quality (x_{st}). This quality–temperature pair is checked in Eq. 7-15, and the search is continued until these equations are simultaneously satisfied. A quality greater than 1 indicates that the equilibrium water condition is superheated, and this search technique fails. This result is unlikely for realistic reactor containment conditions.

Table 7-2 Conditions for containment examples

Fluid	Heat addition during blowdown (Joules)	Volume (m³)	Pressure (MPa)	Temperature (°K)	Quality (x_{st}) or relative humidity (ϕ)
Example 7-1: saturated water in equilibrium with air as final state					
Primary coolant water (initial)		$V_p = 354$	15.5	617.9	Assumed saturated liquid
Containment vessel air (initial)		$V_c = 50,970$	0.101	300.0	$\phi = 80\%$
Mixture (final)	$Q = 0$	$V_T = 51,324$	0.523	415.6	$x_{st} = 50.5\%$
Example 7-2: superheated water in equilibrium with air as final state					
Secondary coolant water (initial)		$V_s = 89$	6.89	558	Assumed saturated liquid
Containment vessel air (initial)		$V_c = 50,970$	0.101	300	$\phi = 80\%$
Mixture (final)	$Q = 10^{11}$	$V_T = 51,059$	0.446 (64.7 psia)	478	$\phi = 17\%$

Let us now elaborate this procedure. To start, assume that the final temperature (T_2) is 415°K. The first value needed is the mass of air in the containment (m_a), which can be found by:

1. Using Eq. 7-19 to find the water vapor partial pressure (p_{w_1}):

$$p_{w_1} = \phi p_{sat}(T_1) = 0.8(3498 \text{ Pa}) = 2798 \text{ Pa}$$

2. Using Eq. 7-20 to find the air partial pressure (p_{a_1}):

$$p_{a_1} = p_1 - p_{w_1} = 101,378 - 2798 = 98,580 \text{ Pa}$$

3. Using Eq. 7-21 to find the air mass (m_a):

$$m_a = \frac{p_{a_1} V_c}{R_a T_{a_1}} = \frac{(98,580)\text{Pa}(50,970)\text{m}^3}{(286) \text{ J/kg°K } (300)°\text{K}} = 5.9 \ (10^4)\text{kg}$$

Using Eq. 7-17, we find the mass of water initially in the containment:

$$m_{wa} = \frac{V_c}{v_{wa_1}} = \frac{50,970 \text{ m}^3}{50.02 \text{ m}^3/\text{kg}} = 1019 \text{ kg}$$

because v_{wa_1} is the specific volume of superheated water vapor at p_{w_1} and T_{a_1}, which numerically equals 50.02 m³/kg.

Now we can find the quality (x_{st}) at state 2 from Eq. 7-18 as:

$$x_{st} = \frac{\dfrac{V_T}{m_w} - v_{f_2}}{v_{fg_2}} = \frac{\dfrac{(51324) \text{ m}^3}{2.11(10^5) \text{ kg}} - 0.00108 \text{ m}^3/\text{kg}}{0.485 \text{ m}^3/\text{kg}} = 0.499$$

where by the steam tables:

v_{f_2} = sat. liquid v at 415°K = 0.00108 m³/kg

v_{fg_2} = volume of vaporization v at 415°K = 0.485 m³/kg

v_{wp} = sat. liquid v at primary system pressure (15.5 MPa) = 0.00168 m³/kg

$$m_{wp} = \frac{V_p}{v_{wp}} = \frac{354 \text{ m}^3}{1.68 \ (10^{-3}) \text{ m}^3/\text{kg}} = 2.1 \ (10^5) \text{ kg}$$

$m_w = m_{wp} + m_{wa} = 2.1(10^5) + 0.01(10^5) = 2.11(10^5) \text{ kg}$

and where the failed system is the primary so that the subscript sys is written as p.

The quality (x_{st}) is checked by using Eq. 7-15 rewritten to express the water conditions separately as primary water and water in air.

$$m_{wp}(u_{f_2} + x_{st}u_{fg_2} - u_{wp_1}) + m_{wa}(u_{f_2} + x_{st}u_{fg_2} - u_{wa_1}) + m_a c_{va}(T_2 - T_{a_1}) = 0$$

2.1(10⁵) [595380 + x_{st}1.95(10⁶) − 1.6(10⁶)] + 1.02(10³)

· [595,380 + x_{st}1.95(10⁶) − 2.41(10⁶)] + 5.9(10⁴) (719) [415 − 300] = 0

Solving for x_{st} we get:

$$x_{st} = 0.505$$

where:

u_{wp_1} = sat. liquid u at 15.5 MPa = $1.6(10^6)$ J/kg

u_{f_2} = sat. liquid u at 415°K = 595,380 J/kg

u_{fg_2} = heat of vaporization u at 415°K = $1.95(10^6)$ J/kg

u_{wa_1} = superheated vapor u at 300°K, 2798 Pa = $2.41(10^6)$ J/kg

Continuing the iteration on x_{st}, the final result is:

$$T_2 = 415.6°K, \, p_{w_2} = 0.386 \text{ MPa}$$

Now the final air partial pressure (p_{a_2}) can be calculated using Eq. 7-18 or 7-22, yielding, respectively:

$$p_{a_2} = \frac{m_a R_a T_a}{V_T} = \frac{5.9(10^4)\text{kg}(286 \text{ J/kg°K})(415°K)}{51,324 \text{ m}^3} = 1.37(10^5)\text{Pa}$$

$$p_{a_2} = p_{a_1}(T_{a_2}/T_{a_1}) = 0.099 \frac{(415.6)}{300} = 0.137 \text{ MPa}$$

This gives us the total pressure from Eq. 7-16 as:

$$p_2 = p_{w_2} + p_{a_2} = 0.386 + 0.137 = 0.523 \text{ MPa}$$

PROBLEM B *Find the containment volume given a design limit for the peak pressure* (p_2).

SOLUTION For this problem we would know the initial air pressure (p_{a_1}), but the mass of air is dependent on the containment volume (Eq. 7-21). Therefore we can use p_{a_1} from Eq. 7-22 directly in Eq. 7-16 as:

$$p_2 = p_{w_2}(T_2) + p_{a_1} \frac{T_2}{T_1} \tag{7-23}$$

Now we can simply iterate on Eq. 7-23 by assuming T_2 and finding $p_{w_2} = p_{sat}(T_2)$ and then comparing the right-hand side to p_2. Once T_2 is found, water–steam properties are available, i.e., v_2, v_{fg_2}, u_{f_2}, u_{fg_2}. These values can be substituted into Eqs. 7-15 and 7-18, which can be solved simultaneously for the two unknowns x_{st} and V_c (when V_T is expressed as V_c plus V_p).

Proceeding numerically for the design condition of $p_2 = 0.523$ MPa, we find that Eq. 7-23 is satisfied when T_2 equals 415.6°K, i.e.,

$$p_2 = p_{w_2}(T_2) + p_{a_1}\left(\frac{T_2}{T_1}\right) = 0.386 + 0.099\left(\frac{415.6}{300}\right) = 0.523 \text{ MPa}$$

Then Eqs. 7-15 and 7-18 in the unknowns x_{st} and V_c become, respectively:

$$\left(m_{wp} + \frac{V_c}{V_{wa_1}}\right)(u_{f_2} + x_{st}u_{fg_2}) - m_{wp}u_{wp_1} - \frac{V_c}{V_{wa_1}}u_{wa_1} + p_{a_1}\frac{V_c}{R_a T_{a_1}}c_{va}(T_2 - T_{a_1}) = 0$$

$$\tag{7-24}$$

and

$$V_c + V_p = \left(m_{w_p} + \frac{V_c}{v_{wa_1}}\right)(v_{f_2} + x_{st}v_{fg_2}) \qquad (7\text{-}25)$$

where v_{wa_1} has been obtained from superheated steam tables.

Substituting numerical values into these two relations yields

$$\left[2.1(10^5) + \frac{V_c}{50}\right][595,380 + x_{st}1.95(10^6)] - 2.1(10^5)[1.6(10^6)]$$
$$- \frac{V_c}{50}[2.41(10^6)] + \frac{0.099(10^6)V_c}{286(300)}(719)(415 - 300) = 0 \qquad (7\text{-}26)$$

$$V_c + 354 = \left[2.1(10^5) + \frac{V_c}{50}\right][0.00108 + x_{st}0.485] \qquad (7\text{-}27)$$

Upon simultaneous solution, x_{st} and V_c are found as:

$$x_{st} = 0.505, \quad V_c = 51,593 \text{ m}^3$$

Example 7-2 Containment pressurization: superheated water in equilibrium with air resulting from a PWR secondary system rupture

A rupture of a main steam line adds water to the containment while the intact primary system circulates water through the steam generator, transferring energy to the secondary water that is discharging into the containment. The amount of water added is primarily dependent on the size of the steam generator. For this example, typical PWR four-loop plant values have been used for the amount of secondary water added and energy transferred via the steam generator. These assumed values are listed in Table 7-2.

PROBLEM A *Find the peak pressure given the containment volume.*

If the water is superheated we have a situation where the relative humidity has increased, as has the equilibrium temperature (T_2). Eqs. 7-16 and 7-17 are not linked by the steam table specific volumes but are equivalently given by:

$$p_2 = \frac{m_w R_w T_2}{V_T} + \frac{m_a R_a T_2}{V_T} \qquad (7\text{-}28)$$

where $m_w = m_{wa} + m_{ws}$ (i.e., water in containment air and the failed secondary system, respectively). Equation 7-28 treats the superheated water as a perfect gas that deviates from reality by only a few percent.

There are now three unknowns (T_2, p_2, u_{w_2}) and three equations (Eqs. 7-15, 7-28, and superheat steam tables). Assume T_2, use Eq. 7-28 to calculate p_2, and find u_{w_2} by the steam tables. Now use Eq. 7-15 to find u_{w_2} and compare to the previous value and iterate until convergence. As before, the initial water internal energy (u_{w_1})

is composed of two parts: (1) the system's liquid water, and (2) water vapor initially in the air. A condition may exist such that the final equilibrium state is above the critical temperature of the water. However, it does not affect the analysis, and the same procedure is used.

First we assume a final temperature, $T_2 = 450°K$, and substitute into Eq. 7-28.

$$p_2 = \frac{m_w R_w T_2}{V_T} + \frac{m_a R_a T_2}{V_T}$$

$$= \frac{(67,223 \text{ kg})(462 \text{ J/kg°K})(450°K)}{(51,059 \text{ m}^3)} + \frac{[5.9(10^4)\text{kg}](286 \text{ J/kg°K})(450°K)}{(51,059 \text{ m}^3)}$$

$$= 0.274 + 0.149 = 0.42 \text{ MPa}$$

where m_a and m_{wa} are found in a similar fashion as before using an equation of the form of Eq. 7-21 for each component.

From the steam tables for $p_{w_2} = 0.274$ MPa and $T_2 = 450°K$, we find $u_{w_2} = 2.61(10^6)$ J/kg. Using Eq. 7-15 with the initial conditions given in Table 7-2, we can solve for u_{w_2} and compare to the steam table value. Expressing Eq. 7-15 for this case and substituting numerical values, we have:

$$m_{ws}(u_{w_2} - u_{ws_1}) + m_{wa}(u_{w_2} - u_{wa_1}) + m_a c_{va}(T_2 - T_{a_1}) = Q \qquad (7\text{-}29)$$

$$(66,204 \text{ kg})[u_{w_2} - 1.25(10^6)\text{J/kg}] + 1019.0 \text{ kg}[u_{w_2} - 2.41(10^6)\text{J/kg}]$$
$$+ 5.9(10^4) \text{ kg } [719(450 - 300)] = 10^{11}\text{J}$$

or

$$u_{w_2} = 2.66(10^6)\text{J/kg}$$

where:

$$u_{ws_1} = \text{sat liquid } u \text{ at } 6.89 \text{ MPa} = 1.25(10^6) \text{ J/kg}$$

$$u_{wa_1} = \text{superheated vapor } u \text{ at } 2798 \text{ Pa}$$
$$= 2.41(10^6) \text{ J/kg } (T_1 = 300°K)$$

We can iterate on T_2 using the calculated u_{w_2} to adjust the guess on T_2; the final result is:

$$T_2 = 478°K, \quad p_2 = 0.4463 \text{ MPa}$$

The final relative humidity is 17%, i.e.,

$$\phi_2 = \frac{p_{w_2}}{p_{sat}(T_2)} = \frac{0.291 \text{ MPa}}{1.725 \text{ MPa}} = 0.17$$

where:

$$p_{w_2} = \frac{(67,223 \text{ kg})(462 \text{ J/kg°K})(478°K)}{51,059 \text{ m}^3} = 0.291 \text{ MPa}$$

PROBLEM B *Find the containment volume given the peak pressure.*
The same procedure can be used here, except that Eq. 7-28 can be solved in terms of V_c.

$$V_T = \frac{m_a R_a T_2}{p_2} + \frac{m_w R_w T_2}{p_2} \qquad (7-30)$$

where $V_T = V_c + V_s$ and $m_w = m_{wa} + m_{ws}$.

The initial air mass (m_a) is normally not known if the initial volume is unknown (V_c). Therefore we can substitute for m_a and m_{wa} by Eq. 7-21 written for each component, yielding:

$$V_c + V_s = \left[\left(\frac{p_{a_1}}{p_2}\right)\left(\frac{T_2}{T_{a_1}}\right) + \left(\frac{p_{wa}}{p_2}\right)\left(\frac{T_2}{T_{a_1}}\right) \right] V_c + \frac{m_{ws} R_w T_2}{p_2} \qquad (7-31)$$

Now we can assume T_2 and solve for V_c and thus m_a and m_{wa}. These values can be substituted into Eq. 7-15 to solve for u_{w_2} and compare it to $u_{w_2}(T_2, p_{w_2})$ from the steam tables. Now we simply iterate until u_{w_2} from Eq. 7-15 matches the steam table value.

We can proceed in this example for the assumed design condition $p_2 = 0.4463$. Eq. 7-31 can be used to find V_c, assuming $T_2 = 478°K$.

$$V_c + 89 \text{ m}^3 = \left(\frac{0.0986}{0.4463}\frac{478}{300} + \frac{0.0027}{0.4463}\frac{478}{300}\right) V_c + \frac{(66,204 \text{ kg})(462 \text{ J/kg°K})478°K}{0.4463 \times 10^6 \text{ Pa}}$$

giving $V_c = 51,093 \text{ m}^3$.

Using V_c we can find u_{w_2} for $p_{w_2} = 0.291$ MPa and $T_2 = 478°K$ from the steam tables:

$$u_{w_2} = 2.66(10^6) \text{ J/kg}$$

Using Eq. 7-15 and solving for u_{w_2}, we get:

$$m_{ws} (u_{w_2} - u_{ws_1}) + m_{wa} (u_{w_2} - u_{wa_1}) + m_a c_{va} (T_2 - T_{a_1}) = Q$$

where:

$$m_{wa} = \frac{V_c}{V_{w_1}} = \frac{51,093 \text{ m}^3}{50 \text{ m}^3/\text{kg}} = 1022 \text{ kg}$$

$$m_a = \frac{p_{a_1} V_c}{R_a T_{a_1}} = \frac{[0.099(10^6) \text{ Pa}][51,093 \text{ m}^3]}{(286 \text{ J/kg°K})(300°K)} = 5.9(10^4) \text{ kg}$$

and the other properties are known from the steam tables as before.

$$(66,204 \text{ kg}) [u_{w_2} - 1.25(10^6)] + 1022 [u_{w_2} - 2.41(10^6)]$$
$$+ 5.9(10^4)(719)(478 - 300) = 10^{11} \text{ J}.$$
$$u_{w_2} = 2.643(10^6) \text{ J/kg}$$

Thus:

$$T_2 = 478°K, \quad V_c = 51,093 \text{ m}^3$$

III RESPONSE OF A PWR PRESSURIZER TO LOAD CHANGES

The pressurizer vessel employed to control system pressure in a PWR provides another example of the analysis of a transient process by the control volume approach. The pressurizer vessel has an upper steam region and a lower water region. Water surges in and out of the lower region as a result of temperature changes in the system connected to the pressurizer. The lower pressurizer region is typically connected to a hot leg.

Upon an insurge, a water spray, which is typically taken from a cold leg, is actuated in the upper steam region to condense steam. Heaters are actuated to restore the initial saturated condition. In an outsurge, electric heaters in the lower water region are actuated to maintain the system pressure, which is otherwise decreased owing to the departure of liquid and subsequent expansion of the vapor volume. Figure 7-9 summarizes the response of the pressurizer to insurge and outsurge.

A Equilibrium Single-Region Formulation

We start the analysis of pressurizer behavior with a simple formulation in which the entire pressurizer is represented as a single region at equilibrium conditions. The

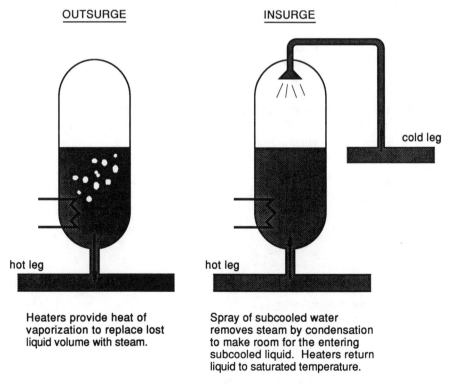

Figure 7-9 Pressurizer operation (without relief valves).

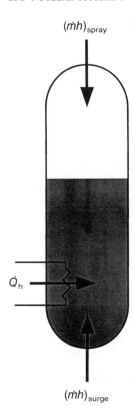

$(\dot{m}h)_{\text{spray}}$

$\dot{Q}_h$

$(\dot{m}h)_{\text{surge}}$

Figure 7-10 Externally supplied mass flow rate, enthalpy, and heat to the one-region pressurizer.

energy quantities being added to or departing the pressurizer owing to inlet spray at the top from the cold leg; insurge and outsurge through the bottom to the hot leg and heater input are illustrated in Figure 7-10. The pressurizer wall is assumed to be perfectly insulated with negligible heat capacity.

The general transient mass and energy equations are as follows:

$$\frac{d}{dt} m = \dot{m}_{\text{surge}} + \dot{m}_{\text{spray}} \tag{7-32}$$

$$\frac{d}{dt} (mu) = \dot{m}_{\text{surge}} h_{\text{surge}} + \dot{m}_{\text{spray}} h_{\text{spray}} + \dot{Q}_h - p \frac{d}{dt} (mv) \tag{7-33}$$

where

$$m = m_v + m_\ell \tag{7-34}$$

$$mu = m_v u_v + m_\ell u_\ell \tag{7-35}$$

$$mv = m_v v_v + m_\ell v_\ell \tag{7-36}$$

Additionally, a constraint exists on the total volume, which is fixed, i.e.,

$$\frac{d}{dt} (mv) = \frac{d}{dt} (m_v v_v + m_\ell v_\ell) = 0 \tag{7-37}$$

There are five prescribed input parameters: $\dot{m}_{spray}$, h_{spray}, $\dot{m}_{surge}$, h_{surge}, and $\dot{Q}_h$.

There are seven unknowns, i.e., p, m_v, u_v, v_v, m_ℓ, u_ℓ, and v_ℓ. These unknowns are so far related by only three equations, (Eqs. 7-32, 7-33, and 7-37). Closure of the problem is obtained by use of four equations of state, reflecting the assumption that all vapor and liquid conditions are maintained at saturation:

$$u_v \equiv u_g = f(p) \tag{7-38}$$

$$u_\ell \equiv u_f = f(p) \tag{7-39}$$

$$v_v \equiv v_g = f(p) \tag{7-40}$$

$$v_\ell \equiv v_f = f(p) \tag{7-41}$$

This set of equations and prescribed inputs is sufficient to perform a transient analysis to determine the unknowns for this simple one-region equilibrium formulation.

B Analysis of Final Equilibrium Pressure Conditions

If we wish to ask simpler questions than the transient nature of pressurizer behavior, the analysis can be simplified to consider only the end states of the transient process. For example, let us determine the size of the pressurizer necessary to accommodate specified insurge and outsurge events based on accommodating the end states only. In this idealized illustration, the initial and final pressurizer states are prescribed as saturated states at a fixed pressure.

In this case the unknowns become the liquid and vapor masses and the heater power, and the equations are those for continuity, energy conservation, and volume constraint. The inputs are spray mass and enthalpy, surge mass and enthalpy, and pressure with the associated saturation internal energy and specific volume properties.

The heaters must always be entirely liquid-covered, which for a fixed pressurizer geometry prescribes the minimum required liquid volume. The insurge is to be accommodated by providing sufficient initial vapor volume that the final state is a liquid-filled pressurizer. In this case the spray completely condenses the initial vapor, leading to a liquid-filled pressurizer, and sufficient heater input is provided to restore the initial pressure. In practice, the limiting condition would be the relief valve pressure set point.

The outsurge is to be accommodated by providing sufficient initial liquid volume that in the final state the heaters remain submerged after having operated to provide enough energy to restore the initial pressure. In practice, the limiting condition for the outsurge could be the minimum allowed primary system pressure.

Figure 7-11 illustrates the initial state of the pressurizer, and Figure 7-12 shows the end states for the insurge and outsurge events. The required total pressurizer volume is the sum of the initial vapor volume to accommodate the insurge, $(V_{g_i})_{insurge}$, and the initial liquid volume to keep the heaters submerged in an outsurge, $(V_{f_i})_{outsurge}$. The sum is required for the pressurizer to operate with sufficient liquid level and vapor space to accommodate either the insurge or the outsurge event.

$$V_T = (V_{g_i})_{insurge} + (V_{f_i})_{outsurge} \tag{7-42}$$

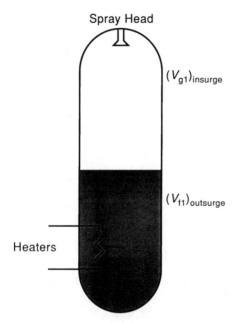

Spray Head

$(V_{g1})_{insurge}$

$(V_{f1})_{outsurge}$

Heaters

Figure 7-11 Initial state of pressurizer awaiting insurge or outsurge.

This example is handled most easily by taking the interior of the entire pressurizer as the control volume. The control volume and process representations for the insurge and outsurge are shown in Figure 7-13.

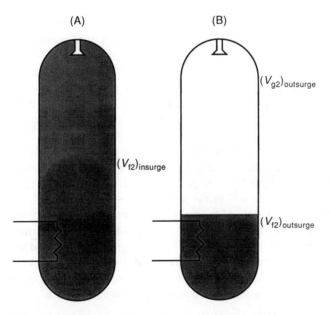

(A)   (B)

$(V_{g2})_{outsurge}$

$(V_{f2})_{insurge}$

$(V_{f2})_{outsurge}$

Figure 7-12 Final states of pressurizer. **A.** Insurge. **B.** Outsurge.

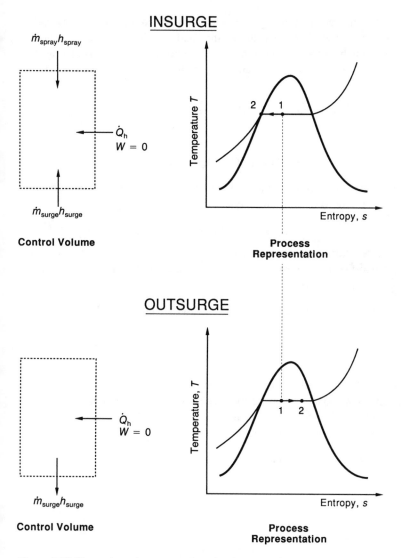

Figure 7-13 Thermodynamic representation of an insurge and an outsurge.

The continuity, energy, and volume constraint equations for the control volume are Eqs. 7-32, 7-33, and 7-37. Integrating these equations between the initial (1) and the final (2) state, assuming the flow rates and surge and spray enthalpies are constant with time, we obtain:

$$m_2 - m_1 = m_{surge} + m_{spray} \tag{7-43}$$

$$m_2 u_2 - m_1 u_1 = m_{surge} h_{surge} + m_{spray} h_{spray} + Q_h \tag{7-44}$$

$$m_2 v_2 = m_1 v_1 \tag{7-45}$$

At this point the insurge and outsurge cases must be treated separately to specialize the equations for each case. For the insurge, the final state is taken to be a liquid-filled pressurizer, which, after the water has been completely mixed, is saturated at the initial pressure. For this case we can express the governing equations in the following form.

$$m_{f_2} = m_{surge} + m_{spray} + m_{f_1} + m_{g_1} \tag{7-46}$$

or

$$m_{f_2} = m_{surge}(1 + f) + m_{f_1} + m_{g_1} \tag{7-47}$$

where the initial mass m_1 is split up into its liquid (m_{f_1}) and steam (m_{g_1}) components, and f is defined as the ratio of spray to surge flow:

$$f m_{surge} = m_{spray} \tag{7-48}$$

Note that, in practice, it is not necessary that $\dot{m}_{surge}$ and $\dot{m}_{spray}$ are related by a time invarient proportionality constant; that is, f in Eq. 7-48 can be a function of time. The energy balance likewise becomes:

$$m_{f_2} u_{f_2} = m_{surge}(h_{surge} + f h_{spray}) + m_{g_1} u_{g_1} + m_{f_1} u_{f_1} + Q_h \tag{7-49}$$

with the volume constraint being:

$$m_{f_2} v_{f_2} = m_{g_1} v_{g_1} + m_{f_1} v_{f_1} \tag{7-50}$$

Because states 1 and 2 are saturated at the same pressure, the properties appear in subsequent equations without state subscripts. The volume constraint (Eq. 7-50) can be substituted into the mass and energy balance (Eqs. 7-47 and 7-49) to eliminate the final mass (m_{f_2}); the resulting equations are:

$$m_{g_1} = \frac{m_{surge}(1 + f)v_f}{v_g - v_f} \tag{7-51}$$

and

$$m_{g_1} = \frac{v_f[m_{surge}(h_{surge} + f h_{spray}) + Q_h]}{v_g u_f - v_f u_g} \tag{7-52}$$

Note that the unknown (m_{f_1}), the initial liquid mass in the pressurizer, does not appear in Eqs. 7-51 or 7-52. This finding is reasonable, as this liquid does not participate in accommodating the insurge. Combining these equations yields the desired solution for Q_h as:

$$(Q_h)_{insurge} = \frac{m_{surge}(1 + f)[v_g u_f - v_f u_g]}{v_g - v_f} - m_{surge}(h_{surge} + f h_{spray}) \tag{7-53}$$

The steam volume, $(V_{g_1})_{insurge}$, is obtained as:

$$(V_{g_1})_{insurge} = m_{g_1} v_g \tag{7-54}$$

To compute the desired total volume, we must now consider the outsurge case and computer $(V_{f_1})_{\text{outsurge}}$. This task is accomplished by specialization of Eqs. 7-43, 7-44, and 7-45 to the outsurge process. In an outsurge, the final state has a liquid level sufficient to cover the pressurizer heaters. The mass balance (Eq. 7-43) becomes:

$$m_{f_2} + m_{g_2} - m_{f_1} - m_{g_1} = -m_{\text{surge}} \tag{7-55}$$

The energy balance equation becomes:

$$m_{f_2} u_{f_2} + m_{g_2} u_{g_2} - m_{f_1} u_{f_1} - m_{g_1} u_{g_1} = -m_{\text{surge}} h_{\text{surge}} + Q_h \tag{7-56}$$

The volume constraint is:

$$m_{f_2} v_{f_2} + m_{g_2} v_{g_2} = m_{f_1} v_{f_1} + m_{g_1} v_{g_1} \tag{7-57}$$

The volume constraint (Eq. 7-57) can be substituted into the mass and energy balances (Eqs. 7-55 and 7-56) to eliminate the final vapor mass (m_{g_2}); the resulting equations are:

$$m_{f_1} = m_{f_2} + m_{\text{surge}} \frac{v_g}{v_g - v_f} \tag{7-58}$$

and

$$m_{f_1} = m_{f_2} + \frac{Q_h - m_{\text{surge}} h_{\text{surge}}}{\dfrac{v_f}{v_g} u_g - u_f} \tag{7-59}$$

Note that the unknown (m_{g_1}), the initial vapor mass in the pressurizer, does not appear in Eq. 7-58 or 7-59. This omission is reasonable, as this vapor does not participate in accommodating the outsurge.

Combining these equations yields the desired solution for Q_h as:

$$(Q_h)_{\text{outsurge}} = m_{\text{surge}} h_{\text{surge}} - \left(u_f - \frac{v_f}{v_g} u_g \right) \left(m_{\text{surge}} \frac{v_g}{v_g - v_f} \right) \tag{7-60}$$

The initial liquid mass (m_{f_1}) can now be obtained from either Eq. 7-58 or Eq. 7-59. Hence:

$$(V_{f_1})_{\text{outsurge}} = m_{f_1} v_f \tag{7-61}$$

The total volume (V_T) can now be obtained by utilizing the results of Eqs. 7-54 and 7-61 in Eq. 7-42, i.e.,

$$V_T = (V_{g_1})_{\text{insurge}} + (V_{f_1})_{\text{outsurge}} \tag{7-42}$$

Example 7-3 Pressurizer sizing example

PROBLEM Determine the size of the pressurizer that can accommodate a maximum outsurge of 14,000 kg and a hot leg insurge of 9500 kg for the conditions of Table 7-3.

Table 7-3 Conditions for pressurizer design problem

Saturation pressure	15.5 MPa	(2250 psia)
Saturation temperature	618.3°K	(652.9°F)
Saturation properties		
u_f	1.60×10^6 J/kg	(689.9 B/lb)
u_g	2.44×10^6 J/kg	(1050.6 B/lb)
v_f	1.68×10^{-3} m³/kg	(0.02698 ft³/lb)
v_g	9.81×10^{-3} m³/kg	(0.15692 ft³/lb)
Mass of maximum outsurge	14,000 kg	
Mass of maximum insurge	9,500 kg	
Hot leg insurge enthalpy	1.43×10^6 J/kg	(612.8 B/lb)
Cold leg spray enthalpy	1.27×10^6 J/kg	(546.8 B/lb)
Cold leg spray expressed as a fraction of hot leg insurge (f)	0.03	
Outsurge enthalpy	1.63×10^6 J/kg	(701.1 B/lb)
Mass of liquid water necessary to cover the heaters (requires an assumption about the pressurizer configuration)	1827 kg	

SOLUTION The value of $(Q_h)_{insurge}$ is obtained from Eq. 7-53 as:

$$(Q_h)_{insurge} = \frac{9500 (1 + 0.03) [(9.81 \times 10^{-3})(1.60 \times 10^6) - (1.68 \times 10^{-3})(2.44 \times 10^6)]}{(9.81 - 1.68) \times 10^{-3}}$$
$$- 9500 [1.43 \times 10^6 + 0.03(1.27 \times 10^6)]$$
$$= 1.06 \times 10^7 \text{ J}$$

The value of m_{g_1} is obtained from Eq. 7-51 as:

$$m_{g_1} = \frac{9500(1 + 0.03)1.68 \times 10^{-3}}{(9.81 - 1.68) \times 10^{-3}} = 2022 \text{ kg}$$

The steam volume needed for the insurge from Eq. 7-54 is:

$$(V_{g_1})_{insurge} = (2.022 \times 10^3)(9.81 \times 10^{-3}) = 19.84 \text{ m}^3$$

Proceeding similarly for the outsurge, obtain $(Q_h)_{outsurge}$ from Eq. 7-60 as:

$$(Q_h)_{outsurge} = 14,000(1.63 \times 10^6)$$
$$- \left(1.60 \times 10^6 - \frac{1.68 \times 10^{-3}}{9.81 \times 10^{-3}}(2.44 \times 10^6)\right)$$
$$\cdot \left(14,000 \left[\frac{9.81 \times 10^{-3}}{(9.81 - 1.68) \times 10^{-3}}\right]\right)$$
$$= 2.2820 \times 10^{10} - (1.1821 \times 10^6)(1.6893 \times 10^4)$$
$$= 2.851 \times 10^9 \text{ J}$$

The value of m_{f_1} from Eq. 7-58 is:

$$m_{f_1} = 1827 + 14,000 \left[\frac{9.81 \times 10^{-3}}{(9.81 - 1.68) \times 10^{-3}} \right] = 1.8720 \times 10^4 \, \text{kg}$$

Hence from Eq. 7-61:

$$(V_{f_1})_{\text{outsurge}} = (1.8720 \times 10^4)(1.68 \times 10^{-3}) = 31.45 \, \text{m}^3$$

Utilizing Eq. 7-42, the total volume is:

$$V_T = (V_{g_1})_{\text{insurge}} + (V_{f_1})_{\text{outsurge}} = 19.84 + 31.45 = 51.29 \, \text{m}^3 \text{ or } 51.3 \, \text{m}^3$$

IV GENERAL ANALYSIS OF TRANSIENT PRESSURIZER BEHAVIOR

Let us next analyze the pressurizer using multiple control volumes and with no restriction on the thermodynamic state within each volume. In general, the upper portion of the pressurizer is a continuous vapor region through which liquid drops fall, and the lower portion is a continuous liquid region through which vapor bubbles rise. This description leads to the identification of four regions to describe this fluid–vapor configuration. With an insurge condition the colder primary fluid that enters the pressurizer may stratify, suggesting use of two control volumes to describe the lower continuous liquid region itself. However, this complexity is not incorporated in the formations presented here.

Section IV.A treats the general four-region case, and the more limited case in which the upper region contains only vapor and the lower region contains only liquid is treated in section IV.B. The latter case reflects the assumption of instantaneous addition of spray and condensate to the lower continuous liquid region and of vapor bubbles to the upper continuous vapor region.

A Transient, Nonequilibrium Four-Region Formulation

The principal characteristics of the general analysis for a transient, nonequilibrium four-region formulation are illustrated in Figures 7-14 and 7-15. Figure 7-14 identifies the liquid–vapor interfaces and the assumed work and heat transfer processes within the pressurizer geometry, whereas Figure 7-15 provides this information in a more compact form. Figure 7-15 also includes the external mass, enthalpy, and heat transfer exchanges that can occur with each region.

In the upper portion of the pressurizer (U) the interface (s'U) is between the continuous vapor phase and discrete liquid, which is in the form of falling liquid drops or wall condensate. In the lower portion of the pressurizer (L) the interface (s'L) is between the continuous liquid phase and the rising vapor bubbles. The interfaces s'' and s''' separate the upper and lower volumes and are defined as follows. The interface s'' is between the continuous vapor phase of the upper volume and the continuous liquid phase of the lower volume. The interface s''' separates the discontinuous phase

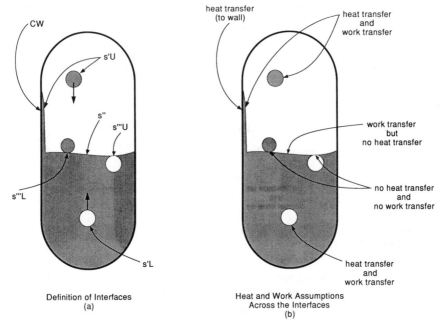

Figure 7-14 Definitions of interfaces and related heat and work transfer.

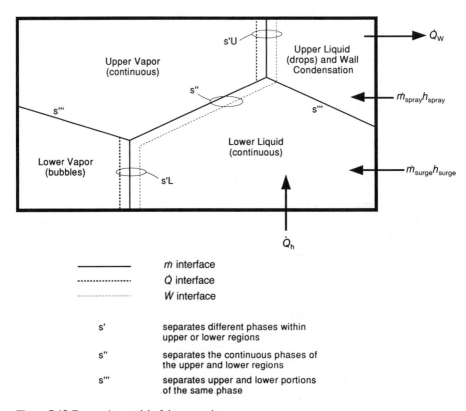

Figure 7-15 Four-region model of the pressurizer.

in either the upper or lower volume from the same (but continuous) phase in the other volume, i.e., upper region liquid drops from the lower region liquid pool and lower region vapor bubbles from the upper region vapor.

Both figures identify the assumed work and heat transfer processes. Because there are numerous drops and bubbles within the upper and lower volumes, respectively, their total surface area can be large, so both processes are allowed across the interfaces s'U and s'L. However, at the interface between the upper and lower volumes (s" + s'''), only work transfer is allowed along that portion (s") that separates the discontinuous phases. Heat transfer occurs across the interface between the condensate and the wall, labeled CW.

The continuity equations and energy equations for the two fluids in each volume are written next. The mass transfer terms are illustrated in Figure 7-16. The mass balance equations, following Eq. 5-83, are obtained by summing the vapor and liquid terms in each of the four volume regions.

$$\text{Upper vapor: } \frac{d}{dt}(m_v)_U = (\dot{m}_v)_{s'U} + (\dot{m}_v)_{s''U} + (\dot{m}_v)_{s'''U} \qquad (7\text{-}62)$$

$$\text{Upper liquid: } \frac{d}{dt}(m_\ell)_U = \dot{m}_{spray} + (\dot{m}_\ell)_{s'U} + (\dot{m}_\ell)_{s''U} \qquad (7\text{-}63)$$

$$\text{Lower vapor: } \frac{d}{dt}(m_v)_L = (\dot{m}_v)_{s'L} + (\dot{m}_v)_{s'''L} \qquad (7\text{-}64)$$

$$\text{Lower liquid: } \frac{d}{dt}(m_\ell)_L = \dot{m}_{surge} + (\dot{m}_\ell)_{s'L} + (\dot{m}_\ell)_{s''L} + (\dot{m}_\ell)_{s'''L} \qquad (7\text{-}65)$$

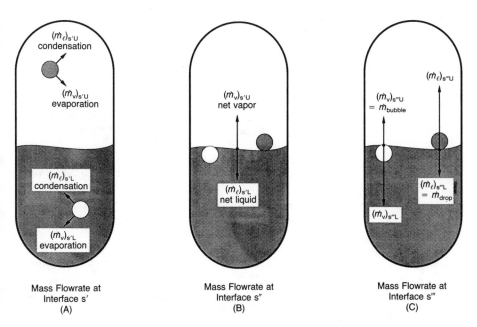

Figure 7-16 Mass balance at interfaces. Mass flow rates at interface s' (**A**), s" (**B**), and s''' (**C**).

The terms appearing in the mass balance equations can be related by introducing the mass jump conditions. The jump conditions are balance equations for mass transfer rates across the interfaces, as discussed in Chapter 5. For all interfaces, no mass sources or sinks exist. Thus for each interface the jump condition is simply that the sum of all mass transfer rates across the interface must be zero. For example, for interface s′ in the upper volume:

$$(\dot{m}_v)_{s'U} + (\dot{m}_\ell)_{s'U} = 0$$

or

$$(\dot{m}_\ell)_{s'U} = -(\dot{m}_v)_{s'U} \tag{7-66}$$

Normally in this region vapor condenses to liquid, which appears in the following forms (the symbols to be used in section IV.B for these condensate forms are noted):

1. As new drops called rainout, i.e., W_{RO}
2. As condensate on drops that have been sprayed into this region, i.e., W_{SC}
3. As condensate on the pressurizer wall, i.e., W_{WC}

Hence in section IV.B, Eq. 7-66 is rewritten as follows:

$$(\dot{m}_\ell)_{s'U} = W_{RO} + W_{SC} + W_{WC} \tag{7-67}$$

Jump conditions for the other interfaces are listed in Table 7-4.

The relevant energy equations are simplified forms of Eq. 5-125. The relevant assumptions are as follows.

Table 7-4 Mass jump conditions

Interface	Volume in which interface is located	Jump condition with nomenclature of section IV.B introduced	Physical significance
s′	Upper	$(\dot{m}_\ell)_{s'U} = -(\dot{m}_v)_{s'U} \equiv W_{RO}$ $+ W_{SC} + W_{WC}$ (Eqs. 7-66 and 7-67)	Vapor is condensed into three forms: rainout drops, condensate film on spray drops, condensate on pressurizer wall.
s″	Upper-lower	$(\dot{m}_v)_{s''U} = -(\dot{m}_\ell)_{s''L}$ (Eq. 7-68)	Liquid from the lower volume flashes across interface s″ to create vapor in the upper volume.
s‴ between vapor–vapor	Upper-lower	$(\dot{m}_v)_{s'''U} = -(\dot{m}_v)_{s'''L}$ (Eq. 7-69)	Vapor bubbles rise from the lower volume and enter the upper volume across interface s‴.
s‴ between liquid–liquid	Upper-lower	$(\dot{m}_\ell)_{s'''L} = -(\dot{m}_\ell)_{s'''U}$ (Eq. 7-70)	Liquid drops fall from the upper volume and enter the lower volume across interface s‴.
s′	Lower	$(\dot{m}_v)_{s'L} = -(\dot{m}_\ell)_{s'L} \equiv W_{FL}$ (Eqs. 7-71 and 7-72)	Liquid flashes to form vapor bubbles.

1. Neglect kinetic energy and gravity:

$$u_v^\circ \to u_v, \; h_v^\circ \to h_v, \; gz \to 0$$
$$u_\ell^\circ \to u_\ell, \; h_\ell^\circ \to h_\ell$$

2. Include volumetric energy generation only in the continuous liquid phase of the lower volume, i.e.,

$$\dot{Q}_{\text{Vol},k} = \dot{Q}_{\ell L} = \dot{Q}_h$$

into the lower liquid region

3. Neglect heat transfer across interfaces s″ and s‴.

4. Include heat transfer across interface s′. In general, then:

$$-(\vec{q}_k'' \cdot \vec{n}_{ks} A_s) \equiv -(\vec{q}_{ks'}'' \cdot \vec{n}_{ks'} A_{s'})_{U \text{ or } L} \equiv \dot{Q}_{ks'U \text{ or } L}$$

(where $k = v$ or ℓ) represents heat transfer from the interface s′ into the region kU or L, as shown in Figure 7-17 for the upper vapor region. It follows the sign convention that heat transfer into a region is positive and the observation that the vectors $\vec{q}_{vs'U}''$ and $\vec{n}_{vs'}$ are oppositely directed.

5. Include heat transfer to the pressurizer wall from only the condensate on the wall in the upper portion of the pressurizer. This heat transfer is:

$$-\sum_j (\vec{q}_k'' \cdot \vec{n}_k)_j A_j \equiv -(\vec{q}_C'' \cdot \vec{n}_C)_W A_W \equiv -\dot{Q}_W$$

where A_W is the area between the wall condensate and the wall.

6. Neglect work transfer across interface s‴.

7. Include work transfer across interface s′ and s″. In general, then, assuming that the phase velocity at the interface, $(\vec{v}_k)_s$, and the interface velocity, $\vec{v}_s$, are equal, the interfacial force term can be written for any general interface s as:

$$-(\overline{\overline{\tau}}_k \cdot \vec{v}_k - p_k \vec{v}_s) \cdot \vec{n}_{ks} A_s = -[(\overline{\overline{\tau}}_k - p_k \overline{\overline{I}}) \cdot \vec{v}_s] \cdot \vec{n}_{ks} A_s$$
$$= -[(\overline{\overline{\tau}}_k - p_k \overline{\overline{I}}) \cdot \vec{n}_{ks}] \cdot \vec{v}_s A_s$$
$$= -\vec{F}_{sk}'' \cdot \vec{v}_s A_s$$
$$\equiv \dot{W}_{ksU \text{ or } L}$$

Figure 7-17 illustrates the work transfer terms $\dot{W}_{vs'U}$ and $\dot{W}_{vs''U}$ for the upper vapor region. By convention, these terms, which represent work done by the vapor, appear as negative signed terms in the energy balance equation. The energy equations become:

Upper vapor:

$$\frac{d}{dt}(m_v u_v)_U = (\dot{m}_v h_v)_{s'U} + (\dot{m}_v h_v)_{s''U} + (\dot{m}_v h_v)_{s'''U} + \dot{Q}_{vs'U} - \dot{W}_{vs''U} - \dot{W}_{vs'U}$$

$$(7\text{-}73)$$

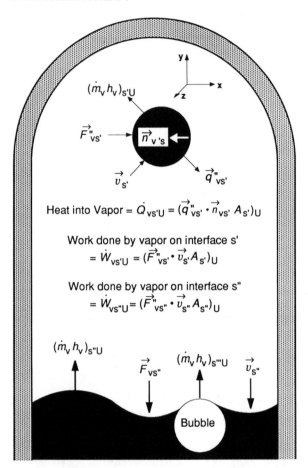

Figure 7-17 Sign convention for terms of the upper vapor region energy (Eq. 7-73). *Note:* All vectors are positive in the positive direction of the coordinate axes.

Upper liquid:

$$\frac{d}{dt}(m_\ell u_\ell)_U = (\dot{m}_\ell h_\ell)_{s'U} + \dot{m}_{spray}h_{spray} + (\dot{m}_\ell h_\ell)_{s'''U} + \dot{Q}_{\ell s'U} - \dot{Q}_W - \dot{W}_{\ell s'U}$$

(7-74)

Lower vapor:

$$\frac{d}{dt}(m_v u_v)_L = (\dot{m}_v h_v)_{s'L} + (\dot{m}_v h_v)_{s'''L} + \dot{Q}_{vs'L} - \dot{W}_{vs'L}$$

(7-75)

Lower liquid:

$$\frac{d}{dt}(m_\ell u_\ell)_L = \dot{m}_{surge}h_{surge} + (\dot{m}_\ell h_\ell)_{s'L} + (\dot{m}_\ell h_\ell)_{s''L}$$
$$+ (\dot{m}_\ell h_\ell)_{s'''L} + \dot{Q}_h + \dot{Q}_{\ell s'L} - \dot{W}_{\ell s''L} - \dot{W}_{\ell s'L}$$

(7-76)

The jump conditions for energy transfer rates across all interfaces are also required. In general, across any interface energy transfer can occur by (1) mass transfer, (2) heat transfer, and (3) work due to interface motion. Furthermore, each phase can contribute to these processes. Therefore for a general interface (s) the jump condition is written as follows using the sign convention of Figure 7-17:

$$(\dot{m}_v h_v + \dot{m}_\ell h_\ell)_s + (\dot{Q}_{vs} + \dot{Q}_{\ell s}) - (\dot{W}_{vs} + \dot{W}_{\ell s}) = 0 \qquad (7\text{-}77a)$$

or in vector form:

$$[\dot{m}_v h_v + \dot{m}_\ell h_\ell]_s - [(\vec{q}''_{vs} - \vec{q}''_{\ell s}) \cdot \vec{n}_{vs} A_s] - [(\vec{F}''_{vs} + \vec{F}''_{\ell s}) \cdot \vec{v}_s A_s] = 0 \qquad (7\text{-}77b)$$

The jump conditions for the interfaces obtained by specializing Eq. 7-77a are listed in Table 7-5. Recall from Figure 7-14 that no heat transfer is assumed to occur across interface s″. It is useful to express the work quantities in terms of the pressure and volume change for each volume. Hence for the upper regions:

$$\dot{W}_{\ell s' U} = p_U \frac{d}{dt} (m_\ell v_\ell)_U \qquad (7\text{-}83)$$

$$\dot{W}_{vs' U} + \dot{W}_{vs'' U} = p_U \frac{d}{dt} (m_v v_v)_U \qquad (7\text{-}84)$$

and for the lower regions:

$$\dot{W}_{vs' L} = p_L \frac{d}{dt} (m_v v_v) \qquad (7\text{-}85)$$

$$\dot{W}_{\ell s' L} + \dot{W}_{\ell s'' L} = p_L \frac{d}{dt} (m_\ell v_\ell)_L \qquad (7\text{-}86)$$

Table 7-5 Energy jump conditions

Interface	Volume in which interface is located	Jump condition	Physical significance
s′	Upper	$(\dot{m}_v h_v + \dot{m}_\ell h_\ell)_{s'U}$ $+ (\dot{Q}_{vs'} + \dot{Q}_{\ell s'})_U$ $- (\dot{W}_{vs'} + \dot{W}_{\ell s'})_U = 0$ (Eq. 7-78)	Mass transfer, heat transfer, and work from interface movement
s′	Lower	$(\dot{m}_v h_v + \dot{m}_\ell h_\ell)_{s'L}$ $+ (\dot{Q}_{vs'} + \dot{Q}_{\ell s'})_L$ $- (\dot{W}_{vs'} + \dot{W}_{\ell s'})_L = 0$ (Eq. 7-79)	Mass transfer, heat transfer, and work from interface movement
s″	Upper-lower	$(\dot{m}_v h_v)_{s''U} + (\dot{m}_\ell h_\ell)_{s''L}$ $- [\dot{W}_{vs''U} + \dot{W}_{\ell s''L}] = 0$ (Eq. 7-80)	Mass transfer and work from interface movement
s‴ between vapor–vapor	Upper-lower	$(\dot{m}_v h_v)_{s'''U} + (\dot{m}_v h_v)_{s'''L} = 0$ (Eq. 7-81)	Mass transfer only
s‴ between liquid–liquid	Upper-lower	$(\dot{m}_\ell h_\ell)_{s'''U} + (\dot{m}_\ell h_\ell)_{s'''L} = 0$ (Eq. 7-82)	Mass transfer only

Although the upper and lower regions may change in size, their sum is constant. This volume constraint can be expressed as:

$$\frac{dV}{dt} = 0 = \frac{d}{dt}(m_v v_v + m_\ell v_\ell)_U + \frac{d}{dt}(m_v v_v + m_\ell v_\ell)_L \qquad (7\text{-}87)$$

The set of Eqs. 7-62 through 7-87 is a general formulation that allows nonequilibrium conditions to exist in each region. The set can be solved if sufficient constitutive relations and state equations are prescribed so that together the number of unknowns is balanced. There are five prescribed input parameters, i.e., $\dot{m}_{spray}$, $\dot{m}_{surge}$, h_{spray}, h_{surge}, and $\dot{Q}_h$.

Note that the outsurge and insurge enthalpies must be specified. However, the outsurge enthalpy can be related to the liquid enthalpy of the lower region of the pressurizer.

There are 45 unknowns for the upper and lower regions.

Upper regions ($n = 23$)	*Lower regions* ($n = 22$)
$(m_v)_U;\ (m_\ell)_U$	$(m_v)_L;\ (m_\ell)_L$
$(\dot{m}_v)_{s'U};\ (\dot{m}_\ell)_{s'U}$	$(\dot{m}_v)_{s'L};\ (\dot{m}_\ell)_{s'L};$
$(\dot{m}_v)_{s''U}$	$(\dot{m}_\ell)_{s''L}$
$(\dot{m}_v)_{s'''U};\ (\dot{m}_\ell)_{s'''U}$	$(\dot{m}_v)_{s'''L};\ (\dot{m}_\ell)_{s'''L}$
$(v_v)_U;\ (v_\ell)_U$	$(v_v)_L;\ (v_\ell)_L$
$(u_v)_U;\ (u_\ell)_U$	$(u_v)_L;\ (u_\ell)_L$
$(h_v)_{s'U};\ (h_\ell)_{s'U}$	$(h_v)_{s'L};\ (h_\ell)_{s'L};$
$(h_v)_{s''U}$	$(h_\ell)_{s''L}$
$(h_v)_{s'''U};\ (h_\ell)_{s'''U}$	$(h_v)_{s'''L};\ (h_\ell)_{s'''L}$
$\dot{Q}_{vs'U};\ \dot{Q}_{\ell s'U}$	$\dot{Q}_{vs'L};\ \dot{Q}_{\ell s'L}$
$\dot{Q}_w$	
$\dot{W}_{vs'U};\ \dot{W}_{\ell s'U}$	$\dot{W}_{vs'L};\ \dot{W}_{\ell s'L}$
$\dot{W}_{vs''U}$	$\dot{W}_{vs''L}$
p_U	p_L

There are 14 equations of state. Using the internal energy and pressure, i.e., $(p, u_v, u_\ell)_U$ and $(p, u_v, u_\ell)_L$, as the independent thermodynamic variables, these relations are as follows.

Upper regions	*Lower regions*	
$(v_v)_U = v_v[p_U, (u_v)_U]$	$(v_v)_L = v_v[p_L, (u_v)_L]$	(7-88, 7-89)
$(v_\ell)_U = v_\ell[p_U, (u_\ell)_U]$	$(v_\ell)_L = v_\ell[p_L, (u_\ell)_L]$	(7-90, 7-91)
$(h_v)_{s'U} = [(h_v)_{s'}(p_U, (u_v)_U]$	$(h_v)_{s'L} = (h_v)_{s'}[p_L, (u_v)_L]$	(7-92, 7-93)
$(h_\ell)_{s'U} = [h_\ell)_{s'}(p_U, (u_\ell)_U]$	$(h_\ell)_{s'L} = (h_\ell)_{s'}[p_L, (u_\ell)_L]$	(7-94, 7-95)
$(h_v)_{s''U} = (h_v)_{s''}[p_U, (u_v)_U]$	$(h_\ell)_{s''L} = (h_\ell)_{s''}[p_L, (u_\ell)_L]$	(7-96, 7-97)
$(h_v)_{s'''U} = (h_v)_{s'''}[p_U, (u_v)_U]$	$(h_v)_{s'''L} = (h_v)_{s'''}[p_L, (u_v)_L]$	(7-98, 7-99)
$(h_\ell)_{s'''U} = (h_\ell)_{s'''}[p_U, (u_\ell)_U]$	$(h_\ell)_{s'''L} = (h_\ell)_{s'''}[p_L, (u_\ell)_L]$	(7-100, 7-101)

There are 37 available equations and constraints so far.

Conservation equations	8	(Eqs. 7-62 through 7-65 and 7-73 through 7-76)
Jump conditions	10	(Eqs. 7-66, and 7-68 through 7-71 and 7-78 through 7-82)
Work–volume change definitions	4	(Eqs. 7-83 and 7-86)
Volume constraint	1	(Eq. 7-87)
Equations of state	14	(Eqs. 7-88 through 7-101)
Total	37	

Eight constitutive equations are required to achieve closure.

$(\dot{m}_\ell)_{s'''U}$ Droplet flow rate	(Eq. 7-102)
$(\dot{m}_v)_{s'''L}$ Bubble flow rate	(Eq. 7-103)
$(\dot{m}_v)_{s''U}$ or $(\dot{m}_\ell)_{s''L}$	(Eq. 7-104)
$\dot{Q}_{vs'U}$; $\dot{Q}_{\ell s'U}$	(Eqs. 7-105, 7-106)
$\dot{Q}_{vs'L}$; $\dot{Q}_{\ell s'L}$	(Eqs. 7-107, 7-108)
$\dot{Q}_W$ (heat transfer from condensate to pressurizer wall)	(Eq. 7-109)

The selection of the specific parameters that are to be prescribed by constitutive laws (Eqs. 7-102 through 7-108) is arbitrary but constrained by the mass and energy jump conditions. From the mass jump equations (Eqs. 7-66, and 7-68 through 7-71), observe that at each interface the vapor and liquid mass transfer rates are equal but opposite. Hence only five mass transfer rates, at most, can be prescribed. This number is further reduced to three by virtue of the energy jump equations at interfaces s'U and s'L (Eqs. 7-78 and 7-79), which would otherwise be overdetermined, as all the interface sensible heat transfer terms $\dot{Q}$ have been chosen to be prescribed by Eqs. 7-105 through 7-108.

This formulation is a general four-region formulation. Many specific analyses have been published using a range of simplifying assumptions. One such simplified case (presented in the next section) is of a two-region nonequilibrium formulation with the liquid and vapor present only in separate regions.

B Nonequilibrium, Two-Region Formulation (Vapor-Only, Liquid-Only Regions)

The general case for a nonequilibrium, two-region formulation can be considerably simplified without loss of accuracy in practical reactor cases. Following the principle lines of the formulation of Kao [2], consider a two-region pressurizer operating at a single pressure that has only vapor present in the upper region and only liquid present in the lower region. The spray (W_{SP}) is thus assumed to instantaneously reach the lower liquid region together with the condensate from the upper region. The condensate comprises the condensed vapor on both the spray during its descent (W_{SC}) and on the pressurizer wall (W_{WC}). Additional vapor and liquid phase transformations are allowed owing to pressure changes. Specifically, flashing of liquid into vapor (W_{FL}) and condensation (also called rainout) of vapor into new liquid drops (W_{RO}) are allowed. The rainout liquid drops and flashed vapor bubbles are assumed to be added instantaneously to the liquid-only and vapor-only regions, respectively. These mass flow rates together with the surge flow rate ($\dot{m}_{surge}$) are illustrated in Figure 7-18. Figure 7-19 compactly provides this information along with the allowed heat and work transfer exchanges between regions and the pressurizer wall.

This two-region model is expressed by the following set of conservation and jump equations where the vapor–liquid interface is labeled s.

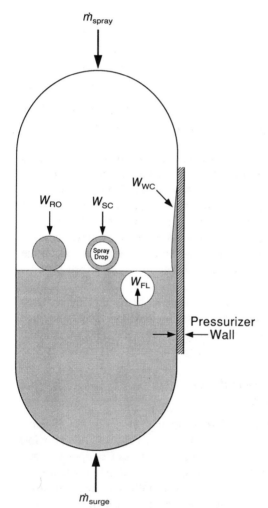

Figure 7-18 Mass flow rates in the two-region formulation.

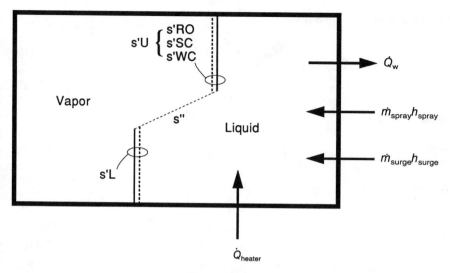

Figure 7-19 Two-region pressurizer model. ────── = $\dot{m}$ interface; ──────── = $\dot{Q}$ interface; ─────────── = $\dot{W}$ interface. *(After Kao [2].)*

Vapor mass:

$$\frac{d}{dt}(m_v) = \dot{m}_{vs} \tag{7-110}$$

Liquid mass:

$$\frac{d}{dt}(m_\ell) = \dot{m}_{\ell s} + \dot{m}_{spray} + \dot{m}_{surge} \tag{7-111}$$

Vapor energy:

$$\frac{d}{dt}(mu)_v = \dot{m}_{vs}h_{vs} + \dot{Q}_{vs} - \dot{W}_{vs} \tag{7-112}$$

Liquid energy:

$$\frac{d}{dt}(mu)_\ell = \dot{m}_{\ell s}h_{\ell s} + \dot{Q}_{\ell s} - \dot{W}_{\ell s} + (\dot{m}h)_{spray} + (\dot{m}h)_{surge} + \dot{Q}_h - \dot{Q}_w \tag{7-113}$$

Mass jump:

$$\dot{m}_{vs} = -\dot{m}_{\ell s} \tag{7-114}$$

Energy jump:

$$\dot{m}_{vs}h_{vs} + \dot{Q}_{vs} - \dot{W}_{vs} + \dot{m}_{\ell s}h_{\ell s} + \dot{Q}_{\ell s} - \dot{W}_{\ell s} = 0 \tag{7-115}$$

Work–volume change definitions:

$$\dot{W}_{vs} = p \frac{d}{dt} (mv)_v \tag{7-116}$$

$$\dot{W}_{\ell s} = p \frac{d}{dt} (mv)_\ell \tag{7-117}$$

Volume constraint:

$$\frac{d}{dt} [(mv)_v + (mv)_\ell] = 0 \tag{7-118}$$

Equations of state:

$$v_v = v_v (p, u_v) \tag{7-119a}$$
$$v_\ell = v_\ell (p, u_\ell) \tag{7-119b}$$
$$h_{vs} = h_{vs} (p, u_v) \tag{7-119c}$$
$$h_{\ell s} = h_{\ell s} (p, u_\ell) \tag{7-119d}$$

The system has the same five prescribed input parameters as did the four-region model: $\dot{m}_{spray}$, $\dot{m}_{surge}$, h_{spray}, h_{surge}, and $\dot{Q}_h$.

This system has 16 unknowns:

Vapor ($n = 7$)	Liquid ($n = 7$)	Others ($n = 2$)
m_v	m_ℓ	p
$\dot{m}_{vs}$	$\dot{m}_{\ell s}$	$\dot{Q}_w$
v_v	v_ℓ	
u_v	u_ℓ	
h_{vs}	$h_{\ell s}$	
$\dot{Q}_{vs}$	$\dot{Q}_{\ell s}$	
$\dot{W}_{vs}$	$\dot{W}_{\ell s}$	

There are 13 available equations and constraints.

Conservation equations	4	(Eqs. 7-110 through 7-113)
Jump conditions	2	(Eqs. 7-114 and 7-115)
Work–volume change definitions	2	(Eqs. 7-116 and 7-117)
Volume constraint	1	(Eq. 7-118)
Equations of state	4	(Eqs. 7-119a through 7-119d)
Total	13	

To achieve closure, three constitutive equations are required among:

$\dot{m}_{vs}$ or $\dot{m}_{\ell s}$ (Eqs. 7-120a and 7-120b)

$\dot{Q}_{vs}$, $\dot{Q}_{\ell s}$, $\dot{Q}_W$ (Eqs. 7-120c through 7-120e)

The specification of these constitutive equations first requires examination of the specific form of the mass and energy jump equations across the single interface s of this model.

1 Characterization of the interface s. This two-region model has a single interface s. However, as illustrated in Figures 7-18 and 7-19, a number of condensing and flashing processes are allowed to occur along portions of this interface. Specifically, the interface s is composed of the following portions from the four-region model:

$$\text{Interface s} = s'L + s'U + s'' = s'L + s'RO + s'SC + s'WC + s'' \quad (7\text{-}121)$$

where the interface portion s''' is neglected because the liquid condensate and the vapor bubbles are assumed to be instantaneously added to the liquid and vapor regions; $s'RO$ = interface between rainout drops and the upper vapor volume; $s'SC$ = interface between the condensate on the spray drops and the upper vapor volume; and $s'WC$ = interface between the condensate on the pressurizer wall and the upper vapor volume.

The corresponding vapor and liquid mass transfer rates across these interface segments are as follows:

$$\dot{m}_{vs} \equiv W_{FL} - W_{RO} - W_{SC} - W_{WC} \quad (7\text{-}122a)$$
$$\dot{m}_{\ell s} \equiv - W_{FL} + W_{RO} + W_{SC} + W_{WC} \quad (7\text{-}122b)$$

where mass transfer across interface s'' has been taken as zero.

The parameters $\dot{Q}_{vs}$ and $\dot{Q}_{\ell s}$ can similarly be partitioned along portions $s'U$ and $s'L$ of the interface and taken zero along s''. The heat transfer partitioning can be based on a reasonable physical assumption, in contrast to the basis that would be necessary to partition the parameters $\dot{W}_{vs}$ and $\dot{W}_{\ell s}$ along interfaces $s'U$ and $s'L$. Thus work terms are taken as zero along $s'U$ and $s'L$ and are assumed to occur totally along interface s''. Although these assumptions affect the values of the individual components of work transfer, they do not affect the net value of work transfer across interface s, which is dictated by the overall framework of the two-region model.

2. Mass and energy jump conditions across all interface segments. Next the relations between mass and energy transfer across the interface segments $s'L$, $s'RO$, $s'SC$, and $s'WC$ are established.

a Liquid region. Within the liquid region, mass transfer and energy transfer occur owing to bulk flashing across the interface $s'L$ (Fig. 7-20). The mass transfer across the interface utilizing the jump condition Eq. 7-71 is:

$$(\dot{m}_v)_{s'L} = -(\dot{m}_\ell)_{s'L} = W_{FL} \quad (7\text{-}123)$$

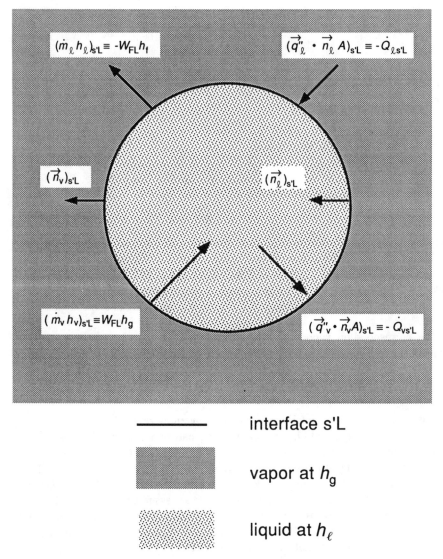

$$(\dot{m}_\ell h_\ell)_{s'L} \equiv -W_{FL}h_f \qquad (\vec{q}''_\ell \cdot \vec{n}_\ell A)_{s'L} \equiv -\dot{Q}_{\ell s'L}$$

$$(\vec{n}_v)_{s'L} \qquad (\vec{n}_\ell)_{s'L}$$

$$(\dot{m}_v h_v)_{s'L} \equiv W_{FL}h_g \qquad (\vec{q}''_v \cdot \vec{n}_v A)_{s'L} \equiv -\dot{Q}_{vs'L}$$

——— interface s'L

vapor at h_g

liquid at h_ℓ

Figure 7-20 Flashing in the lower liquid region by the model of section IV.B (work terms neglected).

In general, the vapor bubbles are created by flashing of lower region liquid at enthalpy h_ℓ to produce a vapor mass per unit time $(\dot{m}_v)_{s'L}$ at enthalpy h_v. The latent heat of evaporation is extracted at the interface s'L due to liquid and vapor heat transfer.

Rearranging the energy jump condition (Eq. 7-79 in Table 7-5) for this interface as:

$$\begin{matrix} \text{Net energy from liquid} \\ \text{region into interface} \end{matrix} = \begin{matrix} \text{net energy from interface} \\ \text{into vapor bubbles} \end{matrix}$$

$$- (\dot{m}_\ell h_\ell)_{s'L} - \dot{Q}_{\ell s'L} + \dot{W}_{\ell s'L} = (\dot{m}_v h_v)_{s'L} + \dot{Q}_{vs'L} - \dot{W}_{vs'L} \qquad (7\text{-}79)$$

where the notation of the heat transfer and work terms has been rearranged for simplification.

It is assumed that flashing occurs only at saturation conditions and that the latent heat of vaporization is supplied only by the liquid region. Thus:

$$\dot{Q}_{vs'L} = 0 \tag{7-124}$$

and

$$(\dot{m}_v h_v)_{s'L} = W_{FL} h_g \tag{7-125}$$
$$(\dot{m}_\ell h_\ell)_{s'L} = - W_{FL} h_f \tag{7-126}$$

Neglecting work terms, the jump condition for energy at s'L (Eq. 7-79) becomes:

$$W_{FL} h_{fg} + \dot{Q}_{\ell s'L} = 0 \tag{7-127}$$

b Vapor region. Within the vapor region, mass and energy transfer occur at three condensation locations, i.e., bulk condensation within the vapor region called rainout (W_{RO}), on the vessel walls, (W_{WC}), and on the spray droplets (W_{SC}). The appropriate mass transfer jump condition is Eq. 7-66, which is rewritten as:

$$(\dot{m}_\ell)_{s'U} = - (\dot{m}_v)_{s'U} = W_{RO} + W_{SC} + W_{WC} \tag{7-128}$$

For energy transfer each of these interface components must be treated separately.

First, *consider the rainout drops,* as illustrated in Figure 7-21. In general, these drops are created by condensation of upper region vapor at enthalpy h_v, which produces a liquid mass per unit time $(\dot{m}_\ell)_{s'RO}$ at enthalpy h_ℓ. The latent heat of condensation is released at the interface s'RO due to vapor and liquid heat transfer. Rearrange the jump condition (Eq. 7-78 in Table 8-5) for this interface s'RO as:

$$\begin{matrix} \text{Net energy from interface} \\ \text{into vapor region} \end{matrix} = \begin{matrix} \text{net energy out of liquid drops} \\ \text{into interface} \end{matrix}$$
$$(\dot{m}_v h_v)_{s'RO} + \dot{Q}_{vs'RO} - \dot{W}_{vs'RO} = - (\dot{m}_\ell h_\ell)_{s'RO} - \dot{Q}_{\ell s'RO} + \dot{W}_{\ell s'RO} \tag{7-129}$$

Assuming that rainout occurs only at saturation conditions and the latent heat of condensation is released only to the vapor region:

$$\dot{Q}_{\ell s'RO} = 0 \tag{7-130}$$
$$(\dot{m}_\ell h_\ell)_{s'RO} = W_{RO} h_f \tag{7-131}$$

and

$$(\dot{m}_v h_v)_{s'RO} = - W_{RO} h_g \tag{7-132}$$

Neglecting work terms, Eq. 7-129 becomes:

$$- W_{RO} h_{fg} + \dot{Q}_{vs'RO} = 0 \tag{7-133}$$

Next, *consider condensation on the spray drops,* which creates the interface s'SC with the vapor region. This process is illustrated in Figure 7-22, which also depicts the interface labeled s'SP between the spray drop and the surrounding condensate shell. Proceeding as in step 3, the energy jump condition (Eq. 7-78) for the interface s'SC can be written as:

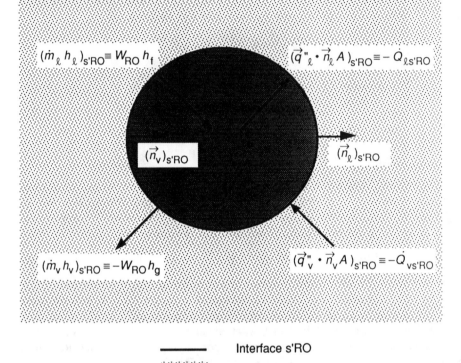

Figure 7-21 Rainout in the upper vapor region by the model of section IV.B (work terms neglected).

Net energy from vapor region into interface $=$ net energy out of interface into condensate

$$- (\dot{m}_v h_v)_{s'SC} - \dot{Q}_{vs'SC} + \dot{W}_{vs'SC} = (\dot{m}_\ell h_\ell)_{s'SC} + \dot{Q}_{\ell s'SC} - \dot{W}_{\ell s'SC} \quad (7\text{-}134)$$

At the interface s'SP only heat transfer is occurring, no mass transfer. Work at this interface is neglected. The energy interface jump condition is simply:

$$- \dot{Q}_{Cs'SP} = \dot{Q}_{Ss'SP} \quad (7\text{-}135)$$

where subscripts C and S = condensate and spray regions, respectively.

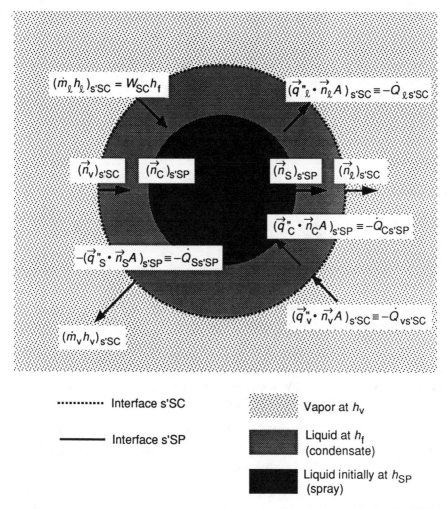

Figure 7-22 Condensation in the upper vapor region on spray drops by the model of section IV.B (work terms neglected).

Assuming that condensation occurs only at saturation conditions and that the latent heat of condensation is released totally to the spray by conduction through the condensate shell:

$$\dot{Q}_{vs'SC} = 0 \qquad (7\text{-}136)$$

$$(\dot{m}_\ell h_\ell)_{s'SC} = W_{SC}h_f \qquad (7\text{-}137)$$

and

$$(\dot{m}_v h_v)_{s'SC} = -W_{SC} h_g \tag{7-138}$$

where the mass jump conditions (Eq. 7-128) for surface s'SC has been utilized to obtain $(\dot{m}_\ell)_{s'SC}$, which equals W_{SC}, and:

$$\dot{Q}_{\ell s'SC} \equiv -\dot{Q}_{Cs'SP} = \dot{Q}_{Ss'SP} \tag{7-139}$$

Neglecting work terms, Eq. 7-134 becomes:

$$W_{SC} h_{fg} = \dot{Q}_{Ss'SP} \tag{7-140a}$$

Finally, if it is assumed that the rate of condensation is just sufficient to raise the spray enthalpy to saturation:

$$\dot{Q}_{Ss'SP} = W_{SP}(h_f - h_{SP}) \tag{7-140b}$$

so Eq. 7-140a becomes:

$$W_{SC} = W_{SP}\left(\frac{h_f - h_{SP}}{h_{fg}}\right) \tag{7-141}$$

i.e., the condensate rate is just sufficient to raise the enthalpy of the spray to saturation as it falls through the upper vapor region. This result for W_{SC} is frequently used in pressurizer models.

Finally, consider condensation on the pressurizer wall, which creates the interface s'WC with the vapor region. This process is illustrated in Figure 7-23, which also depicts the interface labeled s'CW between the condensate and the wall. Proceeding as before, the jump condition (Eq. 7-78) for the interface s'WC is:

$$(\dot{m}_v h_v)_{s'WC} + \dot{Q}_{vs'WC} - \dot{W}_{vs'WC} = -(\dot{m}_\ell h_\ell)_{s'WC} - \dot{Q}_{\ell s'WC} + \dot{W}_{\ell s'WC} \tag{7-142}$$

Assuming that condensation occurs only at saturation conditions and that the latent heat of condensation is released to the condensate:

$$\dot{Q}_{vs'WC} = 0 \tag{7-143}$$
$$(\dot{m}_\ell h_\ell)_{s'WC} = W_{WC} h_f \tag{7-144}$$

and

$$(\dot{m}_v h_v)_{s'WC} = -W_{WC} h_g \tag{7-145}$$

where the mass jump condition (Eq. 7-128) for interface s'WC has been utilized.

Furthermore, it is assumed that the latent heat of condensation that is released to the condensate is transferred by conduction completely to the pressurizer wall. Referring to Figure 7-23 for nomenclature,

$$\dot{Q}_{\ell s'WC} = \dot{Q}_W \tag{7-146}$$

because:

$$(\vec{q}''_\ell \cdot \vec{n}_\ell A)_{s'WC} = (\vec{q}''_C \cdot \vec{n}_C A)_{s'CW} \tag{7-147}$$

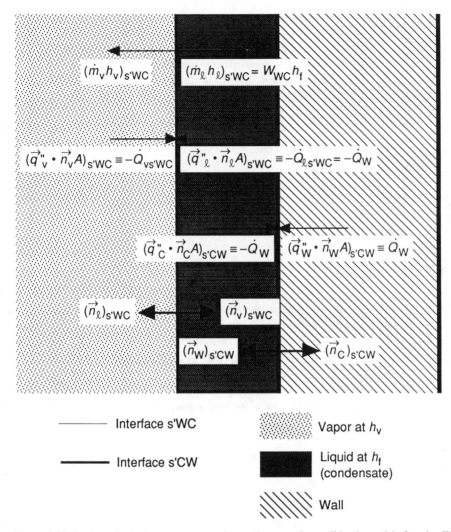

Figure 7-23 Condensation in the upper vapor region on the pressurizer wall by the model of section IV.B (work terms neglected).

Neglecting work terms, Eq. 7-142 becomes:

$$W_{WC} = \frac{\dot{Q}_W}{h_{fg}} \qquad (7\text{-}148)$$

It is important to note that the heat transfer to the wall is treated as heat transfer from liquid (condensate) to the wall even though physically this condensation process is occurring on the pressurizer wall in the upper vapor region. Hence in this model, although $\dot{Q}_W \equiv (\vec{q}''_W \cdot \vec{n}_W A)_{s'W}$ is finite and to be prescribed using a transient heat conduction model of the wall–condensate interaction, the heat transfer to the wall from the bulk vapor or liquid in the pressurizer is assumed negligible.

From the preceding development, the mass jump and energy jump conditions across the segments of interface s are as follows.

$$\text{Mass jump:} \quad \text{s:} \quad \dot{m}_{vs} = -\dot{m}_{\ell s} = W_{FL} - W_{RO} \tag{7-122}$$
$$- W_{SC} - W_{WC}$$

$$\text{Energy jump: s'L:} \quad W_{FL}h_{fg} + \dot{Q}_{\ell s'L} = 0 \tag{7-127}$$

$$\text{s'RO:} \quad -W_{RO}h_{fg} + \dot{Q}_{vs'RO} = 0 \tag{7-133}$$

$$\text{s'SC:} \quad W_{SC} = W_{SP}\left(\frac{h_f - h_{SP}}{h_{fg}}\right) \tag{7-141}$$

$$\text{s'WC:} \quad W_{WC} = \frac{\dot{Q}_W}{h_{fg}} \tag{7-148}$$

$$\text{s'':} \quad \dot{W}_{vs''} + \dot{W}_{\ell s''} = 0 \tag{7-116 through 7-118}$$

Utilizing the jump conditions, the conservation equations (Eqs. 7-110 through 7-113) can be rewritten and are displayed in Table 7-6. The sum of the vapor and liquid mass and energy equations equal the single-region results (Eqs. 7-32 and 7-33) when the volume constraint and Eq. 7-148 are applied.

Table 7-6 Conservation equations for the two-region model

Parameter	Equation
Vapor mass	$\dfrac{d}{dt}(m_v) = W_{FL} - W_{RO} - W_{SC} - W_{WC}$
Liquid mass	$\dfrac{d}{dt}(m_\ell) = -W_{FL} + W_{RO} + W_{SC} + W_{WC} + \dot{m}_{spray} + \dot{m}_{surge}$
Vapor energy	$\dfrac{d}{dt}(mu_v) = (W_{FL} - W_{RO} - W_{SC} - W_{WC})\,h_g + W_{RO}h_{fg} - p\dfrac{d}{dt}(m_v v_v)$
	$\qquad = (W_{FL} - W_{SC} - W_{WC})\,h_g - W_{RO}h_f - p\dfrac{d}{dt}(m_v v_v)$
Liquid energy	$\dfrac{d}{dt}(mu_\ell) = (-W_{FL} + W_{RO} + W_{SC} + W_{WC})\,h_f - W_{FL}h_{fg} + W_{SC}h_{fg} + \dot{Q}_W$
	$\qquad - p\dfrac{d}{dt}(m_\ell v_\ell) + (\dot{m}h)_{spray} + (\dot{m}h)_{surge} + \dot{Q}_h - \dot{Q}_W$
	$\qquad = (W_{RO} + W_{WC})\,h_f + (W_{SC} - W_{FL})\,h_g - p\dfrac{d}{dt}(m_\ell v_\ell)$
	$\qquad + (\dot{m}h)_{spray} + (\dot{m}h)_{surge} + \dot{Q}_h$
	$\qquad = W_{RO}h_f + (W_{SC} - W_{FL} + W_{WC})h_g - p\dfrac{d}{dt}(m_\ell v_\ell)$
	$\qquad + (\dot{m}h)_{spray} + (\dot{m}h)_{surge} + \dot{Q}_h - \dot{Q}_W$

3 Constitutive equations for closure. Return now to the closure requirements of Eqs. 7-120a through 7-120e. Because $\dot{Q}_W$ is externally supplied in this model, two additional relations are required among:

$\dot{Q}_{vs'RO}$, which replaces $\dot{Q}_{vs}$ because the other $\dot{Q}_v$ components are zero
$\dot{Q}_{\ell s'L}$, which replaces $\dot{Q}_{\ell s}$ because the other $\dot{Q}_\ell$ components are zero (i.e., $\dot{Q}_{\ell s'RO}$)
  or known (i.e., $\dot{Q}_{\ell s'SC} = W_{SC}h_{fg}$) (Eqs. 7-139 and 7-140) and $\dot{Q}_{\ell s'WC} = \dot{Q}_W$
  (Eq. 7-146)
$\dot{m}_{vs}$ or $\dot{m}_{\ell s}$

This is accomplished following Moeck and Hinds [3] by assuming that neither phase can exist in a metastable form; i.e., the vapor can be either saturated or superheated but not subcooled, whereas the liquid can be either saturated or subcooled but not superheated.

Table 7-7 categorizes the implications of this assumption for the range of possible combinations of pressure changes and initial conditions. Figure 7-24 illustrates the allowed transitions from the saturated liquid, saturated vapor states ℓ_0 and v_0, respectively. A pressure increases from p_0 to p_2 results in subcooled liquid and superheated vapor at states ℓ_2 and v_2, respectively. As Figure 7-24 illustrates, neither rainout nor flashing occurs during this transition. Hence $W_{RO} = 0$ and $W_{FL} = 0$ so that:

$$\dot{m}_{vs} = -\dot{m}_{\ell s} = -W_{SC} - W_{WC}$$

Table 7-7 Initial conditions and their constraints for the nonequilibrium two-region pressurizer model of section IV.B

Initial liquid conditions		Constraints	
Liquid	Vapor	Decreasing pressure	Increasing pressure
Saturated	Saturated	*Case 1* $u_\ell = u_f(p)$ $u_v = u_g(p)$ $\therefore W_{RO} \neq 0$ $\therefore W_{FL} \neq 0$	*Case 5* $W_{RO} = 0$ $W_{FL} = 0$
Subcooled	Saturated	*Case 2* $W_{FL} = 0$ $u_v = u_g(p)$ $\therefore W_{RO} \neq 0$	*Case 6* $W_{RO} = 0$ $W_{FL} = 0$
Saturated	Superheated	*Case 3* $W_{RO} = 0$ $u_\ell = u_f(p)$ $\therefore W_{FL} \neq 0$	*Case 7* $W_{RO} = 0$ $W_{FL} = 0$
Subcooled	Superheated	*Case 4* $W_{RO} = 0$ $W_{FL} = 0$	*Case 8* $W_{RO} = 0$ $W_{FL} = 0$

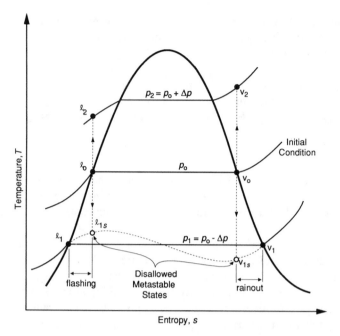

Figure 7-24 Pressurizer pressure transients from saturated initial conditions. *(Adapted from Moeck and Hinds [3].)*

and

$$\dot{Q}_{\ell s'L} = 0$$

by Eq. 7-127, as well as:

$$\dot{Q}_{vs'RO} = 0$$

by Eq. 7-133. Because W_{SC} and W_{WC} are known for a given pressure by Eqs. 7-141 and 7-148, all three parameters $[\dot{m}_{vs}$ (or $\dot{m}_{\ell s}$), $\dot{Q}_{\ell s'L}$, and $\dot{Q}_{vs'RO}]$ are now established. The solution is not overprescribed, however, because the values of these parameters are consistent with each other through the jump conditions.

For a pressure decrease from p_0 to p_1, the metastable superheated liquid and subcooled vapor states ℓ_{1s} and v_{1s}, respectively, which result from an isentropic depressurization for each state, are then assumed transformed to the stable saturated states ℓ_1 and v_1. Between the liquid states ℓ_0 and ℓ_1 some liquid flashes to vapor, whereas between vapor states v_0 and v_1 some vapor condenses or rains out as liquid. In this case, however, the two constraints become:

$$u_\ell = u_f(p)$$
$$u_v = u_g(p)$$

The solution of the equation system yields W_{FL} and W_{RO}, which then through the energy jump conditions dictate $\dot{Q}_{\ell s'L}$ and $\dot{Q}_{vs'RO}$, respectively. Da Silva et al. [1] used this approach for steam generator transient analysis with a procedure in which conditions at the beginning of a time step were used to select the case that would prevail throughout the step. They found that the calculation procedure could be designed so that changes in the selected case did not occur within a time step with great frequency and an inaccurate choice made in one time step was corrected in the next.

REFERENCES

1. Da Silva, Jr., H. C., Lanning, D. D., and Todreas, N. E. Thermit-UTSG: a computer code for thermohydraulic analysis of U-tube steam generators. In: *AIChE Symposium Series, Heat Transfer,* No. 245, Vol. 81, August 1985.
2. Kao, S-P. ScD thesis, Department of Nuclear Engineering, MIT, July 1984.
3. Moeck, E. O., and Hinds, H. W. A mathematical model of steam-drum dynamics. Presented at the 1975 Summer Computer Simulation Conference; San Francisco, July 1975.

PROBLEMS

Problem 7-1 Containment pressure analysis (section II)

For the plant analyzed in Example 7-1, problem A, the peak containment pressure resulting from primary system blowdown is 0.523 MPa. Assume that the primary system failure analyzed in that accident sends an acoustic wave through the primary system that causes a massive failure of steam generator tubes. Although main and auxiliary feedwater to all steam generators are shut off promptly, the entire secondary system inventory of 89 m³ at 6.89 MPa is also now released to containment by blowdown through the primary system. Assume for this case that the secondary system inventory is all at saturated liquid conditions.

(1) What is the new containment pressure? (2) Can it be less than that resulting from only primary system failure?

Answers: 1. 0.6 MPa.

2. Yes, if the secondary fluid properties are such that this fluid acts as an effective heat sink.

Problem 7-2 Ice condenser containment analysis (section II)

Calculate the minimum mass of ice needed to keep the final containment pressure below 0.4 MPa, assuming that the total volume consists of a $5.05(10^4)$ m³ containment volume and a 500 m³ primary volume. Neglect the initial volume of the ice. Additionally, assume the following initial conditions:

Containment pressure = $1.013(10^5)$ Pa (1 atm)
Containment temperature = 300°K
Ice temperature = 263°K (-10°C)
Ice pressure = $1.013(10^5)$ Pa (1 atm)
Primary pressure = 15.5 MPa (saturated liquid conditions)

Relevant properties are:

c_v for air = 719 J/kg°K
R for air = 286 J/kg°K
c for ice = $4.23(10^3)$ J/kg°K
Heat of fusion for water = $3.33(10^5)$ J (at 1 atm)

Answer: $1.77(10^5)$ kg of ice

Problem 7-3 Containment pressure increase due to residual core heat (section II)

Consider the containment system as shown in Figure 7-25 after a loss of coolant accident (LOCA). Assume that the containment is filled with saturated liquid, saturated vapor, and air and nitrogen gas released from rupture of one of the accumulators, all at thermal equilibrium. The containment spray system has begun its recirculating mode, during which the sump water is pumped by the residual heat removal (RHR) pump through the RHR heat exchanger and sprayed into the containment.

Assume that after 1 hour of operation the RHR pump fails and the containment is heated by core decay heat. The decay heat from the core is assumed constant at 1% of the rated power of 2441 MWt.

Find the time when the containment pressure reaches its design limit, 0.827 MPa (121.6 psia). Additional necessary information is given below.

Conditions after 1 hour
Water mixture mass $(m_w) = 1.56(10^6)$ kg
Water mixture quality $(x_{st}) = 0.0249$
Air mass $(m_a) = 5.9(10^4)$ kg
Nitrogen mass $(m_{N_2}) = 1.0(10^3)$ kg
Initial temperature $(T) = 381.6°K$

Thermodynamic properties of gases
$R_{air} = 0.286$ kJ/kg°K
$c_{va} = 0.719$ kJ/kg°K

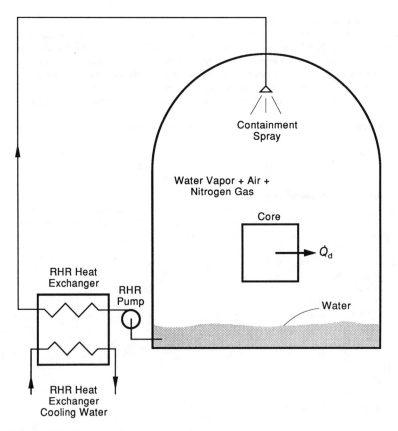

Figure 7-25 Containment.

R_{N_2} = 0.296 kJ/kg°K
c_{vN} = 0.742 kJ/kg°K

Answer: 7.05 hours

Problem 7-4 Loss of heat sink in a sodium-cooled reactor (section II)

A 1000 MW(t) sodium-cooled fast reactor has three identical coolant loops. The reactor vessel is filled with sodium to a prescribed level, with the remainder of the vessel being occupied by an inert cover gas, as shown in Figure 7-26. Under the steady operating condition, the ratio of the volume of cover gas to that of sodium in the primary system is 0.1. At time $t = 0$, the primary system of this reactor suffers a complete loss of heat sink accident, and the reactor power instantly drops to 2% of full power. Using a lumped parameter approach, calculate the pressure of the primary system at time $t = 60$ seconds. At this time, check if the sodium is boiling.

Useful data

Initial average primary system temperature = 980°F
Initial primary system pressure = 50 psi
Total primary system coolant mass = 18,000 lbm
β (volumetric coefficient of thermal expansion of sodium) = $1.6(10^{-4})$/°F
c_p for sodium = 0.3 BTU/lbm °F (at 980°F)
ρ for sodium = 51.4 lbm/ft³ (at 980°F)
Sodium saturated vapor pressure:

$$p = \exp [18.832 - (13113/T) - 1.0948 \ln T + 1.9777 (10^{-4}) T]$$

where p is in atmospheres and T is in °K.

Answer: p = 93.15 psi
T = 1190.7°F, so no boiling occurs

Problem 7-5 Response of a BWR suppression pool to safety/relief valve discharge (section II)

Compute the suppression pool temperature after 5 minutes for a case in which the reactor is scrammed and steam is discharged from the reactor pressure vessel (RPV) into the suppression pool such that the temperature of the RPV coolant is reduced at a specific cooldown rate. During this process, makeup water is supplied to the RPV. Heat input to the RPV is only from long-term decay energy generation.

Numerical parameters applicable to this problem are given in Table 7-8.

Answer: 34.7°C (94.4°F)

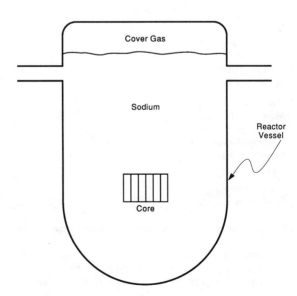

Figure 7-26 Reactor assembly in a sodium-cooled reactor.

Table 7-8 Conditions for suppression pool heat-up analysis

Parameter	Value
Specified cool-down rate	38°C/hr (100°F/hr)
Reactor power level prior to scram	3434 MWt
RPV initial pressure	7 MPa (1015.3 psia)
Saturation properties at 7 MPa	$T = 285.88°C$
	$v_f = 1.3513 (10^{-3})$ m³/kg
	$v_{fg} = 26.0187 (10^{-3})$ m³/kg
	$u_f = 1257.55$ kJ/kg
	$u_{fg} = 1323.0$ kJ/kg
	$s_f = 3.1211$
	$s_{fg} = 2.6922$
Discharge period	5 minutes
Makeup water flow rate	32 kg/s (70.64 lb/s)
Makeup water enthalpy	800 kJ/kg (350 BTU/lb$_m$)
RPV free volume	656.5 m³ [2.3184 (10⁴) ft³]
RPV initial liquid mass	0.303 (10⁶) kg [0.668 (10⁶) lb$_m$]
RPV initial steam mass	9.0264 (10³) kg [19.9 (10³) lb$_m$]
Suppression pool initial temperature	32°C (90°F)
Suppression pool initial pressure	0.1 MPa (15 psia)
Suppression pool water mass	3.44 (10⁶) kg [7.6 (10⁶) lb$_m$]

RHR heat exchanger is actuated at high suppression pool temperature: 43.4°C (110°F).

Problem 7-6 Effect of noncondensable gas on pressurizer response to an insurge (section III)

Compute the pressurizer pressure and heater input resulting from an insurge of liquid from the primary system to a pressurizer containing a mass of air (m_a). Use the following initial, final, and operating conditions.

Initial conditions
Mass of liquid $= m_{f_1}$
Mass of steam $= m_{g_1}$
Mass of air (in steam space) $= m_a$
Total pressure $= p_1$
Equilibrium temperature $= T_1$

Operating conditions
Mass of surge $= m_{surge}$
Mass of spray $= fm_{surge}$
Enthalpy of surge $= h_{surge}$
Enthalpy of spray $= h_{spray}$
Heater input $= Q_h$

Final condition
Equilibrium temperature $= T_2 = T_1$

You may make the following assumptions for the solution:

1. Perfect phase separation
2. Thermal equilibrium throughout the pressurizer
3. Liquid water properties that are independent of pressure

Answers:

$$p_2 = p_w\,(T_{1\ \text{sat}}) + \frac{(m_a RT_1)(v_{g_1} - v_{f_1})}{(v_{g_1})(m_{g_1} v_{g_1} - (1 + f)\,m_{\text{surge}}\,v_{f_1} - m_g v_{f_1})}$$

$$Q_h = \frac{(1 + f)}{v_{fg}}\,m_{\text{surge}}\,(u_f v_g - u_g v_f) - m_{\text{surge}}(h_{\text{surge}} + fh_{\text{spray}})$$

i.e., same as Eq. 7-53.

Problem 7-7 Behavior of a fully contained pressurized pool reactor under decay power conditions (section IV)

A 1600 MW(t) pressurized pool reactor has been proposed in which the entire primary coolant system is submerged in a large pressurized pool of cold water with a high boric acid content. The amount of water in the pool is sufficient to provide for core decay heat removal for at least 1 week following any incident, assuming no cooling systems are operating. In this mode, the pool water boils and is vented to the atmosphere. The vessel geometry is illustrated in Figure 7-27. The core volume can be neglected.

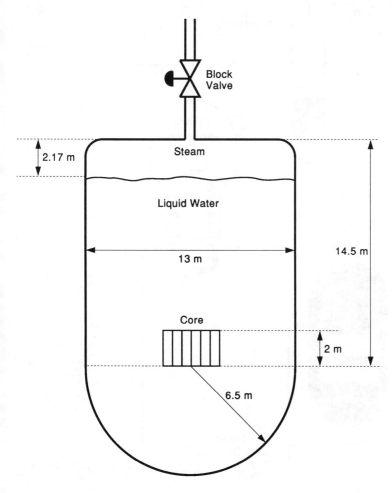

Figure 7-27 Pressurized pool reactor.

Assume that at time $t = 0$ with an initial vessel pressure of 0.10135 MPa (1 atm), the venting to the atmosphere fails. Does water cover the core for all t? Plot the water level measured from the top of the vessel versus time. You may assume that the vessel volume can be subdivided into an upper saturated vapor volume and a lower saturated liquid volume. Also, assume the decay heat rate is constant for the time interval of interest at 25 MW.

Answer: Water always covers the core.

Problem 7-8 Depressurization of a primary system (section IV)

The pressure of the primary system of a PWR is controlled by the pressurizer via the heaters and spray. A simplified drawing of the primary system of a PWR is shown in Figure 7-28. If the spray valve were to fail in the open position, depressurization of the primary system would result.

Calculate the time to depressurize from 15.5 MPa to 12.65 MPa (1835 psia) if the spray rate is 30.60 kg/s with an enthalpy of 1252 kJ/kg (constant with time). Also calculate the final liquid and vapor volumes. You may assume that:

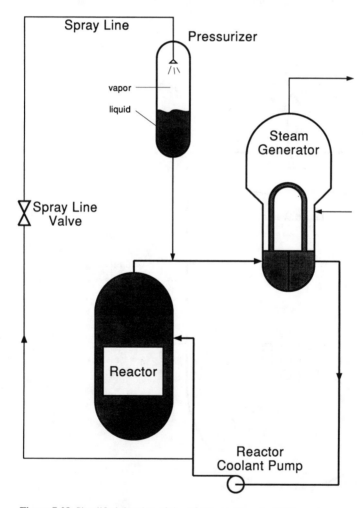

Figure 7-28 Simplified drawing of the primary system of a PWR.

1. Heaters do not operate.
2. Pressurizer wall is adiabatic.
3. Pressurizer vapor and liquid are in thermal equilibrium and occupy initial volumes of 20.39 and 30.58 m^3, respectively.
4. Complete phase separation occurs in the pressurizer.
5. The subcooled liquid in the primary system external to the pressurizer is incompressible so that the spray mass flow rate is exactly balanced by the pressurizer outsurge.

> *Answers:* 230.2 seconds
> 22.16 m^3 vapor
> 28.81 m^3 liquid

Problem 7-9 Three-region pressurizer model (section IV)

For pressurizer insurge cases, an expansion of the two-region model (section IV.B) to a three-region model is of interest. The additional region is a liquid-only region in the lower region of the pressurizer into and out of which liquid surges occur. This additional liquid-only region has the following characteristics.

1. It is labeled "liquid region 2."
2. It communicates with the primary piping and is bounded from above by liquid region 1, which in turn is bounded from above by the vapor region.
3. The pressurizer heaters are immersed in liquid region 2.
4. There is no heat transfer to the wall from liquid region 2.
5. Liquid regions 1 and 2 exchange heat and work only across their common interface.

Maintaining all characteristics of the model of section IV.B, except for those specifically changed by the addition of this liquid-only region:

1. Draw a revision of Figure 7-19.
2. Write the mass and energy equations for all three regions.
3. Write the jump condition(s) for the boundary between liquid regions 1 and 2.
4. Write the volume constraint applicable to the three-region model.

EIGHT

THERMAL ANALYSIS OF FUEL ELEMENTS

I INTRODUCTION

An accurate description of the temperature distributions in the fuel elements and the reactor structures is essential to the prediction of the lifetime behavior of these components. The temperature gradients, which control the thermal stress levels in the materials, together with the mechanical loads contribute to determination of the potential for plastic deformation at high temperatures or cracking at low temperatures. The temperature level at coolant–solid surfaces controls the chemical reactions and diffusion processes, thus profoundly affecting the corrosion process. Furthermore, the impact of the fuel and coolant temperatures on the neutron reaction rates provides an incentive for accurate modeling of the temperature behavior under transient as well as steady-state operating conditions. In this chapter the focus is on the steady-state temperature field in the fuel elements. Many of the principles applied are also useful for describing the temperature field in the structural components.

The temperature in the fuel material depends on the heat-generation rate, the fuel material properties, and the coolant and cladding temperature conditions. The rate of heat generation in a fuel pin depends on the neutron slowing rates near the fuel pin and the neutron reaction rates within the fuel, as described in Chapter 3. In return, the neutron reaction rate depends on the fuel material (both the initial composition and burnup level) and the moderator material (if present) and their temperatures. Hence an exact prediction of the fuel material temperature requires simultaneous determination of the neutronic and temperature fields, although for certain conditions it is possible to decouple the two fields. Thus it is assumed that the heat-generation rate is fixed as we proceed to obtain the fuel temperature field.

Table 8-1 Thermal properties of fuel materials

Property	U	UO$_2$	UC	UN
Theoretical density at room temp (kg/m^3)	19.04 × 10^3	10.97 × 10^3	13.63 × 10^3	14.32 × 10^3
Metal density* (kg/m^3)	19.04 × 10^3	9.67 × 10^3	12.97 × 10^3	13.60 × 10^3
Melting point (°C)	1133	2800	2390	2800
Stability range	Up to 665°C[†]	Up to m.p.	Up to m.p.	Up to m.p.
Thermal conductivity average 200–1000°C (W/m°C)	32	3.6	23 (UC$_{1.1}$)	21
Specific heat, at 100°C (J/kg °C)	116	247	146	—
Linear coefficient of expansion (/°C)		10.1 × 10^{-6} (400–1400°C)	11.1 × 10^{-6} (20–1600°C)	9.4 × 10^{-6} (1000°C)
Crystal structure	Below 655°C: α, orthorhombic Above 770°C: γ, body-centered cubic	Face-centered cubic	Face-centered cubic	Face-centered cubic
Tensile strength, (MPa)	344–1380[‡]	110	62	Not well defined

*Uranium metal density in the compound at its theoretical density.
[†]Addition of a small amount of Mo, Nb, Ti, or Zr extends stability up to the melting point.
[‡]The higher values apply to cold-worked metal.

Uranium dioxide (UO$_2$) has been used exclusively as fuel material in light-water power reactors ever since it was used in the Shippingport PWR in 1955. Uranium metal and its alloys have been used in research reactors. Early liquid-metal-cooled reactors relied on plutonium as a fuel and more recently on a mixture of UO$_2$ and PuO$_2$. The mid-1980s saw a resurgence in the interest in the metal as fuel in U.S.-designed fast reactors. The properties of these materials are highlighted in this chapter. UO$_2$ use in LWRs has been marked with satisfactory chemical and irradiation tolerance. This tolerance has overshadowed the disadvantages of low thermal conductivity and uranium atom density relative to other materials, e.g., the nitrides and carbides or even the metal itself. The carbides and nitrides, if proved not to swell excessively under irradiation, may be used in future reactors.

The general fuel assembly characteristics are given in Chapter 1. Tables 8-1 and 8-2 compare the thermal properties of the various fuel and cladding materials, respectively.

Table 8-2 Thermal properties of cladding materials

Property	Zircaloy 2	Stainless steel 316
Density (kg/m^3)	6.5 × 10^3	7.8 × 10^3
Melting point (°C)	1850	1400
Thermal conductivity (W/m°C)	13 (400°C)	23 (400°C)
Specific heat (J/kg°C)	330 (400°C)	580 (400°C)
Linear thermal expansion coefficient (/°C)	5.9 × 10^{-6}	18 × 10^{-6}

II HEAT CONDUCTION IN FUEL ELEMENTS

A General Equation of Heat Conduction

The energy transport equation (Eq. 4-123) describes the temperature distribution in a solid (which is assumed to be an incompressible material with negligible thermal expansion as far as the effects on temperature distribution are concerned). If Eq. 4-123 is written with explicit dependence on the variables $\vec{r}$ and t, it becomes:

$$\rho c_p(\vec{r},T) \frac{\partial T(\vec{r},t)}{\partial t} = \nabla \cdot k(\vec{r},T)\nabla T(\vec{r},t) + q'''(\vec{r},t) \tag{8-1}$$

Note that for incompressible materials $c_p = c_v$.

At steady state Eq. 8-1 reduces to:

$$\nabla \cdot k(\vec{r},T)\nabla T(\vec{r}) + q'''(\vec{r}) = 0 \tag{8-2}$$

Because by definition the conduction heat flux is given by $\vec{q}'' \equiv -k\nabla T$, Eq. 8-2 can be written as:

$$-\nabla \cdot \vec{q}''(\vec{r},T) + q'''(\vec{r}) = 0 \tag{8-3}$$

B Thermal Conductivity Approximations

In a medium that is isotropic with regard to heat conduction, k is a scalar quantity that depends on the material, temperature, and pressure of the medium. In a nonisotropic medium, thermal behavior is different in different directions. Highly oriented crystalline-like materials can be significantly anisotropic. For example, thermally deposited pyrolytic graphite can have a thermal conductivity ratio as high as 200:1 in directions parallel and normal to basal planes. For anisotropic and nonhomogeneous materials, k is a tensor, which in Cartesian coordinates can be written as:

$$\overline{\overline{k}} = \begin{pmatrix} k_{xx} & k_{xy} & k_{xz} \\ k_{yx} & k_{yy} & k_{yz} \\ k_{zx} & k_{zy} & k_{zz} \end{pmatrix} \tag{8-4}$$

For anisotropic homogeneous solids the tensor is symmetric, i.e., $k_{ij} = k_{ji}$. In most practical cases k can be taken as a scalar quantity. We restrict ourselves to this particular case of a scalar k for the remainder of this text.

As mentioned before, thermal conductivity is different for different media and generally depends on the temperature and pressure. The numerical value of k varies from practically zero for gases under extremely low pressures to about 4000 W/m°K or 7000 BTU/ft hr °F for a natural copper crystal at very low temperatures.

The change of k with pressure depends on the physical state of the medium. Whereas in gases there is a strong pressure effect on k, in solids this effect is negligible. Therefore the conductivity of solids is mainly a function of temperature, $k = k(T)$, and can be determined experimentally. For most metals the empirical formula [1]:

$$k = k_0[1 + \beta_0(T - T_0)] \tag{8-5}$$

gives a good fit to the data in a relatively large temperature range. The values of k_0 and β_0 are constants for the particular metal. It is evident that k_0 corresponds to the reference temperature (T_0). The value of β_0 can be positive or negative. In general, β_0 is negative for pure homogeneous metals, whereas for metallic alloys β_0 becomes positive.

In the case of nuclear fuels, the situation is more complicated because k also becomes a function of the irradiation as a result of change in the chemical and physical composition (porosity changes due to temperature and fission products).

Even when k is assumed to be a scalar, it may be difficult to solve Eq. 8-2 analytically because of its nonlinearity. The simplest way to overcome the difficulties is to transform Eq. 8-2 to a linear one, which can be done by four techniques:

1. In the case of small changes of k within a given temperature range, assume k is constant. In this case Eq. 8-2 becomes:

$$k\nabla^2 T(\vec{r}) + q'''(\vec{r}) = 0 \qquad (8\text{-}6)$$

2. If the change in k over the temperature range is large, define a mean $\bar{k}$ as follows:

$$\bar{k} = \frac{1}{T_2 - T_1} \int_{T_1}^{T_2} k\, dT \qquad (8\text{-}7)$$

and use Eq. 8-6 with $\bar{k}$ replacing k.

3. If an empirical formula for k exists, it may be used to obtain a single variable differential equation, which in many cases can be transformed to a relatively simple linear differential equation. For example, Eq. 8-5 can be used to write:

$$T - T_0 = \frac{k - k_0}{\beta_0 k_0} \qquad (8\text{-}8)$$

so that

$$k\nabla T = \frac{k\nabla k}{\beta_0 k_0} = \frac{\nabla k^2}{2\beta_0 k_0}$$

and

$$\nabla \cdot (k\nabla T) = \frac{\nabla^2 k^2}{2\beta_0 k_0} \qquad (8\text{-}9)$$

Substituting from Eq. 8-9 into Eq. 8-2, we get:

$$\nabla^2 k^2 + 2\,\beta_0 k_0 q''' = 0 \qquad (8\text{-}10)$$

which is a linear differential equation in k^2.

4. Finally, the heat conduction equation can be linearized by Kirchoff's transformation, as briefly described here: In many cases it is useful to know the integral:

$$\int_{T_1}^{T_2} k(T)dT$$

where $T_2 - T_1$ is the temperature range of interest. Kirchoff's method consists of finding such integrals by solving a modified heat conduction equation. Define:

$$\theta \equiv \frac{1}{k_0} \int_{T_0}^{T} k(T)dT \qquad (8\text{-}11)$$

The new variable θ can be used to give:

$$\nabla\theta = \frac{1}{k_0} \nabla \int_{T_0}^{T} k(T)dT = \frac{1}{k_0} \left[\nabla T \frac{d}{dT} \int_{T_0}^{T} k(T)dT \right] = \frac{k(T)}{k_0} \nabla T \qquad (8\text{-}12)$$

From this equation we find:

$$k \nabla T = k_0 \nabla\theta \qquad (8\text{-}13)$$

and

$$\nabla \cdot [k\nabla T] = k_0 \nabla^2\theta \qquad (8\text{-}14)$$

Then at steady state Eq. 8-2 becomes:

$$k_0 \nabla^2\theta + q''' = 0 \qquad (8\text{-}15)$$

which is a linear differential equation that can generally be solved more easily than Eq. 8-2.

In practice, the nuclear industrial computer programs have allowed common use of temperature-dependent conductivity in the numerical solutions. Eq. 8-2 can be readily solved in one-dimensional geometries, as illustrated later in this chapter. Therefore the above approaches are useful only if one is interested in analytic solutions of multidimensional problems.

III THERMAL PROPERTIES OF UO$_2$

The LWR fuel is composed of UO$_2$; hence the focus of this section is on the properties of this most widely used fuel material. In particular, the thermal conductivity, melting point, specific heat, and fractional gas release are discussed.

A Thermal Conductivity

Many factors affect the UO$_2$ thermal conductivity. The major factors are temperature, porosity, oxygen to metal atom ratio, PuO$_2$ content, pellet cracking, and burnup. A brief description of the change in k with each of these factors is discussed.

1 Temperature effects. It has been experimentally observed that $k(T)$ decreases with increasing temperature until $T = 1750°C$ and then starts to increase (Fig. 8-1). The often-used integral of $k(T)$ is also given in Figure 8-1. The polynomials representing

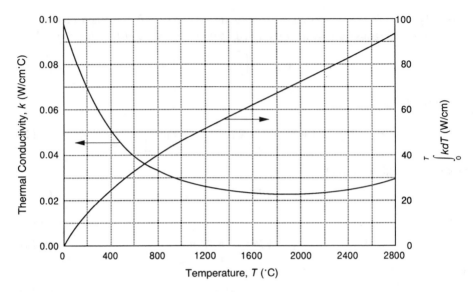

Figure 8-1 Thermal conductivity of UO_2 at 95% density from Lyon. *(From Hann et al. [9].)*

$k(T)$ proposed by the reactor vendors are given in Table 8-3. All these formulas give the value of the integral:

$$\int_{0°C}^{melting} kdT$$

as 93.5 W/cm.

For an ionic solid, thermal conductivity can be derived by assuming that the solid is an ideal gas whose particles are the quantized elastic wave vibrations in a crystal (referred to as *phonons*). It can be shown that the behavior of the UO_2 conductivity with temperature can be predicted by such a model [20].

2 Porosity (density) effects. The oxide fuel is generally fabricated by sintering pressed powdered UO_2 or mixed oxide at high temperature. By controlling the sintering conditions, material of any desired density, usually around 90% of the maximum possible or theoretical density of the solid, can be produced.

Generally, the conductivity of a solid decreases with increasing presence of voids (pores) within its structure. Hence low porosity is desirable to maximize the conductivity. However, fission gases produced during operation within the fuel result in internal pressures that may swell, and hence deform, the fuel. Thus a certain degree of porosity is desirable to accommodate the fission gases and limit the swelling potential. It is particularly true for fast reactors where the specific power level is higher, and hence the rate at which gases are produced per unit fuel volume is higher, than in thermal reactors.

Table 8-3 Formulations for UO$_2$ thermal conductivity at 95% theoretical density (used by reactor vendors) [25]

Temperature-dependent thermal conductivity

1. Derived from Lyon's $\int kdT$ and shown on Figure 8-1 (used by Combustion Engineering)

$$k = \frac{38.24}{402.4 + T} + 6.1256 \times 10^{-13} (T + 273)^3 \qquad \text{(Eq. 8-16a)}$$

where:
  k = UO$_2$ conductivity (W/cm °C)
  T = local UO$_2$ temperature (°C)

2. Composite from a variety of sources (used by Westinghouse)

$$k = \frac{1}{11.8 + 0.0238T} + 8.775 \times 10^{-13} T^3 \qquad \text{(Eq. 8-16b)}$$

where:
  k = W/cm °C
  T = °C

Polynomial representation of Lyon's $\int kdT$ (used by Babcock and Wilcox)

$$\int_{32}^{T} kdT = -170.9124 + 5.597256T - 3.368695 \times 10^{-3}T^2$$
$$+ 1.962784 \times 10^{-6}T^3 - 8.391225 \times 10^{-10}T^4 \qquad \text{(Eq. 8-16c)}$$
$$+ 2.404192 \times 10^{-13}T^5 - 4.275284 \times 10^{-17}T^6$$
$$+ 4.249043 \times 10^{-21}T^7 - 1797017 \times 10^{-25}T^8$$

where:
  T = °F
  $\int kdT$ = Btu/hr ft

Let us define the porosity (P) as:

$$P = \frac{\text{volumes of pores } (V_p)}{\text{total volumes of pores } (V_p) \text{ and solids } (V_s)} = \frac{V_p}{V} = \frac{V - V_s}{V}$$

or

$$P = 1 - \frac{\rho}{\rho_{TD}} \qquad (8\text{-}17)$$

where ρ_{TD} is the theoretical density of the poreless solid. The effect of porosity on $\int kdT$ for mixed oxides is shown in Figure 8-2.

By considering the linear porosity to be $P^{1/3}$ and the cross-sectional porosity to be $P^{2/3}$, Kampf and Karsten [13] derived an equation for negligible pore conductance:

$$k = (1 - P^{2/3})k_{TD} \qquad (8\text{-}18)$$

Earlier, the analysis of Loeb [15] was used by Francl and Kingery [7] to derive the equation referred to as the Loeb equation for this condition of negligible pore conductance:

$$k = (1 - P)k_{TD} \qquad (8\text{-}19)$$

Equation 8-19 was found to underestimate the porosity effect.

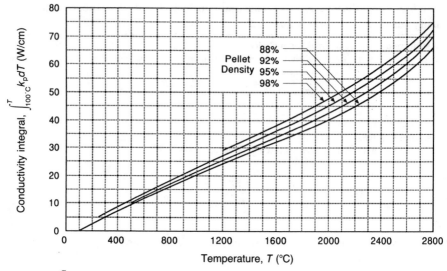

Figure 8-2 $\int_{100°C}^{T} k_p dT$ versus temperature for mixed oxide fuel. Note: 32.8 W/cm = 1 kW/ft.

A modified Loeb equation is often used to fit the UO_2 conductivity measurements as:

$$k = (1 - \alpha_1 P)k_{TD} \tag{8-20}$$

where α_1 is between 2 and 5 [20].

Biancharia [2] derived the following formula for the porosity effect, which accounts for the shape of the pores:

$$k = \frac{(1 - P)}{1 + (\alpha_2 - 1)P} k_{TD} \tag{8-21}$$

where $\alpha_2 = 1.5$ for spherical pores. For axisymmetric shapes (e.g., ellipsoids), α_2 is larger. This formula is often used in LMFBR applications.

The following fit for the temperature and porosity effects on conductivity is used in the MATPRO fuel analysis package for unirradiated fuels [22]:

For $0 \le T \le 1650°C$:

$$k = \eta \left[\frac{B_1}{B_2 + T} + B_3 \exp (B_4 T) \right] \tag{8-22a}$$

For $1650 \le T \le 2940°C$:

$$k = \eta[B_5 + B_3 \exp (B_4 T)] \tag{8-22b}$$

where k is in W/cm °C, T is in °C, and η = porosity correction factor given by:

$$\eta = \frac{1 - \beta(1 - \rho/\rho_{TD})}{1 - \beta(1 - 0.95)} \tag{8-22c}$$

Table 8-4 Values of the constants in the MATPRO correlation for thermal conductivity*

Fuel	B_1 (W/cm)	B_2 (°C)	B_3 (W/cm °C)	B_4 (°C^{-1})	B_5 (W/cm °C)
UO_2	40.4	464	1.216×10^{-4}	1.867×10^{-3}	0.0191
$(U,Pu)O_2$	33.0	375	1.540×10^{-4}	1.710×10^{-3}	0.0171

*Equations 8-22a through 8-22d.

where:

$$\beta = 2.58 - 0.58 \times 10^{-3} T \tag{8-22d}$$

and the constants B_1 through B_5 are given in Table 8-4.

3 Oxygen-to-metal atomic ratio. The oxygen-to-metal ratio of the uranium and plutonium oxides can vary from the theoretical (or stoichiometric) value of 2. This variation affects almost all the physical properties of the fuel. The departure from the initial stoichiometric condition occurs during burnup of fuel. In general, the effect of both the hyper- and hypostoichiometry is to reduce the thermal conductivity, as shown in Figure 8-3.

4 Plutonium content. Thermal conductivity of the mixed oxide fuel decreases as the plutonium oxide content increases, as can be seen in Figure 8-4.

5. Effects of pellet cracking. Fuel pellet cracking and fragment relocation into the pellet–cladding gap during operation alters the fuel thermal conductivity and the gap conductance. A series of tests at the Idaho National Engineering Laboratory have led to an empirical formula for the decrease in the UO_2 thermal conductivity due to cracking. For a fresh, helium-filled LWR fuel rod with cracked and broken fuel pellets, this relation is [16]:

$$k_e = k_{UO_2} - (0.0002189 - 0.050867 X + 5.6578 X^2) \tag{8-23a}$$

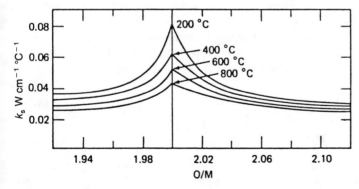

Figure 8-3 Thermal conductivity of $UO_{0.8}Pu_{0.2}O_{2\pm x}$ as a function of the O/(U + Pu) ratio. *(From Schmidt and Richter [24].)*

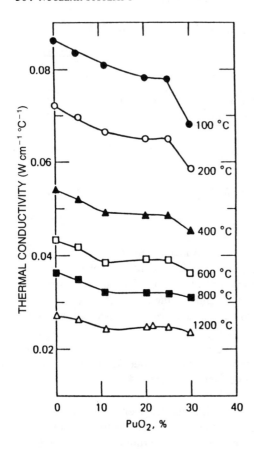

Figure 8-4 Thermal conductivity of $(U,Pu)O_2$ solid solutions as a function of PuO_2 content. *(From Gibby [8].)*

where

$$X = (\delta_{hot} - 0.014 - 0.14\, \delta_{cold}) \left(\frac{0.0545}{\delta_{cold}}\right) \left(\frac{\rho}{\rho_{TD}}\right)^{8} \qquad (8\text{-}23b)$$

where k is in kW/m°K; δ_{hot} = calculated hot gap width (mm) for the uncracked fuel; δ_{cold} = cold gap width (mm); ρ_{TD} = theoretical density of UO_2.

The effect of cracking on fuel conductivity is illustrated in Figure 8-5.

A semiempirical approach by MacDonald and Weisman [17] yielded the following relation between the effective conductivity and the theoretical one:

$$k_e = \frac{k_{UO_2}}{\left[\dfrac{\delta_{hot} - A}{D_{fo}(B\, k_{gas}/k_{UO_2} + C)}\right] + 1} \quad (\text{W/m °K}) \qquad (8\text{-}24)$$

where A, B, and C = constants; D_{fo} = hot pellet diameter in meters; k_{gas} = thermal conductivity of the gas in the gap; δ_{hot} = hot gap in meters. The constants recommended are as follows:

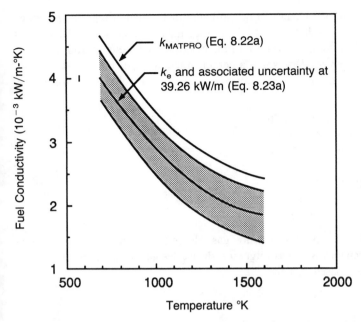

Figure 8-5 Representative comparison between MATPRO fuel thermal conductivity and calculated effective fuel thermal conductivity, with estimated uncertainty, for 2.2% initial gap helium-filled rods at a power of 39.26 kW/m. *(From MacDonald and Smith [16].)*

$A = 6.35 \times 10^{-5}$ m
$B = 0.077$
$C = 0.015$

6 Burnup The irradiation of fuel induces several changes in the porosity, composition, and stoichiometry of the fuel. These changes, however, are generally small in LWRs, where the burnup is only on the order of 3% of the initial uranium atoms. In fast reactors, this effect would be larger, as the expected burnup is on the order of 10% of the initial uranium and plutonium atoms.

Introduction of fission products into the fuel with burnup leads to a slight decrease in the conductivity. Fuel material cracking under thermal cycling also reduces the fuel effective conductivity.

Finally, the oxide material operating at temperatures higher than a certain temperature, about 1400°C, undergoes a sintering process that leads to an increase in the fuel density. This increase in density, which occurs in the central region of the fuel, affects the conductivity and the fuel temperature distribution, which is discussed in detail later.

B Fission gas release.

It is important for the design of the fuel pin to calculate the gas released to the fuel pin plenum. Some of the fission gases are released from the UO_2 pellet at low temperature. Accompanying the change in the structure of the fuel at high temperatures is a significant release of fission gases to the fuel boundaries. The accumulated gases within the cladding lead to pressurization of the cladding. With engineering analysis, a simple scheme is often used in which a certain fraction (f) of the gas is assumed to be released depending on the fuel temperature. An empirically based formula is given as [19]:

$$
\begin{aligned}
f &= 0.05 & T &< 1400°C \\
f &= 0.10 & 1500 &> T > 1400°C \\
f &= 0.20 & 1600 &> T > 1500°C \\
f &= 0.40 & 1700 &> T > 1600°C \\
f &= 0.60 & 1800 &> T > 1700°C \\
f &= 0.80 & 2000 &> T > 1800°C \\
f &= 0.98 & T &> 2000°C
\end{aligned}
\tag{8-25}
$$

More complicated approaches to fission gas release have been proposed based on various physical mechanisms of gas migration. Most designers, however, still prefer the simple empirically based models.

C Melting Point

The melting point of UO_2 is in the vicinity of 2840°C (5144°F). The melting process for the oxide starts at a solidus temperature but is completed at a higher temperature called the *liquidus point*. The melting range is affected by the oxygen-to-metal ratio

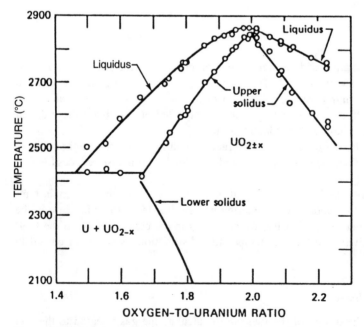

Figure 8-6 Partial phase diagram for uranium from $UO_{1.5}$ to $UO_{2.23}$ should coincide for $UO_{2.0}$. *(From Latta and Fryxell [14].)*

(Fig. 8-6) and by the Pu content. Thus in LWR designs the conservatively low value of 2600°C (4700°F) is often used. Olsen and Miller [21] fitted the melting point for MATPRO as:

$$T(\text{solidus}) = 3113 - 5.414\xi + 7.468 \times 10^{-3}\, \xi^2 \; ^\circ\text{K} \qquad (8\text{-}26)$$

where ξ = mole percent of PuO_2 in the oxide. The effect of increased PuO_2 content is shown in Figure 8-7.

D Specific Heat

The specific heat of the fuel plays a significant role in determining the sequence of events in many transients. It varies greatly over the temperature range of the fuel (Fig. 8-8).

Example 8-1 Effect of cracking on fuel conductivity

PROBLEM Evaluate the effective conductivity of the fuel after cracking at 1000°C for the geometry and operating conditions given below.

1. *Geometry and materials:* BWR fuel rod with UO_2 solid fuel pellet and zircaloy clad. Cold fuel rod dimensions (at 27°C) are:
   a. Clad outside diameter = 12.52 mm
   b. Clad thickness = 0.86 mm
   c. Diametral gap width = 230 μm
2. *Assumptions*
   a. Initial fuel density = 0.88 ρ_{TD}

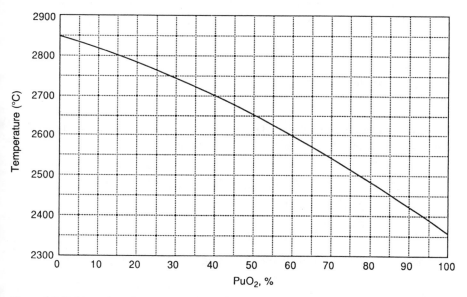

Figure 8-7 Melting points of mixed uranium–plutonium oxides. *(Adapted from Zebroski et al. [27].)*

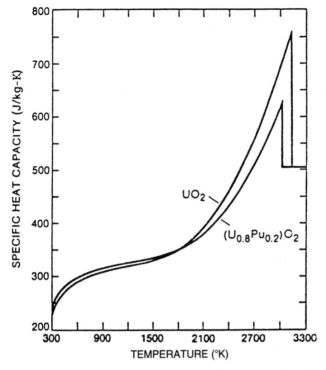

Figure 8-8 Temperature dependence of the specific heat capacity of UO_2 and $(U,Pu)O_2$ *(From Olsen and Miller [21].)*

  b. UO_2 conductivity is predicted by the Westinghouse correlation in Table 8-3.
  c. Porosity correction factor for the conductivity is given by Eq. 8-21 assuming spherical pores.
  d. Fuel conductivity of the cracked fuel is given by Eq. 8-23a.
3. *Operating condition*
  a. Fuel temperature $= 1000°C$
  b. Clad temperature $= 295°C$

SOLUTION Consider first the conductivity of the uncracked fuel pellet. From Eq. 8-16b (Table 8-3), the UO_2 conductivity at $1000°C$ is:

$$
\begin{aligned}
k_{0.95} &= \frac{1}{11.8 + 0.0238(1000)} + 8.775 \times 10^{-13}\,(1000)^3 \\
&= 0.0281 + 0.00088 \\
&= 0.02898 \text{ W/cm°C} \\
&= 0.002898 \text{ kW/m°C}
\end{aligned}
$$

Applying the porosity correction factor of Eq. 8-21:

$$
\frac{k}{k_{TD}} = \frac{1 - P}{1 + 0.5P} = \frac{\rho/\rho_{TD}}{1 + 0.5(1 - \rho/\rho_{TD})}
$$

Because $\rho/\rho_{TD} = 0.88$ and 0.95 for the 88% and 95% theoretical density fuels, respectively, we get:

$$\frac{k_{0.88}}{k_{0.95}} = \frac{0.88}{1 + 0.5(0.12)} \frac{1 + 0.5(0.05)}{0.95} = (0.83)(1.079) = 0.896$$

Therefore the conductivity of the 88% theoretical density uncracked fuel pellet is:

$$k_{0.88} = 0.896 \, (0.002898) = 2.597 \times 10^{-3} \, \text{kW/m°C}$$

Consider the cracked fuel effective conductivity. From Eqs. 8-23a and 8-23b:

$$k_{eff} = k_{UO_2} - (0.0002189 - 0.050867X + 5.6578X^2) \, \text{kW/m°C} \quad \text{(8-23a)}$$

$$X = (\delta_{hot} - 0.014 - 0.14\delta_{cold}) \left(\frac{0.0545}{\delta_{cold}}\right) \left(\frac{\rho}{\rho_{TD}}\right)^8 \quad \text{(8-23b)}$$

For the cold gap, $\delta_{cold} = 0.23$ mm as given. To evaluate δ_{hot}, we must evaluate the change in the radius of the fuel and the cladding. If the fuel pellet radius is R_{fo} and the cladding inner radius is R_{ci}:

$$\begin{aligned}
\delta_{hot} &= (R_{ci})_{hot} - (R_{fo})_{hot} \\
&= (R_{ci})_{cold}[1 + \alpha_c(T_c - 27)] - (R_{fo})_{cold}[1 + \alpha_f(T_f - 27)]
\end{aligned}$$

where $\alpha =$ linear thermal expansion coefficient.
However,

$$(R_{ci})_{cold} = \frac{12.52}{2} - 0.86 = 5.40 \, \text{mm}$$

and

$$(R_{fo})_{cold} = \frac{12.52}{2} - 0.86 - 0.23 = 5.17 \, \text{mm}$$

From Table 8-1, $\alpha_f = 10.1 \times 10^{-6}$ per°C
From Table 8-2, $\alpha_c = 5.9 \times 10^{-6}$ per°C
Therefore $\delta_{hot} = 5.40[1 + 5.9 \times 10^{-6}(295 - 27)] - 5.17[1 + 10.1 \times 10^{-6}(1000 - 27)] = 5.40854 - 5.22081 = 0.18773$ mm

Now we can determine the parameter X from Eq. 8-23b:

$$X = [0.18773 - 0.014 - 0.14(0.23)] \left[\frac{0.0545}{0.23}\right] [0.88]^8$$

$$= 0.14153(0.23696)(0.35963)$$

$$= 0.01206$$

Hence the cracked fuel conductivity for the 88% theoretical density fuel is given by Eq. 8-23a as:

$$\begin{aligned}
k_{eff} &= 0.896(0.002898) - [0.0002189 - 0.050867(0.01206) + 5.6578(0.01206)^2] \\
&= 0.0026 - (0.0002189 - 0.000613 + 0.000823) \\
&= 0.0026 - (0.000439) \\
&= 0.002161 \, \text{kW/m °C}
\end{aligned}$$

Thus the effect of cracking is to reduce the fuel effective thermal conductivity in this fuel from 2.60 to 2.16 W/m °C.

IV TEMPERATURE DISTRIBUTION IN PLATE FUEL ELEMENTS

A General Conduction Equation in Cartesian Coordinates

Assume a fuel plate is operating with a uniform heat-generation rate (q'''). The fuel is clad in thin metallic sheets, with perfect contact between the fuel and the cladding as shown in Figure 8-9.

If the fuel plate is thin and extends in the y and z directions considerably more than it does in the x direction, the heat conduction equation (Eq. 8-2):

$$\frac{\partial}{\partial x} k \frac{\partial T}{\partial x} + \frac{\partial}{\partial y} k \frac{\partial T}{\partial y} + \frac{\partial}{\partial z} k \frac{\partial T}{\partial z} + q''' = 0 \qquad (8\text{-}27)$$

can be simplified by assuming the heat conduction in the y and z directions to be negligible, i.e.,

$$k \frac{\partial T}{\partial y} \simeq k \frac{\partial T}{\partial z} \simeq 0 \qquad (8\text{-}28)$$

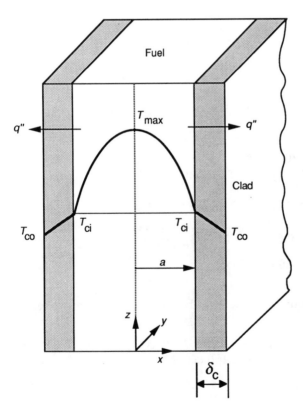

Figure 8-9 Plate fuel element.

Hence, we need only to solve the one-dimensional equation:

$$\frac{d}{dx} k \frac{dT}{dx} + q''' = 0 \qquad (8\text{-}29)$$

By integrating once, we get:

$$k \frac{dT}{dx} + q''' x = C_1 \qquad (8\text{-}30)$$

B Application to a Fuel Plate

Because q''' is uniform, and if the temperatures at both interfaces between cladding and fuel are equal, the fuel temperature should be symmetric around the center plane. In the absence of any heat source or sink at $x = 0$, no heat flux should cross the plane at $x = 0$. Hence:

$$k \frac{dT}{dx}\bigg|_{x=0} = 0 \qquad (8\text{-}31)$$

Applying the condition of Eq. 8-31 to Eq. 8-30 leads to:

$$C_1 = 0 \qquad (8\text{-}32)$$

and

$$k \frac{dT}{dx} + q'''x = 0 \qquad (8\text{-}33)$$

Integrating Eq. 8-33 between $x = 0$ and any position x, and applying the condition that $T = T_{max}$ at $x = 0$, we get:

$$\int_{T_{max}}^{T} kdT + q''' \frac{x^2}{2}\bigg|_0^x = 0$$

or

$$\int_{T}^{T_{max}} k\, dT = q''' \frac{x^2}{2} \qquad (8\text{-}34)$$

Three conditions may exist, depending on whether q''', T_{ci}, or T_{max} are known.

1. q''' specified: A relation between T_{max} and T_{ci} can be determined when q''' is specified by substituting $x = a$ in Eq. 8-34 to get:

$$\int_{T_{ci}}^{T_{max}} kdt = q''' \frac{a^2}{2} \qquad (8\text{-}35)$$

If $k = $ constant,

$$k(T_{max} - T_{ci}) = q''' \frac{a^2}{2} \qquad (8\text{-}36)$$

2. T_{ci} specified: If T_{ci} is known, Eq. 8-34 can be used to specify the relation between q''' and T_{max} in the form:

$$q''' = \frac{2}{a^2} \int_{T_{ci}}^{T_{max}} k dT \tag{8-37}$$

3. T_{max} specified: Equation 8-37 can be used to specify the relation between q''' and T_{ci} given the value of T_{max}.

C Heat Conduction in Cladding

The heat conduction in the cladding can also be assumed to be a one-dimensional problem, so that Eq. 8-29 also applies in the cladding. Furthermore, the heat generation in the cladding is negligible (mainly owing to absorption of γ rays and inelastic scattering of neutrons). Hence in the cladding the heat conduction equation is given by:

$$\frac{d}{dx} k_c \frac{dT}{dx} = 0 \tag{8-38}$$

Integrated once, it leads to the equation:

$$k_c \frac{dT}{dx} = B_1 = \text{constant} \tag{8-39}$$

which implies that the heat flux is the same at any position in the cladding. Let q'' be the heat flux in the cladding in the outward x direction. Therefore:

$$- k_c \frac{dT}{dx} = q'' \tag{8-40}$$

Integrating Eq. 8-40 between $x = a$ and any position x leads to:

$$\int_{T_{ci}}^{T} k_c dT = - q''(x - a) \tag{8-41}$$

or

$$\bar{k}_c(T - T_{ci}) = - q''(x - a) \tag{8-42}$$

where $\bar{k}_c$ = average clad conductivity in the temperature range.

Thus the external clad surface temperature is given by:

$$\bar{k}_c(T_{co} - T_{ci}) = -q''(a + \delta_c - a) \tag{8-43}$$

or

$$T_{co} = T_{ci} - \frac{q''\delta_c}{\bar{k}_c} \tag{8-44}$$

where δ_c = cladding thickness.

D Thermal Resistances

Note that q'' is equal to the heat generated in one-half of the fuel plate. Thus:

$$q'' = q''' a \tag{8-45}$$

Therefore the fuel temperature drop may also be obtained by substituting for $q'''a$ from Eq. 8-45 into Eq. 8-36 and rearranging the result:

$$T_{ci} = T_{max} - q'' \frac{a}{2k} \tag{8-46}$$

Substituting for T_{ci} from Eq. 8-46 into Eq. 8-44, we get:

$$T_{co} = T_{max} - q'' \left(\frac{a}{2k} + \frac{\delta_c}{k_c} \right) \tag{8-47}$$

By simple manipulation of Eq. 8-47, the heat flux q'' can be given as:

$$q'' = \frac{T_{max} - T_{co}}{\dfrac{a}{2k} + \dfrac{\delta_c}{k_c}} \tag{8-48}$$

Thus the temperature difference acts analogously to an electrical potential difference that gives rise to a current (q'') whose value is dependent on two thermal resistances in series (Fig. 8-10). This concept of resistances proves useful for simple transient fuel temperature calculations.

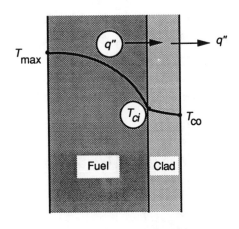

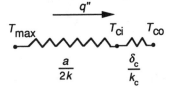

Figure 8-10 Electrical current equivalence with the heat flux.

E Conditions for Symmetric Temperature Distributions

The temperature field symmetry in the preceding discussion enabled us to solve for the temperature by considering only one-half the plate. It is useful to reflect on the required conditions to produce such symmetry.

Consider the general case for a plate fuel element with internal heat generation that is cooled on both sides (Fig. 8-11). In this case some of the heat is removed from the right-hand side, and the rest is removed from the left-hand side. Therefore a plane exists within the fuel plate through which no heat flux passes. Let the position of this plane be x_0. Thus:

$$q''|_{x_0} = k \left.\frac{\partial T}{\partial x}\right|_{x_0} = 0 \qquad (8\text{-}49)$$

The value of x_0 is zero if symmetry of the temperature distribution exists. This symmetry can be a priori known under specific conditions, all of which should be present simultaneously. These conditions are the following:

1. Symmetric distribution of heat generation in the fuel plate
2. Equal resistances to heat transfer on both sides of the plate, which translates into similar material and geometric configurations on both sides (i.e., uniform fuel, fuel–clad gap, and clad material thicknesses)
3. Equal temperatures of the outermost boundary of the plate on both sides, i.e., $T_{co|A} = T_{co|B}$

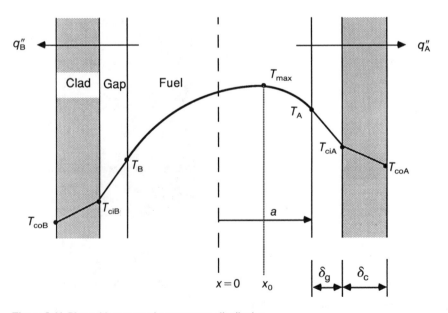

Figure 8-11 Plate with asymmetric temperature distribution.

If the three conditions exist in a fuel element, symmetry of the temperature field can be ascertained and the condition of zero heat flux at the midplane can be supplied, i.e., $x_0 = 0$ in Eq. 8-49:

$$k \frac{\partial T}{\partial x}\bigg|_0 = 0 \qquad (8\text{-}50)$$

Two examples can be cited as violating the conditions of symmetry mentioned above.

1. Nonuniform heat generation due to absorption of an incident flux of γ rays, as occurs in a core barrel or thermal shield (see Chapter 3, Section VII)
2. Nonidentical geometry of the region surrounding the fuel as happens in an off-center fuel element

In a general coordinate system the condition of symmetry is as follows.

$$\begin{aligned} \nabla T = 0 \ &\text{at } x = 0 \ (\text{plate}) \\ &\text{at } r = 0 \ (\text{cylinder}) \\ &\text{at } r = 0 \ (\text{sphere}) \end{aligned} \qquad (8\text{-}51)$$

For condition 8-51 to be valid, it is implied that the fuel elements are solid. If there is a central void, condition 8-51 is not useful for solution of the temperature profile within the fuel region. When an inner void exists, and if a symmetric condition of temperature field applies, no heat flux would exist at the void boundary. This situation is illustrated for cylindrical pins in the following section.

V TEMPERATURE DISTRIBUTION IN CYLINDRICAL FUEL PINS

We derive first the basic relations for cylindrical (solid and annular) fuel pellets. Then some conclusions are made with regard to: (1) the maximum fuel temperature (T_{max}) for a given linear heat rate (q'); and (2) the maximum possible heat rate (q'_{max}) for a given T_{max}.

Cylindrical fuel pellets are nearly universally used as the fuel form in power reactors. Dimensions of the fuel pellet and cladding are given in Table 1-3.

A General Conduction Equation for Cylindrical Geometry

If the neutron flux is assumed to be uniform within the fuel pellet, the heat-generation rate can be assumed uniform. The coolant turbulent flow conditions in a fuel assembly of a pin pitch-to-diameter ratio of more than 1.2 is such that the azimuthal flow conditions can be taken to be essentially the same around the fuel rod. (More information on the azimuthal heat flux distribution is available in Chapter 7, Vol. II.) The above two conditions lead to the conclusion that no significant azimuthal temperature gradients exist in the fuel pellet. Also, for a fuel pin of a length-to-diameter ratio of more than 10, it is safe to neglect the axial heat transfer within the fuel relative to the radial heat transfer for most of the pin length. However, near the top and bottom ends, axial heat conduction plays a role in determining the temperature field.

Thus at steady state the heat conduction equation reduces to a one-dimensional equation in the radial direction:

$$\frac{1}{r}\frac{d}{dr}\left(kr\frac{dT}{dr}\right) + q''' = 0 \tag{8-52}$$

Integrating Eq. 8-52 once, we get:

$$kr\frac{dT}{dr} + q'''\frac{r^2}{2} + C_1 = 0 \tag{8-53}$$

which can be written as:

$$k\frac{dT}{dr} + q'''\frac{r}{2} + \frac{C_1}{r} = 0 \tag{8-54}$$

For an annular fuel pellet with an internal cavity radius (R_v) (Fig. 8-12), no heat flux exists at R_v. For a solid pellet, $R_v = 0$, no heat flux exists at $r = 0$. Hence the general heat flux condition that can be applied is:

$$q''|_{r=R_v} = -k\frac{dT}{dr}\bigg|_{r=R_v} = 0 \tag{8-55}$$

Applying this condition to Eq. 8-54 leads to:

$$C_1 = -\frac{q'''R_v^2}{2} \tag{8-56}$$

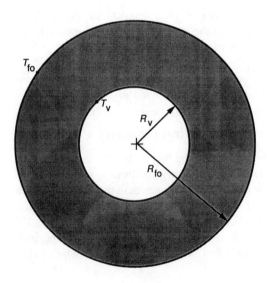

Figure 8-12 Cross section of an annular pellet.

Equation 8-54 can be integrated between r and R_v and r to yield, after rearrangement:

$$- \int_{T_{max}}^{T} kdT = \frac{q'''}{4} [r^2 - R_v^2] + C_1 \ell n \left(\frac{r}{R_v} \right) \tag{8-57}$$

B Solid Fuel Pellet

For a solid fuel pellet, applying Eq. 8-56:

$$R_v = 0 \quad \text{and} \quad C_1 = 0 \tag{8-58}$$

Hence:

$$\int_{T}^{T_{max}} kdT = \frac{q''' r^2}{4} \tag{8-59}$$

A relation between T_{max}, T_{fo}, and R_{fo} can be obtained by Eq. 8-59 at $r = R_{fo}$ to get:

$$\int_{T_{fo}}^{T_{max}} kdT = \frac{q''' R_{fo}^2}{4} \tag{8-60}$$

The linear heat rate is given by:

$$q' = \pi R_{fo}^2 q''' \tag{8-61}$$

Therefore:

$$\int_{T_{fo}}^{T_{max}} kdT = \frac{q'}{4\pi} \tag{8-62}$$

It is interesting to note that the temperature difference across a solid fuel pellet is fixed by q' and is independent of the pellet radius (R_{fo}). Thus a limit on q' is directly implied by a design requirement on the maximum fuel temperature.

It should also be mentioned, and the student can verify on his or her own, that for a constant conductivity the average temperature in a fuel pin is given by:

$$T_{ave} - T_{fo} = \frac{1}{2}(T_{max} - T_{fo}) = \frac{q'}{8\pi k}$$

C Annular Fuel Pellet

For an annular pellet, C_1 from Eq. 8-56 can be substituted for in Eq. 8-57 to get:

$$- \int_{T_{max}}^{T} kdT = \frac{q'''}{4} [r^2 - R_v^2] - \frac{q''' R_v^2}{2} \ell n \, (r/R_v)$$

The above equation can be rearranged into:

$$\int_{T}^{T_{max}} kdT = \frac{q''' r^2}{4} \left\{ \left[1 - \left(\frac{R_v}{r} \right)^2 \right] - \left(\frac{R_v}{r} \right)^2 \ell n \left(\frac{r}{R_v} \right)^2 \right\} \tag{8-63}$$

Equation 8-63 can be used to provide a relation between T_{max}, T_{fo}, R_v, and R_{fo} when the condition $T = T_{fo}$ at $r = R_{fo}$ is applied. Thus we get:

$$\int_{T_{fo}}^{T_{max}} kdT = \frac{q''' R_{fo}^2}{4} \left\{ \left[1 - \left(\frac{R_v}{R_{fo}} \right)^2 \right] - \left(\frac{R_v}{R_{fo}} \right)^2 \ell n \left(\frac{R_{fo}}{R_v} \right) \right\}$$ (8-64)

Note that the linear heat rate is given by:

$$q' = \pi (R_{fo}^2 - R_v^2) q'''$$ (8-65)

so that:

$$q''' R_{fo}^2 = \frac{q'}{\pi \left[1 - \left(\frac{R_v}{R_{fo}} \right)^2 \right]}$$ (8-66)

Substituting for $q''' R_{fo}^2$ from Eq. 8-66 into Eq. 8-64, we get:

$$\int_{T_{fo}}^{T_{max}} kdT = \frac{q'}{4\pi} \left[1 - \frac{\ell n (R_{fo}/R_v)^2}{\left(\frac{R_{fo}}{R_v} \right)^2 - 1} \right]$$ (8-67)

If a void factor is defined as:

$$F_v(\alpha, \beta) \equiv 1 - \frac{\ell n(\alpha^2)}{\beta^2(\alpha^2 - 1)}$$ (8-68)

Equation 8-67 can be written as:

$$\int_{T_{fo}}^{T_{max}} kdT = \frac{q'}{4\pi} \left[F_v \left(\frac{R_{fo}}{R_v}, 1 \right) \right]$$ (8-69)

The value of $\beta = 1$ is encountered in fuel elements of uniform power density. For nonuniform power density $\beta \neq 1$. If the power density for an inner region is higher than that of the outer region, $\beta > 1$. This situation is encountered in restructured fuel pellet analysis (section VI).

Figure 8-13 provides a plot of F_v in terms of α and β. Note that F_v is always less than 1. Hence when comparing the solid and annular pellets, the following conclusions can be made by observing Eqs. 8-62 and 8-67.

1. For the same temperature limit T_{max}:

$$q'_{annular} F_v = q'_{solid}$$ (8-70)

provided that T_{fo} and k are the same in the annular element as in the solid. Hence:

$$q'_{annular} > q'_{solid}$$ (8-71)

That is, an annular pellet can operate at a higher linear heat rate than a solid pellet if T_{max}, T_{fo}, and k are the same.

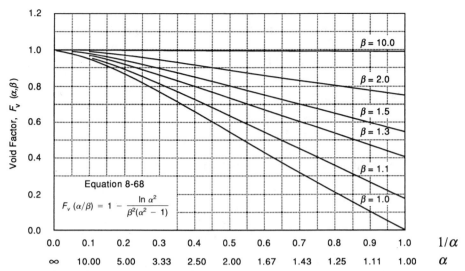

Figure 8-13 Void factor function. Note that for an annular region α is the ratio of the outer to the inner radius, and β is a function of the radii as well as the heat-generation profile in the pellet.

2. For the same heat rate (q') the temperature integrals of the conductivity are related by:

$$\left(\int_{T_{fo}}^{T_{max}} k_{annular} \, dT \right) = \left(\int_{T_{fo}}^{T_{max}} k_{solid} \, dT \right) F_v \tag{8-72}$$

so that if the fuel material conditions are the same, i.e., the same $k(T)$, and T_{fo} is the same, we get:

$$T_{max}|_{annular} < T_{max}|_{solid} \tag{8-73}$$

In this case the maximum operating temperature of the annular fuel is less than that of the solid fuel pellet.

Note that the conductivity integral is a function of the fuel pellet density, which depends on the initial fuel manufacturing conditions and irradiation conditions in the reactor. This dependence was discussed in section III. The conditions of Eqs. 8-71 and 8-73 do not necessarily apply if $k(T)$ of the annular fuel does not equal $k(T)$ of the solid fuel.

Example 8-2 Linear power of a cylindrical fuel pellet

PROBLEM For the geometry of the BWR pellet described in Example 8-1, evaluate the linear power of the pin when the fuel outer temperature is 495°C and the centerline fuel temperature is 1400°C. Assume that the fuel density is 88% ρ_{TD}, and ignore the effect of cracking on fuel conductivity.

SOLUTION

The linear power is related to the temperature difference by Eq. 8-62:

$$\int_{T_{fo}}^{T_{max}} k dT = \frac{q'}{4\pi}$$ (8-62)

which can be used to write:

$$q' = 4\pi \left(\int_{100}^{T_{max}} k dT - \int_{100}^{T_{fo}} k dT \right)$$

Applying the conditions specified in the problem, we get:

$$
\begin{aligned}
q' &= 4\pi \left(\int_{100}^{1400°C} k_{0.88} dT - \int_{100}^{495} k_{0.88} dT \right) \\
&= 4\pi(27.5 - 8.5) \text{ W/cm} \\
&= 239 \text{ W/cm} \\
&= 23.9 \text{ kW/m}
\end{aligned}
$$

VI TEMPERATURE DISTRIBUTION IN RESTRUCTURED FUEL ELEMENTS

Operation of an oxide fuel material at a high temperature leads to alterations of its morphology. The fuel region in which the temperature exceeds a certain sintering temperature experiences loss of porosity. In a cylindrical fuel pellet the inner region is restructured to form a void at the center, surrounded by a dense fuel region. In fast reactors, where the fuel may have a higher temperature near the center, restructuring was found to lead to three distinct regions (Fig. 8-14). In the outermost ring, where no sintering (i.e., no densification) occurs, the fuel density remains equal to the original (as fabricated) density, whereas the intermediate and inner regions have densities of 95 to 97% and 98 to 99%, respectively. It should be noted that most of the restructuring occurs within the first few days of operation, with slow changes afterward. In LWRs, where the fuel temperature is not as high as in liquid-metal-cooled reactors, two-region pellet restructuring may occur in the core regions operating at high powers. The sintering temperatures as well as the density in each fuel structure are not universally agreed on, as seen in Table 8-5.

Table 8-5 Fuel sintering temperature and densities

	Columnar grains		Equiaxed grains	
Recommendation source	T_1 (°C)	ρ_1/ρ_{TD}	T_2 (°C)	ρ_2/ρ_{TD}
Atomics International	1800	0.98	1600	0.95
General Electric	2150	0.99	1650	0.97
Westinghouse	2000	0.99	1600	0.97

Source: From Marr and Thompson [18].

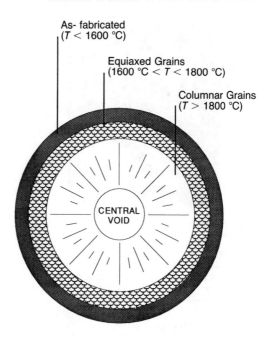

As- fabricated
$(T < 1600 \ °C)$

Equiaxed Grains
$(1600 \ °C < T < 1800 \ °C)$

Columnar Grains
$(T > 1800 \ °C)$

CENTRAL
VOID

Figure 8-14 Restructuring of an oxide fuel pellet during high-temperature irradiation.

In this section, the heat conduction problem in cylindrical fuel elements that have undergone some irradiation, and hence developed sintered (densified) regions, is solved. Temperature distributions are obtained on the assumption that the fuel element may be represented by three zones (Fig. 8-15). The two-zone fuel temperature distribution is obtained by inference from the three-zone treatment.

A Mass Balance

From conservation of mass across a section in the fuel rod before and after restructuring, we conclude that the original mass is equal to the sum of the fuel mass in the three rings. Hence when the pellet length is assumed unchanged:

$$\pi R_{fo}^2 \rho_0 = \pi (R_1^2 - R_v^2)\rho_1 + \pi(R_2^2 - R_1^2)\rho_2 + \pi(R_{fo}^2 - R_2^2)\rho_3 \qquad (8\text{-}74)$$

However, the initial density ρ_0 is equal to ρ_3, so that an explicit expression for R_v can be obtained from Eq. 8-74 as:

$$R_v^2 = \left(\frac{\rho_1 - \rho_2}{\rho_1}\right) R_1^2 + \left(\frac{\rho_2 - \rho_3}{\rho_1}\right) R_2^2 \qquad (8\text{-}75)$$

B Power Density Relations

For a uniform neutron flux the heat-generation density is proportional to the mass density:

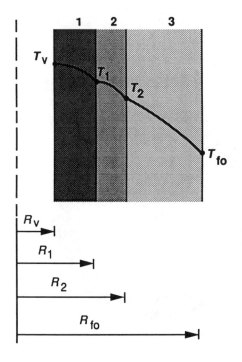

Figure 8-15 Temperature profile in a three-zone restructured fuel pellet.

$$q_2''' = \frac{\rho_2}{\rho_3} q_3''' \tag{8-76}$$

and

$$q_1''' = \frac{\rho_1}{\rho_3} q_3''' \tag{8-77}$$

However, the linear heat rate is given by the summation of the heat-generation rate in the three rings. Therefore the linear heat-generation rate in the restructured fuel (q_{res}') is given by:

$$q_{res}' = q_3''' \, \pi(R_{fo}^2 - R_2^2) + q_2''' \, \pi(R_2^2 - R_1^2) + q_1''' \, \pi(R_1^2 - R_v^2) \tag{8-78}$$

or

$$q_{res}' = q_3''' \left[\pi(R_{fo}^2 - R_2^2) + \frac{\rho_2}{\rho_3} \pi(R_2^2 - R_1^2) + \frac{\rho_1}{\rho_3} \pi(R_1^2 - R_v^2) \right] \tag{8-79}$$

Combining Eqs. 8-74 and 8-79, we obtain:

$$q_{res}' = \pi R_{fo}^2 \, q_3''' \tag{8-80}$$

i.e.,

$$q_3''' = \frac{q_{res}'}{\pi R_{fo}^2} \tag{8-81}$$

That is, the power density in the as-fabricated region can be obtained from the assumption that the mass is uniformly distributed in the fuel pellet. It is equivalent to expecting the power density in the outer region not to be affected by the redistribution of the fuel within the other two zones.

C Heat Conduction in Zone 3

The differential equation is:

$$\frac{1}{r}\frac{d}{dr}\left(rk_3\frac{dT}{dr}\right) = -q_3''' \tag{8-82}$$

By integrating once we get:

$$k_3\frac{dT}{dr} = -q_3'''\frac{r}{2} + \frac{C_3}{r} \tag{8-83}$$

and integrating again between the zone boundaries:

$$\int_{T_2}^{T_{fo}} k_3 dT = -q_3'''\frac{R_{fo}^2 - R_2^2}{4} + C_3\,\ell n\left(\frac{R_{fo}}{R_2}\right) \tag{8-84}$$

Using Eq. 8-81 we can evaluate the temperature gradient at R_{fo} from the heat flux as follows:

$$q_{R_{fo}}'' = -k_3\frac{dT}{dr}\bigg|_{R_{fo}} = \frac{q_{res}'}{2\pi R_{fo}} = q_3'''\frac{\pi R_{fo}^2}{2\pi R_{fo}} = \frac{q_3''' R_{fo}}{2} \tag{8-85}$$

Equation 8-85 provides a boundary condition to be satisfied by Eq. 8-83. This condition leads to:

$$C_3 = 0 \tag{8-86}$$

Hence Eq. 8-84 reduces to:

$$\int_{T_{fo}}^{T_2} k_3 dT = \frac{q_3'''}{4} R_{fo}^2\left[1 - \left(\frac{R_2}{R_{fo}}\right)^2\right] \tag{8-87}$$

which, using Eq. 8-81, can also be written as:

$$\int_{T_{fo}}^{T_2} k_3 dT = \frac{q_{res}'}{4\pi}\left[1 - \left(\frac{R_2}{R_{fo}}\right)^2\right] \tag{8-88}$$

D Heat Conduction in Zone 2

The heat conduction equation integration for zone 2 leads to an equation similar to Eq. 8-83 but applicable to zone 2:

$$k_2\frac{dT}{dr} = -q_2'''\frac{r}{2} + \frac{C_2}{r} \tag{8-89}$$

At R_2, continuity of heat flux leads to:

$$-k_2 \frac{dT}{dr}\bigg|_{R_2} \text{ From zone 2 } = -k_3 \frac{dT}{dr}\bigg|_{R_2} \text{ From zone 3} \qquad (8\text{-}90)$$

From zone 2   From zone 3

Substituting from Eqs. 8-83 (with $C_3 = 0$) and 8-89 into Eq. 8-90 leads to:

$$+ q_2''' \frac{R_2}{2} - \frac{C_2}{R_2} = q_3''' \frac{R_2}{2}$$

or

$$C_2 = \frac{R_2^2}{2} [q_2''' - q_3'''] \qquad (8\text{-}91)$$

By integrating Eq. 8-89 between $r = R_1$ and $r = R_2$, we obtain:

$$\int_{T_1}^{T_2} k_2 dT = -\frac{q_2'''}{4}(R_2^2 - R_1^2) + \frac{R_2^2}{2}(q_2''' - q_3''') \, \ell n \left(\frac{R_2}{R_1}\right)$$

which, using Eqs. 8-76 and 8-81, can be written as:

$$\int_{T_2}^{T_1} k_2 dT = \frac{q_{\text{res}}'}{4\pi} \left(\frac{\rho_2}{\rho_3}\right) \left[1 - \left(\frac{R_1}{R_2}\right)^2\right] \left(\frac{R_2}{R_{\text{fo}}}\right)^2$$
$$- \frac{q_{\text{res}}'}{4\pi} \left(\frac{\rho_2}{\rho_3}\right) \left(\frac{R_2}{R_{\text{fo}}}\right)^2 \left(1 - \frac{\rho_3}{\rho_2}\right) \ell n \left(\frac{R_2}{R_1}\right)^2$$

Hence:

$$\int_{T_2}^{T_1} k_2 dT = \frac{q_{\text{res}}'}{4\pi} \left(\frac{\rho_2}{\rho_3}\right) \left(\frac{R_2}{R_{\text{fo}}}\right)^2 \left[1 - \left(\frac{R_1}{R_2}\right)^2 - \left(\frac{\rho_2 - \rho_3}{\rho_2}\right) \ell n \left(\frac{R_2}{R_1}\right)^2\right] \qquad (8\text{-}92a)$$

Using the notation of Eq. 8-68, Eq. 8-92a can be recast in the form:

$$\int_{T_2}^{T_1} k_2 dT = \frac{q_{\text{res}}'}{4\pi} \left(\frac{\rho_2}{\rho_3}\right) \left(\frac{R_2^2 - R_1^2}{R_{\text{fo}}^2}\right) F_v \left[\frac{R_2}{R_1}, \frac{R_1}{R_2} \left(\frac{\rho_2}{\rho_2 - \rho_3}\right)^{1/2}\right] \qquad (8\text{-}92b)$$

E Heat Conduction in Zone 1

Again the integration of the heat conduction equation once leads to:

$$k_1 \frac{dT}{dr} = -q_1''' \frac{r}{2} + \frac{C_1}{r} \qquad (8\text{-}93)$$

To obtain the value of C_1, we can either use the zero heat flux condition at R_v or the continuity of heat flux at R_1. Applying the boundary condition:

$$k_1 \frac{dT}{dr}\bigg|_{R_v} = 0 \qquad (8\text{-}94)$$

leads to

$$C_1 = q_1''' \frac{R_v^2}{2}$$

(8-95)

By integrating Eq. 8-93 again, we get:

$$\int_{T_v}^{T_1} k_1 dT = \frac{-q_1'''}{4} (R_1^2 - R_v^2) + C_1 \ln \left(\frac{R_1}{R_v}\right)$$

(8-96)

Substituting from Eqs. 8-77 and 8-81 for q_1''' and q_3''' and Eq. 8-95 for C_1 we get:

$$\int_{T_1}^{T_v} k_1 dT = \frac{q_{res}'}{4\pi} \left(\frac{\rho_1}{\rho_3}\right) \left(\frac{R_1}{R_{fo}}\right)^2 \left[1 - \left(\frac{R_v}{R_1}\right)^2 - \left(\frac{R_v}{R_1}\right)^2 \ln \left(\frac{R_1}{R_v}\right)^2\right]$$

(8-97a)

Using the notation of Eq. 8-68, the last equation can be written as:

$$\int_{T_1}^{T_v} k_1 dT = \frac{q_{res}'}{4\pi} \left(\frac{\rho_1}{\rho_3}\right) \frac{R_1^2 - R_v^2}{R_{fo}^2} \left[1 - \frac{\ln(R_1/R_v)^2}{\left(\frac{R_1}{R_v}\right)^2 - 1}\right]$$

$$= \frac{q_{res}'}{4\pi} \left(\frac{\rho_1}{\rho_3}\right) \left(\frac{R_1^2 - R_v^2}{R_{fo}^2}\right) F_v \left(\frac{R_1}{R_v}, 1\right)$$

(8-97b)

F Solution of the Pellet Problem

Equations 8-75, 8-88, 8-92a (or 8-92b), and 8-97a (or 8-97b) provide a set of four equations in R_1, R_2, R_v, T_{fo}, T_v, and q'. Thus if any two are specified, the rest can be evaluated. Note that the values of T_1 and T_2 are assumed known from Table 8-5.

Generally, two conditions are of interest.

1. A linear heat rate (q_{res}') and the outer surface temperature (T_{fo}) are specified; so that T_v, along with R_v, R_1, and R_2 can be evaluated.
2. The maximum temperatures (T_v and T_{fo}) are specified; q_{res}' along with R_v, R_1, and R_2 are to be evaluated.

G Two-Zone Sintering

For two-zone representation of the fuel, $R_1 = R_2 = R_s$, $\rho_1 = \rho_2 = \rho_s$, and $T_1 = T_2 = T_s$; Eq. 8-75 reduces to:

$$R_v^2 = \frac{\rho_s - \rho_3}{\rho_s} R_s^2$$

(8-98)

Equation 8-88 takes the form:

$$\int_{T_{fo}}^{T_s} k_3 dT = \frac{q_{res}'}{4\pi} \left[1 - \left(\frac{R_s}{R_{fo}}\right)^2\right]$$

(8-99)

Equation 8-92a is not needed. Equation 8-97a takes the form:

$$\int_{T_s}^{T_v} k_s dT = \frac{q'_{res}}{4\pi} \left(\frac{\rho_s}{\rho_3}\right) \left(\frac{R_s}{R_{fo}}\right)^2 \left\{1 - \left(\frac{R_v}{R_s}\right)^2 \left[1 + \ell n \left(\frac{R_s}{R_v}\right)^2\right]\right\} \quad (8\text{-}100)$$

Example 8-3 Linear power of a two-zone pellet under temperature constraint

PROBLEM Consider an initially solid LMFBR fuel pellet under fixed maximum temperature constraint. Evaluate the linear power (q'_{res}) for the sintered pellet for the same temperature constraint. The sintered pellet may be represented by two zones. Given:

Solid pellet:
$$q' = 14.4 \text{ kW/ft} = 472.8 \text{ W/cm}$$
$$\rho_0 = 88\% \; \rho_{TD}$$
$$T_{fo} = 960°C$$

Two-zone sintered pellet:
$$\rho_s = 98\% \; \rho_{TD}$$
$$T_s = 1800°C$$
$$T_{fo} = 960°C$$

SOLUTION

1. Evaluate the maximum temperature of the solid pellet using Eq. 8-62 for $k = k_{0.88}$:

$$\int_{T_{fo}}^{T_{max}} k_{0.88} \, dT = \frac{q'}{4\pi} = \frac{14.4}{4\pi} = 1.15 \text{ kW/ft} = 37.6 \text{ W/cm}$$

Expressing the conductivity integral in the form represented in Figure 8-2:

$$\int_{100°C}^{T_{max}} k_{0.88} \, dT - \int_{100°C}^{T_{fo}=960°C} k_{0.88} \, dT = 37.6 \text{ W/cm}$$

which yields T_{max} as follows:

$$\int_{100°C}^{T_{max}} k_{0.88} \, dT = 37.6 + 19 = 56.6 \text{ W/cm} \rightarrow T_{max} = 2600°C$$

2. Obtain the relation between q'_{res} and R_s for the restructured pellet. Using Eq. 8-99 for $k_3 = k_{0.88}$, we get:

$$\int_{T_{fo}=960°C}^{T_s=1800°C} k_{0.88} \, dT = \frac{q'_{res}}{4\pi} \left[1 - \left(\frac{R_s}{R_{fo}}\right)^2\right]$$

or

$$17 \text{ W/cm} = \frac{q'_{res}}{4\pi} \left[1 - \left(\frac{R_s}{R_{fo}}\right)^2\right] \quad (8\text{-}101)$$

3. Obtain the relation between q'_{res}, R_{fo}, R_s, and R_v from Eq. 8-100 for $k_s = k_{0.98}$:

$$\int_{T_s=1800°C}^{T_{max}=2600°C} k_{0.98}\, dT = \frac{q'_{res}}{4\pi} \left(\frac{\rho_s}{\rho_3}\right) \left(\frac{R_s}{R_{fo}}\right)^2 \left\{ 1 - \left(\frac{R_v}{R_s}\right)^2 \left[1 + \ell n \left(\frac{R_s}{R_v}\right)^2\right]\right\}$$

Hence:

$$23.3 = \frac{q'_{res}}{4\pi} \frac{0.98}{0.88} \left(\frac{R_s}{R_{fo}}\right)^2 \left\{ 1 - \left(\frac{R_v}{R_s}\right)^2 \left[1 + \ell n \left(\frac{R_s}{R_v}\right)^2\right]\right\} \qquad (8\text{-}102)$$

4. Eliminate q'_{res} between Eqs. 8-101 and 8-102 and rearrange to obtain:

$$0.8125 \left\{ 1 - \left(\frac{R_v}{R_s}\right)^2 \left[1 + \ell n \left(\frac{R_s}{R_v}\right)^2\right]\right\} = \left(\frac{R_{fo}}{R_s}\right)^2 - 1 \qquad (8\text{-}103)$$

5. From the mass balance equation (Eq. 8-98) in the solid and sintered pellets:

$$R_v^2 = \frac{\rho_s - \rho_o}{\rho_s} R_s^2 = \frac{0.98 - 0.88}{0.98} R_s^2 = 0.102\, R_s^2 \qquad (8\text{-}98)$$

Thus

$$\left(\frac{R_s}{R_v}\right)^2 = 9.8 \qquad (8\text{-}104)$$

6. Substituting the value of $(R_v/R_s)^2$ from Eq. 8-104 into Eq. 8-103:

$$0.8125 \left\{1 - 0.102[1 + \ell n(9.8)]\right\} = \left(\frac{R_{fo}}{R_s}\right)^2 - 1$$

or

$$R_s/R_{fo} = 0.806 \qquad (8\text{-}105)$$

7. Substituting the value of R_s/R_{fo} from Eq. 8-105 into Eq. 8-101, we get:

$$q'_{res} = \frac{4\pi(17)}{1 - (0.806)^2} = 609\ \text{W/cm} = 18.57\ \text{kW/ft}$$

i.e., $q'_{res} > q'$. Hence under the same temperature limits, the linear power of a sintered pellet is higher than that of a solid pellet.

Example 8-4 Comparison between three-zone and two-zone fuel pellet maximum temperature

PROBLEM Consider an LMFBR fuel pin operating at a fixed linear power q' of 14.4 kW/ft. Obtain the maximum fuel temperature if the fuel pin is in one of the following restructured conditions.

Three-zone condition
Columnar: $\rho_1 = 98\% \; \rho_{TD}$, $T_1 = 1800°C$
Equiaxed: $\rho_2 = 95\% \; \rho_{TD}$, $T_2 = 1600°C$
Unrestructured: $\rho_3 = 88\% \; \rho_{TD}$, $T_{fo} = 1000°C$

Two-zone condition
Sintered: $\rho_s = 98\% \; \rho_{TD}$, $T_s = 1700°C$
Unrestructured: $\rho_{fo} = 88\% \; \rho_{TD}$, $T_{fo} = 1000°C$

SOLUTION
We note that $q'_{res} = q' = 14.4$ kW/ft $= 472.8$ W/cm.

Case 1

1. We first obtain the value of (R_2/R_{fo}) from Eq. 8-88:

$$\int_{T_{fo}}^{T_2} k_3 dT = \frac{q'}{4\pi}\left[1 - \left(\frac{R_2}{R_{fo}}\right)^2\right]$$ (8-88)

where $k_3 = k_{0.88 \; TD}$
From Figure 8-2 we get:

$$\int_{1000°C}^{1600°C} k_{0.88} \, dT = 32 - 20 = 12 \text{ W/cm}$$

Hence:

$$\left(\frac{R_2}{R_{fo}}\right)^2 = 1 - \frac{4\pi}{q'} 12 = 1 - \frac{48\pi}{472.8} = 1 - 0.32 = 0.68$$

Therefore $R_2 = 0.825 \; R_{fo}$
2. Obtain a value for R_1/R_2 from the equation:

$$\int_{T_2}^{T_1} k_2 \, dT = \frac{q'}{4\pi}\left(\frac{\rho_2}{\rho_3}\right)\left(\frac{R_2}{R_{fo}}\right)^2\left[1 - \left(\frac{R_1}{R_2}\right)^2 - \left(1 - \frac{\rho_3}{\rho_2}\right)\ell n\left(\frac{R_2}{R_1}\right)^2\right]$$ (8-92a)

where $k_2 = k_{0.95 \; TD}$.
From Figure 8-2:

$$\int_{1600°C}^{1800°C} k_{0.95} \, dT = 40.2 - 35.7 = 4.5 \text{ W/cm}$$

Substituting in Eq. 8-92a:

$$4.5 = \frac{472.8}{4\pi}\left(\frac{0.95}{0.88}\right)(0.681)\left[1 - \left(\frac{R_1}{R_2}\right)^2 - \left(1 - \frac{0.88}{0.95}\right)\ell n\left(\frac{R_2}{R_1}\right)^2\right]$$

$$4.5 = 27.66\left[1 - \left(\frac{R_1}{R_2}\right)^2 - 0.074 \, \ell n \left(\frac{R_2}{R_1}\right)^2\right]$$

$$0.837 = \left(\frac{R_1}{R_2}\right)^2 + 0.074 \, \ell n \left(\frac{R_2}{R_1}\right)^2$$

Solving iteratively, we get:

$$\left(\frac{R_1}{R_2}\right)^2 = 0.8226; \text{ and } \left(\frac{R_1}{R_{fo}}\right)^2 = \left(\frac{R_1}{R_2}\right)^2 \left(\frac{R_2}{R_{fo}}\right)^2 = (0.8226)(0.681) = 0.560$$

or $R_1 = 0.907 R_2$ and $R_1 = 0.748 R_{fo}$.

3. Determine R_v from the mass balance equation:

$$R_v^2 = \left(\frac{\rho_1 - \rho_2}{\rho_1}\right) R_1^2 + \left(\frac{\rho_2 - \rho_3}{\rho_1}\right) R_2^2 \qquad (8\text{-}75)$$

or

$$\left(\frac{R_v}{R_2}\right)^2 = \frac{0.98 - 0.95}{0.98}(0.8226) + \frac{0.95 - 0.88}{0.98}$$

$$= 0.02518 + 0.07143 = 0.09661$$

or

$$R_v = 0.311; \; R_2 = 0.343; \; R_1 = 0.256 \, R_{fo}$$

4. Determine T_v from the integral of the heat conduction equation over zone 1 (Eq. 8-97a):

$$\int_{T_1}^{T_v} k_1 \, dT = \frac{q'}{4\pi}\left(\frac{\rho_1}{\rho_3}\right)\left(\frac{R_1}{R_{fo}}\right)^2 \left[1 - \left(\frac{R_v}{R_1}\right)^2 - \left(\frac{R_v}{R_1}\right)^2 \ell n \left(\frac{R_1}{R_v}\right)^2\right]$$

$$= \frac{472.8}{4\pi}\left(\frac{0.98}{0.88}\right)(0.56)$$

$$\left[1 - (0.343)^2 - (0.343)^2 \ell n \left(\frac{1}{0.343}\right)^2\right]$$

$$= 23.44 \, [1 - 0.1176 - 0.2518]$$

$$= 14.78 \text{ W/cm}$$

Because $k_1 = k_{0.98}$, we now have:

$$\int_{1800}^{T_v} k_{0.98} \, dT = 14.78 \text{ W/cm}$$

or, using Figure 8-2:

$$\int_{100°C}^{T_v} k_{0.98} \, dT = 14.78 + \int_{100°C}^{T_1 = 1800°C} k_{0.98} \, dT = 14.78 + 42.5$$

$$= 57.28 \text{ W/cm} \rightarrow T_v = 2340°C$$

Case 2

Consider the two-zone sintered fuel, under the condition $q'_{res} = q'$, and $k_3 = k_{0.88}$.

1. Obtain R_s from the integral of the heat conduction equation over the unrestructured zone (Eq. 8-99):

$$\int_{T_{fo} = 1000°C}^{T_s = 1700°C} k_{0.88} dT = \frac{q'}{4\pi} \left[1 - \left(\frac{R_s}{R_{fo}} \right)^2 \right] = 13.8 \text{ W/cm}$$

Hence:

$$\left(\frac{R_s}{R_{fo}} \right)^2 = 1 - \frac{4\pi}{472.8} (13.8) = 1 - 0.367 = 0.633$$

$$\therefore R_s = 0.796 R_{fo}$$

2. Obtain R_v from the mass balance equation (Eq. 8-98):

$$R_v^2 = \left(\frac{\rho_s - \rho_o}{\rho_s} \right) R_s^2 = \frac{0.98 - 0.88}{0.98} R_s^2$$

i.e.,

$$R_v^2 = 0.102 R_s^2$$
$$R_v = 0.319 R_s = 0.2543 R_{fo}$$

3. Obtain T_v from the heat conduction integral (Eq. 8-100) when $k_s = k_{0.98}$:

$$\int_{1700°C}^{T_v} k_s \, dT = \frac{q'}{4\pi} \left(\frac{\rho_s}{\rho_o} \right) \left(\frac{R_s}{R_{fo}} \right)^2 \left[1 - \left(\frac{R_v}{R_s} \right)^2 - \left(\frac{R_v}{R_s} \right)^2 \ell n \left(\frac{R_s}{R_v} \right)^2 \right] \quad (8\text{-}100)$$

or

$$\int_{100°C}^{T_v} k_{0.98} \, dT = \int_{100°C}^{1700°C} k_{0.98} \, dT + \frac{472.8}{4\pi} \left(\frac{0.98}{0.88} \right) (0.633)$$

$$\left[1 - 0.102 - 0.102 \, \ell n \left(\frac{1}{0.102} \right) \right]$$

$$= 40 + 26.50 (1 - 0.102 - 0.232)$$
$$= 40 + 26.50 (0.666)$$
$$= 57.65 \text{ w/cm}$$
$$\rightarrow T_v = 2360°C$$

Although the three region model is a more exact one, there is only a small advantage in using the three-zone approach, in that T_v from the three-zone model is 20°C lower than that from the two-zone approach.

H Comments on Design Implications of Restructured Fuel

The result of fuel sintering is to reduce the effective thermal resistance between the highest fuel temperature (T_{max}) and the pellet outer temperature (T_{fo}). Thus two operational options exist after restructuring.

1. If the fuel maximum temperature is kept constant, the linear power can be increased. This condition is applicable to the LMFBRs, as the heat flux to the

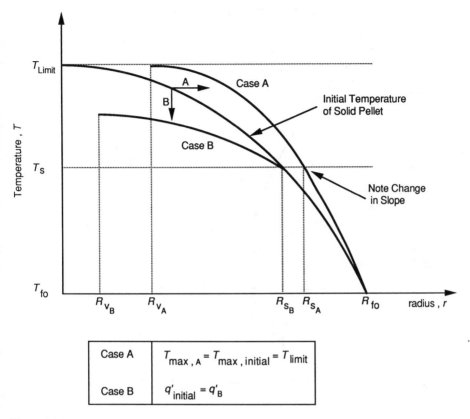

Figure 8-16 Design and irradiation strategies for initially solid fuel pellets.

coolant is not a limiting factor. This case is depicted graphically as case A in Figure 8-16. The assumption of a constant T_{fo} in the figure implies that the coolant conditions have been adjusted.

2. If the linear power is kept constant, the fuel maximum temperature is reduced. This case may be applicable to LWRs, as other considerations limit the operating heat flux of a fuel pin. This case is depicted graphically as case B in Figure 8-16. If the fuel maximum temperature is calculated ignoring the sintering process, a conservative value is obtained.

VII THERMAL RESISTANCE BETWEEN FUEL AND COOLANT

The overall thermal resistance between the fuel and coolant consists of (1) the resistance of the fuel itself, (2) the resistance of the gap between the fuel and the cladding, (3) the resistance of the cladding, (4) the resistance of the coolant.

For typical LWR and LMFBR fuel rods, the resistance of the UO_2 fuel is by far the largest, as can be inferred from the temperature profile of Figure 8-17. The next

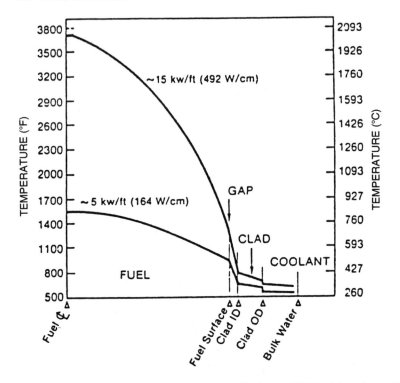

Figure 8-17 Typical PWR fuel rod temperature profile for two LHGRs. *(From Jordan [12].)*

largest resistance is that of the gap. Hence models for the gap conductance with increasing sophistication have been developed over the years. In this section, the gap resistance models are presented first, and then the overall resistance is discussed.

A Gap Conductance Models

It is usually assumed that the gap consists of an annular space occupied by gases. The gas composition is initially the fill gas, which should be an inert gas such as helium, but is gradually altered with burnup by the addition of gaseous fission product such as xenon and krypton. (Sodium-cooled reactors have also considered sodium bonding.) However, this simple picture does not reflect the real conditions of the fuel pin after some irradiation. The fuel pellets usually crack upon irradiation, as shown in Figure 8-18, and this situation leads to circumferential variation in the gap. In addition, thermal expansions of the fuel and cladding are often different, the result being substantial pellet–cladding contact at the interface. This contact reduces the thermal resistance and hence effectively increases the "gap" conductance of the burnup. It occurs despite the lower conductivity of the fission gas products, compared to the initial conductivity of helium. A typical change of gap conductance with burnup is shown in Figure 8-19.

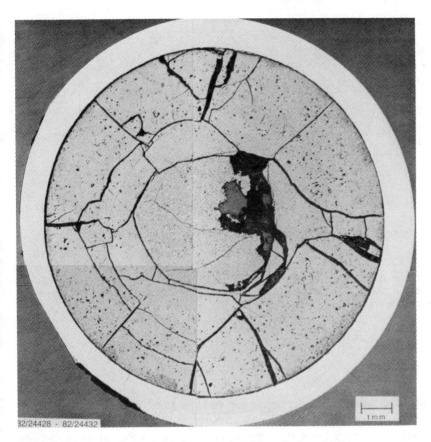

Figure 8-18 Example of a cracked fuel cross section. *(From Clark et al. [4].)*

1 As-fabricated gap. The gap conductance at the as-fabricated condition of tne fuel can be modeled as due to conduction through an annular space as well as to radiation from the fuel. Thus the heat flux at an intermediate gap position can be given by:

$$q_g'' = h_g(T_{fo} - T_{ci}) \qquad (8\text{-}106)$$

where for an open gap:

$$h_{g,open} = \frac{k_{gas}}{\delta_{eff}} + \frac{\sigma}{\dfrac{1}{\epsilon_f} + \dfrac{1}{\epsilon_c} - 1} \frac{T_{fo}^4 - T_{ci}^4}{T_{fo} - T_{ci}} \qquad (8\text{-}107a)$$

where $h_{g,open}$ = heat transfer coefficient for an open gap; T_{fo} = fuel surface temperature; T_{ci} = clad inner surface temperature; k_{gas} = thermal conductivity of the gas; δ_{eff} = effective gap width; σ = Stefan-Boltzman constant; ϵ_f, ϵ_c = surface emissivities of the fuel and cladding, respectively.

Equation 8-107a can often be approximated by:

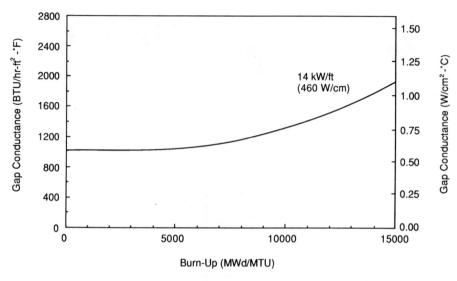

Figure 8-19 Variations of gap conductance with burnup for a PWR fuel rod (pressurized with helium) and operating at 14 kW/ft (460 W/cm). *(From Fenech [6].)*

$$h_{g,open} = \frac{k_{gas}}{\delta_{eff}} + \frac{\sigma T_{fo}^3}{\dfrac{1}{\epsilon_f} + \dfrac{1}{\epsilon_c} - 1} \qquad (8\text{-}107b)$$

It should be noted that the effective gap width is larger than the real gap width because of the temperature discontinuities at the gas–solid surface. The temperature discontinuities arise near the surface owing to the small number of gas molecules present near the surface. Thus it is possible to relate δ_{eff} to the real gap width (δ_g), as illustrated in Figure 8-20, by:

$$\delta_{eff} = \delta_g + \delta_{jump1} + \delta_{jump2} \qquad (8\text{-}108)$$

At atmospheric pressure, $\delta_{jump1} + \delta_{jump2}$ were found to equal 10 μm in helium and 1 μm in xenon [23].

The gas conductivity of a mixture of two gases is given by [13]:

$$k_{gas} = (k_1)^{x_1} (k_2)^{x_2} \qquad (8\text{-}109)$$

where x_1 and x_2 are the mole fractions of gases 1 and 2, respectively. For rare gases the conductivity dependence on temperature is given by [26]:

$$k(\text{pure gas}) = A \times 10^{-6} T^{0.79} \text{ W/cm } ^\circ\text{K} \qquad (8\text{-}110)$$

where T is in $^\circ$K and $A = 15.8$ for helium, 1.97 for argon, 1.15 for krypton, and 0.72 for xenon.

2 Gap closure effects. Calza-Bini et al. [3] observed that the model of Eqs. 8-106 to 8-108 provides a reasonable estimate for the gap conductance on the first rise to power. Subsequently, the measured gap conductance increased and was attributed to

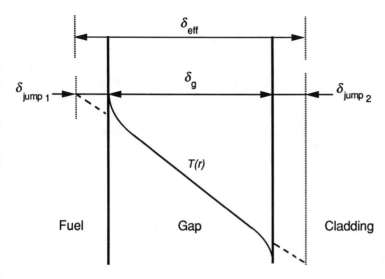

Figure 8-20 Temperature profile across the fuel–clad gap.

a portion of the fuel contacting the cladding and, after fuel cracking, remaining in contact. Additionally, cracking has a negative effect on thermal conductivity of the fuel (see section III.A.5).

When gap closure occurs because of fuel swelling and thermal expansion, the contact area with the cladding is proportional to the surface contact pressure between the fuel and cladding (Fig. 8-21). Thus the contact-related heat transfer coefficient can be given by:

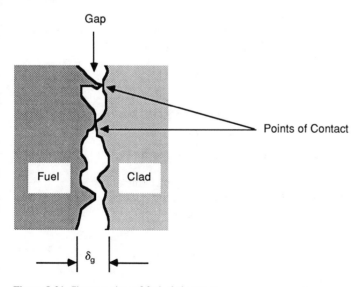

Figure 8-21 Close-up view of fuel–clad contact.

$$h_{contact} = C \frac{2k_f k_c}{k_f + k_c} \frac{p_i}{H\sqrt{\delta_g}} \text{ (Btu/ft}^2 \text{ hr)} \tag{8-111}$$

where C = a constant (= $10 \text{ ft}^{-1/2}$); p_i = surface contact pressure (psi) (usually calculated based on the relative thermal expansion of the fuel and the cladding, but ignoring the elastic deformation of the cladding); H = Meyer's hardness number of the softer material (typical values are, for steel, 13×10^4 psi and, for zircaloy, 14×10^4 psi); δ_g = mean thickness of the gas space (feet) (calculated based on the roughness of the materials in contact); k_f and k_c = thermal conductivities of fuel and cladding (Btu/hr ft °F).

The total gap conductance upon contact may be given by:

$$h_g = h_{g,open} + h_{contact} \tag{8-112}$$

Ross and Stoute [23] have shown that the general features of Eqs. 8-111 and 8-112 can be observed in experiments. From their data they concluded that δ_{jump} is 1 to 10 μm at atmospheric pressure, which is 10 to 30 times the gas molecule mean free path. Jacobs and Todreas [11] used the work of Cooper et al. [5] to formulate a better model by recognizing the nonuniform distribution of the surface protrusions. Yet even their improved model could not explain all the experimental observations. Thus inadequacies in modeling remain to be resolved. The effect of increased swelling at higher linear powers is to increase the gap conductance unless the initial gap thickness is so small as to lead to saturation of the contact effect, as can be observed in Figure 8-22. The reader should consult the empirically based relations available in the database of MATPRO for additional information [21].

B Overall Resistance

The linear power of a cylindrical fuel pin can be related to the temperature drop $(T_{max} - T_m)$ by considering the series of thermal resistances posed by the fuel, the gap, the cladding, and the coolant. Consider, for simplicity, a solid fuel pellet with a constant thermal conductivity $(\bar{k}_f)$. From Eq. 8-62, we get:

$$T_{max} - T_{fo} = \frac{q'}{4\pi \bar{k}_f} \tag{8-113}$$

Across the gap the temperature drop is predicted from Eq. 8-106 by:

$$T_{fo} - T_{ci} = \frac{q''_g}{h_g} = \frac{2\pi R_g q''_g}{2\pi R_g h_g} = \frac{q'}{2\pi R_g h_g} \tag{8-114}$$

where R_g = mean radius in the gap; h_g = effective gap conductance.

For a thin cladding a linear temperature drop across the cladding may be assumed. Hence the temperature drop across the cladding is given by:

$$T_{ci} - T_{co} = \frac{q''}{k_c/\delta_c} = \frac{q'}{2\pi R_c k_c/\delta_c} \tag{8-115a}$$

where R_c = mean radius in the cladding.

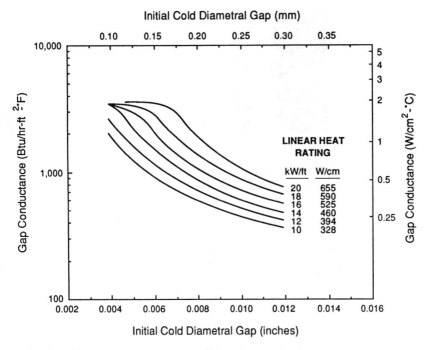

Figure 8-22 Calculated gap conductance as a function of cold diametral gap in a typical LWR fuel rod. *(From Horn and Panisko, [10].)*

The temperature drop across a thick clad can be determined as:

$$T_{ci} - T_{co} = \frac{q'}{2\pi k_c} \ell n \left(\frac{R_{co}}{R_{ci}}\right) \tag{8-115b}$$

The heat flux emerging from the cladding is given by

$$q''_{co} = h(T_{co} - T_m) \tag{8-116}$$

where T_m = mean coolant temperature at the cross section. The heat transfer coefficient (h) is dependent on the coolant flow conditions. The linear power is then given by:

$$q' = 2\pi R_{co} q''_{co} = 2\pi R_{co} h(T_{co} - T_m) \tag{8-117}$$

Equation 8-117 can be recast to give the temperature drop ($T_{co} - T_m$):

$$T_{co} - T_m = \frac{q'}{2\pi R_{co} h} \tag{8-118}$$

Hence the linear power of the pin can be related to an overall thermal resistance and the temperature drop $T_{max} - T_m$ (from Eqs. 8-113, 8-114, 8-115b, and 8-118):

$$T_{max} - T_m = q' \left[\frac{1}{4\pi \bar{k}_f} + \frac{1}{2\pi R_g h_g} + \frac{1}{2\pi k_c} \ell n \left(\frac{R_{co}}{R_{ci}}\right) + \frac{1}{2\pi R_{co} h}\right] \tag{8-119}$$

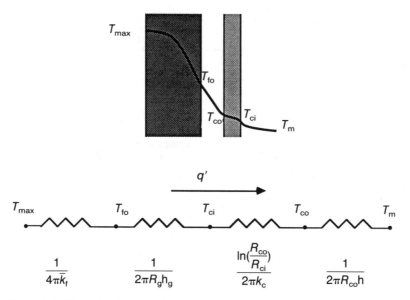

Figure 8-23 Thermal resistance analogy circuit for a cylindrical fuel pin.

The linear power is then analogous to an electrical current driven by the potential $T_{max} - T_m$ across a series of resistances, as depicted in Figure 8-23. This concept can be used to analyze the rate of temperature change in the fuel during a transient, provided the transient is sufficiently slow to allow the quasi-steady-state treatment of the thermal resistance.

REFERENCES

1. Arpaci, V. S. *Conduction Heat Transfer*. Menlo Park, CA: Addison-Wesley, 1966.
2. Biancharia, A. The effect of porosity on thermal conductivity of ceramic bodies. *Trans. ANS* 9:15, 1966.
3. Calza-Bini, A., et al. In-pile measurement of fuel cladding conductance for pelleted and vipac zircaloy-2 sheathed fuel pin. *Nucl. Technol.* 25:103, 1975.
4. Clark, P. A. E., et al. *Post Irradiation Examination of Two High Burnup Fuel Rods Irradiated in the Halden BWR*. AEREG3207, AERE Harwell Report. May 1985.
5. Cooper, M. G., Mikic, B. B., and Yovanovich, M. M. Thermal contact conductance. *Int. J. Heat Mass Transfer* 12:279, 1969.
6. Fenech, H. General considerations on thermal design and performance requirements of nuclear reactor cores. In H. Fenech (ed.), *Heat Transfer and Fluid Flow in Nuclear Systems*. Oxford: Pergamon Press, 1981.
7. Francl, J., and Kingery, W. D. Thermal conductivity. IX. Experimental investigation of effect of porosity on thermal conductivity. *J. Am. Ceram. Soc.* 37:99, 1954.
8. Gibby, R. L. The effect of plutonium content on the thermal conductivity of (U, Pu)O$_2$ solid solutions. *J. Nucl. Mater.* 38:163, 1971.
9. Hann, C. R., et al. *GAPCON-Thermal-7: A Computer Program for Calculating the Gap Conductance in Oxide Fuel Pins*. BNWL-1778, September 1973.
10. Horn, G. R., and Panisko, F. E. HEDL-TME72-128, September 1979.

11. Jacobs, G., and Todreas, N. Thermal contact conductance in reactor fuel elements. *Nucl. Sci. Eng.* 50:282, 1973.

12. Jordan, R. MIT Reactor Safety Course, 1979.

13. Kampf, H., and Karsten, G. Effects of different types of void volumes on the radial temperature distribution of fuel pins. *Nucl. Appl. Technol.* 9:288, 1970.

14. Latta, R. E., and Fryxell, R. E. Determination of solidus–liquidus temperatures in the UO_{2+n} system ($-0.50 < n < 0.20$). *J. Nucl. Mater.* 35:195, 1970.

15. Loeb, A. L. Thermal conductivity. VIII. A theory of thermal conductivity of porous material. *J. Am. Ceram. Soc.* 37:96, 1954.

16. MacDonald, P. E., and Smith, R. H. An empirical model of the effects of pellet cracking on the thermal conductivity of UO_2 light water reactor fuel. *Nucl. Eng. Design* 61:163, 1980.

17. MacDonald, P. E., and Weisman, J. Effect of pellet cracking on light water reactor fuel temperatures. *Nucl. Technol.* 31:357, 1976.

18. Marr, W. M., and Thompson, D. H. *Trans. Am. Nucl. Soc.* 14:150, 1971.

19. Notley, M. J. F. A computer model to predict the performance of UO_2 fuel element irradiated at high power output to a burnup of 10,000 MWD/MTU. *Nucl. Appl. Technol.* 9:195, 1970.

20. Olander, D. R. *Fundamental Aspects of Nuclear Reactor Fuel Elements.* T1D-26711-P1, 1976.

21. Olsen, C. S., and Miller, R. L. *MATPRO, Vol. II: A Handbook of Materials Properties for Use in the Analysis of Light Water Reactor Fuel Behavior.* NUREG/CR-0497, USNRC, 1979.

22. Reymann, G. A. *MATPRO* (Vol. II).

23. Ross, A. M., and Stoute, R. L. *Heat Transfer Coefficient between UO_2 and Zircaloy-2.* AECL-1552, 1962.

24. Schmidt, H. E., and Richter, J. Presented at the Symposium on Oxide Fuel Thermal Conductivity, Stockholm, 1967.

25. Todreas, N. E. Pressurized subcooled light water systems. In H. Fenech (ed.), *Heat Transfer and Fluid Flow in Nuclear Systems.* Oxford: Pergamon Press, 1981.

26. Von Ubisch, H., Hall, S., and Srivastov, R. Thermal conductivities of mixtures of fission product gases with helium and argon. Presented at the 2nd U.N. International Conference on Peaceful Uses of Atomic Energy, Sweden, 1958.

27. Zebroski, E. L., Lyon, W. L., and Bailey, W. E. Effect of stoichiometry on the properties of mixed oxide U-Pu fuel. In: *Proceedings of the Conference on Safety, Fuels, and Core Design in Large Fast Power Reactors,* 11–14 October, 1965. USAEC Report ANL-7120, p. 382, Argonne National Laboratory, 1965.

PROBLEMS

Problem 8-1 Application of Kirchoff's law to pellet temperature distribution (section II)

For PWR cylindrical solid fuel pellet operating at a heat flux equal to 1.7 MW/m^2 and a surface temperature of 400°C, calculate the maximum temperature in the pellet for two assumed values of conductivities.

1. $k = 3$ W/m °C independent of temperature
2. $k = 1 + 3e^{-0.0005T}$ where T is in °C

   UO_2 pellet diameter = 10.0 mm
   UO_2 density 95% theoretical density
   Answers: 1. $T_{max} = 1817$°C
            2. $T_{max} = 1974$°C

Problem 8-2 Effect of cracking on UO_2 conductivity (section III)

For the conditions given in Example 8-1, evaluate the effective conductivity of the UO_2 pellet after cracking using the empiric relation of Eq. 8-24. Assume the gas is helium at a temperature $T_{gas} = 0.7 T_{ci} + 0.3 T_{co}$. Compare the results to those obtained in Example 8-1.

Answer: $k_{eff} = 1.73$ W/m°K

Problem 8-3 Comparison of UO$_2$ and UC fuel temperature fields (section IV)

A fuel plate is of half width $a = 10$ mm and is clad in a zircaloy sheet of thickness $\delta_c = 2$ mm. The heat is generated uniformly in the fuel.

Compare the temperature drop across the fuel plate when the fuel is UO$_2$ with that of a UC fuel for the same heat-generation rate (i.e., calculate the ratio of $T_{max} - T_{co}$ for the UC plate to that of the UO$_2$ plate).

Answer: $\Delta T_{UO_2}/\Delta T_{UC} = 4.15$

Problem 8-4 Temperature fields in fresh and irradiated fuel (section V)

Consider two conditions for heat transfer in the pellet and the pellet–clad gap of a BWR fuel pin.

• Initial uncracked pellet with no relocation
• Cracked and relocated fuel

1. For each combination, find the temperatures at the clad inner surface, the pellet outer surface, and the pellet centerline.
2. Find for each case the volume-weighted average temperature of the pellet.

Geometry and material information
Clad outside diameter = 12.52 mm
Clad thickness = 0.86 mm
Fuel–clad diametral gap = 230 μm
Initial solid pellet with density = 88%

Basis for heat transfer calculations
Clad conductivity is constant at 17 W/m °K
Gap conductance
  Without fuel relocation, 4300 W/m^2 °K,
  With fuel relocation, 31,000 W/m^2 °K;
Fuel conductivity (average) at 95% density
  Uncracked, 2.7 W/m °K,
  Cracked, 2.4 W/m °K;
Volumetric heat deposition rate: uniform in the fuel and zero in the clad

Do not adjust the pellet conductivity for restructuring.
Use Biancharia's porosity correction factor (Eq. 8-21).

Operating conditions
Clad outside temperature = 295°C
Linear heat-generation rate = 44 kW/m

Answers:

	Uncracked	Cracked
1. T_{max} (°C)	2112	2027
T_{fo} (°C)	664	398
T_{ci} (°C)	355.9	355.9
2. T_{ave} (°C)	1395.9	1213.9

Problem 8-5 Temperature field in a restructured fuel pin (section VI)

Using the conditions of Problem 8-4 for the uncracked fuel, calculate the maximum fuel temperature for the given operating conditions. Assume two-zone sintering, with $T_{sintering} = 1700$°C and $\rho_{sintered} = 98\%$ TD.

Answer: $T_v = 2140$°C

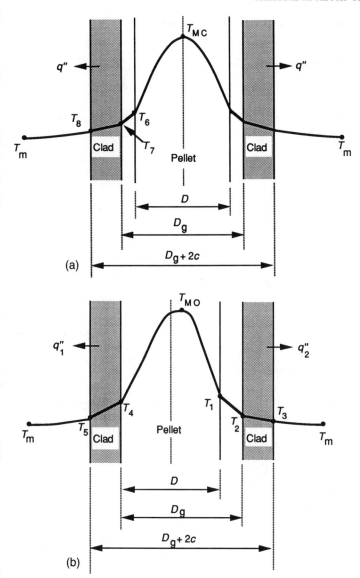

Figure 8-24 Effect of fuel offset. (*a*) Centered fuel pellet. (*b*) Effect of fuel pellet offset.

Problem 8-6 Eccentricity effects in a plate-type fuel (section VII)

A nuclear fuel element is of plate geometry (Fig. 8-24). It is desired to investigate the effects of fuel offset within the clad. For simplicity, assume uniform heat generation in the fuel, temperature-independent fuel conductivity, and heat conduction only in the gap. Calculate:

1. The temperature difference between the offset fuel and the concentric fuel maximum temperatures:

$$T_{MO} - T_{MC}$$

2. The temperature difference between the cladding maximum temperatures:

$$T_7 - T_4$$

3. The ratio of heat fluxes to the coolant.

$$\frac{q''}{q_1''} \quad \text{and} \quad \frac{q''}{q_2''}$$

$T_{coolant}$ and heat transfer coefficient to coolant may be assumed constant on both sides and for both cases. Neglect interface contact resistance for fuel and clad.

$k_f = 3.011 \text{ W/m°C (PuO}_2 - \text{UO}_2)$
$k_g = 0.289 \text{ W/m°C (He)}$
$k_c = 21.63 \text{ W/m°C (SS)}$
$D = 6.352 \text{ mm}$
$D_g = 6.428 \text{ mm}$
$c = 0.4054 \text{ mm}$
$q''' = 9.313 \times 10^5 \text{ kW/m}^3$
$h_{coolant} = 113.6 \text{ kW/m}^2\text{°C (Na)}$

> Answer: 1. $T_{MO} - T_{MC} = -23.9°C$
> 2. $T_7 - T_4 = -8.83°C$
> 3. $\dfrac{q_1''}{q''} = 1.109$ and $\dfrac{q_2''}{q''} = 0.8914$

Problem 8-7 Determining the linear power given a constraint on the fuel average temperature (section VII)

For a PWR fuel pin with pellet radius of 4.7 mm, clad inner radius of 4.89 mm, and outer radius of 5.46 mm, calculate the maximum linear power that can be obtained from the pellet such that the mass average temperature in the fuel does not exceed 1204°C (2200°F). Take the bulk fluid temperature to be 307.5°C and the coolant heat transfer coefficient to be 28.4 kW/m² °C. In the gap, consider only conduction heat transfer.

Fuel conductivity $(k_f) = 3.011 \text{ W/m °C}$
Clad conductivity $(k_c) = 18.69 \text{ W/m °C}$
Helium gas conductivity $(k_g) = 0.277 \text{ W/m °C}$

> Answer: $q' = 22.9 \text{ kW/m}$

SINGLE-PHASE FLUID MECHANICS

I APPROACH TO SIMPLIFIED FLOW ANALYSIS

The objective of the fluid mechanics analysis is to provide the velocity and pressure distributions in a given geometry for specified boundary and initial conditions. The transport equations of mass, momentum, and energy were derived for both a control volume and at a local point in Chapter 4. Theoretically, we need to solve these detailed equations simultaneously to obtain the velocity, pressure, and temperature distributions in the flow system of interest. Practically, however, we first simplify the equations to be solved by eliminating insignificant terms for the situation of interest.

Furthermore, our objective can often be achieved by applying the accumulated engineering experience in a manner that empirically relates macroscopic quantities, e.g., the pressure drop and the flow rate through a tube, without obtaining the detailed distribution of the fluid velocity or density in the tube. This engineering approach can be used whenever the flow characteristics fall within the range of previously established empiric relations.

The analytic approach and the empiric engineering relations for single-phase flow analysis are discussed in this chapter. The heat transport analysis is dealt with in Chapter 10, and the fluid mechanics of two-phase flow are considered in Chapter 11.

A Solution of the Flow Field Problem

Determination of the velocity field in a moving fluid requires simultaneous solution of the mass, momentum, and energy equations:

$$\text{Mass: } \frac{\partial}{\partial t} \rho + \nabla \cdot \rho \vec{v} = 0 \qquad (4\text{-}73)$$

Linear momentum: $\dfrac{\partial}{\partial t}\, \rho\vec{v} + \nabla \cdot \rho\vec{v}\vec{v} = \nabla \cdot (\overline{\overline{\tau}} - p\,\overline{\overline{I}}) + \rho\vec{g}$ (4-80)

Energy, in one of its various forms, e.g.,

$$\dfrac{\partial}{\partial t}\, \rho u^{\circ} + \nabla \cdot \rho u^{\circ}\vec{v} = -\nabla \cdot \vec{q}'' + \nabla \cdot (\overline{\overline{\tau}} - p\,\overline{\overline{I}}) \cdot \vec{v} + q''' + \rho\vec{v} \cdot \vec{g} \quad (4\text{-}96)$$

or equivalently:

$$\rho c_{\mathrm{p}} \dfrac{DT}{Dt} = -\nabla \cdot \vec{q}'' + q''' + \beta T \dfrac{Dp}{Dt} + \Phi \qquad (4\text{-}122a)$$

There are six unknowns in the above equations: ρ, $\vec{v}$, u° (or T), p, $\overline{\overline{\tau}}$, and $\vec{q}''$. Two additional quantities are a priori given, q''' and $\vec{g}$. Therefore the three transport equations need to be supplemented with three additional equations: the equation of state for the fluid:

$$\rho = \rho(p,T) \qquad (5\text{-}61)$$

and two constitutive equations that relate shear stress and heat flux to the unknown or given quantities (in magnitude as well as spatial gradients):

$$\overline{\overline{\tau}} = \overline{\overline{\tau}}(\rho,\vec{v},T) \qquad (5\text{-}62)$$

$$\vec{q}'' = \vec{q}''(\rho,\vec{v},T) \qquad (5\text{-}63)$$

Note that T in Eqs. 5-61 through 5-63 can be replaced by u° if the energy equations in the form 4-96 is used.

Finally, initial and boundary conditions necessary for completely specifying the solution of the differential transport equations should be supplied.

Thus given q''' and $\vec{g}$, the above six equations can be solved to obtain the six variables ρ, $\vec{v}$, T (or u°), p, $\overline{\overline{\tau}}$, $\vec{q}''$.

Often, however, we tend to use simplified forms that represent acceptable approximations to the physical conditions. Thus the complexity of the problem can be reduced, and even analytic forms for all the variables may be obtained.

B Possible Simplifications

1. The most significant of these approximations is the assumption of temperature-independent physical properties, which leads to decoupling of the velocity field solution from the energy equation, as both ρ and $\overline{\overline{\tau}}$ no longer depend on T. Thus Eqs. 4-73 and 4-80 can then be solved along with the simplified equations:

$$\rho = \rho(p) \qquad (5\text{-}61)$$

$$\overline{\overline{\tau}} = \overline{\overline{\tau}}(\rho,\vec{v}) \qquad (5\text{-}62)$$

to obtain the unknowns ρ, $\vec{v}$, p, $\overline{\overline{\tau}}$. This assumption removes the energy equation from the system of equations to be solved, as illustrated in Figure 9-1. This assumption is good in flow fields that do not span a wide range of temperatures, provided the

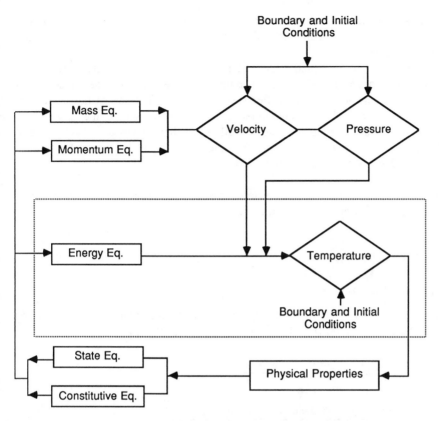

Figure 9-1 Fluid mechanics analysis schemes. The region within the dotted line is considered separately when the physical properties are assumed temperature independent. Note that in a transient situation the initial conditions for each time period (one time step) are the values of the variable at the end of the preceding time period.

selection for $\rho(p)$ and $\overline{\overline{\tau}}(\rho,\vec{v})$ is made to represent their values at an appropriate temperature within the range of interest.

Other possible simplifying assumptions are:

2. The density (ρ) is constant, which is a valid assumption when the effect of pressure as well as temperature is small. This is an incompressible flow problem with constant-temperature properties. It is a reasonable assumption for nearly all practical problems involving liquids. At high pressure it may also be applied to gases if the pressure variation within the system is small.

3. The effects of viscosity are negligible, so that $\nabla \cdot \overline{\overline{\tau}}$ is a negligible term in the momentum equation. This is called an inviscid flow problem and is appropriate for flows where the momentum effects dominate, such as the case at high flow velocities in large compartments or even with open channel flow. In some nuclear reactor components, e.g., large pipes, or within a large reactor vessel plenum, the flow may be considered inviscid.

4. The problem can be solved in the fewest reasonable number of dimensions. For example, a problem can be solved as one-dimensional if the flow in the other

dimensions is either nonexistent or very small. In that case the point equation is to be integrated over the directions that are to be eliminated, which results in the one-dimensional problem.

5. Finally, if the information desired is for the endpoint conditions (e.g., the pressure drop across a channel), the momentum integral equation can be solved, rather than the local momentum equation. For steady-state one-dimensional flows, this approach is often used, and the change in momentum flux between the inlet and the exit is related to the static pressure change as well as the shear forces and gravity forces. The laws representing shear forces depend on the flow velocity and material properties in a manner described in the next few sections.

The relevant categories of single-flow situations are given in Figure 9-2. Some of the categories that appear are further explained in this chapter.

II INVISCID FLOW

A Dynamics of Inviscid Flow

In the absence of viscosity (or shear forces), the momentum equation (Eq. 4-80) and the mass balance equation (Eq. 4-73) can be combined to yield:

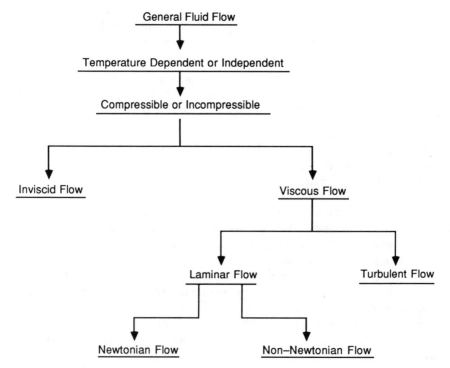

Figure 9-2 Categorization of fluid flow situations.

$$\rho \frac{\partial \vec{v}}{\partial t} + \rho \vec{v} \cdot \nabla \vec{v} = -\nabla p + \rho \vec{f} \qquad (4\text{-}91)$$

The above equation can be rewritten in the form of gradient and curl operations with the help of the definition of $\vec{v} \cdot \nabla \vec{v}$ as:

$$\rho \vec{v} \cdot \nabla \vec{v} = \frac{\rho}{2} \nabla v^2 - [\rho \vec{v} \times (\nabla \times \vec{v})] \qquad (9\text{-}1)$$

so that the inviscid flow momentum equation takes the form:

$$\frac{\partial \vec{v}}{\partial t} + \nabla \left(\frac{v^2}{2}\right) - [\vec{v} \times (\nabla \times \vec{v})] = -\frac{1}{\rho} \nabla p + \vec{f} \qquad (9\text{-}2a)$$

The term $\nabla \times \vec{v}$ represents an angular rotation of the fluid element and is referred to as the *vorticity* of the fluid ($\vec{\omega}$). It can be shown [2] that the vorticity is twice the angular velocity of any two perpendicular lines intersecting at a point (Fig. 9-3). Thus the inviscid flow momentum equation takes the form:

$$\frac{\partial \vec{v}}{\partial t} + \nabla \left(\frac{v^2}{2}\right) - \vec{v} \times \vec{\omega} = -\frac{1}{\rho} \nabla p + \vec{f} \qquad (9\text{-}2b)$$

B Bernoulli's Integral

1 Time-Dependent Flow. The body forces can often be expressed as the gradient of a scalar function called the potential function (ψ):

$$\vec{f} = -\nabla \psi \qquad (4\text{-}21)$$

For example, in a gravity field $\psi = gz$, where g = the gravitational constant, and z = the coordinate in the vertical direction:

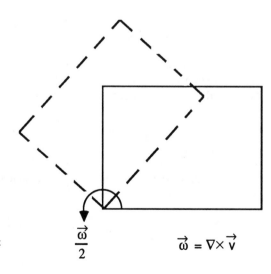

Figure 9-3 Rotation of a fluid element about a point.

$$\frac{\vec{\omega}}{2}$$

$$\vec{\omega} = \nabla \times \vec{v}$$

$$\vec{f} = -\left(\frac{\partial}{\partial z} gz\right) \vec{k} = -g\vec{k} = \vec{g} \qquad (9\text{-}3)$$

When the body forces are derived from a field potential, they are said to be conservative. For conservative body forces, the inviscid fluid momentum (Eq. 9-2) can be written as:

$$\frac{\partial \vec{v}}{\partial t} + \frac{1}{\rho} \nabla p + \nabla \left(\psi + \frac{v^2}{2}\right) = \vec{v} \times \vec{\omega} \qquad (9\text{-}4)$$

Let $d\vec{r}$ be an elementary displacement (dx, dy, dz). Note that for an arbitrary scalar quantity (X), $\nabla X \cdot d\vec{r} = dX$, as:

$$\begin{aligned}
\nabla X \cdot d\vec{r} &= \left(\vec{i}\, \frac{\partial X}{\partial x} + \vec{j}\, \frac{\partial X}{\partial y} + \vec{k}\, \frac{\partial X}{\partial z}\right) \cdot (\vec{i}\, dx + \vec{j}\, dy + \vec{k}\, dz) \\
&= \frac{\partial X}{\partial x}\, dx + \frac{\partial X}{\partial y}\, dy + \frac{\partial X}{\partial z}\, dz = dX
\end{aligned} \qquad (9\text{-}5)$$

Thus when the terms in Eq. 9-4 are multiplied by $d\vec{r}$, we get:

$$\frac{\partial \vec{v}}{\partial t} \cdot d\vec{r} + \frac{dp}{\rho} + d\left(\psi + \frac{v^2}{2}\right) = (\vec{v} \times \vec{\omega}) \cdot d\vec{r} \qquad (9\text{-}6)$$

The right-hand side term is zero if $\vec{\omega} = 0$, i.e., if the flow is irrotational, or if $d\vec{r} = d\vec{s}$, when $d\vec{r}$ is directed along a stream line $d\vec{s}$ and hence in the same direction as $\vec{v}$.

In the second case it is true because of the identity:

$$(\vec{v} \times \vec{\omega}) \cdot d\vec{r} = (\vec{\omega} \times d\vec{r}) \cdot \vec{v} = (d\vec{r} \times \vec{v}) \cdot \vec{\omega} = 0 \qquad (9\text{-}7)$$

as $d\vec{r}$ is parallel to $\vec{v}$ and $d\vec{r} \times \vec{v} = 0$. Thus either when the flow is irrotational or when we are considering changes along a stream line:

$$\frac{\partial \vec{v}}{\partial t} \cdot d\vec{r} + \frac{dp}{\rho} + d\left(\psi + \frac{v^2}{2}\right) = 0 \qquad (9\text{-}8)$$

For a barotropic fluid, where ρ is only a function of p, we can integrate Eq. 9-8 along a stream line, or between any two points if the flow is irrotational, to obtain the general form of the Bernoulli integral equation:

$$\int \frac{\partial \vec{v}}{\partial t} \cdot d\vec{r} + \int \frac{dp}{\rho} + \left(\psi + \frac{v^2}{2}\right) = f(t) \qquad (9\text{-}9)$$

Note that $f(t)$, the constant of integration, is the same for any two points at a given time but might change with time. Eq. 9-9 is generally written as the integral between two positions along the stream line:

$$\int_1^2 \frac{\partial \vec{v}}{\partial t} \cdot d\vec{r} + \int_1^2 \frac{dp}{\rho} + \Delta\left(\psi + \frac{v^2}{2}\right) = 0 \qquad (9\text{-}10)$$

where

$$\Delta \left(\psi + \frac{v^2}{2} \right) = \left(\psi + \frac{v^2}{2} \right)_2 - \left(\psi + \frac{v^2}{2} \right)_1$$

2 Steady-state flow. At steady state the Bernoulli equation is given by:

$$\int_1^2 \frac{dp}{\rho} + \Delta \left(\psi + \frac{v^2}{2} \right) = 0 \qquad (9\text{-}11)$$

which for an incompressible flow is given by:

$$\Delta \left(\frac{p}{\rho} + \psi + \frac{v^2}{2} \right) = 0 \qquad (9\text{-}12)$$

In most cases the body forces consist only of gravitational forces, $\psi = gz$; and a total head (z°) is usually defined as the actual height (z) plus the static head and the velocity head:

$$z^\circ = \frac{p}{\rho g} + \frac{v^2}{2g} + z \qquad (9\text{-}13)$$

Thus the total head of an incompressible and inviscid flow in a channel is constant along a stream line or at any location if the flow is also irrotational.

3 Flow through a nozzle. The flow of an incompressible fluid through a nozzle can be predicted by application of Eq. 9-12. Consider a nozzle of diameter d_2 attached to a pipe of diameter d_1 (Fig. 9-4). From Equation 9-12, when gravity is the only body force, we get:

$$\left(\frac{p}{\rho g} + \frac{v^2}{2g} + z \right)_1 = \left(\frac{p}{\rho g} + \frac{v^2}{2g} + z \right)_2 \qquad (9\text{-}14)$$

For a horizontal nozzle, however,

$$z_1 = z_2 \qquad (9\text{-}15)$$

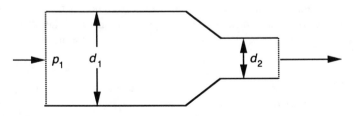

Figure 9-4 Nozzle.

From the mass balance consideration, we get:

$$\rho v_1 \frac{\pi d_1^2}{4} = \rho v_2 \frac{\pi d_2^2}{4}$$

which leads to:

$$v_1 = v_2 \left(\frac{d_2}{d_1}\right)^2 \tag{9-16}$$

Thus substituting from Eqs. 9-15 and 9-16 into 9-14, we get:

$$\frac{v_2^2}{2}\left[\left(\frac{d_2}{d_1}\right)^4 - 1\right] = \frac{p_2 - p_1}{\rho}$$

or

$$v_2 = \sqrt{\frac{2(p_1 - p_2)}{\rho[1 - (d_2/d_1)^4]}} \tag{9-17a}$$

It should be stated that under real conditions the velocity in the nozzle is less than that predicted by Eq. 9-17a. The departure from ideal conditions due to fluid shear forces is accounted for by introducing a nozzle coefficient (C_D) such that the velocity v_2 is given by:

$$v_2 = C_D \sqrt{\frac{2(p_1 - p_2)}{\rho[1 - (d_2/d_1)^4]}} \tag{9-17b}$$

The values of C_D are typically 0.3 to 0.7 depending on the nozzle geometry.

The flow rate is given by

$$\dot{m} = \rho v_2 \left(\frac{\pi d_2^2}{4}\right) = C_D \left(\frac{\pi d_2^2}{4}\right) \sqrt{\frac{2\rho(p_1 - p_2)}{[1 - (d_2/d_1)^4]}} \tag{9-18}$$

Example 9-1 Venturi meter for flow measurement

PROBLEM A venturi meter is inserted into one flow loop of a PWR as shown in Figure 9-5. The dimensions are $d_1 = 28$ in. (0.711 m) and $d_2 = 27$ in. (0.686 m). The venturi meter is mostly filled with stagnant water that is separated from the primary flow by an air bubble. This setup enables visual determination of the difference in water elevation levels (h) and hence the pressure difference between the contraction and the loop. Given that the height (h) is 3 ft (0.914 m) and the water density is approximately $\rho_w \simeq 1000$ kg/m^3, what are the velocity and the mass flow rate in the loop?

SOLUTION From Figure 9-5, the loop velocity (v_1) is related to throat velocity (v_2) from a mass balance such that:

$$v_1 = v_2 \left(\frac{d_2}{d_1}\right)^2$$

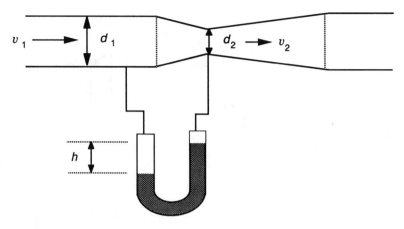

Figure 9-5 Venturi meter.

Using this equation and Eq. 9-17a, v_1 can be found from:

$$v_1 = \sqrt{\frac{2(p_1 - p_2)}{\rho_w[(d_1/d_2)^4 - 1]}}$$

Because $p_1 - p_2 = \rho_w gh$,

$$v_1 = \sqrt{\frac{2gh}{[(d_1/d_2)^4 - 1]}} = \sqrt{\frac{2(9.81 \text{ m/s}^2)(0.914 \text{ m})}{\left[\left(\dfrac{0.711}{0.686}\right)^4 - 1\right]}} = 10.71 \text{ m/s}$$

The mass flow rate through the loop is:

$$\dot{m} = \rho_w v_1 A_1 = (1000 \text{ kg/m}^3)(10.71 \text{ m/s})\left[\frac{\pi}{4}(0.711 \text{ m})^2\right] = 4252 \text{ kg/s}$$

C Incompressible Inviscid Flow through a Variable-Geometry Channel

Consider the time-dependent flow of an incompressible, inviscid fluid created by the pressure difference between two ends of a channel. To make this case as general as possible, we consider a channel consisting of several sections of variable geometry. Typical transient situations arise during normal reactor operation, such as the startup of a hydraulic loop or incidents such as inadvertent closing of a valve. The inviscid-flow assumption in this case must be interpreted as meaning that the frictional pressure drop is negligible compared to pressure drops created by temporal and spatial acceleration of the fluid. Figure 9-6 illustrates the system under consideration.

We seek to determine the variation of the flow rate, $\dot{m}(t)$, through the system of N sections for an arbitrary pressure drop between the inlet and the exit, $p_{in} - p_{out}$.

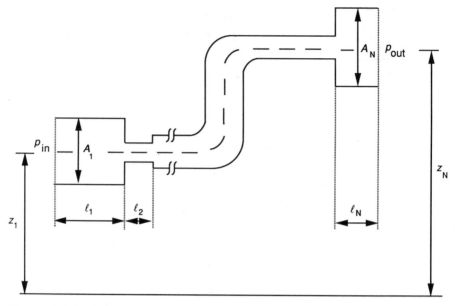

Figure 9-6 Hydraulic system configuration.

Assuming that the flow can be considered incompressible and one-dimensional (irrotational assumption implied) everywhere in the flow duct, Eq. 9-10 yields:

$$\int_1^N \frac{\partial \vec{v}}{\partial t} \cdot d\vec{r} + \frac{p_{\text{out}} - p_{\text{in}}}{\rho} + \Delta\psi + \Delta\frac{v^2}{2} = 0$$

The integral can be performed by recalling that for incompressible flow the flow rate ($\dot{m}$) is not space-dependent:

$$
\begin{aligned}
\int_1^N \frac{\partial \vec{v}}{\partial t} \cdot d\vec{r} &= \frac{\partial}{\partial t} \int_1^N \vec{v} \cdot d\vec{r} = \frac{\partial}{\partial t} \int_1^N \frac{\dot{m}}{\rho A} d\ell \\
&= \frac{1}{\rho} \frac{d\dot{m}}{dt} \int_1^N \frac{d\ell}{A} = \frac{1}{\rho} \frac{d\dot{m}}{dt} \sum_{n=1}^N \frac{\ell_n}{A_n}
\end{aligned}
\tag{9-19}
$$

The term $\sum_{n=1}^N \frac{\ell_n}{A_n}$ represents an equivalent inertia length $\left(\frac{\ell}{A}\right)_T$ for the system and can be computed knowing the system dimensions. Thus after multiplying Eq. 9-19 by ρ, the Bernoulli integral gives:

$$\left(\frac{\ell}{A}\right)_T \frac{d\dot{m}}{dt} + (p_{\text{out}} - p_{\text{in}}) + \rho g(z_N - z_1) + \frac{\dot{m}^2}{2\rho}\left(\frac{1}{A_N^2} - \frac{1}{A_1^2}\right) = 0 \tag{9-20}$$

Note that for incompressible flow only the inertia term (first term alone) involves a channel integral quantity, whereas all other terms represent conditions at the endpoints. If $p_{\text{in}} - p_{\text{out}}$ is prescribed, this equation can be solved to determine the variation of the flow rate in time.

Example 9-2 Pump start-up in an inviscid flow loop

PROBLEM Consider a reactor flow loop, shown in Figure 9-7. We are interested in determining the time it takes the coolant, initially at rest, to reach a steady-state flow level once the pump is turned on. From a hydraulic standpoint, the loop can be modeled as a one-dimensional flow path where a pressure head $\Delta p = p_{in} - p_{out}$ is provided by the pump. Assume the pump inlet and outlet are at the same elevation.

Starting from Eq. 9-20, derive an expression for $\dot{m}(t)$. Note in particular what happens when $A_5 = A_1$. Does it seem reasonable? What has been neglected in deriving Eq. 9-20 that, if included, would give a more reasonable answer for the case $A_1 = A_5$?

Given that the pump provides a pressure head equivalent to 85.3 m of water, and that the density of water is approximately $\rho_w \simeq 1000 \text{ kg/m}^3$, evaluate the various parameters appearing in the expression for $\dot{m}(t)$. What is the expected steady-state value of the flow rate? How long does it take the system to reach 90% of this value?

(Note that in Example 9-1 we estimated the flow rate through a venturi meter located at the pump outlet piping of a PWR system with hydraulic characteristics similar to those of this reactor loop.)

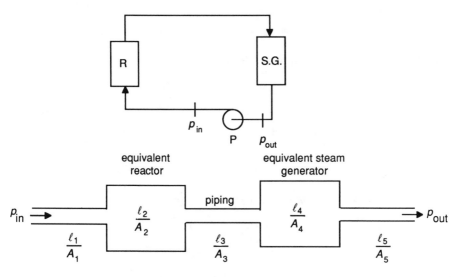

Table of Characteristic Dimensions

Component No.	Component Name	ℓ(m)	A(m^2)
1	Pump Outlet Piping	8.0	0.4
2	Reactor	14.5	20.9
3	Reactor Outlet Piping	17.0	0.4
4	Steam Generator	16.5	1.5
5	Pump Inlet Piping	10.0	0.35

Figure 9-7 Simplified reactor loop with a table of characteristic dimensions.

SOLUTION First, because the pump inlet and outlet are at the same elevation there is no net change in the gravity head. Thus the gravity term can be dropped from Eq. 9-20. Then, because $\Delta p = p_{in} - p_{out}$, we have:

$$\left(\frac{\ell}{A}\right)_T \frac{d\dot{m}}{dt} - \Delta p + \frac{\dot{m}^2}{2\rho}\left(\frac{1}{A_5^2} - \frac{1}{A_1^2}\right) = 0$$

$$\left(\frac{\ell}{A}\right)_T \frac{d\dot{m}}{dt} = \Delta p - \dot{m}^2\left[\frac{1}{2\rho}\left(\frac{1}{A_5^2} - \frac{1}{A_1^2}\right)\right]$$

Then

$$\frac{\left(\dfrac{\ell}{A}\right)_T \dfrac{1}{\Delta p} d\dot{m}}{1 - \dot{m}^2 C^2} = dt$$

where

$$C^2 = \frac{1}{2\rho\Delta p}\left(\frac{1}{A_5^2} - \frac{1}{A_1^2}\right)$$

Hence

$$\frac{d\dot{m}}{(1 - \dot{m}C)(1 + \dot{m}C)} = \Delta p \left(\frac{A}{\ell}\right)_T dt$$

Expanding the left-hand side in partial fractions:

$$\frac{1/2\, d\dot{m}}{1 - \dot{m}C} + \frac{1/2\, d\dot{m}}{1 + \dot{m}C} = \Delta p \left(\frac{A}{\ell}\right)_T dt$$

Integrating the last equation we get:

$$-\frac{1}{2C}\ell n\,(1 - \dot{m}C) + \frac{1}{2C}\ell n\,(1 + \dot{m}C) = \Delta p \left(\frac{A}{\ell}\right)_T t + C_o$$

Because at $t = 0$, $\dot{m} = 0$, we find that $C_o = 0$, and:

$$\frac{1 + \dot{m}C}{1 - \dot{m}C} = \exp\left[2C\,\Delta p \left(\frac{A}{\ell}\right)_T t\right]$$

or

$$\dot{m} = \frac{1}{C}\left\{\frac{\exp\left[2C\Delta p \left(\dfrac{A}{\ell}\right)_T t\right] - 1}{\exp\left[2C\Delta p \left(\dfrac{A}{\ell}\right)_T t\right] + 1}\right\}$$

Thus we see that as $t \to \infty$, $\dot{m} \to 1/C$. Hence, at steady state there is a balance between the pressure driving force and the acceleration pressure drop.

We can evaluate C using the information given above.

$$\Delta p = \rho g \, \Delta z$$
$$\Delta p = (1000 \text{ kg/m}^3)(9.81 \text{ m/s}^2)(85.3 \text{ m})$$
$$= 8.37 \times 10^5 \text{ kg/m} \cdot \text{s}^2$$

$$C^2 = \frac{1}{2(1000 \text{ kg/m}^3)(8.37 \times 10^5 \text{ kg/m} \cdot \text{s}^2)}$$
$$\cdot \left(\frac{1}{(0.35 \text{ m}^2)^2} - \frac{1}{(0.4 \text{ m}^2)^2} \right) = 1.143 \times 10^{-9} \text{ s}^2/\text{kg}^2$$

$\dot{m}$ at steady state $= \dfrac{1}{C} = 2.96 \times 10^4 \text{ kg/s}$

To obtain the transient time constant we evaluate the term appearing in the exponent:

$$\left(\frac{\ell}{A} \right)_T = \sum_{n=1}^{5} \left(\frac{\ell}{A} \right)_n$$

$$= \left(\frac{8}{0.4} \right) + \left(\frac{14.5}{20.9} \right) + \left(\frac{17.0}{0.4} \right) + \left(\frac{16.5}{1.5} \right) + \left(\frac{10.0}{0.35} \right)$$

$$= 102.8 \text{ m}^{-1}$$

$$2C \, \Delta p \left(\frac{A}{\ell} \right)_T = 2(1.143 \times 10^{-9})^{1/2} \text{ s/kg} \, (8.37 \times 10^5) \text{ kg/m} \cdot \text{s}^2 \left(\frac{1}{102.8} \text{ m} \right)$$

$$= 0.551 \text{ s}^{-1}$$

To find the time it takes for the system to reach 90% of its full flow value, let:

$$0.9 = \frac{e^{0.551t} - 1}{e^{0.551t} + 1}$$
$$0.9(e^{0.551t} + 1) = e^{0.551t} - 1$$
$$1.9 = 0.1e^{0.551t}$$
$$t = \left(\frac{1}{0.551} \, \ell n \, 19 \right) = 5.34 \text{ s}$$

We note some interesting observations here. First, in Example 9-1 the flow rate through the PWR loop (measured by a venturi meter located at the pump outlet piping) was approximately 4250 kg/s. However, the result of this calculation shows that the inviscid flow rate would be 29,000 kg/s and that the flow rate is sensitive to the values of A_5 and A_1.

When $A_5 = A_1$, we can no longer use the above expressions for $\dot{m}(t)$ because $C \to 0$. The equation governing the flow rate is still Eq. 9-20 but with both the gravity and acceleration terms set to zero (remember the friction term does not show up because the flow is inviscid):

$$\left(\frac{\ell}{A} \right)_T \frac{d\dot{m}}{dt} - \Delta p = 0$$

or

$$\dot{m} = \Delta p \left(\frac{A}{\ell} \right)_T t$$

Hence there is no limit to the flow rate, as there is no resistance to the applied force in such a system. In reality, form and friction forces are encountered in a flow loop with a viscous fluid. This subject is discussed in Example 9-3. In addition, the pressure head of the pump decreases with the flow rate. Thus if $\Delta p = A - B \dot{m}$, the maximum achievable flow rate is limited to $\dot{m}_{max} = A/B$ even in the absence of friction forces. Pump head representation is described in Chapter 3, Vol. II.

D Real Flow through a Variable-Geometry Channel

In real fluids, the effects of compressibility and viscosity may have to be considered.

1 Compressibility. Let us qualitatively discuss the effects of compressibility. If the pressure at one end of the system changes abruptly, this effect is not experienced immediately at the other end, as pressure waves are transmitted with finite velocities in compressible fluids. Thus the flow rate at the other end of the system cannot change until the pressure wave arrives there.

Because the sonic wave propagation velocity is high in liquids (more than 1000 m/s), it is generally acceptable to assume that pressure changes propagate instantaneously and that the flow rate responds instantaneously to pressure transients. This assumption is of course equivalent to assuming incompressible flow. The assumption of incompressible flow, however, breaks down as the time scale of interest becomes comparable to the sonic transit time in the system. (Sonic effects on flow transient calculations are discussed in Chapter 2, Vol. II in some detail.)

At the other end of the scale, as the transients become slower, the transient (or inertia) term in Eq. 9-20, $(\ell/A)_T \, d\dot{m}/dt$, diminishes in magnitude, and for slow enough transients it can be completely neglected. The problem can then be solved in a quasi-steady-state fashion, i.e., by calculating the instantaneous flow rate using the steady-state relation between momentum and the driving pressure difference.

2 Viscous effects. Let us now consider the effects of viscosity, which introduces shear within the fluid (internal friction) as well as friction with the confining wall. A difficulty arises because we do not know how the shear stress varies under transient conditions, e.g., due to a change in the velocity profile. For engineering applications this difficulty is often ignored, and the "quasi-steady-state" values are assumed to hold for the transient conditions. This approximation is strictly valid for relatively slow transients for which it can be argued that dynamic effects in the governing equations are small.

The shear effects lead to a loss of the driving pressure, so that a pressure loss term must be added to the Bernoulli equation. Eq. 9-20 takes then the form:

$$\left(\frac{\ell}{A}\right)_T \frac{d\dot{m}}{dt} + p_{out} - p_{in} + \rho g(z_N - z_1) + \frac{\dot{m}^2}{2\rho}\left(\frac{1}{A_N^2} - \frac{1}{A_1^2}\right) + \Delta p_{loss} = 0 \quad (9\text{-}21)$$

where Δp_{loss} is to be evaluated either from a viscous flow analysis or empirically. It is usually decomposed into the pressure losses due to wall friction ($\Delta p_{friction}$) and flow form losses (Δp_{form}).

Note that Eq. 9-21 can be rearranged into the form:

$$p_{in} - p_{out} = \Delta p_{inertia} + \Delta p_{acc} + \Delta p_{gravity} + \Delta p_{friction} + \Delta p_{form} \qquad (9\text{-}22)$$

where: $\Delta p_{inertia} = \left(\dfrac{\ell}{A}\right)_T \dfrac{d\dot{m}}{dt}$

$$\Delta p_{acc} = \frac{\dot{m}^2}{2\rho}\left(\frac{1}{A_N^2} - \frac{1}{A_1^2}\right)$$

$$\Delta p_{gravity} = \rho g(z_N - z_1)$$

The pressure loss due to an abrupt change in flow direction and/or geometry is usually called a *form loss*. The pressure head loss due to form losses is, in practice, related to the kinetic pressure, so the pressure loss is given by:

$$\Delta p_{form} \equiv K\left(\frac{\rho v_{ref}^2}{2}\right) \qquad (9\text{-}23)$$

The reference velocity in Eq. 9-23 is usually the higher of the two velocities on both sides of the abrupt flow area change.

The frictional pressure losses within a pipe can be written in the form of Eq. 9-23. The frictional pressure drop coefficient is, however, conveniently taken to be proportional to the length of the flow channel and inversely proportional to the channel diameter, so the pressure drop is, in practice, given by:

$$\Delta p_{friction} \equiv \bar{f}\,\frac{L}{D}\left(\frac{\rho v_{ref}^2}{2}\right) \qquad (9\text{-}24)$$

The average friction factor $\bar{f}$ depends on the channel geometry and flow velocity as is discussed in sections III and IV for laminar flow and turbulent flow, respectively. The reference velocity in Eq. 9-24 is the average velocity in the channel.

The modified Bernoulli's equation (Eq. 9-21) can be used to obtain the effective pressure loss between points 1 and 2 at steady state:

$$\left(\frac{p}{\rho} + gz + \frac{v^2}{2}\right)_1 - \left(\frac{p}{\rho} + gz + \frac{v^2}{2}\right)_2 = K_T\frac{v_{ref}^2}{2} + \bar{f}\,\frac{L}{D}\frac{v_{ref}^2}{2} \qquad (9\text{-}25)$$

where:

$$K_T\frac{v_{ref}^2}{2} = \sum_i K_i\frac{(v_{ref}^2)_i}{2} \qquad (9\text{-}26)$$

and L and D = length and diameter, respectively, of the pipe connecting the two points 1 and 2.

For incompressible fluids and steady-state flow, the velocity varies inversely to the flow area; thus:

$$\frac{K_T}{A_{ref}^2} = \sum_i \frac{K_i}{A_i^2} \qquad (9\text{-}27)$$

Table 9-1 Form loss coefficients for various flow restrictions*

Parameter		K	Reference velocity
Pipe entrance from a plenum			
Well rounded entrance to pipe		0.04	In pipe
Slightly rounded entrance to pipe		0.23	In pipe
Sharp-edged entrance		0.50	In pipe
Projecting pipe entrance		0.78	In pipe
Pipe exit to a plenum			
Any pipe exit		1.0	In pipe
Sudden changes in cross-sectional area			
Sudden contraction		$0.5 (1 - \beta^2)$	Downstream
Sudden expansion		$(1 - \beta)^2$	Upstream
where $\beta \equiv \dfrac{\text{small cross-sectional area}}{\text{large cross-sectional area}}$			
	$(L/D)_{\text{equiv}}$		
Fittings[†]			
90° Standard elbow	30	0.35–0.9	
90° Long-radius elbow	20	0.2 –0.6	
45° Standard elbow	16	0.17–0.45	
Standard tee (flow through run)	20	0.2 –0.6	
Standard tee (flow through branch)	60	0.65–1.70	
Valves (various types)			
Fully open		0.15–15.00	
Half-closed		13 –450	

*Approximate values; consult Idelchik [13] for extensive tabulation. Also, section VII gives an accounting for the theoretical basis for obtaining K.

[†]Values of K depend on the pipe diameter.

and

$$K_T = \sum_i K_i \left(\frac{A_{\text{ref}}}{A_i}\right)^2 \tag{9-28}$$

The relation in Eq. 9-28 allows all the partial pressure losses to be referred to the velocity at some reference cross section (A_{ref}).

Loss coefficients for typical flow geometry changes are given in Table 9-1. The methodology that may be applied to theoretically derive such coefficients is described in section VII.

Example 9-3 Pump start-up for a viscous flow loop

PROBLEM Let us return to the problem of Example 9-2 and solve for the time-dependent flow rate, taking frictional losses into account. Eq. 9-21 is now specified for our horizontal system, so it becomes:

$$\left(\frac{\ell}{A}\right)_T \frac{d\dot{m}}{dt} - \Delta p + \frac{\dot{m}^2}{2\rho}\left(\frac{1}{A_5^2} - \frac{1}{A_1^2}\right) + K_R \frac{\rho V_R^2}{2} + K_{SG} \frac{\rho V_{SG}^2}{2} + \sum_i f \frac{L_i}{D_i} \frac{\rho v_i^2}{2} = 0$$

where K_R and K_{SG} = form pressure loss coefficients for the reactor and steam generator, respectively, and the friction pressure loss term is a summation over the different pipes in the system assuming a constant friction factor (f). Writing it in terms of the total mass flow rate, we find:

$$\left(\frac{\ell}{A}\right)_T \frac{d\dot{m}}{dt} - \Delta p + \frac{\dot{m}^2}{2\rho}\left(\frac{1}{A_5^2} - \frac{1}{A_1^2} + \frac{K_R}{A_R^2} + \frac{K_{SG}}{A_{SG}^2} + \sum_i f \frac{L_i}{D_i} \frac{1}{A_i^2}\right) = 0$$

Taking $K_R = 18$, $K_{SG} = 52$, $f = 0.015$, $\rho = 1000$ kg/m^3, and the pump head as 85.3 m, and using the information on lengths and diameters from the previous problem, what is the asymptotic value of the flow rate?

SOLUTION Note that this differential equation is of the same form as the one solved in the previous problems. The solution to the equation is still:

$$\dot{m}(t) = \frac{1}{C}\left[\frac{e^{2C \Delta p\left(\frac{A}{\ell}\right)_T t} - 1}{e^{2C \Delta p\left(\frac{A}{\ell}\right)_T t} + 1}\right]$$

where now:

$$C^2 = \frac{1}{2\rho\Delta p}\left(\frac{1}{A_5^2} - \frac{1}{A_1^2} + \frac{K_R}{A_R^2} + \frac{K_{SG}}{A_{SG}^2} + \sum_i f \frac{L_i}{D_i} \frac{1}{A_i^2}\right)$$

Evaluating C, we find:

$$\Delta p = 85.3 \text{ (m)} \times 9.81 \text{ (m/s}^2) \times 1000 \text{ (kg/m}^3)$$
$$= 8.37 \times 10^5 \text{ kg/m s}^2$$

$$C^2 = \frac{1}{2(1000)(8.37 \times 10^5)}$$
$$\cdot \left\{\left[\frac{1}{(0.35)^2} - \frac{1}{(0.4)^2} + \frac{18}{(20.9)^2} + \frac{52}{(1.5)^2} + 0.015\right.\right.$$
$$\cdot \left[\frac{8}{0.714}\frac{1}{(0.4)^2} + \frac{17}{0.714}\left(\frac{1}{0.4}\right)^2 + \frac{10}{0.668}\left(\frac{1}{0.35}\right)^2\right]\right\}$$
$$= 1.803 \times 10^{-8} \text{ s}^2/\text{kg}^2$$

$$\dot{m} \text{ (steady state)} = \frac{1}{C} = 7447 \text{ kg/s}$$

which is in more reasonable agreement with the practice in real plants.

Example 9-4 Work requirement of a pump

PROBLEM For the system of Example 9-3, how much power is required for the pump in steady state if the pump efficiency (η) is 85%? Does it constitute a significant fraction of the work produced by a power plant?

SOLUTION The pumping power for a 100% efficient pump was introduced in Chapter 2, section V as:

$$\text{Pump power} = (\Delta p)(A)(V) = (\Delta p)(\dot{m}/\rho) \tag{2-8}$$

For an 85% efficient pump we can estimate the pumping power from:

$$\text{Pump power} = \frac{(\Delta p)(\dot{m})}{\eta\rho} = \frac{(8.37 \times 10^5 \text{ kg/m} \cdot \text{s}^2)(7447 \text{ kg/s})}{(0.85)(1000 \text{ kg/m}^2)}$$
$$= 7.33 \times 10^6 \text{ W} = 7.33 \text{ MW}$$

The typical electrical output of a large PWR power plant is on the order of 1000 MW, so the work required to run the pump is essentially negligible.

III VISCOUS FLOW FUNDAMENTALS

A Viscosity

The shear stress $\bar{\bar{\tau}}$ in the momentum balance Eq. 4-80 arises from the resistance of a fluid to move with a uniform velocity when only a portion of it is subjected to an externally imposed velocity. Consider as an example a fluid layer initially at rest between two plates. If the upper plate begins to move to the right at a constant velocity (V_x), the rest of the fluid, with time, acquires a finite velocity. Thus the x momentum is said to be transportable in the y direction, as indicated in Figure 9-8.

When eventually a steady-state velocity profile is established, a force (F_x) is required to maintain the motion of the upper plate. The value of F_x is generally a function of the velocity (V_x), the distance Y between the two plates, and the fluid properties, so that:

$$F_x = F_x(V_x, Y, \text{fluid properties}) \tag{9-29}$$

For most ordinary fluids, the velocity profile is linear in y, and the force can be expressed as:

$$F_x = \tau_{yx}A = \mu\left(\frac{V_x}{Y}\right)A \tag{9-30}$$

where A = plate area, τ = shear force per unit area (i.e., shear stress), and μ = dynamic viscosity of the fluid. Such fluid behavior is called Newtonian, and the stress tensor components can be written as (see Chapter 4 for the convention on direction and sign of the shear stresses):

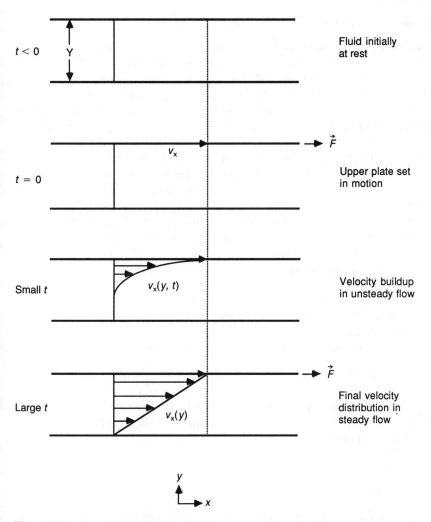

Figure 9-8 Laminar velocity profile in a fluid between two plates when the upper plate moves at a constant velocity (V_x).

$$\tau_{ij} = \mu \left(\frac{\partial v_i}{\partial x_j} + \frac{\partial v_j}{\partial x_i} \right) \tag{4-85}$$

Fluids that do not follow the Newtonian mechanics include slurries and aerosol-carrying gases.

The kinematic viscosity (ν) is defined as:

$$\nu \equiv \frac{\mu}{\rho}$$

B Viscosity Changes with Temperature and Pressure

Molecular movement can be used to predict the dynamic viscosity of gases and liquids as well as the dependence of viscosity on the pressure and temperature [2]. The data on the dynamic viscosity of ordinary fluids, e.g., water and air, can be generalized as a function of the critical (thermodynamic) pressure (p_c) and temperature (T_c). In Figure 9-9, the reduced viscosity (μ/μ_c; where μ_c = viscosity at the critical point) is plotted against the reduced temperature (T/T_c) and the reduced pressure (p/p_c). It is seen from Figure 9-9 that at a given temperature the viscosity of a gas approaches a limit as the pressure approaches zero. For ideal gases, the dynamic viscosity is independent of pressure (note that gases can be considered ideal at pressures much lower than their critical pressures). Also seen are an increase in gas viscosity and a decrease in liquid viscosity with increasing temperature. This picture is also clear in the plot of viscosity of various materials given in Figure 9-10. In Figure 9-11 the trend of dependence of the kinematic viscosity (ν) on temperature is found to be similar to that of the dynamic viscosity (μ).

In the absence of empiric data, μ_c can be estimated from:

$$\mu_c = 7.7\ M^{1/2}\ p_c^{2/3} T_c^{-1/6} \tag{9-31}$$

where μ_c is in micropoise, p_c is in atmospheres, and T_c is in °K; M = molecular weight of the material.

For a mixture of gases, pseudocritical properties can be used:

$$p_c = \sum_n x_n p_{cn}; \quad T_c = \sum_n x_n T_{cn}; \quad \mu_c = \sum_n x_n \mu_{cn} \tag{9-32}$$

where x_n = mole fraction of the component n. These pseudocritical properties can be used along with Figure 9-9 to determine the mixture viscosity, following the same procedure as that for pure fluids.

C Boundary Layer

For most practical flow conditions, the effects of viscosity on the flow over a surface can be assumed as confined to a "thin" region close to the surface. This region is called the *boundary layer*. The velocity of the fluid at the surface is taken to be zero. For external flows, the flow away from the layer can be treated as inviscid (Fig. 9-12). Thus by definition the boundary layer thickness is taken as the region in which the velocity changes from a free stream (inviscid) velocity to zero at the surface. In reality, the velocity at the fluid side of the boundary layer is taken to be about 99% of the free stream velocity to account for the presence of a weak effect of the viscosity even in the bulk of the flow.

In the case of internal flows, e.g., flow inside a tube, the boundary layers are assumed to start developing at the entrance of the channel and grow from the surface until they meet the lines of symmetry in the channel. Thus for a tube the ultimate boundary layer thickness is the tube radius (Fig. 9-13).

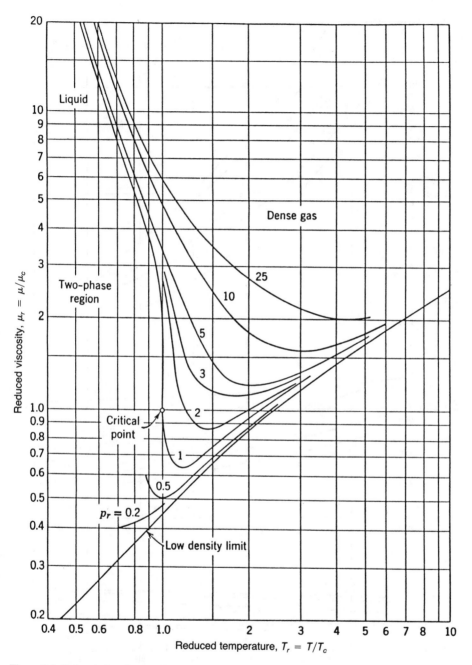

Figure 9-9 Reduced viscosity as a function of temperature for various values of reduced pressure. *(From Bird et al. [2].)*

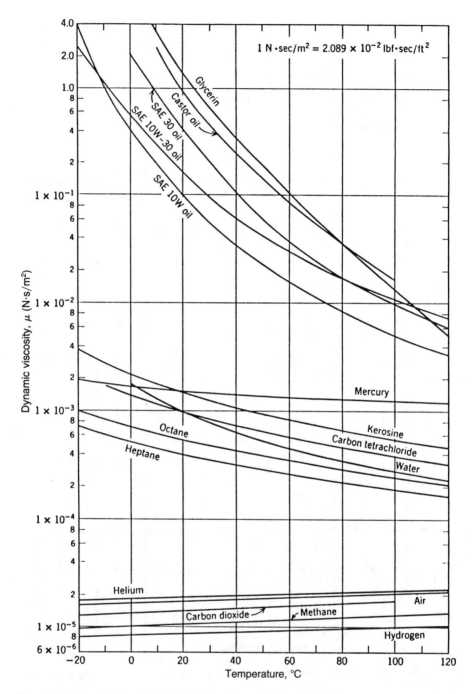

Figure 9-10 Dynamic viscosity of fluids. *(From Fox and McDonald [10].)*

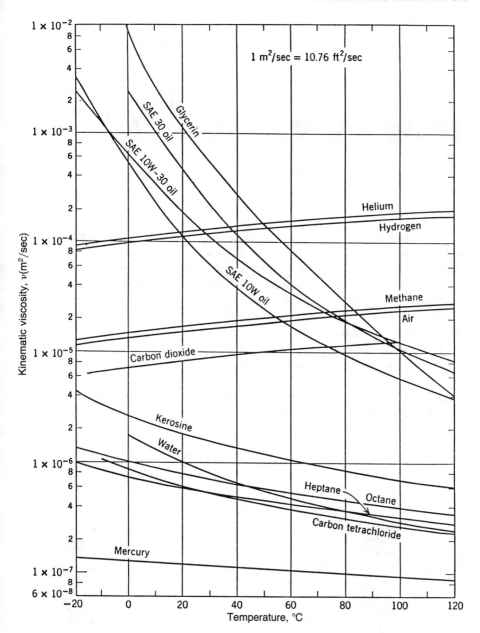

Figure 9-11 Kinematic viscosity of fluids. *(From Fox and McDonald [10].)*

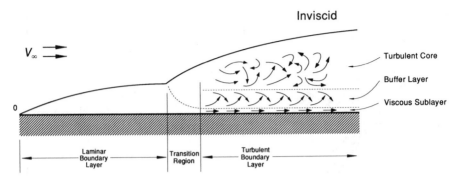

Figure 9-12 Boundary layer velocity distribution for flow on an external surface.

The length from the channel entrance required for the boundary layers to grow to occupy the entire flow area is called the *hydrodynamic entrance* or *developing length*. The flow is deemed fully developed once it is past the entrance length.

When heat is transferred between the flowing fluid, externally or internally, and the surface, the temperature gradient can be similarly treated as occurring mostly across a thermal boundary layer. The hydrodynamic and thermal layers are not necessarily of the same dimensions.

The boundary layer concept is of great value because it allows for simplification of the flow equations, as is shown later in this section. The greatest simplification arises from the ability to assume that the flow in the boundary layer is predominantly tangential to the surface. Thus the pressure gradient in the perpendicular directions to the flow can be taken as zeroe. For a flow in a tube, within the boundary layer the pressure gradient conditions are:

$$\frac{\partial p}{\partial r} \simeq 0 \quad \text{and} \quad \frac{\partial p}{r \partial \theta} = 0 \tag{9-33}$$

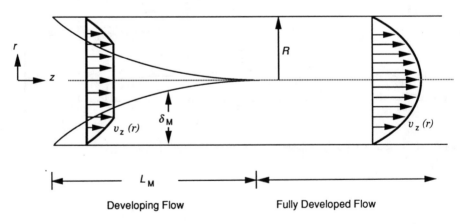

Figure 9-13 Boundary layer development in a tube.

The velocity conditions are:

$$v_z \gg v_r$$

and

$$\frac{\partial v_z}{\partial r} \gg \frac{\partial v_z}{\partial z} \tag{9-34}$$

The second condition arises because the velocity changes from zero at the wall to the free stream value across the boundary.

D Turbulence

At low velocities, the flow within a boundary layer proceeds along stream lines and hence is of laminar type (Fig. 9-12). However, with sufficient length, disturbances within the flow appear, leading to variable velocity (in direction as well as magnitude) at any position; thus the flow becomes turbulent. The higher the flow velocity, the shorter is the purely laminar flow length. With turbulent flow, eddies are formed that have a random velocity, which destroys the laminar flow lines. However, even for turbulent flow, the flow near the wall (where the velocity is small) appears to have a laminar region or sublayer, as illustrated in Figure 9-12.

The eddies substantially increase the rate at which momentum and energy can be transferred across the main (or mean) flow direction. The enhancement, by the eddies, of the transport processes above the level possible by molecular effects leads to the practical preference for the use of turbulent flow in most heat transfer equipment. The turbulent flow equations are more complicated than the laminar flow equations. Experimentally based correlations for the macroscopic turbulent flow characteristics have been developed for most applications.

E Dimensionless Analysis

It is useful to relate the hydraulic characteristics of flow to the ratio of the various forces encountered in the flow. A list of dimensionless groups that have been applied to fluid mechanics analysis appears in the Appendix G.

These groupings originate from the various ways by which the flow equations can be nondimensionalized. As an example, the fluid equations for Newtonian fluids of constant density and viscosity can be nondimensionalized as follows.

Let

$$\vec{v}^* \equiv \frac{\vec{v}}{V}$$

where V = some characteristic velocity

$$x^*, y^*, z^* \equiv \frac{x}{D_e}, \frac{y}{D_e}, \frac{z}{D_e}$$

where D_e = a reference length

$$t^* \equiv \frac{tV}{D_e} \quad \text{and} \quad p^* \equiv \frac{p}{\rho V^2}$$

The Navier Stokes equation:

$$\rho \frac{D\vec{v}}{Dt} = -\nabla p + \mu \nabla^2 \vec{v} + \rho \vec{g} \tag{4-90b}$$

can then be written as:

$$\rho \frac{V}{D_e} \frac{D}{Dt^*} (V\vec{v}^*) = -\frac{\nabla}{D_e} (\rho V^2 p^*) + \mu \frac{\nabla^{*2}}{D_e^2} (V\vec{v}^*) + \rho \vec{g}$$

which can be rearranged as:

$$\frac{D\vec{v}^*}{Dt^*} = -\nabla^* p^* + \left(\frac{\mu}{\rho V D_e}\right) \nabla^{*2}\vec{v}^* + \left(\frac{D_e\, g}{V^2}\right) \frac{\vec{g}}{g} \tag{9-35}$$

where g = the absolute value of gravity.

Equation 9-35 can be written, using the definitions of the Reynolds number (Re) and Froude number (Fr), as:

$$\frac{D\vec{v}^*}{Dt^*} = -\nabla^* p^* + \frac{1}{Re} \nabla^{*2}\vec{v}^* + \frac{1}{Fr} \frac{\vec{g}}{g} \tag{9-36}$$

Thus when two flow systems with the same Reynolds and Froude numbers and initial and boundary conditions are found, they can be described by the same dimensionless velocity and pressure. Such systems are said to be dynamically similar. This similarity is used for identifying flow patterns in large systems by constructing less expensive, smaller prototypes.

IV LAMINAR FLOW INSIDE A CHANNEL

In this section, we examine the velocity and pressure relations for laminar flow inside a channel. The approach to the solution of laminar flow problems is illustrated via a detailed analysis for a channel of circular geometry. However, results for other geometries are also presented. As mentioned before, the analysis can be split into two regions:

1. The developing flow region, where subdivision between an inviscid core flow and a laminar boundary layer is needed to determine the velocity variation across the entire channel
2. The fully developed flow region, where the viscous laminar flow analysis can be applied over the entire flow region

A Fully Developed Laminar Flow in a Circular Tube

Consider steady-state viscous flow in a circular tube of radius R. The flow enters the tube, at $z = 0$, with a uniform velocity. The boundary layer is zero at the inlet but eventually grows to the centerline, and the layer occupies the entire flow channel. Because of symmetry in the θ direction, the velocity $(\vec{v})$ and pressure (p) are independent of θ. Fully developed laminar flow implies that the velocity at any point is only in the main flow direction (z). Hence,

$$v_r = v_\theta = 0 \qquad (9\text{-}37)$$

Let the flow be for constant ρ and μ. Thus the continuity equation leads to:

$$\nabla \cdot \vec{v} = \frac{\partial v_z}{\partial z} = 0 \qquad (9\text{-}38)$$

Applying the results of Eqs. 9-37 and 9-38 to the momentum equation in the z direction, we get:

$$\underset{\rho v_z\, \partial v_z/\partial z}{\overset{=\,0}{}} + \underset{\rho v_r\, \partial v_z/\partial r}{\overset{=\,0}{}} = -\frac{\partial p}{\partial z} + \underset{\mu\, \partial^2 v_z/\partial z^2}{\overset{=\,0}{}} + \frac{\mu}{r}\frac{\partial}{\partial r}\left(r\frac{\partial v_z}{\partial r}\right)$$

or

$$\frac{\mu}{r}\frac{\partial}{\partial r}\left(r\frac{\partial v_z}{\partial r}\right) = \frac{\partial p}{\partial z} \qquad (9\text{-}39)$$

Because from the conditions of Eq. 9-33 $\partial p/\partial r = 0$ and $\partial p/\partial \theta = 0$, however,

$$\frac{\partial p}{\partial z} = \frac{dp}{dz} \qquad (9\text{-}40)$$

Hence the equation to be solved for v_z as a function of r is:

$$\frac{\mu}{r}\frac{d}{dr}\left(r\frac{dv_z}{dr}\right) = \frac{dp}{dz} \qquad (9\text{-}41)$$

Note that the total derivatives are used because v_z and p are only functions of r and z, respectively.

Equation 9-41 can be directly integrated twice over r, as the pressure is not a function of r and dp/dz is not a function of r. Applying the boundary conditions

at $\qquad\qquad r = R \quad v_z = 0 \qquad\qquad\qquad (9\text{-}42b)$

$$r = 0 \quad \frac{\partial v_z}{\partial r} = 0 \text{ (i.e., } v_z \text{ is maximum)} \qquad (9\text{-}42b)$$

we get:

$$v_z = \frac{R^2}{4\mu}\left(-\frac{dp}{dz}\right)\left(1 - \frac{r^2}{R^2}\right) \qquad (9\text{-}43)$$

Hence, when dp/dz is negative, the velocity is positive in the axial direction.

It is useful to obtain the mean (mass weighted) velocity (V_m):

$$V_m = \frac{\int_0^R \rho v_z (2\pi r)\, dr}{\int_0^R \rho (2\pi r)\, dr}$$

For a constant density:

$$V_m = \frac{\int_0^R v_z (2\pi r)\, dr}{\pi R^2}$$

or

$$V_m = \frac{R^2}{8\mu}\left(-\frac{dp}{dz}\right) \tag{9-44}$$

From Eqs. 9-43 and 9-44, the local velocity can be written as:

$$\frac{v_z}{V_m} = 2\left(1 - \frac{r^2}{R^2}\right) \tag{9-45}$$

The parabolic nature of the velocity profile leads to a linear profile of the shear stress (τ), as,

$$\tau = +\mu \frac{dv_z}{dr} = -4\mu \frac{V_m}{R}\left(\frac{r}{R}\right) \tag{9-46}$$

Note that τ is opposite in sign to V_m, which means that it acts opposite to the flow direction, as expected. The wall shear stress is given by specifying the value of $r = R$ in Eq. 9-46 to get:

$$\tau_w = -4\mu \frac{V_m}{R} = \frac{R}{2}\left(+\frac{dp}{dz}\right) \tag{9-47}$$

Note that the wall shear stress could have been obtained from an integral balance over an infinitesimal δz (Fig. 9-14) in the pipe such that the net pressure force is balanced by the wall shear stress force so that:

$$\tau_w (2\pi R)(\delta z) = \frac{dp}{dz}(\pi R^2)(\delta z)$$

or

$$\tau_w = \frac{R}{2}\left(\frac{dp}{dz}\right) \tag{9-47}$$

Note both τ_w and dp/dz have negative values for flow in the positive z direction.

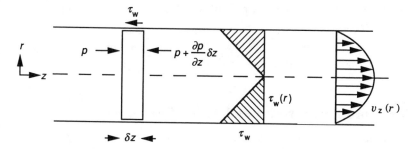

Figure 9-14 Shear stress distribution and velocity profile in a pipe.

Of great practical importance is the definition of a friction factor (f) such that the pressure gradient is related to the kinetic head based on the average velocity and the diameter of the pipe. From Eq. 9-44 we can see that:

$$-\frac{dp}{dz} = \frac{8\mu}{R^2} V_m = \frac{64\mu}{\rho D^2 V_m} \left(\frac{\rho V_m^2}{2} \right)$$

(9-48)

which can be recast in the form originally proposed by Darcy as:

$$-\frac{dp}{dz} = \frac{f}{D} \frac{\rho V_m^2}{2}$$

(9-49)

where f, the friction factor, in this case is given by:

$$f = \frac{64}{\rho D V_m / \mu} = \frac{64}{Re}$$

(9-50)

The result of this analysis leads to a condition usually observed for laminar flow, i.e., that the product fRe is a constant dependent only on the geometry of the flow. Experience shows that laminar flow in a tube exists up to a Reynolds number of about 2100. By combining Eqs. 9-47 and 9-49, we get a relation for the wall shear stress and the friction factor f:

$$\tau_w = -\frac{R}{2} \frac{f}{2R} \frac{\rho V_m^2}{2} = -\frac{f}{4} \frac{\rho V_m^2}{2}$$

(9-51)

Unfortunately, there is another friction factor that appears in the literature, the Fanning friction factor, which is defined in terms of the shear stress (τ_w) relation to the kinetic pressure, such that:

$$\tau_w = -f' \frac{\rho V_m^2}{2}$$

(9-52)

Thus the Darcy (or Moody) factor f is related to the Fanning factor f' as:

$$f = 4f'$$

(9-53)

The friction pressure drop across a pipe of length L, when the developing flow region can be ignored, is given by:

$$\Delta p_{\text{friction}} = \int_{z_{\text{in}}}^{z_{\text{out}}} \left(-\frac{dp}{dz} \right) dz = p_{\text{in}} - p_{\text{out}} = \frac{fL}{D} \frac{\rho V_{\text{m}}^2}{2} \tag{9-54}$$

B Fully Developed Laminar Flow with Other Geometries

By following a procedure similar to the one outlined above, solutions for the velocity distribution and wall friction factor coefficient have been obtained for a variety of flow geometries. The values of $f'\text{Re}$ for fully developed flow in rectangular and annular channels are given in Figures 9-15 and 9-16. In those cases the Reynolds number is defined by:

$$\text{Re} = \frac{\rho V_{\text{m}} D_{\text{e}}}{\mu} \tag{9-55}$$

where D_{e} = the equivalent hydraulic diameter defined by:

$$D_{\text{e}} = \frac{4A}{P_{\text{w}}} \tag{9-56}$$

where A = flow area, and P_{w} = wetted perimeter (or surface per unit length) of the channel.

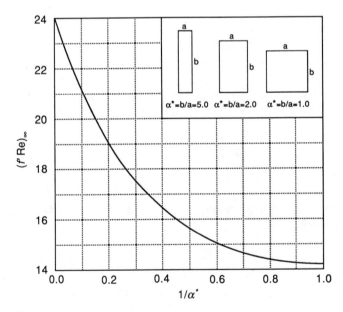

Figure 9-15 Product of laminar friction factor and Reynolds number for fully developed flow with rectangular geometry. (*From Kays* [14].)

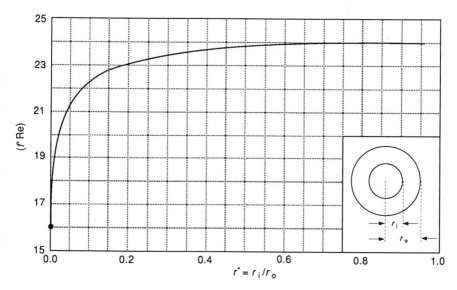

Figure 9-16 Product of laminar friction factor and Reynolds number for fully developed flow in an annular channel. *(From Kays [14].)*

Although the values of f'Re for these cases are expressed in terms of the equivalent hydraulic diameter, these results are not equivalent to simply transforming the circular tube results utilizing the equivalent diameter concept. It can be seen directly by noting from Eq. 9-50 that such a transformation would yield:

$$f'\mathrm{Re} = 16 \quad (\text{or } f\mathrm{Re} = 64) \tag{9-50}$$

where Re is defined in Eq. 9-55, and f' is used in place of f. Note that the plots in Figures 9-15 and 9-16 yield the same value of f'Re $= 24$ for infinite parallel plates (i.e., for $1/\alpha^* = 0$ and $r_i/r_o = 1$). On the other hand, Figures 9-15 and 9-16 illustrate that values of f'Re are not constant. This result is to be expected with laminar flow, where the molecular shear effects are significant throughout the flow cross section so that the governing equations have to be solved for each specific geometry. As discussed in Chapter 10, it is also the case for heat transfer in laminar flow.

For rod bundle analysis the reader is referred to section VI.

C Laminar Developing Flow Length

In the developing (or entrance) region of flow, the steady-state velocity distribution and friction factor coefficients can be obtained from the equation:

$$\frac{\mu}{r} \frac{\partial}{\partial r} \left(r \frac{\partial v_z}{\partial r} \right) = \rho v_z \frac{\partial v_z}{\partial z} + \frac{dp}{dz} \tag{9-57}$$

because $\partial v_z/\partial z$ is not zero in that region. However, v_r remains approximately zero and $\partial p/\partial r$ also is approximately zero. Momentum integral solutions to this problem

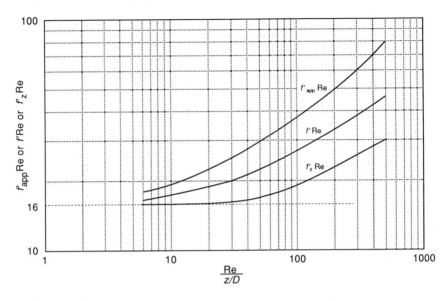

Figure 9-17 Developing laminar flow friction factor. *(From Langhaar and No [17].)*

lead to determination of the dependence of $f\text{Re}$ on $\text{Re}/(z/D)$ as the nondimensional parameter of significance. Langhaar and No [17] calculated the value of both the local value of the friction factor f_z (actually f'_z) as well as the effective value over any length for a circular tube. Their results are shown in Figure 9-17. It is seen that the local value for $f'_z\text{Re}$ approaches 16 as $\text{Re}/(z/D)$ approaches 20. Hence it can be said that the flow becomes fully developed at that distance, such that:

$$\frac{z}{D} \simeq \frac{\text{Re}}{20} \tag{9-58}$$

Three friction coefficients are indicated in Figure 9-17: f', f'_z, and f'_{app}. The local friction coefficient is described as f'_z and is based on the actual local wall shear stress at z. The mean friction coefficient from $z = 0$ to z is described by f'. Part of the pressure drop in the entrance region of a tube is attributable to an increase in the total fluid momentum flux, which is associated with the development of the velocity profile. The combined effects of surface shear and momentum flux have led to identifying an apparent friction factor (f'_{app}), which is the mean friction factor from $z = 0$ to z. If the momentum flux variation is accounted for explicitly in the momentum equations, f' (not f'_{app}) should be used to calculate the pressure drop due to friction alone.

It should be noted that the friction factor, or pressure gradient, is higher in the developing region than in the fully developed region. Note that for large values of z/D, f'_z and f' have essentially the same value. Thus in channels with multiple flow obstructions the pressure drop is higher than that of unobstructed open channels for two reasons: (1) the form losses imposed by the obstruction; and (2) the destruction of the fully developed flow pattern that occurs both upstream and downstream of the obstruction. The solution of the entrance region flow for concentric annuli, including flow between parallel plates, can be found in Heaton et al. [12] and Fleming and Sparrow [9].

Example 9-5 Laminar flow characteristics in a steam generator tube

PROBLEM During a shutdown condition in a PWR, the flow is driven through the loop by natural circulation at a rate corresponding to about 1% of the flow rate provided by the pumps. Assuming that the total flow rate is $\dot{m}_T = 4686$ kg/s in the full flow condition and that there are approximately 3800 tubes of 0.0222 m ($\frac{7}{8}$ in.) inside diameter in the steam generator of average length 16.5 m, determine:

1. Whether the flow is turbulent or laminar
2. The value of the friction factor
3. The friction pressure loss between the inlet and outlet of one tube

Use $\rho = 1000$ kg/m^3 and $\mu = 0.001$ kg/m $\cdot$ s.

SOLUTION
1. The value of the Reynolds number is:

$$\text{Re} = \frac{\rho V_m D}{\mu} = \frac{\dot{m}D}{\mu A} = \frac{4\dot{m}}{\pi\mu D} = \frac{4(0.01)(4686/3800) \text{ kg/s}}{\pi(0.001 \text{ kg/m·s})(0.0222 \text{ m})} = 706.5$$

Flow in pipes is laminar below an Re of about 2100, so the flow is laminar.
2. For laminar flow, it is appropriate to use Eq. 9-50 to evaluate f.

$$f = \frac{64}{\text{Re}} = \frac{64}{706.5} = 0.0906$$

3. The pressure drop due to friction in one tube is then:

$$\Delta p = f\left(\frac{L}{D}\right)\frac{\rho V_m^2}{2}$$

$$V_m = \frac{\dot{m}}{\rho A} = \frac{(0.01)4686/3800 \text{ kg/s}}{(1000 \text{ kg/m}^3)\frac{\pi}{4}(0.0222 \text{ m})^2} = 0.0318 \text{ m/s}$$

$$\Delta p = 0.0906\left(\frac{16.5 \text{ m}}{0.0222 \text{ m}}\right)\frac{(1000 \text{ kg/m}^3)(0.0318 \text{ m/s})^2}{2} = 34.0 \text{ Pa}$$

V TURBULENT FLOW INSIDE A CHANNEL

A Turbulent Diffusivity

As mentioned in section III.D, at sufficiently long distances or high velocities (i.e., high Re values) the smooth flow of the fluid is disturbed by the irregular appearance of eddies. For flow in pipes, the laminar flow becomes unstable at Re values above 2100. However, the value of Re that is needed to stabilize a fully turbulent flow is about 10,000. For Re values between 2100 and 10,000, the flow is said to be in transition from laminar to turbulent flow.

The turbulent enhancement of the momentum and energy lateral transport above the rates possible by molecular effects alone flattens the velocity and temperature profiles (Fig. 9-18).

In Chapter 4 it was shown that the instantaneous velocity can be divided into a time-averaged component and a fluctuating component:

$$\vec{v} = \bar{\vec{v}} + \vec{v}\,' \tag{4-129b}$$

The time-averaged mass and momentum equations (Eqs. 4-135 and 4-136) can be reduced for negligible density fluctuations to the form:

$$\nabla \cdot [\rho \bar{\vec{v}}] = 0 \tag{9-59}$$

$$\frac{\partial \rho \bar{\vec{v}}}{\partial t} + \nabla \cdot [\rho \bar{\vec{v}}\bar{\vec{v}}] = -\nabla p + \nabla \cdot [\bar{\bar{\tau}} - \rho \overline{\vec{v}'\vec{v}'}] + \rho \vec{g} \tag{9-60}$$

From Eq. 9-60 it is seen that the lateral momentum transport rate can be considered to be composed of two parts: one due to the molecular effects and the other due to the fluctuations of the eddies, i.e.,

$$\bar{\bar{\tau}}_{\text{eff}} = \bar{\bar{\tau}} - \rho \overline{\vec{v}'\vec{v}'} \tag{9-61}$$

Thus for flow inside a pipe, the effective shear in the flow direction can be given by:

$$(\tau_{zr})_{\text{eff}} = \tau_{zr} - \rho \overline{v_z' v_r'} \tag{9-62}$$

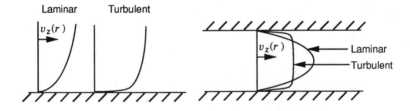

Figure 9-18 Velocity profiles of laminar and turbulent flows.

The most common approach to analyzing turbulent flow problems is to relate the additional term to the time-averaged behavior by defining a new momentum diffusivity (ϵ_M) in the plane of interest. For example:

$$(\epsilon_M)_{zr} \equiv \frac{-\overline{v_z' v_r'}}{\left(\dfrac{\partial \overline{v}_z}{\partial r} + \dfrac{\partial \overline{v}_r}{\partial z}\right)} \tag{9-63}$$

so that from Eqs. 9-62 and 9-63 the effective shear can be given by:

$$(\tau_{zr})_{\text{eff}} = (\mu + \rho \epsilon_M) \left(\frac{\partial \overline{v}_z}{\partial r} + \frac{\partial \overline{v}_r}{\partial z}\right) \tag{9-64}$$

(Under steady-state flow conditions in a fully developed pipe flow $\overline{v}_r = 0$.) In general, the turbulent diffusivity is assumed to be independent of the orientation of flow, so that:

$$(\epsilon_M)_{zr} = (\epsilon_M)_{r\theta} = (\epsilon_M)_{\theta z} \tag{9-65}$$

The quantity $\rho \epsilon_M$ is clearly analogous to the molecular viscosity (μ); but whereas μ is a property of the fluid, $\rho \epsilon_M$ depends on the velocity and geometry as well.

Other approaches have been introduced to determine the effect of turbulence on the momentum transport and energy transport. A brief discussion of two common approaches is found in Chapter 10, section IV.

B Turbulent Velocity Distribution

It has been found useful to express the turbulent velocity profile in a dimensionless form that depends on the wall shear:

$$v_z^+ = \frac{v_z}{\sqrt{\tau_w/\rho}} \tag{9-66}$$

v_z^+ is often called the universal turbulent velocity. The term $\sqrt{\tau_w/\rho}$ is referred to as the "shear velocity." Also, the distance from the wall (y) can be nondimensionalized as:

$$y^+ = y \frac{\sqrt{\tau_w/\rho}}{\nu} \tag{9-67}$$

If a linear shear stress distribution is assumed near the wall where molecular effects dominate, we have:

$$\tau_w \approx \rho \nu \frac{dv_z}{dy} \tag{9-68}$$

where y = distance from the wall (in a pipe, $y = R - r$). After integrating Eq. 9-68:

$$v_z = \frac{\tau_w}{\rho \nu} y + C \tag{9-69}$$

Because at $y = 0$, $v_z = 0$ $\qquad$ (9-70)

we get:

$$v_z = \frac{\tau_w}{\rho \nu} y$$

or

$$v_z^+ = y^+ \qquad (9\text{-}71)$$

Various investigators have developed expressions for the universal velocity. Often the boundary layer is subdivided into a laminar sublayer near the wall (where $\epsilon_M = 0$), an intermediate or buffer sublayer, and a fully turbulent sublayer (where $\mu \ll \rho\epsilon_M$). Martinelli [19] described the resulting distribution by:

$$
\begin{align}
y^+ < 5 \qquad & v_z^+ = y^+ & (9\text{-}72\mathrm{a}) \\
5 < y^+ < 30 \qquad & v_z^+ = -3.05 + 5.00 \, \ell n y^+ & (9\text{-}72\mathrm{b}) \\
y^+ > 30 \qquad & v_z^+ = 5.5 + 2.5 \, \ell n y^+ & (9\text{-}72\mathrm{c})
\end{align}
$$

An important equation for ϵ_M (for flow in a pipe) in the turbulent sublayer is that reported by Reichart [26]. He proposed that:

$$\frac{\epsilon_M}{\nu} = \frac{kR^+}{6}\left[1 - \left(\frac{r}{R}\right)^2\right]\left[1 + 2\left(\frac{r}{R}\right)^2\right] \qquad (9\text{-}73)$$

where

$$R^+ = R\sqrt{\frac{\tau_w/\rho}{\nu}} \qquad (9\text{-}74)$$

and $k = $ a constant (≈ 0.4). This equation leads to the following expression for the velocity in the turbulent sublayer:

$$v_z^+ = 5.5 + 2.5 \, \ell n \left[y^+ \frac{1.5(1 + r/R)}{1 + 2(r/R)^2}\right] \qquad (9\text{-}75)$$

This equation appears to satisfy experiments at all values of r, including at the center line.

A fully developed turbulent velocity profile for a pipe is given in Figure 9-19. Note that the velocity is flatter than the laminar flow. The velocity profile can be approximated by:

$$\frac{v_z}{V_{\underline{\mathfrak{C}}}} = \left(\frac{y}{R}\right)^{1/7} = \left(\frac{R - r}{R}\right)^{1/7} \qquad (9\text{-}76)$$

Consequently the average velocity is given by:

$$V_m = 0.817 \, V_{\underline{\mathfrak{C}}} \qquad (9\text{-}77)$$

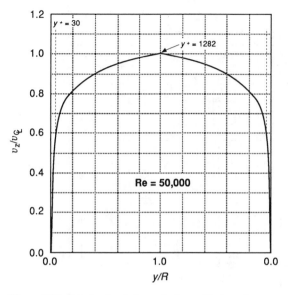

Figure 9-19 Calculated turbulent velocity profile in a pipe.

C Turbulent Friction Factor

A commonly encountered expression for the friction factor is the Karman-Nikuradse equation:

$$\frac{1}{\sqrt{f}} = -0.8 + 0.87 \, \ell n \, (\text{Re}\sqrt{f}) \tag{9-78}$$

This transcendental equation is difficult to use in practice, and simplified relations are often applied. An approximate equation for a smooth tube where $30{,}000 < \text{Re} < 1{,}000{,}000$ is the McAdams relation:

$$f = 0.184 \, \text{Re}^{-0.2} \tag{9-79}$$

For $\text{Re} < 30{,}000$, the turbulent friction factor for a smooth tube may be given by the Blasius relation:

$$f = 0.316 \, \text{Re}^{-0.25} \tag{9-80}$$

The tube roughness, characterized by the ratio of depth of surface protrusions to the tube diameter (λ/D), may increase the effective friction factor. The most commonly available factor is that of Moody [20] and is given in Figure 9-20. In Moody's chart, the effect of the roughness depends on the pipe diameter (D), as is reasonable to expect. The Moody diagram is a graphic representation of the empiric Colebrook equation [4], given by:

$$\frac{1}{\sqrt{f}} = -2 \log_{10} \left[\frac{\lambda/D}{3.70} + \frac{2.51}{\text{Re}\sqrt{f}} \right] \tag{9-81}$$

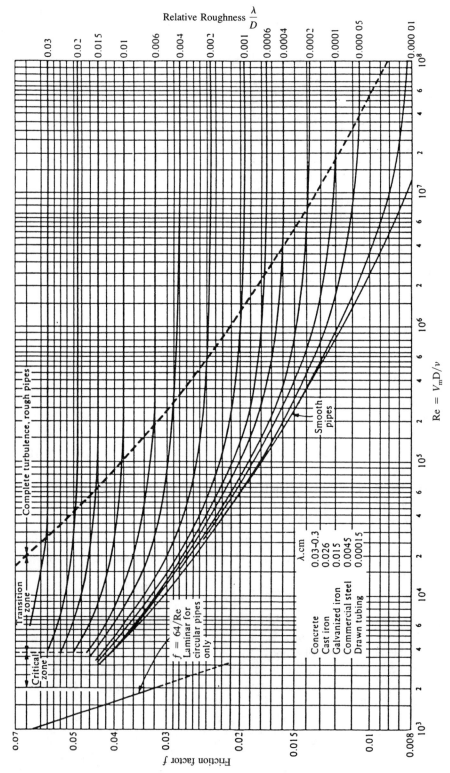

Figure 9-20 Moody's chart for friction factors: Friction factor for use in the relation Δp for pressure drop for flow inside circular pipes. (*From Moody* [20].)

D Fully Developed Turbulent Flow with Noncircular Geometries

The relations developed for the friction factor in circular tubes can similarly be derived for other geometries. However, in the turbulent flow case the velocity gradient is principally near the wall. Hence the flow channel geometry does not have as important an influence on the friction factor. Therefore the hydraulic diameter concept can be more accurate in predicting the friction factor.

E Turbulent Developing Flow Length

The entrance length of laminar flow was found to extend to a maximum axial distance z such as:

$$\frac{z}{D} = 0.05 \text{ Re} \quad \text{for Re} < 2000 \tag{9-58}$$

For turbulent flow the boundary layer can develop faster than in the high Re laminar region, so that:

$$\frac{z}{D} = 25 \text{ to } 40$$

Example 9-6 Turbulent flow in a steam generator tube

PROBLEM For the condition of full flow through a U-tube steam generator, find the same quantities asked for in Example 9-5:

1. Determine whether the flow is turbulent or laminar.
2. Find the value of the friction factor.
3. Find the average pressure loss between the inlet and outlet of a tube.

SOLUTION
1. Because all quantities are the same as in Example 9-5, except the velocity, which is now increased 100 times, the Reynolds number is:

$$\text{Re} = 70,650$$

which is clearly in the turbulent range (Re > 2100).
2. The appropriate expression for the friction factor is now Eq. 9-79:

$$f = 0.184 \text{ Re}^{-0.2} = 0.184 \, (70650)^{-0.2} = 0.0197$$

3. The friction pressure drop in one of the tubes is now:

$$\Delta p = f \left(\frac{L}{D}\right) \frac{\rho V_m^2}{2} = 0.0197 \left(\frac{16.5}{0.0222}\right) \frac{1000 \text{ kg/m}^3 (3.18 \text{ m/s})^2}{2} = 74 \text{ kPa}$$

The friction factor is lower by a factor of 5, but the velocity has increased by a factor of 100, and the pressure drop has therefore increased by a factor of 2000.

VI PRESSURE DROP IN ROD BUNDLES

The total pressure drop along a reactor core includes (1) entrance and exit pressure losses between the vessel plena and the core internals, (2) the friction pressure drop along the fuel rods, and (3) the form losses due to the presence of spacers. The entrance and exit losses are those (described earlier) due to a sudden change in flow area (discussed in sections II.D.3 and VII). Hence attention here is focused on the friction along the rod bundles and the effect of the spacers.

A Friction Along Bare Rod Bundles

1 Laminar flow. In section IV.B it was demonstrated that friction factors for non-circular geometries in laminar flow could not be obtained by transforming circular tube results using the equivalent diameter concept. However, an alternate approximate method follows from the observation that the coolant region in rod arrays can be represented as an array of equivalent annuli around the rods. As Figure 9-21 suggests, the equivalent annulus approximation improves as the rod spacing increases.

The solution of the momentum equations for the exact bare rod array geometry and for the equivalent annulus approximation have been obtained for laminar and turbulent flow. The solution procedure and a number of significant special cases are presented in Chapter 7, Vol. II. Here the most commonly encountered cases are summarized.

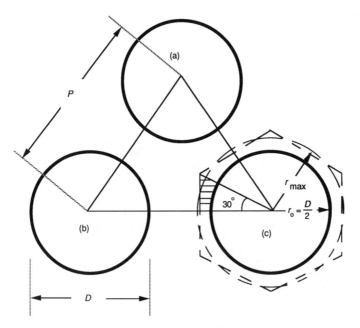

Figure 9-21 Definition of an equivalent annulus in a triangular array. *Cross-hatched area* represents an elemental coolant flow section. Circle of radius r_{max} represents the equivalent annulus with equal flow area.

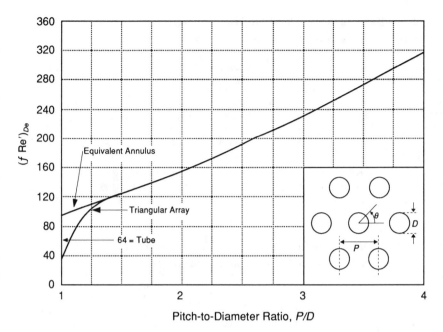

Figure 9-22 Product of laminar friction factors and Reynolds number for parallel flow in a rod bundle. *(From Sparrow and Loeffler [28].)*

In this section we use the superscripted symbol Re′ to refer to the Reynolds number in a bare (spacerless) rod bundle. In section VI.B we use Re to refer to the Reynolds number in a bundle with spacers.

Figure 9-22 presents the product $f\mathrm{Re}'_{De}$ for fully developed laminar flow in a triangular array derived by Sparrow and Loeffler [28] and for the equivalent annulus. The equivalent annulus approximation is seen as satisfactory for pitch-to-diameter (P/D) ratios greater than about 1.3. A complete set of laminar results are available from Rehme [22]. His results have been fit by Cheng and Todreas [3] with polynomials for each subchannel type. The polynomials have the form:

$$C'_{fiL} = a + b_1(P/D - 1) + b_2(P/D - 1)^2 \tag{9-82}$$

where:

$$f_{iL} \equiv \frac{C'_{fiL}}{\mathrm{Re}'^n_{iL}} \tag{9-83}$$

where $n = 1$ for laminar flow. When Eq. 9-82 is used for edge and corner subchannels, P/D is replaced by W/D, where W = rod diameter plus gap between rod and bundle wall. The effect of P/D (or W/D) was separated into two regions; $1.0 \le P/D \le 1.1$ and $1.1 \le P/D \le 1.5$. Tables 9-2 and 9-3 present the coefficients a, b_1, and b_2 for the subchannels of hexagonal and square arrays.

Bundle average friction factors are obtained from the subchannel friction factors by assuming that the pressure difference across all subchannels is equal and applying

Table 9-2 Coefficients in Eqs. 9-82 and 9-87 for bare rod subchannel friction factor constants C'_{fi} in hexagonal array

Subchannel	$1.0 \leq P/D \leq 1.1$			$1.1 < P/D \leq 1.5$		
	a	b_1	b_2	a	b_1	b_2
Laminar flow						
Interior	26.00	888.2	−3334	62.97	216.9	−190.2
Edge	26.18	554.5	−1480	44.40	256.7	−267.6
Corner	26.98	1636.	−10,050	87.26	38.59	−55.12
Turbulent flow						
Interior	0.09378	1.398	−8.664	0.1458	0.03632	−0.03333
Edge	0.09377	0.8732	−3.341	0.1430	0.04199	−0.04428
Corner	0.1004	1.625	−11.85	0.1499	0.006706	−0.009567

the mass balance condition for total bundle flow in terms of the subchannel flow. This procedure, which is demonstrated in Chapter 4, Vol. II, yields:

$$C'_{bL} = D'_{eb} \left[\sum_{i=1}^{3} S_i \left(\frac{D'_{ei}}{D'_{eb}} \right)^{\frac{n}{2-n}} \left(\frac{C'_{fi}}{D'_{ei}} \right)^{\frac{1}{n-2}} \right]^{n-2} \tag{9-84}$$

where S_i = the ratio of the total flow area of subchannels of type i to the bundle flow area. Figure 9-23 compares Eq. 9-84 to the available data for laminar flow ($n = 1$) in a 37-pin triangular array.

2 Turbulent flow. Early work in the area of turbulent flow includes that of Deissler and Taylor [5], who derived friction factors that depend on a universal velocity profile obtained from early measurements. Their approach was compared, along with the equivalent diameter concept, to measurements in square and triangular rod bundles with $P/D = 1.12$, 1.20, and 1.27. The results show that the circular tube prediction of Eq. 9-79 for Re > 100,000 provides an answer that lies within the scatter of the

Table 9-3 Coefficients in Eqs. 9-82 and 9-87 for bare rod subchannel friction factor constants C'_{fi} in square array

Subchannel	$1.0 \leq P/D \leq 1.1$			$1.1 < P/D \leq 1.5$		
	a	b_1	b_2	a	b_1	b_2
Laminar flow						
Interior	26.37	374.2	−493.9	35.55	263.7	−190.2
Edge	26.18	554.5	−1480	44.40	256.7	−267.6
Corner	28.62	715.9	−2807	58.83	160.7	−203.5
Turbulent flow						
Interior	0.09423	0.5806	−1.239	0.1339	0.09059	−0.09926
Edge	0.09377	0.8732	−3.341	0.1430	0.04199	−0.04428
Corner	0.09755	1.127	−6.304	0.1452	0.02681	−0.03411

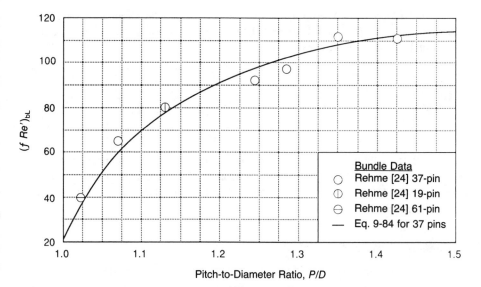

Figure 9-23 Laminar flow results in triangular array bare rod bundles. *(From Cheng and Todreas [3].)*

data. However, the data show a dependence on P/D that cannot be reproduced by the equivalent diameter concept and the circular tube correlation. LeTourneau et al. [18] tested rod bundles of square lattice with P/D ratios of 1.12 and 1.20 and of triangular lattice with a P/D ratio of 1.12. These data fall within a band between the smooth tube prediction and a curve 10% below that for the Re $= 3 \times 10^3$ to 3×10^5.

Later, Trupp and Azad [30] obtained velocity distributions, eddy diffusivities, and friction factors with airflow in triangular array bundles. These data indicated friction factors somewhat higher than Deissler and Taylor's predictions. For Reynolds numbers between 10^4 and 10^5, their data at $P/D = 1.2$ were about 17% higher than the circular tube data. The data at $P/D = 1.5$ were about 27% higher than the circular tube data.

For the turbulent flow situation, solution of both the exact and the equivalent annulus geometry require assumptions about the turbulent velocity distribution. For a triangular array, Rehme [24] obtained the following equivalent annulus solutions:

For $\mathrm{Re}'_{De} = 10^4$:

$$\frac{f}{f_{\text{c.t.}}} = 1.045 + 0.071(P/D - 1) \qquad (9\text{-}85)$$

For $\mathrm{Re}'_{De} = 10^5$:

$$\frac{f}{f_{\text{c.t.}}} = 1.036 + 0.054(P/D - 1) \qquad (9\text{-}86)$$

where $f_{\text{c.t.}}$ = circular tube friction factor.

Rehme [25] also proposed a method for solving the turbulent flow case in the actual geometry. Cheng and Todreas [3] fitted results of this method with the polynomial of Eq. 9-82, where now:

$$C'_{fiT} = a + b_1(P/D - 1) + b_2(P/D - 1)^2 \tag{9-87}$$

where

$$f_{iT} \equiv \frac{C'_{fiT}}{(Re'_{iT})^n} \tag{9-88}$$

and $n = 0.18$. Tables 9-2 and 9-3 list these coefficients for subchannels of triangular and square arrays. The turbulent bundle friction factor constant (C'_{bT}) can be obtained from Eq. 9-84. Figure 9-24 compares this bundle friction factor for a 37-pin triangular array with available data.

B Pressure Loss at Spacers

Pressure losses across spacer grids or wires (Fig. 9-25) are form drag-type pressure losses that can be calculated using pressure-loss coefficients. The spacers' pressure drop can be comparable in magnitude to the friction along the bare rod bundle.

DeStordeur [6] measured the pressure drop characteristics of a variety of spacers and grids. He correlated his results in terms of a drag coefficient (C_s). The pressure drop (Δp_s) across the grid or spacer is given by:

$$\Delta p_s = C_s(\rho V_s^2/2)(A_s/A_v) \tag{9-89}$$

where A_v = unrestricted flow area away from the grid or spacer; V_s = velocity in the spacer region; and A_s = projected frontal area of the spacer.

The grid drag coefficient is a function of the Reynolds number for a given spacer or grid type. At high Reynolds number (Re $\simeq 10^5$) honeycomb grids showed drag

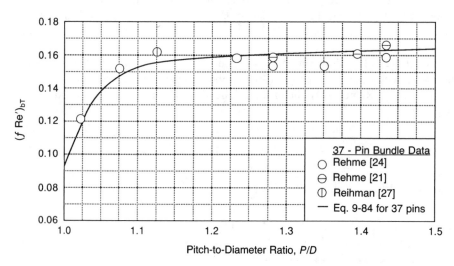

Figure 9-24 Turbulent flow results in triangular array bare rod bundles. *(From Cheng and Todreas [3].)*

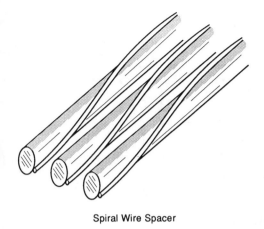

Spiral Wire Spacer

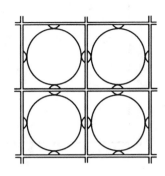

Honeycomb Grid Spacer

Figure 9-25 Rod bundle fuel element spacers.

coefficients of ≈ 1.65. The pressure drop across a crossed circular wire grid was $\approx 10\%$ lower (Fig. 9-26).

On the basis of tests of several grids, Rehme [23] found that the effect of the ratio A_s/A_v is more pronounced than was indicated by deStordeur. Rehme concluded that grid pressure drop data are better correlated by:

$$\Delta p_s = C_v \, (\rho V_v^2/2)(A_s/A_v)^2 \tag{9-90}$$

where C_v = modified drag coefficient, and V_v = average bundle fluid velocity.

The drag coefficient (C_v) is a function of the average bundle, unrestricted area. Reynolds number. Rehme's data indicated that for square arrays $C_v = 9.5$ at Re = 10^4 and $C_v = 6.5$ at Re = 10^5 (Fig. 9-27).

Rehme [23] also correlated the total pressure losses in wire-wrapped bundles. More recently Cheng and Todreas [3] correlated wire-wrapped pressure losses utilizing the much larger database existing in the literature up to 1984. Their correlations covered the laminar, transition, and turbulent flow regimens and reduced smoothly to the bare rod correlations presented in section VI.A. Friction factors and flow split factors are presented for each subchannel for the hexagonal array. The friction factor for the rod bundle is also presented, as given below. It predicts most of the data within 10% except at the extreme ends of the P/D and H/D range (where H = axial lead of the wire wrap).

Turbulent region ($\text{Re}_T \leq \text{Re}$):

$$f = \frac{C_{fT}}{\text{Re}^{0.18}} \tag{9-91a}$$

Transition region ($\text{Re}_L < \text{Re} < \text{Re}_T$):

$$f = \left(\frac{C_{fT}}{\text{Re}^{0.18}}\right) \psi^{1/3} + \left(\frac{C_{fL}}{\text{Re}}\right) (1 - \psi)^{1/3} \tag{9-91b}$$

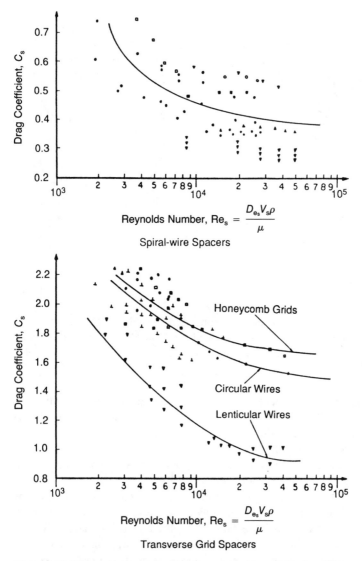

Figure 9-26 Drag coefficients for rod bundle spacers. *(From deStordeur [6].)*

Laminar region ($Re_L \geq Re$):

$$f = \frac{C_{fL}}{Re} \tag{9-91c}$$

where

$$\psi = \log_{10}(Re/Re_L)/\log_{10}(Re_T/Re_L) \tag{9-92}$$
$$= [\log_{10}(Re) - (1.7P/D + 0.78)]/(2.52 - P/D)$$

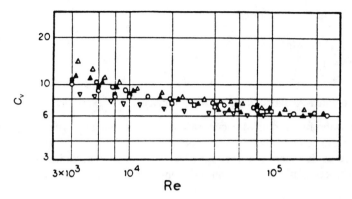

Triangular array

▽ Spacer coils   △ Triangular-type spacer
◻ Honey-comb-type spacer,n-1   ▲ Rhombus-type spacer
◼ Honey-comb-type spacer,n-2   ○ Ring-type spacer

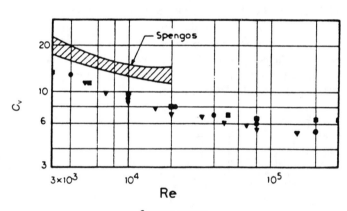

Square array

Rehme ⎨ ● Tube spacer transversally connected
        ▼ Tube spacer axially connected
        ◼ Honey-comb type spacer

Spengos ▨▨▨

Figure 9-27 Modified drag coefficients. *(From Rehme [23].)*

$$C_{fT} = [0.8063 - 0.9022(\log_{10}(H/D)] \tag{9-93a}$$
$$+ 0.3256[\log_{10}(H/D)]^2(P/D)^{9.7}(H/D)^{1.78 - 2.0 \ (P/D)}$$
$$C_{fL} = [-974.6 + 1612.0(P/D) - 598.5(P/D)^2](H/D)^{0.06 - 0.085(P/D)} \tag{9-93b}$$

and all parameters are bundle average values.

The range of applicability of the correlation is:

$$1.025 \le P/D \le 1.42$$
$$8.0 \le H/D \le 50.0$$

The flow region boundary definitions are:

Turbulent $(Re \geq Re_T) = 10^{(0.7P/D+3.3)}$
Laminar $(Re < Re_L) = 10^{(1.7P/D+0.78)}$

C Lateral Flow Resistance Across Bare Rod Arrays

Lateral flow (normal to the fuel rod axis) occurs to some extent in a core where there are significant pressure differences between assemblies. In the lateral direction, the flow can be considered as flow across a large tube bank. A simple correlation is that of Zukauskas [31], who correlated laminar and turbulent flow pressure drop by equations of the form:

$$\Delta p = f \frac{N G_{max}^2}{2\rho} Z \qquad (9\text{-}94)$$

where f = friction factor, G_{max} = maximum mass flux, N = number of tube rows in the direction of flow, Z = a correction factor depending on the array arrangement.

Figures 9-28 and 9-29 provide the values of f and Z for various flow conditions as a function of Re = $G_{max} D/\mu$, where D = rod diameter.

A large number of more comprehensive but also more complex correlations exist. The various types of in-line and staggered arrays that have been investigated, together with identification of their salient geometric characteristics, are illustrated in Figure 9-30. From the defining expression for the friction factor:

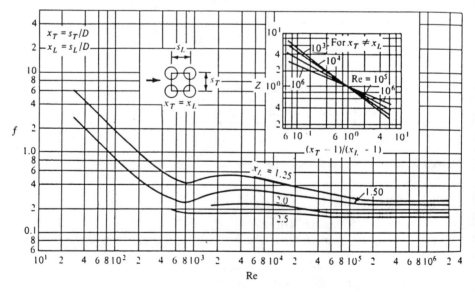

Figure 9-28 Friction factor (f) and the correction factor (Z) for use in Eq. 9-94 for in-line tube arrangement. *(From Zukauskas [31].)*

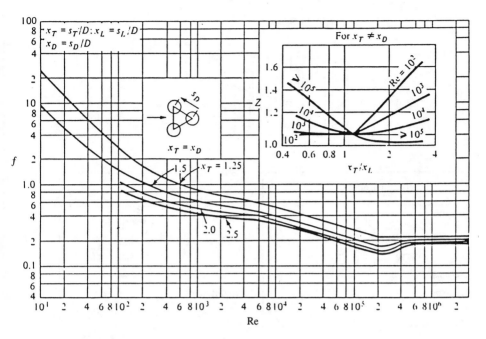

Figure 9-29 Friction factor (f) and the correction factor (Z) for use in Eq. 9-94 for staggered tube arrangement. *(From Zukauskas [31].)*

$$f \equiv \frac{2 \, \Delta p}{\rho} \left(\frac{1}{V_{ref}^2}\right) \left(\frac{D_{ref}}{L}\right) \tag{9-95}$$

The correlations must adopt definitions for V_{ref}, D_{ref}, and D_{ref}/L; and in the literature on cross flow pressure drop, a wide range of definition sets have been adopted. For example, the velocity can be taken as any of the following velocities: V_{min}, V_{avg}, or V_{max}.

The reference length D_{ref} can be taken as: D (rod diameter), D_e, D_c (gap spacing), or D_V, where:

$$D_V \equiv \text{volumetric hydraulic diameter} = \frac{4(\text{free volume of rod bundle})}{\text{friction surface area of rods}}$$

and the ratio D/L can be taken as: N (the number of major restrictions encountered by the flow) or with D or L individually identified with the selected characteristic **length.**

Figure 9-31 compares a number of correlations for the specific case of an equi-lateral triangular array of dimension $S_T/D = 1.25$. The parameter set V_{max}, D_V, and N, of the Gunter-Shaw correlation [11], is utilized to put all correlations on a consistent basis. The difference among these correlations exhibited in Figure 9-31 is typically less than the scatter among data from different experimenters. The Zukauskas corre-lation [32], which is a power law fit, appears the most comprehensive and also includes

Coordinate System

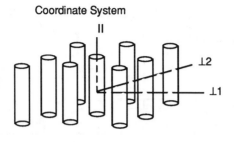

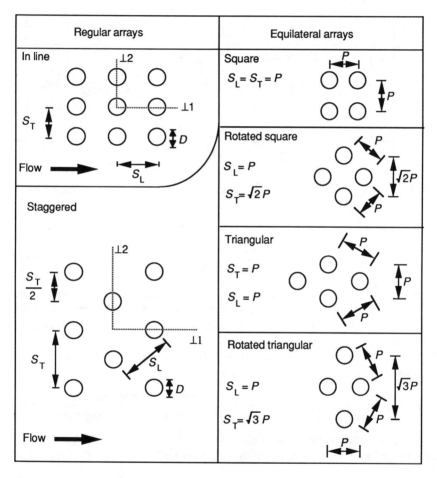

Figure 9-30 Regular rod array geometry and coordinate definitions. *(Adapted from Ebeling-Koning et al. [8].)*

a correction for entrance effects. The Gunter-Shaw correlation [11] is reasonably simple and covers a variety of geometries. It is fully detailed in Chapter 5, Vol. II where it is utilized to evaluate the distributed resistance for an in-line array in cross flow.

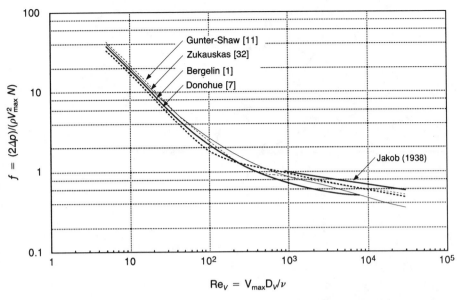

Figure 9-31 Friction factor correlations for equilateral-triangular-array bare tube (or rod) bundle with $S_T/D = 1.25$.

D Lateral Flow Resistance Across Wire-Wrapped Rod Bundles

A correlation has been presented by Suh and Todreas [29] based on their experimental data for cross flow over rod arrays with wire separators (also called displacers) parallel to the rod axes. Whereas in the wire-wrapped arrays the wires are arranged in helical fashion, numerical analyses of these arrays treat the wires as parallel to the rod axes in each axial control volume. For typical cases of control volumes with axial lengths small with respect to the wire lead length, this assumed parallel configuration of the wire and the rod is satisfactory. Figure 9-32 illustrates the geometry of the displacer and the rod in terms of the angle θ, which defines the position of the displacer with respect to the gap through which the cross flow passes. Because the displacer restricts the cross flow, the associated pressure drop is expected to be greater than that for a bare array and strongly dependent on the value of the angle θ.

The correlation for the friction factor is:

$$f_{\text{displacer}} = \frac{f_{\text{bare}}}{E(\theta)} \tag{9-96}$$

The Reynolds number is:

$$\text{Re}_V = \frac{V_{\max}D_V}{\nu} \tag{9-97}$$

and the parameter $E(\theta)$ is:

$$E(\theta) = \sum_{i=0}^{8} a_i^i \theta_i \tag{9-98}$$

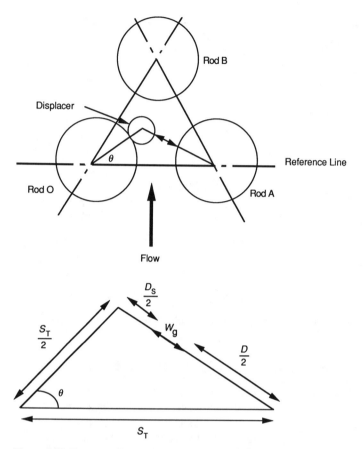

Figure 9-32 Geometry of displacer-rod bundle of equilateral triangular array.

where j = laminar, transition, turbulent for $Re_V \leq 500$; $500 \leq Re_V \leq 1000$; $Re_V \geq 1000$ respectively. The regression coefficients a_i^j are given in Table 9-4. It should be noted that this result has been evaluated for only one geometry, an equilateral trian-

Table 9-4 Regression coefficients for polynomial Eq. 9-98 (for θ in radians)

i	a_i^j		
	Laminar	Transition	Turbulent
0	−0.048629	−0.089567	−0.060247
1	−0.050047	−0.064236	−0.085772
2	2.893754	2.948909	2.534854
3	0.075660	0.086888	0.140045
4	−4.245446	−4.263047	−3.604788
5	−0.035919	−0.037395	−0.072517
6	2.213902	2.216716	1.869185
7	0.005437	0.004490	0.010781
8	−0.368285	−0.367479	−0.306346

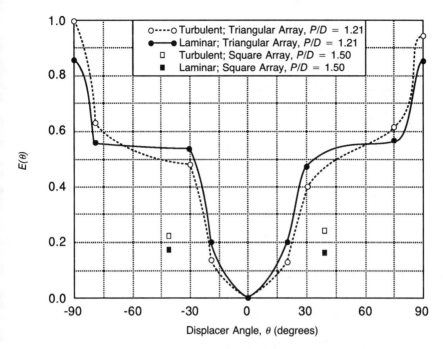

Figure 9-33 Parameter $E(\theta)$ as a function of displacer angle.

gular array of $S_T/D = S_L/D = 1.21$. The shape of $E(\theta)$, shown in Figure 9-33, illustrates that the displacer bundle friction factor increases as the displacer angular position approaches the gap.

Example 9-7 Pressure loss at spacers

PROBLEM For the PWR fuel assembly conditions of Tables 1-2, 1-3, and 2-3, calculate the pressure drop across the spacers. Use both deStordeur's and Rehme's correlations for spacer pressure drop and compare the results.

Assume the spacer type is honeycomb, and there are eight spacers along the fuel length. Take the grid thickness (t) as one-half the pin-to-pin clearance. Figure 9-34 illustrates the unit cell to be considered. Relevant data:

$L = 4$ m
$D = 9.5$ mm
$P = 12.62$ mm
Inlet temp. $= 286°C$
Outlet temp. $= 324°C$
Inlet pressure $= 15.5$ MPa
$\rho_{in} = 739$ kg/m^3
$\rho_{out} = 657$ kg/m^3
$\nu_{avg} = 1.26 \times 10^{-7}$ m^2/s

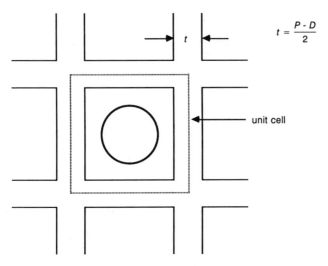

$$t = \frac{P - D}{2}$$

unit cell

Figure 9-34 Unit cell analysis in Example 9-7.

From Table 2-3: 193 assemblies, 17 × 17 rod array, total core flow 17,400 kg/s.

SOLUTION Using the deStordeur model of Eq. 9-89, we get:

$$\Delta p_{\text{spacers}} = 8\rho \frac{C_s V_s^2 A_s}{2 A_v} \tag{9-99}$$

First determine the velocity in the spacer region. It requires the flow area in one cell: $A_{\text{flow}} = A_v - A_s$. However, A_v is the cell unobstructed flow area:

$$A_v = P^2 - \frac{\pi D^2}{4} = 8.84 \times 10^{-5} \, \text{m}^2$$

and A_s is the spacer projected area per unit cell:

$$A_s = 2 \left[P \left(\frac{t}{2} \right) + (P - t) \frac{t}{2} \right] = 2 \left(Pt - \frac{t^2}{2} \right) = 3.69 \times 10^{-5} \, \text{m}^2$$

Then

$$A_{\text{flow}} = 5.15 \times 10^{-5} \, \text{m}^2$$

Also

$$\rho = \frac{\rho_{\text{in}} + \rho_{\text{out}}}{2} = 698 \, \text{kg/m}^3$$

Assuming the flow area of the interior cell can be used as a core-wide cell area (see Chapter 1 for details of edge and corner cell area formulas), we determine the average velocity at the spacer region as:

$$V_s = \frac{\dot{m}}{\rho A_{\text{flow}}} = \frac{17{,}400 \text{ kg/s}}{(698 \text{ kg/m}^3)(5.15 \times 10^{-5}) \, 193(17 \times 17) \text{ m}^2} = 8.678 \text{ m/s}$$

Now:

$$Re_s = \frac{D_{es} V_s}{\nu}$$

$$D_{es} = \frac{4 A_{\text{flow}}}{P_{w_s}}$$

$$\therefore D_{es} = 4 \left[\frac{\left(P^2 - \frac{\pi D^2}{4} \right) - 2 \left(Pt - \frac{t^2}{2} \right)}{\pi D + 4(P - t)} \right]$$

$$= 4 \left[\frac{5.15 \times 10^{-5} \text{ m}^2}{\pi(0.0095) + 4(0.01262 - 0.00156) \text{ m}} \right]$$

$$= 2.78 \times 10^{-3} \text{ m}$$

$$Re_s = \frac{(2.78 \times 10^{-3} \text{ m})(8.678 \text{ m/s})}{1.26 \times 10^{-7} \text{ m}^2/\text{s}} = 1.915 \times 10^5$$

$$\therefore C_s = 1.75 \text{ (by extrapolation from Fig. 9-26)}$$

Because $A_s/A_v = 3.69/8.84 = 0.417$
From Eq. 9-99:

$$\Delta p_s = 8(698 \text{ kg/m}^3) \frac{(1.75)(8.678 \text{ m/s})^2(0.417)}{2} = 0.153 \text{ MPa}$$

Now using Rehme's model (Eq. 9-90):

$$\Delta p_s = 8 C_v \, \rho \, \frac{V_v^2}{2} \left(\frac{A_s}{A_v} \right)^2$$

From a mass balance:

$$V_v = V_s \frac{(A_v - A_s)}{A_v} = 8.678 \left(\frac{5.15 \times 10^{-5}}{8.84 \times 10^{-5}} \right) = 5 \text{ m/s}$$

$$D_e = \frac{4[P^2 - \pi D^2/4]}{\pi D} = \frac{4 \times 8.84 \times 10^{-5}}{\pi \times 9.5 \times 10^{-3}} = 0.0118 \text{ m}$$

$$Re = \frac{D_e V_v}{\nu} = \frac{(0.0118 \text{ m})(5 \text{ m/s})}{1.26 \times 10^{-7} \text{ m}^2/\text{s}} = 4.68 \times 10^5$$

$$\therefore C_v \approx 6 \text{ from Figure 9-27}$$

$$\therefore \Delta p_s = 8(6) \, (698 \text{ kg/m}^3) \frac{(5 \text{ m/s})^2}{2} (0.417)^2 = 0.073 \text{ MPa}$$

Rehme's model leads to a lower spacer pressure drop.

VII PRESSURE LOSS COEFFICIENT AT ABRUPT AREA CHANGES

The purpose of this section is to present in some detail the approach to calculation of acceleration, form, and friction pressure drops in single-phase flows. Compressible and incompressible flows are considered. The equations that apply under special conditions are organized in Table 9-5. The major pressure drop components for a channel with an inlet contraction and an exit expansion are shown in Figure 9-35.

To describe the flow variables in a channel, the subscript 1 denotes an upstream value and the subscript 2 a downstream value. Unsubscripted variables are constants in the flow. First, a uniform velocity profile across the flow area is assumed for the evaluation of the acceleration pressure drop. Note that this approximation is not made when determining the friction pressure drop; in fact, the velocity profile effect is automatically included in the well known treatment, by Kays and London [15], of form pressure drop between a large area and a restricted flow area, where velocity profile effects can be important. A more general discussion is presented later.

Table 9-5 Pressure changes due to abrupt area changes

Flow	Incompressible	Compressible
Constant area channels*		
Frictionless	$\Delta p_{acc} = 0$	$\Delta p_{acc} = \rho_2 V_2^2 - \rho_1 V_1^2$
	$\Delta p_{form} = 0$	$\Delta p_{form} = 0$
	$\Delta p_{fric} = 0$	$\Delta p_{fric} = 0$
	(section VII.A.1)	(section VII.A.4)
Viscous	$\Delta p_{acc} = 0$	$\Delta p_{acc} = \rho_2 V_2^2 - \rho_1 V_1^2$
	$\Delta p_{form} = 0$	$\Delta p_{form} = 0$
	$\Delta p_{fric} = f \dfrac{L}{D_e} \rho \dfrac{V^2}{2}$	$\Delta p_{fric} \approx \displaystyle\int_0^L \dfrac{f(\ell)}{2\,D_e} \rho(\ell) V^2(\ell) d\ell$
	(section VII.A.1)	(section VII.A.4)
Short channels with an abrupt area change†		
Frictionless	$\Delta p_{acc} = \rho \left(\dfrac{V_2^2}{2} - \dfrac{V_1^2}{2} \right)$	$\Delta p_{acc} \approx \rho \left(\dfrac{V_2^2}{2} - \dfrac{V_1^2}{2} \right)$
	$\Delta p_{form} = 0$	$\Delta p_{form} = 0$
	$\Delta p_{fric} = 0$	$\Delta p_{fric} = 0$
	(section VII.A.2)	(section VII.A.5)
Viscous	$\Delta p_{acc} = \rho \left(\dfrac{V_2^2}{2} - \dfrac{V_1^2}{2} \right)$	$\Delta p_{acc} \approx \rho \left(\dfrac{V_2^2}{2} - \dfrac{V_1^2}{2} \right)$
	$\Delta p_{form} = \begin{cases} + K_c \, \rho \, \dfrac{V_2}{2} \\ + K_e \, \rho \, \dfrac{V_1^2}{2} \end{cases}$	$\Delta p_{form} = \begin{cases} + K_c \, \rho \, V_2^2/2 \\ + K_e \, \rho \, V_1^2/2 \end{cases}$
	$\Delta p_{fric} = 0$	$\Delta p_{fric} = 0$
	(section VII.B.3)	(section VII.B.5)

*$\Delta p_{form} = 0$ by definition in this part.

†$\Delta p_{fric} = 0$ in this part for viscous flow because the effect of friction is included in Δp_{form}.

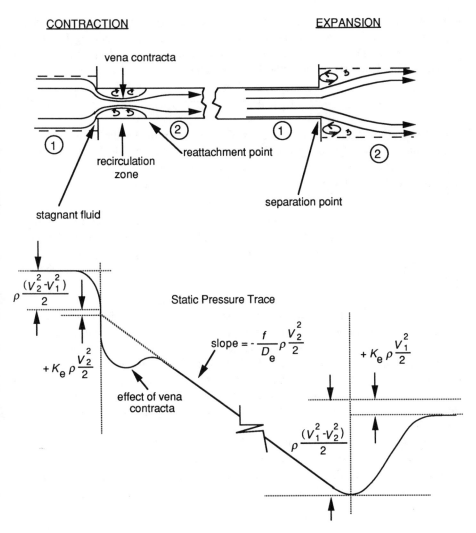

Figure 9-35 Flow of a viscous, incompressible fluid (which could be laminar or turbulent) through a sudden expansion or contraction.

A Assumption of Uniform Velocity Profile

1 Incompressible flow in a constant-area duct. A constant density fluid in a constant area has a constant velocity (because the velocity profile is uniform). Hence there are no form or acceleration pressure drops. Additionally, the frictionless flow case has no friction pressure drop, whereas the viscous flow case has the familiar friction pressure drop.

2 Frictionless, incompressible flow through sudden expansion or contraction. The detailed solution of the velocity field of a frictionless fluid can be found in a straightforward way by conformal mapping. We can find the pressure drop (in the expansion case it is a pressure rise or pressure recovery) by integrating Bernoulli's equation (Eq. 9-11) along a stream line from some arbitrary point up to station 1 or station 2. For steady-state horizontal flow we get:

$$p_1 - p_2 = \rho \frac{V_2^2 - V_1^2}{2} \tag{9-100}$$

Note that stations 1 and 2 need to be just far enough from the abrupt change that $\vec{v}$ is essentially parallel to the wall. This formula is valid for ideal contractions or expansions. The flow field for this case is far removed from the flow of real fluids. Real fluids can exhibit separation, reattachment, and recirculation.

3 Viscous, incompressible flow through a sudden expansion or contraction. To evaluate the pressure drop in these flows, a correction is made in the frictionless pressure drop results; the result is the form pressure drop. A static pressure trace is given in Figure 9-35 to show explicitly the effect of the form pressure drop or recovery. Table 9-5 has the applicable equations.

It is important to realize that the entire effect of friction in the developing region is usually included in the form pressure drop coefficients: There is no need to separately evaluate the familiar friction pressure drop in the entrance length. Values for contraction and expansion coefficients K_c and K_e generally depend on the Reynolds number, the ratio of the upstream and downstream areas, and the flow geometry. A detailed discussion of entrance and exit coefficients for single tubes is found in section II.D. From Kays' original papers, it can be concluded that the effect of decreasing Re is to decrease K_e and increase K_c. There is also a large change in the coefficients from turbulent to laminar flow.

It is possible to have a negative exit "loss" coefficient (K_e) for channels that may have substantially different velocity profiles. For example, when considering parallel flow in a rod array or through connected channels into a large plenum, it is possible to have a negative loss coefficient (K_e). This subject is discussed at length in section VII.B.

4 Compressible flow in a constant-area duct. Compressible flow in a constant-area duct is of interest in a nuclear reactor with a gaseous coolant that is being heated as it flows through the core. For the frictionless flow case, we use Bernoulli's equation (Eq. 9-11), which when applied to horizontal flow leads to:

$$\int \frac{dp}{\rho} + \frac{1}{2} (\vec{v} \cdot \vec{v}) = \text{constant along a stream line} \tag{9-101}$$

The pressure, density, and velocity are all functions of ℓ, where ℓ = distance down the duct in the direction of the flow. In a constant-area duct the stream lines are parallel to the walls, so $\vec{v} \rightarrow V(\ell)$. Generally, the gas has a temperature, $T(\ell)$, a pressure, $p(\ell)$, and a density, $\rho(\ell)$, none of which has to be linear. Integrating from an arbitrary $\ell = 0$ to $\ell = \ell$ gives:

$$\int_0^\ell \frac{dp(\ell)}{\rho(\ell)} = \frac{1}{2}[V^2(0) - V^2(\ell)] \qquad (9\text{-}102)$$

which is to be solved for $p(\ell)$. At first glance it looks difficult because ρ and V depend on p. A simple trick is to differentiate Eq. 9-102 with respect to ℓ. The left-hand side can be differentiated as:

$$\frac{d}{d\ell} \int_0^\ell \frac{dp(\ell)}{\rho(\ell)} = \frac{d}{d\ell} \int_0^\ell \frac{1}{\rho(\ell)} \frac{dp(\ell)}{d\ell} d\ell = \frac{1}{\rho(\ell)} \frac{dp(\ell)}{d\ell} \qquad (9\text{-}103)$$

Thus we can cast the differential of Eq. 9-102 as:

$$\frac{1}{\rho(\ell)} \frac{dp(\ell)}{d\ell} = -V(\ell) \frac{dV(\ell)}{d\ell} \qquad (9\text{-}104)$$

Then:

$$\frac{dp(\ell)}{d\ell} = -\rho(\ell)V(\ell) \frac{dV(\ell)}{d\ell} \qquad (9\text{-}105)$$

For a constant-area duct, however, $\rho(\ell)V(\ell) = $ constant, so when we integrate between station 1 and 2 we obtain:

$$p_1 - p_2 = \rho_2 V_2^2 - \rho_1 V_1^2 \qquad (9\text{-}106)$$

Note that the factor of $1/2$ has disappeared from the result compared to Eq. 9-100 for the frictionless incompressible case in section VII.A.2.

For flow with friction, the usual approximation is to simply introduce an integrated friction pressure drop. This treatment implicitly assumes that the friction pressure drop is small enough not to perturb the frictionless solutions for $\rho(\ell)$ and $V(\ell)$.

5 Compressible flow through a sudden expansion or contraction. The problem of compressible flow through a sudden expansion or contraction is complicated. Two simplifying assumptions can be made: (1) the flow has a small Mach number (so the pressure field does not greatly affect the density), and (2) the expansion or contraction region is short enough that there is little heating or cooling of the gas for the case of adiabatic change in pressure (hence the temperature field does not greatly affect the density). In this case the results of sections VII.A.2 and VII.A.3 are applicable without modification.

B Assumption of Nonuniform Velocity Profile

In 1950 Kays outlined an analytic procedure for calculation of pressure loss coefficients for abrupt changes in flow cross section. His results, commonly referenced in textbooks dealing with the subject, are applicable to entrance and exit losses between a large flow area and a restricted flow area such as the case at the inlet and exit of the fuel assemblies, but they are not applicable to flow between two channels with similar geometry. Hence, as shown in Figure 9-36, the loss coefficients do not ap-

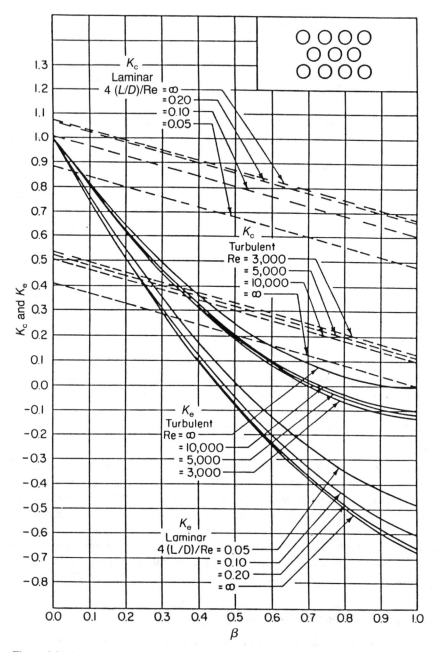

Figure 9-36 Loss coefficients for multitubular systems connected to headers. *(From Kays and Perkins [16].)*

proach zero if no area change is involved (i.e., an area ratio of 1), as should be the case at the intersection of identical geometry channels.

The reason for Kays' results is that the velocity distributions in the flow areas are not identical on both sides of the abrupt change and hence indicate a pressure differential even when the area ratio is unity. For less drastic flow geometry changes, similar velocity distributions can be assumed on both sides of the geometry change. An outline of the method and illustrative results are given here.

1 Method of calculation. Consider a sudden expansion in the flow area (Fig. 9-37). The behavior of this flow can be predicted by considering a momentum–force balance if the pressure on the downstream end of section 2 in Figure 9-37 is known. The pressure on the downstream face may be taken as equal to the static pressure in the stream just prior to the expansion. Thus if we neglect wall friction, a force balance between sections 2 and 3 in Figure 9-37 should yield the momentum change between sections 2 and 3:

$$\int \rho v^2 dA_3 - \int \rho v^2 dA_2 = p_2 A_2 - p_3 A_3 \qquad (9\text{-}107)$$

The pressure p_2 acts on all of A_2 because at the throat there is no change in the jet area. Note that:

$$\int \rho v^2 dA_2 = \int \rho v^2 dA_1 \text{ (upstream momentum)} \qquad (9\text{-}108)$$

and

$$A_2 = A_3 \qquad (9\text{-}109)$$

Furthermore, let K_d be defined as:

$$K_d = \frac{1}{V_m^2 A} \int v^2 dA \qquad (9\text{-}110)$$

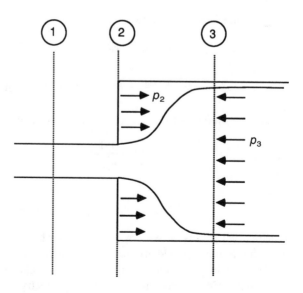

Figure 9-37 Flow past an expansion area.

where:

$$V_m = \frac{1}{A} \int v \, dA \qquad (9\text{-}111)$$

Assuming ρ is constant, Eqs. 9-108 to 9-110 can be used to recast Eq. 9-107 in the form:

$$K_{d_3} \rho V_{m3}^2 A_3 - K_{d_1} \rho V_{m1}^2 A_1 = p_2 A_3 - p_3 A_3 \qquad (9\text{-}112)$$

Now, let:

$$\beta = A_1/A_3 \qquad (9\text{-}113)$$

From the mass continuity equation:

$$V_{m3} = \beta V_{m1} \qquad (9\text{-}114)$$

Equation 9-112 can then be written in the form:

$$\frac{p_2 - p_3}{\rho} = - (\beta K_{d_1} - \beta^2 K_{d_3}) V_{m1}^2 \qquad (9\text{-}115)$$

Ideally, for conservation of the mechanical energy in the stream and uniform fluid velocity, the acceleration part of the pressure drop can be obtained from:

$$\frac{V_{m1}^2}{2} + \frac{p_2}{\rho} = \frac{V_{m3}^2}{2} + \frac{p_3}{\rho} \qquad (9\text{-}116)$$

or

$$\left(\frac{p_2 - p_3}{\rho} \right)_{\substack{\text{ideal} \\ \text{acceleration}}} = - (1 - \beta^2) \frac{V_{m1}^2}{2} \qquad (9\text{-}117)$$

The pressure loss is given by:

$$(\Delta p)_{\text{loss}} = (p_2 - p_3) - (p_2 - p_3)_{\substack{\text{ideal} \\ \text{acceleration}}} \qquad (9\text{-}118)$$

$$\therefore \frac{(\Delta p)_{\text{loss}}}{\rho} = [1 - 2\beta K_{d_1} + \beta^2 (2K_{d_3} - 1)] \frac{V_{m1}^2}{2} \qquad (9\text{-}119)$$

From the definition of the loss coefficient at an expansion:

$$K_e = \frac{(\Delta p)_{\text{loss}}}{\rho V_{m1}^2/2} \qquad (9\text{-}120)$$

Thus:

$$K_e = 1 - 2\beta K_{d_1} + \beta^2 (2K_{d_3} - 1) \qquad (9\text{-}121)$$

Equation 9-121 was derived by Kays. He further simplified the expression for K_e by assuming a uniform velocity in the larger area ($K_{d_3} = 1$). This assumption is appropriate for a large plenum connected to small channels but is not justified for the interactions of two channels with similar geometry.

Example 9-8 Application of form loss derivation

PROBLEM For turbulent flow, a semiempiric velocity distribution in circular tubes is given by:

$$v = V_m \left\{ \sqrt{f} \left[2.15 \log \left(\frac{y}{R} \right) + 1.43 \right] + 1 \right\} \tag{9-122}$$

where y = distance from the wall of a tube of radius R:

$$y = R - r$$

Use this velocity distribution to determine K_d and K_e for abrupt expansion of a circular tube for cases of laminar flow, Re $= 10^4$ and Re $= \infty$.

SOLUTION Substituting from Eq. 9-122 in Eq. 9-110 yields:

$$K_d = 1.09068(f) + 0.05884\sqrt{f} + 1 \tag{9-123}$$

Equation 9-123 together with the equations:

$$f = 0.184 \, \text{Re}^{-0.2} \tag{9-79}$$
$$\text{Re}_3 = \text{Re}_1\sqrt{\beta} \tag{9-124}$$

can be used to substitute for the respective parameters in Eq. 9-121 to obtain the value of K_e in turbulent flow. For laminar flow, a parabolic velocity distribution can be assumed, and hence:

$$K_d = 1.333, \text{ independent of Re} \tag{9-125}$$

So, for laminar flow from Eq. 9-121:

$$K_e = 1 - 2.666 \, \beta + 1.666 \, \beta^2 \tag{9-126}$$

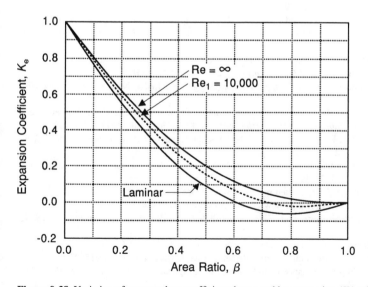

Figure 9-38 Variation of pressure loss coefficient due to sudden expansion (K_e) with the area ratio (β).

Laminar and turbulent cases have been plotted in Figure 9-38. It is clear that these relations accurately predict no pressure loss if no area change is involved between channels of similar geometry. Note that because the pressure recovery due to acceleration has been accounted for, the negative values of K_e for $\beta = 0.6$ to 1.0 for the laminar case implies that there is additional pressure gain due to the velocity profile alone. This was also the case for Kays' original results, shown in Figure 9-36. Also note that the kinetic head $(v^2/2)$ is almost completely lost when the flow is discharged from a tube into a large area $(\beta = 0)$.

REFERENCES

1. Bergelin, O. P., et al. Heat transfer and fluid friction during flow across banks of tubes-IV. *Trans. ASME* 74:953, 1952.
2. Bird, R. B., Stewart, W. E., and Lightfoot, E. N. *Transport Phenomena.* New York: Wiley, 1960.
3. Cheng, S. K., and Todreas, N. E. Hydrodynamic models and correlations for wire-wrapped LMFBR bundles and subchannel friction factors and mixing parameters. *Nucl. Eng. Design* 92:227, 1985.
4. Colebrook, C. F. Turbulent flow in pipes with particular reference to the transition region between the smooth and rough pipe laws. *Proc. Inst. Civil Eng.* 11:133, 1939.
5. Deissler, R. G., and Taylor, M. F. Analysis of axial turbulent flow and heat transfer through banks of rods or tubes. TID-7529. In: *Reactor Heat Transfer Conference.* Part 1, Book 2. USAEC, 1957.
6. DeStordeur, A. M. Drag coefficients for fuel elements spacers. *Nucleonics* 19(6):74, 1961.
7. Donohue, D. A. Heat transfer and pressure drop in heat exchangers. *Ind. Eng. Chem.* 41:2499, 1949.
8. Ebeling-Koning, D., et al. Hydrodynamics of Single- and Two-Phase Flow in Inclined Rod Arrays. MIT Nuclear Engineering Department Report DOE/ER/12075-4FR, 1985.
9. Fleming, D. P., and Sparrow, E. M. Flow in the hydrodynamic entrance regions of ducts of arbitrary cross sections. *Trans. ASME J. Heat Transfer* 91:345, 1969.
10. Fox, R. W., and McDonald, A. T. *Introduction to Fluid Mechanics* (2nd ed.). New York: Wiley, 1978.
11. Gunter, A. Y., and Shaw, W. A. A general correlation of friction factors for various types of surfaces in crossflow. *Trans. ASME* 67:643, 1945.
12. Heaton, H. S., Reynolds, W. C., and Kays, W. M. Heat transfer in annular passages: simultaneous development of velocity and temperature fields in laminar flow. *Int. J. Heat Mass Transfer* 7:763, 1964.
13. Idelchik, I. E. *Handbook of Hydraulic Resistance* (2nd ed.). New York: Hemisphere, 1986.
14. Kays, W. M. *Convective Heat and Mass Transfer.* New York: McGraw-Hill, 1966.
15. Kays, W. M., and London, A. L. *Compact Heat Exchangers* (2nd ed.). New York: McGraw-Hill, 1964.
16. Kays, W. M., and Perkins, H. C. Forced convection, internal flow in ducts. In W. M. Rohsenow and J. P. Hartnett (eds.), *Handbook of Heat Transfer.* New York: McGraw-Hill, 1971.
17. Langhaar, H. L., and No, P. H. Steady flow in the transition length of a straight tube. *J. Appl. Mech.* Sect. 3, Vol. 9, June 1942.
18. LeTourneau, B. W., Grimble, R. E., and Zerke, J. E. Pressure drop for parallel flow through rod bundles. *Trans. ASME* 79:483, 1957.
19. Martinelli, R. C. Heat transfer to molten metals. *Trans. ASME* 69:947, 1947.
20. Moody, L. F. Friction factors for pipe flow. *Trans. ASME* 66:671, 1944.
21. Rehme, K. Widerstandsbeiwerte von Gitterabstandshaltern fur Reaktorbrennelemente. *ATKE* 15 (2):127–130 (1970).
22. Rehme, K. Laminarstromung in Stabbundden. *Chem. Ingenieur Technik* 43:17, 1971.
23. Rehme, K. Pressure drop correlations for fuel elements spacers. *Nucl. Technol.* 17:15, 1973.
24. Rehme, K. Pressure drop performance of rod bundles in hexagonal arrangements. *Int. J. Heat Mass Transfer* 15:2499, 1972.

25. Rehme, K. Simple method of predicting friction factors of turbulent flow in non-circular channels. *Int. J. Heat Mass Transfer* 16:933, 1973.

26. Reichardt, H., ZAMM, Vollstandige Darsteilung der turbulenten geshwin digkeitsverteilung in glatten Leitungen. 31:208, 1951.

27. Reihman, T. C. *An Experimental Study of Pressure Drop in Wire Wrapped FFTF Fuel Assemblies.* BNWL-1207, September 1969.

28. Sparrow, E. M., and Loeffler, A. L., Jr. Longitudinal laminar flow between cylinders arranged in regular array. *A.I.Ch.E. J.* 5:325, 1959.

29. Suh, K., and Todreas, N. E. An experimental correlation of cross-flow pressure drop for triangular array wire-wrapped rod assemblies. *Nucl. Technol.* 76:229, 1987.

30. Trupp, A. C., and Azad, R. S. Structure of turbulent flow in triangular rod bundles. *Nucl. Eng. Design* 32:47, 1975.

31. Zukauskas, A. Heat transfer from tubes in crossflow. *Adv. Heat Transfer* 8:93, 1972.

32. Zukauskas, A., and Ulinkskas, R. Banks of plain and finned tubes. In: *Heat Exchanger Design Handbook* (Vol. 2). New York: Hemisphere, 1983, pp. 2.2.2.4-1–2.2.4-17.

PROBLEMS

Problem 9-1 Emptying of a liquid tank (section II)

Consider an emergency water tank, shown in Figure 9-39, that is supposed to deliver water to a reactor following a loss of coolant event. The tank is prepressurized by the presence of nitrogen at 1.0 MPa. The water is discharged through an 0.2 m (I.D.) pipe. What is the maximum flow rate delivered to the reactor if the water is considered inviscid and the reactor pressure is:

1. 0.8 MPa
2. 0.2 MPa

Answers: 1. $\dot{m} = 827.6$ kg/s
       2. $\dot{m} = 1367.2$ kg/s

Problem 9-2 Laminar flow velocity distribution and pressure drop in parallel plate geometry (section IV)

For flow between parallel flat plates:

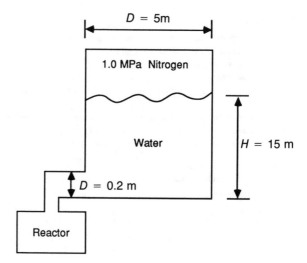

Figure 9-39 Emergency water tank.

1. Show that the momentum equation for fully developed, steady-state, constant density and viscosity flow takes the form:

$$\frac{dp}{dx} = \mu \frac{d^2 v_x}{dy^2}$$

2. For plate separation of $2y_0$ show that the velocity profile is given by:

$$v_x(y) = \frac{3}{2} V_m \left[1 - \left(\frac{y}{y_0} \right)^2 \right]$$

3. Show that:

$$-\frac{dp}{dx} = \frac{96}{Re} \cdot \frac{1}{4y_0} \cdot \frac{\rho V_m^2}{2}$$

Problem 9-3 Velocity distributions in single-phase flow (section V)

Consider a smooth circular flow channel of diameter 13.5 mm (for a hydraulic simulation of flow through a PWR assembly).

Operating conditions: Assume two flow conditions, with fully developed flow patterns but with no heat addition:
a. High flow (mass flow rate = 0.5 kg/s)
b. Low flow (mass flow rate = 1 g/s).
Properties (approximately those of pressurized water at 300°C and 15.5 MPa)
Density = 720 kg/m³
Viscosity = 91 μPa · s

1. For each flow condition draw a quantitative sketch to show the velocity distribution based on Martinelli's formalism (Eqs. 9-72a to 9-72c).
2. Find the positions of interfaces between the laminar sublayer, the buffer zone, and the turbulent core.

Answers: 2a. Laminar layer: $0 \leq y \leq 3.2$ μm
Buffer zone: 3.2 μm $\leq y < 19.2$ μm
Turbulent core: 19.2 μm $\leq y \leq 6.75$ mm
2b. Laminar flow

Problem 9-4 Sizing of an orificing device

In a hypothetical reactor an orificing scheme is sought such that the core is divided into two zones. Each zone produces one-half of the total reactor power. However, zone 1 contains 100 assemblies, whereas zone 2 contains 80 assemblies.

It is desired to obtain equal average temperature rises in the two zones. Therefore the flow in the assemblies of lower power production is to be constricted by the use of orificing blocks, as shown in Figure 9-40.

Determine the appropriate diameter (D) of the flow channels in the orificing block. Assume negligible pressure losses in all parts of the fuel assemblies other than the fuel rod bundle and the orifice block. The flow in all the assemblies may be assumed fully turbulent. All coolant channels have smooth surfaces.

Data

Pressure drop across assemblies	$\Delta p_A = 7.45 \times 10^5$ N/m²
Total core flow rate	$\dot{m}_T = 17.5 \times 10^6$ kg/hr
Coolant viscosity	$\mu = 2 \times 10^{-4}$ N s/m²
Coolant density	$\rho = 0.8$ g/cm³

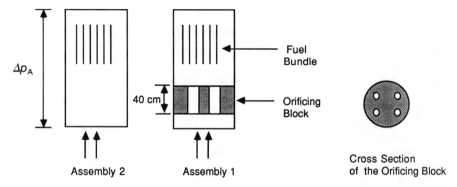

Figure 9-40 Orificing technique.

| Contraction pressure loss coefficient | $K_c = 0.5$ |
| Expansion pressure loss coefficient | $K_e = 1.0$ |

Answer: $D = 2.15$ cm

Problem 9-5 Pressure drop features of a PWR core (section VI)

Consider a PWR core containing 38,000 fuel rods cooled with a total flow rate of 15 Mg/s. Each rod has a total length of 3.7 m and a smooth outside diameter of 11.2 mm. The rods are arranged in a square array with pitch = 14.7 mm. The lower-end and upper-end fittings are represented as a honeycomb grid spacer, with the thickness of individual grid elements being 1.5 mm. There are also five intermediate honeycomb grid spacers with thickness of 1 mm. Consider the upper and lower plenum regions to be entirely open.

1. What is the plenum-to-plenum pressure drop?
2. What are the components of this pressure drop (entrance, exit, end fixtures, wall friction, elevation, grids)?

Properties
Water density = 720 kg/m^3
Water viscosity = 91 μPa s

$\qquad$ *Answers:* $\Delta p_{gravity} = 26.11$ kPa
$\qquad\qquad\quad$ $\Delta p_{friction} = 29.29$ kPa
$\qquad\qquad\quad$ $\Delta p_{entrance} + \Delta P_{exit} = 11.73$ kPa
$\qquad\qquad\quad$ $\Delta p_{spacer} = 13.68$ kPa
$\qquad\qquad\quad$ $\Delta p_{fittings} = 11.90$ kPa
$\qquad\qquad\quad$ $\Delta p_T = 92.71$ kPa

Problem 9-6 Comparison of laminar and turbulent friction factors of water and sodium heat exchangers (section VI)

Consider square arrays of vertical tubes utilized in two applications: a recirculation PWR steam generator and an intermediate heat exchanger for LMFBR service. In each case primary system liquid flows through the tubes, and secondary system liquid flows outside the tubes within the shell side.

The operating and geometric conditions of both units are given in the table. Assume the wall heat flux is axially constant.

Parameter	PWR steam generator	LMFBR intermediate heat exchanger
System characteristics		
Primary fluid	Water	Sodium
Secondary fluid	Water	Sodium
P/D	1.5	1.5
D (cm)	1.0	1.0
Nominal shell-side properties		
Pressure (MPa)	5.5	0.202
Temperature (°C)	270	480
Density (kg/m³)	767.9	837.1
Thermal conductivity (W/m °C)	0.581	88.93
Viscosity (kg/m s)	1.0×10^{-4}	2.92×10^{-4}
Heat capacity (J/kg °K)	4990.0	1195.8

For the shell side of the tube array in each application answer the following questions:

1. Find the friction factor (f) for fully developed laminar flow at $Re_{De} = 10^3$.
2. Find the friction factor (f) for fully developed turbulent flow at $Re_{De} = 10^5$.
3. Can either of the above friction factors be found from a circular tube using the equivalent diameter concept? Demonstrate and explain.
4. What length is needed to achieve fully developed laminar flow?
5. What length is needed to achieve fully developed turbulent flow?

SINGLE-PHASE HEAT TRANSFER

I FUNDAMENTALS OF HEAT TRANSFER ANALYSIS

A Objectives of the Analysis

The objectives of heat transfer analysis for single-phase flows are generally: (1) determination of the temperature field in a coolant channel so as to ensure that the operating temperatures are within the specified limits; and (2) determination of the parameters governing the heat-transport rate at the channel walls. These parameters can then be used to choose materials and flow conditions that maximize heat transport in the process equipment.

Attainment of the first objective leads directly to the second one, as knowledge of the temperature field in the coolant leads to determination of the heat flux, $\vec{q}''$ (W/m^2), at the solid wall via Fourier's law for heat transfer. At any surface this law states that:

$$\vec{q}'' = -k \frac{\partial T}{\partial n} \vec{n} \qquad (10\text{-}1)$$

where k = thermal conductivity of the coolant (W/m °K), and $\vec{n}$ = unit vector perpendicular to the surface, so that $\partial T/\partial n$ = temperature gradient in the direction of heat transfer (°K/m). However, in engineering analyses, where only the second objective is desired, the heat flux is related to the bulk or mean temperature of the flow (T_b), via Newton's law for heat transfer:

$$\vec{q}'' \equiv h(T_w - T_b) \vec{n} \qquad (10\text{-}2)$$

where T_w = wall temperature (°K), and h = the heat transfer coefficient (W/m^2 °K). Equation 10-2 is applied in the engineering analysis when it is possible to determine h for the flow conditions based on prior engineering experience. Often the heat transfer coefficient is a semiempirical function of the coolant properties and velocity as well as the flow channel geometry.

B Approximations to the Energy Equation

The general energy equation for single-phase flow is used in its temperature form:

$$\rho c_p \frac{DT}{Dt} = - \nabla \cdot \vec{q}'' + q''' + \beta T \frac{Dp}{Dt} + \Phi \tag{4-122a}$$

When the velocity $\vec{v}(\vec{r},t)$, pressure $p(\vec{r},t)$, and heat-generation rate, q''', are known a priori, Eq. 4-122a is used to specify the temperature field $T(\vec{r},t)$. The equations of state— $\rho(p,T)$, $c_p(p,T)$, and $\beta(T)$—as well as the constitutive relations for $q''(\rho,v,T)$ and $\Phi(\rho,v,T)$ are needed for the solution. Also needed are initial values of $T(\vec{r},o)$ and $p(\vec{r},o)$ as well as the appropriate number of boundary conditions.

In general, as discussed in Chapter 9, section I, the equation is first simplified to a form that is an acceptable approximation of the situation at hand. The common approximations are twofold:

1. The pressure term is negligible (in effect considering the phase incompressible).
2. The material properties are temperature- and pressure-independent.

The above two approximations are acceptable in forced convection flow analysis but should not be applied indiscriminately in natural-flow and mixed-flow analyses. When natural flow is important, the Boussinesq approximation is applied where the material properties are assumed to be temperature-independent with the exception of the density in the gravity or buoyancy term in the momentum equation, which is assumed to vary linearly with temperature.

It is possible to ignore radiation heat transfer within the single-phase liquids and high-density gases so that the heat flux is due to molecular conduction alone. Using Eq. 4-114, we get:

$$\nabla \cdot \vec{q}'' = - \nabla \cdot k \nabla T$$

Thus for incompressible fluids and purely conductive heat flux, the energy equation is written as:

$$\rho c_p \frac{DT}{Dt} = \nabla \cdot k \nabla T + q''' + \Phi \tag{10-3a}$$

For rectangular geometry and a fluid with constant μ and k properties, Eq. 10-3a takes a form given in Table 4-12 when dh is taken as $c_p dT$.

$$\rho c_p \left(\frac{\partial T}{\partial t} + v_x \frac{\partial T}{\partial x} + v_y \frac{\partial T}{\partial y} + v_z \frac{\partial T}{\partial z} \right) =$$

$$k \left(\frac{\partial^2 T}{\partial x^2} + \frac{\partial^2 T}{\partial y^2} + \frac{\partial^2 T}{\partial z^2} \right) + q'''$$

$$+ 2\mu \left[\left(\frac{\partial v_x}{\partial x} \right)^2 + \left(\frac{\partial v_y}{\partial y} \right)^2 + \left(\frac{\partial v_z}{\partial z} \right)^2 \right] \tag{10-3b}$$

$$+ \mu \left[\left(\frac{\partial v_x}{\partial y} + \frac{\partial v_y}{\partial x} \right)^2 + \left(\frac{\partial v_x}{\partial z} + \frac{\partial v_z}{\partial x} \right)^2 + \left(\frac{\partial v_y}{\partial z} + \frac{\partial v_z}{\partial y} \right)^2 \right]$$

If the dissipation energy (Φ) is also negligible, which is generally true unless the velocity gradients are very large, the terms between brackets [ ] are neglected.

C Dimensional Analysis

Consider the case of spatially and temporally constant pressure ($Dp/Dt = 0$) and no heat generation ($q''' = 0$). Also, let the heat flux be due to conduction alone, and let the material properties be temperature-independent. Then Eq. 10-3a takes the form:

$$\rho c_p \frac{DT}{Dt} = k\nabla^2 T + \mu\phi \tag{10-4}$$

where the viscosity has been factored out of the dissipation function Φ (i.e., $\Phi = \mu\phi$); Φ is given in Table 4-12 as an explicit function of the velocity components.

Using the nondimensional parameters:

$\vec{v}* = \vec{v}/V$
$x* = x/D_e$
$t* = tV/D_e$
$T* = (T - T_o)/(T_1 - T_o)$

where V, D_e, and $T_1 - T_o$ = convenient characteristic velocity, length, and temperature differences in the system. Then Eq. 10-4 can be cast in the form:

$$\rho c_p \left[(T_1 - T_o) \frac{V}{D_e} \frac{\partial T*}{\partial t*} + (T_1 - T_o) \frac{V}{D_e} \vec{v}* \cdot \nabla*T* \right] = \tag{10-5}$$

$$\frac{k(T_1 - T_o)}{D_e^2} \nabla*^2 T* + \mu \left(\frac{V}{D_e} \right)^2 \phi*$$

or, by rearranging Eq. 10-5:

$$\frac{\partial T*}{\partial t*} + \vec{v}* \cdot \nabla*T* = \frac{1}{Re\ Pr} \nabla*^2 T* + \frac{Br}{Re\ Pr} \phi* \tag{10-6}$$

where $\nabla*$ and $\nabla*^2$ involve differentiation with respect to $x*$:

$$Re \equiv \frac{\rho V D_e}{\mu}, \text{ the Reynolds number} \tag{9-55}$$

$$Pr \equiv \frac{\mu c_p}{k}, \text{ the Prandtl number} \tag{10-7}$$

$$Br \equiv \frac{\mu V^2}{k(T_1 - T_o)}, \text{ the Brinkmann number} \tag{10-8}$$

The Prandtl number signifies the ratio of molecular diffusivity of momentum to that of heat in a fluid. The Brinkmann number is the ratio of heat production by viscous dissipation to heat transfer by conduction. In some analyses the Eckert number (Ec) is used instead of the ratio (Br/Pr). Because:

$$Ec = \frac{Br}{Pr} = \frac{V^2/c_p}{T_1 - T_o}$$

it signifies the ratio of the dynamic temperature due to motion to the static temperature difference. The physical significance of several nondimensional groups of importance in single-phase heat transfer is given in Appendix G.

D Thermal Conductivity

Thermal conductivity is the property relating the heat flux (rate of heat transfer per unit area) in a material to the temperature spatial gradient in the absence of radiation effects.

The thermal conductivities of engineering materials vary widely (Fig. 10-1). The

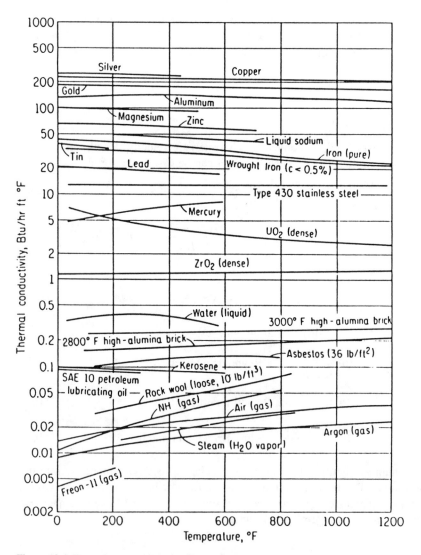

Figure 10-1 Thermal conductivity of engineering materials. *(From Ozisik [35].)*

highest conductivities belong to the metals and the lowest to the gases. Even for the same material, the conductivity is a function of temperature and, in the case of gases, also of pressure. Figure 10-2 illustrates the pressure and temperature dependence of monoatomic substances. It may also be used to approximate the behavior of polyatomic substances.

For gases, an approximate relation for predicting the conductivity is given by the Eucken formula [3]:

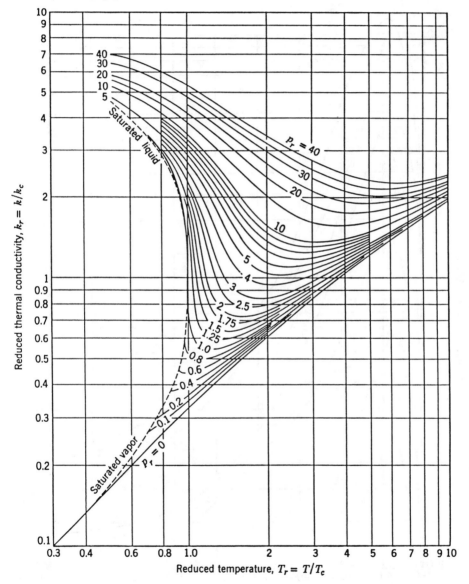

Figure 10-2 Thermal conductivity of monoatomic substances as a function of pressure and temperature where $p_r = \dfrac{p}{p_c}$ *(From Bird et al. [3].)*

$$k = \left(c_p + \frac{5R}{4M}\right)\mu \qquad (10\text{-}9)$$

where R = the universal gas constant, and M = gas molecular weight.

For liquids and solids, the conductivity is more difficult to predict theoretically; however, for metals, thermal conductivity (k) is related to electrical conductivity (k_e) by:

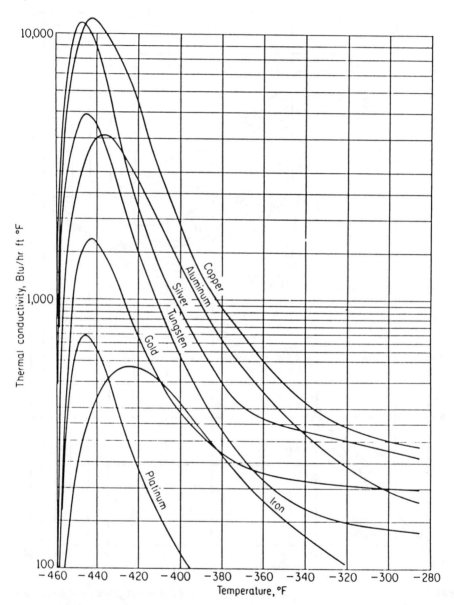

Figure 10-3 Thermal conductivity of super cold metals. *(From Ozisik [35].)*

$$\frac{k}{k_e T} = L = \text{constant} \tag{10-10}$$

where L = Lorentz number, and T = absolute temperature (L is about 25×10^{-9} volt^2/°K^2 for pure metals at 0°C and increases only at about 10% per 100°C). At very low temperatures (below -200°C), the metals become superconductors of electricity and heat, as can be seen in Figure 10-3.

E Engineering Approach to Heat Transfer Analysis

For engineering analyses, the difference between the wall temperature and the bulk flow temperature is obtained by defining the heat transfer coefficient (h) through the nondimensional Nusselt number:

$$\text{Nu} = f(\text{Re, Pr, Gr, } \mu_w/\mu_b) \tag{10-11}$$

where:

$$\text{Nu} \equiv \frac{hD_H}{k} \tag{10-12}$$

and D_H = an appropriate length or lateral dimension. (For external flows the length dimension is usually used, whereas a lateral dimension is used for internal flows.)

The general form of Nu is obtained by a boundary layer analysis similar to that for the momentum equation (see Chapter 9). Experiments are often needed, particularly for turbulent flow, to define the numerical constants of Eq. 10-11. The form of the Nu relation depends on the flow regime (laminar versus turbulent, external versus internal) and the coolant (metallic versus nonmetallic). At high values of Re, the heat transfer is aided by the presence of turbulent eddies, resulting in an increased heat-transfer rate over the case of purely laminar flow. For metallic liquids, the molecular thermal conductivity is so high that the relative effect of turbulence is not as significant as in the case of nonmetallic flows.

Typical values of the heat-transfer parameters associated with various coolants and processes are given in Tables 10-1 and 10-2, respectively.

Table 10-1 Representative heat transfer parameters

Material	T (°C)	p (MPa)	k (W/m °K) (Btu/hr ft°F)	Pr	Nu in a tube Re = 10,000	Nu in a tube Re = 100,000
Water	275	7.0	0.59 (0.35)	0.87	34.8	219.5
Gases						
Helium	500	4.0	0.31 (0.18)	0.67	31.9	195
CO_2	300	4.0	0.042 (0.024)	0.76	33.2	209
Sodium	500	0.3	52 (31)	0.004	5.44	7.77

Table 10-2 Typical values of the heat-transfer coefficient for various processes

Process	Heat-transfer coefficient (h)	
	Btu/hr ft² °F	W/m² °K
Natural convection		
Low pressure gas	1–5	6–28
Liquids	10–100	60–600
Boiling water	100–2000	60–12,000
Forced convection in pipes		
Low pressure gas	1–100	6–600
Liquids		
Water	50–2000	250–12,000
Sodium	500–5000	2,500–25,000
Boiling water	500–10,000	2,500–50,000
Condensation of steam	1,000–20,000	5,000–100,000

Example 10-1 Importance of terms in the energy equation under various flow conditions

PROBLEM Consider the following two flow conditions in a pressurized water reactor steam generator on the primary side.

	Forced flow	Natural circulation
Flow per tube	1.184 kg/s	0.01184 kg/s
Characteristic temperature difference	15°C	25°C

For a tube of inner diameter $\frac{7}{8}$ in. and an average temperature of 305°C, evaluate the various dimensionless parameters in Eq. 10-6 and determine which terms are important under both flow conditions.

SOLUTION The energy equation (Eq. 10-6) is:

$$\frac{\partial T^*}{\partial t^*} + \vec{v}^* \cdot \nabla^* T^* = \frac{1}{\text{RePr}} \nabla^{*2} T^* + \frac{\text{Br}}{\text{RePr}} \phi^*$$

where:

$$\text{Re} = \frac{\rho V D_e}{\mu}$$

$$\text{Pr} = \frac{\mu c_p}{k}$$

$$\text{Br} = \frac{\mu V^2}{k(T_1 - T_o)}$$

Evaluating ρ, μ, c_p, and k for saturated water at 305°C we find:

$\rho = 701.9 \text{ kg/m}^3$
$\mu = 8.9 \times 10^{-5} \text{ kg/m s}$
$k = 0.532 \text{ W/m°C}$
$c_p = 5969 \text{ J/kg°C}$

For the forced flow condition:

$$V = \frac{\dot{m}}{\rho A} = \frac{1.184 \text{ kg/s}}{(701.9 \text{ kg/m}^3) \frac{\pi}{4} \left[(0.875 \text{ in.}) \frac{0.0254 \text{ m}}{\text{in.}} \right]^2} = 4.348 \text{ m/s}$$

so that:

$$\text{Re} = \frac{(701.9 \text{ kg/m}^3)(4.348 \text{ m/s})(0.0222 \text{ m})}{8.90 \times 10^{-5} \text{ kg/m s}} = 7.613 \times 10^5$$

$$\text{Pr} = \frac{(8.90 \times 10^{-5} \text{ kg/m s})(5969 \text{ J/kg°C})}{0.532 \text{ W/m°C}} = 1.00$$

$$\text{Br} = \frac{(8.90 \times 10^{-5} \text{ kg/m s})(4.348 \text{ m/s})^2}{(0.532 \text{ W/m°C})(15°C)} = 2.11 \times 10^{-4}$$

Then:

$$\frac{1}{\text{RePr}} = \frac{1}{\text{Pe}} = \frac{1}{(7.613 \times 10^5)(1.00)} = 1.314 \times 10^{-6}$$

$$\frac{\text{Br}}{\text{RePr}} = \frac{\text{Br}}{\text{Pe}} = (2.11 \times 10^{-4})(1.314 \times 10^{-6}) = 2.773 \times 10^{-9}$$

Thus in the forced-flow condition, these parameter values are sufficiently low that the conduction term and the dissipative term in the energy equation can often be ignored in the energy balance.

In the natural circulation condition, the velocity is two orders of magnitude lower than in the forced flow case, so that:

$$\text{Re} = 7.613 \times 10^3$$
$$\text{Pr} = 1.00$$
$$\text{Br} = (2.11 \times 10^{-4})(1 \times 10^{-4}) \left(\frac{15}{25} \right) = 1.266 \times 10^{-8}$$

Then:

$$\frac{1}{\text{Pe}} = \frac{1}{\text{Re Pr}} = 1.314 \times 10^{-4}$$

$$\frac{\text{Br}}{\text{Pe}} = \frac{\text{Br}}{\text{Re Pr}} = (1.266 \times 10^{-8})(1.314 \times 10^{-4}) = 1.664 \times 10^{-12}$$

Again, neither term on the right-hand side is found to be large. However, the conduction term parameter has increased in value, whereas the dissipation term has decreased. The dissipation energy term can always be ignored, and the conduction term can often be ignored. It is not necessarily the case for all working fluids, however. In the case of sodium-cooled breeder reactors, for example, the high thermal conductivity of sodium results in a Pr number of the order 0.004. In natural circulation conditions, $1/\text{Pe} \approx 1$, and conduction may become important.

II LAMINAR HEAT TRANSFER IN A PIPE

The development of the temperature profile in a boundary layer parallels the development of the velocity profile, as discussed in Chapter 9. For external flows, the thermal boundary layer is taken as the distance over which the temperature changes from the wall temperature to the stream temperature. For internal flows, the thermal boundary developing on the wall merges (at the center for symmetric channels) and provides thereafter the thermally developed region of the flow in the pipe. The temperature profiles for external and internal flows are illustrated in Figures 10-4 and 10-5, respectively.

A Fully Developed Flow in a Circular Tube

Let us consider the fully developed flow region in a cylinder of radius R, with azimuthal symmetry. The applicable equations and boundary conditions in this case are shown in Table 10-3. Two surface boundary conditions are of major importance— constant heat flux and constant wall temperature—because the solutions of these cases can be used to solve the cases of arbitrary wall temperature and heat flux distributions by superposition [48].

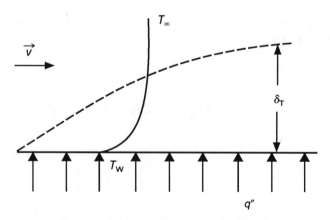

Figure 10-4 Thermal boundary layer for external flows.

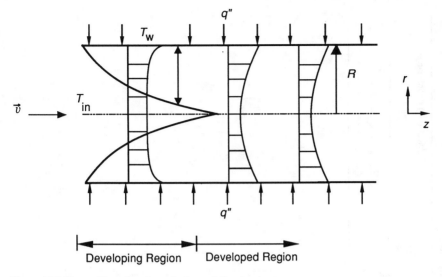

Figure 10-5 Thermal boundary layer for internal flow in pipe.

In the constant heat flux case the wall and fluid temperatures increase linearly with length at the same rate. Because of the symmetry of the circular tube, the wall temperature is circumferentially constant at every cross section, though increasing axially, and the thermal properties of the wall material do not enter into any circumferential considerations. For a geometry that lacks circular symmetry, the analogous constant heat flux condition is more complex. Although the axial profile of the heat flux at any circumferential position can still be specified as constant, the conditions of circumferentially constant heat flux and temperature may be incompatible, as in the case of rod bundles.

The velocity characteristics of the fully developed flow conditions (shown in Table 10-3) are:

$$v_r = 0 \text{ and } \frac{\partial v_z}{\partial z} = 0$$

Let T_m be the fluid mean temperature defined by:

$$T_m(z) \equiv \frac{\int_o^R \rho v_z T \, 2\pi r dr}{\int_o^R \rho v_z \, 2\pi r dr} \tag{10-13}$$

The temperature lateral profile in a flow with fully developed heat transfer is independent of the axial distance, which implies the general condition:

$$\frac{\partial}{\partial z}\left(\frac{T_w - T}{T_w - T_m}\right) = 0 \tag{10-14a}$$

Table 10-3 Steady-state fully developed laminar flow equations: momentum and energy

Parameter	Rectangular (x,y) coordinates	Cylindrical (r,z) coordinates
Momentum		
Region characteristics	$v_y = 0;\ \dfrac{\partial v_x}{\partial x} = 0$	$v_r = 0;\ \dfrac{\partial v_z}{\partial z} = 0$
Governing equations	$-\dfrac{\partial p}{\partial x} + \mu\,\dfrac{\partial^2 v_x}{\partial y^2} = 0$ $\dfrac{\partial p}{\partial y} = 0$	$-\dfrac{\partial p}{\partial z} + \mu\left(\dfrac{\partial^2 v_z}{\partial r^2} + \dfrac{1}{r}\dfrac{\partial v_z}{\partial r}\right) = 0$ $\dfrac{\partial p}{\partial r} = 0$
Boundary conditions	$v_x = 0$ at $y = y_0$ $\dfrac{\partial v_x}{\partial y} = 0$ at $y = 0$	$v_z = 0$ at $r = R$ $\dfrac{\partial v_z}{\partial r} = 0$ at $r = 0$
Resultant solution	$\dfrac{v_x}{V_m} = \dfrac{3}{2}\left[1 - \left(\dfrac{y}{y_0}\right)^2\right]$	$\dfrac{v_z}{V_m} = 2\left[1 - \left(\dfrac{r}{R}\right)^2\right]$
Friction factor	$f = \dfrac{96}{Re}$	$f = \dfrac{64}{Re}$
Characteristic dimension	$D_e = 4\,y_0$	$D_e = 2R$

Energy

	q_w'' uniform $0 \le y < y_o$	T_w uniform $0 \le y \le y_o$	q_w'' uniform $0 < r \le R$	T_w uniform $0 \le r \le R$
Region characteristics	$\dfrac{\partial T}{\partial x} = \dfrac{\partial T_m}{\partial x}$	$\dfrac{\partial T}{\partial x} = f\left(\dfrac{y}{y_o}\right)\dfrac{\partial T_m}{\partial x}$	$\dfrac{\partial T}{\partial z} = \dfrac{\partial T_m}{\partial z}$	$\dfrac{\partial T}{\partial z} = f\left(\dfrac{r}{R}\right)\dfrac{\partial T_m}{\partial z}$
Governing equations (neglecting axial conduction)	$\rho c_p v_x \dfrac{\partial T}{\partial x} = \dfrac{\partial}{\partial y}\left(k\,\dfrac{\partial T}{\partial y}\right) + q''' + \mu\left(\dfrac{\partial v_x}{\partial y}\right)^2$		$\rho c_p v_z \dfrac{\partial T}{\partial z} = \dfrac{1}{r}\dfrac{\partial}{\partial r}\left(rk\,\dfrac{\partial T}{\partial r}\right) + q''' + \mu\left(\dfrac{\partial v_z}{\partial r}\right)^2$	
Boundary conditions	$\dfrac{\partial T}{\partial y} = 0$ at $y = 0$ $T = T_w(x)$ at $y = y_o$		$\dfrac{\partial T}{\partial r} = 0$ at $r = 0$ $T = T_w(z)$ at $r = R$	
Nusselt number (parabolic velocity)	8.23	7.54	4.36	3.66
Nusselt number (slug velocity)	12	—	8.00	5.75
Characteristic dimension used in Nu	$D_H = 4y_o$		$D_H = 2R$	

or

$$\frac{T_w - T}{T_w - T_m} = f\left(\frac{r}{R}\right)$$ (10-14b)

Hence:

$$\frac{\partial T_w}{\partial z} - \frac{\partial T}{\partial z} = f\left(\frac{r}{R}\right)\left(\frac{\partial T_w}{\partial z} - \frac{\partial T_m}{\partial z}\right)$$ (10-14c)

If the heat flux is constant, $\partial q_z''/\partial z = 0$ and the radial profile is constant, so that:

$$\frac{\partial T_w}{\partial z} = \frac{\partial T_m}{\partial z} = \frac{\partial T}{\partial z}; \; q_w''(z) = \text{constant}$$ (10-15)

whereas if the wall temperature is axially constant,

$$\frac{\partial T}{\partial z} = f\left(\frac{r}{R}\right)\frac{\partial T_m}{\partial z}; \; T_w(z) = \text{constant}$$ (10-16)

In this case, $\partial q_z''/\partial z \neq 0$.

The steady-state temperature profile in the pipe is now derived for the constant heat flux case. As shown in Chapter 9, the velocity profile in this case is given by:

$$v_z = 2V_m\left[1 - \left(\frac{r}{R}\right)^2\right]$$ (9-45)

where:

$$V_m = \frac{\int_0^R \rho v_z \, 2\pi r dr}{\int_0^R \rho \, 2\pi r dr}$$

Neglecting the internal energy generation and the dissipation energy, at steady state the cylindrical (r,z) energy equation for fully developed conditions is given by:

$$\rho c_p v_z \frac{\partial T}{\partial z} = \frac{\partial}{\partial z} k \frac{\partial T}{\partial z} + \frac{1}{r}\frac{\partial}{\partial r}\left(kr\frac{\partial T}{\partial r}\right)$$ (10-17)

In most applications, even a small velocity is sufficient to make the axial conduction heat transfer negligible, as:

$$\frac{\partial}{\partial z} k \frac{\partial T}{\partial z} << \rho c_p v_z \frac{\partial T}{\partial z}$$ (10-18)

Substituting for v_z from Eq. 9-45 and applying the condition 10-18, Eq. 10-17 can be written as:

$$2\rho c_p V_m\left[1 - \left(\frac{r}{R}\right)^2\right]\frac{\partial T}{\partial z} = \frac{1}{r}\frac{\partial}{\partial r}\left(kr\frac{\partial T}{\partial r}\right)$$ (10-19)

However, because $\partial T/\partial z$ is not a function of r, Eq. 10-19 can be integrated to yield:

$$2\rho c_p \, V_m \frac{\partial T}{\partial z} \left(\frac{r^2}{2} - \frac{r^4}{4R^2} \right) = kr \frac{\partial T}{\partial r} + C_1 \tag{10-20}$$

Applying the symmetry condition that at:

$$r = 0 \quad \frac{\partial T}{\partial r} = 0 \tag{10-21}$$

we get $C_1 = 0$.

Integrating Eq. 10-20 over r again, we get:

$$2\rho c_p \, V_m \frac{\partial T}{\partial z} \left(\frac{r^2}{4} - \frac{r^4}{16R^2} \right) = kT + C_2 \tag{10-22}$$

Because at $r = R$, $T = T_w$ \hfill (10-23)

$$C_2 = -kT_w + 2\rho c_p V_m \frac{\partial T}{\partial z} \left(\frac{3R^2}{16} \right) \tag{10-24}$$

and Eq. 10-22 can be rearranged to give:

$$T = T_w + \frac{2\rho c_p}{k} V_m \frac{\partial T}{\partial z} \left(\frac{r^2}{4} - \frac{r^4}{16R^2} - \frac{3R^2}{16} \right) \tag{10-25}$$

The wall heat transfer q_w'' is given by:

$$q_w'' = -k \frac{\partial T}{\partial r} \bigg|_R = -2\rho c_p \, V_m \frac{\partial T}{\partial z} \left(\frac{R}{2} - \frac{R}{4} \right) \tag{10-26}$$

or

$$q_w'' = - \left(2\rho c_p \, V_m \frac{\partial T}{\partial z} \right) \left(\frac{R}{4} \right) \tag{10-27}$$

Thus:

$$\frac{\partial T}{\partial z} = - \frac{2q_w''}{\rho \, c_p \, V_m R} \tag{10-28}$$

Realizing that for the axially constant heat flux case $\partial T/\partial z = \partial T_m/\partial z$, Eq. 10-28 could have been obtained from the energy balance for the cross section:

$$\rho V_m \pi R^2 \, c_p \left(\frac{\partial T_m}{\partial z} \right) = (-q_w'') 2\pi R \tag{10-29}$$

Note that the outward heat flux is given by $-k \, (\partial T/\partial r)$; hence for a heated channel, $-q_w''$ is a positive number.

The mean temperature (T_m) can be evaluated using Eq. 10-13 and Eq. 9-45 as:

$$T_m - T_w = \frac{\displaystyle\int_0^R (T - T_w) \, 2V_m \left[1 - \left(\frac{r}{R} \right)^2 \right] 2\pi r dr}{V_m \pi R^2}$$

$$= \frac{8\rho c_p \, V_m (\partial T/\partial z)}{R^2 k} \int_0^R \left[\frac{r^2}{4} - \frac{r^4}{16R^2} - \frac{3R^2}{16} \right] \left[1 - \left(\frac{r}{R} \right)^2 \right] r \, dr \quad (10\text{-}30)$$

$$= -\frac{11}{48} \frac{\rho c_p V_m}{k} \left(\frac{\partial T}{\partial z} \right) R^2$$

By combining Eqs. 10-29 and 10-30, we get:

$$T_m - T_w = \frac{11}{24} \frac{R}{k} q_w'' \quad (10\text{-}31)$$

From the definitions of h and Nu, we get:

$$h = \frac{q_w''}{T_m - T_w} = \frac{24}{11} \frac{k}{R} \quad (10\text{-}32)$$

$$Nu = \frac{hD}{k} = \frac{h(2R)}{k} = \frac{48}{11} = 4.364 \quad (10\text{-}33)$$

The Nusselt number (Nu) derived above is applicable to the constant heat flux case in a circular tube. The value of Nu depends on the boundary conditions as well as the geometry (Table 10-4). It can be observed that for the case of a constant wall temperature heated tube, the Nusselt number is a constant of a different value, i.e., 3.66. In general, laminar flow conditions lead to Nusselt numbers that are constants, independent of flow velocity (or Re) and Pr.

B Developed Flow in Other Geometries

In a manner similar to the tube flow with constant heat flux, the laminar flow heat transfer coefficient and the Nusselt numbers can be computed for other geometries. In all cases the laminar Nusselt number is a constant independent of the flow velocity or Prandtl number. This case is also true for a constant axial wall temperature.

For geometries other than the round tube, a Nusselt number is defined using the concept of equivalent heated diameter where:

$$D_H \equiv \frac{4 \times \text{flow area}}{\text{heated perimeter}} = \frac{4A_f}{P_h} \quad (10\text{-}34)$$

In many cases the heated perimeter and the wetted perimeter are the same, resulting in equal equivalent diameters. Table 10-4 provides the Nu values for various geometries.

For flow parallel to a bundle of circular tubes Sparrow et al. [47] solved the laminar problem for axially constant q_w'' and circumferentially constant T_w (boundary condition A). Dwyer and Berry [15] solved the laminar case for axially and circumferentially constant q_w'' (boundary condition B). Their results are given in Figure 10-6. For sufficiently large P/D values, Figure 10-6 demonstrates that the equivalent annulus approximation is good relative to the exact solution. The exact solution for the azimuthal variation of the temperature when P/D is less than 1.5 as described in Chapter 7, Vol. II.

**Table 10-4 Nusselt number for laminar fully developed velocity
and temperature profiles in tubes of various cross sections**

Cross-sectional shape	b/a	Nu* $q'' = $ constant	Nu $T_w = $ constant
circle	—	4.364	3.66
square	1.0	3.63	2.98
rectangle	1.4	3.78	
rectangle	2.0	4.11	3.39
rectangle	3.0	4.77	
rectangle	4.0	5.35	4.44
rectangle	8.0	6.60	5.95
parallel plates	∞	8.235	7.54
parallel plates (insulated)	∞	5.385	4.86
triangle	—	3.00	2.35

Source: From Kays [22].

*The constant-heat-rate solutions are based on constant *axial* heat rate but with constant *temperature* around the tube periphery. Nusselt numbers are averages with respect to tube periphery.

C Developing Laminar Flow Region

In a pipe entry region, the heat transfer is more involved, as both the velocity and temperature profiles may vary axially. When the Prandtl number is higher than 5 [22], the velocity profile develops faster than the temperature profile. Hence, as in the early work of Graetz, reported in 1886 [17], the parabolic velocity profile may be assumed to exist at the tube entrance with a uniform fluid temperature. When the Prandtl number is low, a tube with limited length may fully develop a thermal boundary layer while the velocity is still developing; hence a slug (flat) velocity profile may be assumed.

Several solutions exist for simultaneous momentum and thermal laminar boundary layer development. Sparrow et al. [46] found the heat-transfer coefficient for various Pr numbers and constant heat flux. Kays [23] solved the developing laminar flow problem, for Pr = 0.7, in a tube with uniform heat flux and uniform wall temperature (Fig. 10-7). It is seen that in this region the Nu value is higher than the asymptotic value.

In general, the developing region (or thermal entry region) may be assumed to extend to a length ξ_T, such that:

$$\frac{\xi_T}{D_e} = 0.05 \text{ Re Pr} \tag{10-35}$$

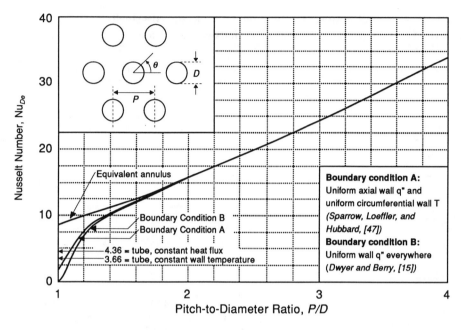

Figure 10-6 Nusselt numbers for fully developed laminar flow parallel to an array of circular tubes.

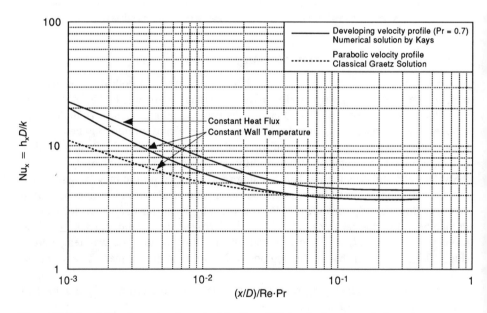

Figure 10-7 Local Nusselt number determined by Kays [23] for simultaneous velocity and temperature development for laminar flow in a circular tube ($Pr = 0.7$).

Bhatti and Savery [1] found that the thermal laminar developing length may be predicted by:

$$\frac{\xi_T}{D_e} = 0.1 \, \text{Re} \, \text{Pr} \qquad \text{for } 0.7 < \text{Pr} < 1$$

$$\approx 0.004 \, \text{Re} \qquad \text{for } \text{Pr} = 0.01 \qquad (10\text{-}36)$$

$$\approx 0.15 \, \text{RePr} \qquad \text{for } \text{Pr} > 5$$

III TURBULENT HEAT TRANSFER: MIXING LENGTH APPROACH

A Equations for Turbulent Flow in Circular Coordinates

It was shown in Chapter 4 that the instantaneous values of velocity and temperature can be expanded into a time-averaged component and a fluctuating component. The steady-state incompressible fluid transport equations for mass, momentum, and energy can be given by inspection of Eqs. 4-135, 4-136, and 4-137, respectively:

$$\nabla \cdot [\rho \vec{v}] = 0 \tag{9-59}$$

$$\nabla \cdot [\rho \vec{v} \, \vec{v}] = -\nabla p + \nabla \cdot [\overline{\overline{\tau}} - \rho \overline{\vec{v}' \vec{v}'}] + \rho \vec{g} \tag{9-60}$$

$$\nabla \cdot [\rho u^\circ \vec{v}] = -\nabla \cdot \overline{q''} + [\overline{q'''} - p\nabla \cdot \vec{v} + \Phi] - \nabla \cdot \rho \overline{u^{\circ\prime} \vec{v}'} \tag{10-37a}$$

or

$$\nabla \cdot [\rho c_p \vec{v} \, \overline{T}] = -\nabla \cdot \overline{q''} + \overline{q'''} + \overline{\Phi} - \nabla \cdot \rho c_p \overline{T' \vec{v}'} \tag{10-37b}$$

For a fluid with constant properties, the axial momentum equation can be expanded for the cylindrical coordinates as:

$$\rho \left(\bar{v}_r \frac{\partial \bar{v}_z}{\partial r} + \frac{\bar{v}_\theta}{r} \frac{\partial \bar{v}_z}{\partial \theta} + \bar{v}_z \frac{\partial \bar{v}_z}{\partial z} \right) = -\frac{\partial p}{\partial z} + \frac{1}{r} \frac{\partial}{\partial r} r (\tau_{rz})_{\text{eff}}$$

$$+ \frac{1}{r} \frac{\partial}{\partial \theta} (\tau_{\theta z})_{\text{eff}} + \frac{\partial (\tau_{zz})_{\text{eff}}}{\partial z} + \rho \vec{g} \tag{10-38}$$

For fully developed flow, the r and θ direction momentum balances reduce to the condition of constant pressure in these directions. Also $\bar{v}_r = \bar{v}_\theta = 0$.

The momentum fluxes $(\tau_{rz})_{\text{eff}}$, $(\tau_{\theta z})_{\text{eff}}$, and $(\tau_{zz})_{\text{eff}}$ incorporate contributions due to viscous effects and turbulent velocity fluctuations. In general, the effective stresses can be given by:

$$(\tau_{rz})_{\text{eff}} = + \mu \left(\frac{\partial \bar{v}_z}{\partial r} + \frac{\partial \bar{v}_r}{\partial z} \right) - \rho \overline{v_r' v_z'} \tag{10-39}$$

$$(\tau_{\theta z})_{\text{eff}} = + \mu \left(\frac{\partial \bar{v}_\theta}{\partial z} + \frac{1}{r} \frac{\partial \bar{v}_z}{\partial \theta} \right) - \rho \overline{v_z' v_\theta'} \tag{10-40}$$

$$(\tau_{zz})_{\text{eff}} = + \mu \left(2 \frac{\partial \bar{v}_z}{\partial z} \right) - \rho \overline{v_z' v_z'} \tag{10-41}$$

where $\bar{v}_z$, $\bar{v}_r$, $\bar{v}_\theta$ = time-averaged flow velocities, and v_z', v_r', v_θ' = fluctuating components of v_z, v_r, and v_θ. The turbulent stress terms $\overline{\rho v_r' v_z'}$ and $\overline{\rho v_z' v_\theta'}$ = one-point correlations of mutually perpendicular velocity fluctuations and therefore have negative values. These correlation terms are called *Reynolds stresses*. The fluctuating components are typically small in magnitude relative to the time-averaged flow velocities.

For axial flow in axisymmetric geometries under fully developed conditions, the time-averaged transverse velocities $\bar{v}_r$ and $\bar{v}_\theta$ are zero. However, in nonaxisymmetric geometries these components exist even under fully developed conditions. They are caused by nonuniformities in wall turbulence and are called *secondary flows of the second kind*. They are in contrast to *secondary flows of the first kind*, which are produced by turning or skewing the primary flow, e.g., flow in curved ducts. The secondary flow effects are small and much less studied than turbulent effects and are usually inferred from the pattern of shear stress distribution [49].

Neglecting the dissipation energy and internal heat generation, the energy equation (Eq. 10-37b) can also be expressed as:

$$\rho c_p \left(\bar{v}_r \frac{\partial \bar{T}}{\partial r} + \bar{v}_\theta \frac{1}{r} \frac{\partial \bar{T}}{\partial \theta} + \bar{v}_z \frac{\partial \bar{T}}{\partial z} \right) = - \left[\frac{1}{r} \frac{\partial r (q_r'')_{\text{eff}}}{\partial r} + \frac{1}{r} \frac{\partial (q_\theta'')_{\text{eff}}}{\partial \theta} + \frac{\partial (q_z'')_{\text{eff}}}{\partial z} \right]$$

(10-42)

The heat fluxes q_r'', q_θ'', and q_z'' can be given by:

$$(q_r'')_{\text{eff}} = - \rho c_p \alpha \frac{\partial \bar{T}}{\partial r} + \rho c_p \overline{T' v_r'}$$

(10-43a)

$$(q_\theta'')_{\text{eff}} = - \rho c_p \alpha \frac{1}{r} \frac{\partial \bar{T}}{\partial \theta} + \rho c_p \overline{T' v_\theta'}$$

(10-43b)

$$(q_z'')_{\text{eff}} = - \rho c_p \alpha \frac{\partial \bar{T}}{\partial z} + \rho c_p \overline{T' v_z'}$$

(10-43c)

where $\bar{T}$ = time-averaged temperature, and T' = its fluctuating component.

The foregoing equations can be developed further only when expressions are available relating the turbulent flux terms to the mean flow properties. Here, however, we treat the simpler case of fully developed flow in circular geometry where secondary flows do not exist and the velocity fluctuations can be related to the velocity gradients by the turbulent or eddy diffusivity for momentum (ϵ_M) and for heat (ϵ_H) such that:

$$\overline{v_z' v_r'} = - \epsilon_{M,r} \frac{\partial \bar{v}_z}{\partial r}$$

(10-44)

$$\overline{v_z' v_\theta'} = - \epsilon_{M,\theta} \frac{1}{r} \frac{\partial \bar{v}_z}{\partial \theta}$$

(10-45)

$$\overline{T' v_r'} = - \epsilon_{H,r} \frac{\partial \bar{T}}{\partial r}$$

(10-46)

$$\overline{T' v_\theta'} = - \epsilon_{H,\theta} \frac{1}{r} \frac{\partial \bar{T}}{\partial \theta}$$

(10-47)

With the restrictions of fully developed flow:

$$\bar{v}_\theta = \bar{v}_r = 0 \tag{10-48}$$

$$\frac{\partial \bar{v}_z}{\partial z} = 0 \tag{10-49}$$

$$\tau_{zz} = 0 \tag{10-50}$$

For fully developed heat transfer, in general (see Eq. 10-14a):

$$\frac{\partial}{\partial z} \left(\frac{T_w - \bar{T}(r)}{T_w - \bar{T}_m} \right) = 0 \tag{10-51}$$

To evaluate $\partial \bar{T} / \partial z$ we need to specify the axial boundary conditions at the wall.

Applying Eqs. 10-48 through 10-50 and neglecting gravity, Eqs. 10-38 and 10-42 become, respectively:

$$\frac{1}{r} \left(\frac{\partial(\tau_{rz}r)}{\partial r} + \frac{\partial \tau_{\theta z}}{\partial \theta} \right) = \frac{\partial p}{\partial z} \tag{10-52}$$

$$\frac{1}{r} \left[\frac{\partial}{\partial r} (q_r'' r) + \frac{\partial q_\theta''}{\partial \theta} \right] + \frac{\partial q_z''}{\partial z} = -\rho c_p \bar{v}_z \frac{\partial \bar{T}}{\partial z} \tag{10-53}$$

where the subscript eff has been dropped for brevity.

For an axially constant wall heat flux, the fully developed flow features require that (see Table 10-3):

$$\frac{\partial T_w}{\partial z} = \frac{\partial \bar{T}_m}{\partial z} \tag{10-54}$$

With this condition, Eq. 10-51 reduces to a form similar to Eq. 10-15 for laminar flow:

$$\frac{\partial \bar{T}}{\partial z} = \frac{\partial T_w}{\partial z} = \frac{\partial \bar{T}_m}{\partial z} \tag{10-55}$$

independent of r; $q_w''(z) = $ constant. The magnitude of $\partial \bar{T} / \partial z$ can be determined from an energy balance as:

$$\frac{\partial \bar{T}}{\partial z} = \frac{\partial \bar{T}_m}{\partial z} = \frac{4q_w''}{V_m \rho c_p D_H} = \frac{q'}{\dot{m} c_p} = \text{constant} \tag{10-56}$$

For Eq. 10-56 to be true, note that the axial heat conduction is neglected.

$$-k \frac{\partial^2 \bar{T}}{\partial z^2} = \frac{\partial q_z''}{\partial z} = 0 \tag{10-57}$$

This condition is acceptable for practically all flows with reasonable velocities, so that $\rho c_p v_z > k (\partial \bar{T} / \partial z)$.

On the other hand, for a constant axial wall temperature boundary condition, $T_w(z) = $ constant:

$$\frac{\partial T_w}{\partial z} = 0 \tag{10-58}$$

and Eq. 10-51 reduces to a form similar to Eq. 10-16 for laminar flow:

$$\frac{\partial \bar{T}}{\partial z} = f\left(\frac{r}{R}\right) \frac{\partial \bar{T}_m}{\partial z} \tag{10-59}$$

Applying the conditions of Eqs. 10-44 through 10-47 to Eqs. 10-39 and 10-40, we get the shear stress in terms of the kinematic viscosity (ν) and eddy diffusivity of momentum ($\epsilon_{M,r}$) and ($\epsilon_{M,\theta}$):

$$(\tau_{rz})_{eff} = + \rho(\nu + \epsilon_{M,r}) \frac{\partial \bar{v}_z}{\partial r} \tag{10-60}$$

$$(\tau_{\theta z})_{eff} = + \rho(\nu + \epsilon_{M,\theta}) \frac{\partial \bar{v}_z}{r\partial \theta} \tag{10-61}$$

Similarly, we get the heat flux in terms of the molecular thermal diffusivity (α) and the eddy diffusivity of heat, ($\epsilon_{H,r}$) and ($\epsilon_{H,\theta}$):

$$(q_r'')_{eff} = -\rho c_p [\alpha + \epsilon_{H,r}] \frac{\partial \bar{T}}{\partial r} \tag{10-62}$$

$$(q_\theta'')_{eff} = -\rho c_p [\alpha + \epsilon_{H,\theta}] \frac{\partial \bar{T}}{r\partial \theta} \tag{10-63}$$

Of course, with circular geometry with azimuthal symmetry, the angular dependence is eliminated. Thus in round tubes the applicable equations for axial momentum (Eq. 10-52) and energy (Eq. 10-53) are:

$$\frac{1}{r} \frac{\partial}{\partial r} \left[\rho r (\nu + \epsilon_{M,r}) \frac{\partial \bar{v}_z}{\partial r} \right] = \frac{\partial p}{\partial z} \tag{10-64}$$

$$\frac{1}{r} \frac{\partial}{\partial r} \left[r(\alpha + \epsilon_{H,r}) \frac{\partial \bar{T}}{\partial r} \right] = \bar{v}_z \frac{\partial \bar{T}}{\partial z} \tag{10-65}$$

B Relation between ϵ_M, ϵ_H, and Mixing Lengths

The momentum and energy equations for fully developed flow (Eqs. 10-64 and 10-65) have the same form. The eddy diffusivity of momentum has the same dimensions as the eddy diffusivity of heat (both units are square meters per second or square feet per hour). There are no eddies of momentum and energy as such, but these diffusivities are related to the turbulent fluid eddy properties. Thus there is reason to believe that these diffusivities are related. Indeed, in laminar flow both are zero, but in highly turbulent flow they may be so much higher than molecular diffusivities so as to control the heat transfer.

For ordinary fluids, with Pr $\simeq$ 1, it was often assumed that $\epsilon_H = \epsilon_M$, with a reasonable measure of success. However, it was noted by Jenkins [20] and Deissler

[11] that in liquid metal flow an eddy may give up or gain heat during its travel owing to the high conductivity value. Hence $\epsilon_H < \epsilon_M$, but the exact value depends on the flow geometry and Reynold's number.

The mixing length theory has been extensively used to define ϵ_H and ϵ_M. The basic idea of the mixing length is similar to that of the mean free path of molecules in the kinetic theory of gases. Thus a fluid eddy is assumed to travel a certain length (ℓ_M) perpendicular to the flow stream before losing its momentum. Thus:

$$v'_z = \ell_M \frac{d\bar{v}_z}{dy} \tag{10-66}$$

where $y = R - r$.

Prandtl first suggested that, for the parallel plate situation, v'_z is of the same order as v'_y, although by continuity consideration they should be of opposite sign. Thus he obtained:

$$\overline{v'_z v'_y} = -\ell_M^2 \left| \frac{d\bar{v}_z}{dy} \right| \frac{d\bar{v}_z}{dy} \tag{10-67}$$

Hence by inspection of Eq. 10-44 we get:

$$\epsilon_M = \ell_M^2 \left| \frac{d\bar{v}_z}{dy} \right| \tag{10-68}$$

Prandtl proposed that the mixing length is proportional to the distance from the wall, so that:

$$\ell_M = Ky \tag{10-69}$$

where $K =$ a universal constant, which was empirically found to be equal to 0.42 in the area adjacent to the wall.

Schlichting [40] reviewed the theories advanced for the mixing lengths since the original work of Prandtl. He suggested that for fully developed flow in ducts and pipes the mixing length is well characterized by the Nikuradse formula:

$$\frac{\ell_M}{R} = 0.14 - 0.08 \left(1 - \frac{y}{R} \right)^2 - 0.06 \left(1 - \frac{y}{R} \right)^4 \tag{10-70}$$

From Dwyer's work [14] the following relation is recommended for liquid metals (Pr $<<$ 1):

$$\frac{\epsilon_H}{\epsilon_M} \equiv \bar{\psi} = 1 - \frac{1.82}{Pr \left(\frac{\epsilon_M}{\nu} \right)_{max}^{1.4}} \tag{10-71}$$

where $(\epsilon_M/\nu)_{max} =$ maximum value in channel flow and for each geometry is a unique function of the Reynolds number. Various values of $(\epsilon_M/\nu)_{max}$ are given for rod bundles in Figure 10-8. At low Reynolds numbers, $\bar{\psi}$ approaches zero because hot eddies lose heat in transit to the surrounding liquid metal. Equation 10-71 may produce

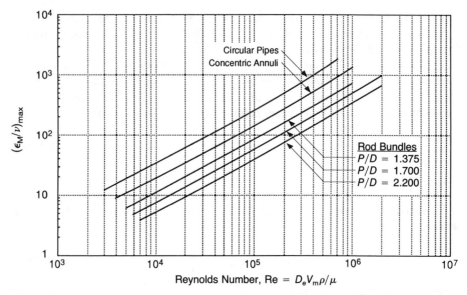

Figure 10-8 Values of $(\epsilon_M/\nu)_{max}$ for fully developed turbulent flow of liquid metals through circular tubes, annuli, and rod bundles with equilateral triangular spacing. *(From Dwyer [14].)*

negative values for $\overline{\psi}$ at very low values of Pr. Negative values are not permissible so in this case $\overline{\psi}$ should be set at zero. The contribution of turbulence to heat transfer would then be negligible, and a laminar heat transfer treatment would be adequate. At high Reynolds numbers, $\overline{\psi}$ approaches unity, as the eddies move so fast that they lose an insignificant amount of heat in transit.

The selection of ϵ_M and ϵ_H was excellently reviewed by Nijsing [34]. The mixing length theory was used extensively to define ϵ_M and ϵ_H and to solve for the temperature distribution. The advent of computers, however, has shifted the treatment of turbulence to a more involved set of equations, for which more than one turbulent parameter (ϵ_M or ϵ_H) can exist in each transport equation. A brief description is given in section IV.

C Turbulent Temperature Profile

Analytic predictions of the temperature profile in turbulent flow date back to the simplest treatment of Reynolds (1874). Many refined treatments followed, notably those of Prandtl [36,37], von Karman [50], Martinelli [31], and Diessler [10,11]. These treatments described, using various approximations, the velocity profile and the eddy diffusivities ϵ_M and ϵ_H for fully developed flow.

Reynolds assumed the entire flow field to consist of a single zone that is highly turbulent so that molecular diffusivities can be neglected, i.e., $\epsilon_M \gg \nu$ and $\epsilon_H \gg \alpha$. From Eqs. 10-60 and 10-62 we get in circular geometry where there is no azimuthal variation:

$$\frac{(\tau_{rz})_{eff}}{(q''_r)_{eff}} = -\frac{\rho\epsilon_{M,r}}{\rho c_p \epsilon_{H,r}} \frac{d\bar{v}_z}{d\bar{T}}$$ (10-72)

Reynolds also assumed that $\epsilon_{M,r}/\epsilon_{H,r} = 1.0$ and that $(\tau_{rz}/q''_r)_{eff}$ is constant (i.e., equal to τ_w/q''_w) throughout the field. Therefore integration of Eq. 10-72 between the wall and the bulk or mean values gives:

$$\int_{T_w}^{T_m} d\bar{T} = -\frac{q''_w}{\tau_w c_p} \int_o^{V_m} d\bar{v}_z$$

or

$$T_w - T_m = \frac{q''_w V_m}{\tau_w c_p}$$ (10-73)

By definition, the heat transfer coefficient is given by:

$$h = \frac{q''_w}{T_w - T_m}$$ (10-74)

and the wall shear stress is related to the friction factor by (see Eq. 9-51):

$$\tau_w = \frac{f}{4}\rho\frac{V_m^2}{2}$$ (10-75)

Equations 10-74 and 10-75, when substituted in Eq. 10-73, lead to:

$$St = \frac{h}{\rho c_p V_m} = \frac{f}{8}$$ (10-76)

This equation is known as the Reynolds analogy for momentum and heat transfer for fully developed turbulent flow and is valid only if $Pr \simeq 1.0$.

Prandtl assumed the flow consists of two zones: a laminar sublayer where molecular effects dominate ($\epsilon_M \ll \nu$ and $\epsilon_H \ll \alpha$) and a turbulent layer where eddy diffusivities dominate ($\epsilon_M \gg \nu$ and $\epsilon_H \gg \alpha$). Again assuming $\epsilon_M/\epsilon_H = 1$ and a constant ratio of q''/τ in the turbulent layer, Prandtl obtained:

$$St = \frac{f}{8}\frac{1}{1 + 5\sqrt{f/8}\,(Pr - 1)}$$ (10-77)

Note that for $Pr = 1$, the Prandtl analogy, reduces to the Reynolds analogy.

Von Karman extended earlier treatments by assuming that three zones exist: a laminar sublayer, a buffer zone, and the turbulent core. He made assumptions similar to those of Prandtl but allowed molecular and eddy contributions in the buffer zone.

Martinelli [31] obtained solutions for the temperature profile in various geometries by using the assumption that the momentum and heat fluxes near the wall vary linearly with the distance from the wall (y). Thus:

$$\frac{q''_w}{\rho c_p}\left(1 - \frac{y}{R}\right) = (\alpha + \epsilon_H)\frac{d\bar{T}}{dy}$$ (10-78)

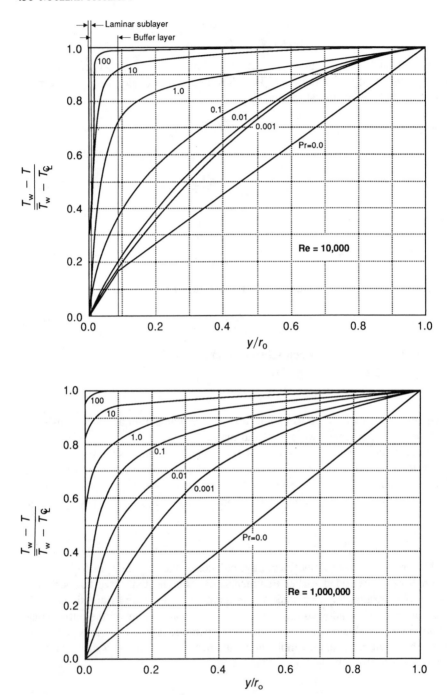

Figure 10-9 Martinelli's solution for the temperature distribution in a uniformly heated round tube. *(From Martinelli [31].)*

$$\frac{\tau_w}{\rho}\left(1 - \frac{y}{R}\right) = (\nu + \epsilon_M)\frac{d\bar{v}_z}{dy} \qquad (10\text{-}79)$$

If the velocity profile is empirically known, ϵ_M can be obtained from Eq. 10-79. For a given relation of $\bar{\psi} = \epsilon_H/\epsilon_M$, Eq. 10-78 can then be solved to obtain $\bar{T}(r)$.

Martinelli's results [31] for the temperature distribution in the three sublayers and for $\bar{\psi} = 1$ are shown in Figure 10-9 for Re $= 10,000$ and $1,000,000$ for various Prandtl numbers. It is interesting to note that for Pr ≥ 1.0, most of the temperature drop occurs in the laminar sublayer. For Pr $\ll 1$ (e.g., in liquid metals) the temperature drop is more evenly distributed throughout the cross section. The higher Reynolds numbers appear to lead to more uniform temperature in the turbulent core sublayer.

numerical results for the Nusselt number dependence on Re in a uniformly heated round tube. Their results are shown in Figure 10-10. It is seen that for Pr ≥ 0.5 the change of Nu with Re is logarithmically linear. For Pr ≤ 0.3, the change of Nu with Re is more gradual, reflecting the influence of the high molecular thermal diffusivity.

Sleicher and Tribus [45] presented results for low Prandtl heat transfer solution with a constant wall temperature. Their results were used by Kays [22] to plot the ratio of Nu for a constant heat flux, $(Nu)_H$, to Nu for a constant wall temperature, $(Nu)_T$ (Fig. 10-11). It is seen that the Nusselt number for constant q''_w is higher than that for constant T_w but that the difference is significant only for Pr < 0.7. Thus for metallic fluids the wall boundary condition significantly affects the turbulent Nu number, but for nonmetallic fluids the Nu number is practically independent of the boundary condition.

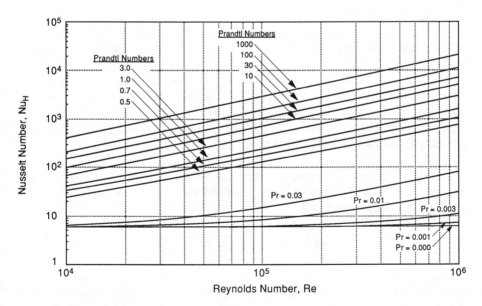

Figure 10-10 Nusselt number dependence on Reynolds number as predicted by Kays and Leung [24] for a uniformly heated round tube.

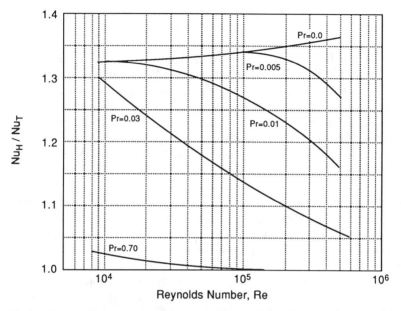

Figure 10-11 Ratio of Nusselt number for constant heat rate (Nu_H) to Nusselt number for constant surface temperature (Nu_T) for fully developed conditions in a circular tube. *(From Sleicher and Tribus [45].)*

Example 10-2 Turbulent heat transfer in a steam generator tube

PROBLEM Consider a typical tube in a steam generator. We are interested in determining the linear heat transfer rate from the primary side (bulk temperature T_p) to the single-phase region of the secondary side (bulk temperature T_s). The typical conditions of this region are described below. The tubes are made of Inconel ($k = 35$ W/m °C) with I.D. $= \frac{7}{8}$ in. and O.D. $= 1$ in.; the flow rate is 1.184 kg/s per tube. The average reactor coolant temperature is approximately 305°C, and the average secondary side temperature in the single-phase region is 280°C. Assume that the heat transfer coefficient on the outside of the tube is the same as that on the inside (not generally true—used here for simplicity).

SOLUTION Using the concept of thermal resistances in series (see section VII.C of Chapter 8) it can be seen that for a cylindrical geometry the heat-transfer rate per unit length may be written as:

$$q' = \frac{2\pi(T_p - T_s)}{\dfrac{1}{r_i h_i} + \dfrac{\ell n(r_o/r_i)}{k} + \dfrac{1}{r_o h_o}}$$

From Example 10-1 we found that, under approximately the same conditions, Re $\approx 7.613 \times 10^5$, Pr $= 1$. We can therefore find an approximate value for the Nusselt number from Figure 10-10. (Note that it is not exactly correct, as there is not uniform heat rejection along the length of a tube in a steam generator.) However, Figure 10-10 indicates that, for Re $= 7.613 \times 10^5$ and Pr $= 1$, Nu ≈ 1200.

$$\therefore \text{h} = \text{Nu}\,\frac{k}{D} = 1200\,\frac{0.532 \text{ W/m }^\circ\text{C}}{(0.875 \text{ in.})(0.0254 \text{ m/in.})} = 2.87 \times 10^4 \text{ W/m}^2\,^\circ\text{C}$$

Then the linear heat transfer rate is:

$$q' = \frac{2\pi(25^\circ\text{C})}{\dfrac{1}{(0.0111 \text{ m})(2.87 \times 10^4 \text{ W/m}^2\,^\circ\text{C})} + \dfrac{\ell n(1/0.875)}{35 \text{ W/m }^\circ\text{C}}}$$

$$+ \frac{1}{(0.0127 \text{ m})(2.87 \times 10^4 \text{ W/m}^2\,^\circ\text{C})}$$

$$q' = \frac{157.1}{3.14 \times 10^{-3} + 3.82 \times 10^{-3} + 2.74 \times 10^{-3}} = 1.62 \times 10^4 \text{ W/m}$$

IV TURBULENT HEAT TRANSFER: DIFFERENTIAL APPROACH

A Basic Models

The discussion in section III focuses on the representation of the fluctuation terms by algebraic relations involving the gradient of the mean velocity and temperature. It is the oldest and probably most popular approach to solving the time-averaged transport equations. The Prandtl mixing length hypothesis is the best known method for the algebraic connection between the turbulent terms and the mean flow properties. However, this approach implies that the generation and dissipation of turbulence are in balance at each point. Therefore it does not allow for flow at a point to be influenced by turbulence generated at other points. In channel flow, turbulence is produced mainly near the walls and is transported to the bulk flow by diffusion. Turbulence generated by an obstruction is transported downstream, where the local velocity gradients are much smaller. The mixing length (and turbulent diffusivities) approach neglects the transport and diffusion of turbulence.

The differential approach to modeling turbulence was started during the 1940s and became intensive after 1960. Many reviews have appeared in this area, and the reader is referred to works by Launder and Spaulding [27], Rodi [39], and Bradshaw et al. [7] for a thorough review of the various possible modeling approaches. Here only brief descriptions of two popular models are presented: a one-equation model and a two-equation model. The mixing length approach of section III is referred to as a zero-equation model, as it does not introduce any new transport equation.

It should be expected that in complex flow situations, e.g., rapidly developing flow or recirculating flows, the mixing length approach is not suitable. However, it remains a useful tool for simple flows where ℓ_M can be well specified.

Prandtl was first to suggest a one-equation model, which involves a transport equation for the turbulent kinetic energy per unit mass (k_t). The turbulent energy is defined by:

$$k_t^2 \equiv \tfrac{1}{2}(\overline{v_x'^2} + \overline{v_y'^2} + \overline{v_z'^2})$$ (10-80a)

or, using the tensor form:

$$k_t^2 \equiv \tfrac{1}{2}(\overline{v_i' v_i'}) = \tfrac{1}{2}(\overline{v_i'^2})$$ (10-80b)

Other one-equation models are based on the addition of a transport equation for the shear stresses or the turbulent viscosity.

Two-equation models for various parameters have also been introduced. The most popular model is the $k_t - \epsilon_t$ model, where ϵ_t is the dissipation rate of the turbulent viscosity given by:

$$\epsilon_t = \nu \overline{\left(\frac{\partial v_i'}{\partial x_j}\right)^2}$$ (10-81)

where ν = kinematic viscosity. In this model, the transport equations of both k_t and ϵ_t are solved to obtain a value for the turbulent viscosity (μ_t), defined by:

$$\mu_t = \frac{C\rho k_t^2}{\epsilon_t}$$ (10-82)

where C = a constant with a recommended value of 0.09 [18,21]. It may be important to note that the subscripts t are not normally included in the literature on turbulence. Hence the model above is referred to as simply the $k - \epsilon$ model.

B Transport Equations for the $k_t - \epsilon_t$ Model

The time-averaged transport equations for mass and momentum may be written in tensor form as:

$$\frac{\partial \rho}{\partial t} + \frac{\partial \rho \bar{v}_i}{\partial x_i} = 0$$ (10-83)

$$\rho\left(\frac{\partial \bar{v}_i}{\partial t} + \bar{v}_j \frac{\partial \bar{v}_i}{\partial x_j}\right) = -\frac{\partial p}{\partial x_i} + \frac{\partial \tau_{ji}}{\partial x_j} - \frac{\partial \overline{\rho v_i' v_j'}}{\partial x_j} + \rho g_i$$ (10-84)

$$
\begin{aligned}
\frac{Dk_t}{Dt} = &\frac{\partial}{\partial x_i}\left[\overline{v_i'\left(\frac{v_j' v_j'}{2} + \frac{p}{\rho}\right)}\right] && \text{Diffusion term} \\[2mm]
&- \overline{v_i' v_j'}\,\frac{\partial \bar{v}_i}{\partial x_j} && \text{Production by shear} = P_k \\[2mm]
&- \beta g_i\, \overline{v_i' T'} && \text{Production due to buoyancy} = G_k \\[2mm]
&- \nu \overline{\left(\frac{\partial v_i'}{\partial x_j}\right)^2} && \text{Viscous dissipation} = -\epsilon_t
\end{aligned}
$$

where:

$$\beta = -\frac{1}{\rho}\frac{\partial\rho}{\partial T} \tag{10-85}$$

The term P_k represents the transfer of kinetic energy from the mean to the turbulent motion, and G_k represents an exchange between the kinetic and potential energies. In stable stratification G_k is negative, whereas for unstable stratification it is positive. The term ϵ_t is always a sink term.

This energy equation is of little use given the number of new unknowns. To overcome this problem, it is assumed that the gradient of k_t controls the diffusion process:

$$\overline{v_i'\left(\frac{v_j'v_j'}{2} + \frac{p}{\rho}\right)} = \frac{\nu_t}{\sigma_k}\frac{\partial k_t}{\partial x_i} \tag{10-86}$$

where σ_k = empirical diffusion constant with a recommended value of 1.0 for high Reynolds numbers and $\nu_t = \mu_t/\rho$.

Recalling that

$$-\overline{v_i'v_j'} = \nu_t\left(\frac{\partial\overline{v}_i}{\partial x_j} + \frac{\partial\overline{v}_j}{\partial x_i}\right) - \frac{2}{3}\delta_{ij}k_t \qquad \delta_{ij} \begin{array}{l} = 1 \quad i = j \\ = 0 \quad i \neq j \end{array} \tag{10-87a}$$

$$-\overline{v_i'T'} = \frac{\nu_t}{\sigma_t}\frac{\partial\overline{T}}{\partial x_i} \tag{10-87b}$$

where σ_t = a turbulent Prandtl number ($= 1/\overline{\psi}$). Eq. 10-85 can now be rewritten as:

$$\frac{Dk_t}{Dt} = \frac{\partial}{\partial x_i}\left(\frac{\nu_t}{\sigma_k}\frac{\partial k_t}{\partial x_i}\right) + \nu_t\left(\frac{\partial\overline{v}_i}{\partial x_j} + \frac{\partial\overline{v}_j}{\partial x_i}\right)\frac{\partial\overline{v}_i}{\partial x_j} - \frac{2}{3}k_t\delta_{ij}\frac{\partial\overline{v}_i}{\partial x_j} + \beta g_i\frac{\nu_t}{\sigma_t}\frac{\partial\overline{T}}{\partial x_i} - \epsilon_t \tag{10-88}$$

An exact equation can be derived for ϵ_t from the Navier-Stokes equation, which again has been found to be of little use. A more useful semiempirical relation for ϵ_t may be given by:

$$\rho\frac{D\epsilon_t}{Dx_j} = C_1\frac{\epsilon_t}{k_t}(P_k + G_k) - C_2\frac{\rho\epsilon_t^2}{k_t} + \frac{\partial}{\partial x_j}\left[\left(\frac{\mu_t}{\sigma_t} + \mu\right)\frac{\partial\epsilon_t}{\partial x_j}\right] \tag{10-89}$$

Here the values of C_1 and C_2 are determined empirically. For grid turbulence, diffusion and $P_k + G_k$ are not important; thus C_2 can be determined alone and is usually found to be between 1.8 and 2.0. Launder et al. [28] recommended that $C_1 = 1.92$. The value of the turbulent Prandtl number is $\sigma_t = 1.3$ [8]. Equations 10-82, 10-88, and 10-89 comprise the full $k_t - \epsilon_t$ model for turbulence.

C One-Equation Model

When only the k_t equation is used, ϵ_t is calculated using the relation:

$$\epsilon_t = \frac{C^{3/4} k_t^{3/2}}{\ell_M} \qquad (10\text{-}90)$$

where $\ell_M = 0.42y$; $y = $ distance from the wall; and C is that of Eq. 10-82 with a recommended value of 0.09.

D Effect of Turbulence on the Energy Equation

In the energy equation the transport of enthalpy due to turbulence can be described by the addition of a turbulent diffusivity (ϵ_H) such that:

$$\epsilon_H = \mu_t/\rho\sigma_h \qquad (10\text{-}91)$$

where $\sigma_h = $ turbulent Prandtl number for thermal energy transfer of a recommended value of 0.9 [8].

V HEAT TRANSFER CORRELATIONS IN TURBULENT FLOW

A large amount of experimental and theoretical work has been done for the purpose of assessing the influence of various parameters on the heat-transfer coefficient in the case of relatively simple geometries. These studies have shown the following.

   1. The $h(x)$ value changes significantly in the entrance region (i.e., the regions where the velocity and temperature profiles are still developing). When the entrance region is a small percentage of the whole channel (typical of LWR reactor channels), it can be neglected. If the entrance region is a large percentage of a channel (typical of LMFBR reactor channels), proper averaging is required. In any event, we can represent the heat-transfer ability of a channel by the overall heat-transfer coefficient of the channel (averaged for the whole length).
   2. For the nonmetallic fluids, where Pr is ≥ 1, the laminar layer is very thin compared with the turbulent region (i.e., turbulent mechanism of heat transfer predominates) and the heat-transfer coefficient is not very sensitive to boundary conditions. In the liquid metals case (Pr < 0.4), where heat transfer by conduction is important, the heat transfer coefficient is sensitive to boundary conditions and channel shape. The prediction of heat-transfer coefficients for these two ranges of Pr values is different, which makes it necessary to examine them separately.

A Nonmetallic Fluids

1 Fully developed turbulent flow. The application of greatest interest is the fully developed turbulent flow, mainly in long channels. In this section we designate the asymptotic value (i.e., that of the fully developed flow) of the Nusselt number by Nu_∞.

Both experiment and theory show that for almost all nonmetallic fluids Nu_∞ is given by the equation:

$$\mathrm{Nu}_\infty = C \, \mathrm{Re}^\alpha \, \mathrm{Pr}^\beta \left(\frac{\mu_w}{\mu}\right)^\kappa \tag{10-92a}$$

where μ_w = fluid viscosity at $T = T_w$; μ = fluid viscosity at $T = T_m$; C, α, β, κ = constants that depend on the fluid and the geometry of the channel.

When $T_w - T_m$ is not very large, $\mu_w \approx \mu$ and Eq. 10-92a reduces to:

$$\mathrm{Nu}_\infty = C \, \mathrm{Re}^\alpha \, \mathrm{Pr}^\beta \tag{10-92b}$$

Let us now specialize these equations to geometries of interest.

a. Circular tubes For ordinary fluids, extensive use has been made of three equations.

i Seider and Tate [43] equation The equation is:

$$\mathrm{Nu}_\infty = 0.023 \, \mathrm{Re}^{0.8} \, \mathrm{Pr}^{0.4} \left(\frac{\mu_w}{\mu}\right)^{0.14} \tag{10-93}$$

Whereas μ_w is evaluated at the wall temperature, all other fluid properties are evaluated at the arithmetic mean bulk temperature (i.e., the average of the bulk inlet and outlet temperature). This equation is applicable for $0.7 < \mathrm{Pr} < 120$ and $\mathrm{Re} > 10,000$ and $L/D > 60$.

ii Dittus–Boelter [13] equation For cases when $\mu \approx \mu_w$ the Dittus–Boelter equations are the most universally used correlations:

$$\mathrm{Nu}_\infty = 0.023 \, \mathrm{Re}^{0.8} \, \mathrm{Pr}^{0.4} \quad \text{when the fluid is heated} \tag{10-94a}$$
$$\mathrm{Nu}_\infty = 0.023 \, \mathrm{Re}^{0.8} \, \mathrm{Pr}^{0.3} \quad \text{when the fluid is cooled} \tag{10-94b}$$

for $0.7 < \mathrm{Pr} < 100$, $\mathrm{Re} > 10,000$, and $L/D > 60$. All fluid properties are evaluated at the arithmetic mean bulk temperature.

iii Colburn [9] equation For fluids with high viscosity, Colburn attempted to unify the exponent of Pr for heating and cooling. The result is:

$$\mathrm{St} \, \mathrm{Pr}^{2/3} = 0.023 \, \mathrm{Re}^{-0.2} \tag{10-95a}$$

where the Stanton number (St) $= \mathrm{Nu}/\mathrm{RePr}$. It is equivalent to:

$$\mathrm{Nu} = 0.023 \, \mathrm{Re}^{0.8} \, \mathrm{Pr}^{1/3} \tag{10-95b}$$

The validity range is the same as for the Dittus–Boelter equation and all properties except for c_p in St are used at the mean film temperature. The c_p value for St is used at the bulk fluid temperature.

For the specific case of organic liquids, Silberberg and Huber [44] recommended:

$$\mathrm{Nu}_\infty = 0.015 \, \mathrm{Re}^{0.85} \, \mathrm{Pr}^{0.3} \tag{10-96}$$

For gases for which compressibility effects can be neglected (i.e., Mach number $<<1$) the preceding equations can be used. If the Mach number is on the order of 1 or higher, the above equations are used, with h defined as [2]:

$$h \equiv \frac{q''_w}{T_w - T^\circ} \tag{10-97}$$

where:

$$T^\circ \equiv \left(1 + \frac{\gamma - 1}{2} M^2\right) T_m \tag{10-98}$$

is called the stagnation temperature, and $M =$ the Mach number (see Appendix G).

b Annuli and noncircular ducts The same relations as in the case of the circular tube are used for annuli and noncircular ducts by employing the concept of the hydraulic diameter:

$$D_e \equiv \frac{4A_f}{P_w}$$

for the Reynolds number and D_H, where:

$$D_H \equiv \frac{4A_f}{P_h}$$

for the Nusselt number relation. Of course, use of the previous equations is justified only if the sections do not vary appreciably from circular. Such channels are square, rectangular not too far from square, and probably equilateral or nearly equilateral triangles. For geometries far from circular this method may not be satisfactory [19].

c Rod bundles For fully turbulent flow along rod bundles, Nu values may significantly deviate from the circular geometry because of the strong geometric nonuniformity of the subchannels. For this reason Nu and h are expected to depend on the position of the rod within the bundle.

However, it is important to point out that, as stated earlier (section III.C), the value of Nu is insensitive to the boundary conditions for $Pr > 0.7$. Furthermore, if the coolant area per rod is taken to define an equivalent annulus with zero shear at its outer boundary, it is found that the Nu predictions are accurate to within $\pm$ 10% for $P/D \geq 1.12$. This point is illustrated for an interior pin in triangular geometry in Figure 10-12 for $Pr > 0.7$ and in Figure 10-13 for $Pr < 0.1$.

The usual way to represent the relevant correlation is to express the Nusselt number for fully developed conditions (Nu_∞) as a product of $(Nu_\infty)_{c.t.}$ for a circular tube multiplied by a correction factor:

$$Nu_\infty = \psi (Nu_\infty)_{c.t.} \tag{10-99}$$

where $(Nu_\infty)_{c.t.}$ is usually given by the Dittus-Boelter equation unless otherwise stated.

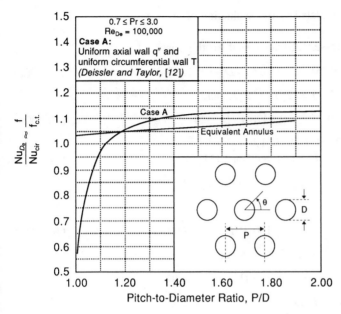

Figure 10-12 Fully developed turbulent flow parallel to a bank of circular tubes or rods. Reynolds number influence is small, and Nusselt number behavior is virtually the same as friction behavior.

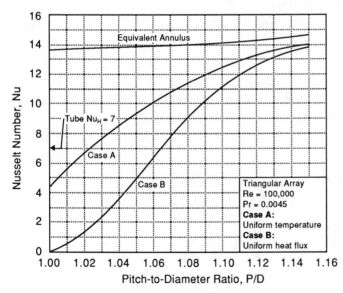

Figure 10-13 Variation of Nusselt number for fully developed turbulent flow with rod spacing for Prandtl number < 0.01. *(From Nijsing [34].)*

The problem is then formulated as the evaluation of ψ.

i Infinite array Presser [38] suggested for this case:

$$\psi = 0.9090 + 0.0783 \, P/D - 0.1283 \, e^{-2.4(P/D-1)} \qquad \text{(10-100a)}$$

for triangular array and $1.05 \leq P/D \leq 2.2.$, and

$$\psi = 0.9217 + 0.1478 \, P/D - 0.1130 \, e^{-7(P/D-1)} \qquad \text{(10-100b)}$$

for square array and $1.05 \leq P/D \leq 1.9$.
   In the particular case of water, Weisman [51] gave:

$$(Nu_\infty)_{\text{c.t.}} = 0.023 \, Re^{0.8} \, Pr^{0.333}$$

and

$$\psi = 1.130 \, P/D - 0.2609 \qquad \text{(10-101a)}$$

for triangular array and $1.1 < P/D < 1.5$ and

$$\psi = 1.826 \, P/D - 1.0430 \qquad \text{(10-101b)}$$

for square array and $1.1 \leq P/D \leq 1.3$

ii Finite array Markoczy [30] gave a general expression for ψ valid for every rod within a finite lattice. The concept is as follows: Consider a fuel rod (R), as those in Figure 10-14, surrounded by J subchannels. Let the cross section of the flow area and the wetted perimeter of the j-th subchannel be A_j and P_{wj}, respectively.
   The general expression for ψ is:

$$\psi = 1 + 0.9120 Re^{-0.1} \, Pr^{0.4} \, (1 - 2.0043 \, e^{-B}) \qquad \text{(10-102)}$$

where:

1. The Reynolds number (Re) is based on the hydraulic diameter (D_e) and velocity characteristic of the subchannels surrounding the subject fuel rod (R). The hydraulic diameter is evaluated as:

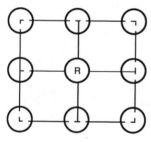

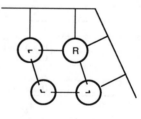

a) Interior Rod   b) Edge (or Corner) Rod

Figure 10-14 Interior rod and edge (or corner) rod.

$$D_e = 4 \frac{\displaystyle\sum_{j=1}^{J} A_j}{\displaystyle\sum_{j=1}^{J} P_{wj}} \tag{10-103a}$$

2. The exponent B is given by:

$$B = \frac{D_e}{D} \tag{10-103b}$$

where D is the diameter of the rod.

Note that for an interior rod the surrounding subchannels are those of an infinite array; and B, evaluated by Eqs. 10-103a and 10-103b, becomes (Fig. 10-14):

$$B = \frac{2\sqrt{3}}{\pi} (P/D)^2 - 1 \tag{10-104a}$$

for triangular arrays, and

$$B = \frac{4}{\pi} (P/D)^2 - 1 \tag{10-104b}$$

for square arrays. The value of ψ from Eq. 10-102 successfully fits the experimental data for the following conditions (an average deviation of 12.7% with a probable error of 8.58%).

$3 \times 10^3 \leq Re \leq 10^6$
$0.66 \leq Pr \leq 5.0$
$1 \leq P/D \leq 2.0$ for interior rods in triangular arrays
$1 \leq P/D \leq 1.8$ for interior rods in square arrays

2 Entrance region effect. In reality, h is not constant throughout the channel because (1) the entrance region where the temperature profile is still developing, and (2) there are changes of the properties due to the change in temperature. The effect of the changing temperature can be approximated by evaluating the physical constants at a representative temperature, which would be the mean value of the inlet and exit temperatures in the channel.

The case of the entrance region is more complicated because the entrance length is sensitive to Re, Pr, heat flux shape, and entrance conditions for the fluid. Within the entrance region, h decreases dramatically from a theoretical infinite value to the asymptotic value (Fig. 10-15). In such a case it is always possible to conservatively approximate the overall value of $\overline{Nu}$ by the fully developed value Nu_∞.

Generally, for circular geometry, at

$$\frac{L}{D_e} \gtrsim 60, \; Nu_z \approx Nu_\infty \qquad \text{for } Pr \ll 1.0$$

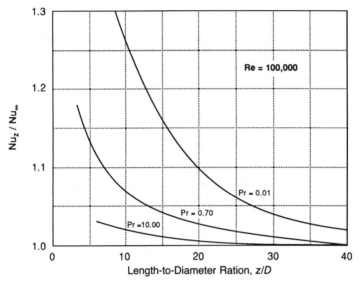

Figure 10-15 Local heat transfer coefficient variation with length to diameter of a tube at Re $= 10^5$. *(From Kays [22].)*

and

$$\frac{L}{D_e} \gtrsim 40, \; Nu_z = Nu_\infty \qquad \text{for } Pr \gtrsim 1.0$$

a Overall heat-transfer coefficient Several correlations have been proposed to include the entrance effects, which depend on the type of entrance condition. In general, the heat-transfer rate is much higher in the entrance region.

1. For Re > 10,000 and 0.7 < Pr < 120 and square-edged entries, McAdams [32] recommended

$$\overline{Nu} = Nu_\infty[1 + (D_e/L)^{0.7}] \tag{10-105}$$

2. For tubes with bell-mouthed entry, McAdams [32] cited the relation of Latzko [26] for $L/D_e < 0.693 \, Re^{1/4}$:

$$\overline{Nu} = 1.11 \left[\frac{Re^{1/5}}{(L/D_e)^{4/5}} \right]^{0.275} Nu_\infty \tag{10-106}$$

and for $L/D_e > 0.693 \, Re^{1/4}$:

$$\overline{Nu} = \left(1 + \frac{A}{L/D_e}\right) Nu_\infty \tag{10-107}$$

where $A =$ a function of Re:

$$A = 0.144 \, Re^{1/4} \tag{10-108}$$

Latzko's prediction was acceptable for $26{,}000 < \mathrm{Re} < 56{,}000$, but with $A = 1.4$ compared with 1.83 to 2.22 predicted by Eq. 10-107.

Correlations of the form of Equation 10-108 for gases are given for various types of entries in Table 10-5. For the particular case of superheated steam, McAdams et al. [33] predicted:

$$\overline{\mathrm{Nu}} = 0.0214\ \mathrm{Re}^{0.8}\ \mathrm{Pr}^{1/3} \left(1 + \frac{2.3}{L/D_e} \right) \tag{10-109}$$

Table 10-5 Nusselt number for gas flow in tubes*

Type of entrance	Illustration	A
Bellmouth		0.7
Bellmouth and one screen		1.2
Long calming section		1.4
Short calming section		~3
45° Angle-bend entrance		~5
90° Angle-bend entrance		~7
Large orifice entrance (ratio of pipe I.D. to orifice diameter = 1.19)		~16
Small orifice entrance (ratio of pipe I.D. to orifice diameter = 1.789)		~7

Source: From Boelter et al. [4].

*$\mathrm{Nu} = \mathrm{Nu}_\infty \left(1 + \dfrac{AD_e}{L} \right)$; for $\dfrac{L}{D_e} > 5$ and $\dfrac{D_e G}{\mu} > 17{,}000$.

b Entrance region heat transfer coefficient For short tubes, where the region of fully developed flow is a small percentage of the total length, the local value of the Nusselt number must be sought. Experimental data for gases have shown that [5]:

1. For uniform velocity and temperature profile in the entrance:

$$Nu(z) = 1.5 \left(\frac{z}{D_e}\right)^{-0.16} Nu_\infty \quad \text{for } 1 < \frac{z}{D_e} < 12 \tag{10-110a}$$

$$Nu(z) = Nu_\infty \quad \text{for } \frac{z}{D_e} > 12 \tag{10-110b}$$

2. For abrupt entrance:

$$Nu(z) = \left(1 + \frac{1.2}{z/D_e}\right) Nu_\infty \quad \text{for } 1 < \frac{z}{D_e} < 40 \tag{10-111a}$$

$$Nu(z) = Nu_\infty \quad \text{for } \frac{z}{D_e} > 40 \tag{10-111b}$$

B Metallic Fluids

The behavior of Nusselt (Nu) for liquid metals follows the relation:

$$Nu_\infty = A + B \, (Pe)^C \tag{10-112}$$

where Pe = the Peclet number, i.e., Pe = RePr.

$A, B, C,$ are constants that depend on the geometry and the boundary conditions. The constant C is a number close to 0.8. The constant A reflects the fact that significant transfer in liquid metals occurs even as Re goes to zero.

1 Circular tube. The following relations hold for the boundary conditions cited and fully developed flow conditions.

1. Constant heat flux along and around the tube [29]:

$$Nu_\infty = 7 + 0.025 \, Pe^{0.8} \tag{10-113}$$

2. Uniform axial wall temperature and uniform radial heat flux [42]:

$$Nu_\infty = 5.0 + 0.025 \, Pe^{0.8} \tag{10-114}$$

2 Parallel plates. For fully developed flow [41]:

1. Constant heat flux through one wall only (the other is adiabatic):

$$Nu_\infty = 5.8 + 0.02 \, Pe^{0.8} \tag{10-115}$$

2. For constant heat flux through both walls, a graphic correction factor for the heat-transfer coefficient was supplied by Seban [41].

3 Concentric annuli For fully developed flow and the boundary condition of uniform heat flux in the inner wall when $D_2/D_1 > 1.4$:

$$\text{Nu}_\infty = 5.25 + 0.0188 \text{ Pe}^{0.8} \left(\frac{D_2}{D_1}\right)^{0.3} \tag{10-116}$$

If D_2/D_1 is close to unity, the use of Eq. 10-115 was recommended by Seban [41].

4 Rod bundles. As noted before, the nonuniformity of the subchannel shape creates substantial azimuthal variation of Nu. Also in finite rod bundles the turbulent effects in a given subchannel affect adjacent subchannels differently depending on the location of the subchannels with respect to the duct boundaries. Therefore the value of Nu is a function of position within the bundle. However, we provide here overall heat-transfer correlations and defer the discussion of the variation of heat transfer with angle and location to Chapter 7, Vol. II. Again, as shown in Figure 10-13, for $P/D > 1.1$ the equivalent diameter concept provides an acceptable answer.

The correlations considered here are as follows:
Westinghouse [25]:

$$\text{Nu} = 4.0 + 0.33(P/D)^{3.8} \, (\text{Pe}/100)^{0.86} + 0.16(P/D)^{5.0} \tag{10-117}$$

for $1.1 \leq P/D \leq 1.4$ and $10 \leq \text{Pe} \leq 5000$.
Schad—Modified [25]:

$$\text{Nu} = [-16.15 + 24.96(P/D) - 8.55(P/D)^2] \, \text{Pe}^{0.3} \tag{10-118a}$$

for $1.1 \leq P/D \leq 1.5$ and $150 \leq \text{Pe} \leq 1000$.

$$\text{Nu} = 4.496[-16.15 + 24.96(P/D) - 8.55(P/D)^2] \tag{10-118b}$$

for $\text{Pe} \leq 150$.
Graber and Rieger [16]:

$$\text{Nu} = [0.25 + 6.2(P/D) + 0.32(P/D) - 0.007] \, (\text{Pe})^{0.8 - 0.024(P/D)} \tag{10-119}$$

for $1.25 \leq P/D \leq 1.95$ and $150 \leq \text{Pe} \leq 3000$.
Borishanskii et al. [6]:

$$\text{Nu} = 24.15 \log[-8.12 + 12.76(P/D) - 3.65(P/D)^2] \\ + 0.0174[1 - \exp(6 - 6 \, P/D)][\text{Pe} - 200]^{0.9} \tag{10-120a}$$

for $1.1 \leq P/D \leq 1.5$ and $200 \leq \text{Pe} \leq 2000$.

$$\text{Nu} = 24.15 \log[-8.12 + 12.76(P/D) - 3.65(P/D)^2] \tag{10-120b}$$

for $1.1 < P/D < 1.5$ $\text{Pe} \leq 200$.

Experimental and predicted values of Nu in rod bundles with P/D values of 1.3, and 1.15 are shown in Figures 10-16 and 10-17, respectively.

The comparison indicates that the correlations of Borishanskii and Schad—modified yield the best agreement over the entire range of P/D values. The Graber and

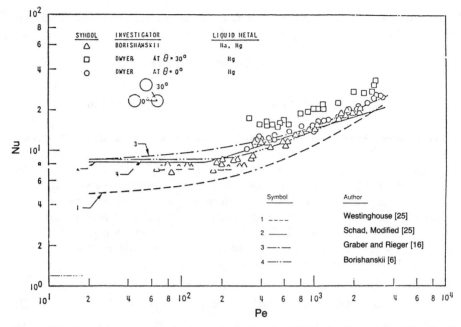

Figure 10-16 Comparison to predicted and experimental results of Nu for liquid metals in rod bundles for $P/D = 1.3$. *(From Kazimi and Carelli [25].)*

Rieger correlation appears to significantly overpredict the heat-transfer coefficient if extended beyond the published range of applicability $P/D \leq 1.15$. The Westinghouse correlation underestimates Nu at high values of P/D.

Example 10-3 Turbulent heat transfer in a steam generator—more exact calculation

PROBLEM In Example 10-2 the linear heat-transfer rate of a tube in the single-phase region of the secondary side of the steam generator was based on the assumption that the shell side film coefficient was the same as the tube side coefficient. Using the information in Example 10-2 and the information below, determine the shell side heat-transfer coefficient with one of the empirical relations presented in section V.

Shell side information
Lower shell shroud I.D. = 1.8 m
Secondary flow rate = 480 kg/s
Flow characteristics = assume saturated water at secondary side temperature of 280°C
Triangular tube array (not common practice for PWRs)
Total number of tubes = 3800

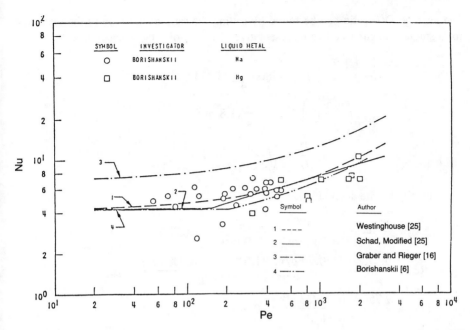

Figure 10-17 Comparison of predicted and experimental results of Nu for metals in rod bundles for P/D = 1.15. *(From Kazimi and Carelli [25].)*

SOLUTION We choose a correlation based on flow conditions that most closely approximate those of the steam generator. We must be sure, in particular, that the parameter space falls within the range of the correlation. To find the Re and Pr, first find the properties of water in the shell side:

Saturated water at 280°C:
$\rho = 750.7 \text{ kg/m}^3$
$\mu = 9.75 \times 10^{-5} \text{ kg/m s}$
$k = 0.574 \text{ W/m °C}$
$c_p = 5307 \text{ J/kg °C}$

Now calculate the equivalent dimensions needed for the empirical formulas. The area and hydraulic diameter on the shell side are:

$$A = \frac{\pi}{4}(1.8 \text{ m})^2 - 3800\frac{\pi}{4}(0.0254 \text{ m})^2 = 0.619 \text{ m}^2$$

$$D_e = \frac{4A}{P_w} = \frac{4(0.619)}{\pi(1.8) + 3800\,\pi(0.0254)} = 0.0080 \text{ m}$$

$$D_H = \frac{4A}{P_h} = \frac{4(0.619)}{3800\,\pi(0.0254)} = 0.0082 \text{ m}$$

The average P/D ratio can be found by setting the ratio of the tube to the total area for a single triangular cell as representative of the whole steam generator.

$$\frac{\frac{1}{2}\left(\frac{\pi}{4}D^2\right)}{\frac{1}{2}\left(\frac{\sqrt{3}}{2}P^2\right)} = \frac{3800\,\frac{\pi}{4}(0.0254\text{ m})^2}{\frac{\pi}{4}(1.8\text{ m})^2}$$

$$\frac{P}{D} = \sqrt{\frac{2\pi}{4\sqrt{3}}\,\frac{1}{3800}}\left(\frac{1.8}{0.0254}\right) = 1.1$$

The velocity on the shell side is:

$$V = \frac{\dot{m}}{\rho A} = \frac{480\text{ kg/s}}{(750.7\text{ kg/m}^3)(0.619\text{ m}^2)} = 1.033\text{ m/s}$$

$$\text{Re} = \frac{\rho V D_e}{\mu} = \frac{(750.7\text{ kg/m}^3)(1.033\text{ m/s})(0.0080\text{ m})}{9.75\times10^{-5}\text{ kg/m s}} = 6.363\times10^4$$

$$\text{Pr} = \frac{9.75\times10^{-5}\,(5307)}{0.574} = 0.901$$

Using the Weisman correlation (Eq. 10-101) for water in triangular arrays, we find:

$$(\text{Nu}_\infty)_{\text{c.t.}} = 0.023(6.363\times10^4)^{0.8}\,(0.901)^{0.333} = 154.7$$

$$\psi = 1.130(1.1) - 0.2609 = 0.9821$$

Therefore:

$$\text{Nu}_\infty = 154.7(0.9281) = 151.9$$

The shell side heat-transfer coefficient is now:

$$h = \frac{\text{Nu}k}{D_\text{H}} = \frac{151.9(0.574)}{0.0082}\text{ W/m}^2\,°\text{C} = 10.63\times10^3\text{ W/m}^2\,°\text{C}$$

and the linear heat transfer rate is:

$$q' = \frac{2\pi(305 - 280)}{\dfrac{1}{(0.0111)(2.87\times10^4)} + \dfrac{\ell n(1/0.875)}{35} + \dfrac{1}{(0.0127)(10.63\times10^3)}}$$

$$q' = \frac{157.1}{3.14\times10^{-3} + 3.82\times10^{-3} + 7.41\times10^{-3}} = 1.093\times10^4\text{ W/m}$$

So the refinement in the calculation of the linear heat-transfer rate, in this case, resulted in a decrease of about 33%.

Example 10-4 Typical heat transfer coefficients of water and sodium

PROBLEM Liquid flow velocity along rod bundles is limited to a maximum of 10 m/s to avoid cavitation at obstructions. At that velocity, compare the heat-transfer coefficient of water (at average coolant temperature for a PWR) to the heat-transfer coefficient of sodium (at average temperatures for an LMFBR). Use geometric data for the PWR and LMFBR in Table 1-3. How do the coefficients compare at a velocity of 1 m/s?

Parameter	Water (315°C)	Sodium (538°C)	
k	0.5	62.6	W/m °C
ρ	704	817.7	kg/m^3
μ	8.69×10^{-5}	2.28×10^{-4}	kg/m s
c_p	6270	1254	J/kg °C

SOLUTION

For water at 10 m/s

$$\text{Nu}_\infty = \psi(\text{Nu}_\infty)_{\text{c.t.}}$$

$$(\text{Nu}_\infty)_{\text{c.t.}} \text{ Presser} = 0.023 \, \text{Re}^{0.8} \, \text{Pr}^{0.4} = 0.023 \left(\frac{\rho V D_e}{\mu}\right)^{0.8} \left(\frac{c_p \mu}{k}\right)^{0.4}$$

$$D_e = \frac{4 A_f}{P_w} = \frac{4\left(P^2 - \dfrac{\pi D^2}{4}\right)}{\pi D} = 11.78 \text{ mm}$$

$$(\text{Nu}_\infty)_{\text{c.t.}} \text{ Presser} = 0.023 \left[\frac{(704)(10)(11.78 \times 10^{-3})}{8.69 \times 10^{-5}}\right]^{0.8}$$
$$\cdot \left[\frac{(6270) \, 8.69 \times 10^{-5}}{0.5}\right]^{0.4} = 1447$$

$$(\text{Nu}_\infty)_{\text{c.t.}} \text{ Weisman} = 0.023 \, \text{Re}^{0.8} \, \text{Pr}^{0.333} = 1438$$

$$\psi \text{ Presser (Eq. 10-100b)} = 0.9217 + 0.1478 \left(\frac{0.0126}{0.0095}\right)$$
$$- 0.1130 \, e^{-7\left(\frac{0.0126}{0.0095} - 1\right)} = 1.11$$

$$\psi \text{ Weisman (Eq. 10-101b)} = 1.826 \left(\frac{0.0126}{0.0095}\right) - 1.0430 = 1.38$$

$$\text{Nu}_\infty \text{ Presser} = 1447 \times 1.11 = 1606$$

$$\text{Nu}_\infty \text{ Weisman} = 1438 \times 1.38 = 1984$$

However,

$$\text{Nu} = \frac{h D_H}{k}$$

and

$$D_H = \frac{4A_f}{P_h} = D_e$$

$$\therefore h \text{ Presser} = \frac{k \text{ Nu}}{D_H} = \frac{0.5(1606)}{11.78 \times 10^{-3}} = 68.2 \text{ kW/m}^2 \text{ °C}$$

$$h \text{ Weisman} = \frac{0.5(1984)}{11.78 \times 10^{-3}} = 84.2 \text{ kW/m}^2 \text{ °C}$$

For sodium at 10 m/s

$$\text{Pe} = \text{Re Pr} = \left(\frac{\rho V D_e}{\mu}\right)\left(\frac{c_p \mu}{k}\right)$$

For a triangular array interior channel,

$$D_e = 4 \frac{\left(\frac{\sqrt{3}}{4}P^2 - \pi\frac{D^2}{8}\right)}{\frac{\pi D}{2}}$$

from Table 1-3, $P = 9.7$ mm and $D = 8.65$ mm:

$$\therefore D_e = 4 \frac{[0.433(9.7)^2 - \pi(8.65)^2/8]}{\frac{\pi(8.65)}{2}} = 3.344 \text{ mm}$$

$$\text{Pe} = \left[\frac{(817.7)(10)(3.344 \times 10^{-3})}{2.28 \times 10^{-4}}\right]\left[\frac{(1254)(2.28 \times 10^{-4})}{62.6}\right] = 548$$

Using Eq. 10.117:

$$\text{Nu} = 4 + 0.33\left(\frac{9.7}{8.65}\right)^{3.8}\left(\frac{548}{100}\right)^{0.86} + 0.16\left(\frac{9.7}{8.65}\right)^{5.0} = 6.49$$

$$h = \frac{k \text{ Nu}}{D_H}; D_H = D_e$$

$$\therefore h = \frac{(62.6)(6.49)}{3.344 \times 10^{-3}} = 121.5 \text{ kW/m}^2 \text{ °C}$$

For water at 1 m/s
Note that Re is still in the turbulent region.

$$h_{\text{Presser}} = (0.1)^{0.8}(68.2) = 10.8$$
$$h_{\text{Weisman}} = (0.1)^{0.8}(84.2) = 13.3$$

For sodium at 1 m/s

$$Pe = 0.1 \times 548 = 54.8$$

$$Nu = 4 + 0.33 \left(\frac{9.7}{8.65}\right)^{3.8} \left(\frac{54.8}{100}\right)^{0.86} + 0.16 \left(\frac{9.7}{8.65}\right)^{5.0} = 4.59$$

$$h = kNu/D_H = \frac{(62.6)(4.59)}{3.344 \times 10^{-3}} = 85.9 \text{ kW/m}^2 \text{ °C}$$

In summary, at velocities of 1 and 10 m/s, the heat-transfer coefficients (kW/m² °C) are:

Velocity (m/s)	Water		Sodium
	Presser	Weisman	
10	68.2	84.2	121.5
1	10.8	13.3	85.9

REFERENCES

1. Bhatti, M. S., and Savery, C. W. *Heat Transfer in the Entrance Region of a Straight Channel: Laminar Flow with Uniform Wall Heat Flux.* ASME 76-HT-20, 1976. (Also in condensed form: *J. Heat Transfer* 99:142, 1972.
2. Bialokoz, I. G., and Saunders, O. A. Heat transfer in pipe flow at high speed. *Proc. Inst. Mech. Eng. (Lond.)* 170:389, 1956.
3. Bird, R. B., Stewart, W. E., and Lightfoot, E. N. *Transport Phenomena.* New York: Wiley, 1960.
4. Boelter, L. M. K., Young, G., and Iverson, H. W. *An Investigation of Aircraft Heaters.* XXVII, NACA-TN-1451, 1948.
5. Bonilla, C. F. *Heat Transfer.* New York: Interscience, 1964, Ch. 2.
6. Borishanskii, V. M., Gotorsky, M. A., and Firsova, E. V. Heat transfer to liquid metal flowing longitudinally in wetted bundles of rods. *Atomic energy* 27:549, 1969.
7. Bradshaw, P., Cebeci, T., and Whitelaw, J. H. *Engineering Calculation Methods for Turbulent Flow.* New York: Academic Press, 1981.
8. Chen, F. F., Domanus, H. M., Sha, W. T., and Shah, V. L. *Turbulence Modeling in the COMMIX Computer Code.* NUREG/CR-3504, ANL-83-65, April 1984.
9. Colburn, A. P. A method of correlating forced convection heat transfer data and a comparison with liquid friction. *Trans. AIChE* 29:170, 1933.
10. Deissler, R. G. Investigation of turbulent flow and heat transfer in smooth tubes including the effects of variable physical properties. *Trans. ASME* 73:101, 1951.
11. Deissler, R. G. Turbulent heat transfer and friction in the entrance regions of smooth passages. *Trans. ASME* 77:1221, 1955.
12. Deissler, R. G., and Taylor, M. F. Analysis of axial turbulent flow and heat transfer through banks of rods or tubes. In: *Reactor Heat Transfer Conference,* 1956. TID-7529, Book 2, 1957, p. 416.
13. Dittus, F. W., and Boelter, L. M. K. University of California, Berkeley, *Publ. Eng.* 2:443, 1930.
14. Dwyer, O. E. Eddy transport in liquid metal heat transfer. *AIChE J.* 9:261, 1963.
15. Dwyer, O. E., and Berry, H. C. Laminar flow heat transfer for in-line flow through unbaffled rod bundles. *Nucl. Sci. Eng.* 42:81, 1970.

16. Graber, H., and Rieger, M. Experimental study of heat transfer to liquid metals flowing in-line through tube bundles. *Prog. Heat Mass Transfer* 7:151, 1973.
17. Graetz, L. Uber die Warmeleitfahigkeit von Flussingeiten. *Ann. Physik.* 25:337, 1885.
18. Harlow, F. H., and Nakayama, P. I. *Transport of Turbulence Energy Decay.* Los Alamos Scientific Laboratory, Report LA-3854-UC-34, TID-4500, 1968.
19. Irvine, T. R. Non-circular convective heat transfer. In W. Ible (ed.), *Modern Developments in Heat Transfer.* New York: Academic Press, 1963.
20. Jenkins, R. *Heat Transfer and Fluid Mechanics Institute.* Stanford University Press, 1951, pp. 147–158.
21. Jones, W. P., and Launder, B. E. The prediction of laminarization with a 2-equation model of turbulence. *Int. J. Heat Mass Transfer* 15:301, 1972.
22. Kays, W. M. *Convective Heat and Mass Transfer.* New York: McGraw-Hill, 1966.
23. Kays, W. M. Numerical solutions for laminar flow heat transfer in circular tubes. *Trans. ASME* 77:1265, 1955.
24. Kays, W., and Leung, E. Y. Heat transfer on annular passages. *Int. J.H.M.T.* 10:1533, 1963.
25. Kazimi, M. S., and Carelli, M. D. *Heat Transfer Correlation for Analysis of CRBRP Assemblies.* Westinghouse Report, CRBRP-ARD-0034, 1976.
26. Latzko, H. Der Warmeubergang an einer Turbulenten Flussigkeits-Oder Gasstrom. *ZAMM* 1:268, 1921.
27. Launder, B. E., and Spaulding, D. B. *Mathematical Models of Turbulence.* New York: Academic Press, 1972.
28. Launder, B. E., Morse, A., Rode, W., and Spaulding, D. B. The prediction of free shear flows—a comparison of the performance of six turbulent models. In: *Proceedings of NASA Conference on Free Shear Flows,* Langberg, 1972.
29. Lyon, R. N. Liquid metal heat transfer coefficients. *Chem. Eng. Prog.* 47:75, 1951.
30. Markoczy, G. Convective heat transfer in rod clusters with turbulent axial coolant flow. 1. Mean value over the rod perimeter. *Warme Stoffubertragung S* 204, 1972.
31. Martinelli, R. B. Heat transfer in molten metals. *Trans. ASME* 69:947, 1947.
32. McAdams, W. H. *Heat Transmission* (3rd ed.). New York: McGraw-Hill, 1954.
33. McAdams, W. H., Kennel, W. E., and Emmons, J. N. Heat transfer to superheated steam at high pressures addendum. *Trans ASME* 72:421, 1950.
34. Nijsing R. *Heat Exchange and Heat Exchangers with Liquid Metals.* AGRD-Ls-57-12; AGRD Lecture Series No. 57 on Heat Exchangers by J. J. Ginoux, Lecture Series Director, 1972.
35. Ozisik, M. N. *Basic Heat Transfer.* New York: McGraw-Hill, 1977.
36. Prandtl, L. *Z. Phys.* 11:1072, 1910.
37. Prandtl, L. Uber die Ausgebildete Turbulenz. *ZAMM* 5:136, 1925.
38. Presser, K. H. *Warmeubergang und Druckverlust und Reaktorbrennelementen in Form Langsdurch-stromter Rundstabbundel.* Jul-486-RB, KFA Julich, 1967.
39. Rodi, W. *Turbulence Models and Their Application to Hydraulics—A State of the Art Review.* International Association for Hydraulic Research, 1980.
40. Schlichting, H. *Boundary Layer Theory* (6th ed.). New York: McGraw-Hill, 1968.
41. Seban, R. A. Heat transfer to a fluid flowing turbulently between parallel walls and asymmetric wall temperatures. *Trans. ASME* 72:789, 1950.
42. Seban, R. A., and Shimazaki, T. Heat transfer to a fluid flowing turbulently in a smooth pipe with walls at constant temperature. ASME Paper 50-A-128, 1950.
43. Seider, E. N., and Tate, G. E. Heat transfer and pressure drop of liquids in tubes. *Ind. Eng. Chem.* 28:1429, 1936.
44. Silberberg, M., and Huber, D. A. *Forced Convection Heat Characteristics of Polyphenyl Reactor Coolants.* AEC Report NAA-SR-2796, 1959.
45. Sleicher, C. A., and Tribus, M. Heat transfer in a pipe with turbulent flow and arbitrary wall-temperature distribution. *Trans. ASME* 79:789, 1957.
46. Sparrow, E. M., Hallman, T. M., and Siegel, R. Turbulent heat transfer in the thermal entrance region of a pipe with uniform heat flux. *Appl. Sci. Res.* A7:37, 1957.
47. Sparrow, E. M., Loeffler Jr., A. L., and Hubbard, H. A. Heat transfer to longitudinal laminar flow between cylinders. *J. Heat Transfer* 83:415, 1961.

48. Tribus, M., and Klein, S. J. *Forced Convection from Nonisothermal Surfaces in Heat Transfer: A Symposium*. Engineering Research Institution, University of Michigan, 1953.

49. Trup, A. C., and Azad, R. S. The structure of turbulent flow in triangular array rod bundles. *Nucl. Eng. Design* 32:47, 1975.

50. Von Karman. The analogy between fluid friction and heat transfer. *Trans. ASME* 61:701, 1939.

51. Weisman, J. Heat transfer to water flowing parallel to tube bundles. *Nucl. Sci. Eng.* 6:79, 1959.

PROBLEMS

Problem 10-1 Derivation of Nusselt number for laminar flow in rectangular geometry (section II)
Prove that the asymptotic Nusselt number for flow of a coolant with constant properties between two flat plates of infinite width, heated with uniform heat flux on both walls, is 8.235 for laminar flow (i.e., parabolic velocity distribution) and 12 for slug flow (i.e., uniform velocity distribution).

Problem 10-2 Derivation of Nusselt number for laminar flow in circular geometry (section II)
Show that for a round tube, uniform wall temperature, and slug velocity conditions the asymptotic Nusselt number for a round tube is 5.75, whereas for flow between two flat plates it is 9.87.

Problem 10-3 Derivation of Nusselt number for laminar flow in an equivalent annulus (section II)
Derive the Nusselt number for slug flow in the equivalent annulus of an infinite rod array by solving the differential energy equation subject to the appropriate boundary conditions, i.e.,

$$\frac{\partial^2 T}{\partial r^2} + \frac{1}{r}\frac{\partial T}{\partial r} = \frac{\upsilon \rho c_p}{k}\frac{\partial T}{\partial z}$$

Answer:

$$Nu = \frac{2(r_o^2 - r_i^2)^3}{r_i^2[r_o^4 \ell n(r_o/r_i) - (r_o^2 - r_i^2)(3r_o^2 - r_i^2)/4]}$$

Problem 10-4 Determining the temperature of the primary side of a steam generator (section III)
Consider the flow of high pressure water through the U-tubes of a PWR steam generator. There are 5700 tubes with outside diameter 19 mm, wall thickness 1.2 mm, and average length 16.0 m.

The steady-state operating conditions are:

Total primary flow through the tubes = 5100 kg/s
Total heat transfer from primary to secondary = 820 MW
Secondary pressure = 5.6 MPa (272°C saturation)

1. What is the primary temperature at the tube inlet?
2. What is the primary temperature at the tube outlet?

Use a Dittus–Boelter equation for the primary side heat-transfer coefficient. Assume that the tube wall surface temperature on the secondary side is constant at 276°C.

Properties
For water at 300°C and 15 MPa
  Density = 726 kg/m³
  Specific heat = 5.7 kJ/kg °K
  Viscosity = 92 μPa · s
  Thermal conductivity = 0.56 W/m °K
For tube wall
  Thermal conductivity = 26 W/m °K

Hint: Consideration of the axial variation of the primary coolant bulk temperature is required.

    Answer: 1. $T_{p,in} = 307.2°C$

           2. $T_{p,out} = 279.0°C$

Problem 10-5 Comparison of heat transfer characteristics of water and helium (section V)

Consider a new design of a thermal reactor that requires square arrays of fuel rods. Heat is being generated uniformly along the fuel rods. Water and helium are being considered as single-phase coolants. The design condition is that the maximum cladding surface temperature should remain below 600°F.

1. Find the minimum mass flow rate of water to meet the design requirement.
2. Would the required mass flow rate of helium be higher or lower than that of water?

Geometry of square array
  P = pitch = 0.55 in.
  D = fuel rod diameter = 0.43 in.
  H = fuel height = 12 ft

Operating conditions
  Heat flux = 250,000 Btu/hr ft^2
  Coolant inlet temperature = 500°F

Coolant properties

Coolant	ρ(lb/ft^3)	c_p(Btu/lb °F)	μ(lb/hr ft)	k(Btu/ft hr °F)
Water	45.9	1.270	0.231	0.326
Helium	0.0540	1.248	0.072	0.133

    Answers: 1. $\dot{m}_{water} = 3610$ lbm/hr

           2. $\dot{m}_{He} = 3950$ lbm/hr

Problem 10-6 Estimating the effect of turbulence on heat transfer in LMFBR fuel bundles

Consider the fuel bundle of an LMFBR whose geometry is described in Table 1-3. Using Dwyer's recommendations for the values of ϵ_m and ϵ_H (Eq. 10-71), estimate the ratio of ϵ_H/ϵ_M for the sodium velocities in the bundles.

1. $V = 10$ m/s
2. $V = 1$ m/s

    Answers:
1. For $V = 10$ m/s:

$$\frac{\epsilon_H}{\epsilon_M} \text{ in core} = 0.510; \left(\frac{\epsilon_M}{\nu}\right)_{max} = 120$$

$$\frac{\epsilon_H}{\epsilon_M} \text{ in blanket} = 0.722; \left(\frac{\epsilon_M}{\nu}\right)_{max} = 180$$

2. For $V = 1$ m/s:

$$\frac{\epsilon_H}{\epsilon_M} \text{ in both core and blanket} = 0$$

TWO-PHASE FLOW DYNAMICS

I INTRODUCTION

In the present chapter the focus is on the hydraulics of two-phase flow, which includes phase configuration, pressure drop relations, and critical flow. The discussion in Chapter 5 indicates the wide range of possibilities for the selection of the sets of equations to be solved. The appropriate set of equations (more commonly called the *model*) for a particular two-phase flow system is influenced by the conditions of the system. The more detailed models can be expensive to solve because of the requirement for a large computing facility. The less detailed models introduce certain simplifying assumptions that are not always correct but can be solved more readily. Thus, it is appropriate to start the process of analyzing the two-phase flow systems by asking the following questions.

1. What is the number of flow dimensions that need to be represented? The difficulty of solving a multidimensional problem greatly exceeds that of a one-dimensional flow problem. In fact, there is relatively little information regarding the multidimensional effects on the flow hydraulics, and most of the existing information is related to one-dimensional channel flow.

2. What is the expected degree of mechanical equilibrium between the phases? Under conditions of high mass fluxes, the two phases can be expected to move at the same velocity. However, the relative velocity of one phase to the other is a complex function of the flow conditions and geometry under consideration.

3. What is the expected degree of thermal nonequilibrium in the flow? Thermal nonequilibrium is generally more important under transient conditions than under steady-state conditions. Even in steady-state conditions, however, the effects of non-equilibrium often must be included, as for subcooled boiling of liquids.

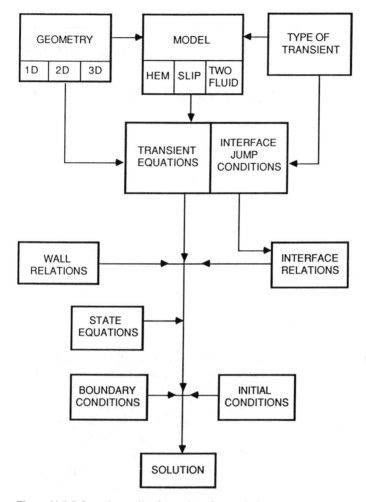

Figure 11-1 Information needs of two-phase flow analysis.

Once a model is selected, constitutive relations are needed to describe the rate of exchange of mass, momentum, and energy among the wall and the two phases. These relations depend on the flow configuration, or *flow regimes*. The state equations and boundary and initial conditions are also needed to close the problem. The needed information for solving the two-phase flow equations is depicted in Figure 11-1.

II FLOW REGIMES

A Regime Identification

The void distribution patterns of gas–liquid flows have been characterized by many researchers. They did not always refer to the flow regimes with consistent terminology.

However, the main flow regimes, the only ones addressed here, are sufficiently distinct. The void distribution depends on the pressure, channel geometry, gas and liquid flow rates, and the orientation of the flow with respect to gravity. Of major significance to the nuclear engineer are the cases of vertical upward co-current flow (as in a BWR fuel assembly) and horizontal co-current flow (as in the piping of an LWR during transients).

The flow regimes identified in vertical flows are the bubbly, slug, churn, and annular regimes (Fig. 11-2). The bubbly regime is distinguished by the presence of dispersed vapor bubbles in a continuous liquid phase. The bubbles can be of variable size and shape. Bubbles of 1 mm or less are spherical, but larger bubbles have variable shapes. Slug flow is distinguished by the presence of gas plugs (or large bubbles) separated by liquid slugs. The liquid film surrounding the gas plug usually moves downward. Several small bubbles may also be dispersed within the liquid. The churn flow is more chaotic but of the same basic character as the slug flow. Annular flow is distinguished by the presence of a continuous core of gas surrounded by an annulus of the liquid phase. If the gas flow in the core is sufficiently high, it may be carrying liquid droplets. In this case an annular-dispersed flow regime is said to exist. Hewitt and Roberts [19] also suggested that the droplets can gather in clouds forming an annular–wispy regime. The liquid droplets are torn from the wavy liquid film, get entrained in the gas core, and can be de-entrained to join the film downstream of the point of their origin.

With horizontal flow, the flow may stratify, creating additional patterns, as shown in Figure 11-3.

B Flooding and Flow Reversal

Flooding and flow reversal are basic phenomena encountered in several conditions in nuclear reactor thermohydraulics, including flow regime transitions (described in sec-

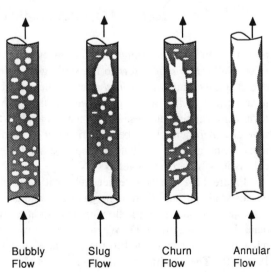

Figure 11-2 Flow patterns in vertical flow.

| Bubbly Flow | Slug Flow | Churn Flow | Annular Flow |

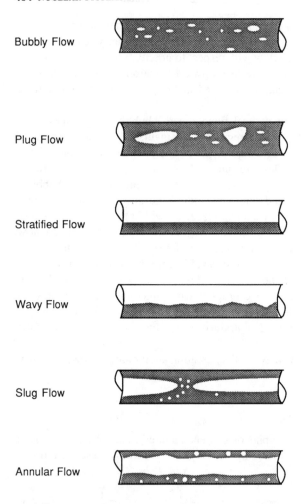

Bubbly Flow

Plug Flow

Stratified Flow

Wavy Flow

Slug Flow

Annular Flow

Figure 11-3 Flow patterns in horizontal flow.

tion II.C) and the rewetting of hot surfaces after a loss of coolant accident (LOCA) in an LWR. *Flooding* refers to the stalling of a liquid downflow by a sufficient rate of gas upflow, and *flow reversal* refers to the change in flow direction of liquid initially in co-current upflow with a gas as the gas flow rate is sufficiently decreased. Here we review information on these phenomena as applied to simple tube geometry. Although several experiments have been carried out to predict the onset of flooding and flow reversal, no general adequate theories are available so far. However, some empirical correlations have been developed.

Figure 11-4 shows a vertical tube with a liquid injection device composed of a porous wall. Initially there is no gas flow, and a liquid film flows down the tube at a constant flow rate. If gas is flowing upward at a low rate, a countercurrent flow takes place in the tube (tube 1). When the gas flow rate is increased, the film thickness remains constant and equal to the thickness of a laminar film derived by Nusselt in 1916 [20]. The nondimensional thickness is given by:

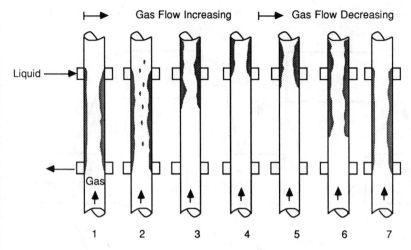

Figure 11-4 Flooding and flow reversal experiment.

$$\delta^* \equiv \delta \left[\frac{g \sin \theta}{\nu_\ell^2} \right]^{1/3} = 1.442 \, \mathrm{Re}_\ell^{1/3} \tag{11-1}$$

where:

$$\mathrm{Re}_\ell \equiv \frac{4Q_\ell}{\pi D \nu_\ell} \tag{11-2}$$

and where θ = inclination angle from the horizontal, ν_ℓ = liquid kinematic viscosity, Q_ℓ = liquid volumetric flow rate, D = pipe diameter, and g = acceleration due to gravity. This relation is valid for liquid Reynolds number (Re_ℓ) up to 2000. For turbulent films the thickness may be given by [3]:

$$\delta^* = 0.304 \, \mathrm{Re}_\ell^{7/12} \tag{11-3}$$

For a higher gas flow rate, the liquid film becomes unstable, large-amplitude waves appear, and the effective film thickness is increased. Droplets are torn from the crests of the waves and are entrained with the gas flow above the liquid injection level. Some of the droplets impinge on the wall and give rise to a liquid film (tube 2). Meanwhile, the pressure drop in the tube above the liquid injection level increases sharply. When the gas flow rate is increased further, the tube section below the liquid injection dries out progressively (tube 3), until the liquid downflow is completely prevented at a given gas flow rate. This point is called the *flooding condition* (tube 4). At higher gas flow rates, churn flow or annular flow appears in the upper tube.

Suppose we decrease the gas flow rate. For a given value the liquid film becomes unstable, and large-amplitude waves appear on the interface. The pressure drop increases, and the liquid film tends to fall below the liquid injection port. This point is called the *flow reversal phenomenon* (tube 5). When the gas flow rate decreases, the

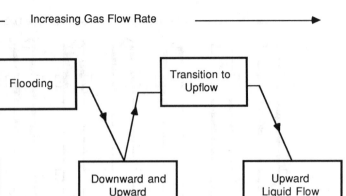

Figure 11-5 Flooding and flow reversal.

liquid film flows downward, and the upper tube dries out for a given gas flow rate (tube 6).

The experiment shown in Figure 11-4 is summarized in Figure 11-5. It is noted that the change from upward to downward liquid flow and vice versa is preceded by a transition period of partial flow in both directions.

Two nondimensional numbers have been used extensively to correlate the transition conditions for two-phase systems: the Wallis number and the Kutateladze number. The former represents the ratio of inertial force to hydrostatic force on a bubble or drop of diameter D. The latter replaces the length scale D with the Laplace constant $[\sigma/g(\rho_\ell - \rho_v)]^{1/2}$.

The *Wallis number* is defined as:

$$\{j_k^+\} \equiv \{j_k\} \left[\frac{\rho_k}{gD(\rho_\ell - \rho_v)} \right]^{0.5} \tag{11-4}$$

where $k = \ell$ for liquid and v for gas.

The *Kutateladze number* (Ku) is defined as:

$$\text{Ku}_k \equiv \{j_k\} \left[\frac{\rho_k}{[g\sigma(\rho_\ell - \rho_v)]^{0.5}} \right]^{0.5} \tag{11-5}$$

Wallis [38] had proposed correlations for flooding in vertical tubes that depend on the size of the channel:

$$\{j_v^+\}^{0.5} + m \{j_\ell^+\}^{0.5} = C \tag{11-6}$$

where m and $C =$ constants that depend on the channel exit conditions. It is often found that $m = 1.0$ and $C = 0.9$ for round-edged tubes and 0.725 for sharp-edged tubes.

Porteous [31] gave the limit of flooding in a semitheoretical treatment as:

$$\frac{\{j\}}{\sqrt{gD}} = 0.105 \sqrt{\frac{\rho_\ell - \rho_v}{\rho_v}} \tag{11-7}$$

For flow reversal, Wallis [38] recommended:

$$\{j_v^+\} = 0.5 \tag{11-8}$$

However, later Wallis and Kuo [39] modified this relation for flow reversal to reconcile it with the data, which showed less dependence on the tube diameter.

When analyzing their result for air–water flow in tubes of 6 to 309 mm diameter, Pushkina and Sorokin [32] proposed that flow reversal occurs when the Kutateladze number is 3.2:

$$Ku_v = 3.2 \tag{11-9}$$

They argued that this condition is also close to the flooding condition.

The flooding phenomenon has been described analytically by several authors. The onset of flooding has been explained by four mechanisms:

1. Occurrence of a standing wave on the liquid film
2. Interfacial instabilities between the two phases due to their relative velocities
3. No net flow in the liquid film
4. Inception of entrainment from the liquid film

The flow reversal phenomenon has been shown to be strongly connected to the hydrodynamics of hanging films. A review of the available experimental and analytic studies of countercurrent flow phenomena associated with flooding and flow reversal in vertical and horizontal flow has been given by Bankoff and Lee [3]. Because of the various definitions used for the onset of each phenomenon, the empirical relations should always be reviewed to ascertain their experimental base prior to application.

Experimental information on the pressure drop characteristics in vertical flow is given in Figure 11-6.

C Flow Regime Maps

Most of the flow regimes have been defined in one-dimensional channels. The flow regime transition points (or the flow regime maps) have generally been related to the average flow properties in the channel, e.g., the flow quality. In the more advanced reactor safety analysis codes, multidimensional effects have necessitated the use of

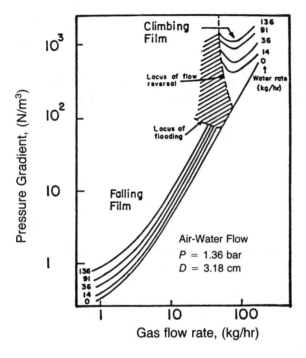

Figure 11-6 Pressure drop characteristics in vertical countercurrent flow transitions. *(From Hewitt et al. [18].)*

flow regime maps based on static quantities, e.g., the void fraction. The existing multidimensional flow maps are not as extensively tested as the one-dimensional maps.

1 Vertical flow. The one-dimensional flow map of Hewitt and Roberts [19] (Fig. 11-7) was developed on the basis of air–water data, obtained in a pipe of 31.2 mm diameter and at pressures varying from 0.14 to 0.54 MPa. It was found suitable for steam–water data in a pipe of 12.7 mm diameter at pressures of 3.45 to 6.90 MPa. It is based on the superficial liquid and vapor momentum fluxes:

$$\rho_\ell \{j_\ell\}^2 = \frac{G_m^2 (1 - x)^2}{\rho_\ell} \tag{11-10}$$

and

$$\rho_v \{j_v\}^2 = \frac{G_m^2 x^2}{\rho_v} \tag{11-11}$$

Taitel and co-workers [36] compared several flow regime maps and found some discrepancies among them. They also developed their own map based on a theoretical analysis for the mechanisms contributing to the transition between regimes.

a Bubbly to slug The transition from bubbly to slug or churn flow was related to a maximum possible void fraction of 0.25 to prevent bubble coalescence to form slugs. The liquid and vapor velocities are related by:

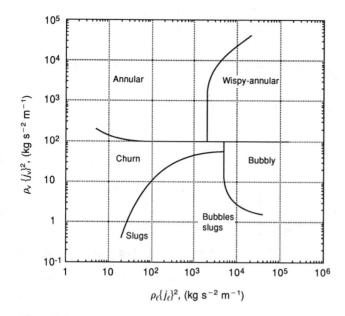

Figure 11-7 Hewitt and Roberts map for vertical upward flow. *(From Hewitt and Roberts, [19].)*

$$V_\ell = V_v - V_o \tag{11-12}$$

where V_o = rise velocity of large bubbles (5 mm $< d <$ 20 mm), which was shown to be insensitive to bubble size [17] and is given by:

$$V_o = 1.53 \left(\frac{g(\rho_\ell - \rho_v)\sigma}{\rho_\ell^2} \right)^{1/4} \tag{11-13}$$

where σ = surface tension.

Using the definition of the volumetric fluxes and assuming uniform velocity distributions, we can obtain from Eqs. 11-12 and 11-13 the relation:

$$\frac{\{j_\ell\}}{\{1 - \alpha\}} = \frac{\{j_v\}}{\{\alpha\}} - 1.53 \left(\frac{g(\rho_\ell - \rho_v)\sigma}{\rho_\ell^2} \right)^{1/4} \tag{11-14}$$

For $\{\alpha\} = 0.25$, Eq. 11-14 yields:

$$\frac{\{j_\ell\}}{\{j_v\}} = 3 - 1.15 \frac{[g(\rho_\ell - \rho_v)\sigma]^{1/4}}{j_v \rho_\ell^{1/2}} \tag{11-15}$$

Equation 11-15 is shown as line A in Figure 11-8. It should be noted that in an earlier paper by Taitel and Dukler [35] $\{\alpha\} = 0.3$ was chosen as the maximum permissible packing of bubbles to prevent coalescence, which yielded the transition limit of:

$$\frac{\{j_\ell\}}{\{j_v\}} = 2.34 - 1.07 \frac{[g(\rho_\ell - \rho_v)\sigma]^{1/4}}{j_v \rho_\ell^{1/2}} \tag{11-16}$$

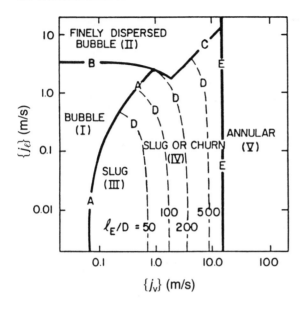

Figure 11-8 Flow regime map of Taitel et al. [36] for air–water at 25°C and 0.1 MPa in 50 mm diameter tubes.

For small pipes, the rising bubbles can catch up to form a slug. Thus for very small channels the bubbly regime may not exist, in which case only slug flow would exist at low liquid and gas velocities. This situation is shown in Figure 11-9 for air–water flow in a 25-mm diameter tube under the same pressure and temperature conditions as Figure 11-8.

The criterion for eliminating bubbly flow is that a deformable bubble rising at a velocity V_o approaches and coalesces with the slug or Taylor bubble rising at velocity V_b, i.e.,

$$V_o \geq V_b$$

where

$$V_b = 0.35\sqrt{gD} \text{ (for } \rho_v \ll \rho_\ell) \tag{11-17}$$

Because V_o is given by Eq. 11-13, this criterion yields the following relation for a small-diameter (no bubbly regime) system:

$$\left[\frac{\rho_\ell^2 gD^2}{(\rho_\ell - \rho_v)\sigma}\right]^{1/4} \leq 4.36 \tag{11-18}$$

b Finely dispersed bubbly to bubbly/slug flow For high liquid volumetric fluxes, bubble breakup would occur due to turbulent forces so that coalesence would also be prevented above a limit on the sum of $\{j_\ell\}$ and $\{j_v\}$ derived by Taitel et al. [36] as:

$$\{j_\ell\} + \{j_v\} = 4\left\{\frac{D^{0.429}(\sigma/\rho_\ell)^{0.089}}{\nu_\ell^{0.072}}\left[\frac{g(\rho_\ell - \rho_v)}{\rho_\ell}\right]^{0.446}\right\} \tag{11-19}$$

where ν_ℓ = liquid kinematic viscosity, and D = tube diameter. Equation 11-19 is shown as line B in Figures 11-8 and 11-9.

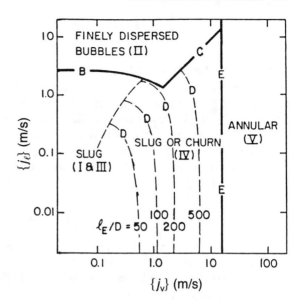

Figure 11-9 Flow regime map of Taitel et al. [36] for air–water at 25°C and 0.1 MPa in 25 mm diameter tube.

When small bubbles are packed together, the maximum allowable void fraction (which occurs when they touch each other) is equal to 0.52. At high liquid velocities (when bubble breakup can occur), however, the relative velocity is zero, so from Table 5-2 or the one-dimensional relations of Figure 5-3 we get:

$$\frac{\{\alpha\}}{1 - \{\alpha\}} = \frac{\{\beta\}}{1 - \{\beta\}} \frac{v_\ell}{v_v}$$

which, using Eq. 5-59, reduces to

$$\{\alpha\} = \frac{\{j_v\}}{\{j_v\} + \{j_\ell\}} \quad \text{for } v_\ell \simeq v_v \tag{11-20}$$

Thus the line:

$$0.52 = \frac{\{j_v\}}{\{j_v\} + \{j_\ell\}} \tag{11-21}$$

defined as line C in Figures 11-8 and 11-9, is another limiting line on the existence of the bubbly regime.

c Slug to churn flow The slug flow pattern develops from a bubbly flow pattern if enough small bubbles can coalesce to form Taylor bubbles. If the liquid slug between two Taylor bubbles is too small to be stable, the churn regime develops. Several mechanisms have been proposed for the onset of churn flow including flooding, which was suggested by Nicklin and Davidson [30] (see section II.B).

Taitel et al. [36] presented the alternative view that the churn regime is essentially a developing length region for slug flow. Thus they derived a maximum length (ℓ_e) of the churn regime, given by:

$$\frac{\ell_\epsilon}{D} = 40.6 \left(\frac{\{j\}}{\sqrt{gD}} + 0.22 \right) \tag{11-22}$$

This equation shows the developing length to be dependent on $\{j\}/\sqrt{gD}$.

For conditions not allowing the existence of bubbly or annular flow, their criterion can be used to determine the length over which a churn regime can exist before changing to a slug flow. If a position along the tube length is shorter than the developing length, churn or slug flow may be observed. If the developing length is short relative to the position of interest along the tube, slug flow alone exists. This situation is depicted in Figures 11-8 and 11-9 as lines D for given ℓ_ϵ/D.

In an earlier paper, Taitel and Dukler [35] had suggested that for $\ell_\epsilon/D > 50$ churn flow would exist if:

$$\{j_v\}/\{j\} \geq 0.85 \tag{11-23}$$

It should be noted that Eq. 11-23 implies that churn flow exists at $\{j_v\}/\{j_\ell\} \geq 5.5$ whenever $\ell_\epsilon/D > 50$. However, the more recent approach allows churn flow to exist even at lower values of $\{j_v\}/\{j_\ell\}$. For example, at $\{j_v\} = 1.0$ m/s, line D in Figure 11-9 for $\ell_\epsilon/D = 100$ implies that churn flow can exist for $\{j_\ell\} \simeq 0.5$ m/s or up to $\{j_v\}/\{j_\ell\} \simeq 2.0$. Therefore there is a substantial difference between the two approaches for defining the boundary between slug flow and churn flow.

d Slug/churn to annular flow For transition from slug/churn flow to annular flow, Taitel et al. [36] argued that the gas velocity should be sufficient to prevent liquid droplets from falling and bridging between the liquid film. The minimum gas velocity to suspend a drop is determined from a balance between the gravity and drag forces:

$$\frac{1}{2} C_d \left(\frac{\pi d^2}{4} \right) \rho_v V_v^2 = \left(\frac{\pi d^3}{6} \right) g(\rho_\ell - \rho_v) \tag{11-24}$$

or

$$V_v = \frac{2}{\sqrt{3}} \left[\frac{g(\rho_\ell - \rho_v)d}{\rho_v C_d} \right]^{1/2} \tag{11-25}$$

The drop diameter is determined from the criteria shown by Hinze [21] for maximum stable droplet sizes:

$$d = \frac{K\sigma}{\rho_v V_v^2} \tag{11-26}$$

where K = the critical Weber number, which has a value between 20 and 30. Combining Eqs. 11-25 and 11-26, we get:

$$V_v = \left(\frac{4K}{3C_d} \right)^{1/4} \frac{[\sigma g(\rho_\ell - \rho_v)]^{1/4}}{\rho_v^{1/2}} \tag{11-27}$$

If K is taken as 30 and C_d as 0.44 (note that V_v is insensitive to the exact values because of the 1/4 power), and it is assumed that in annular flow the liquid film is thin so that $j_v \simeq V_v$, Eq. 11-27 yields:

$$\frac{\{j_v\}\rho_v^{1/2}}{(\sigma g(\rho_\ell - \rho_v))^{1/4}} = 3.1 \qquad (11\text{-}28)$$

Thus the transition to annular flow is independent of the tube diameter and the liquid flow rate (depicted as line E in Figure 11-8).

The left-hand side group of Eq. 11-28 is the vapor Kutateladze number (Ku), which represents the ratio of the gas kinetic head to the inertial forces acting on liquid capillary waves with a dimension of:

$$\sqrt{\frac{\sigma}{g(\rho_\ell - \rho_v)}}$$

It is interesting that a similar criterion would be obtained for the initiation of annular flow if it were related to the onset of flow reversal (see section II.B).

2 Horizontal flow. For horizontal flow, Mandhane et al. [26] proposed a map on the basis of the gas and liquid volumetric fluxes as shown in Figure 11-10. The range of data used to produce their results appears in Table 11-1. Taitel and Dukler [34] found from a mechanistic approach similar map boundaries.

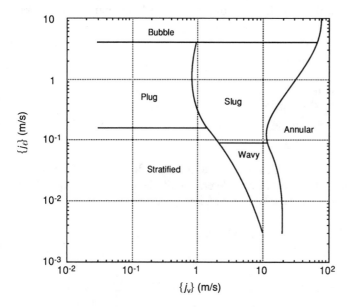

Figure 11-10 Flow map of Mandhane et al. [26] for horizontal flow.

Table 11-1 Parameter range for the flow map of Mandhane et al. [26]

Property	Parameter range
Pipe inner diameter	12.7–165.1 mm
Liquid density	705–1009 kg/m^3
Gas density	0.80–50.5 kg/m^3
Liquid viscosity	3×10^{-4} to 9×10^{-2} kg/m · s
Gas viscosity	10^{-5} to 2.2×10^{-5} kg/m · s
Surface tension	24–103 mN/m
Liquid superficial velocity	0.9–7310 mm/s
Gas superficial velocity	0.04–171 m/s

3 Regime maps in multidimensional flow. For three-dimensional flows, there is more than a single velocity component at each location. To use the one-dimensional flow maps required new approaches. Two can be considered: (1) The use of a static quantity such as the void fraction, to characterize the flow. One such flow map has been used in the RELAP-5 [33] analysis code (Fig. 11-11). (2) The use of the total flow velocity as the characteristic velocity in the flow maps. Neither approach is without limitations. For example, the effect of the heat flux from the walls on the existence of a liquid or vapor region near the wall is not included.

Example 11-1 Flow regime calculations

PROBLEM Consider a vertical tube of 17 mm I.D. and 3.8 m length. The tube is operated at steady state with the following conditions, which approximate the BWR fuel bundle condition.

$p \equiv$ operating pressure $= 7.44$ MPa
$T_{in} \equiv$ inlet temperature $= 275°C$
$T_{sat} \equiv$ saturation temperature $\simeq 290°C$
$G_m \equiv$ mass velocity $= 1700$ kg/m^2 · s
$q'' \equiv$ heat flux (constant) $= 670$ kW/m^2
$x \equiv$ outlet quality $= 0.185$

1. Using the Hewitt and Roberts flow map, determine the flow regime at the exit.
2. Using the approach of Taitel et al., determine if there is a transition from slug/churn flow to annular flow before the exit.

SOLUTION Using $T_{sat} = 290°C$, the following fluid properties pertain:

$\rho_\ell = 732.33$ kg/m^3
$\rho_v = 39.16$ kg/m^3
$\sigma = 0.0167$ N/m

1. The coordinates for the Hewitt and Roberts flow map (Fig. 11-7) are:

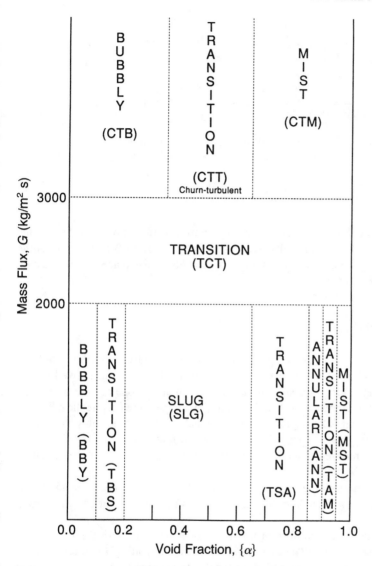

Figure 11-11 Vertical flow regime map of RELAP-5 [33].

$$\rho_\ell\{j_\ell\}^2 = \frac{[G_m(1 - x)]^2}{\rho_\ell} = \frac{[1700(1 - 0.185)]^2}{732.33} = 2621.2 \text{ kg/m} \cdot \text{s}^2$$

$$\rho_v\{j_v\}^2 = \frac{(G_m x)^2}{\rho_v} = \frac{[1700(0.185)]^2}{39.16} = 2525.8 \text{ kg/m} \cdot \text{s}^2$$

Using these values with Figure 11-7, we find that the flow at the outlet is on the border between annular and wispy-annular.

2. The transition from slug/churn to annular flow in the approach of Taitel et al. is predicted by Eq. 11-28:

$$\frac{\{j_v\}\rho_v^{1/2}}{[\sigma g(\rho_\ell - \rho_v)]^{1/4}} = Ku_v = 3.1$$

At the outlet,

$$\frac{\sqrt{2525.8}}{[(0.0167)(9.8)(732.33 - 39.16)]^{1/4}} = 15.4 > 3.1$$

Thus at the exit the flow is already well into the annular regime.

The expression in the numerator of Eq. 11-28 can be written as:

$$\{j_v\}\rho_v^{1/2} = \frac{G_m x}{\rho_v^{1/2}}$$

Because the flow quality is linearly increasing from 0 at saturation condition to the exit value of $x = 0.185$, we see that Ku_v takes on values from 0 to 15.4. Thus the transition from slug/churn flow to annular flow should take place at the location corresponding to $Ku_v = 3.1$.

III FLOW MODELS

As described in Chapter 5, many approaches can be applied to describe two-phase flow. The two degrees of freedom that can be either allowed or disallowed by the various models are: (1) *thermal nonequilibrium*, which allows one or both of the two phases to have temperatures other than the saturation temperature despite the existence of the other phase; and (2) *unequal velocities*, which allows for not only potentially higher vapor velocity but the possibility of countercurrent flows. The equal velocity assumption is often referred to as the homogeneous flow assumption or the condition of mechanical equilibrium.

Table 5-1 summarizes the possible models and the implied need for additional constitutive relations for each case. The main models can be summarized as follows

1. The homogeneous equilibrium mixture model (HEM). Here the two phases move with the same velocity. The two phases also exist only at the same temperature (i.e., they are at the saturation temperature for the prevailing pressure). The mixture can then be treated as a single fluid. This model is particularly useful for high pressure and high flow rate conditions.

2. A thermal equilibrium mixture model with an algebraic relation between the velocities (or a slip ratio) of the two phases. A widely used such model is the drift flux model. This model is different from the HEM model only in allowing the two phases to have different velocities that are related via a predetermined relation. This model is useful for low pressure and/or low flow rate flows under steady state or near-steady-state conditions.

3. The two-fluid (separate flow) model allows the phases to have thermal non-equilibrium as well as unequal velocities. In this case, each phase has three inde-

pendent conservation equations. The two-fluid model is needed for very fast transients when nonequilibrium conditions can be expected to be significant.

IV VOID–QUALITY–SLIP RELATIONS

It is possible to relate the area-averaged void fraction at any axial position to the flow quality (x). It is accomplished by defining a relation between the vapor velocity and the liquid velocity in the form of relative velocity or slip ratio. Here we discuss only the slip ratio approach. The slip ratio is the ratio of the time- and area-averaged phase velocities in direction i:

$$S_i = \frac{\{\bar{v}_{v,i}\}_v}{\{\bar{v}_{\ell,i}\}_\ell} \tag{5-48}$$

A General One-Dimensional Relations

When time fluctuations in $\{\alpha\}$ can be ignored, for one-dimensional flow Eq. 5-48 can be used to establish the relation:

$$S = \frac{x}{1 - x} \frac{\rho_\ell}{\rho_v} \frac{\{1 - \alpha\}}{\{\alpha\}} \tag{11-29}$$

An explicit value of $\{\alpha\}$ can also be obtained as:

$$\{\alpha\} = \frac{1}{1 + \dfrac{1 - x}{x} \dfrac{\rho_v}{\rho_\ell} S} \tag{5-55}$$

From Eq. 5-55 it is seen that the void fraction decreases with a higher slip ratio and higher ρ_v/ρ_ℓ. For steam–water flow, the void fraction at 6.9 MPa (1000 psi) is shown for various slip ratios in Figure 11-12. It is seen that with $S = 1$ the void fraction approaches 50% even when the quality is only 5%, whereas for higher values of S the void fraction is less. For lower pressures the density ratio ρ_v/ρ_ℓ decreases, and the void fraction is higher for the same quality and slip ratio.

The slip ratio is itself affected by the pressure (or density ratio) as well as by the profile of the void fraction within the cross-sectional area. As is shown later, the macroscopic slip ratio can be obtained from a knowledge of the local phase velocity ratio (or microscopic slip) and the void distribution profile. The slip can be obtained as a specified value or on the basis of a flow-regime-dependent approach.

For the homogeneous equilibrium model the slip ratio is taken as 1.0. Therefore the relation of the void to the quality is independent of the flow regimen and is simply given by:

$$\{\alpha\} = \frac{1}{1 + \dfrac{1 - x}{x} \dfrac{\rho_g}{\rho_f}} \quad \text{(for HEM model)} \tag{11-30}$$

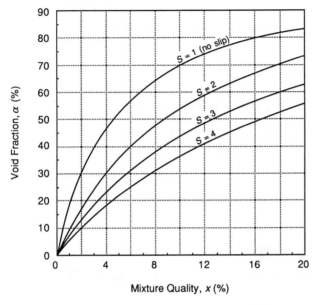

Figure 11-12 Variation of the void fraction with the quality for steam–water flow at 6.9 MPa (1000 psi).

Note that $\{\beta\}$ for saturation conditions can be given by Eq. 5-60 as:

$$\{\beta\} = \frac{x/\rho_g}{x/\rho_g + (1 - x)/\rho_f} = \frac{1}{1 + \frac{1 - x}{x} \frac{\rho_g}{\rho_f}} \tag{11-31}$$

It is clear that for the HEM model $\{\alpha\} = \{\beta\}; S = 1$.

B Drift Flux Model

A more general approach to obtain the slip ratio was suggested by Zuber and Findlay [40] by considering the average velocity of the vapor in the channel. First, let us express the local vapor velocity as the sum of the two-phase local volumetric velocity (j) and the local drift velocity of the vapor (v_{vj}). Thus:

$$v_v = j + v_{vj} \tag{11-32}$$

Hence:

$$j_v = \alpha v_v = \alpha j + \alpha(v_v - j) \tag{11-33}$$

When averaging Eq. 11-33 over the flow area, we get:

$$\{j_v\} = \{\alpha j\} + \{\alpha(v_v - j)\} \tag{11-34}$$

Note that the second term on the right-hand side is defined as the drift flux. It physically represents the rate at which vapor passes through a unit area (normal to the channel axis) that is already traveling with the flow at a velocity j. The superficial velocity, $\{j_v\}$, is obtained from:

$$\{j_v\} = C_o\{\alpha\}\{j\} + \{\alpha\} V_{vj} \tag{11-35}$$

where the concentration parameter C_o is defined by:

$$C_o \equiv \frac{\{\alpha j\}}{\{\alpha\}\{j\}} \tag{11-36}$$

and the effective drift velocity is:

$$V_{vj} \equiv \{\alpha(v_v - j)\}/\{\alpha\} \tag{11-37}$$

The value of $\{\alpha\}$ can now be obtained from Eq. 11-35 as:

$$\{\alpha\} = \frac{\{j_v\}}{C_o\{j\} + V_{vj}} \tag{11-38}$$

The void fraction can be seen as due to two effects if Eq. 11-38 is rewritten as:

$$\frac{\{\alpha\}}{\{\beta\}} = \frac{1}{C_o + \dfrac{V_{vj}}{\{j\}}} \tag{11-39}$$

The C_o terms represents the global effect due to radial nonuniform void and velocity profiles. The $V_{vj}/\{j\}$ term represents the local relative velocity effect. At high total flow rate, the local effect term is negligible, as the relative velocity is negligible (i.e., $V_{vj} \approx 0$). This condition is valid for a steam–water system, which is implied by the relations of Armand and Treschev [1] and Bankoff [2], which are of the form:

$$\{\alpha\} = K\{\beta\} \tag{11-40}$$

where the constant K is then defined by:

$$K = \frac{1}{C_o} \tag{11-41}$$

For homogeneous flow, neither local slip nor concentration profile effects are considered; therefore $C_o = 1$ and:

$$\{\alpha\} = \{\beta\} \text{ for HEM.} \tag{11-42}$$

When the local slip (or drift) is small, Armand and Treschev [1] suggested:

$$\frac{1}{C_o} = K = 0.833 + 0.05\ell n(10p) \tag{11-43}$$

where p is in MPa. Bankoff [2] suggested that:

$$K = 0.71 + 0.0001p \text{ (psi)} \tag{11-44}$$

Dix [9] suggested a general expression for C_o for all flow regimens, given by:

$$C_o = \{\beta\} \left[1 + \left(\frac{1}{\{\beta\}} - 1 \right)^b \right] \tag{11-45}$$

where:

$$b = \left(\frac{\rho_v}{\rho_\ell} \right)^{0.1} \tag{11-46}$$

It is also useful to express Eq. 11-38 in terms of the mass flow rate and quality:

$$\{\alpha\} = \frac{x\dot{m}/\rho_v A}{C_o \left(\frac{x}{\rho_v} + \frac{1-x}{\rho_\ell} \right) \frac{\dot{m}}{A} + V_{vj}} \tag{11-47}$$

where $\{j_v\}$ and $\{j\}$ have been expressed in terms of the mass flow rate and the quality as given in Eqs. 5-46a and 5-46b, or:

$$\{\alpha\} = \frac{1}{C_o \left(1 + \frac{1-x}{x} \frac{\rho_v}{\rho_\ell} \right) + \frac{V_{vj}\rho_v}{xG}} \tag{11-48}$$

Comparing Eqs. 11-48 and 5-55, we obtain:

$$S = C_o + \frac{(C_o - 1)x\rho_\ell}{(1-x)\rho_v} + \frac{V_{vj}\rho_\ell}{(1-x)G} \tag{11-49}$$

Due to nonuniform   Due to local
void distribution   velocity differential
between vapor and liquid

For uniform void distribution in the flow area, $C_o = 1$. Zuber and Findlay [40] suggested that C_o and V_{vj} are functions of the flow regime; $C_o = 1.2$ for bubbly and slug flow, and $C_o = 0$ for near zero void fraction and 1.0 for a high void fraction. The drift velocity (V_{vj}) for the bubbly and slug flow regimens can be given as:

$$V_{vj} = (1 - \{\alpha\})^n V_\infty; \quad 0 < n < 3 \tag{11-50}$$

where V_∞ is the bubble rise terminal velocity in the liquid, as given in Table 11-2. It should be noted that in the bubbly regime the drift velocity is smaller than the terminal single bubble velocity because of the presence of the other bubbles. In the slug and churn flow regimes V_{vj} and V_∞ are equal.

Ishii [23] extended the drift flux relations to annular flow. However, in annular flow there is little difference between the vapor volumetric flow rate and the total volumetric flow rate, and V_{vj} is insignificant.

Example 11-2 Void–quality–slip calculations

PROBLEM Consider water flow in a two-phase channel under saturated conditions for two cases:

Table 11-2 Values of n and V_∞ for various regimes

Regimen	n	V_∞
Small bubbles ($d < 0.5$ cm)	3	$\dfrac{g(\rho_\ell - \rho_v) d^2}{18\mu_\ell}$
Large bubbles ($d < 2$ cm)	1.5	$1.53 \left[\dfrac{\sigma g(\rho_\ell - \rho_v)}{\rho_\ell^2} \right]^{1/4}$
Churn flow	0	$1.53 \left[\dfrac{\sigma g(\rho_\ell - \rho_v)}{\rho_\ell^2} \right]^{1/4}$
Slug flow (in tube of diameter D)	0	$0.35 \sqrt{g \left(\dfrac{\rho_\ell - \rho_v}{\rho_\ell} \right) D}$

$T_{\text{sat}} = 100°C$
$T_{\text{sat}} = 270°C$

If the flow quality $x = 0.1$, for two cases of slip ratio ($S = 1$ and $S = 2$) calculate:

1. Void fraction $\{\alpha\}$
2. Vapor flux $\{j_v\}$ in terms of the mass flux G
3. Volumetric fraction $\{\beta\}$
4. Average density $\{\rho\}$ (i.e., ρ_m)

SOLUTION There are four conditions to consider ($x = 0.1$ for all cases):

Case	T_{sat} (°C)	S
1	100	1
2	100	2
3	270	1
4	270	2

Fluid properties:
For $T_{\text{sat}} = 100°C$: $p_{\text{sat}} = 0.1$ MPa, $\rho_v = 0.5978$ kg/m³, $\rho_\ell = 958.3$ kg/m³
For $T_{\text{sat}} = 270°C$: $p_{\text{sat}} = 5.5$ MPa, $\rho_v = 28.06$ kg/m³, $\rho_\ell = 767.9$ kg/m³

1. To find $\{\alpha\}$, let us use:

$$\{\alpha\} = \cfrac{1}{1 + \cfrac{1 - x}{x} \cfrac{\rho_v}{\rho_\ell} S} \tag{5-55}$$

$$\text{Case 1: } \{\alpha\} = \cfrac{1}{1 + \cfrac{1 - 0.1}{0.1} \cfrac{0.5978}{958.3}} \tag{1} = 0.9944$$

$$\text{Case 2: } \{\alpha\} = \cfrac{1}{1 + \cfrac{1 - 0.1}{0.1}\cfrac{0.5978}{958.3}(2)} = 0.9889$$

$$\text{Case 3: } \{\alpha\} = \cfrac{1}{1 + \cfrac{1 - 0.1}{0.1}\cfrac{28.06}{767.9}(1)} = 0.7525$$

$$\text{Case 4: } \{\alpha\} = \cfrac{1}{1 + \cfrac{1 - 0.1}{0.1}\cfrac{28.06}{767.9}(2)} = 0.6032$$

2. To find $\{j_v\}$:

$$\{j_v\} = \{\alpha V_v\} = \frac{x G_m}{\rho_v} \tag{5-46a}$$

Thus:

$$\text{Cases 1 and 2: } j_v = \frac{0.1}{0.5978} G_m = 0.1673 G_m$$

$$\text{Cases 3 and 4: } j_v = \frac{0.1}{28.06} G_m = 0.0036 G_m$$

3. To find $\{\beta\}$, use:

$$\{\beta\} = \cfrac{1}{1 + \cfrac{1 - x}{x}\cfrac{\rho_v}{\rho_\ell}} \tag{5-60}$$

This equation is the same as Eq. 5-55 for α with $S = 1$. Therefore we have, from solution 1:

$$\text{Cases 1 and 2 } \{\beta\} = 0.9944$$
$$\text{Cases 3 and 4 } \{\beta\} = 0.7525$$

4. To find $\{\rho\}$, use:

$$\{\rho\} = \rho_m = \{\alpha \rho_v\} + \{(1 - \alpha)\rho_\ell\} = \alpha \rho_v + (1 - \alpha)\rho_\ell \tag{5-50b}$$

Case 1: $\{\rho\} = (0.9944)0.5978 + (1 - 0.9944)958.3 = 5.961 \text{ kg/m}^3$
Case 2: $\{\rho\} = (0.9889)0.5978 + (1 - 0.9889)958.3 = 11.23 \text{ kg/m}^3$
Case 3: $\{\rho\} = (0.7525)28.06 + (1 - 0.7525)767.9 = 211.17 \text{ kg/m}^3$
Case 4: $\{\rho\} = (0.6032)28.06 + (1 - 0.6032)767.9 = 321.63 \text{ kg/m}^3$

Example 11-3 Comparison of void fraction correlations

PROBLEM Consider vertical upflow of two-phase water in a 20 mm diameter tube. Assume thermal equilibrium at 290°C (7.45 MPa). Find and plot the void fraction, $\{\alpha\}$, versus flow quality, $\{x\}$, for $0.1 \le x \le 0.9$. Use three values of mass velocity:

$G_1 = 4000 \text{ kg/m}^2 \cdot \text{s}$
$G_2 = 400 \text{ kg/m}^2 \cdot \text{s}$
$G_3 = 40 \text{ kg/m}^2 \cdot \text{s}$

Use three correlations for the void fraction:

Homogeneous, $\{\alpha\} = \{\beta\}$
Martinelli-Nelson (from Fig. 11-17, below)
Drift-flux (slug flow)

How appropriate are these correlations for the flow conditions considered?

SOLUTION Fluid properties: $v_g = 2.554 \times 10^{-2} \text{ m}^3/\text{kg}$
$v_f = 1.366 \times 10^{-3} \text{ m}^3/\text{kg}$

Homogeneous void fraction correlation

Use Eq. 11-30:

$$\{\beta\} = \{\alpha\} = \cfrac{1}{1 + \cfrac{1-x}{x} \cfrac{\rho_g}{\rho_f}} = \cfrac{1}{1 + \cfrac{1-x}{x} \cfrac{v_f}{v_g}}$$

Independent of G, we can find the following values for $\{\alpha\}$:

x	$\{\alpha\}$
0.1	0.68
0.3	0.89
0.5	0.949
0.7	0.978
0.9	0.994

Martinelli-Nelson

For $T_{sat} = 290°C$ and $p = 74.4$ bar, we can read $\{\alpha\}$ versus x directly from Figure 11-17 (below).

Independent of G:

x	$\{\alpha\}$
0.1	0.58
0.3	0.78
0.5	0.87
0.7	0.93
0.9	0.98

Drift flux

Use Eq. 11-48 expressed for saturated conditions:

$$\{\alpha\} = \cfrac{1}{C_o\left[1 + \cfrac{1-x}{x}\cfrac{\rho_g}{\rho_f}\right] + \cfrac{V_{vj}\rho_g}{xG}} = \cfrac{1}{C_o\left[1 + \cfrac{1-x}{x}\cfrac{v_f}{v_g}\right] + \cfrac{V_{vj}}{v_g\, xG}}$$

For slug flow from Eq. 11-50 and for saturated conditions, $C_o = 1.2$.

$$V_{vj} = 0.35\sqrt{\frac{g(\rho_f - \rho_g)D}{\rho_f}} = 0.35\sqrt{\cfrac{g\left(\cfrac{1}{v_f} - \cfrac{1}{v_g}\right)D}{\cfrac{1}{v_f}}} = 0.15 \text{ m/s}$$

Using $g = 9.8 \text{ m/s}^2$ and $D = 0.02$ m, we find the following values of $\{\alpha\}$:

$\{\alpha\}$	$x = 0.1$	$x = 0.3$	$x = 0.5$	$x = 0.7$	$x = 0.9$
G_1	0.56	0.74	0.79	0.81	0.83
G_2	0.52	0.71	0.77	0.80	0.82
G_3	0.31	0.54	0.64	0.70	0.73

These results are plotted in Figure 11-13.

To determine how appropriate these correlations are, we need to consider the flow conditions and the flow regimes associated with G_1, G_2, and G_3.

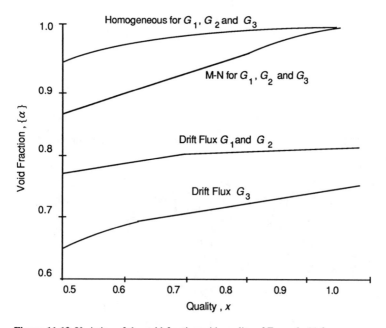

Figure 11-13 Variation of the void fraction with quality of Example 11-3.

Homogeneous model: This model is good for high-pressure, high-flow-rate conditions. Therefore the $G_1 = 4000$ kg/m² s case is more appropriate for use with this model than G_2 or G_3, as this model is best used when $G > 2000$ kg/m² s.

Martinelli-Nelson model: This model is good for $500 < G$ (kg/m² s) < 1000. Therefore the results for $G_1 = 400$ kg/m² s should be reasonably better than the other two models.

Drift-flux (slug flow) model: From Figure 11-7, based on the values of the two coordinates:

$$\rho_v\{j_v\}^2 = v_v(Gx)^2, \quad \rho_\ell\{j_\ell\}^2 = v_\ell[G(1 - x)]^2$$

we find that for G_1 and G_2 most of the calculation region gives annular flow. Hence the slug flow assumptions used are not appropriate. However, for G_3 the values fall in the bubbly and churn flow regions, and therefore the slug flow assumptions are more appropriate. It is interesting to note that it is the only model that accounts for the effect of the mass flux G on the void fraction.

V PRESSURE–DROP RELATIONS

A Pressure Gradient and Drop Components

Evaluation of the pressure drop in a two-phase flow channel can be accomplished using several models. Here we address the pressure drop prediction on the basis of the mixture models first HEM then allowing for unequal phase velocities. (This format follows the historical development of the insight into this area.) Furthermore, it should not be surprising that a large degree of empiricism is involved, as the flow is inherently chaotic.

The pressure gradient can be calculated using the momentum equation for a two-phase mixture flowing in the z direction of a one-dimensional channel:

$$\frac{\partial}{\partial t}(G_m A_z) + \frac{\partial}{\partial z}\left(\frac{G_m^2 A_z}{\rho_m^+}\right) = -\frac{\partial}{\partial z}(pA_z) - \int_{P_z} \tau_w dP_z - \rho_m g \cos \theta A_z \quad (5\text{-}66)$$

where $\theta = $ the flow angle with the upward vertical.

For steady state in a constant area channel, it is possible to simplify Eq. 5-66 to:

$$-\frac{dp}{dz} = \frac{d}{dz}\left(\frac{G_m^2}{\rho_m^+}\right) + \frac{1}{A_z}\int_{P_z} \tau_w dP_z + \rho_m g \cos \theta \quad (11\text{-}51)$$

In this equation the radial variation of p within the cross section is assumed negligible.

Equation 11-51 expresses the rate of change of the static pressure in the channel as the sum of three components due to acceleration, friction, and gravity:

$$-\frac{dp}{dz} = \left(\frac{dp}{dz}\right)_{acc} + \left(\frac{dp}{dz}\right)_{fric} + \left(\frac{dp}{dz}\right)_{gravity} \quad (11\text{-}52)$$

where:

$$\left(\frac{dp}{dz}\right)_{\text{acc}} = \frac{d}{dz}\left(\frac{G_m^2}{\rho_m^+}\right) \tag{11-53}$$

$$\left(\frac{dp}{dz}\right)_{\text{fric}} = \frac{1}{A_z}\int_{P_z} \tau_w \, dP_z = \frac{\bar{\tau}_w P_z}{A_z} \tag{11-54}$$

$$\left(\frac{dp}{dz}\right)_{\text{gravity}} = \rho_m g \cos\theta \tag{11-55}$$

where $\bar{\tau}$ is the circumferentially averaged wall shear stress. Note that dp/dz is negative for flow in the positive z direction, and $(dp/dz)_{\text{fric}}$ is always negative. The other two terms depend on the channel conditions. For heated channels ρ_m decreases as z increases, and $(dp/dz)_{\text{acc}}$ is positive. The $(dp/dz)_{\text{gravity}}$ term is positive if $\cos\theta$ is positive.

To obtain the pressure drop, we integrate the gradient equation:

$$\Delta p \equiv p_{\text{in}} - p_{\text{out}} = \int_{z_{\text{in}}}^{z_{\text{out}}} \left(-\frac{dp}{dz}\right) dz \tag{11-56}$$

or

$$\Delta p = \Delta p_{\text{acc}} + \Delta p_{\text{fric}} + \Delta p_{\text{gravity}} \tag{11-57}$$

where:

$$\Delta p_{\text{acc}} = \left(\frac{G_m^2}{\rho_m^+}\right)_{\text{out}} - \left(\frac{G_m^2}{\rho_m^+}\right)_{\text{in}} \tag{11-58}$$

$$\Delta p_{\text{fric}} = \int_{z_{\text{in}}}^{z_{\text{out}}} \frac{\bar{\tau}_w P_z}{A_z} dz \tag{11-59}$$

$$\Delta p_{\text{gravity}} = \int_{z_{\text{in}}}^{z_{\text{out}}} \rho_m g \cos\theta \, dz \tag{11-60}$$

The dynamic density ρ_m^+ can be written in terms of the flow quality if we write the momentum flux of each phase in terms of the mass flux. From Eq. 5-67 we have:

$$\frac{G_m^2}{\rho_m^+} = \{\rho_v \alpha v_v^2\} + \{\rho_\ell(1 - \alpha)v_\ell^2\} \tag{5-67}$$

Because the flow quality can be given by

$$xG_m = \{\rho_v \alpha v_v\} \text{ and } (1 - x)G_m = \{\rho_\ell(1 - \alpha)v_\ell\} \tag{11-61}$$

then:

$$\frac{G_m^2}{\rho_m^+} = \frac{x^2 G_m^2}{c_v\{\rho_v \alpha\}} + \frac{(1 - x)^2 G_m^2}{c_\ell\{\rho_\ell(1 - \alpha)\}} \tag{11-62}$$

where:

$$c_v \equiv \frac{\{\rho_v \alpha v_v\}^2}{\{\rho_v \alpha v_v^2\}\{\rho_v \alpha\}} \tag{11-63a}$$

and

$$c_\ell \equiv \frac{\{\rho_\ell(1 - \alpha)v_\ell\}^2}{\{\rho_\ell(1 - \alpha)v_\ell^2\}\{\rho_\ell(1 - \alpha)\}} \tag{11-63b}$$

For radially uniform velocity of each phase in the channel, $c_v = c_\ell = 1.0$. In that case Eq. 11-62 leads to:

$$\frac{1}{\rho_m^+} = \frac{x^2}{\{\rho_v \alpha\}} + \frac{(1 - x)^2}{\{\rho_\ell(1 - \alpha)\}} \tag{11-64}$$

for uniform velocities. We ignore the radial distribution of the velocity in the development of the acceleration term in this section.

The friction pressure gradient for two-phase flow can be expressed in a general form similar to the single-phase flow:

$$\left(\frac{dp}{dz}\right)_{\text{fric}} = \frac{\tilde{\tau}_w P_w}{A_z} \equiv \frac{f_{\text{TP}}}{D_e}\left[\frac{G_m^2}{2\,\rho_m^+}\right] \tag{11-65}$$

where:

$$D_e = \frac{4A_z}{P_w} \tag{9-56}$$

is the hydraulic equivalent diameter.

The general approach for formulating the two-phase friction factor or the friction pressure gradient, $\left(\dfrac{dp}{dz}\right)_{\text{fric}}^{\text{TP}}$, is to relate them to friction factors and multipliers defined for a single phase (either liquid or vapor) flowing at the same mass flux as the total two-phase mass flux. If the single phase is liquid, the relevant parameters are $f_{\ell o}$ and $\phi_{\ell o}^2$, whereas if the single phase is vapor the parameters are f_{vo} and ϕ_{vo}^2. These parameters are related as:

$$\left(\frac{dp}{dz}\right)_{\text{fric}}^{\text{TP}} = \phi_{\ell o}^2\left(\frac{dp}{dz}\right)_{\text{fric}}^{\ell o} = \phi_{vo}^2\left(\frac{dp}{dz}\right)_{\text{fric}}^{vo} \tag{11-66}$$

so that

$$\phi_{\ell o}^2 = \frac{\rho_\ell}{\rho_m^+}\frac{f_{\text{TP}}}{f_{\ell o}} \tag{11-67a}$$

$$\phi_{vo}^2 = \frac{\rho_v}{\rho_m^+}\frac{f_{\text{TP}}}{f_{vo}} \tag{11-67b}$$

Typically, the "liquid-only" parameters are utilized in boiling channels, and the "vapor-only" multipliers are utilized for condensing channels. Therefore in a two-phase boiling channel the friction pressure gradient is given by:

$$\left(\frac{dp}{dz}\right)_{\text{fric}}^{\text{TP}} = \phi_{\ell o}^2\frac{f_{\ell o}}{D_e}\left[\frac{G_m^2}{2\rho_\ell}\right] \tag{11-68}$$

B HEM Model Pressure Drop

The HEM model implies that the velocity of the liquid equals that of the vapor, that the two velocities are uniform within the area, and that the two phases are in thermodynamic equilibrium. If:

$$V_m \equiv \frac{G_m}{\rho_m} = \{\rho_g \alpha v_g + \rho_f (1 - \alpha) v_f\} / \{\rho_g \alpha + \rho_f (1 - \alpha)\} \qquad (11\text{-}69)$$

we can see that for the HEM model (or any model with equal phase velocities):

$$v_g = v_f = V_m \qquad (11\text{-}70)$$

Then, substituting from Eqs. 11-60, 11-61, and 11-62 into Eq. 11-64, we get:

$$\frac{1}{\rho_m^+} = \frac{x\{\rho_g \alpha\} V_m}{\{\rho_g \alpha\} G_m} + \frac{(1 - x)\{\rho_f (1 - \alpha)\} V_m}{\{\rho_f (1 - \alpha)\} G_m}$$
$$= \frac{x V_m + (1 - x) V_m}{G_m} = \frac{V_m}{G_m} = \frac{1}{\rho_m} \qquad (11\text{-}71)$$

Equation 11-71 indicates that in the HEM model $\rho_m^+ = \rho_m$.

It is useful to note that we can use the equivalence of ρ_m^+ and ρ_m in the homogeneous model to write:

$$\frac{1}{\rho_m^+} = \frac{V_m}{G_m} = \frac{\alpha V_m + (1 - \alpha) V_m}{G_m} \qquad (11\text{-}72)$$

so that using the definition of x in Eq. 11-61 we get:

$$\frac{1}{\rho_m^+} = \frac{x G_m / \rho_g + (1 - x) G_m / \rho_f}{G_m} \qquad (11\text{-}73a)$$

and

$$\frac{1}{\rho_m^+} = \frac{x}{\rho_g} + \frac{1 - x}{\rho_f} \quad \text{(HEM flow)} \qquad (11\text{-}73b)$$

From Eqs. 11-53 and 11-73b, because G_m is constant for a constant area channel, the HEM acceleration pressure gradient is given by:

$$\left(\frac{dp}{dz}\right)_{acc} = G_m^2 \frac{d}{dz} \left[\frac{1}{\rho_f} + \left(\frac{1}{\rho_g} - \frac{1}{\rho_f} \right) x \right] \qquad (11\text{-}74a)$$

or

$$\left(\frac{dp}{dz}\right)_{acc} = G_m^2 \left[\frac{dv_f}{dz} + x \left(\frac{dv_g}{dz} - \frac{dv_f}{dz} \right) + (v_g - v_f) \frac{dx}{dz} \right] \qquad (11\text{-}74b)$$

If v_g and v_f are assumed independent of z, i.e., both the liquid and the gas are assumed incompressible:

$$\left(\frac{dp}{dz}\right)_{acc} = G_m^2(v_g - v_f)\frac{dx}{dz} = G_m^2 v_{fg}\frac{dx}{dz} \tag{11-75}$$

When only the liquid compressibility is ignored:

$$\left(\frac{dp}{dz}\right)_{acc} = G_m^2 \left(x\frac{\partial v_g}{\partial p}\frac{dp}{dz} + v_{fg}\frac{dx}{dz}\right) \tag{11-76}$$

For HEM, Eq. 11-65 for the friction pressure gradient reduces to:

$$\left(\frac{dp}{dz}\right)_{fric} = \frac{f_{TP}}{D_e}\left(\frac{G_m^2}{2\rho_m}\right) \tag{11-77}$$

Thus the total pressure drop when the gas compressibility is accounted for is obtained by substituting in Eq. 11-52 from Eqs. 11-55, 11-76, and 11-77 and rearranging to get:

$$-\left(\frac{dp}{dz}\right)_{HEM} = \frac{\dfrac{f_{TP}}{D_e}\left(\dfrac{G_m^2}{2\rho_m}\right) + G_m^2 v_{fg}\dfrac{dx}{dz} + \rho_m g \cos\theta}{\left(1 + G_m^2 x\dfrac{\partial v_g}{\partial p}\right)} \tag{11-78}$$

To evaluate the friction multiplier $\phi_{\ell o}^2$ in the HEM model, two approximations are possible for the two-phase friction factor:

1. f_{TP} is equal to the friction factor for liquid single-phase flow at the same mass flux G_m as the total two-phase mass flux:

$$f_{TP} = f_{\ell o} \tag{11-79a}$$

2. f_{TP} has the same Re dependence as the single-phase $f_{\ell o}$, so that:

$$\frac{f_{TP}}{f_{\ell o}} = \frac{C_1/Re_{TP}^n}{C_1/Re_{\ell o}^n} = \left(\frac{\mu_{TP}}{\mu_f}\right)^n \tag{11-79b}$$

Turbulent flow conditions are usually assumed so that $C_1 = 0.316$ and $n = 0.25$; or $C_1 = 0.184$ and $n = 0.2$. Several formulas for the two-phase viscosity μ_{TP} have been proposed in terms of saturation properties:

$$\text{McAdams et al. [28]:}\quad \frac{\mu_{TP}}{\mu_f} = \left[1 + x\left(\frac{\mu_f}{\mu_g} - 1\right)\right]^{-1} \tag{11-80a}$$

$$\text{Cichitti et al. [7]:}\quad \frac{\mu_{TP}}{\mu_f} = \left[1 + x\left(\frac{\mu_g}{\mu_f} - 1\right)\right] \tag{11-80b}$$

$$\text{Dukler et al. [10]:}\quad \frac{\mu_{TP}}{\mu_f} = \left[1 + \beta\left(\frac{\mu_g}{\mu_f} - 1\right)\right] \tag{11-80c}$$

Note that all the above equations reduce to the proper single-phase viscosity in the extreme cases of $x = \beta = 1$ or $x = \beta = 0$.

The friction pressure drop, $\phi_{\ell o}^2$, has been defined by Eq. 11-67a. For homogeneous flow, the condition of Eq. 11-79a can now be given as:

$$\phi_{\ell o}^2 = \frac{\rho_\ell}{\rho_m} \frac{f_{TP}}{f_{\ell o}} = \frac{\rho_\ell}{\rho_m} \tag{11-81}$$

For thermodynamic equilibrium Eq. 11-81 reduces to:

$$\phi_{\ell o}^2 = \left[1 + x \left(\frac{\rho_f}{\rho_g} - 1 \right) \right] \tag{11-82}$$

Alternatively, if Eqs. 11-79b and 11-80a were used to define $f_{TP}/f_{\ell o}$, we get:

$$\phi_{\ell o}^2 = \frac{\rho_\ell}{\rho_m} \frac{f_{TP}}{f_{\ell o}} = \left[1 + x \left(\frac{\rho_f}{\rho_g} - 1 \right) \right] \left[1 + x \left(\frac{\mu_f}{\mu_g} - 1 \right) \right]^{-n} \tag{11-83}$$

Other similar relations can be obtained for the conditions 11-80b and 11-80c.

Table 11-3 gives representative values of $\phi_{\ell o}^2$ for various pressures and qualities. It is seen that $\phi_{\ell o}^2$ increases with decreasing pressure and higher quality, and the inclusion of the viscosity effect is particularly important for high quality conditions.

C Separate Flow Pressure Drop

In the general two-fluid case, the velocities and the temperatures of the two phases may be different. However, in the separate flow model, thermodynamic equilibrium conditions (and hence equal temperatures) are assumed. The model is different from the homogeneous equilibrium case only in that the velocities are not equal. Thus simple relations for the friction pressure gradient may be derived.

Table 11-3 Two-phase multiplier ($\phi_{\ell o}^2$) of various models

| p(psia) | $\phi_{\ell o}^2$ at various qualities (x) | | | | | | Source |
	$x = 0.0$	$x = 0.1$	$x = 0.2$	$x = 0.5$	$x = 0.8$	$x = 1.0$	
1020	1	2.73	4.27	8.30	11.81	13.98	Eq. 11-83, n = 0.25
1020	1	2.07	4.14	10.35	16.6	20.7	Eq. 11-82
1020	1	5.4	8.6	17.0	22.9	15.0	Martinelli-Nelson [27], Figure 11-15
738	1	3.9	6.4	12.9	18.5	21.9	Eq. 11-83, n = 0.25
738	1	2.98	5.96	14.9	23.8	29.8	Eq. 11-82
738	1	7.1	12.4	25.5	35	22.5	Martinelli-Nelson [27], Figure 11-15
291	1	8.25	14.4	29.7	42.9	51.0	Eq. 11-83, n = 0.25
291	1	8.5	17.0	42.5	67.0	85.0	Eq. 11-82
291	1	18.4	36.2	90	132	80.0	Martinelli-Nelson [27], Figure 11-15

Ignoring the liquid compressibility and assuming radially uniform phasic densities and velocities, the acceleration pressure drop for a constant mass flux is defined by substituting from Eq. 11-64 into Eq. 11-53 to get:

$$
\begin{aligned}
\left(\frac{dp}{dz}\right)_{acc} &= G_m^2 \frac{d}{dz}\left[\frac{(1-x)^2 \, v_f}{\{1-\alpha\}} + \frac{x^2 v_g}{\{\alpha\}}\right] \\
&= G_m^2 \left[-\frac{2(1-x)v_f}{\{1-\alpha\}} + \frac{2xv_g}{\{\alpha\}}\right]\left(\frac{dx}{dz}\right) \\
&\quad + G_m^2 \left[\frac{(1-x)^2 \, v_f}{\{1-\alpha\}^2} - \frac{x^2 v_g}{\{\alpha\}^2}\right]\left(\frac{d\alpha}{dz}\right) \\
&\quad + G_m^2 \frac{x^2}{\{\alpha\}} \frac{\partial v_g}{\partial p}\left(\frac{dp}{dz}\right)
\end{aligned}
\tag{11-84}
$$

Thus the total static pressure drop is given by substituting from Eqs. 11-55, 11-68, and 11-84 in Eq. 11-52 and rearranging to get:

$$
\begin{aligned}
-\left(\frac{dp}{dz}\right)_{SEP} &= \left[1 + G_m^2 \frac{x^2}{\{\alpha\}} \frac{\partial v_g}{\partial p}\right]^{-1} \\
&\quad \cdot \left\{\phi_{\ell o}^2 \frac{f_{\ell o}}{D_e}\left[\frac{G_m^2}{2\rho_f}\right] + G_m^2 \left[\frac{2xv_g}{\{\alpha\}} - \frac{2(1-x)v_f}{\{1-\alpha\}}\right]\frac{dx}{dz}\right. \\
&\quad \left. + G_m^2\left[\frac{(1-x)^2}{\{1-\alpha\}^2}v_f - \frac{x^2}{\{\alpha\}^2}v_g\right]\frac{d\alpha}{dz} + \rho_m g \cos\theta\right\}
\end{aligned}
\tag{11-85}
$$

Comparing Eqs. 11-78 and 11-85, it is seen that, because in the separate flow model the change in α is not uniquely related to the change in x, an additional term for the explicit dependence of the acceleration pressure drop on $d\alpha/dz$ appears in Eq. 11-85.

Note that under high pressures the compressibility of the gas may also be ignored, i.e.,

$$
G_m^2 \frac{x^2}{\{\alpha\}} \frac{\partial v_g}{\partial p} \ll 1
$$

This adjustment greatly simplifies the integration of Eqs. 11-78 and 11-85.

D Evaluation of Two-Phase Friction Multipliers

1 Method of Lockhart-Martinelli [25].
There are two basic assumptions here:

1. The familiar single-phase pressure drop relations can be applied to each of the phases in the two-phase flow field.
2. The pressure gradients of the two phases are equal at any axial position.

In the previous discussion the friction pressure drop was related to that of a single-phase flow at a mass flux equal to the total mixture mass flux. An alternative assumption is that the friction pressure gradient along the channel may also be predicted from the flow of each phase separately in the channel. Thus:

$$\left(\frac{dp}{dz}\right)^{\text{TP}}_{\text{fric}} \equiv \phi_\ell^2 \left(\frac{dp}{dz}\right)^{\ell}_{\text{fric}} \equiv \phi_v^2 \left(\frac{dp}{dz}\right)^{v}_{\text{fric}} \tag{11-86}$$

where $\left(\dfrac{dp}{dz}\right)^{\ell}_{\text{fric}}$ and $\left(\dfrac{dp}{dz}\right)^{v}_{\text{fric}}$ = pressure drops obtained when the liquid phase and the gas phase are assumed to flow alone in the channel at their actual flow rate, respectively.

By definition:

$$\left(\frac{dp}{dz}\right)^{\ell}_{\text{fric}} = \frac{f_\ell}{D_e} \left[\frac{G_m^2(1-x)^2}{2\rho_\ell}\right]; \left(\frac{dP}{dz}\right)^{v}_{\text{fric}} = \frac{f_v}{D_e}\left[\frac{G_m^2 x^2}{2\rho_v}\right] \tag{11-87}$$

Hence:

$$\phi_\ell^2 = \frac{f_{\text{TP}}}{f_\ell} \frac{\rho_\ell}{\rho_m^+} \frac{1}{(1-x)^2} \tag{11-88a}$$

$$\phi_v^2 = \frac{f_{\text{TP}}}{f_v} \frac{\rho_v}{\rho_m^+} \frac{1}{x^2} \tag{11-88b}$$

It is seen from Eqs. 11-67a and 11-88a that:

$$\phi_{\ell o}^2 = \frac{f_{\text{TP}}}{\rho_m^+} \frac{\rho_\ell}{f_{\ell o}} = \phi_\ell^2 \frac{f_\ell}{\rho_\ell} \frac{\rho_\ell}{f_{\ell o}} (1-x)^2 \tag{11-89}$$

but

$$\text{Re}_{\ell o} = G_m D_e / \mu_\ell \text{ and } \text{Re}_\ell = G_m(1-x)D_e/\mu_\ell$$

The relation between the friction factor and the respective Re number can be given as:

$$f_\ell \sim \left(\frac{\mu_\ell}{D_e G_\ell}\right)^n, f_v \sim \left(\frac{\mu_v}{D_e G_v}\right)^n, f_{\ell o} \sim \left(\frac{\mu_\ell}{D_e G_m}\right)^n \tag{11-90}$$

where $n = 0.25$ or 0.2 depending on the correlation used to define the relation of Eq. 11-79. Substituting from Eq. 11-90 into Eq. 11-89, we get:

$$\phi_{\ell o}^2 = \phi_\ell^2 \frac{[G_m(1-x)D_e/\mu_\ell]^{-n}}{(G_m D_e/\mu_\ell)^{-n}} (1-x)^2 = \phi_\ell^2 (1-x)^{2-n} \tag{11-91}$$

Lockhart and Martinelli defined the parameter X such that

$$X^2 \equiv \frac{(dp/dz)^{\ell}_{\text{fric}}}{(dp/dz)^{v}_{\text{fric}}} \tag{11-92}$$

Note that from Eq. 11-86, $X^2 = \phi_v^2/\phi_\ell^2$. Substituting from Eqs. 11-87 and 11-90 into Eq. 11-92, we obtain X^2 under thermal equilibrium condition. When f is taken proportional to $\text{Re}^{-0.25}$:

$$X^2 = \left(\frac{\mu_f}{\mu_g}\right)^{0.25} \left(\frac{1-x}{x}\right)^{1.75} \left(\frac{\rho_g}{\rho_f}\right) \tag{11-93}$$

for $n = 0.25$. If f is taken to be proportional to $\text{Re}^{-0.2}$, we get:

$$X^2 = \left(\frac{\mu_f}{\mu_g}\right)^{0.2} \left(\frac{1-x}{x}\right)^{1.8} \left(\frac{\rho_g}{\rho_f}\right) \tag{11-94}$$

for $n = 0.2$.

Lockhart and Martinelli [25] suggested that ϕ_ℓ and ϕ_v can be correlated uniquely as a function of X. The graphic relation is shown in Figure 11-14. Their results were obtained from data on horizontal flow of adiabatic two-component systems at low pressure. These curves can be represented by the relations:

$$\phi_\ell^2 = 1 + \frac{C}{X} + \frac{1}{X^2} \tag{11-95a}$$

$$\phi_v^2 = 1 + CX + X^2 \tag{11-95b}$$

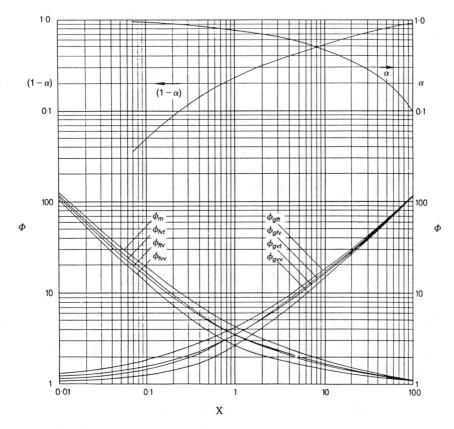

Figure 11-14 Martinelli model for pressure gradient ratios and void fractions *(From Martinelli and Nelson [27].)*

Table 11-4 Values of constant C

Liquid–gas	C
Turbulent–turbulent (tt)	20
Viscous–turbulent (vt)	12
Turbulent–viscous (tv)	10
Viscous–viscous (vv)	5

$$1 - \alpha = \frac{X}{\sqrt{X^2 + CX + 1}} \tag{11-95c}$$

where C is given in Table 11-4, dependent on the phase flow being laminar (viscous) or turbulent. In practice, the scatter of the experimental results is as large if not larger than the variation in ϕ_ℓ^2 and ϕ_v^2 owing to the flow regime assumptions. Hence only the value of $C = 20$ is usually used for all regimes (i.e., a turbulent–turbulent regime is used).

Equations 11-95a and 11-95c imply that:

$$\phi_\ell^2 = (1 - \alpha)^{-2} \tag{11-96}$$

which was theoretically derived by Chisholm [5].

2 Martinelli-Nelson [27]. Martinelli and Nelson dealt with steam–water data. Their basic assumption was that $\phi_{\ell o}^2$ can be related to the flow quality at any given pressure. Thermodynamic equilibrium and a turbulent-turbulent flow regime were assumed to exist. Using X_{tt} (turbulent-turbulent) from Lockhart-Martinelli's work, values of $\phi_{\ell o}^2$ were established (Fig. 11-15).

An analytic value of $\phi_{\ell o}^2$ for steam–water may be calculated from the expression suggested by Jones [24]:

$$\phi_{\ell o}^2 = 1.2 \left[\frac{\rho_f}{\rho_g} - 1 \right] x^{0.824} + 1.0 \tag{11-97}$$

Note that this approach assumes, as with the homogeneous approach, that the flow rate does not affect $\phi_{\ell o}^2$.

When calculating total pressure drop over a heated channel, the friction pressure drop needs to be calculated from an averaged $\phi_{\ell o}^2$ of the form:

$$\overline{\phi_{\ell o}^2} = \frac{1}{x} \int_0^x \phi_{\ell o}^2 \, dx = \frac{1}{x} \int_0^x \phi_\ell^2 (1 - x)^{1.75} \, dx$$

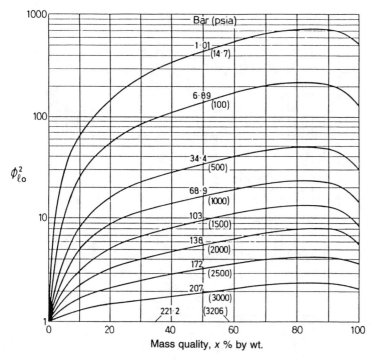

Figure 11-15 Martinelli-Nelson's $\overline{\phi_{\ell o}^2}$ as a function of quality and pressure. *(From Martinelli and Nelson [27].)*

or

$$\overline{\phi_{\ell o}^2} = \frac{1}{x} \int_0^x \left(1 + \frac{C}{X} + \frac{1}{X^2} \right) (1 - x)^{1.75} \, dx \qquad (11\text{-}98)$$

The Martinelli-Nelson results for $\overline{\phi_{\ell o}^2}$ using X_{tt} ($C = 20$) are shown in Figure 11-16.

Integrating Eq. 11-85 and ignoring the vapor compressibility leads to:

$$\Delta p = \frac{f_{\ell o}}{D_e} \frac{G_m^2}{2\rho_\ell} \int_0^L \overline{\phi_{\ell o}^2} \, dz + G_m^2 \left[\frac{(1 - x)^2}{(1 - \alpha)\rho_\ell} + \frac{x^2}{\alpha \rho_v} \right]_{\alpha,x_{z=0}}^{\alpha,x_{z=L}}$$

$$+ \int_0^L g \cos \theta \, [\rho_v \alpha + \rho_\ell (1 - \alpha)] \, dz \qquad (11\text{-}99)$$

For $\alpha = x = 0$ at $z = 0$ and constant heat addition rate over a length L, the vapor quality increases at a constant rate with distance z:

$$\frac{dx}{dz} = \text{constant} = \frac{x_{\text{out}}}{L}$$

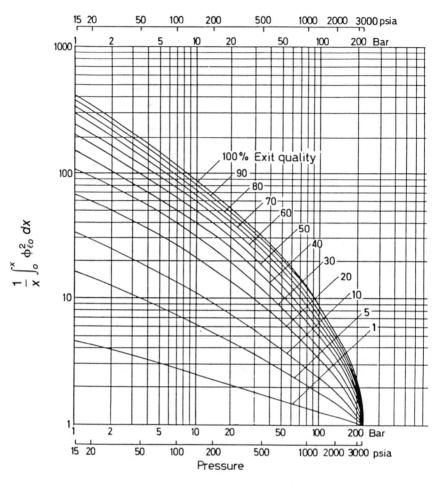

Figure 11-16 Martinelli-Nelson average $\phi_{\ell o}^2$ from zero quality to given exit quality ($C = 20$). *(From Martinelli and Nelson [27].)*

Hence Eq. 11-99 can be written as:

$$\Delta p = \frac{f_{\ell o}}{D_e}\frac{G_m^2}{2\rho_\ell}L\left[\frac{1}{x_{out}}\int_o^{x_{out}}\phi_{\ell o}^2\,dx\right] + \frac{G_m^2}{\rho_\ell}\left[\frac{(1-x_{out})^2}{(1-\alpha_{out})} + \frac{x_{out}^2\rho_\ell}{\alpha_{out}\rho_v} - 1\right]$$
$$+ \frac{L\rho_\ell\, g\cos\theta}{x_{out}}\int_o^{x_{out}}\left[1 - \left(1 - \frac{\rho_v}{\rho_\ell}\right)\alpha\right]dx \qquad (11\text{-}100)$$

or

$$\Delta p = \frac{f_{\ell o}G_m^2 L}{D_e 2\rho_\ell}(r_3) + \frac{G_m^2}{\rho_\ell}(r_2) + L\rho_\ell g\cos\theta\,(r_4) \qquad (11\text{-}101)$$

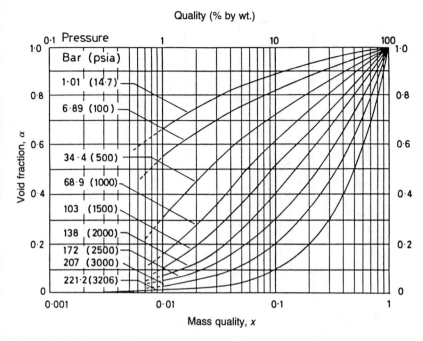

Quality (% by wt.)

Figure 11-17 Martinelli-Nelson void fraction. *(From Martinelli and Nelson [27].)*

The parameter r_3 is a length-averaged two-phase multiplier based on liquid flow. For this case of uniform axial heat addition rate, r_3 is equal to $\overline{\phi_{\ell o}^2}$, which is given by Eq. 11-98. Values of α are shown as a function of quality in Figure 11-17. Values of r_2 can be obtained using the values of α given exit quality conditions. The value of r_2 ranges from nearly 2.3 to 1500 for exit quality between 1% and 100% at atmospheric pressure and from 0.2 to 20 for 7.0 MPa pressure.

The form of Eq. 11-101 is useful for representing the pressure drop components as predicted under more general heating conditions and using other models. However, the values of r_2, r_3, and r_4 are different for nonconstant heating situations and for different models for $\phi_{\ell o}^2$ and α.

It is important to realize some shortcomings in the Martinelli-Nelson and HEM models:

1. They assume negligible effects for the mass flow rate (G_m) at a given quality, whereas experimentally $\phi_{\ell o}^2$, ϕ_ℓ^2, and ϕ_v^2 are found to depend on G_m.
2. They do not account for surface tension effects, which are important at high pressures (near the critical point).

Various investigators have found the Martinelli-Nelson results to be better than the homogeneous model for $500 < G_m < 1000$ kg/m^2 s, whereas the homogeneous

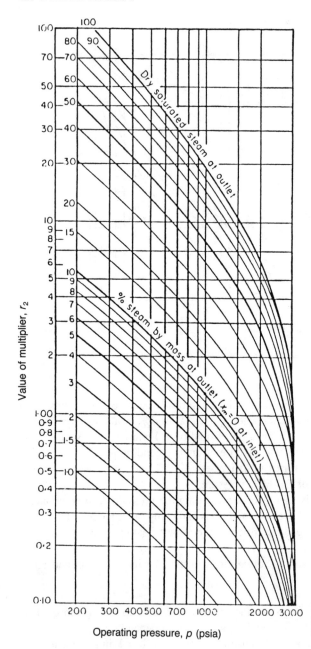

Figure 11-18 Thom's acceleration pressure drop multiplier (r_2). *(From Thom [37].)*

model appears to be better for $G_m \simeq 2000 \text{ kg/m}^2 \text{ s}$. This finding may be due to the fact that at a given quality higher mass flow rates may lead to a well mixed (dispersed) flow pattern instead of the annular (separated) flow assumed in the Martinelli-Nelson and Thom approaches.

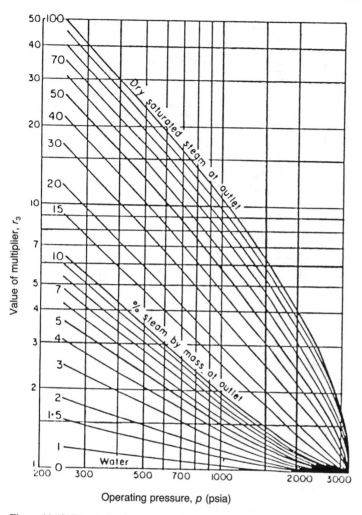

Figure 11-19 Thom's friction pressure drop multiplier (r_3). *(From Thom [37].)*

3 Thom method. Thom [37] evaluated the friction multiplier using additional results but the same representation as Martinelli-Nelson. Values of r_2, r_3, and r_4 are shown in Figures 11-18, 11-19, and 11-20, respectively.

4 Baroczy correlation. Baroczy [4] attempted to correct for the influence of G_m on $\phi_{\ell o}^2$ for fluids other than steam–water. He generated two sets of curves. The first defines $\phi_{\ell o}^2$ at a reference $G_m = 1356$ kg/m^2 s (or 10^6 lbm/ft^2 hr) as a function of fluid property index and at a given quality (Fig. 11-21). The other set provides a multiplier correction factor (Ω) for various values of G as a function of the same property index (Fig. 11-22).

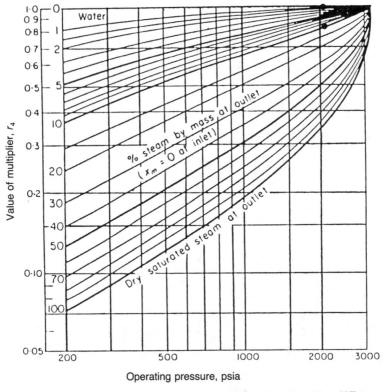

Figure 11-20 Thom's gravitational pressure drop multiplier (r_4). *(From Thom [37].)*

$$\phi_{\ell o}^2(G_m) = \Omega \, \phi_{\ell o}^2(G_{ref}) \tag{11-102}$$

5 Chisholm. Chisholm and Sutherland [6] proposed the following procedure to account for mass flow rate effects for steam–water flow in pressure tubes at pressures above 3 MPa (435 psia).

For $G_m \leq G^*$ (G^* is a reference mass flux as specified below):

$$\phi_\ell^2 = 1 + \frac{C}{X} + \frac{1}{X^2}$$

where:

$$C = \left[\lambda + (C_2 - \lambda) \left(\frac{v_{fg}}{v_g} \right)^{0.5} \right] \left[\left(\frac{v_g}{v_f} \right)^{0.5} + \left(\frac{v_f}{v_g} \right)^{0.5} \right]$$

$$\lambda = 0.5 \, [2^{(2-n)} - 2] \tag{11-103a}$$

n = power coefficient of Re in the friction factor relation

$C_2 = G^*/G_m$

For $G_m > G^*$: Let:

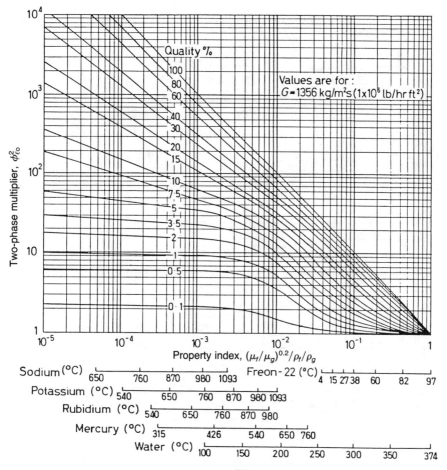

Figure 11-21 Baroczy two-phase friction multiplier ($\overline{\phi_{\ell o}^2}$). *(From Baroczy [4].)*

$$\phi_\ell^2 = \left[1 + \frac{\overline{C}}{X} + \frac{1}{X^2} \right] \psi$$

$$\overline{C} = \left(\frac{v_g}{v_f} \right)^{0.5} + \left(\frac{v_f}{v_g} \right)^{0.5}$$

$$\psi = \left[1 + \frac{C}{T} + \frac{1}{T^2} \right] \Big/ \left[1 + \frac{\overline{C}}{T} + \frac{1}{T^2} \right] \qquad (11\text{-}103b)$$

$$T = \left(\frac{x}{1-x} \right)^{\frac{2-n}{2}} \left(\frac{\mu_f}{\mu_g} \right)^{n/2} \left(\frac{v_f}{v_g} \right)^{1/2}$$

C is obtained from Eq. 11-103a using the same procedure as for $G_m \leq G^*$.

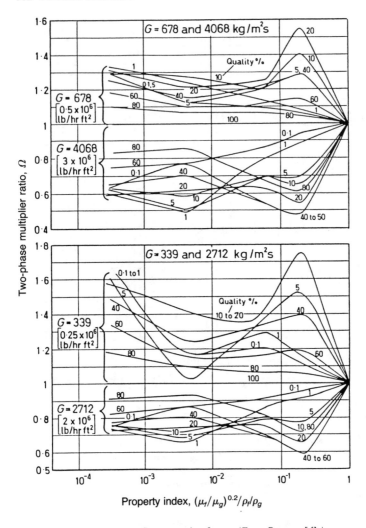

Figure 11-22 Baroczy mass flux correction factor. *(From Baroczy [4].)*

For rough tubes:

$$G^* = 1500 \text{ kg/m}^2\text{s}, \ \lambda = 1.0, \ n = 0.0$$

For smooth tubes:

$$G^* = 2000 \text{ kg/m}^2\text{s}, \ \lambda = 0.75, \ n = 0.2$$

6 Jones. Jones [24] developed an empirical approximation to the Martinelli-Nelson $\phi_{\ell o}^2$ for water. He included a synthesized flow rate effect. His proposed correlation is:

$$\phi_{\ell o}^2 = \Omega(p\ G_m) \left\{ 1.2 \left[\left(\frac{\rho_\ell}{\rho_v} \right) - 1 \right] x^{0.824} \right\} + 1.0 \qquad (11\text{-}104)$$

where: $\Omega(p, G_m)$

$$\equiv \begin{cases} 1.36 + 0.0005p + 0.1 \left(\dfrac{G_m}{10^6} \right) - 0.000714p \left(\dfrac{G_m}{10^6} \right) \text{ for } \left(\dfrac{G_m}{10^6} \right) \le 0.7 \\ \qquad\qquad\qquad\qquad\qquad\qquad\qquad\qquad\qquad\qquad (11\text{-}105a) \\ 1.26 - 0.0004p + 0.119 \left(\dfrac{10^6}{G_m} \right) + 0.00028p \left(\dfrac{10^6}{G_m} \right) \text{ for } \left(\dfrac{G_m}{10^6} \right) > 0.7 \\ \qquad\qquad\qquad\qquad\qquad\qquad\qquad\qquad\qquad\qquad (11\text{-}105b) \end{cases}$$

Equation 11-104 is shown graphically in Figure 11-23. In the above Jones' relations, G_m is in lb/hr ft^2, and p is in psia.

7 Armand-Treschev. The model proposed by Armand and Treschev [1] does not account for the mass flux effect but attempts to describe the effect of the void fraction in a more precise way. The database of the model is steam–water flowing in rough pipes 25.5 to 56 mm in diameter at pressures between 1.0 and 18 MPa. The proposed correlation for the void fraction is:

$$\frac{\{\alpha\}}{\{\beta\}} = 0.833 + 0.05 \ell n\ (10p) \qquad (11\text{-}43)$$

where p is in MPa.

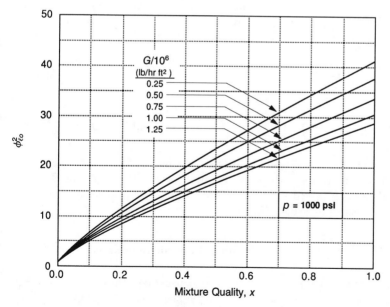

Figure 11-23 Jones' mass flux correction to Martinelli-Nelson model.

The friction factor is given by:

1. $\qquad$ For $\{\beta\} < 0.9$ and $\{\alpha\} < 0.5$, $\quad \phi_{\ell o}^2 = \dfrac{(1 - x)^{1.75}}{(1 - \{\alpha\})^{1.2}}$ $\qquad$ (11-106a)

2. $\qquad$ For $\{\beta\} < 0.9$ and $\{\alpha\} > 0.5$, $\quad \phi_{\ell o}^2 = \dfrac{0.48(1 - x)^{1.75}}{(1 - \{\alpha\})^n}$ $\qquad$ (11-106b)

where $n = 1.9 + 1.48 \times 10^{-2} p$ (in MPa).

3. $\qquad$ For $\{\beta\} > 0.9$, $\quad \phi_{\ell o}^2 = \dfrac{0.025p + 0.055}{(1 - \{\beta\})^{1.75}} (1 - x)^{1.75}$ $\qquad$ (11-106c)

8 Comparison of various models. Eighteen two-phase friction pressure drop models and correlations were tested against about 2220 experimental steam–water pressure drop measurements under adiabatic conditions and about 1230 of diabatic flow conditions by Idsinga et al. [22]. The data represented several geometries and flow regimes and had the following property ranges:

Pressure   1.7–10.3 MPa (250–1500 psia)
Mass velocity   270–4340 kg/m²s (0.2×10^6 to 3.2×10^6 lbm/hr-ft²)
Quality   Subcooled to 1.0
Equivalent diameters   2.3–33.0 mm (0.09–1.30 in.)
Geometry   Tube, annular, rectangular channel, rod array

The four models and correlations that were found to have the best performance were the Baroczy correlation, the Thom correlation, and the homogeneous model two-phase friction multipliers.

HEM 1: $\phi_{\ell o}^2 = 1 + x \left(\dfrac{v_{fg}}{v_f} \right)$ $\qquad$ (11-82)

HEM 2: $\phi_{\ell o}^2 = \left[1 + x \left(\dfrac{v_{fg}}{v_f} \right) \right] \left[1 + x \left(\dfrac{\mu_g}{\mu_f} - 1 \right) \right]^{0.25}$ (with $n = 0.25$) $\qquad$ (11-83)

For geometries with equivalent diameter of about 13 mm (0.5 in.) i.e., BWR conditions, the Baroczy correlation performed the best for $x > 0.6$ while the Armand-Treschev correlation performed the best for $x < 0.3$. Other comparison studies of various models are listed in Table 11-5.

Example 11-4 Pressure drop calculations in condensing units

PROBLEM To predict the pressure drop in condensing equipment it is possible to relate the friction pressure drop to an all-gas (single-phase) pressure drop by defining a new two-phase multiplier ϕ_{vo}^2 from Eqs. 11-76 and 11-77:

Table 11-5 Comparison of two-phase pressure drop correlations with steam–water data

Model	ESDU [11] Upflow, downflow, and horizontal flow			Friedel [16] Upflow only			Idsinga [22] Upflow and horizontal			Harwell Upflow and horizontal		
	n	e	σ	n	e	σ	n	e	σ	n	e	σ
HEM 1	—	—	—	—	—	—	2238	−9.2	26.7	—	—	—
HEM 2	1709	−13.0	32.2	2705	−19.9	42.0	2238	−26.0	22.8	4313	−23.1	34.6
Baroczy [4]	1447	4.2	30.5	2705	−11.6	36.7	2238	−8.8	29.7	4313	−2.2	30.8
Chisholm-Sutherland [6]	1536	19.0	36.0	2705	−3.8	36.0	2238	0.5	40.5	4313	13.9	34.4
Martinelli-Nelson [27]	1422	16.3	36.6	—	—	—	2238	47.8	43.7	—	—	—

Source: Adapted from Collier [8].

Correlations: n = number of data points analyzed; e = mean error (%) = $(\Delta p_{cal} - \Delta p_{exp}) \times 100/\Delta p_{exp}$; σ = standard deviation of errors about the mean (%).

$$\left(\frac{dp}{dz}\right)_{friction} = \frac{f_{TP}}{D_e}\frac{G_m^2}{2\rho_m} = \phi_{vo}^2 \frac{f_{vo}}{D_e}\frac{G_m^2}{2\rho_v} \tag{11-107}$$

1. Using the HEM model, determine the multiplier in terms of the vapor density. Assume that the two-phase mixture viscosity is equal to the vapor viscosity.
2. Evaluate the pressure drop across a horizontal tube of length L and diameter D using the HEM approach. Assume axially uniform heat flux and the following conditions:

D = 20 mm
L = 2 m
$f_{TP} = f_{go}$ = 0.005
$p_{in} \approx 1.0$ MPa (150 psi)
$\dot{m}$ = 0.1 kg/s
Inlet equilibrium quality = 0.05
Exit equilibrium quality = 0.00

SOLUTION 1. From the problem statement and recognizing that $\rho_v = \rho_g$, we can write:

$$\phi_{go}^2 = \frac{f_{TP}}{f_{go}}\frac{\rho_g}{\rho_m}$$

Using the HEM model, $\rho_m = \rho_m^+$, so Eq. 11-69 gives:

$$\frac{1}{\rho_m} = \frac{x}{\rho_g} + \frac{1-x}{\rho_f} = xv_g + (1-x)v_f$$

Thus:

$$\phi_{go}^2 = \frac{f_{TP}}{f_{go}} \rho_g [x v_g + (1 - x)v_f]$$

or

$$\phi_{go}^2 = \frac{f_{TP}}{f_{go}} \left[\frac{v_f}{v_g} + x \left(1 - \frac{v_f}{v_g} \right) \right]$$

2. The pressure gradient is found from:

$$-\frac{dp}{dz} = \left(\frac{dp}{dz} \right)_{acc} + \left(\frac{dp}{dz} \right)_{fric} + \left(\frac{dp}{dz} \right)_{gravity} \tag{11-52}$$

For a horizontal tube $(dp/dz)_{gravity} = 0$.

The pressure drop equation can be written in a manner similar to Eq. 11-100, as:

$$\Delta p = p_{in} - p_{out} = G_m^2 \left[\left(\frac{1}{\rho_m} \right)_{out} - \left(\frac{1}{\rho_m} \right)_{in} \right] + \overline{\phi_{go}^2} \frac{f_{go} L}{D} \frac{G_m^2}{2 \rho_g} \tag{11-108}$$

when the gravity term has been neglected. Now:

$$\alpha_{in} = \frac{1}{1 + \frac{1 - x}{x} \frac{\rho_g}{\rho_g}} = \frac{1}{1 + \frac{1 - 0.05}{0.05} \frac{1.1274 \times 10^{-3}}{0.1943}} = 0.901$$

$$\rho_{in} = \alpha_{in} \rho_g + (1 - \alpha)\rho_f = \frac{\alpha_{in}}{v_g} + \frac{1 - \alpha}{v_f} = \frac{0.901}{0.1943} + \frac{1 - 0.901}{1.1274 \times 10^{-3}}$$

$$\rho_{in} = 92.45 \text{ kg/m}^3$$

$$\rho_{out} = \rho_f = \frac{1}{v_f} = \frac{1}{1.1274 \times 10^{-3}} = 887 \text{ kg/m}^3$$

To find $\overline{\phi_{go}^2}$ we integrate the expression for ϕ_{go}^2, noting that $f_{TP} = f_{go}$, $x_{in} = 0.05$ to $x_{out} = 0$.

$$\overline{\phi_{go}^2} = \frac{1}{x_{in} - x_{out}} \int_{x_{in}}^{x_{out}} \phi_{go}^2 \, dx$$

$$= \frac{1}{x_{in}} \left[\frac{v_f}{v_g} x_{in} + \frac{x_{in}^2}{2} \left(1 - \frac{v_f}{v_g} \right) \right]$$

$$= \frac{v_f}{v_g} + \frac{x_{in}}{2} \left(1 - \frac{v_f}{v_g} \right)$$

$$= \frac{1.1274 \times 10^{-3}}{0.1943} + \frac{0.05}{2} \left(1 - \frac{1.1274 \times 10^{-3}}{0.1943} \right) = 0.031$$

$$G_m = \frac{\dot{m}}{A} = \frac{\dot{m}}{\frac{\pi}{4} D^2} = \frac{0.1}{\frac{\pi}{4} (0.02)^2} = 318.3 \text{ kg/m}^2 \cdot \text{s}$$

Substituting in Eq. 11-108:

$$\Delta p = (318.3)^2 \left(\frac{1}{887} - \frac{1}{92.45} \right)$$
$$+ (0.031) \frac{(0.005)(2)}{(0.02)} 0.1943 \frac{(318.3)^2}{2} = -829.1 \text{ Pa}$$

Thus for condensation the pressure at the exit is larger than at the inlet.

VI CRITICAL FLOW

A Background

The critical flow rate is the maximum flow rate that can be attained by a compressible fluid as it passes from a high-pressure region to a low-pressure region. Although the flow rate of an incompressible fluid from the high-pressure region can be increased by reducing the receiving end pressure, a compressible fluid flow rate reaches a maximum for a certain (critical) receiving end pressure. This situation occurs for both single flow of gases as well as two-phase gas–liquid flows. The observed velocity and pressure of a compressible fluid through a pipe from an upstream pressure (p_o) to a downstream (or back) pressure (p_b) for various levels of p_b are illustrated in Figure 11-24. Note that the discharge velocity for all values of p_b below the critical pressure p_{cr} is constant.

The critical flow phenomenon has been studied extensively in single-phase and two-phase systems. It plays an important role in the design of two-phase bypass systems in steam turbine plants and of venting valves in the chemical and power industries. The critical flow conditions during a loss of coolant event from a nuclear power plant comprised the motivation behind much of the experimental and theoretical studies of two-phase flows in recent years.

B Single-Phase Critical Flow

It may be useful to illustrate the analysis of critical flow by first considering single-phase flow in a one-dimensional horizontal tube.

The mass and momentum transport equations can be written as:

$$\dot{m} = \rho V A \qquad (11\text{-}109)$$

$$\frac{\dot{m}}{A} \frac{dV}{dz} = -\frac{dp}{dz} - \left(\frac{dp}{dz} \right)_{\text{friction}} \qquad (11\text{-}110)$$

If no heat is being added and friction is ignored, the flow becomes ideal (i.e., reversible), adiabatic, and hence isentropic.

The critical flow implies that $\dot{m}$ is maximum, irrespective of change in the pressure downstream, i.e.,

$$\frac{d\dot{m}}{dp} = 0 \qquad (11\text{-}111)$$

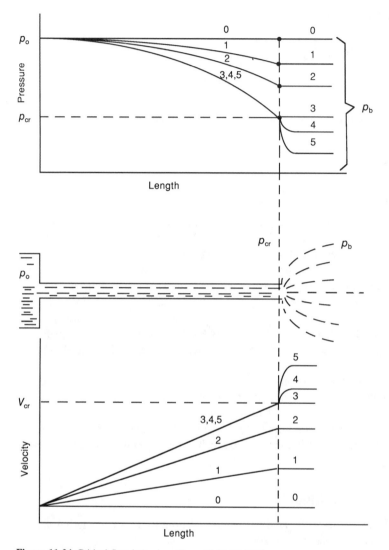

Figure 11-24 Critical flow behavior. (From El-Wakil [13])

However, from the mass equation (Eq. 11-109):

$$\frac{d\dot{m}}{dp} = VA\frac{d\rho}{dp} + \rho A\frac{dV}{dp} \tag{11-112}$$

Therefore at the critical condition:

$$\left(\frac{dV}{dp}\right)_{cr} = -\frac{V}{\rho}\frac{d\rho}{dp} \tag{11-113}$$

From the momentum equation (Eq. 11-110) for isentropic flows (no friction):

$$\frac{dV}{dp} = -\frac{A}{\dot{m}}$$

(11-114)

Therefore the critical mass flow rate is determined from:

$$\frac{\dot{m}_{cr}}{A} = G_{cr} = \frac{\rho}{V}\frac{dp}{d\rho}$$

(11-115)

or

$$G_{cr}^2 = \rho^2 \frac{dp}{d\rho}$$

(11-116)

The critical mass flux can also be given in terms of the specific volume as follows. Because:

$$\rho = \frac{1}{v}$$

(11-117)

$$\frac{dp}{d\rho} = -v^2 \frac{dp}{dv}$$

(11-118)

and from Equations 11-116 and 11-118 we get

$$G_{cr}^2 = -\frac{dp}{dv}$$

(11-119)

Note that for a single phase the speed of sound under isentropic conditions is given by:

$$c^2 = \left(\frac{dp}{d\rho}\right)_s$$

(11-120)

Thus the critical flow is identical to the mass flow at the sonic speed for isentropic conditions. This identity does not hold for two-phase flows.

Equations 11-116 and 11-120 are not easily usable for applications because they require definition of the local conditions (temperature and pressure) at the throat of a nozzle or the exit of a pipe. An alternative approach relates the velocity at the throat (V) to the thermodynamic enthalpy of the fluid at the upstream point of entry into the pipe (or the orifice) (h_o). Usually the fluid is stagnant in a tank of known temperature and pressure, and hence enthalpy. Then at the entrance of the nozzle or the discharge pipe, the stagnation enthalpy ($\overset{\circ}{h}_o$) is equal to the thermodynamic enthalpy (h_o). Along the pipe the stagnation enthalpy is constant in the absence of heat addition and friction. Thus:

$$\overset{\circ}{h}_o = h_o = h + \frac{V^2}{2}$$

(11-121)

Thus the flow rate in the pipe is given by:

$$G = \rho V = \rho \sqrt{2(h_o - h)} \tag{11-122}$$

For isentropic expansion of an ideal gas, however:

$$\frac{\rho}{\rho_o} = \left(\frac{p}{p_o}\right)^{1/\gamma} \tag{11-123}$$

and

$$\frac{T}{T_o} = \left(\frac{p}{p_o}\right)^{(\gamma-1)/\gamma} \tag{11-124}$$

where $\gamma = c_p/c_v$.
For an ideal gas:

$$dh = c_p dT \tag{4-120b}$$

Therefore, using Eqs. 11-123, 11-124, and 4-120b, Eq. 11-122 can be written as:

$$G = \rho_o \sqrt{2c_p T_o \left(1 - \frac{T}{T_o}\right)\left(\frac{p}{p_o}\right)^{2/\gamma}}$$

$$G = \rho_o \sqrt{2c_p T_o \left[\left(\frac{p}{p_o}\right)^{2/\gamma} - \left(\frac{p}{p_o}\right)^{(\gamma+1)/\gamma}\right]} \tag{11-125}$$

The mass flux varies with the ratio p/p_o, with $p = p_b$ so long as p_b is higher or equal to a critical pressure (p_{cr}). For lower values of p_b, the mass flux stays constant. The value of the mass flux can be obtained by differentiating G of Eq. 11-125 with respect to pressure.

$$\frac{\partial G}{\partial p} = 0 \tag{11-126a}$$

which leads to:

$$\left(\frac{p_b}{p_o}\right)_{cr} = \left(\frac{2}{\gamma + 1}\right)\frac{\gamma}{\gamma - 1} \tag{11-126b}$$

C Two-Phase Critical Flow

The two-phase critical flow condition can be obtained from the requirement that $\partial p/\partial z$ in the momentum equation becomes infinite. Hence for the HEM model, Eq. 11-78 implies that the denominator should be zero or:

$$(G_m^2)_{cr} = -\frac{1}{x}\frac{dp}{dv_g} \tag{11-127}$$

Note that Eq. 11-127 reduces to Eq. 11-119 for purely gas flow (i.e., $x = 1.0$).

For the separate flow model, Eq. 11-85 implies that the critical flow condition occurs when:

Table 11-6 Relaxation length observed in various critical flow experiments with flashing liquids

Source	D (mm)	L/D	L (mm)
Fauske (water)	6.35	~16	~100
Sozzi & Sutherland (water)	12.7	~10	~127
Flinta (water)	35	~3	~100
Uchida & Nariai (water)	4	~25	~100
Fletcher (freon 11)	3.2	~33	~105
Van Den Akker et al. (freon 12)	4	~22	90
Marviken data (water)	500	>0.33	<166

Source: From Fauske [14].

$$(G_m^2)_{cr} = -\frac{\{\alpha\}}{x^2}\frac{dp}{dv_g} \tag{11-128}$$

The difference in the results between the above two models is due to the slip ratio in the separate flow model.

Nonequilibrium conditions, in terms of both the velocity and the temperature differences between the two fluids play a significant role in determining the flow rate at the exit. In flashing fluids, a certain length of flow in a valve (or pipe) is needed before thermal equilibrium is achieved. In the absence of subcooling and noncondensable gases the length to achieve equilibrium appears to be on the order of 0.1 m, as illustrated in Table 11-6 [14]. For flow lengths of less than 0.1 m, the discharge rate increases strongly with decreasing length as the degree of nonequilibrium increases, and more of the fluid remains in a liquid state. Figure 11-25 provides a

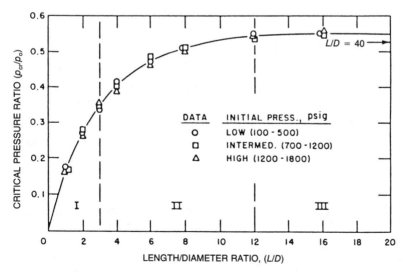

Figure 11-25 Critical or choked pressure ratio as a function of *L/D*. I, II, III = the three regions mentioned in the text of the nonequilibrium models. *(From Fauske [15].)*

demonstration of the critical pressure dependence on the length to diameter ratio (L/D).

1 Equilibrium models. Following treatment of the single-phase gas in Eq. 11-121, the enthalpy of the two-phase mixture undergoing isentropic expansion under thermal equilibrium conditions can be related by:

$$h_o = x h_g + (1 - x)h_f + x \frac{V_g^2}{2} + (1 - x) \frac{V_f^2}{2} \tag{11-129}$$

where $x =$ the flow quality under the assumption of thermodynamic equilibrium.

The entropy can also be written as:

$$s_o = x s_g + (1 - x)s_f \tag{11-130}$$

so that:

$$x = \frac{s_o - s_f}{s_g - s_f} \tag{11-131}$$

It is straightforward to determine the flow rate by relating the velocities V_g and V_f in Eq. 11-129 to the mass flux (G) and the quality, to get:

$$G_{cr} = (\rho''') \sqrt{2[h_o - xh_g - (1 - x)h_f]} \tag{11-132}$$

where:

$$\rho''' = \left\{ \left[\frac{x}{\rho_g} + \frac{(1 - x)S}{\rho_f} \right] \left[x + \frac{1 - x}{S^2} \right]^{1/2} \right\}^{-1} \tag{11-133}$$

and $S = V_g/V_f$. Therefore $G_{cr} = G(h_o, p_o, p_{cr}, S)$. Thus if the critical pressure is known, ρ_g, ρ_f, h_g, and h_f are known. Knowing s_o and p_{cr}, x can be determined using Eq. 11-131. Hence the only remaining unknown is the slip ratio (S).

The more widely used values of S are the:

HEM model:	$S = 1.0$	(11-134a)
Moody model:	$S = (\rho_f/\rho_g)^{1/3}$	(11-134b)
Fauske model:	$S = (\rho_f/\rho_g)^{1/2}$	(11-134c)

The Moody [29] model is based on maximizing the specific kinetic energy of the mixture with respect to the slip ratio:

$$\frac{\partial}{\partial S} \left[\frac{xV_g^2}{2} + \frac{(1 - x)V_f^2}{2} \right] = 0 \tag{11-135}$$

The Fauske [15] model is based on maximizing the flow momentum with respect to the slip ratio:

$$\frac{\partial}{\partial S} [xV_g + (1 - x)V_f] = 0 \tag{11-136}$$

The water discharge rate is provided in Figure 11-26 at various values of p_o and h_o, with S given by the Fauske model.

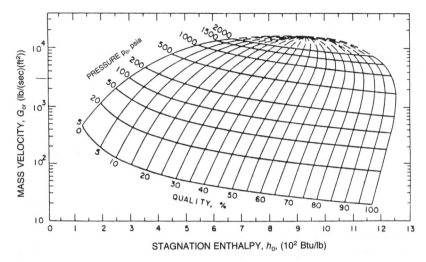

Figure 11-26 Choked flow mass velocity as a function of stagnation enthalpy and pressure for steam and water, with S given by the Fauske model.

An alternative approach to determining G_{cr} is to require that the throat conditions would maximize the flow rate with respect to changes in the throat pressure:

$$\left(\frac{\partial G}{\partial p}\right)_S = 0 \qquad \left(\frac{\partial^2 G}{\partial p^2}\right)_S < 0 \qquad (11\text{-}137)$$

This condition has been applied in the single-phase gas flow case (see Eq. 11-126a). The application of the condition in Eq. 11-137 provides a relation that, along with a specified slip ratio (any of Eqs. 11-134), also allows determination of the discharge rate for equilibrium conditions knowing only h_o and p_o.

A comparison of the predicted flow rate from various models suggests that the homogeneous flow prediction is good for pipe lengths greater than 300 mm and at pressures higher than 2.0 MPa [12]. Moody's model overpredicts the data by a factor of 2, whereas Fauske's model falls in between. When the length of the tube is such that $L/D > 40$, the HEM model appears to do better than the other models. Generally, the predictability of critical two-phase flow remains uncertain. The results of one model appear superior for one set of experiments but not others.

2 Thermal nonequilibrium cases. For orifices ($L/D = 0$) the experimental data show that the discharge rate can be given by:

$$G_{cr} = 0.61 \sqrt{2\rho_f(p_o - p_b)} \qquad (11\text{-}138)$$

For $0 < L/D < 3$ (region I in Fig. 11-25), it is possible to use the equation:

$$G_{cr} = 0.61 \sqrt{2\rho_f(p_o - p_{cr})} \qquad (11\text{-}139)$$

where p_{cr} is obtained from Figure 11-25.

For $3 < L/D < 12$ (region II), the flow is less than that predicted by Eq. 11-139 using p_{cr} of Figure 11-25.

For $40 > L/D > 12$ (region III), the flow can be predicted from Eq. 11-139 and Figure 11-25.

In the absence of significant frictional losses, Fauske [14] proposed the following correlations in SI units:

$$G_{cr} = \frac{h_{fg}}{v_{fg}} \sqrt{\frac{1}{NTc_f}} \tag{11-140}$$

where h_{fg} = vaporization enthalpy (J/kg); v_{fg} = change in specific volume (m³/kg); T = absolute temperature (°K); c_f = specific heat of the liquid (J/kg °K); and N = a nonequilibrium parameter given by:

$$N = \frac{h_{fg}^2}{2\Delta p \, \rho_f K^2 \, v_{fg}^2 \, Tc_f} + 10L \tag{11-141}$$

where $\Delta p = p_o - p_b$, Pa; K = discharge coefficient (0.61 for sharp edge); and L = length of tube, ranging from 0 to 0.1 m. For large values of L ($L \geq 0.1$ m), N = 1.0, and Eq. 11-140 becomes:

$$G_{cr} = \frac{h_{fg}}{v_{fg}} \sqrt{\frac{1}{Tc_f}} \tag{11-142}$$

When the properties are evaluated at p_o, the value of G_{cr} predicted by Eq. 11-142 is called the *equilibrium rate model* (ERM). The ERM model is compared to experimental data and other models in Figure 11-27.

The effect of subcooling on the discharge rate is simply obtained by accounting for the increased single-phase pressure drop $[p_o - p(T_o)]$ resulting from the subcooling, where the subscript o refers to stagnation conditions. For example, for flow geometries where equilibrium rate conditions prevail for saturated inlet conditions ($L \geq 0.1$ m), the critical flow rate can be stated as:

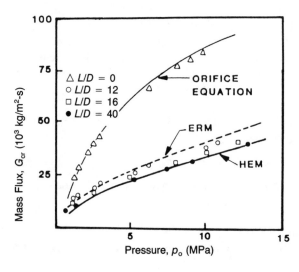

Figure 11-27 Typical flashing discharge data of initially saturated water and comparison with analytic models. *(From Fauske [14].)*

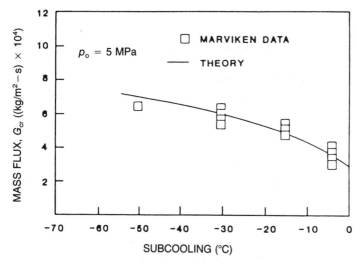

Figure 11-28 Comparison of typical Marviken data (*D* ranging from 200 to 509 mm and *L* ranging from 290 to 1809 mm) and calculated values based on Eq. 11-143. *(From Fauske [14].)*

$$G_{cr} \simeq \sqrt{2[p_o - p(T_o)]\rho_\ell + G_{ERM}^2} \qquad (11\text{-}143)$$

If the subcooling is zero [$p(T_o) = p_o$], the critical flow rate is approximated by Eq. 11-142. Good agreement is illustrated between Eq. 11-143 and various data including the large-scale Marviken data (Figs. 11-28 and 11-29).

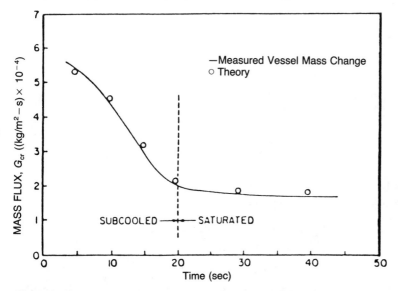

Figure 11-29 Comparison of Marviken test 4 (*D* = 509 mm and *L/D* = 3.1) and calculated values based on Eq. 11-143. *(From Fauske [14].)*

Example 11-5 Critical flow in a tube of $L/D \simeq 3$

PROBLEM Find the critical mass flux (G_{cr}) for water discharge from a pressurized vessel (5 MPa) through a tube with a diameter $(D) = 509$ mm, and length $(L) = 1580$ mm. Consider two cases for the water temperature: (1) saturated and (2) 20°C, subcooled. Use Figure 11-26 and the prescriptions of Eqs. 11-142 and 11-143. Compare the values obtained to those in Figure 11-27 and comment on the results.

SOLUTION First, find L/D:

$$\frac{L}{D} = \frac{1580}{509} = 3.1$$

Therefore the nonequilibrium case is appropriate here.

Saturated condition

From Figure 11-25: $p_{cr} \simeq 0.37 \, p_o = 0.37 \, (5 \text{ MPa})$.

$$\therefore p_{cr} = 1.85 \text{ MPa} = 268.3 \text{ psia}$$

At $p = 5$ MPa, the saturated liquid enthalpy:

$$h°(5\text{MPa}) = 1154.23 \text{ kJ/kg} = 496.26 \text{ Btu/lb}$$

From Figure 11-26, we find for $h° = 496$ Btu/lb and $p_{cr} = 268$ psi:

$$G_{cr} \simeq 3.5 \times 10^3 \text{ lb/ft}^2 \cdot \text{s} = 17,070 \text{ kg/m}^2 \cdot \text{s}$$

Now we try the more recent equations. For saturated liquid, use Eq. 11-142:

$$G_{cr} = \frac{h_{fg}}{v_{fg}} \sqrt{\frac{1}{T \, c_f}}$$

$h_{fg} = 1640.1 \text{ kJ/kg}$

$v_{fg} = 0.03815 \text{ m}^3/\text{kg}$

$T_{sat} = 263.99°\text{C} = 537.14°\text{K}$

$c_f \simeq 5.0 \text{ kJ/kg °K}$

$$\rightarrow G_{cr} = \frac{1640.1 \times 10^3}{0.03815} \sqrt{\frac{1}{(537.14)(5.0 \times 10^3)}} = 26,233 \text{ kg/m}^2 \text{ s}$$

Subcooled condition

For 20°C subcooled, use Eq. 11-143:

$$G_{cr} \simeq \sqrt{2[p_o - p(T_o)]\rho_\ell + G_{ERM}^2}$$
$$G_{ERM}^2 = (26,233 \text{ kg/m}^2 \cdot \text{s})^2$$
$$p_o = 5 \text{ MPa}$$

From $T_o = 243.99$ °C, we get $p(T_o) = 3.585$ MPa.

$$\rho_\ell \simeq 809.4 \text{ kg/m}^3 \text{ (at } T_o \text{ and } p_o)$$
$$G_{cr} \simeq \sqrt{2[5 \times 10^6 - 3.585 \times 10^6]809.4 + (26,233)^2}$$
$$= \sqrt{2290 \times 10^6 + 688 \times 10^6}$$
$$\therefore G_{cr} = 54,578 \text{ kg/m}^2 \cdot \text{s}$$

For $L/D = 3.1$, the value obtained using Figures 11-25 and 11-26 is lower than the value obtained using Eq. 11-142. The flow rate for the subcooled water is higher than that for the saturated water owing to the larger fraction of single phase in the flow. The calculated values agree with those in Figure 11-28.

REFERENCES

1. Armand, A. A., and Treshchev, G. G. *Investigation of the Resistance during the Movement of Steam–Water Mixtures in Heated Boiler Pipe at High Pressures.* AERE Lib/Trans. 81, 1959.
2. Bankoff, S. G. A variable density single fluid model for two-phase flow with particular reference to steam-water flow. *J. Heat Transfer* 82:265, 1960.
3. Bankoff, S. G., and Lee, S. C. *A Critical Review of the Flooding Literature.* NUREG/CR-3060 R2, 1983.
4. Baroczy, C. J. A systematic correlation for two-phase pressure drop. AIChE reprint no. 37. Paper presented at 8th National Heat Transfer Conference, Los Angeles, 1965.
5. Chisholm, D. *A Theoretical Basis for the Lockhart-Martinelli Correlation for Two-Phase Flow.* NEL Report No. 310, 1967.
6. Chisholm, D., and Sutherland, L. A. Prediction of pressure gradients in pipeline systems during two-phase flow. Paper 4, Presented at Symposium on Fluid Mechanics and Measurements in Two-Phase Flow Systems, University of Leeds, 1969.
7. Cichitti, A., et al. Two-phase cooling experiments—pressure drop, heat transfer and burnout measurements. *Energia Nucl.* 7:407, 1960.
8. Collier, J. G. Introduction to two-phase flow problems in the power industry. In Bergles et al. (eds.), *Two-Phase Flow and Heat Transfer in the Power and Process Industries.* New York: Hemisphere, 1981.
9. Dix, G. E. *Vapor Void Fractions for Forced Convection with Subcooled Boiling at Low Flow Rates.* NEDO-10491. General Electric Company, 1971.
10. Dukler, A. E., et al. Pressure drop and hold-up in two-phase flow: Part A—A comparison of existing correlations. Part B—An approach through similarity analysis. Paper Presented at the AIChE Meeting, Chicago, 1962. Also *AIChE J.* 10:38, 1964.
11. EDSU. *The Frictional Component of Pressure Gradient for Two-Phase Gas or Vapor/Liquid Flow through Straight Pipes.* Eng. Sci., Data Unit (ESDU), London, 1976.
12. Edwards, A. R. *Conduction Controlled Flashing of a Fluid, and the Prediction of Critical Flow Rates in One Dimensional Systems.* UKAEA Report AHSB (5) R147, Risley, England, 1968.
13. El-Wakil, M. M. *Nuclear Heat Transport.* International Textbook Co., 1971.
14. Fauske, H. K. Flashing flows—some practical guidelines for emergency releases. *Plant Operations Prog.* 4:132, 1985.
15. Fauske, H. K. The discharge of saturated water through tubes. *Chem. Eng. Sym. Series* 61:210, 1965.
16. Friedel, L. Mean void fraction and friction pressure drop: comparison of some correlations with experimental data. Presented at the European Two-Phase Flow Group Meeting, Grenoble, 1977.
17. Harmathy, T. Z. Velocity of large drops and bubbles in media of infinite or restricted extent. *AIChE J.* 6:281, 1960.
18. Hewitt, G. F., Lacey, P. M. C., and Nicholls, B. *Transitions in Film Flow in a Vertical Flow.* AERE-R4614, 1965.

19. Hewitt, G. F., and Roberts, D. N. *Studies of Two-Phase Flow Patterns by Simultaneous X-Ray and Flash Photography.* AERE-M2159, 1969.
20. Hewitt, G. F., and Wallis, G. B. *Flooding and Associated Phenomena in Falling Film Flow in a Tube.* AERE-R, 4022, 1963.
21. Hinze, J. V. Fundamentals of the hydrodynamic mechanism of splitting in dispersion processes. *AIChE J.* 1:289, 1955.
22. Idsinga, W., Todreas, N. E., and Bowring, R. An assessment of two-phase pressure drop correlations for steam–water systems. *Int. J.* Multiphase Flow *3,* 401–413 (1977).
23. Ishii, M. *One-Dimensional Drift Flux Model and Constitutive Equations for Relative Motion between Phases in Various Two-Phase Flow Regimens.* ANL-77-47, 1977.
24. Jones, A. B. *Hydrodynamic Stability of a Boiling Channel.* KAPL-2170, Knolls Atomic Power Laboratory, 1961.
25. Lockhart, R. W., and Martinelli, R. C. Proposed correlation of data for isothermal two-phase two-component flow in pipes. *Chem. Eng. Prog.* 45:no. 39, 1949.
26. Mandhane, J. M., Gregory, G. A., and Aziz, K. A flow pattern map for gas–liquid flow in horizontal pipes. *Int. J. Multiphase Flow* 1:537, 1974.
27. Martinelli, R. C., and Nelson, D. B. Prediction of pressure drop during forced circulation boiling of water. *Trans. ASME* 70:695, 1948.
28. McAdams, W. H., et al. Vaporization inside horizontal tubes. II. Benzene–oil mixtures. *Trans. ASME* 64:193, 1942.
29. Moody, F. J. Maximum two-phase vessel blowdown from pipes. *J. Heat Transfer* 88:285, 1966.
30. Nicklin, D. J., and Davidson, J. F. The onset of instability on two phase slug flow. In: *Proceedings Symposium on Two Phase Flow.* Paper 4, Institute Mechanical Engineering, London. 1962.
31. Porteous, A. Prediction of the upper limit of the slug flow regime. *Br. Chem. Eng.* 14:117, 1969.
32. Pushkin, O. L., and Sorokin, Y. L. Breakdown of liquid film motion in vertical tubes. *Heat Transfer Soviet Res.* 1:56, 1969.
33. *RELAP-5 MOD1 Code Manual.* EGG-270, 1980.
34. Taitel, Y., and Dukler, A. E. A model for predicting flow regimes transition in horizontal and near horizontal gas–liquid flow. *AIChE J.* 22:47, 1976.
35. Taitel, Y., and Dukler, A. E. Flow regime transitions for vertical upward gas–liquid flow: a preliminary approach. Presented at the AIChE 70th Annual Meeting, New York, 1977.
36. Taitel, Y., Bornea, D., and Dukler, A. E. Modelling flow pattern transitions for steady upward gas–liquid flow in vertical tubes. *AIChE J.* 26:345, 1980.
37. Thom, J. R. S. Prediction of pressure drop during forced circulation boiling of water. *Int. J. Heat Mass Transfer* 7:709, 1964.
38. Wallis, G. B. *One Dimensional, Two-Phase Flow.* New York: McGraw-Hill, 1969.
39. Wallis, G. B., and Kuo, J. T. The behavior of gas–liquid interface in vertical tubes. *Int. J. Multiphase Flow* 2:521, 1976.
40. Zuber, N., and Findlay, J. A. Average volumetric concentration in two-phase flow systems. *J. Heat Transfer* 87:453, 1965.

PROBLEMS

Problem 11-1 Methods of describing two-phase flow (section II)

Consider vertical flow through a subchannel formed by fuel rods (12.5 mm O.D.) arranged in a square array (pitch 16.3 mm). Neglect effects of spacers and unheated walls and assume the following:

1. Saturated water at 7.2 MPa (288°C); liquid density = 740 kg/m³; vapor density = 38 kg/m³.
2. For vapor fraction determination in the slug flow regime, use the drift flux representation (substituting subchannel hydraulic diameter for tube diameter).
3. Consider a simplified flow regime representation, including bubbly flow, slug flow, and annular flow. The slug-to-bubbly transition occurs at a vapor fraction of 0.15. The slug-to-annular transition occurs at a vapor fraction of 0.75.

Calculate and draw on a graph with axes: the superficial vapor velocity, $\{j_v\}$ (ordinate, from 0 to 40 m/s); and the superficial liquid velocity, $\{j_\ell\}$ (abscissa, from 0 to 3 m/s):

1. Flow regime map (indicating transitions of item 3 above)
2. Locus of all points with a mass velocity of 1300 kg/m² · s
3. Location of a point with mass velocity = 1300 kg/m² · s and with a flowing quality of 14%

Answers: 1. Slug-to-bubbly $\{j_v\} = 0.0235 + 0.22 \{j_\ell\}$; slug-to-annular $\{j_v\} = 0.966 + 9.0 \{j_\ell\}$
        2. $\{j_v\} = 34.2 - 19.5 \{j_\ell\}$
        3. $\{j_\ell\} = 1.51$ m/s
           $\{j_v\} = 4.78$ m/s

Problem 11-2: Flow regime transitions (section II)

Line B of the flow regime map of Taitel et al. [36] is given as Eq. 11-19. Derive this equation assuming that, because of turbulence, a Taylor bubble cannot exist with a diameter larger than:

$$d_{max} = k \left(\frac{\sigma}{\rho_\ell}\right)^{3/5} (\epsilon)^{-2/5}$$

where: $k = 1.14$

$$\epsilon = \left|\frac{dp}{dz}\right| \frac{j}{\rho_m}$$

Use the friction factor $(f) = 0.046 \, (jD/\nu_\ell)^{-0.2}$ and the fact that for small bubbles the critical diameter that can be supported by the surface tension is given by:

$$d_{cr} = \left[\frac{0.4\sigma}{(\rho_\ell - \rho_g)g}\right]^{0.5}$$

Problem 11-3 Regime map for horizontal flow (section II)

Construct a flow regime map based on coordinates $\{j_v\}$ and $\{j_\ell\}$ using the transition criteria of Taitel et al. [36] for a horizontal steam generator.

Assume that the length of the steam generator is 3.7 m and the characteristic hydraulic diameter is the volumetric hydraulic diameter:

$$D_v = \frac{4(\text{net free volume})}{\text{friction surface}} = 0.134 \text{ m}$$

Operation conditions
Saturated water at 282°C
Liquid density = 747 kg/m³
Vapor density = 34 kg/m³
Surface tension = 17.6×10^{-3} N/m

Answers:

Bubbly-to-slug:	$\{j_\ell\} = 3.0 \{j_v\} - 0.14$
Bubbly-to-dispersed bubbly:	$\{j_\ell\} + \{j_v\} = 5.554$
Disperse bubble-to-slug:	$\{j_\ell\} = 0.923 \{j_v\}$
Slug-to-churn:	$\{j_\ell\} + \{j_v\} = 0.5344$
Churn-to-annular:	$\{j_v\} = 1.78$

Problem 11-4 Comparison of correlations for flooding (section II)

1. It was experimentally verified that for tubes longer than about 2.5 in. flooding is independent of diameter for air–water mixtures at low-pressure conditions. Test this assertion against the Wallis [38] correlation and the Pushkin and Sorokin [32] correlation (Eq. 11-9) by finding the diameter at which they are equal at atmospheric pressure.

2. Repeat question 1 for a saturated steam–water mixture at 1000 psia.

> *Note:* For the $j_\ell = 0$ condition, the terms flooding and flow reversal are effectively synonymous.
> *Answers:* 1. $D = 1.25$ in.
>   2. $D = 0.8$ in.

The Taylor bubble diameter should be less than the tube diameter if the tube diameter is not to influence flooding. At extremely low pressure situations, the limiting value of bubble diameter can reach 2.5 in. Thus the assertion that for $D > 2.5$ in. flooding is independent of diameter is applicable.

Problem 11-5 Impact of slip model on the predicted void fraction (section IV)

In a water channel, the flow conditions are such that:

Mass flow rate: $\dot{m} = 0.29$ kg/s
Flow area: $A = 1.5 \times 10^{-4}$ m^2
Flow quality: $x = 0.15$
Operating pressure: $p = 7.2$ MPa

Calculate the void fraction using (1) the HEM model, (2) Bankoff's slip correlation, and (3) the drift flux model using Dix's correlation for C_o and V_{vj} calculated assuming churn flow conditions.

> *Answers:* 1. $\{\alpha\} = 0.775$
>   2. $\{\alpha\} = 0.631$
>   3. $\{\alpha\} = 0.703$

Problem 11-6 Calculation of a pipe's diameter for a specific pressure drop (section V)

For the vertical riser shown in Figure 11-30, calculate the steam-generation rate and the riser diameter necessary for operation at the flow conditions given below:

Geometry: downcomer height = riser height = 15.125 ft

Flow conditions:
T(steam) = 544.6°F = 284.8°C
T(feed) = 440.0°F = 226.7°C

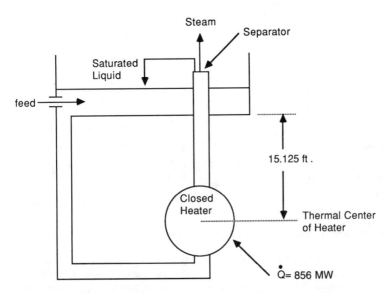

Figure 11-30 Vertical user schematic for Problem 11-6. Closed heater primary side not shown.

Steam pressure = 1000 psia = 6.7 MPa
Thermal power = 856 MW (from heater primary side to fluid in riser)

Assumptions
Homogeneous flow model is applicable.
Friction losses in the downcomer, upper plenum, lower plenum, and heater are negligible.
Riser and downcomer are adiabatic.
Quality at heater exit (x_{out}) is 0.10.

Answers: D = 3.5 ft
Steam flow = 3.8×10^6 lbm/hr = 476 kg/s

Problem 11-7 Level swell in a vessel due to two-phase conditions
Compute the level swell in a cylindrical vessel with volumetric heat generation under thermodynamic equilibrium and steady-state conditions. The vessel is filled with vertical fuel rods such that D_e = 0.04 ft (0.0122 m). The collapsed water level is 7 ft (2.134 m).

Operating conditions
No inlet flow to the vessel
p = 800 psia = 5.516 MPa
Q''' (volumetric heat source) = 4.0×10^5 BTU/hr ft^3 = 4.14×10^6 W/m^3

Water properties
σ = 0.07 N/m
h_{fg} = 1.6×10^6 J/kg
ρ_g = 1.757 lbm/ft^3 = 28.14 kg/m^3
ρ_f = 47.9 lbm/ft^3 = 767.5 kg/m^3

Assumptions

Flow regime	Selected values from Table 11-2 (n, V_∞)
Bubbly	Large bubbles
Slug flow	Slug bubbles
Churn flow	—

Select appropriate transition for flow regimes so that there is a continuous shape for the α versus z curve. For these conditions, the following result for α versus z is known:

$$\{\alpha\}V_\infty \, (1 - \{\alpha\})^{n-1} = \frac{Q'''}{h_{fg}\rho_g} z$$

Answer: Level = 3.02 m

TWO-PHASE HEAT TRANSFER

I INTRODUCTION

Boiling heat transfer is the operating mode of heat transfer in the cores of BWRs and is allowable under more restricted conditions in PWR cores. Furthermore, it is present in the steam generators and steam equipment of practically all nuclear plants. Therefore it is of interest to the analysis of normal operating thermal conditions in reactor plants. Additionally, the provision of sufficient margin between the anticipated transient heat fluxes and the critical boiling heat fluxes is a major factor in the designs of LWR cores, as are the two-phase coolant thermal conditions under accidental loss of coolant events.

In this chapter the heat transfer characteristics of two-phase flow in heated channels are outlined for a variety of flow conditions, and the mechanisms controlling the occurrence of critical heat fluxes are discussed. Where useful, boiling heat transfer under nonflow conditions, so-called pool boiling, are discussed.

II NUCLEATION SUPERHEAT

It is instructive to review some of the thermodynamic limits and basic processes involved in boiling prior to discussion of the heat-transfer relations. The change of phase from liquid to vapor can occur via homogeneous or heterogeneous nucleation. In the former, no foreign bodies aid in the change of phase, but a number of liquid molecules that attain sufficiently high energy come together to form a nucleus of vapor. Therefore homogeneous nucleation takes place only when the liquid temperature is substantially above the saturation temperature. Thus in the pressure–volume relation of a material (Fig. 12-1) the heating of the substance at a constant pressure from a subcooled state (point A) to a highly superheated state (point B') is needed before homogeneous nucleation occurs. However, if a foreign surface is available to aid in the formation of a vapor nucleus, it is possible to form the vapor at the saturated liquid state (point B). Obviously, the solid surfaces present in a core or in a steam generator become the first sites for vapor formation within the heated water.

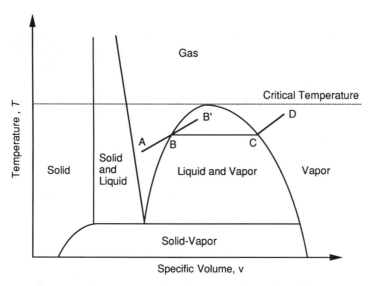

Figure 12-1 Typical temperature–volume relation for a pure substance. A–B–C–D = constant pressure line; B = saturated liquid; C = saturated vapor; and B′ = superheated liquid.

The existence of a spherical vapor bubble of radius r^* in a liquid at constant temperature (T_ℓ) and pressure (p_ℓ) requires that the vapor bubble pressure (p_b) be high enough to overcome the surface tension forces. Thus for mechanical equilibrium:

$$(p_b - p_\ell)\pi r^{*2} = \sigma 2\pi r^*$$

or

$$p_b - p_\ell = \frac{2\sigma}{r^*} \tag{12-1}$$

To relate the vapor pressure to the vapor temperature, we use the Clausius–Clapeyron relation between saturation pressure and temperature:

$$\left(\frac{dp}{dT}\right)_{sat} = \frac{h_{fg}}{T_{sat}(v_g - v_f)} \tag{12-2}$$

Assuming that the vapor exists at saturation within the bubble and that $v_g \gg v_f$, we get:

$$\frac{dp_g}{dT_g} = \frac{h_{fg}}{T_g(v_g)} \tag{12-3}$$

Applying the perfect gas law, $p_g v_g = RT_g$, Eq. 12-3 becomes:

$$\frac{dp_g}{p_g} = \frac{h_{fg}}{R T_g^2} dT_g \tag{12-4}$$

Integration between the limits of bubble pressure (p_b) and liquid pressure (p_ℓ) yields:

$$\ell n\left(\frac{p_b}{p_\ell}\right) = -\frac{h_{fg}}{R}\left(\frac{1}{T_b} - \frac{1}{T_{sat}}\right) \tag{12-5}$$

where T_b and T_{sat} = saturation temperatures at p_b and p_ℓ, respectively. Hence:

$$T_b - T_{sat} = \frac{RT_bT_{sat}}{h_{fg}} \ell n \left(\frac{p_b}{p_\ell} \right) \tag{12-6}$$

Combining Eqs. 12-6 and 12-1, we get:

$$T_b - T_{sat} = \frac{RT_bT_{sat}}{h_{fg}} \ell n \left(1 + \frac{2\sigma}{p_\ell r^*} \right) \tag{12-7}$$

Therefore a certain superheat equal to $T_b - T_{sat}$ is required to maintain a bubble. From Eq. 12-1, when $2\sigma/p_\ell r^* \ll 1$, $p_b \simeq p_\ell$. Also $RT_b/p_b = v_b \simeq v_{fg}$. Therefore the liquid superheat required to maintain a bubble of radius r^* is:

$$T_b - T_{sat} \simeq \frac{RT_bT_{sat}2\sigma}{h_{fg} \, p_b r^*} \simeq \frac{2\sigma T_{sat}v_{fg}}{h_{fg}} \left(\frac{1}{r^*} \right) \tag{12-8}$$

Equation 12-8 can be used only if the pressure is below the critical thermodynamic pressure; $p_\ell \ll p_c$. At system pressures near the critical pressure, the dependence of the saturation pressure on the temperature cannot be well represented by Eq. 12-3.

For homogeneous nucleation, r^* is on the order of molecular dimensions, so that $T_b - T_{sat}$ must be very high. For water at atmospheric pressure the calculated superheat to initiate homogeneous nucleation is on the order of 220°C, a value that is much higher than what is measured in practice. A major reduction in the superheat requirement is realized when dissolved gases are present, which reduces the required vapor pressure for bubble mechanical equilibrium. Equation 12-1 can then be generalized into the form:

$$p_v + p_{gas} - p_\ell = \frac{2\sigma}{r^*} \tag{12-9}$$

Further discussion on homogeneous nucleation can be found in Blander and Katz [9].

For nucleation at solid surfaces or on suspended bodies, microcavities (size $\sim 10^{-3}$ mm) at the surfaces act as gas storage volumes. This arrangement allows the vapor to exist in contact with subcooled liquid, provided the angular opening of the crack is small (microcavity). A solid surface contains a large number of microcavities with a distribution in sizes. Therefore the boiling process at the surface can begin if the coolant temperature near the surface is high enough that the preexisting vapor at the cavity site may attain sufficient pressure to initiate the growth of a vapor bubble at that site.

III HEAT-TRANSFER REGIMES

A Pool Boiling

The first determination of the heat-transfer regimes of pool boiling was that of Nukiyama in 1934. The boiling curve (Fig. 12-2), which remains useful for outlining the general features of pool boiling, is generally represented on a log–log plot of heat flux versus wall superheat $(T_w - T_{sat})$.

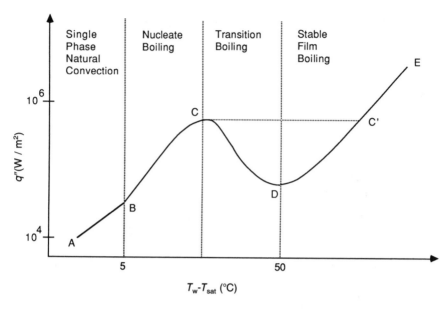

Figure 12-2 Typical pool boiling curve for water under atmospheric pressure.

As can be seen starting with point B, the formation of bubbles leads to a more effective heat transfer, so the nucleate boiling heat flux can be one to two orders of magnitude higher than that of the single-phase natural convection heat flux. However, above a given initial heat flux (point C) the nucleation rate becomes high enough to lead to formation of a continuous vapor film at the surface. Film boiling can also be established at lower heat fluxes if the surface temperature is sufficiently high, as in the region C–D–C'. However, at low wall superheat (between C and D) the formation of the film is unstable, and the region is often called *transition boiling* (from nucleate to film boiling). Point C defines the critical heat flux or the departure from nucleate boiling condition. Point D is called the minimum stable *film boiling* temperature.

It may be useful to mention that the critical heat flux (q''_{cr}) has been linked to the onset of fluidization of the pool (suspension of the liquid by the vapor streams). This form of hydrodynamic instability has been the basis of numerous correlations since it was first suggested by Kutateladze in 1952 [45]. The vapor velocity leading to the onset of fluidization is given by:

$$j_g = C_1 \left[\frac{\sigma(\rho_f - \rho_g)g}{\rho_g^2} \right]^{1/4} \tag{12-10}$$

where $C_1 = 0.16$ [45], ≈ 0.13 [64], or 0.18 [52]. The critical heat flux for saturated boiling is then given by $q''_{cr} = \rho_g j_g \, h_{fg}$. The critical heat flux values of the three correlations are shown in Figure 12-3. As evident from the figure, the water critical heat flux seems to have a peak value at a pressure near 800 psi. At low pressure, the effect of an increase in pressure is to increase the vapor density, thus reducing the velocity of the vapor and leaving the film for a given heat flux. Therefore more vaporization is possible before the vapor flux reaches such a magnitude as to prevent

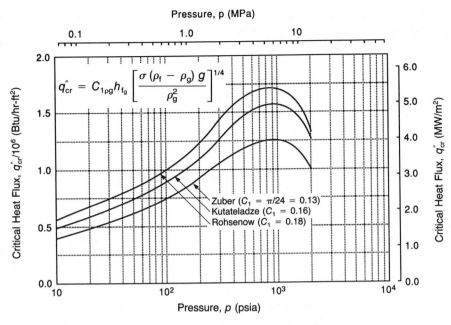

Figure 12-3 Effect of pressure on pool boiling CHF.

the liquid from reaching the surface. On the other hand, because the heat of vaporization of water decreases with pressure, the heat flux associated with this critical vapor flux for fluidization starts to decrease at high pressures, when the relative changes in the vapor density with pressure become smaller.

For minimum stable film boiling, the hydrodynamic stability analysis of Berenson [6] leads to:

$$j_g = C_2 \left[\frac{\sigma(\rho_f - \rho_g)g}{(\rho_\ell + \rho_g)^2} \right]^{1/4} \tag{12-11}$$

where $C_2 = 0.09$. Berenson went on to derive a formula for the minimum wall temperature for a stable film (T^M) that depends only on the properties of the fluid, as given in Table 12-1. Henry [33] later suggested that the interface temperature upon sudden contact between the liquid and the surface equals Berenson's minimum stable film boiling condition. Henry's correlation is recommended for highly subcooled water and for liquid metals. The correlation of Spiegler [57], on the other hand, is based on the assumption that T^M should be related to the thermodynamic critical temperature. The Kalinin et al. [40] correlation requires the contact temperature between the solid surface and the liquid to reach a certain value relative to the thermodynamic critical temperature.

B Flow Boiling

The regimes of heat transfer in a flowing system depend on a number of variables: mass flow rate, fluids employed, geometry of the system, heat flux magnitude, and

Table 12-1 Summary of correlations for prediction of minimum wall temperature to sustain film boiling (T^M)*

Author	Correlation
Berenson [6]	$T_B^M - T_{sat} = 0.127 \dfrac{\rho_{vf} h_{fg}}{k_{vf}} \left[\dfrac{g(\rho_f - \rho_g)}{\rho_f + \rho_g} \right]^{2/3} \left[\dfrac{g_c \sigma}{g(\rho_f - \rho_g)} \right]^{1/2} \left[\dfrac{\mu_{vf}}{g_c(\rho_\ell - \rho_v)} \right]^{1/3}$
Spiegler et al. [57]	$T_S^M = 0.84 \, T_c$
Kalinin et al. [40]	$\dfrac{T_K^M - T_{sat}}{T_c - T_\ell} = 0.165 + 2.48 \left[\dfrac{(\rho k c)_\ell}{(\rho k c)_w} \right]^{0.25}$
Henry [33]	$\dfrac{T_H^M - T_B^M}{T_B^M - T_\ell} = 0.42 \left[\sqrt{\dfrac{(\rho k c)_\ell}{(\rho k c)_w}} \dfrac{h_{fg}}{c_w(T_B^M - T_{sat})} \right]^{0.6}$

*(1) The subscripts given to T^M in the correlations refer to the originator(s) of the correlation. (2) The British system of units is to be used in Berenson's correlation. Absolute temperatures are to be used in the correlation of Spiegler et al. The other correlations include only dimensionless parameters. (3) Properties with the subscript vf are to be evaluated at the average temperature in the vapor film.

distribution. Figure 12-4 gives an example of some flow and heat transfer regimes. The example is for a tube with vertical upflow forced convection subjected to axially uniform heat addition. Here, as the temperature graph indicates, the independent variable for the thermal analysis is the heat flux. In another situation a fluid could be flowing through a tube with a fixed wall temperature. In that case, the heat flux would vary along the length of the tube.

Figure 12-5 illustrates the effect of the flow rate on the heat flux relation with $T_w - T_{sat}$. Compared with the pool boiling heat transfer coefficient, the forced convection heat transfer coefficient is higher, because of the contribution of forced convection to heat removal. It should be noted that while in single phase flow the heat flux is proportional to $T_w - T_{bulk}$, the nucleate boiling heat flux is proportional to $(T_w - T_{sat})^m$ where m $\simeq$ 3. The locus of the point of the boiling inception is approximately proportional to $(T_w - T_{sat})^{m-1}$.

Returning to Figure 12-4, the fluid enters the channel in a subcooled state ($T_{in} < T_{sat}$), and rises in temperature due to heat addition. Then, at a certain height, the fluid near the wall becomes superheated and can nucleate a vapor bubble while the bulk liquid temperature may still be subcooled (the flow thermodynamic quality is still negative). When subcooled boiling starts, the boiling process and the turbulence caused by the boiling enhance heat transfer and the wall temperature ceases to rise as fast as in the single phase entry region. The bulk liquid temperature continues to rise until it reaches T_{sat}, thus starting a region of saturated nucleate boiling. Here, the bubbles become numerous and as they detach may start to agglomerate into larger bubbles thus changing the flow pattern into slug or churn flow. In this regime, the wall temperature does not rise because the bulk fluid temperature is the saturation temperature, T_{sat}. In fact, due to the ever increasing turbulence near the wall because of nucleation, the wall temperature decreases slightly. As boiling continues, the bubbles may merge totally into a vapor core in the tube while the liquid remains partially in a film on the walls and partially as entrained droplets in the vapor core. This annular flow pattern leads to a forced convection heat transfer mechanism through the liquid

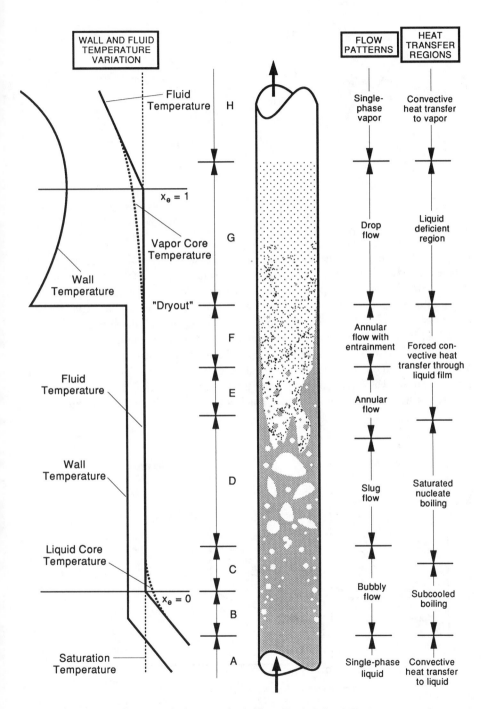

Figure 12-4 Regions of heat transfer in convective boiling. *(From Collier [17].)*

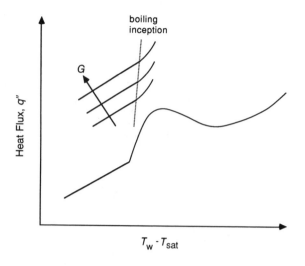

Figure 12-5 Effect of flow rate on heat flux.

film. The film may become too thin to maintain sufficient wall superheat to form bubbles. Evaporation may continue only at the liquid–vapor interface (referred to as the *nucleation suppression phenomenon*). At some point, the liquid film on the wall becomes depleted owing to vapor entrainment and evaporation, and a "dryout" condition occurs. Above that point, the flow is mostly vapor, with dispersed liquid droplets. This liquid-deficient heat transfer causes the wall temperature to rise abruptly, as vapor heat transfer is less efficient than liquid heat transfer. Thus dryout is a mechanism for the occurrence of a critical heat flux condition. The impingement of droplets on the wall may reduce the wall temperature right after the dryout position. Finally, as the thermodynamic quality reaches unity, superheated vapor exists along

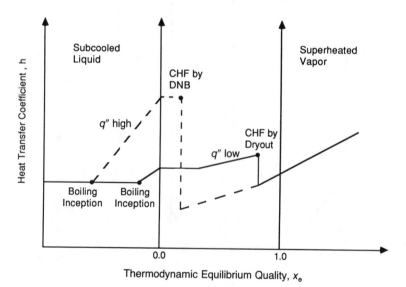

Figure 12-6 Possible variation of the heat-transfer coefficient with quality. Pressure and flow rate are fixed.

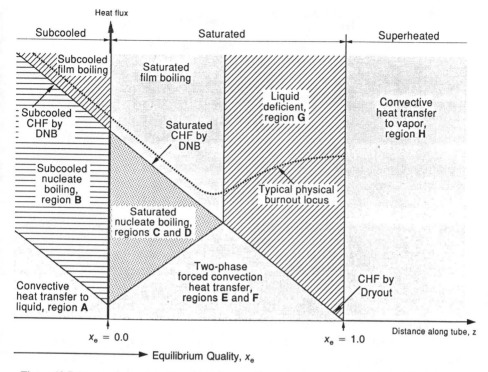

Figure 12-7 Dependence of two-phase forced convective heat-transfer regimes on quality and heat flux. A–H = regions associated with axial regions of Figure 12-4 for conditions leading to critical heat flux by dryout. *(From Collier [17].)*

with liquid droplets. With convective heat transfer mainly to vapor, the wall temperature again increases.

If the imposed heat flux is high, it is possible that the vapor-generation rate in the nucleate boiling regimen becomes so high as to establish a gas film that separates the liquid from the wall. This situation leads to departure from nucleate boiling (DNB) at the wall, similar to the critical heat flux condition in pool boiling. The maximum heat flux that can be tolerated without establishing a vapor film is the critical heat flux for these conditions.

The dependence of the heat transfer coefficient on the vapor quality for both high flux and low flux conditions is qualitatively depicted in Figure 12-6. The effect of the level of the imposed heat flux is shown in Figure 12-7. It is seen that the higher the heat flux, the lower is the thermodynamic quality at which the boiling inception and the critical heat flux occur.

IV SUBCOOLED BOILING

The boundaries and void fraction behavior in the subcooled boiling regions are now discussed using the subregions defined in Figure 12-8. The parameters of interest are the locations of the transition points (Z_{NB}, Z_D, Z_B, and Z_E) as well as predictions of

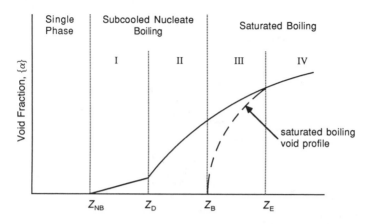

Figure 12-8 Development of area-averaged void fraction in a heated channel. Region I: $\{\alpha\}$ is small and may be neglected. Region II: Bubbles are significant; they are ejected from the wall into the bulk and collapse there. Region III: Bubbles do not collapse, as thermal equilibrium exists in the channel. Region IV: Void fraction loses the subcooling history.

the quality and void fraction. The short summary that follows gives different models for the determination of these parameters.

The subcooled boiling region begins with the onset of nucleate boiling (at $z = Z_{NB}$) while the mean (or bulk) temperature is below the saturated temperature. However, for nucleation to occur, the fluid temperature near the wall must be somewhat higher than T_{sat}. Therefore vapor bubbles begin to nucleate at the wall. Because most of the liquid is still subcooled, the bubbles do not detach but grow and collapse while attached to the wall, giving a small nonzero void fraction (region I in Fig. 12-8) that may be neglected. As the bulk of the coolant heats up, the bubbles can grow larger, and the possibility that they will detach from the wall surface into the flow stream increases. In region II ($z > Z_D$), the bubbles detach regularly and condense slowly as they move through the fluid, and the vapor voidage penetrates to the fluid bulk. As Figure 12-8 indicates, the void fraction increases significantly. The next stage, region III, is initiated when the bulk liquid becomes saturated at $z = Z_B$. The void fraction continues to increase, approaching the thermal equilibrium condition (at $z = Z_E$). This point marks the beginning of region IV, where the thermodynamic nonequilibrium history is completely lost.

A Boiling Incipience

A criterion for boiling inception (i.e., $z = Z_{NB}$) in forced flow was developed by Bergles and Rohsenow [7] based on the suggestion of Hsu and Graham [35]. Their analyses were later confirmed in a more general derivation by Davis and Anderson [21]. The basic premise is that the liquid temperature due to the heat flux near the wall must be equal to the temperature associated with the required superheat for bubble stability (Fig. 12-9). The first possible equality occurs when the two temperatures tangentially make contact, which assumes that the wall has cavities of various sizes

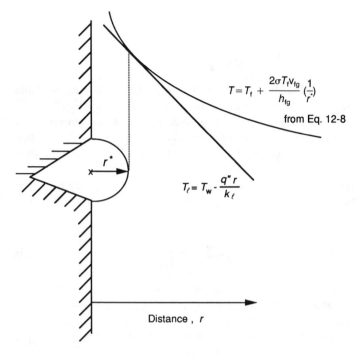

$$T = T_f + \frac{2\sigma T_f v_{fg}}{h_{fg}}\left(\frac{1}{r}\right)$$

from Eq. 12-8

$$T_\ell = T_w - \frac{q'' r}{k_\ell}$$

r^*

x

Distance , r

Figure 12-9 Critical cavity size for nucleation at a wall.

and the bubble grows at the cavity of radius r^*. The liquid temperature and the heat flux near the wall are related by:

$$q'' = -k_\ell \frac{\partial T}{\partial r} \quad \text{or} \quad \frac{\partial T}{\partial r} = -\frac{q''}{k_\ell}$$

This gradient is equal to the gradient of the required superheat (given in Eq. 12-8) at a cavity radius given by:

$$r^*_{tang} = \sqrt{\frac{2\sigma T_{sat} v_{fg} k_\ell}{h_{fg} q''}}$$

To support bubble nucleation, the liquid superheated boundary layer was assumed to extend to a thickness twice the critical cavity radius. This assumption, after some manipulation, leads to:

$$(q'')_i = \frac{k_\ell h_{fg}}{8\sigma T_{sat} v_{fg}} (T_w - T_{sat})_i^2 \tag{12-12}$$

At the point of incipience the heat flux is also given by the convection law, so that:

$$(q'')_i = h_c (T_w - T_{bulk})_i \tag{12-13}$$

Hence the incipient wall superheat is given by:

$$\frac{(T_w - T_{sat})_i^2}{(T_w - T_{bulk})_i} = \frac{1}{\Gamma} \tag{12-14}$$

where:

$$\Gamma \equiv \frac{k_\ell h_{fg}}{8\sigma T_{sat} v_{fg} h_c}$$

At low heat fluxes the point of tangency may be at a radius larger than the available size cavities in the wall. In this case, as in natural convection, Eq. 12-12 would underpredict the required superheat and $(q'')_i$. Bjorge et al. [10] recommended that a maximum cavity for most surfaces in contact with water be $r_{max} = 10^{-6}$ m. When the liquid has a good surface wetting ability, the apparent cavity size is smaller than the actual size.

Rohsenow [51] showed that when the single-phase point of tangency occurs at the maximum available cavity radius the convective heat-transfer coefficient (h_c) is given by:

$$h_c = \frac{k_\ell/r_{max}}{1 + \sqrt{1 + 4\Gamma(T_{sat} - T_{bulk})}} \qquad (12\text{-}15)$$

If h_c is less than this magnitude, Eq. 12-12 would not be reliable. When the liquid temperature T_{bulk} is at saturation, the value of h_c is $k_\ell/2r_{max}$, consistent with the assumption of a superheat boundary layer thickness equal to twice the cavity radius.

For water at a pressure between 0.1 and 13.6 MPa, Bergles and Rohsenow [7] provided the following empirical nucleation criterion:

$$(q'')_i = 15.6\, p^{1.156}\, (T_w - T_{sat})_i^{2.3/p^{0.0234}} \qquad (12\text{-}16)$$

where p is in psi; T_w and T_{sat} are in °F; and q'' is in BTU/hr ft². This correlation assumes the availability of large wall microcavities and ignores the surface finish effects. The incipient heat flux from Eq. 12-16 is relatively low, as this relation was based on visual observation of the first nucleation occurrence rather than on observation of wall temperature response, which requires a significant number of nucleation bubbles.

B Net Vapor Generation (Bubble Departure)

The point at which bubbles can depart from the wall before they suffer condensation (Z_D) has been proposed to be either hydrodynamically controlled or thermally controlled. Among the early proposals for thermally controlled departure are those by Griffith et al. [27], Bowring [12], Dix [23], and Levy [47]. Griffith et al. [27] proposed:

$$T_{sat} - T_{bulk} = \frac{q''}{5h_{\ell o}} \qquad (12\text{-}17a)$$

where $h_{\ell o}$ = the heat-transfer coefficient of single-phase liquid flowing at the same total mass flow rate.

Bowring [12] proposed:

$$T_{sat} - T_{bulk} = \frac{\eta \, q''}{G/\rho_f} \tag{12-17b}$$

where $\eta = 0.94 + 0.00046 \, p$ ($156 < p < 2000$ psia); T is in °F, G is in $lb_m/hr \, ft^2$, ρ_f is in lb_m/ft^3, and q'' is in Btu/hr ft^2.

Dix [23] proposed:

$$T_w - T_{sat} = 0.00135 \, \frac{q''}{k_\ell} (Re_\ell)^{1/2} \tag{12-17c}$$

The above three criteria are based on the assumption that at Z_D the wall heat flux is balanced by heat removal due to liquid subcooling.

Levy [47] introduced a hydrodynamically based model, assuming that the bubble detachment is primarily the result of drag (or shear) force overcoming the surface tension force. Staub [58] added the effect of buoyancy to the Levy model.

More recently, Saha and Zuber [53] postulated that both the hydrodynamic and the heat-transfer mechanisms may apply. Thus in the low mass flow region, the heat diffusion controls the condensation process and the departure process is heat-transfer-limited, signified by the Nusselt number:

$$Nu = \frac{q'' \, D_e}{k_\ell (T_{sat} - T_{bulk})} \tag{12-18}$$

whereas for high flow rates both the heat transfer and the hydrodynamics are controlling, signified by the Stanton number:

$$ST = \frac{q''}{G \, c_{p\ell}(T_{sat} - T_{bulk})} \tag{12-19}$$

The data from various sources were plotted against the Peclet number (Fig. 12-10), where:

$$Pe = \frac{Nu}{St} = \frac{G \, D_e \, c_{p\ell}}{k_\ell} \tag{12-20}$$

in rectangular, annular, and circular tubes as well as for some freon data. They developed the following criteria:

For $Pe < 7 \times 10^4$:

$$(Nu)_{Dep} = 455 \quad \text{or} \quad T_{sat} - T_{bulk} = 0.0022 \left(\frac{q'' D_e}{k_\ell}\right) \tag{12-21a}$$

For $Pe > 7 \times 10^4$:

$$(St)_{Dep} = 0.0065 \quad \text{or} \quad T_{sat} - T_{bulk} = 154 \left(\frac{q''}{G \, c_{p\ell}}\right) \tag{12-21b}$$

The data used by Saha and Zuber [53] covered the following range of parameters for water: $p = 0.1$ to 13.8 MPa; $G = 95$ to 2760 kg/m^2 s; and $q'' = 0.28$ to 1.89 MW/m^2.

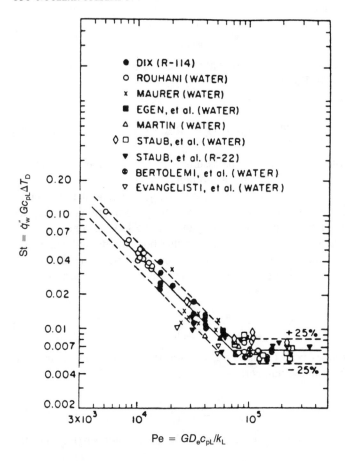

Figure 12-10 Bubble departure conditions as a function of the Peclet number. *(From Saha and Zuber [53].)*

C Subcooled Flow Quality and Void Fraction

Attempts to mechanistically determine the flow quality in the subcooled region have been reviewed by Lahey and Moody [46]. No completely satisfactory approach has been found, although the qualitative aspects of this region are well understood. The currently accepted approach for modeling this region is the profile-fit approach. An example of such approaches is the profile suggested by Levy [47], in which the flow quality $x(z) = 0$ at Z_D and approaches the equilibrium quality asymptotically:

$$x(z) = x_e(z) - x_e(Z_D) \exp\left[\frac{x_e(z)}{x_e(Z_D)} - 1\right]$$
(12-22)

The void fraction may then be predicted from the void drift model:

$$\{\alpha(z)\} = \cfrac{1}{C_o \left\{ 1 + \cfrac{1 - x(z)}{x(z)} \cfrac{\rho_g}{\rho_\ell} \right\} + \cfrac{V_{gj}}{xG} \rho_g} \tag{11-48}$$

where C_o can be obtained from the suggestion of Dix [23]:

$$C_o = \{\beta\} \left[1 + \left(\frac{1}{\{\beta\}} - 1 \right)^b \right] \tag{11-45}$$

$$b = \left(\frac{\rho_g}{\rho_\ell} \right)^{0.1} \tag{11-46}$$

and β = volumetric flow fraction of vapor.

Lahey and Moody [46] suggested that a general expression for V_{gj} may be given by:

$$V_{gj} = 2.9 \left[\left(\frac{\rho_\ell - \rho_g}{\rho_\ell^2} \right) \sigma g \right]^{0.25} \tag{12-23}$$

which is about twice the value of the bubble terminal velocity (V_∞) in a churn flow regime (see Table 11-2).

V SATURATED BOILING

The heat transfer in the early region of saturated boiling depends on the degree of bubble formation at the solid wall. So long as nucleation is present, it dominates the heat-transfer rate. Hence the correlations developed for estimating the heat flux associated with incipient boiling are still applicable in this region. Thus the heat flux associated with subcooled nucleate boiling and low quality saturated nucleate boiling are equal. However, when the flow quality is high, the liquid film becomes thin owing to evaporation and entrainment of droplets. The heat removal from the liquid film to the vapor core becomes efficient so that the nucleation within the film may be suppressed. Evaporation occurs mainly at the liquid film–vapor interface. Most correlations used for the annular flow heat-transfer coefficient are found empirically.

The heat-transfer rate in the saturated boiling region is expressed as:

$$q'' = h_{2\phi}(T_w - T_{sat}) \tag{12-24}$$

because the bulk fluid temperature is at saturation conditions. The two-phase heat-transfer coefficient ($h_{2\phi}$) is commonly formulated as the sum of a term due to nucleate boiling (h_{NB}) and a term due to convection heat transfer (h_c):

$$h_{2\phi} = h_{NB} + h_c \tag{12-25}$$

A number of authors have utilized this two-term approach by formulating the two-phase heat transfer coefficient as a multiple of $h_{\ell o}$, the single-phase liquid heat-transfer coefficient for the same total mass flux, i.e.,

Table 12-2 Values of constants for saturated flow boiling heat-transfer coefficient in Eq. 12-26

Author	a_1	a_2	b
Dengler and Addoms* [22]	0	3.5	0.5
Bennett et al.* [4]	0	2.9	0.66
Schrock and Grossman [54]	7400	1.11	0.66
Collier and Pulling [20]	6700	2.34	0.66

*Correlations appropriate only in the annular flow regimen; hence $a_1 = 0$.

$$\frac{h_{2\phi}}{h_{\ell o}} = a_1 \frac{q''}{G \, h_{fg}} + a_2 \, X_{tt}^{-b} \tag{12-26}$$

where a_1, a_2, b = empirical constants whose values in various correlations are summarized in Table 12-2
and

$$\frac{1}{X_{tt}} = \left(\frac{x}{1-x}\right)^{0.9} \left(\frac{\rho_f}{\rho_g}\right)^{0.5} \left(\frac{\mu_g}{\mu_f}\right)^{0.1} \tag{11-94}$$

Two simpler frequently used correlations for the nucleate boiling region (subcooled as well as saturated) for water at a pressure between 500 and 1000 psi are those of Jens and Lottes [39] and Thom et al. [59]. They are, respectively:

$$\frac{q''}{10^6} = \frac{\exp{(4p/900)}}{(60)^4} (T_w - T_{sat})^4 \tag{12-27a}$$

and

$$\frac{q''}{10^6} = \frac{\exp{(2p/1260)}}{(72)^2} (T_w - T_{sat})^2 \tag{12-28a}$$

where q'' is in BTU/hr ft^2, p is in psi, and T is in °F. In SI units the respective equations are:

$$q'' = \frac{\exp{(4p/6.2)}}{(25)^4} (T_w - T_{sat})^4 \tag{12-27b}$$

and

$$q'' = \frac{\exp{(2p/8.7)}}{(22.7)^2} (T_w - T_{sat})^2 \tag{12-28b}$$

where q'' is in MW/m^2, p is in MPa, and T is in °C.

It should be noted that the lower slope of Thom et al.'s correlation is due to the limitation of the data they used to the early part of the boiling curve usually encountered in conventional boilers.

A popular composite correlation to cover the entire range of saturated boiling is that of Chen [14]. His correlation, which is widely used, is expressed in the form of Eq. 12-25.

The convective part, h_c, is a modified Dittus–Boelter correlation given by:

$$h_c = 0.023 \left(\frac{G(1 - x)D_e}{\mu_f} \right)^{0.8} (Pr_f)^{0.4} \frac{k_f}{D_e} F \qquad (12\text{-}29)$$

The factor F accounts for the enhanced flow and turbulence due to the presence of vapor. F was graphically determined, as shown in Figure 12-11. It can be approximated by:

$$F = 1 \qquad \text{for } \frac{1}{X_{tt}} < 0.1$$

$$F = 2.35 \left(0.213 + \frac{1}{X_{tt}} \right)^{0.736} \qquad \text{for } \frac{1}{X_{tt}} > 0.1 \qquad (12\text{-}30)$$

The nucleation part is based on the Forster–Zuber [24] equation with a suppression factor S:

$$h_{NB} = S(0.00122) \left[\frac{(k^{0.79} c_p^{0.45} \rho^{0.49})_f}{\sigma^{0.5} \mu_f^{0.29} h_{fg}^{0.24} \rho_g^{0.24}} \right] \Delta T_{sat}^{0.24} \Delta p^{0.75} \qquad (12\text{-}31)$$

where: $\Delta T_{sat} = T_w - T_{sat}$; $\Delta p = p(T_w) - p(T_{sat})$; and S is a function of the total Reynolds number as shown in Figure 12-12. It can be approximated by:

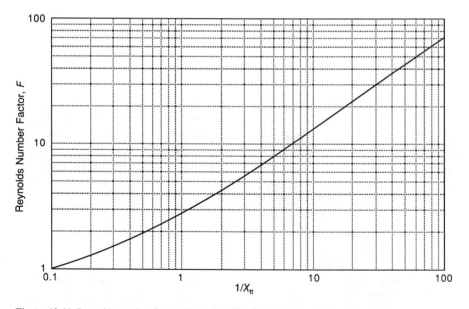

Figure 12-11 Reynolds number factor (F) used in Chen's correlation. *(From Chen [14].)*

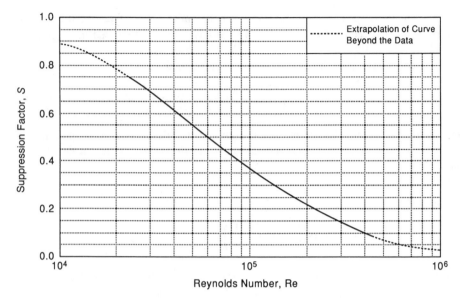

Figure 12-12 Suppression factor (S) used in Chen's correlation. *(From Chen [14].)*

$$S = \frac{1}{1 + 2.53 \times 10^{-6} \, \text{Re}^{1.17}} \qquad (12\text{-}32)$$

where $\text{Re} = \text{Re}_\ell \, F^{1.25}$

The range of water conditions of the original data was:

Pressure $= 0.17$ to 3.5 MPa
Liquid inlet velocity $= 0.06$ to 4.5 m/s
Heat flux up to 2.4 MW/m^2
Quality $= 0$ to 0.7

Other tested fluids include methanol, cyclohexane, pentane, and benzene.

Chen's correlation has the advantage of being applicable over the entire boiling region. It also has a lower deviation error (11%) than the earlier correlations of Dengler and Addoms (38%) [22], Bennett et al. (32.6%) [5], and Shrock and Grossman (31.7%) [54] when tested for the range indicated above [15]. Subsequent data have extended the pressure range of the application to 6.9 MPa. A comparison of the Chen correlation to the Dengler and Addoms correlation is shown in Figure 12-13.

Collier [18] discussed the possibility of using the Chen correlation in the sub-cooled region by adding a temperature difference weighting to each of its components:

$$q'' = h_{NB}(T_w - T_{sat}) + h_c(T_w - T_{bulk}) \qquad (12\text{-}33)$$

For subcooled boiling, F can be set to unity, and S can be calculated with the quality x being set to zero. This method was found acceptable against experimental data of water and ammonia.

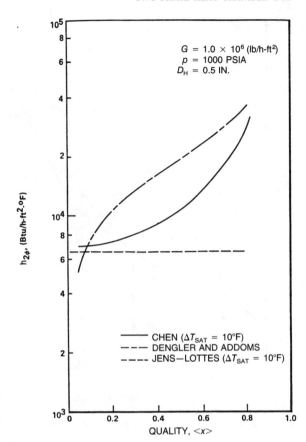

Figure 12-13 Comparison of some convective boiling heat-transfer correlations. *(From Lahey and Moody [46].)*

Bjorge et al. [10] also proposed that superposition of nucleate boiling and convection heat transfer may be used to produce a heat flux relation in the low quality two-phase region as:

$$q''^2 = q_c''^2 + (q_{NB}'' - q_i'')^2 \qquad (12\text{-}34)$$

where q_i'' is subtracted to make $q'' = q_c''$ at the incipience of boiling. Figure 12-14 illustrates this superposition approach. The fully developed nucleate boiling curve has a slope of about 3, i.e., $q_{NB}'' = \Delta T_{sat}^3$, so that Eq. 12-34 can be rewritten as:

$$q''^2 = q_c''^2 + q_{NB}''^2 \left\{ 1 - \left[\frac{(T_w - T_{sat})_i}{(T_w - T_{sat})} \right]^3 \right\}^2 \qquad (12\text{-}35)$$

where (Eq. 12-12):

$$(T_w - T_{sat})_i = \frac{8\sigma T_{sat} v_{fg} h_c}{k_\ell h_{fg}} \qquad (12\text{-}36)$$

and

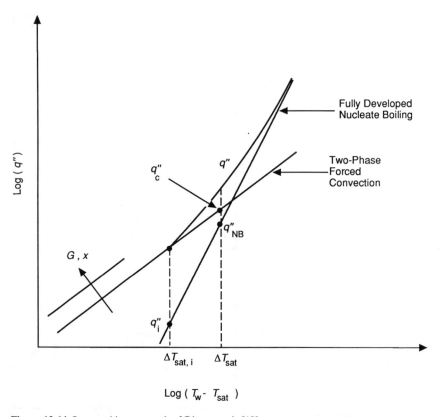

Figure 12-14 Superposition approach of Bjorge et al. [10].

$$q''_{NB} = B_M \frac{(g/g_c)^{1/2} h_{fg}^{1/8} k_\ell^{1/2} \rho_\ell^{17/8} C_\ell^{19/8} \rho_g^{1/8}}{\sigma^{9/8} (\rho_\ell - \rho_g)^{5/8} T_{sat}^{1/8}} (T_w - T_{sat})^3 \qquad (12\text{-}37)$$

where $B_M = 1.89 \times 10^{-14}$ in SI units or 2.13×10^{-5} in British units.

For high quality flow ($\alpha > 80\%$) Bjorge et al. [10] recommended a different correlation, given by:

$$q'' = q''_c + q''_{NB} \left\{ 1 - \left[\frac{(T_w - T_{sat})_i}{(T_w - T_{sat})} \right]^3 \right\} \qquad (12\text{-}38)$$

where:

$$q''_c = \frac{F(X_{tt})}{F_2} k_\ell \, \mathrm{Re}_\ell^{0.9} \, \mathrm{Pr}_\ell \, (T_w - T_{sat})$$

$$F(X_{tt}) = 0.15 \left[\frac{1}{X_{tt}} + 2.0 \left(\frac{1}{X_{tt}} \right)^{0.32} \right], \quad \mathrm{Re}_\ell = \frac{(1 - x)GD}{\mu_\ell}$$

and

$$F_2 = \begin{cases} \text{Pr}_\ell + 5\ ln(1 + 5\text{Pr}_\ell) + 2.5\ ln(0.0031\ \text{Re}_\ell^{0.812}); & \text{Re}_\ell > 1125 \\ 5\text{Pr}_\ell + 5\ ln[1 + \text{Pr}_\ell(0.0964\ \text{Re}_\ell^{0.585} - 1)]; & 60 < \text{Re}_\ell < 1125 \\ 0.707\ \text{Pr}_\ell \text{Re}_\ell^{0.5} \end{cases}$$

Example 12-1 Heat flux calculation for saturated boiling

PROBLEM Water is boiling at 7.0 MPa (1000 psi) in a tube of an LMFBR steam generator. Using the Chen correlation, determine the heat flux at a position in the tube where the quality $x = 0.2$ and the wall temperature is 290°C.
Tube flow conditions

Diameter of tube $(D) = 25$ mm
Mass flow rate $(\dot{m}) = 800$ kg/hr

SOLUTION
Required water properties

$\mu_f = 96 \times 10^{-6}$ N s/m^2
$\mu_g = 18.95 \times 10^{-6}$ N s/m^2
$c_{pf} = 5.4 \times 10^3$ J/kg°K
$\rho_f = 740$ kg/m^3
$\rho_g = 36.5$ kg/m^3
$\sigma = 18.03 \times 10^{-3}$ N/m
$h_{fg} = 1513.6 \times 10^3$ J/kg
T_{sat} (7.0 MPa) $= 284.64$°C
$T_w = 290$°C
$\Delta T_{sat} = 5.36$°K
$k_f = 0.567$ W/m °K

Pressure and mass flux

$$\Delta p_{sat} = p_{sat}\ (290°C) - p_{sat}(284.64°C) = 5.66 \times 10^5 \text{ (Pa)}$$

$$G = \frac{\dot{m}}{\frac{\pi}{4} D^2} = \frac{800\ \dfrac{1}{3600}}{\frac{\pi}{4} (0.025)^2} = 452.7 \text{ kg/m}^2 \text{ s}$$

Chen correlation

$$h_{2\phi} = h_{NB} + h_c \tag{12-25}$$

$$h_c = 0.023 \left[\frac{G(1 - x)D}{\mu_f}\right]^{0.8} \left[\frac{\mu\ c_p}{k}\right]_f^{0.4} \left(\frac{k_f}{D}\right) F \tag{12-29}$$

$$= 0.023 \left[\frac{(452.7)(1 - 0.2)(0.025)}{96 \times 10^{-6}} \right]^{0.8}$$

$$\cdot \left[\frac{(96 \times 10^{-6})(5.4 \times 10^3)}{0.567} \right]^{0.4} \left(\frac{0.567}{0.025} \right) F$$

$$F = 2.35 \left(\frac{1}{X_{tt}} + 0.213 \right)^{0.736} \quad \text{for } \frac{1}{X_{tt}} > 0.1 \tag{12-30}$$

$$X_{tt} = \left(\frac{1 - x}{x} \right)^{0.9} \left(\frac{\rho_g}{\rho_f} \right)^{0.5} \left(\frac{\mu_f}{\mu_g} \right)^{0.1}$$

$$= \left(\frac{1 - 0.2}{0.2} \right)^{0.9} \left(\frac{36.54}{740.0} \right)^{0.5} \left(\frac{96 \times 10^{-6}}{18.95 \times 10^{-6}} \right)^{0.1} = 0.9$$

$$\therefore F = 2.87$$

$$\therefore h_c = 13{,}783 \text{ W/m}^2 \text{ °K}$$

$$h_{NB} = 0.00122 \frac{(k^{0.79} c_p^{0.45} \rho^{0.49})_f}{\sigma^{0.5} \mu_f^{0.29} h_{fg}^{0.24} \rho_g^{0.24}} \Delta T_{sat}^{0.24} \Delta p_{sat}^{0.75} S \tag{12-31}$$

$$= 0.00122 \frac{(0.567)^{0.79}(5.4 \times 10^3)^{0.45}(740.0)^{0.49}}{(18.03 \times 10^{-3})^{0.5}(96 \times 10^{-6})^{0.29} \cdot} (1513.6 \times 10^3)^{0.24}(36.5)^{0.24}$$

$$\cdot (5.36)^{0.24}(5.66 \times 10^5)^{0.75} S$$

To find S, use Figure 12-12:

$$\text{Re}_\ell = \frac{G(1 - x)D}{\mu_f} = \frac{(452.7)(1 - 0.2)(0.025)}{96 \times 10^{-6}} = 94{,}313$$

$$\text{Re} = \text{Re}_\ell F^{1.25} = 9.43 \times 10^4 (2.87)^{1.25} = 3.52 \times 10^5$$

$$\therefore S \approx 0.12 \text{ and } h_{NB} = 5309 \text{ W/m}^2 \text{ °K}$$

$$\text{So } h_{2\phi} = 5309 + 13{,}783 = 19{,}092 \text{ W/m}^2 \text{ °K}$$

and

$$q'' = h_{2\phi} \Delta T_{sat} = (19{,}092)(5.36) = 102.3 \text{ kW/m}^3$$

Example 12-2 B-H-R (Bjorge et al) superposition method

PROBLEM Consider a coolant flowing in forced convection boiling inside a round tube. The following relations have been calculated from the liquid and vapor properties and flow conditions:

Forced convection:

$$q_c'' = h \Delta T_{sat}$$
$$h = 1000 \text{ Btu/hr ft}^2 \text{ °F } (5.68 \text{ kW/m}^2 \text{ °K})$$

Fully developed nucleate boiling

$$q''_{NB} = \gamma_B \, \Delta T^3_{sat}$$
$$\gamma_B = 10 \text{ Btu/hr ft}^2 \text{ °F}^3 \, (0.18 \text{ kW/m}^2 \text{ °K}^3)$$

Incipient boiling

$$q''_i = \gamma_i \, \Delta T^2_{sat,i}$$
$$\gamma_i = 167 \text{ Btu/hr ft}^2 \text{ °F}^2 \, (1.71 \text{ kW/m}^2 \text{ °K}^2)$$

1. Determine the heat flux q'' when $\Delta T_{sat} = T_w - T_{sat} = 12°F \, (6.67°C)$.
2. The above equations apply at the particular value of G_1. Keeping the heat flux q'' the same as above, how much must G be increased in order to stop nucleate boiling? What is the ratio of G_2/G_1? Assume the flow is turbulent.

SOLUTION

1. Determine the heat flux q'' for $T_w - T_{sat} = 12°F$. Assuming that this superheat is sufficient to cause nucleate boiling at low vapor quality, the heat flux is given by:

$$q''^2 = (q''_c)^2 + (q''_{NB})^2 \left\{ 1 - \left[\frac{(\Delta T_{sat,i})}{\Delta T_{sat}} \right]^3 \right\}^2 \tag{12-35}$$

Now we have to find $\Delta T_{sat,i}$ to calculate the heat flux. This quantity is the temperature difference between the wall and the saturated temperature for incipient boiling. For incipient boiling, the nucleate boiling heat flux should equal that by forced convection; therefore by considering the energy balance:

$$q''_i = q''_c$$
$$\gamma_i (\Delta T_{sat,i})^2 = h \Delta T_{sat}$$
$$\Delta T_{sat,i} = \frac{h}{\gamma_i} = \frac{1000}{167} \frac{(\text{Btu/hr ft}^2 \text{ °F})}{(\text{Btu/hr ft}^2 \text{ °F}^2)} = 5.98°F$$
$$\Delta T_{sat,i} \approx 6°F$$

The heat flux is therefore given by

$$q'' = \left\{ [1000(12)]^2 + [10(12)^3]^2 \left[1 - \left(\frac{6}{12} \right)^3 \right]^2 \right\}^{1/2}$$

$$q'' = 19300 \text{ Btu/hr ft}^2$$

2. The above heat-transfer coefficient (h_c) is for a particular flow rate. We want to look for the flow rate that has a high enough forced convection heat-transfer rate such that the wall temperature would be too low to allow incipient boiling for the given heat flux. Let the wall temperature in this case be T_{w2}.

To calculate the temperature $\Delta T_2 = T_{w2} - T_{sat}$

$$q''_i = \gamma_i \Delta T^2_{2,i}$$

$$\Delta T_{2,i} = \sqrt{\frac{q''_i}{\gamma_i}}$$

$$= \sqrt{\frac{19300}{167}}$$

$$= 10.75°F$$

The heat-transfer coefficient of forced convection corresponding to ΔT_2 is:

$$h_c = \frac{q''}{\Delta T_2}$$

$$= \frac{19300}{10.75}$$

$$= 1795 \text{ Btu/hr ft}^2 \text{ °F}$$

From Equation 12-26, the forced convection heat transfer coefficient is expressed as:

$$h_c = 0.023 \left[\frac{(1-x)D}{\mu_\ell} \right]^{0.8} G^{0.8} \text{Pr}_\ell^{0.4} \frac{k_\ell}{D} F = \text{const. } G^{0.8}$$

So the new flow rate is obtained from:

$$\frac{h_{c2}}{h_{c1}} = \frac{\text{const. } G_2^{0.8}}{\text{const. } G_1^{0.8}} = \left(\frac{G_2}{G_1} \right)^{0.8}$$

Hence:

$$\frac{G_2}{G_1} = \left(\frac{h_{c2}}{h_{c1}} \right)^{1/0.8} = \left(\frac{1795}{1000} \right)^{1/0.8} = 2.08$$

VI POST-CRITICAL-HEAT-FLUX HEAT TRANSFER

Knowledge of the heat-transfer rate in the liquid-deficient flow regime beyond the critical heat flux (CHF) is important in many reactor applications. In LWRs a loss of coolant accident or possibly an overpower accident may result in exposure of at least part of the fuel elements to CHF and post-CHF conditions. Once-through steam generators routinely operate with part of the length of their tubes in this heat-transfer region.

Mechanisms and correlations of CHF are discussed in section VII.

A Observed Post-CHF Regimens

It is important to point out that the post-CHF behavior in flow boiling is somewhat dependent on the manner in which the wall heat flux and wall temperature are controlled.

For a uniformly applied heat flux in a vertical tube, as in Figure 12-15a, either increasing the heat flux or reducing the flow rate causes CHF to occur first at the exit of the tube (condition A). Further increasing the heat flux drives the CHF point upstream and causes the downstream section to enter the post-CHF region (condition

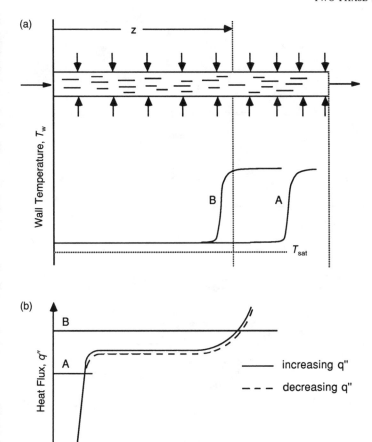

Figure 12-15 Burnout in a uniformly heated tube. (a) Heat flux increase from condition A to B. (b) Lack of hysteresis for heat flux change at position z. *(Adapted from Collier [19].)*

B). Plotting the variation of surface temperature at a given position (z) as the heat flux is first increased and then decreased, one obtains the curve shown in Figure 12-15b without any hysteresis [5]. This finding is in contrast to the situation for pool boiling, where the return from film boiling conditions to nucleate boiling occurs at a considerably lower heat flux than the CHF (Fig. 12-16) [2].

Hysteresis has been observed in a tube with a short length over which the heat flux can be controlled independently from that in the remainder of the tube (Fig. 12-17b). CHF is initiated at this downstream location for conditions in which the two sections are operated at equal heat fluxes (condition A). Post-CHF behavior can be studied on the downstream section by progressively decreasing the value of the heat flux, q_2'' (condition B). This technique was used by Bailey [1] with water. The wall temperature decreased smoothly until rewetting occurred and the wall temperature

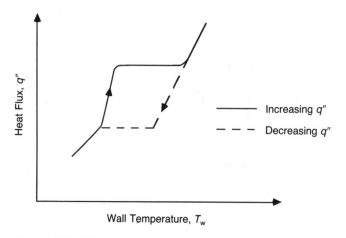

Figure 12-16 CHF in pool boiling.

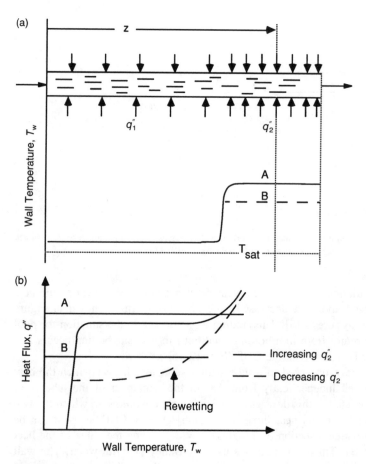

Figure 12-17 CHF in a two-section heated tube. (a) Heat flux decreases from condition A to B. (b) Hysteresis for the heat flux change at position z. *(Adapted from Collier [19].)*

dropped sharply. Figure 12-17b depicts this behavior, which exhibits a similarity to the pool boiling situation of Figure 12-16. This behavior can be induced by initiating CHF at a heat flux spike and observing the postdryout response on the downstream section at a lower heat flux. Groeneveld [28] utilized this arrangement with freon and Plummer et al. [49] and Iloeje et al. [36] with nitrogen. It can be concluded that significant hysteresis can occur in situations where rewetting by an advancing liquid front is prevented. Collier [18] suggested that if a surface remains partially wet rewetting of the areas where CHF previously occurred happens soon after the local surface heat flux falls below the CHF value. The time to rewet, however, is limited by the conduction-controlled velocity of the "quench" front. On the other hand, if a surface has "dried out" completely during a transient, the surface heat flux must fall significantly below the CHF value (to 10 to 20% of the CHF heat flux) before rewetting can be initiated.

Transition boiling can be established using a test section with a controlled temperature rather than controlled heat flux. Consider a section consisting of two parts: a long, uniformly heated upstream section and a short, thick-walled (high heat capacity) downstream section (Fig. 12-18a). CHF is established at the downstream end

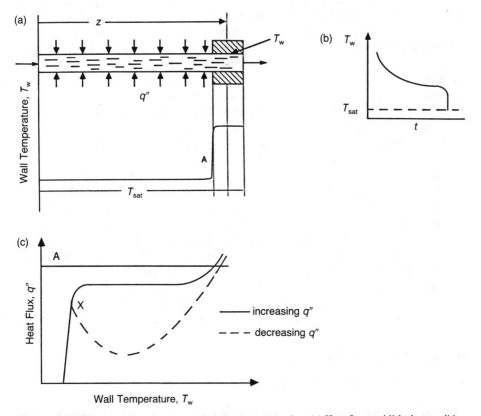

Figure 12-18 CHF in a temperature-controlled downstream section. (a) Heat flux established at condition A. (b) Temperature during quenching at position z. (c) Transition boiling region established at position z. *(Adapted from Collier [19].)*

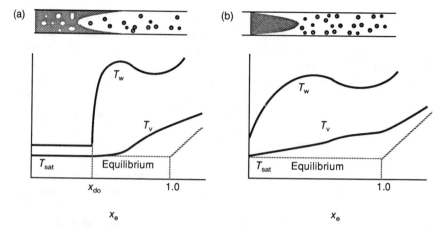

Figure 12-19 Post-CHF temperature distribution. (a) Annular flow. (b) Inverted annular flow.

of the long section as the short section is heated to a high initial temperature. As the short section is cooled to the saturation temperature (Fig. 12-18b), a different quench curve is seen. The resulting curve of the heat flux versus wall temperature is shown in Figure 12-18c. Note that the point of return to efficient flow boiling (X) does not coincide with the CHF point because, in general, both the flow in the liquid film and the droplet deposition flux contribute to the CHF. Measurements by Hewitt [34] showed that point X may normally be about 80% of the CHF value, rising on some occasions to as much as 100%, in which case X and the CHF value coincide.

Thus the post-CHF heat transfer can be affected by two major factors:

1. The independent boundary condition is one factor. Is it the wall heat flux or the wall temperature? In the first case transition boiling is not obtained, whereas in the second case transition boiling may be realized.
2. The type of CHF (DNB versus dryout) affects the post-CHF heat transfer even in the controlled heat flux case. Perhaps more accurately, the flow regime of the post-DNB region allows for the existence of a vapor film that separates the wall from the bulk liquid (inverted annular). In this case, film boiling occurs, and the flow quality could be relatively low. In the postdryout (postannular) case, the liquid exists only as droplets carried by the vapor, in which case the vapor receives the heat from the wall and only some of the droplets impinge on the wall. Most of the droplets receive heat from the vapor core. In this case, the vapor temperature may exceed T_{sat} (Fig. 12-19). It is seen that in the inverted annular flow (i.e., film boiling) T_w is high even at low values of the thermodynamic quality.

B Post-CHF Correlations

Three types of correlation have been identified by Collier [19].

1. Empirical correlations which make no assumptions about the mechanism but attempt to relate the heat-transfer coefficient (assuming the coolant is at the saturation temperature) and the independent variables.

2. Mechanistic correlations which recognize that departure from a thermodynamic equilibrium condition and attempt to calculate the "true" vapor quality and vapor temperature. A conventional single-phase heat-transfer correlation is then used to calculate the heated wall temperature.
3. Semitheoretical correlations which attempt to model individual hydrodynamic and heat-transfer processes in the heated channel and relate them to the wall temperature.

Groeneveld [29] compiled a bank of selected data from a variety of experimental postburnout studies in tubular, annular, and rod bundle geometries for steam–water flows. He recommended the following correlation, which is of type 1.

$$ \mathrm{Nu_g} = a \left\{ \mathrm{Re_g} \left[x + \frac{\rho_g}{\rho_f} (1 - x) \right] \right\}^b \mathrm{Pr_g^c} \, Y \qquad (12\text{-}39) $$

where $\mathrm{Re_g} = GD/\mu_g$

$$ Y = \left[1 - 0.1 \left(\frac{\rho_f - \rho_g}{\rho_g} \right)^{0.4} (1 - x)^{0.4} \right]^d \qquad (12\text{-}40) $$

The coefficients a, b, c, and d are given in Table 12-3.

Slaughterbeck et al. [55] improved on the correlation, particularly at low pressure. They recommended that the parameter Y be changed to:

$$ Y = (q'')^e \left(\frac{k_g}{k_{cr}} \right)^f \qquad (12\text{-}41) $$

where k_{cr} = conductivity at the thermodynamic critical point. Their recommended constants are also included in Table 12-3.

Table 12-3 Empirical post-CHF correlations*

Author	a	b	c	d	e	f	No. of points	% rms error[†]
Groeneveld								
Tubes	1.09×10^{-3}	0.989	1.41	-1.15			438	11.5
Annuli	5.20×10^{-2}	0.688	1.26	-1.06			266	6.9
Slaughterback								
Tubes	1.16×10^{-4}	0.838	1.81		0.278	-0.508		12.0

*Range of data:

Parameter	Tubes (vertical and horizontal)	Annuli (vertical)
D_e (mm)	2.5–25.0	1.5–6.3
p (MPa)	6.8–21.5	3.4–10.0
G (kg/m²s)	700–5300	800–4100
q'' (kW/m²)	120–2100	450–2250
x	0.1–0.9	0.1–0.9

[†]rms = root mean square.

1 Transition boiling. The existence of a transition boiling region in the forced convection condition as well as in pool boiling, under controlled wall temperature conditions, has been demonstrated experimentally.

Attempts have been made to produce correlations for the transition boiling region. Groeneveld and Fung [31] tabulated those available for forced convective boiling of water. In general, the correlations are valid only for the range of conditions of the data on which they were based. Figure 12-20 provides an examples of the level of uncertainty in this region.

One of the earliest experimental studies of forced convection transition boiling was that by McDonough et al. [48], who measured heat-transfer coefficients for water over the pressure range 5.5 to 13.8 MPa inside a 3.8 mm I.D. tube heated by NaK. They proposed the following correlation:

$$\frac{q''_{cr} - q''(z)}{T_w(z) - T_{cr}} = 4.15 \exp \frac{3.97}{p} \tag{12-42}$$

where q''_{cr} = CHF heat flux (kW/m²); $q''(z)$ = transition region heat flux (kW/m²); T_{cr} = wall temperature at CHF (°C); $T_w(z)$ = wall temperature in the transition region (°C); and p = system pressure (MPa).

Tong [62] suggested the following equation for combined transition and stable film boiling at 6.9 MPa.

$$h_{tb} = 39.75 \exp\left(-0.0144\Delta T\right) + 2.3 \times 10^{-5} \frac{k_g}{D_e} \exp\left(-\frac{105}{\Delta T}\right) Re_f^{0.8} \, Pr_f^{0.4} \tag{12-43}$$

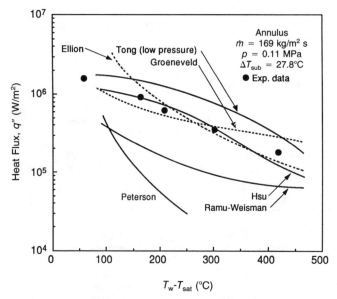

Figure 12-20 Comparison of various transition boiling correlations with Ellion's data. *(From Groeneveld and Fung [31].)*

where h_{tb} = the heat-transfer coefficient for the transition region (kW/m^2°K); ΔT = $T_w - T_{sat}$ (°C); D_e = equivalent diameter (m); k_g = vapor conductivity; Pr_f = liquid Pr; and Re = $D_e G_m / \mu_f$.

Ramu and Weisman [50] later attempted to produce a single correlation for post-CHF and reflood situations. They proposed the transition boiling heat-transfer coefficient:

$$h_{tb} = 0.5 \, S \, h_{cr} \{ \exp \left[-0.0140(\Delta T - \Delta T_{cr}) \right] + \exp \left[-0.125(\Delta T - \Delta T_{cr}) \right] \}$$

(12-44)

where h_{cr} (kW/m^2 °C) and ΔT_{cr} (°C) = the heat-transfer coefficient and wall superheat $(T_w - T_{sat})$, respectively, corresponding to the pool boiling CHF condition; and S = the Chen nucleation suppression factor.

Cheng et al. [16] suggested a simple correlation of the form:

$$\frac{q''_{tb}}{q''_{cr}} = \left(\frac{T_w - T_{sat}}{\Delta T_{cr}} \right)^{-n}$$

(12-45)

Cheng et al. found that $n = 1.25$ fitted their low-pressure data acceptably well. A similar approach was adopted by Bjornard and Griffith [11], who proposed:

$$q''_{tb} = \delta q''_{cr} + (1 - \delta) q''_{min}$$

(12-46)

$$\delta = \left(\frac{T^M - T_w}{T^M - T_{cr}} \right)^2$$

(12-47)

where q''_{min} and T^M = heat flux and wall temperature, respectively, corresponding to the minimum heat flux for stable film boiling in the boiling curve; and q''_{cr} and T_{cr} = heat flux and wall temperature, respectively, at CHF. Some of the correlations for wall temperature for stable film boiling are given in Table 12-1.

2 Postdryout region. It has been observed experimentally that vapor can become substantially superheated in the postdryout region. Thus the thermodynamic quality of the flow may exceed unity, whereas the actual quality is substantially less than unity. This was modeled by dividing the heat flux into two parts, a vapor heating part and a liquid heating part. Because the liquid is at saturation, the liquid heat flux is actually consumed in causing the evaporation of droplets. The correlations in this region are somewhat involved and can be found in Collier's work [18]. Kumamaru et al. [44] have tested several postdryout correlations against data obtained in a 5 × 5 rod bundle at a pressure of 3 MPa. They found the best wall temperature predictions over the mass flux range of 80 to 220 kg/m^{2}s, and heat fluxes from 30 to 260 kW/m^2 are those by the Varone–Rohsenow [63] correlation. That correlation is semimechanistic, requiring knowledge of the dryout thermodynamic quality and average liquid drop diameter at dryout locations. Earlier versions of this correlation as well as other correlations can be found in Rohsenow's work [51].

3 Film boiling (post-DNB). Film boiling has been modeled analytically by allowing the vapor film to exist continuously on the surface and considering various film layer conditions. For laminar flow of vapor in the film, it results in a correlation for a vertical tube given by:

$$Nu = C(Pr^* \, Gr)^{1/4}$$
$$C = 0.943 \text{ for zero shear at the vapor–liquid interface} \qquad (12\text{-}48)$$
$$C = 0.707 \text{ for zero velocity at the interface}$$

where:

$$Nu = \frac{hz}{k_g}$$

$$Pr^* = \frac{\mu_g h_{fg}}{k_g \, \Delta T}$$

$$Gr = \frac{\rho_g g (\rho_\ell - \rho_g) z^3}{\mu_g^2}$$

z = the distance from the point at which film boiling starts. Thus:

$$h = \frac{k_g}{z} \left[\frac{\rho_g g (\rho_\ell - \rho_g) z^3 h_{fg}}{4 k_g \, \mu_g \, \Delta T} \right]^{1/4} \qquad (12\text{-}49)$$

This equation is similar to the Berenson [6] correlation for film boiling on a horizontal flat plate given by:

$$h = 0.425 \frac{k_g}{\lambda_c} \left[\frac{\rho_g g (\rho_\ell - \rho_g) \lambda_c^3 h_{fg}}{k_g \, \mu_g \, \Delta T} \right]^{1/4} \qquad (12\text{-}50)$$

where:

$$\lambda_c = 2\pi \left[\frac{\sigma}{g(\rho_\ell - \rho_g)} \right]^{1/2} \qquad (12\text{-}51)$$

is the spacing between the bubbles leaving the film, which is the wavelength of a Taylor-type instability at the vapor–liquid interface.

VII CRITICAL HEAT FLUX

The CHF is used here to denote the conditions at which the heat-transfer coefficient of the two-phase flow substantially deteriorates. For given flow conditions, it occurs at a sufficiently high heat flux or wall temperature. In a system in which the heat flux is independently controlled, the consequences of the CHF occurrence is the rapid rise in the wall temperature. For systems in which the wall temperature is independently controlled, the occurrence of CHF implies a rapid decrease in the heat flux.

Many terms have been used to denote the CHF conditions, including "boiling crisis" (mostly in non-English speaking countries) and "burnout" (preferred in Britain). Departure from nucleate boiling (DNB) is the term originally used to describe

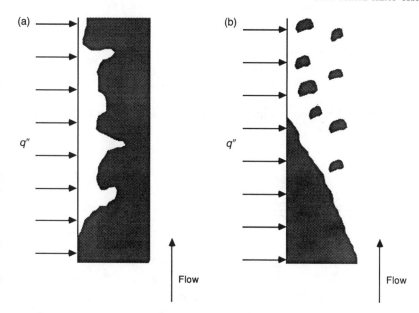

Figure 12-21 CHF mechanisms. (a) DNB. (b) Dryout.

the CHF conditions in pool boiling; it can also be encountered in flow boiling, when bubble formation is rapid enough to cause a continuous vapor film to form at the wall (Fig. 12-21). "Dryout" is the term that is now reserved for liquid film dryout in annular flow; "boiling transition" is the term preferred by some to represent the fact that the local heat flux condition is not the only factor influencing the deterioration of the heat-transfer coefficient. "CHF" is used here because it is more widely used than any of the other terms.

A CHF at Low Quality (Departure from Nucleate Boiling)

1 Pool boiling. The earliest studies of CHF involved positioning a heating surface in a static pool of liquid. The observed heat flux–wall temperature behavior clearly showed a CHF condition leading to transition boiling. This pool boiling CHF provides lower heat fluxes than does flow boiling, as the critical flux of the DNB-type increases with mass flow. This CHF mechanism (DNB) can occur for both saturated and subcooled liquid conditions.

The effect of pressure on the pool boiling heat flux is evident from Figure 12-3, in which three well known correlations are compared.

2 Flow boiling. The nucleate boiling regime exists at low-quality flow conditions. If the heat flux is very high, vapor blanketing of the surface may occur. It is possible to set two limits to the CHF (q''_{cr}). At the low end, it must be sufficiently high to cause the wall temperature to reach saturation ($T_w = T_{sat}$) whereas the bulk flow is still subcooled. Hence:

$$(q''_{cr})_{min} = h_{\ell o}(T_w - T_{bulk}) = h_{\ell o}(T_{sat} - T_{bulk}) \tag{12-52}$$

For axially uniform heat flux:

$$(q''_{cr})_{min}\pi DL = \frac{G\pi D^2}{4} c_p(T_{bulk} - T_{in}) \tag{12-53}$$

where L = tube length. Therefore eliminating T_{bulk} from Eqs. 12-52 and 12-53 yields:

$$(q''_{cr})_{min} = \frac{(T_{sat} - T_{in})}{\dfrac{1}{h_{\ell o}} + \dfrac{4L}{GDc_p}} \tag{12-54}$$

At the high end, q''_{cr} should be sufficiently high to cause an equilibrium quality (x_e) = 1.0. Hence:

$$(q''_{cr})_{max} = \frac{GD}{4L} [c_p(T_{sat} - T_{in}) + h_{fg}] \tag{12-55}$$

From Eqs. 12-54 and 12-55, it is clear that the higher the inlet temperature (or thermodynamic quality) the lower is the value of q''_{cr} in a uniformly heated tube. Generally, it would be expected that:

$$q''_{cr} = q''_{cr}[p,G,D,L,(\Delta T_{sub})_{in},q''(z)] \tag{12-56}$$

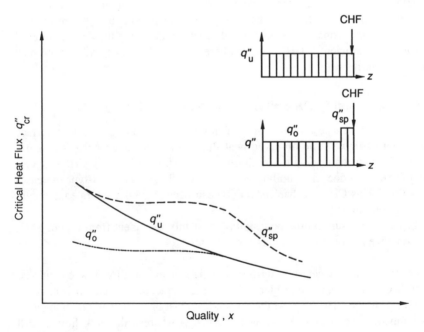

Figure 12-22 Effect of heat flux spike on q''_{cr}.

The dependence on the inlet subcooling and the heat flux profile may be replaced by a dependence on the local value of the quality at the point of CHF (i.e., at L):

$$q''_{cr} = q''_{cr}[p,G,D,L,x(L)] \tag{12-57}$$

It is noted that vapor blanketing is mostly affected by the local vaporization rate; hence the effect of length of heating on DNB is small. However, for the dryout form of CHF, hydrodynamic effects on the liquid film behavior are pronounced. The development of a particular form of hydrodynamic behavior is affected by the length of the flow, which has been demonstrated by Groeneveld [30]. He observed the effect of a heat flux spike at the end of tube on q''_{cr} (Fig. 12-22), and found that at low and negative exit qualities the spike heat flux did not change the critical heat flux: $(q''_{sp})_{cr} = (q''_{u})_{cr}$. However, for high quality conditions, the value $(q''_{sp})_{cr}$ was higher than $(q''_{u})_{cr}$, but the value of the average heat flux $(q'')_{cr}$ was about equal to the uniform flux condition $(q''_{u})_{cr}$. Hence, unlike DNB, dryout is controlled more by the total heat input in a channel than by the local heat flux.

B CHF at High Quality (Dryout)

At high quality flow conditions, the vapor mostly exists in the core. Both the shear action of the vapor and the local vaporization rate may lead to the liquid film being stripped from the wall. In uniformly heated tubes, the value of q''_{cr} at high flow qualities is, generally speaking, lower than that at low flow qualities. It is also observed that q''_{cr} is lower for longer tubes. However, the overall critical power input increases with length. Therefore the critical quality (x_{cr}) at the tube exit also increases with length.

Because of the difference between the DNB and the dryout mechanisms, they exhibit a reversed dependence on the mass flux. As shown in Figure 12-23, the higher mass flux leads to a higher q''_{cr} for low quality flow but decreases the q''_{cr} for high

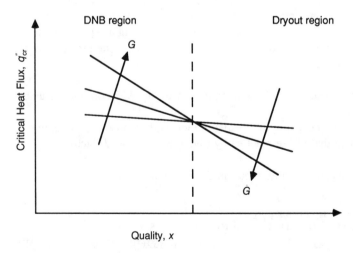

Figure 12-23 Effect of mass flux on critical heat flux.

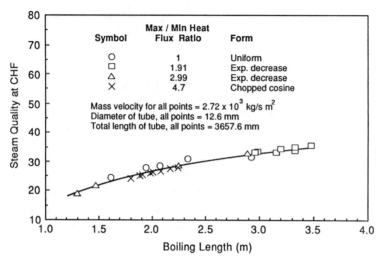

Figure 12-24 Effect of flux shape on critical quality *(From Keeys et al. [42].)*

quality flow. The overall critical power in a uniformly heated tube of fixed length (L), however, increases with the increase in mass flux.

For high quality CHF, the critical channel power is not sensitive to the distribution of the heat flux along a fixed length, as demonstrated in Figure 12-24 [42].

C Correlation of CHF by U.S. Reactor Vendors

1 DNB. The most widely used correlation for evaluation of DNB conditions for PWRs is the W-3 correlation developed by Tong [60, 61]. The correlation may be applied to circular, rectangular, and rod-bundle flow geometries. The correlation has been developed for axially uniform heat flux, with a correcting factor for nonuniform flux distribution. Also local spacer effects can be taken into account by specific factors.

For predicting the DNB condition in a nonuniform heat flux channel, the following two steps are to be followed:

1. The uniform critical heat flux (q''_{cr}) is computed with the W-3 correlation, using the local reactor conditions.
2. The nonuniform DNB heat flux ($q''_{cr,n}$) distribution is then obtained (assuming a flux shape similar to that of the reactor) by dividing q''_{cr} by the F factor.

For a channel with axially uniform heat flux in British units:

$$q''_{cr}/10^6 = \{(2.022 - 0.0004302\,p)$$
$$+ (0.1722 - 0.0000984\,p)$$
$$\exp\,[(18.177 - 0.004129\,p)x_e]\}[(0.1484 - 1.596x_e$$
$$+ 0.1729x_e|x_e|)G/10^6 + 1.037](1.157 - 0.869x_e)[0.2664 \qquad (12\text{-}58)$$
$$+ 0.8357\,\exp\,(-3.151\,D_h)][0.8258 + 0.000794(h_f - h_{in})]$$

where $q_{cr}'' =$ DNB flux (BTU/hr ft^2); $p =$ pressure (psia); $x_e =$ local steam thermodynamic quality; $D_h =$ equivalent heated diameter (in.); $h_{in} =$ inlet enthalpy (BTU/lb); $G =$ mass flux (lb/hr ft^2); $h_f =$ saturated liquid enthalpy (BTU/lb).

The correlation is valid in the ranges:

$p =$ 800 to 2000 psia
$G/10^6 =$ 1.0 to 5.0 lb/hr ft^2
$D_h =$ 0.2 to 0.7 in.
$x_e =$ −0.15 to 0.15
$L =$ 10 to 144 in.

The axially nonuniform heat flux $(q_{cr,n}'')$ is obtained by applying a corrective F factor to the uniform critical heat flux:

$$q_{cr,n}'' = q_{cr}''/F \qquad (12\text{-}59)$$

where:

$$F = \frac{C \int_0^\ell q''(z') \exp\left[-C(\ell - z')\right]dz'}{q''(\ell)[1 - \exp(-C\ell)]} \qquad (12\text{-}60)$$

where $q_{cr,n}'' = q''$ local at DNB position; $\ell =$ distance to DNB as predicted by the uniform q_{cr}'' model; and:

$$C = \frac{0.44[1 - x_e(\ell)]^{7.9}}{(G/10^6)^{1.72}} \text{ in.}^{-1} \qquad (12\text{-}61)$$

In SI units the critical heat flux (kW/m^2) for uniformly heated channels is given by:

$$
\begin{aligned}
q_{cr}'' = &[(2.022 - 0.06238p) + (0.1722 - 0.001427p)\exp(18.177 \\
&- 0.5987p)x_e][(0.1484 - 1.596x_e + 0.1729x_e|x_e|)2.326G \\
&+ 3271][1.157 - 0.869x_e][0.2664 + 0.837 \\
&\exp(-124.1 D_h)][0.8258 + 0.0003413(h_f - h_{in})]
\end{aligned}
\qquad (12\text{-}62)
$$

where p is in MPa; G is in kg/m^{2}s; h is in kJ/kg; and D_h is in meters (m).

The F factor can still be used with:

$$C = \frac{4.23 \times 10^6[1 - x_e(\ell)]^{7.9}}{G^{1.72}} \text{ m}^{-1} \qquad (12\text{-}63)$$

Smith et al. [56] suggested that for the limited conditions of $p =$ 515 psi (3.5 MPa), $D_h =$ 0.2285 in. (5.8 mm), and $G =$ 0.5 to 1.1 × 10^6 lb/ft^{2}hr (750 to 1500 kg/m^{2}s), the value of C is given by:

$$C \simeq \frac{0.135}{D_h} \text{ in.}^{-1} \qquad (12\text{-}64)$$

Because $F \geq 1$, the axial nonuniformity in the heat flux reduces the CHF. The parameter C decreases with higher quality so that F also decreases and there is a less pronounced effect for the heat flux shape. In the limit of sufficiently high quality that dryout occurs, the value of F becomes unity. For low $x(z')$ (or high C), the upstream flux shape effects are more pronounced.

As discussed in Chapter 2, reactors operate at a safety margin called the *CHF ratio* (CHFR) (or departure of nucleate boiling ratio), which is defined as:

$$\text{CHFR} = \frac{q''_{cr,n}}{q''(z)} = \frac{q''_{cr}}{Fq''(z)} \tag{12-65}$$

For PWRs using the W-3 correlation, the margin of safety has been to require the minimum value of CHFR (called MCHFR) at any location in the core to be at least 1.3 at full power. The correlations adopted by the reactor vendors have undergone several modifications over the years. Generally, the newer correlations have less uncertainty and are therefore associated with smaller MCHFR values.

2 Dryout. In the high vapor quality regions of interest to BWRs, two approaches have been taken to establish the required design margin. The first approach was to develop a limit line. That is a conservative lower envelope to the appropriate CHF data, such that virtually no data points fall below this line. The first set of limit lines used by the General Electric Company was based largely on single-rod annular boiling transition data having uniform axial heat flux. These design lines were known as the Janssen–Levy limit lines [38]. They are considered valid for mass fluxes from 0.4×10^6 to 6.0×10^6 lb/hr ft², hydraulic diameters between 0.245 to 1.25 in., and system pressure from 600 to 1450 psia. For 1000 psia and hydraulic diameter less than 0.60 in., these limit line heat fluxes in Btu/hr ft² are expressed in terms of mass flux (G) and equilibrium quality (x_e) as:

$$(q''_{cr}/10^6) = 0.705 + 0.237 \, (G/10^6) \tag{12-66a}$$

for $x_e < 0.197 - 0.108 \, (G/10^6)$;

$$(q''_{cr}/10^6) = 1.634 - 0.270 \, (G/10^6) - 4.71 \, x_e \tag{12-66b}$$

for $0.197 - 0.108 \, (G/10^6) < x_e < 0.254 - 0.026 \, (G/10^6)$; and

$$(q''_{cr}/10^6) = 0.605 - 0.164 \, (G/10^6) - 0.653 \, x_e \tag{12-66c}$$

for $x_e > 0.254 - 0.026 \, (G/10^6)$.

For hydraulic diameters greater than 0.60 in., these three equations should be modified by subtracting:

$$2.19 \times 10^6 (D_h^2 - 0.36) \left[x_e - 0.0714 \left(\frac{G}{10^6} \right) - 0.22 \right] \tag{12-67}$$

At system pressures other than 1000 psia, the following pressure correction (in psi) was recommended:

$$q''_{cr}(p) = q''_{cr}(1000) + 400 \, (1000 - p) \tag{12-68}$$

An experimental program in four- and nine-rod uniform axial heat flux bundles was conducted later. Some adjustment to the Janssen–Levy limit lines were found to be needed, and thus the Hench–Levy limit lines were developed [32]. The Hench–Levy limit lines are considered valid for mass fluxes of 0.2×10^6 to 1.6×10^6 lb/hr ft^2, hydraulic diameters of 0.324 to 0.485 in., system pressures of 600 to 1450 psia, and rod-to-rod and rod-to-wall spacings greater than 0.060 in.

The mathematic expressions for the CHF (in Btu/hr ft^2) predictions by this correlation at 1000 psia are given by:

$$(q_{cr}''/10^6) = 1.0 \tag{12-69a}$$

for $x_e \le (x_e)_1$;

$$(q_{cr}''/10^6) = 1.9 - 3.3\, x_e - 0.7 \tanh^2 (3G/10^6) \tag{12-69b}$$

for $(x_e)_1 < x_e < (x_e)_2$; and

$$(q_{cr}''/10^6) = 0.6 - 0.7\, x_e - 0.09 \tanh^2 (2G/10^6) \tag{12-69c}$$

for $x_e \ge (x_e)_2$, where:

$$(x_e)_1 = 0.273 - 0.212 \tanh^2 (3G/10^6)$$
$$(x_e)_2 = 0.5 - 0.269 \tanh^2 (3G/10^6) + 0.0346 \tanh^2 (2G/10^6)$$

Values at various mass fluxes are shown in Figure 12-25. At system pressures other than 1000 psia, the following pressure correction is recommended:

$$q_{cr}''(p) = q_{cr}''(1000) \left[1.1 - 0.1 \left(\frac{p - 600}{400} \right)^{1.25} \right] \tag{12-70}$$

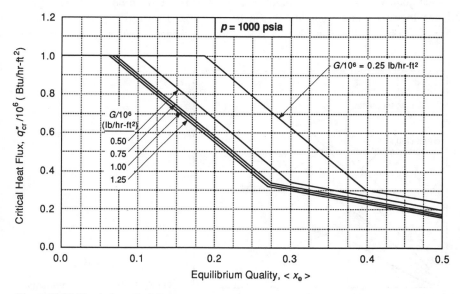

Figure 12-25 Hench–Levy limit lines.

These design curves were also constructed such that they fell below virtually all the data at each mass flux. As with the W-3 correlation, they have always been applied in BWR design with a margin of safety, in terms of an MCHFR. For the Hench–Levy limit lines the MCHFR was 1.9 for limiting transients. Obviously, although the limit line concept can be used for design purposes, it does not capture the axial heat flux effect; thus the axial CHF location is normally predicted incorrectly. It also is based on the assumption of local control of CHF conditions, which is invalid for high quality flow.

In order to eliminate the undesirable features inherent in the local CHF hypothesis, a new correlation, known as the General Electric critical quality-boiling length (GEXL) correlation, was developed [26]. The GEXL correlation is based on a large amount of boiling transition data taken in General Electric's ATLAS Heat Transfer Facility, which includes full-scale 49- and 64-rod data. The generic form of the GEXL correlation is:

$$x_{cr} = x_{cr}(L_B, D_H, G, L, p, R)$$

where x_{cr} = bundle average critical quality; L_B = boiling length (i.e., downstream distance from the position at which the equilibrium quality equals zero); D_H = heated diameter (i.e., four times the ratio of total flow area to heated rod perimeter); G = mass flux; L = total heated length; p = system pressure; and R = a parameter that characterizes the local peaking pattern with respect to the most limiting rod.

Unlike the limit line approach, the GEXL correlation is a "best fit" to the experimental data and is said to be able to predict a wide variety of data with a standard deviation of about 3.5% [26]. Additionally, the GEXL correlation is a relation between parameters that depend on the total heat input from the channel inlet to a position within the channel. The correlation line is plotted in terms of critical bundle radially averaged quality versus boiling length (Fig. 12-26). For a given bundle power ($\dot{Q}_1$ or $\dot{Q}_2$), the locus of bundle conditions can be represented on this figure. The critical power of a bundle is the value that leads to a point of tangency between the correlation

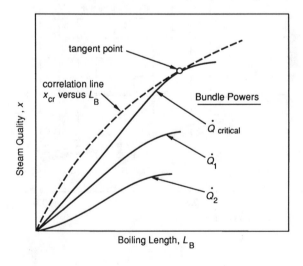

Figure 12-26 Correlation of critical conditions under dryout.

line and a bundle operating condition (Fig. 12-26). The critical quality–boiling length correlation leads to identification of a new margin of safety in BWR designs, i.e., the *critical power ratio* (CPR) of a channel:

$$CPR = \frac{\text{critical power}}{\text{operating power}}$$

The GEXL correlation has been kept as proprietary information.

D General CHF Correlations

The CHF correlations considered here are the round-tube correlations of Biasi et al. [8], Bowring [13], and CISE-4 [25], and the annulus correlation of Barnett [3]. The databases for all correlations are summarized in Table 12-4. Figure 12-27 illustrates the pressure and mass flux range of the databases.

1 Biasi correlation. The Biasi [8] correlation is a function of pressure, mass flux, flow quality, and tube diameter. The root-mean-square (rms) error of the correlation is 7.26% for more than 4500 data points, and 85.5% of all points are within $\pm$ 10%. The correlation is capable of predicting both DNB and dryout CHF conditions. The correlation is given by the following equations.

Use Eq. 12-71b, below, for $G < 300$ kg/m^{2}s; for higher G, use the larger of the two values. Hence:

Table 12-4 CHF correlation databases

Author	Database
Biasi [8]	$D = 0.0030$–0.0375 m $L = 0.2$–6.0 m $p = 0.27$–14 MPa $G = 100$–6000 kg/m$^2 \cdot$ s $x = 1/(1 + \rho_l/\rho_g)$ to 1
CISE-4 [25]	$D = 0.0102$–0.0198 m $L = 0.76$–3.66 m $p = 4.96$–6.89 MPa $G = 1085$–4069 kg/m$^2 \cdot$ s
Bowring [13]	$D = 0.002$–0.045 m $L = 0.15$–3.7 m $p = 0.2$–19.0 MPa $G = 136$–$18{,}600$ kg/m$^2 \cdot$ s
Barnett [3]	$D_l = 0.0095$–0.0960 m $D_s = 0.014$–0.102 m $L = 0.61$–2.74 m $p = 6.9$ MPa $G = 190$–8409 kg/m$^2 \cdot$ s

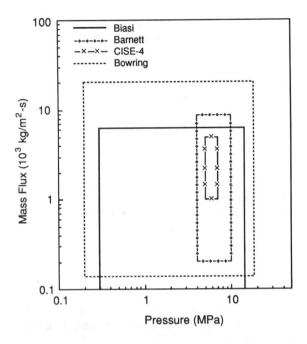

Figure 12-27 Comparison of CHF correlation databases.

$$q''_{\text{Biasi}} = (2.764 \times 10^7)(100D)^{-n}G^{-1/6}[1.468F(p_{\text{bar}})G^{-1/6} - x] \text{ W/m}^2 \quad (12\text{-}71a)$$

$$q''_{\text{Biasi}} = (15.048 \times 10^7)(100D)^{-n} G^{-0.6}H(p_{\text{bar}}) [1 - x] \text{ W/m}^2 \quad (12\text{-}71b)$$

where:

$$F(p_{\text{bar}}) = 0.7249 + 0.099p_{\text{bar}} \exp(-0.032p_{\text{bar}}) \quad (12\text{-}71c)$$

$$H(p_{\text{bar}}) = -1.159 + 0.149p_{\text{bar}} \exp(-0.019p_{\text{bar}}) \quad (12\text{-}71d)$$
$$+ 9p_{\text{bar}} (10 + p_{\text{bar}}^2)^{-1}$$

Note: $p_{\text{bar}} = 10p$, when p is in MPa; and:

$$n = \begin{cases} 0.4 & D \geq 0.01 \text{ m} \\ 0.6 & D < 0.01 \text{ m} \end{cases} \quad (12\text{-}71e)$$

2 Bowring correlation. The Bowring correlation [13] contains four optimized pressure parameters. The rms error of the correlation is 7% for its 3800 data points. The correlation probably has the widest range of applicability in terms of pressure and mass flux (Fig. 12-27). The correlation is described by the following equations in SI units:

$$q''_{\text{cr}} = \frac{A - Bh_{\text{fg}}x}{C} \text{ W/m}^2 \quad (12\text{-}72a)$$

where:

$$A = \frac{2.317(h_{\text{fg}}DG/4)F_1}{1 + 0.0143F_2D^{1/2}G} \quad (12\text{-}72b)$$

$$B = \frac{DG}{4} \tag{12-72c}$$

$$C = \frac{0.077 F_3 \, DG}{1 + 0.347 F_4 \left(\dfrac{G}{1356}\right)^n} \tag{12-72d}$$

$$p_R = 0.145p \text{ (where } p \text{ is in MPa)} \tag{12-72e}$$

$$n = 2.0 - 0.5 p_R \tag{12-72f}$$

For $p_R < 1$ MPa:

$$\left. \begin{aligned}
F_1 &= \{p_R^{18.942} \exp [20.89(1 - p_R)] + 0.917\}/1.917 \\
F_2 &= F_1/(\{p_R^{1.316} \exp [2.444(1 - p_R)] + 0.309\}/1.309) \\
F_3 &= \{p_R^{17.023} \exp [16.658(1 - p_R)] + 0.667\}/1.667 \\
F_4 &= F_3 p_R^{1.649}
\end{aligned} \right\} \tag{12-72g}$$

For $p_R > 1$ MPa:

$$\left. \begin{aligned}
F_1 &= p_R^{-0.368} \exp [0.648(1 - p_R)] \\
F_2 &= F_1/\{p_R^{-0.448} \exp [0.245(1 - p_R)]\} \\
F_3 &= p_R^{0.219} \\
F_4 &= F_3 p_R^{1.649}
\end{aligned} \right\} \tag{12-72h}$$

3 CISE-4 correlation. The CISE-4 correlation [25] is a modification of CISE-3 in which the critical flow quality (x_c) approaches 1.0 as the mass flux decreases to 0. Unlike others, this correlation is based on the quality-boiling length concept and is restricted to BWR applications. The correlation was optimized in the flow range of $1000 < G < 4000$ kg/m² s. It has been suggested that the application of the CISE correlation in rod bundles is possible by including the ratio of heated to wetted perimeters in the correlation. The correlation is expressed by the following equations:

$$x_{cr} = \frac{D_h}{D_e} \left(a \frac{L_{cr}}{L_{cr} + b} \right) \tag{12-73a}$$

where:

$$a = \frac{1}{1 + 1.481 \times 10^{-4}(1 - p/p_c)^{-3}G} \quad \text{if } G \le G^* \tag{12-73b}$$

and

$$a = \frac{1 - p/p_c}{(G/1000)^{1/3}} \quad \text{if } G \ge G^* \tag{12-73c}$$

where $G^* = 3375(1 - p/p_c)^3$; p_c = critical pressure (MPa); L_{cr} = boiling length to CHF (m); and:

$$b = 0.199(p_c/p - 1)^{0.4} GD^{1.4} \tag{12-73e}$$

where G is in kg/m²s and D is in meters (m).

4 Barnett correlation. The Barnett correlation [3] has the same basic form as MacBeth's correlation. The rms error of the correlation is 5.9% based on 724 data points obtained in uniformly heated annuli at a pressure of 1000 psi. The application of the correlation is extended to rod bundles using the equivalent diameter concept. The correlation consists of the following equations in SI units:

$$q''_{cr} = 3.1546 \times 10^6 \frac{3.584 \, Ah_{fg} + 4.3 \times 10^{-4} B(h_f - h_i)}{C + 39.37L} \tag{12-74a}$$

where:

$$L = \text{heated length (m)}$$
$$A = 230.7 \, D_h^{0.68} \, G^{0.192}[1 - 0.744 \exp(-0.3477D_eG)] \tag{12-74b}$$
$$B = 0.1206D_h^{1.415}G^{0.817} \tag{12-74c}$$
$$C = 8249D_e^{1.415}G^{0.212} \tag{12-74d}$$

For annuli, the wetted and heated equivalent diameters (D_e and D_h) are given by:

$$D_e = (D_s - D_i) \tag{12-74e}$$
$$D_h = (D_s^2 - D_i^2)/D_i \tag{12-74f}$$

where D_s = diameter of the shroud; and D_i = diameter of the inner rod.

Example 12-3 Comparison of critical heat flux correlations

PROBLEM A vertical test tube in a high pressure water boiling channel has the following characteristics:

$p = 6.89$ MPa $= 68.9$ bar
$D = 10.0$ mm
$L = 3.66$ m
$T_{in} = 204°C$

Using the Biasi and CISE-4 correlations, find the critical channel power for uniform heating at $G = 2000$ kg/m² s.

SOLUTION

Biasi correlation

Because $G > 300$ kg/m² s, check Eqs. 12-71a and 12-71b. Use the larger q''_{Biasi}. In this case Eq. 12-71b is the larger.

$$q''_{Biasi} = 15.048 \times 10^7 \, (100 \, D)^{-n} \, G^{-0.6} \, H(p_{bar})(1 - x)$$

$$H(p_{bar}) = -1.159 + 0.149 \, p_{bar} \exp(-0.019 \, p_{bar})$$
$$+ 9p_{bar}/(10 + p_{bar}^2) \tag{12-71b}$$
$$= -1.159 + 0.149(68.9) \exp[(-0.019(68.9)]$$
$$+ 9 (68.9)/[10 + (68.9)^2]$$
$$= 1.744$$

Because $D = 0.01$ m, $n = 0.4$.
An expression is needed for x, which is obtained from a heat balance:

$$q_{cr} = \dot{m} x \, h_{fg} + \dot{m} \, \Delta h_{sub}$$

Hence

$$x(z) = \frac{q_{cr}}{\dot{m} h_{fg}} - \frac{\Delta h_{sub}}{h_{fg}} = \frac{4q_{cr}'' \, z}{GD \, h_{fg}} - \frac{\Delta h_{sub}}{h_{fg}}$$

For uniform heating, we want $z = L$ and $q_{Biasi}'' = q_{cr}''$.

$$\Delta h_{sub} = 0.389 \times 10^6 \text{ J/kg}$$
$$h_{fg} = 1.51 \times 10^6 \text{ J/kg}$$

Thus:

$$q_{cr}'' = 15.048 \times 10^7 \, [100(0.010)]^{-0.4} \, (2000)^{-0.6} \, (1.744)$$
$$* \left[1 - \left(\frac{4q_{cr}''(3.66)}{2000(0.01)(1.51 \times 10^6)} - \frac{0.386 \times 10^6}{1.51 \times 10^6} \right) \right]$$
$$q_{cr}'' = 3.423 \times 10^6 - 1.301 \, q_{cr}'' \rightarrow q_{cr}'' = 1.488 \times 10^6 \text{ W/m}^2$$
$$q_{cr} = q_{cr}'' \, \pi D L = (1.488 \times 10^6)\pi(0.01)(3.66)$$
$$q_{cr} = 173.8 \text{ kW}$$

CISE-4 correlation

We need expressions for x_{cr} and L_{cr} to use in Eq. 12-73a. They come from energy balance considerations:

$$q_{cr}'' \, \pi D(L - L_{cr}) = G \frac{\pi D^2}{4} \, \Delta h_{sub}$$

$$\therefore L_{cr} = L - \frac{GD\Delta h_{sub}}{4q_{cr}''}$$

$$q_{cr}'' \, \pi D L_{cr} = G \frac{\pi D^2}{4} \, x_{cr} \, h_{fg}$$

$$\therefore x_{cr} = \frac{4q_{cr}'' L_{cr}}{GD h_{fg}}$$

Thus Eq. 12-73a becomes:

$$\frac{4q_{cr}''}{GD h_{fg}} \left(L - \frac{GD\Delta h_{sub}}{4q_{cr}''} \right) = \frac{D_h}{D_e} \, a \left[\frac{\left(L - \frac{GD\Delta h_{sub}}{4q_{cr}''} \right)}{L - \frac{GD\Delta h_{sub}}{4q_{cr}''} + b} \right]$$

Table 12-5 Comparison of General Electric nine-rod bundle CHF predictions*

Run number	G_2 (kg/m² · s)	x_{CHF}	CISE-4 (MCPR)		Biasi (MCHFR)		Bowring (MCHFR)		Barnett (MCHFR)	
			Subchannel method	Bundle-averaged method	Subchannel method	Bundle-averaged method	Subchannel method	Bundle-averaged method	Subchannel method	Bundle-averaged method
278	681	0.4885	0.854	0.867	1.192	1.548	1.634	2.109	0.708	0.794
279	678	0.4640	0.883	0.903	1.222	1.511	1.676	2.061	0.713	0.808
280	678	0.4242	0.932	0.957	1.272	1.490	1.735	2.024	0.770	0.864
271	1024	0.3749	0.893	0.929	1.174	1.357	1.351	1.633	0.760	0.828
272	1024	0.3518	0.929	0.963	1.148	1.377	1.356	1.587	0.763	0.835
273	1020	0.3328	0.948	0.985	1.156	1.376	1.393	1.577	0.775	0.850
266	1367	0.2957	0.935	0.970	1.194	1.428	1.206	1.425	0.754	0.855
267	1358	0.2582	1.019	1.060	1.300	1.450	1.307	1.456	0.807	0.893
268	1362	0.2349	1.030	1.061	1.282	1.368	1.304	1.404	0.841	0.901
297	1690	0.2038	1.056	1.090	1.327	1.418	1.231	1.216	0.856	0.914
298	1691	0.1783	1.025	1.108	1.225	1.280	1.157	1.217	0.807	0.890
299	1687	0.1510	1.043	1.132	1.228	1.258	1.183	1.215	0.584	0.901

*Pressure = 6.9 MPa.

or

$$\frac{4q''_{cr}}{GDh_{fg}} = \frac{D_h}{D_e} \left(\frac{a}{L - \dfrac{GD\Delta h_{sub}}{4q''_{cr}} + b} \right)$$

where $D_h = D_e = D = 0.01$ m; $p = 68.9$ bar $= 6.89$ MPa; $p_{cr} = 22.04$ MPa; and $G^* = 3375(1 - 6.89/22.04)^3 = 1096.17$ kg/m$^2 \cdot$ s.

$$\therefore G > G^*$$

$$\therefore a = \frac{1 - p/p_c}{(G/1000)^{1/3}} = \frac{1 - 6.89/22.04}{(2000/1000)^{1/3}} = 0.54558$$

$$b = 0.199 \, (p_c/p - 1)^{0.4} \, GD^{1.4}$$

$$= 0.199(22.04/6.89 - 1)^{0.4} \, (2000)(0.01)^{1.4} = 0.8839$$

Thus:

$$\frac{4q''_{cr}}{(2000)(0.01)(1.51 \times 10^6)} = \frac{0.54558}{3.66 - \dfrac{(2000)(0.01)(0.389 \times 10^6)}{4q''_{cr}} + 0.8839}$$

$$1.30365 \times 10^{-7} \, q''_{cr} = \frac{0.54558}{4.544 - 1.976 \times 10^6/q''_{cr}}$$

$$5.92366 \times 10^{-7} \, q''_{cr} = 0.54558 + 0.2576$$

$$q''_{cr} = 1{,}356 \text{ kW/m}^2$$

$$q_{cr} = q''_{cr} \, \pi DL = (1{,}355{,}890.0)\pi(0.01)(3.66)$$

$$q_{cr} = 158.4 \text{ kW}$$

E Special Considerations in Rod Bundles

The CHF in rod bundles can be predicted by considering the average flow conditions in the bundle as well as by considering the flow conditions on a subchannel analysis. Kao and Kazimi [41] compared four correlations, applied in both approaches, to results from the GE nine-rod bundle tests [37]. The cross-sectional view of the GE bundle is illustrated in Figure 12-28. The rod diameter and the spacing between rods resemble those in a typical BWR rod bundle. The test section was uniformly heated, with a heated length of 3.66 m. The test conditions are also given in Figure 12-28. During the test, CHF was approached by holding the pressure, flow, and subcooling constant while gradually increasing the power until one or more thermocouples indicated CHF conditions (rapid temperature excursion). It was observed that CHF occurred usually in the corner rods, at the end of the heated section or immediately upstream of the spacer at the end of the heated section.

With the subchannel method, the traditional coolant-centered subdivision was used (Fig. 12-28). The THERMIT code was used to predict the flow distribution [43]. Table 12-5 shows the predicted minimum critical power ratios (MCPR) and minimum CHF ratios (MCHFR) with both the subchannel and bundle-averaged methods. The

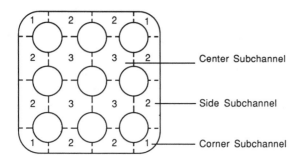

Geometrical Details (mm)

Rod diameter	14.3
Rod-rod gap	4.42
Rod wall gap	3.51
Radius of corner	10.2
Heated length	1829

Test Conditions

Pressure	6.9 MPa
Mass flux	339 to 1695 kg/m^{2}s
Inlet subcooling	16.5 to 800 kJ/kg

Figure 12-28 Cross-sectional view of General Electric nine-rod bundle.

MCPR is defined on a subchannel basis for the subchannel approach and on a bundle basis for the bundle-averaged approach. As seen, the subchannel method consistently predicts a lower MCPR than the bundle-averaged approach. Additionally, the Biasi and Bowring correlations were found to overestimate the MCHFRs by more than 15% and to predict CHF to occur first in the center subchannel. Although Barnett's correlation was the most conservative of all, the CHF location was correctly predicted to be in the corner subchannel.

The CISE-4 and Barnett correlations predicted the CHF to occur within the top axial node, which was 200 mm in height. It should be noted that CISE-4 includes a factor that accounts for the difference between heated and wetted parameters, and Barnett allowed for a heated equivalent diameter different from the wetted equivalent diameter, which may be the reason that they were able to predict the correct CHF location, i.e., the corner rod.

Furthermore, the subchannel and bundle-averaged method results for CISE-4 are the closest together among the correlations studied, as well as the closest to the measured CHF condition.

REFERENCES

1. Bailey, N. A. *The Interaction of Droplet Deposition and Forced Convection in Post-Dryout Heat Transfer at High Subcritical Pressure.* AEEW-R807, 1973.

2. Bailey, N. A., Collier, J. G., and Ralph, J. C. *Post Dryout Heat Transfer in Nuclear and Cryogenic Equipment.* AERE-R7519, 1973.
3. Barnett, P. G. *A Correlation of Burnout Data for Uniformly Heated Annuli and Its Use for Predicting Burnout in Uniformly Heat Rod Bundles.* AEEW-R-463, U.K. Atomic Energy Authority, 1966.
4. Bennett, J. A., et al. Heat transfer to two-phase gas liquid systems. I. Steam/water mixtures in the liquid dispersed region in an annulus. *Trans. Inst. Chem. Eng.* 39:113, 1961; and AERE-R-3519, 1959.
5. Bennett, A. W., et al. *Heat Transfer to Steam-Water Mixture Flowing in Uniformly Heated Tubes in Which the Critical Heat Flux Has Been Exceeded.* AERE-R5373, 1967.
6. Berenson, P. J. Film boiling heat transfer from a horizontal surface. *J. Heat Transfer* 83:3, 1961.
7. Bergles, A. E., and Rohsenow, W. M. The determination of forced convection surface boiling heat transfer. *J. Heat Transfer* 86:363, 1964.
8. Biasi, L., et al. Studies on burnout. Part 3. *Energy Nucl.* 14:530, 1967.
9. Blander, M., and Katz, J. L. Bubble nucleation in liquids. *AIChE J.* 21:833, 1975.
10. Bjorge, R., Hall, G., and Rohsenow, W. Correlation for forced convection boiling heat transfer data. *Int. J. Heat Mass Transfer* 20:763, 1982.
11. Bjornard, T. A., and Griffith, P. PWR blowdown heat transfer. In O. C. Jones and G. Bankoff (eds.), *Symposium on Thermal and Hydraulic Aspects of Nuclear Reactor Safety* (Vol. 1). New York: ASME, 1977.
12. Bowring, R. W. *Physical Model Based on Bubble Detachment and Calculation of Steam Voidage in the Subcooled Region of a Heated Channel.* Report HPR10. Halden, Norway: Halder Reactor Project, 1962.
13. Bowring, R. W. *Simple but Accurate Round Tube, Uniform Heat Flux Dryout Correlation over the Pressure Range 0.7 to 17 MPa.* AEEW-R-789. U.K. Atomic Energy Authority, 1972.
14. Chen, J. C. *A Correlation for Boiling Heat Transfer to Saturated Fluids in Convective Flow.* ASME paper 63-HT-34, 1963.
15. Chen, J. C. A correlation for boiling heat transfer in convection flow. *ISEC Process Design Dev.* 5:322, 1966.
16. Cheng, S. C., Ng, W. W., and Heng, K. T. Measurements of boiling curves of subcooled water under forced convection conditions. *Int. J. Heat Mass Transfer* 21:1385, 1978.
17. Collier, J. G. *Convective Boiling and Condensation* (2nd ed.). New York: McGraw-Hill, 1981.
18. Collier, J. G. Heat transfer in the post dryout region and during quenching and reflooding. In G. Hetsroni (ed.), *Handbook of Multiphase Systems Systems.* New York: Hemisphere, 1982.
19. Collier, J. G. Post dryout heat transfer. In A. Bergles et al. (eds.), *Two-Phase Flow and Heat Transfer in the Power and Process Industries.* New York: Hemisphere, 1981.
20. Collier, J. G., and Pulling, D. J. *Heat Transfer to Two-Phase Gas-Liquid Systems, Part II.* Report AERE-R3809. Harwell: UKAEA, 1962.
21. Davis, E. J., and Anderson, G. H. The incipience of nucleate boiling in forced convection flow. *AIChE J.* 12:774, 1966.
22. Dengler, C. E., and Addoms, J. N. Heat transfer mechanism for vaporization of water in a vertical tube. *Chem. Eng. Prog. Symp. Ser.* 52:95, 1956.
23. Dix, G. E. *Vapor Void Fraction for Forced Convection with Subcooled Boiling at Low Flow Rates.* Report NADO-10491. General Electric, 1971.
24. Forster, K., and Zuber, N. Dynamics of vapor bubbles and boiling heat transfer. *AIChE J.* 1:531, 1955.
25. Gaspari, G. P., et al. A rod-centered subchannel analysis with turbulent (enthalpy) mixing for critical heat flux prediction in rod clusters cooled by boiling water. In: *Proceedings 5th International Heat Transfer Conference,* Tokyo, Japan, 3–7 September 1974, CONF-740925, 1975.
26. *General Electric BWR Thermal Analysis Basis (GETAB) Data: Correlation and Design Applications.* NADO-10958, 1973.
27. Griffith, P. J., Clark, A., and Rohsenow, W. M. *Void Volumes in Subcooled Boiling Systems.* ASME paper 58-HT-19, 1958.
28. Groeneveld, D. C. Effect of a heat flux spike on the downstream dryout behavior. *J. Heat Transfer* 96C:121, 1974.
29. Groeneveld, D. C. *Post Dryout Heat Transfer at Reactor Operating Conditions.* AECL-4513, 1973.

30. Groeneveld, D. C. *The Effect of Short Flux Spikes on the Downstream Dryout Behavior*. AECL-4927, 1975.

31. Groeneveld, D. C., and Fung, K. K. Heat transfer experiments in the unstable post-CHF region. Presented at the Water Reactor Safety Information Exchange Meeting, Washington, D.C., 1977.

32. Healzer, J. M., Hench, et al. *Design Basis for Critical Heat Flux Condition in Boiling Water Reactors*. APED-5286, General Electric, 1966.

33. Henry, R. E. A correlation for the minimum wall superheat in film boiling. *Trans. Am. Nucl. Soc.* 15:420, 1972.

34. Hewitt, G. F. Experimental studies on the mechanisms of burnout in heat transfer to steam-water mixture. In: *Proceedings 4th International Heat Transfer Conference*, paper B6.6, Versailles, France, 1970.

35. Hsu, Y. Y., and Graham, R. W. *Analytical and Experimental Study of Thermal Boundary Layer and Ebullition Cycle*. NASA Technical Note TNO-594, 1961.

36. Iloeje, O. C., Plummer, D. N., Rohsenow, W. M., and Griffith, P. *A Study of Wall Rewet and Heat Transfer in Dispersed Vertical Flow*. MIT, Department of Mechanical Engineering Report 72718-92, 1974.

37. Janssen, E. *Two-Phase Flow and Heat Transfer in Multirod Geometries, Final Report*. GEAP-10347. General Electric, 1971.

38. Janssen, E., and Levy, S. *Burnout Limit Curves for Boiling Water Reactors*. APED-3892, General Electric, 1962.

39. Jens, W. H., and Lottes, P. A. *Analysis of Heat Transfer, Burnout, Pressure Drop and Density Data for High Pressure Water*. ANL-4627, 1951.

40. Kalinin, E. K., et al. *Investigation of the Crisis of Film Boiling in Channels*. Moscow Aviation Institute, USSR, 1968.

41. Kao, S. P., and Kazimi, M. S. Critical heat flux predictions in rod bundles. *Nucl. Tech.* 60:7, 1983.

42. Keeys, R. K. F., Ralph, J. C., and Roberts, D. N. *Post Burnout Heat Transfer in High Pressure Steam-Water Mixtures in a Tube with Cosine Heat Flux Distribution*. AERE-R6411. U.K. Atomic Energy Research Establishment, 1971.

43. Kelly, J. E., Kao, S. P., and Kazimi, M. S. *THERMIT-2: A Two-Fluid Model for Light Water Reactor Subchannel Transient Analysis*. Energy Laboratory Report MIT-EL 81-014. Cambridge: Massachusetts Institute of Technology, 1981.

44. Kumamaru, H., Koizumi, Y., and Tasaka, K. Investigation of pre- and post-dryout heat transfer of steam-water two-phase flow in a rod bundle. *Nucl. Eng. Design* 102:71, 1987.

45. Kutateladze, S. S. *Heat Transfer in Condensation and Boiling*. AEC-TR-3770, 1952.

46. Lahey, R. T., and Moody, F. J. *The Thermal Hydraulic of Boiling Water Reactor*. American Nuclear Society, 1977, p. 152.

47. Levy, S. Forced convection subcooled boiling-prediction of vapor volumetric fraction. *Int. J. Heat Mass Transfer* 10:951, 1967.

48. McDonough, J. B., Milich, W., and King, E. C. An experimental study of partial film boiling region with water at elevated pressures in a round vertical tube. *Chem. Eng. Prog. Sym. Series* 57:197, 1961.

49. Plummer, D. N., Iloeje, O. C., Rohsenow, W. M., and Griffith, P. *Post Critical Heat Transfer to Flowing Liquid in a Vertical Tube*. MIT Department of Mechanical Engineering Report 72718-91, 1974.

50. Ramu, K., and Weisman, J. A method for the correlation of transition boiling heat transfer data. Paper B.4.4. In: *Proceedings 5th International Heat Transfer Meeting*, Tokyo, 1974.

51. Rohsenow, W. M. Boiling. In W. M. Rohsenow, J. P. Hartnett, and E. N. Ganic (eds.), *Handbook of Heat Transfer Fundamentals* (2nd ed.). New York: McGraw-Hill, 1985.

52. Rohsenow, W. M. Pool boiling. In G. Hetsroni (ed.), *Handbook of Multifluid Systems*. New York: Hemisphere, 1982.

53. Saha, P., and Zuber, N. Point of net vapor generation and vapor void fraction in subcooled boiling. In: *Proceedings 5th International Heat Transfer Conference*, Tokyo, 1974, pp. 175–179.

54. Schrock, V. E., and Grossman, L. M. Forced convection boiling in tubes. *Nucl. Sci. Eng.* 12:474, 1962.

55. Slaughterback, D. C., et al. Flow film boiling heat transfer correlations—parametric study with data comparisons. Paper presented at the National Heat Transfer Conference, Atlanta, 1973.

56. Smith, O. G., Tong, L. S., and Rohrer, W. M. *Burnout in Steam-Water Flow with Axially Nonuniform Heat Flux.* ASME paper, WA/HT-33, 1965.

57. Spiegler, P., et al. Onset of stable film boiling and the foam limit. *Int. J. Heat Mass Transfer* 6:987, 1963.

58. Staub, F. The void fraction in subcooled boiling—prediction of the initial point of net vapor generation. *J. Heat Transfer* 90:151, 1968.

59. Thom, J. R. S., et al. Boiling in subcooled water during flow in tubes and annuli. *Proc. Inst. Mech. Eng.* 180:226, 1966.

60. Tong, L. S. *Boiling Crisis and Critical Heat Flux.* USAEC Critical Review Series, Report TID-25887, 1972.

61. Tong, L. S. Heat transfer in water cooled nuclear reactors. *Nucl. Eng. Design* 6:301, 1967.

62. Tong, L. S. Heat transfer mechanisms in nucleate and film boiling. *Nucl. Eng. Design* 21:1, 1972.

63. Varone, A. F., and Rohsenow, W. M. Post dryout heat transfer prediction. Presented at International Workshop on Fundamental Aspects of Post Dryout Heat Transfer, Salt Lake City, Utah, NUREG-0060, 1984.

64. Zuber, N. *Hydrodynamic Aspects of Boiling Heat Transfer.* AECU-4439, 1959.

PROBLEMS

Problem 12-1 Comparison of liquid superheat required for nucleation in water and sodium (section II)

1. Calculate the relation between superheat and equilibrium bubble radius for sodium at 1 atmosphere using Eq. 12-7 and compare it to Eq. 12-8. What are the two relations for water at 1 atmosphere? Does sodium require higher or lower superheat than water?

2. Consider bubbles of 10 μm radius. Evaluate the vapor superheat and the bubble pressure difference for the bubbles of question 1.

Answers:
1. For sodium:

$$T_v - T_{sat} = \frac{1154\psi}{1 - \psi}; \; \psi = 0.11 \; \ell n \left[1 + \frac{0.223 \times 10^{-5}}{r^*} \right]$$

For water:

$$T_v - T_{sat} = \frac{373\psi}{1 - \psi}; \; \psi = 0.0763 \; \ell n \left[1 + \frac{0.116 \times 10^{-5}}{r^*} \right]$$

2. $\quad (T_v - T_{sat})_{sodium} = 25.2°C; \; (T_v - T_{sat})_{water} = 3.25°C$

$$(p_v - p_\ell)_{sodium} = 22.6 \; kN/m^2; \; (p_v - p_\ell)_{water} = 11.8 \; kN/m^2$$

Problem 12-2 Evaluation of pool boiling conditions at high pressures (section III)

1. A manufacturer has free access on the weekends to a supply of 8000 amps of 440 V electric power. If his boiler operates at 3.34 MPa, how many 2.5 cm diameter, 2 m long electric emersion heaters would be required to utilize the entire available electric power? He desires to operate at 80% of critical heat flux.

2. At what heat flux would incipient boiling occur? Assume that the water at the saturation temperature corresponds to 3.35 MPa. The natural convection heat flux is given by:

$$q_{NC}'' = 2.63 \; (\Delta T)^{1.25} \; kW/m^2$$

$p = 3.35$ MPa
$T_{sat} = 240°C$
$h_{fg} = 1803$ kJ/kg
$\rho_\ell = 813$ kg/m^3
$\rho_v = 17.3$ kg/m^3
$\sigma = 0.0204$ N/m
$k = 0.606$ W/m^2 °C

Assume that the maximum cavity radius is very large.

Answers:

1. $N = 6$
2. $q_i'' = 3.04$ kW/m^2

Problem 12-3 Comparison of stable film boiling conditions in water and sodium (section III)
Compare the value of the wall superheat required to sustain film boiling on a horizontal steel wall as predicted by Berenson's correlation to that predicted by Henry's correlation (Table 12-1). For simplicity, calculate the properties with subscript vf as if they were for saturated vapor.

1. Consider the cases of saturated water at (1) atmospheric pressure and (2) $p = 7.0$ MPa.
2. Consider the case of saturated sodium at atmospheric pressure.

Answers:

1. H$_2$O at 0.1 MPa:	$T_B^M - T_{sat} = 176°F$	$T_H^M - T_{sat} = 406°F$
H$_2$O at 7.0 MPa:	$T_B^M - T_{sat} = 194°F$	$T_H^M - T_{sat} = 374°F$
2. Na at 0.1 MPa:	$T_B^M - T_{sat} = 206°F$	$T_H^M - T_{sat} = 1071°F$

Problem 12-4 Factors affecting incipient superheat in a flowing system (section IV)

1. Saturated liquid water at atmospheric pressure flows inside a 20 mm diameter tube. The mass velocity is adjusted to produce a single-phase heat-transfer coefficient equal to 10 kW/m^2°K. What is the incipient boiling heat flux? What is the corresponding wall superheat?
2. Provide answers to the same questions for saturated liquid water at 290°C, flowing through a tube of the same diameter, and with a mass velocity adjusted to produce the same single-phase heat-transfer coefficient.
3. Provide answers to the same questions if the flow rate in the 290°C case is doubled.

Answers:

1. $q_i'' = 1.91 \times 10^4$ W/m^2
   $T_w - T_{sat} = 1.91°C$
2. $q_i'' = 220.9$ W/m^2
   $T_w - T_{sat} = 0.022°C$
3. $q_i'' = 670$ W/m^2
   $T_w - T_{sat} = 0.038°C$

Problem 12-5 Critical heat flux correlation (section VII)
Using the data of Example 12-3, calculate the critical heat flux using the Bowring correlation.
Answer: $q_{cr}'' = 1.37 \times 10^6$ W/m^2

SINGLE HEATED CHANNEL: STEADY-STATE ANALYSIS

I INTRODUCTION

Solutions of the mass, momentum, and energy equations of the coolant in a single channel are presented in this chapter. The channel is generally taken as a representative coolant subchannel within an assembly, which is assumed to receive coolant only through its bottom inlet. The fuel and clad heat transport equations are also solved under steady-state conditions for the case in which they are separable from the coolant equations.

We start with a discussion of one-dimensional transient transport equations of the coolant with radial heat input from the clad surfaces. It is assumed that the flow area is axially uniform, although form pressure losses due to local area changes (e.g., spacers) can still be accounted for. Solutions for the coolant equations are presented for steady state in this chapter and for transient conditions in Chapter 2, Vol. II.

II FORMULATION OF ONE-DIMENSIONAL FLOW EQUATIONS

Consider the coolant as a mixture of liquid and vapor flowing upward, as shown in Figure 13-1. Following the basic concepts discussed in Chapter 5, the radially averaged coolant flow equations can be derived by considering the flow area at any axial position as the control area. For simplicity we consider the mixture equations of a two-phase system rather than deal with each fluid separately.

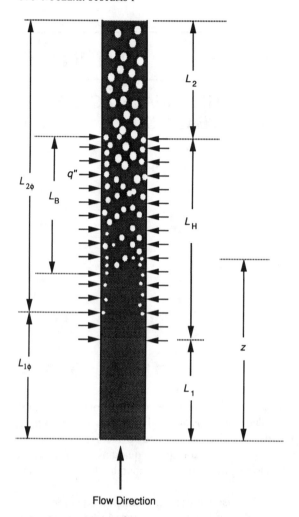

Figure 13-1 Surface heated flow channel.

Flow Direction

A Nonuniform Velocities

The one-dimensional mass, momentum, and energy transport equations have been derived in Chapter 5 and can be written as follows.

Mass:

$$\frac{\partial}{\partial t} (\rho_m A_z) + \frac{\partial}{\partial z} (G_m A_z) = 0 \tag{5-63}$$

where

$$\rho_m = \{\rho_v \alpha\} + \{\rho_\ell (1 - \alpha)\} \tag{5-50b}$$

$$G_m = \{\rho_v \alpha v_{vz}\} + \{\rho_\ell (1 - \alpha) v_{\ell z}\} \tag{5-40c}$$

Momentum:

$$\frac{\partial}{\partial t}(G_m A_z) + \frac{\partial}{\partial z}\left(\frac{G_m^2}{\rho_m^+} A_z\right) = -\frac{\partial(p\,A_z)}{\partial z} - \int_{P_z} \tau_w dP_z - \rho_m g A_z \cos\theta \quad (5\text{-}66)$$

where:

$$\frac{1}{\rho_m^+} \equiv \frac{1}{G_m^2}\{\rho_v \alpha v_{vz}^2 + \rho_\ell(1-\alpha)v_{\ell z}^2\} \quad (5\text{-}67)$$

$$p = p_v \alpha + p_\ell(1-\alpha) \quad (5\text{-}142)$$

$$\theta = \text{angle of the z direction with the upward vertical}$$

$$\frac{1}{A_z}\int_{P_z} \tau_w \, dP_z = \left(\frac{\partial p}{\partial z}\right)_{\text{fric}} \quad (11\text{-}54)$$

Note that in Chapter 11 it was shown that the friction pressure gradient for two-phase flow can be related to the wall shear stress and the momentum flux by terms analogous to the single-phase flow case:

$$\left(\frac{\partial p}{\partial z}\right)_{\text{fric}} = -\frac{\tau_w P_w}{A_z} = f_{TP}\frac{G_m^2}{D_e 2\rho_m} \quad (11\text{-}77)$$

Energy:

$$\frac{\partial}{\partial t}[(\rho_m h_m - p)A_z] + \frac{\partial}{\partial z}(G_m h_m^+ A_z) = q_m''' A_z - q_w'' P_w + \frac{G_m}{\rho_m}\left[F_{wz}''' + \frac{\partial p}{\partial z}\right]A_z \quad (5\text{-}160)$$

where:

$$h_m = \{\alpha\rho_v h_v + (1-\alpha)\rho_\ell h_\ell\}/\rho_m \quad (5\text{-}72)$$

$$h_m^+ = \{\alpha\rho_v h_v v_{vz} + (1-\alpha)\rho_\ell h_\ell v_{\ell z}\}/G_m \quad (5\text{-}73)$$

$$F_{wz}''' = \frac{1}{A_z}\int_{P_z} \tau_w \, dP_z \quad (5\text{-}143)$$

From Eq. 11-77:

$$F_{wz}''' = \left(\frac{\partial p}{\partial z}\right)_{\text{fric}}$$

For a vertical constant area channel, and under the assumption of $p_v \approx p_\ell \approx p$, the mass, momentum, and energy equations take the form:

$$\frac{\partial \rho_m}{\partial t} + \frac{\partial}{\partial z}(G_m) = 0 \quad (13\text{-}1)$$

$$\frac{\partial G_m}{\partial t} + \frac{\partial}{\partial z}\left(\frac{G_m^2}{\rho_m^+}\right) = -\frac{\partial p}{\partial z} - \frac{f\,G_m|G_m|}{2D_e\rho_m} - \rho_m g \cos\theta \quad (13\text{-}2a)$$

$$\frac{\partial}{\partial t}(\rho_m h_m - p) + \frac{\partial}{\partial z}(h_m^+ G_m) = \frac{q_h'' P_h}{A_z} + \frac{G_m}{\rho_m}\left(\frac{\partial p}{\partial z} + \frac{f\,G_m|G_m|}{2D_e\rho_m}\right) \quad (13\text{-}3a)$$

where the friction factor f is not subscripted to allow for both single-phase and two-phase conditions.

The absolute value notation is used with G_m in the momentum equation to account for the friction force change in direction depending on the flow direction.

The rearrangement of Eq. 13-3a leads to:

$$\frac{\partial}{\partial t}(\rho_m h_m) + \frac{\partial}{\partial z}(h_m^+ G_m) = \frac{q'' P_h}{A_z} + \frac{\partial p}{\partial t} + \frac{G_m}{\rho_m}\left(\frac{\partial p}{\partial z} + \frac{f\,G_m|G_m|}{2D_e\rho_m}\right) \quad (13\text{-}3b)$$

In the above equations all parameters are functions of time and axial position.

Although the specific enthalpy energy equations are used here, there is no fundamental difficulty in using the internal energy instead. However, when the pressure (p) can be assumed to be a constant (in both time and space), the energy equation (Eq. 13-3b) can be mathematically manipulated somewhat more easily. For numerical solutions, the particular form of the energy equation may have more significance.

B Uniform and Equal Phase Velocities

The momentum and energy equations can be simplified by combining each with the continuity equation. The mass equation can be written as:

$$\frac{\partial}{\partial t}\rho_m + \frac{\partial}{\partial z}\rho_m V_m = 0 \quad (13\text{-}4)$$

where:

$$V_m \equiv G_m/\rho_m \quad (11\text{-}69)$$

Note that even the two-phase velocities are uniform and equal (for the homogeneous two-phase flow model); then:

$$V_v = V_\ell = V_m$$

For uniform velocity across the channel, Eq. 5-67 yields:

$$\rho_m^+ = G_m^2/(G_v V_v + G_\ell V_\ell) \quad (13\text{-}5)$$

only for uniform velocity. Thus if the uniform phase velocities are also equal:

$$\rho_m^+ = G_m^2/(G_v + G_\ell)V_m = \rho_m \quad (13\text{-}6)$$

The left-hand side of the momentum equation (Eq. 13-2a) can be simplified if it is assumed that the vapor and liquid velocities are equal to the form:

$$\frac{\partial}{\partial t}\rho_m V_m + \frac{\partial}{\partial z}(\rho_m V_m V_m) = \rho_m \frac{\partial V_m}{\partial t} + V_m \frac{\partial \rho_m}{\partial t} + V_m \frac{\partial \rho_m V_m}{\partial z} + \rho_m V_m \frac{\partial V_m}{\partial z}$$

$$= \rho_m \frac{\partial V_m}{\partial t} + \rho_m V_m \frac{\partial V_m}{\partial z} \quad (13\text{-}7)$$

By substituting from Eq. 13-7 into the momentum equation 13-2a, it can be written as:

$$\rho_m \frac{\partial V_m}{\partial t} + \rho_m V_m \frac{\partial V_m}{\partial z} = -\frac{\partial p}{\partial z} - \frac{f \rho_m V_m |V_m|}{2D_e} - \rho_m g \cos \theta \qquad (13\text{-}2b)$$

By using Eq. 11-69, the momentum equation (Eq. 13-2b) can also be written as:

$$\rho_m \frac{\partial V_m}{\partial t} + G_m \frac{\partial V_m}{\partial z} = -\frac{\partial p}{\partial z} - \frac{f G_m |G_m|}{2\rho_m D_e} - \rho_m g \cos \theta \qquad (13\text{-}2c)$$

Similarly, the left-hand side of the energy equation (Eq. 13-3a) for the case of uniform and equal velocities (where $h_m^+ = h_m$) can be written as:

$$\frac{\partial}{\partial t}(\rho_m h_m - p) + \frac{\partial}{\partial z}(\rho_m h_m V_m) = \rho_m \frac{\partial h_m}{\partial t} - \frac{\partial p}{\partial t} + h_m \frac{\partial \rho_m}{\partial t} \qquad (13\text{-}8)$$

$$+ h_m \frac{\partial \rho_m V_m}{\partial z} + \rho_m V_m \frac{\partial h_m}{\partial z}$$

Again, applying the continuity equation to Eq. 13-8 and combining the result with Eq. 13-3b gives:

$$\rho_m \frac{\partial h_m}{\partial t} + \rho_m V_m \frac{\partial h_m}{\partial z} - \frac{\partial p}{\partial t} = q'' \frac{P_h}{A_z} + V_m \left(\frac{\partial p}{\partial z} + \frac{f G_m |G_m|}{2D_e \rho_m} \right) \qquad (13\text{-}9a)$$

By substituting from Eq. 11-69 into Eq. 13-9a:

$$\rho_m \frac{\partial h_m}{\partial t} + G_m \frac{\partial h_m}{\partial z} = \frac{q'' P_h}{A_z} + \frac{\partial p}{\partial t} + \frac{G_m}{\rho_m} \left(\frac{\partial p}{\partial z} + \frac{f G_m |G_m|}{2D_e \rho_m} \right) \qquad (13\text{-}9b)$$

III DELINEATION OF BEHAVIOR MODES

Before illustrating solutions of the equations presented in the last section, it is advantageous to identify the hydrodynamic characteristics of interest. These characteristics tend to influence the applicability of the simplifying assumptions used to reduce the complexity of the equations.

To begin with, it should be mentioned that the underlying assumption of the discussion to follow is that the axial variation in the laterally variable local conditions may be represented by the axial variation of the bulk conditions. It is true only when the flow is fully developed (i.e., the lateral profile is independent of axial location). For single-phase flow, the development of the flow is achieved within a length equal to 10 to 100 times the channel hydraulic diameter. For two-phase flow in a heated channel, the flow does not reach a "developed" state owing to changing vapor quality and distribution along the axial length.

Another important flow feature is the degree to which the pressure field is influenced by density variation in the channel and the connecting system. For "forced convection" conditions, the flow is meant to be unaffected, or only mildly affected,

Table 13-1 Considerations of flow conditions in a single channel

Inlet boundary condition*	Flow conditions[†]	
	Forced convection (buoyancy can be neglected)	Natural convection (buoyancy is dominant)
Pressure	✓	✓
Flow rate or velocity	✓	This boundary condition cannot be applied.

*Exit pressure is prescribed for both cases.

[†]Conditions in which buoyancy effects are neither negligible nor dominant lead to "mixed convection" in the channel.

by the density change along the length of the channel. Hence buoyancy effects can be neglected. For "natural convection" the pressure gradient is governed by density changes with enthalpy, and therefore the buoyancy head should be described accurately. When neither the external pressure head nor the buoyancy head govern the pressure gradient independently, the convection is termed "mixed." Table 13-1 illustrates the relation between the convection state and the appropriate boundary conditions in a calculation.

Lastly, as discussed in Chapter 1, Vol. II, the conservation equations of coolant mass and momentum may be solved for a single channel under boundary conditions of (1) specified inlet pressure and exit pressure, (2) specified inlet flow and exit pressure, or (3) specified inlet pressure and outlet velocity. For case 2 and case 3 boundary conditions, the unspecified boundary pressure is uniquely obtained. However, for case 1 boundary conditions, more than one inlet flow rate may satisfy the equations. Physically, this is possible because density changes in a heated channel can create several flow rates at which the integrated pressure drops are identical, particularly if boiling occurs within the channel. This subject is discussed in section V.B.

IV STEADY-STATE SINGLE-PHASE FLOW IN A HEATED CHANNEL

Consider an upward vertical channel, as shown in Figure 13-1. Under steady-state conditions, Eq. 13-2a may be written as:

$$\frac{d}{dz}\left(\frac{G_m^2}{\rho_m^+}\right) = -\frac{dp}{dz} - f\frac{G_m|G_m|}{2D_e\rho_m} - \rho_m g \qquad (13\text{-}10)$$

Note that the partial derivative terms are changed to the full derivative terms, as z is the only variable in Eq. 13-10.

The exact solution of the momentum equation requires identifying the axial dependence of the properties of the fluid, such as ρ and μ (which are indirectly present in Eq. 13-10 through the Reynolds number dependence of f). Determining the pressure drop is then accomplished by integrating Eq. 13-10, which can be written after rearrangement as:

$$(p_{in} - p_{out}) = \left(\frac{G_m^2}{\rho_m^+}\right)_{out} - \left(\frac{G_m^2}{\rho_m^+}\right)_{in} + \int_{Z_{in}}^{Z_{out}} \frac{fG_m|G_m|}{2D_e\rho_m} dz + \int_{Z_{in}}^{Z_{out}} \rho_m g\, dz \quad (13\text{-}11)$$

Because the acceleration term is an exact differential, it depends on the endpoint conditions only, whereas the friction and gravity terms are path-dependent.

A Single-Phase Pressure Drop

Under conditions of single-phase liquid flow it is possible to assume that the physical property change along the heated channel is negligible, thereby decoupling the momentum equation from the energy equation. If, in addition, the flow area is axially constant, the mass flux (G_m) is constant; and for $\rho_m^+ = \rho_\ell \approx$ constant, the acceleration pressure drop is negligible. Hence Eq. 13-11 can be approximated by:

$$p_{in} - p_{out} = + \frac{f\, G_m|G_m|}{2D_e\rho_\ell}(Z_{out} - Z_{in}) + \Delta p_{form} + \rho_\ell g\,(Z_{out} - Z_{in}) \quad (13\text{-}12)$$

In practice, "average" properties, evaluated at the center of the channel, are used. The accuracy of such an evaluation is good for single-phase liquid fluid flow. However, for gas flow or two-phase flow, the radial and axial variation in fluid properties cannot always be ignored, and a proper average would then have to be defined by performing the integration of Eq. 13-11.

If the flow area varies axially, G_m is not a constant and the first two terms of the right-hand side of Eq. 13-11 no longer sum to zero but are evaluated in accordance with the discussion of Chapter 9 about pressure changes due to flow area changes.

Furthermore, if p_{in} and p_{out} are taken to be the pressures in the plena at the channel extremities, the entrance and exit pressure losses (see Chapter 9, section VII) should be added to the right-hand side of Eq. 13-12.

Example 13-1 Calculation of friction pressure drop in a bare rod PWR assembly

PROBLEM For the PWR conditions described below, determine the pressure drop across the core. You may neglect spacer, inlet, and exit losses; i.e., consider a bare rod PWR fuel assembly. Assume that the flow is evenly distributed across the core. Also ignore subcooled boiling, if it exists.

Design parameters for a 3411 MWt PWR (see Tables 1-2 and 2-3) are as follows:

Pressure = 15.5 MPa
Inlet temperature = 286°C
Exit temperature = 324°C
Linear heat rate of midplane (averaged radially in the core) = 17.8 kW/m
Number of fuel pins = 50,952
Core flow rate = 17.4 Mg/s
Fuel rod outside diameter (D) = 9.5 mm

Clad thickness = 0.57 mm
Gap = 0.08 mm
Pitch = 12.6 mm
Fuel rod height = 4.0 m
Active fuel height = 3.66 m

SOLUTION Equation 13-12 is utilized. Fluid properties are evaluated at the center of the core, where the average temperature is given by:

$$\frac{286 + 324}{2} = 305°C$$

$$\rho_m = \frac{1}{v_f} = \frac{1}{1.4201 \times 10^{-3}} = 704.2 \text{ kg/m}^3$$

$$c_p = 6.143 \text{ kJ/kg °K}$$

$$\mu = 917 \text{ }\mu\text{poise} = 91.7 \times 10^{-6} \text{ kg/m s} = 91.7 \text{ }\mu\text{Pa s}$$

Using the assumption that the flow is evenly distributed across the core, the average mass flow rate per fuel rod is:

$$\dot{m} = \frac{17.4 \text{ Mg/s}}{50,952} = 0.341 \text{ kg/s}$$

Consider a coolant-centered subchannel as defined in Figure 13-2. To obtain the mass flow rate per unit area (G_m) the flow area (A_z) in each subchannel must first be obtained. The rod-centered cell illustrated in Figure 13-2 produces the same results.

$$A_z = P^2 - \frac{\pi}{4}D^2 = (12.6)^2 - \frac{\pi}{4}(9.5)^2 = 87.88 \text{ mm}^2$$

Thus:

$$G_m = \frac{\text{mass flow per subchannel}}{\text{flow area}} = \frac{0.341 \text{ kg/s}}{87.88 \times 10^{-6} \text{ m}^2} = 3880.4 \text{ kg/m}^2\text{s}$$

To determine f, the Reynolds number must first be determined, which is a function of the subchannel equivalent diameter (D_e):

$$D_e = \frac{4(\text{flow area})}{\text{wetted perimeter}} = \frac{4(87.88 \text{ mm}^2)}{\pi(9.5 \text{ mm})} = 11.8 \text{ mm} = 0.0118 \text{ m}$$

Thus:

$$\text{Re} = \frac{G_m D_e}{\mu} = \frac{(3880.4)(0.0118)}{91.7 \times 10^{-6}} = 4.98 \times 10^5$$

Using the Chapter 9 correlations to obtain the friction factor (f):

$$f = 0.184 \text{ (Re)}^{-0.20} = 0.01334$$

Substituting f, D_e, G_m, ρ_m, and z into Eq. 13-12, the change in pressure is then obtained:

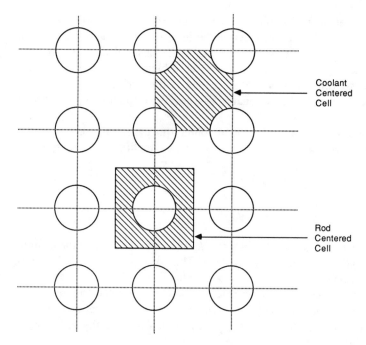

Figure 13-2 Representative interior cells in a PWR fuel assembly.

$$p_{in} - p_{out} = \frac{(0.01334)(3880.4)^2}{2(0.0118)(704.2)} (4) + (704.2)(9.8)(4)$$
$$= 48,428 + 27,605 = 76,033 \text{ Pa} \approx 0.08 \text{ MPa}$$

B Solution of the Energy Equation of Single-Phase Coolant and Fuel Rod

The radial temperature distribution across a fuel rod surrounded by flowing coolant was derived in Chapter 8. Here the axial distribution of the temperature is derived and linked with the radial temperature distribution.

At steady state the energy equation (Eq. 13-9b) in a channel with constant axial flow area (i.e., G_m = constant) reduces to:

$$G_m \frac{d}{dz} h_m = \frac{q'' P_h}{A_z} + \frac{G_m}{\rho_m}\left(\frac{dp}{dz} + f \frac{G_m|G_m|}{2D_e\rho_m}\right) \qquad (13\text{-}13a)$$

Neglecting the energy terms due to the pressure gradient and friction dissipation:

$$G_m A_z \frac{d}{dz} h_m = q'' P_h \qquad (13\text{-}13b)$$

or

$$\dot{m} \frac{d}{dz} h_m = q'(z) \qquad (13\text{-}13c)$$

For a given mass flow rate ($\dot{m}$) the coolant enthalpy rise depends on the axial variation of the heat-generation rate. In nuclear reactors the local heat generation depends on the distribution of both the neutron flux and the fissile material distribution. The neutron flux is affected by the moderator density, absorbing materials (e.g., control rods), and the local concentration of the fissile and fertile nuclear materials. Thus a coupled neutronic-thermal hydraulic analysis is necessary for a complete design analysis.

For the purpose of illustrating the general solution methods and some essential features of the axial distribution, we apply the following simplifying assumptions:

1. The variation of $q'(z)$ is sinusoidal:

$$q'(z) = q'_o \cos \frac{\pi z}{L_e} \tag{13-14}$$

where q'_o = peak linear heat-generation rate, and L_e = length over which the neutron flux has a nonzero value, as seen in Figure 13-3. This axial power profile is representative of what may be the neutron flux shape in a homogeneous reactor core when the effects of neutron absorbers within the core or axially varying coolant/moderator density due to change of phase are neglected. In real reactors the axial neutron flux shape cannot be given by a simple analytic expression but is generally less peaked than the sinusoidal distribution.

Note that whereas the neutron flux extends from $z = - L_e/2$ to $z = + L_e/2$, the heat-generation rate is confined to the actual heated length between $z = - L/2$ to $z = + L/2$.

2. The change in the physical properties of the coolant, fuel, or cladding can be ignored. Thus the convective heat-transfer coefficient, coolant heat capacity, and fuel and clad thermal conductivities are constant, independent of z. In reality, some variation does take place owing to axial variation in temperatures.

3. The coolant remains in the liquid phase.

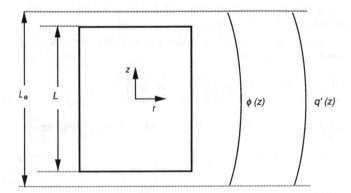

Figure 13-3 Axial profile of the neutron flux (ϕ) and the heat addition rate [$q'(z)$].

1 Coolant temperature. Equation 13-13c can be integrated over the axial length:

$$\dot{m} \int_{h_{in}}^{h_m(z)} dh_m = q'_o \int_{-L/2}^{z} \cos\left(\frac{\pi z}{L_e}\right) dz \qquad (13\text{-}15)$$

Equation 13-15 for single-phase flow may be written as:

$$\dot{m} c_p \int_{T_{in}}^{T_m(z)} dT = q'_o \int_{-L/2}^{z} \cos\left(\frac{\pi z}{L_e}\right) dz \qquad (13\text{-}16)$$

The subscript "in" denotes the inlet coolant condition. Given the single-phase nature of the flow and the temperature-independent physical properties, the result of integrating Eq. 13-16 may be written in the form:

$$T_m(z) - T_{in} = \frac{q'_o}{\dot{m} c_p} \frac{L_e}{\pi} \left(\sin\frac{\pi z}{L_e} + \sin\frac{\pi L}{2L_e}\right) \qquad (13\text{-}17)$$

This equation specifies the bulk temperature of the coolant as a function of height. The exit temperature of the coolant is given by

$$T_{out} \equiv T_m\left(\frac{L}{2}\right) = T_{in} + \left(\frac{q'_o}{\dot{m} c_p}\right)\left(\frac{2L_e}{\pi}\right) \sin\frac{\pi L}{2L_e} \qquad (13\text{-}18)$$

When the neutronic extrapolation height (L_e) can be approximated by the physical core height (L), the equation simplifies to:

$$T_{out} = T_{in} + \frac{2q'_o L}{\pi \dot{m} c_p} \qquad (13\text{-}19)$$

2 Cladding temperature. The axial variation of the outside cladding temperature can be determined by considering the heat flux at the cladding outer surface:

$$h[T_{co}(z) - T_m(z)] = q''(z) = \frac{q'(z)}{P_h} \qquad (13\text{-}20)$$

where $P_h = 2\pi R_{co}$, and h = coolant heat-transfer coefficient.
When combining Eqs. 13-14 and 13-20:

$$T_{co}(z) = T_m(z) + \frac{q'_o}{2\pi R_{co}h} \cos\left(\frac{\pi z}{L_e}\right) \qquad (13\text{-}21)$$

Eliminating $T_m(z)$ with Eq. 13-17 gives:

$$T_{co}(z) = T_{in} + q'_o\left[\frac{L_e}{\pi \dot{m} c_p}\left(\sin\frac{\pi z}{L_e} + \sin\frac{\pi L}{2L_e}\right) + \frac{1}{2\pi R_{co}h}\cos\frac{\pi z}{L_e}\right] \qquad (13\text{-}22)$$

The maximum cladding surface temperature can be evaluated by the condition:

$$\frac{dT_{co}}{dz} = 0 \qquad (13\text{-}23)$$

and

$$\frac{d^2 T_{co}}{dz^2} < 0 \tag{13-24}$$

The condition of Eq. 13-23 leads to:

$$\tan\left(\frac{\pi z_c}{L_e}\right) = \frac{2\pi R_{co} L_e h}{\pi \dot{m} c_p} \tag{13-25a}$$

or equivalently:

$$z_c = \frac{L_e}{\pi} \tan^{-1}\left[\frac{2\pi R_{co} L_e h}{\pi \dot{m} c_p}\right] \tag{13-25b}$$

Because all the quantities in the arc-tangent are positive, z_c is a positive value. Hence the maximum cladding temperature occurs at z_c such that $0 < z_c < L/2$.

The second derivative is given by:

$$\frac{d^2 T_{co}}{dz^2} = -q'_o\left[\frac{\pi}{L_e \dot{m} c_p} \sin\left(\frac{\pi z}{L_e}\right) + \frac{(\pi)^2}{L_e^2 2\pi R_{co} h} \cos\left(\frac{\pi z}{L_e}\right)\right] \tag{13-26}$$

and yields a negative value when z is positive.

3 Fuel centerline temperature. It is possible to extrapolate this approach to determine the maximum rod temperature in the fuel rod itself. First by combining Eqs. 8-119 and 13-22 we find that the fuel centerline temperature, $T_{\text{₵}}$ corresponding to T_{max} in Eq. 8-119) and T_m is given by:

$$T_{\text{₵}}(z) = T_{in} + q'_o\left\{\frac{L_e}{\pi \dot{m} c_p}\left(\sin\frac{\pi z}{L_e} + \sin\frac{\pi L}{2L_e}\right)\right.$$
$$\left. + \left[\frac{1}{2\pi R_{co} h} + \frac{1}{2\pi k_c} \ln\left(\frac{R_{co}}{R_{ci}}\right) + \frac{1}{2\pi R_g h_g} + \frac{1}{4\pi \bar{k}_f}\right] \cos\frac{\pi z}{L_e}\right\} \tag{13-27}$$

where $\bar{k}_f$ and k_c = the thermal conductivities of the fuel and cladding, respectively; h_g = gap conductance; and R_{co} and R_{ci} = outer and inner clad radii, respectively. By differentiating the last equation, we find the position of maximum fuel temperature as:

$$z_f = \frac{L_e}{\pi} \tan^{-1}\left\{\frac{L_e}{\pi \dot{m} c_p\left[\frac{1}{4\pi \bar{k}_f} + \frac{1}{2\pi k_c} \ln\left(\frac{R_{co}}{R_{ci}}\right) + \frac{1}{2\pi R_g h_g} + \frac{1}{2\pi R_{co} h}\right]}\right\} \tag{13-28}$$

The maximum fuel temperature is found by substituting the position z_f into Eq. 13-27. Note that again z_f is expected to be a positive quantity. An illustration of the axial variation of T_m, T_{co}, and $T_{\text{₵}}$ is given in Figure 13-4.

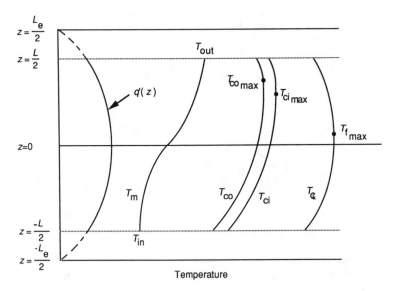

Figure 13-4 Axial variation of the bulk coolant temperature (T_m), the clad temperatures (T_{co} and T_{ci}), and the fuel center temperature (T_{C_L}).

Example 13-2 Determination of fuel maximum temperature

PROBLEM For the PWR described in Example 13-1, determine the coolant temperature as a function of height and the maximum fuel and cladding temperatures. Neglect extrapolation heights. Assume that the heat-transfer coefficients and thermal conductivities remain constant with the following values:

Fuel conductivity (k_f) = 0.002163 kW/m °C
Clad conductivity (k_c) = 0.01385 kW/m °C
Heat-transfer coefficient (h) = 34.0 kW/m² °C
Gap conductance (h_g) = 5.7 kW/m² °C

SOLUTION Given the sinusoidal heat addition over the heated length (L) = 3.66 m, the following values from Example 13-1 may be substituted into Eq. 13-17 to obtain the coolant temperature, $T_m(z)$:

$$T_m(z) = 286 + \left(\frac{17.8}{(0.341)(6.143)} \right) \left(\frac{3.66}{\pi} \right) \left(\sin \frac{\pi z}{3.66} + 1 \right)$$

$$= 295.9 + 9.9 \sin \frac{\pi z}{3.66}$$

Equation 13-25b may be utilized to determine the position at which the maximum cladding temperature occurs:

$$z_c = \frac{L}{\pi} \tan^{-1} \left[\frac{h\pi D\, L}{\pi\, c_p \dot{m}} \right]$$

$$= \frac{3.66}{\pi} \tan^{-1} \left[\frac{(34.0)\, \pi(9.5 \times 10^{-3})(3.66)}{\pi(6.143)(0.341)} \right]$$

$$= 0.6 \text{ m}$$

Substituting this value into Eq. 13-22 gives the cladding surface maximum temperature:

$$T_c(z_c) = 286 + 17.8 \left\{ \frac{3.66 \left[\sin\left(\dfrac{\pi(0.6)}{3.66} \right) + 1 \right]}{\pi(0.341)(6.143)} + \frac{\cos\left(\dfrac{\pi(0.6)}{3.66} \right)}{(\pi)(9.5 \times 10^{-3})(34)} \right\}$$

$$= 286 + 17.8(0.83 + 0.858) = 316.0°C$$

To obtain the maximum fuel centerline temperature, Eq. 13-28 is applied to determine the position (z_f) where this maximum temperature occurs. First we determine the fuel pellet radius and the effective gap radius:

$$R_f = 0.5(9.5) - 0.57 - 0.08 = 4.1 \text{ mm}$$

$$R_g = (4.18 + 4.1)/2 = 4.14 \text{ mm}$$

$$z_f = \frac{3.66}{\pi} \tan^{-1} \left\{ \frac{3.66/\pi(0.341)(6.143)}{\left[\dfrac{1}{4\pi(0.002163)} + \dfrac{ln(4.75/4.18)}{2\pi(0.01385)} + \dfrac{10^3}{2\pi(4.14)(5.7)} + \dfrac{10^3}{2\pi(4.75)(34)} \right]} \right\}$$

$$= \frac{3.66}{\pi} \tan^{-1} \left(\frac{0.556}{36.79 + 1.47 + 6.74 + 0.985} \right) \approx 0.014 \text{ m}$$

Substituting z_f into Eq. 13-27 gives the maximum fuel centerline temperature as:

$$T_{C_L}(z_f) = 286 + 17.8 \left[0.556 \left(\sin \frac{\pi(0.014)}{3.66} + 1 \right) + 45.99 \left(\cos \frac{\pi(0.014)}{3.66} \right) \right]$$

$$= 1114.7°C$$

V STEADY-STATE TWO-PHASE FLOW IN A HEATED CHANNEL UNDER EQUILIBRIUM CONDITIONS

A Solution to the Steady-State Two-Phase Energy Equation

The liquid coolant may undergo boiling as it flows in a heated channel under steady-state conditions, as it does in BWR assemblies. The coolant state undergoes a transition to a two-phase state at a given axial length. Thus the initial channel length (Fig. 13-5) can be called the nonboiling length and is analyzed by the equations described in section IV. The *boiling height* requires the development of new equations in terms

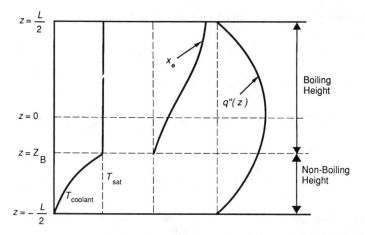

Figure 13-5 Coolant temperature and quality for sinusoidal heat addition in a single channel.

of enthalpy because phase change occurs at isothermal conditions. Starting with Eq. 13-13b and integrating from the inlet to any position along the length:

$$h(z) - h_{in} = \frac{P_h}{G_m A_z} \int_{-L/2}^{z} q''(z)dz$$ (13-29)

If the onset of boiling is determined to occur at a given bulk coolant enthalpy (normally a maximum subcooling level can be specified), Eq. 13-29 can be used to determine Z_B, the demarcation between the boiling height and the nonboiling height. For simplicity we assume thermodynamic equilibrium, in which case $h(Z_B) = h_f$ (the saturated liquid enthalpy) at the local pressure. Subcooled boiling does occur in BWRs; but because the channel is mostly in saturated boiling, the subcooled boiling impact on the pressure drop and the temperature field under steady-state conditions tends to be small.

For the case of sinusoidal heat generation (Eq. 13-14), and $L_e = L$, Eq. 13-29 yields:

$$h_f - h_{in} = \frac{P_h q_o'' L}{G_m A_z \pi} \left(\sin \frac{\pi Z_B}{L} + 1 \right)$$ (13-30a)

However, because:

$$\frac{2}{\pi} q_o'' P_h L = \frac{2}{\pi} q_o' L = \dot{q} \quad \text{and} \quad G_m A_z = \dot{m}$$

Eq. 13-30a takes the form:

$$h_f - h_{in} = \frac{\dot{q}}{2\dot{m}} \left(\sin \frac{\pi Z_B}{L} + 1 \right)$$ (13-30b)

The boiling boundary (Z_B) is determined from Eq. 13-30b as:

$$Z_B = \frac{L}{\pi} \sin^{-1}\left[-1 + \frac{2\dot{m}}{\dot{q}}(h_f - h_{in})\right] \qquad (13\text{-}31a)$$

Applying Eq. 13-29 from the inlet to the outlet also yields:

$$h_{out} - h_{in} = \frac{P_h q_o''}{G_m A_z} \int_{-L/2}^{L/2} \cos\frac{\pi z}{L}\, dz = \frac{2P_h q_o'' L}{G_m A_z \pi} = \frac{\dot{q}}{\dot{m}} \qquad (13\text{-}32)$$

which when combined with Eq. 13-31 yields:

$$Z_B = \frac{L}{\pi} \sin^{-1}\left[-1 + 2\left(\frac{h_f - h_{in}}{h_{out} - h_{in}}\right)\right] \qquad (13\text{-}31b)$$

For a two-phase fluid the average density may be written under homogeneous equilibrium conditions as (see Eqs. 5-50b and 11-73b):

$$\rho_m = \rho_f(p) + \alpha\,\rho_{fg}(p) = \frac{1}{v_f(p) + x_e v_{fg}(p)} \qquad (13\text{-}33)$$

The mixture enthalpy under these conditions may be obtained from Eqs. 5-52 and 5-75 as:

$$h_m = h_f(p) + x_e h_{fg}(p) \qquad (13\text{-}34)$$

where x_e = flow thermodynamic (equilibrium) quality; $\rho_{fg} \equiv \rho_g - \rho_f$; and $h_{fg} \equiv h_g - h_f$.

When the axial variation of pressure is small with respect to the inlet pressure, h_f and h_{fg} can be assumed to be axially constant. Therefore the quality at any axial position can be predicted from Eqs. 13-29 and 13-34, which combine to give:

$$x_e(z) = x_{e_{in}} + \frac{P_h}{G_m A_z h_{fg}} \int_{-L/2}^{z} q''(z)dz \qquad (13\text{-}35a)$$

$$x_e(z) = x_{e_{in}} + \frac{\dot{q}}{2\dot{m} h_{fg}}\left(\sin\frac{\pi z}{L} + 1\right) \qquad (13\text{-}35b)$$

Note that $x_{e_{in}}$ is negative under normal conditions because the liquid enters the core in a subcooled state. Once the quality axial distribution is known, it is possible to predict the axial void fraction distribution, assuming thermal equilibrium, from Eq. 5-55.

$$\alpha = \frac{1}{1 + \dfrac{1 - x}{x}\dfrac{\rho_v V_v}{\rho_\ell V_\ell}} = \frac{1}{1 + \dfrac{1 - x}{x}\dfrac{\rho_v}{\rho_\ell}S} \qquad (5\text{-}55)$$

Example 13-3 Calculation of axial distribution of vapor quality in a BWR

PROBLEM If in a BWR-6 the exit quality of the core is 14.6%, determine the coolant flow rate and the axial profile of the radially averaged enthalpy, $h_m(z)$, and quality $x_e(z)$, in this core. Assume thermodynamic equilibrium and sinusoidal heat generation. Neglect effects of the neutron flux extrapolation heights.

BWR-6 design parameters (see Tables 1-2 and 2-1) are as follows:

Gross thermal output = 3579 MWt
Inlet temperature = 278°C
Pressure = 7.2 MPa
Saturation temperature = 288°C
Fuel height = 3.81 m

The saturation enthalpies may be obtained using the following values for the water properties at the saturation pressure and temperature of 7.2 MPa and 288°C.

$h_f = 1277.2$ kJ/kg
$h_{fg} = 1492.2$ kJ/kg
$c_p = 5.307$ kJ/kg °K

SOLUTION The exit and inlet enthalpies (h_{in} and h_{out}) may be evaluated as follows:

$$h_{in} = h_f - c_p(T_f - T_{in}) = 1277.2 - (5.307)(10) = 1224.1 \text{ kJ/kg}$$
$$h_{out} = h_f + x_{e_{out}} h_{fg} = 1277.2 + (0.146)(1492.2) = 1495.1 \text{ kJ/kg}$$

To obtain Z_B, the location where bulk boiling starts to occur, Eq. 13-31b must be applied, yielding:

$$Z_B = \frac{3.81}{\pi} \sin^{-1} \left[-1 + \frac{2(1277.2 - 1224.1)}{(1495.1 - 1224.1)} \right] = -0.793 \text{ m}$$

Solving Eq. 13-30b for $q/\dot{m}$ with $z = Z_B = -0.793$ m gives:

$$\frac{\dot{q}}{\dot{m}} = \frac{2(h_f - h_{in})}{\left(\sin \dfrac{\pi Z_B}{L} + 1 \right)} = h_{out} - h_{in}$$

$$= 1495.1 - 1224.1 = 271 \frac{\text{kW s}}{\text{kg}}$$

Thus $\dot{m} = q/271 = 3,579,000/271 = 13,207$ kg/s.
In general, one may obtain $h(z)$ by applying Eq. 13-29:

$$h(z) = h_{in} + \frac{\dot{q}}{2\dot{m}} \left(\sin \frac{\pi z}{L} + 1 \right)$$

$$= 1224.1 + \frac{271}{2} \left(\sin \frac{\pi z}{L} + 1 \right)$$

or

$$h(z) = 1359.6 + 135.5 \sin \left(\frac{\pi z}{3.81} \right)$$

The quality may then be obtained by applying Eq. 13-35b as follows:

$$x_e(z) = x_{e_{in}} + \frac{\dot{q}}{2\dot{m}h_{fg}}\left(\sin\frac{\pi z}{L} + 1\right)$$

$$= \frac{1224.1 - 1277.2}{1492.2} + \frac{135.5}{1492.2}\left(\sin\frac{\pi z}{3.81} + 1\right)$$

$$= -0.0356 + 0.0908\left(\sin\frac{\pi z}{3.81} + 1\right)$$

or

$$x_e(z) = 0.0552 + 0.0908\sin\frac{\pi z}{3.81}$$

The values of $h(z)$ and $x_e(z)$ are plotted on Figures 13-6 and 13-7.

B Two-Phase Pressure Drop Characteristics

The general equation (Eq. 13-11) is in principle sufficient for solving the pressure drop problem. However, the channel pressure drop–flow rate behavior, when extended over the entire flow rate range, is not a simple linearly increasing function. In fact, for a wide range of pressure and heating rates, the $\Delta p - \dot{m}$ characteristic curve has a shape that can lead to multiple solutions and instabilities. Familiarity with these characteristics provides the physical basis essential to understanding and analyzing complex system behavior.

We develop here the channel pressure drop–flow rate relation in the friction-dominated regime (rather than a gravity-dominated regime, which is discussed in

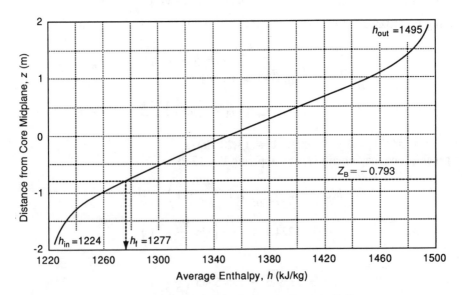

Figure 13-6 Axial distribution of average enthalpy in the core (Example 13-3).

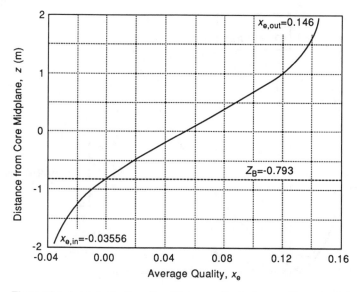

Figure 13-7 Axial distribution of equilibrium quality in the core (Example 13-3).

Chapters 3 and 4 of Vol. II). Because heated channels are considered, at sufficiently low flows, the channel is occupied mostly by the gas phase.

For a constant total power to the channel, the friction-dominated regime characteristic is shown in Figure 13-8. Point A is arbitrarily chosen at a high enough flow rate that the channel is in liquid single-phase flow throughout its length. The region A–B is characterized principally by a pressure drop that decreases as $\dot{m}^{2-n}$ where $n =$ the exponent of the flow rate in the governing expression for the friction factor, i.e.:

$$f = \frac{C}{\mathrm{Re}^n} = \frac{C}{\left(\dfrac{\dot{m}D_e}{A_z\mu}\right)^n} \qquad (13\text{-}36)$$

Additional factors influencing this curve in the region A–B, but not shown on the figure, are through property (ρ and μ) variations with temperature and perhaps a change in f due to transition from turbulent to laminar flow conditions.

Prior but close to point B, boiling is initiated in the channel. At point B, vapor generation is sufficient to reverse the trend of decreasing Δp with decreasing $\dot{m}$. It occurs when the two phase conditions lead to an increase in the friction pressure drop that overcomes the decrease in the gravity pressure drop due to boiling. Therefore this increase in total Δp occurs whenever boiling can be initiated at a relatively large flow rate and hence for relatively high power input. At point C, where the flow rate is sufficiently low, the channel has considerable vapor flow so that the increase in f with decreasing $\dot{m}$ no longer governs the curve. Under these conditions the region C–D is also characterized principally by a pressure drop, decreasing as $\dot{m}^{2-n}$. Here,

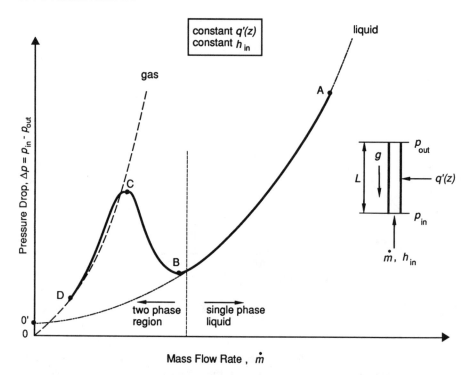

Figure 13-8 Pressure drop-flow rate characteristic for a heated channel. Curve O′–liquid is for a liquid flowing adiabatically in the channel. Curve O–gas is for a gas flowing adiabatically in the region.

however, the flow condition is mainly a single-phase vapor. Beyond point D the gravity-dominated region is encountered.

If the flow rate is extremely low when boiling occurs, the decrease in the gravity head may be larger than the increase in the friction pressure drop, leading to a decrease in the total pressure drop (Fig. 13-9). This behavior was demonstrated experimentally by Rameau et al. [4].

Let us assume that an external pressure drop (Δp_{ex}) boundary condition is imposed; i.e., p_{in} and p_{out} are held constant. Three levels of Δp can be assumed (Fig. 13-10).

$$\Delta p_{ex} > \Delta p_C \tag{13-37a}$$

$$\Delta p_C > \Delta p_{ex} > \Delta p_B \tag{13-37b}$$

$$\Delta p_{ex} < \Delta p_B \tag{13-37c}$$

For the intermediate case only, multiple channel flow rates are possible. In this case, however, not all intersections are stable conditions. The criterion for stability can be developed by the following perturbation analysis. The fluid in a heated channel accelerates owing to the difference between the imposed external (or boundary) pressure drop (Δp_{ex}) and the intrinsic (mostly friction) pressure drop (Δp_f), as given by:

$$I \frac{d\dot{m}}{dt} = \Delta p_{ex} - \Delta p_f \tag{13-38}$$

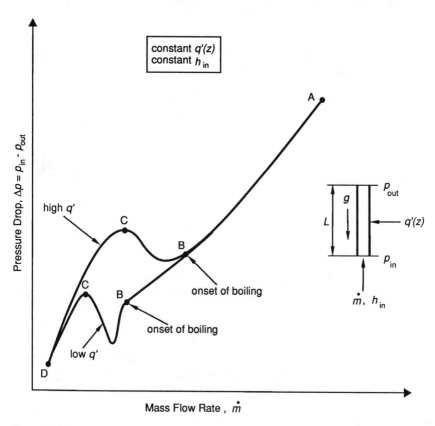

Figure 13-9 Pressure drop characteristics for the gases of high power and low power.

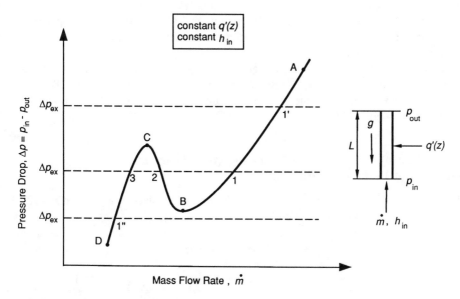

Figure 13-10 Stable and unstable operating conditions for prescribed external pressure drop conditions.

where $I = L/A_z$, the geometric inertia of the fluid in the channel. The small perturbation equation for a small velocity change $\Delta \dot{m}$ is:

$$I \frac{\partial \Delta \dot{m}}{\partial t} = \frac{\partial (\Delta p_{ex})}{\partial \dot{m}} \Delta \dot{m} - \frac{\partial (\Delta p_f)}{\partial \dot{m}} \Delta \dot{m} \qquad (13\text{-}39)$$

Expressing $\Delta \dot{m}$ as:

$$\Delta \dot{m} = \epsilon e^{\omega t} \qquad (13\text{-}40)$$

Equation 13-39 leads to the result that:

$$\omega = \frac{\left[\dfrac{\partial (\Delta p_{ex})}{\partial \dot{m}} \right] - \left[\dfrac{\partial (\Delta p_f)}{\partial \dot{m}} \right]}{I} \qquad (13\text{-}41)$$

The criterion of stability is that small perturbations should not grow with time. Hence ω should be zero or a negative value. Equation 13-41 thus yields the criterion that:

$$\frac{\partial (\Delta p_{ex})}{\partial \dot{m}} < \frac{\partial (\Delta p_f)}{\partial \dot{m}} \qquad (13\text{-}42)$$

The external Δp_{ex} is represented in Figure 13-10 by a horizontal or zero-sloped line. Therefore from Eq. 13-42 a stable operating condition exists only for the positive-sloped region of the friction loss curve, i.e., segments A–B and C–D. We conclude that point 2 is not a stable operating point, whereas points 1', 1, 3, and 1" are stable operating points.

The behavior at point 2 can be explained physically by considering slight changes in $\dot{m}$ about that point. Suppose $\dot{m}$ is slightly decreased (increased). The friction pressure drop increases (decreases), further decelerating (accelerating) the mass flow rate. This decrease (increase) causes the transition to point 3 (1) where the friction pressure drop is once again balanced by the imposed external pressure drop.

Consequently for the condition of Eq. 13-37b, two solutions are admissible (intersections 3 and 1) for a given flow channel. This perturbation analysis of the well-known Leddineg instability was presented by Maulbetsch and Griffith [3]. This instability is pronounced when the system pressure is low and therefore the liquid-to-vapor density ratio is high.

C Two-Phase Pressure Drop Using the Homogeneous Equilibrium Boiling Assumption

The pressure drop across a channel in which boiling takes place at a position $z = Z_B$ (Fig. 13-1) can be obtained through the application of Eq. 13-11. Thus:

$$p_{in} - p_{out} = \left(\frac{G_m^2}{\rho_m^+} \right)_{out} - \left(\frac{G_m^2}{\rho_m^+} \right)_{in} + \int_{Z_{in}}^{Z_B} \rho_\ell \, g \, dz + \int_{Z_B}^{Z_{out}} \rho_m \, g \, dz$$

$$+ \frac{f_{\ell o} G_m |G_m| (Z_B - Z_{in})}{2 D_e \rho_\ell} + \frac{\overline{\phi_{\ell o}^2} f_{\ell o} G_m |G_m| (Z_{out} - Z_B)}{2 D_e \rho_\ell}$$

$$+ \sum_i \left(\phi_{\ell o}^2 \, K \, \frac{G_m |G_m|}{2\rho_\ell} \right)_i \tag{13-43}$$

where the friction pressure drop multiplier depends on the axial variation of the flow quality, system pressure, and flow rate. Note that ρ_ℓ is used in Eq. 13-43 for both the single-phase and two-phase regions—because the value of ρ_ℓ is not significantly different along the channel. To evaluate the boiling height and the two-phase friction multiplier, the energy equation needs to be solved as illustrated earlier. (Note that the multiplier values plotted in Chapter 11 were all for axially uniform heat addition in the channels.)

Equation 13-43 is commonly written as:

$$\Delta p = \Delta p_{acc} + \Delta p_{gravity} + \Delta p_{fric} + \Delta p_{form} \tag{13-44}$$

where

$$\Delta p = p_{in} - p_{out} \tag{13-45a}$$

$$\Delta p_{acc} = \left(\frac{G_m^2}{\rho_m^+} \right)_{out} - \left(\frac{G_m^2}{\rho_m^+} \right)_{in} \tag{13-45b}$$

$$\Delta p_{gravity} = \int_{Z_{in}}^{Z_B} \rho_\ell g \, dz + \int_{Z_B}^{Z_{out}} \rho_m g \, dz \tag{13-45c}$$

$$\Delta p_{fric} = [(Z_B - Z_{in}) + \overline{\phi_{\ell o}^2}(Z_{out} - Z_B)] \frac{f_{\ell o} G_m |G_m|}{2 D_e \rho_\ell} \tag{13-45d}$$

$$\Delta p_{form} = \sum_i \left(\phi_{\ell o}^2 \, K \, \frac{G_m |G_m|}{2\rho_\ell} \right)_i \tag{13-45e}$$

Let us now evaluate these components for the sinusoidal heat input case.

1 ΔP_{acc}. From Eqs. 13-45b and 11-64, we get:

$$\Delta p_{acc} = \left[\left(\frac{(1-x)^2}{(1-\alpha)\rho_f} + \frac{x^2}{\alpha \rho_g} \right)_{out} - \frac{1}{\rho_\ell} \right] G_m^2 \tag{13-46}$$

For high flow rate conditions when boiling does not occur $(\rho_m^+)_{out} \simeq \rho_\ell$ and $\Delta p_{acc} \simeq 0$. For low flow conditions, the term ρ_m^+ decreases, approaching ρ_g when the total flow is evaporated. It implies that as G_m decreases,

$$\left(\frac{1}{\rho_m^+} \right)_{out} - \left(\frac{1}{\rho_\ell} \right)$$

increases and the net effect is for a maximum Δp_{acc} to occur at an intermediate flow rate.

Note that, from Eq. 13-35b, when $x_e = x$ (under homogeneous equilibrium flow conditions):

$$x = x_{in} + \frac{q}{2\dot{m}h_{fg}} + \frac{q}{2\dot{m}h_{fg}} \sin\left(\frac{\pi z}{L}\right)$$

$$= x_{in} + \left(\frac{x_{out} - x_{in}}{2}\right) + \left(\frac{x_{out} - x_{in}}{2}\right) \sin\left(\frac{\pi z}{L}\right)$$

$$= \frac{x_{in} + x_{out}}{2} + \frac{x_{out} - x_{in}}{2} \sin\left(\frac{\pi z}{L}\right)$$

$$= x_{ave} + x_{rise} \sin\left(\frac{\pi z}{L}\right) \tag{13-47}$$

where:

$$x_{ave} \equiv \frac{x_{in} + x_{out}}{2} \tag{13-48a}$$

$$x_{rise} \equiv \frac{x_{out} - x_{in}}{2} \tag{13-48b}$$

Also, the void fraction is given by:

$$\alpha = \frac{x}{x + (1 - x) \dfrac{\rho_g}{\rho_\ell} S} \tag{5-55}$$

Therefore for homogeneous equilibrium conditions, $S = 1$, and the void fraction is given by:

$$\alpha = \frac{x_{ave} + x_{rise} \sin\left(\dfrac{\pi z}{L}\right)}{\left(1 - \dfrac{\rho_g}{\rho_\ell}\right)\left(x_{ave} + x_{rise} \sin\left(\dfrac{\pi z}{L}\right)\right) + \dfrac{\rho_g}{\rho_\ell}}$$

or

$$\alpha = \frac{x_{ave} + x_{rise} \sin\left(\dfrac{\pi z}{L}\right)}{x' + x'' \sin\left(\dfrac{\pi z}{L}\right)} \tag{13-49}$$

where:

$$x' \equiv x_{ave} + \frac{\rho_g}{\rho_\ell}(1 - x_{ave}) \tag{13-50a}$$

$$x'' \equiv \left(1 - \frac{\rho_g}{\rho_\ell}\right) x_{rise} \tag{13-50b}$$

Note that at low pressure $\rho_g/\rho_\ell \ll 1$; therefore $x' \simeq x_{ave}$ and $x'' \simeq x_{rise}$.

2 $\Delta p_{gravity}$. From Eq. 13-45c the gravity pressure drop is given by:

$$\Delta p_{gravity} = \rho_{\ell}g\ (Z_B - Z_{in}) + \int_{Z_B}^{Z_{out}} [\rho_\ell - \alpha(\rho_\ell - \rho_g)]g\,dz$$

$$= \rho_{\ell}g\ (Z_{out} - Z_{in}) - (\rho_\ell - \rho_g)g \int_{Z_B}^{Z_{out}} \frac{x_{ave} + x_{rise}\ \sin\left(\dfrac{\pi z}{L}\right)}{x' + x''\ \sin\left(\dfrac{\pi z}{L}\right)}\,dz \quad (13\text{-}51)$$

If $|x''| > |x'|$, which is generally the case at very low pressure:

$$\Delta p_{gravity} = \rho_\ell\, gL - (\rho_\ell - \rho_g)g \left\{ \frac{x_{rise}}{x''} (Z_{out} - Z_B) \right.$$

$$\left. + \left(x_{ave} - \frac{x_{rise}x'}{x''}\right) \frac{L}{\pi} \frac{1}{(x''^2 - x'^2)^{1/2}}\ \ln \left[\frac{x'\tan\left(\dfrac{\pi z}{2L}\right) + x'' - (x''^2 - x'^2)^{1/2}}{x'\tan\left(\dfrac{\pi z}{2L}\right) + x'' + (x''^2 - x'^2)^{1/2}} \right]_{Z_B}^{Z_{out}} \right\}$$

$$(13\text{-}52)$$

If $|x'| > |x''|$, which is typically the case at BWR pressure (7.0 MPa):

$$\Delta p_{gravity} = \rho_{\ell}gL - (\rho_\ell - \rho_g)g \left\{ \frac{x_{rise}(Z_{out} - Z_B)}{x''} + \left(x_{ave} - \frac{x_{rise}\ x'}{x''}\right) \right.$$

$$\left. \frac{L}{\pi} \frac{1}{(x'^2 - x''^2)^{1/2}}\ \tan^{-1} \left[\frac{x'\tan\left(\dfrac{\pi z}{2L}\right) + x''}{(x'^2 - x''^2)^{1/2}} \right]_{Z_B}^{Z_{out}} \right\}$$

$$(13\text{-}53)$$

3 Δp_{fric}. Using the homogeneous friction pressure drop coefficients when the effect of viscosity is neglected:

$$\phi_{\ell o}^2 = \left(\frac{\rho_\ell}{\rho_g} - 1\right)x + 1.0 \quad (11\text{-}82)$$

The friction pressure drop is obtained from Eq. 13-45d as:

$$\Delta p_{\text{fric}} = \frac{f_{\ell o} G_m |G_m|}{2 D_e \rho_\ell} \left((Z_B - Z_{\text{in}}) + \int_{Z_B}^{Z_{\text{out}}} \left[\left(\frac{\rho_\ell}{\rho_g} - 1 \right) x + 1.0 \right] dz \right)$$

$$= \frac{f_{\ell o} G_m |G_m|}{2 D_e \rho_\ell} \left[(Z_{\text{out}} - Z_{\text{in}}) + \left(\frac{\rho_\ell}{\rho_g} - 1 \right) \int_{Z_B}^{Z_{\text{out}}} \right. \tag{13-54}$$

$$\left. \cdot \left[x_{\text{ave}} + x_{\text{rise}} \sin \left(\frac{\pi z}{L} \right) \right] dz \right]$$

Performing the integration of Eq. 13-54:

$$\Delta p_{\text{fric}} = \frac{f_{\ell o} G_m |G_m|}{2 D_e \rho_\ell} \left\{ L + \left(\frac{\rho_\ell}{\rho_g} - 1 \right) \left[x_{\text{ave}} (Z_{\text{out}} - Z_B) \right. \right.$$

$$\left. \left. + x_{\text{rise}} \frac{L}{\pi} \left(\cos \left(\frac{\pi Z_B}{L} \right) - \cos \left(\frac{\pi Z_{\text{out}}}{L} \right) \right) \right] \right\} \tag{13-55}$$

This approach can also be used for other forms of $\phi_{\ell o}^2$, as demonstrated by Chen and Olson [1].

4 Δp_{form}. The form losses due to abrupt geometry changes, e.g., spacers, can be evaluated using the homogeneous multiplication factor, so that:

$$\Delta p_{\text{form}} = \sum_i \frac{G_m^2 K_i}{2 \rho_\ell} \left[1 + \left(\frac{\rho_\ell}{\rho_g} - 1 \right) x_i \right] \tag{13-56}$$

where K_i = the single-phase pressure loss coefficients at location "i."

Example 13-4 Determination of pressure drop in a bare rod BWR bundle

PROBLEM For the BWR conditions of Example 13-3, calculate the pressure drops Δp_{acc}, $\Delta p_{\text{gravity}}$, Δp_{fric}, and Δp_{total} for a bare rod bundle for the flow rate range from $\dot{m}/4$ to $4\dot{m}$ where $\dot{m}$ is the nominal flow rate, and given these operating parameters:

Gross thermal output = 3579 MWt
Inlet temperature = 278°C
Saturation temperature = 288°C
Pressure = 7.2 MPa
Exit quality = 14.6%
Axial heat flux = sinusoidal

The 8 × 8 rod array has the following parameters:

Pitch = 16.2 mm
Fuel pin diameter = 12.27 mm
Interior channel mass flow rate = 0.29 kg/s

Total fuel pin length $= 4.1$ m
Active fuel pin length $= 3.81$ m

Assumption: Neglect the unheated channel length and consider that the heated length is 3.81 m.

SOLUTION

Channel flow area $(A) = P^2 - \dfrac{\pi}{4}D^2$

$$= \left(\frac{16.2}{10^3}\right)^2 - \frac{\pi}{4}\left(\frac{12.27}{10^3}\right)^2 = 1.442 \times 10^{-4}\, m^2$$

$\therefore$ mass flux $(G_m) = \dfrac{0.29\ kg/s}{1.442 \times 10^{-4}\ m^2} = 2.011 \times 10^3\ kg/m^2\, s$

To determine the friction factor (f) the Reynold's number, which is a function of the subchannel equivalent diameter (D_e), must first be determined:

$$D_e = \frac{4(\text{flow area})}{\text{wetted perimeter}} = \frac{4(1.442 \times 10^{-4}m^2)}{\pi\left(\dfrac{12.27}{10^3}\ m\right)} = 0.01496\ m$$

Thus:

$$Re = \frac{G_m D_e}{\mu} = \frac{(2.0111 \times 10^3)(0.01496)}{96.93 \times 10^{-6}} = 3.105 \times 10^5$$

where μ is calculated at $T_{avg} = 283°C$ and $p = 7.2$ MPa.

$$f = \frac{0.184}{Re^{0.2}} = \frac{0.184}{(4.7039 \times 10^5)^{0.2}} = 0.01467$$

Assuming homogeneous equilibrium conditions (slip ratio $= 1$), the void fraction is given by:

$$\alpha = \frac{1}{1 + \dfrac{(1 - x)\,\rho_g}{x}\dfrac{\rho_g}{\rho_f}}$$

To calculate ρ_g and ρ_f at the exit conditions, i.e., where $p = 7.2$ MPa and $T_{sat} = 288°C$:

$$\rho_g = \frac{1}{26.52 \times 10^{-3}} = 37.71\ kg/m^3$$

and

$$\rho_f = \frac{1}{1.3578 \times 10^{-3}} = 736.49\ kg/m^3$$

then:

$$\alpha = \frac{1}{1 + \dfrac{(1 - 0.146)}{0.146} \dfrac{37.71}{736.49}} = 0.77$$

At the inlet we have subcooled water, and ρ_ℓ should be calculated at $T = 278°C$ and $p = 7.2$ MPa

$$\therefore \rho_\ell \text{ at inlet condition} = 752.56 \text{ kg/m}^3$$

The inlet quality can be calculated from Eq. 13-35b as in Example 13-3:

$$\therefore x_{e_{in}} = -0.0356$$

From Eq. 13-46:

$$\Delta p_{acc} = \left\{ \left[\frac{(1 - x)^2}{(1 - \alpha)\rho_f} + \frac{x^2}{\alpha\rho_g} \right]_{out} - \left(\frac{1}{\rho_\ell} \right)_{in} \right\} G_m^2$$

$$\Delta p_{acc} = \left\{ \left[\frac{(1 - 0.146)^2}{(1 - 0.77)(736.49)} + \frac{(0.146)^2}{0.77(37.71)} \right] - \frac{1}{752.56} \right\} (2.0111 \times 10^3)^2$$

$$\therefore \Delta p_{acc} = 15.0 \text{ kPa}$$

From Eq. 13-55:

$$\Delta p_{fric} = f_{\ell o} \frac{G_m^2}{2D_e\rho_f} \left\{ L + \left(\frac{\rho_f}{\rho_g} - 1 \right) \right.$$

$$\left. \cdot \left[x_{ave}(Z_{out} - Z_B) + x_{rise} \frac{L}{\pi} \left(\cos \frac{\pi Z_B}{L} - \cos \frac{\pi Z_{out}}{L} \right) \right] \right\}$$

where:

$$Z_{out} = \frac{3.81}{2} = 1.905 \text{ m}$$

$$L = 3.81 \text{ m}$$

$$Z_B = -0.793 \text{ m}$$

$$G_m = 2.011 \times 10^3 \text{ kg/m}^2\text{s}$$

$$\rho_g = 37.71 \text{ kg/m}^3; \rho_\ell = 736.49 \text{ kg/m}^3$$

$$x_{ave} = \frac{x_{in} + x_{out}}{2} = \frac{-0.0356 + 0.146}{2} = 0.0552$$

$$x_{rise} = \frac{x_{out} - x_{in}}{2} = \frac{0.146 - (-0.0356)}{2} = 0.0908$$

$$f_{\ell o} = 0.01467; D_e = 0.01496 \text{ m}$$

$$\therefore \Delta p_{fric} = 0.01467 \frac{(2.011 \times 10^3)^2}{2 (0.01496)(736.49)}$$

(*continued*)

$$\cdot \left\{ 3.81 + \left(\frac{736.49}{37.71} - 1 \right) \right.$$

$$\cdot \left[0.0552(1.905 + 0.793) + 0.0908 \left(\frac{3.81}{\pi} \right) \right.$$

$$\left. \left. \cdot \left(\cos \frac{-0.793\pi}{3.81} - \cos \frac{1.905\pi}{3.81} \right) \right] \right\}$$

$$\therefore \Delta p_{\text{fric}} = 2.69 \times 10^3 \{ 3.81 + 18.53[0.1489 + 0.1101(0.794 - 0)] \}$$

$$\therefore \Delta p_{\text{fric}} = 22.05 \text{ kPa}$$

Let us now calculate the gravity term. First determine x' and x''.

$$x' = 0.0552 + \frac{37.71}{736.49}(1 - 0.0552) = 0.1036$$

$$x'' = \left(1 - \frac{37.71}{736.49} \right)(0.0908) = 0.0862$$

$$\therefore x' > x''$$

From Eq. 13-53:

$$\Delta p_{\text{gravity}} = \rho_f g L - (\rho_f - \rho_g)g \left\{ \frac{x_{\text{rise}}(Z_{\text{out}} - Z_{\text{B}})}{x''} \right.$$

$$+ \left(x_{\text{ave}} - \frac{x_{\text{rise}}x'}{x''} \right) \frac{L}{\pi} \frac{1}{(x' - x''^2)^{1/2}}$$

$$\cdot \left[\tan^{-1} \frac{x' \tan \dfrac{\pi Z_{\text{out}}}{2L} + x''}{(x'^2 - x''^2)^{1/2}} \right.$$

$$\left. \left. - \tan^{-1} \frac{x' \tan \dfrac{\pi Z_{\text{B}}}{2L} + x''}{(x'^2 - x''^2)^{1/2}} \right] \right\}$$

$$\therefore \Delta p_{\text{gravity}} = (736.49)(9.81)(3.81) - (736.49 - 37.71)(9.81)$$

$$\left\{ \frac{0.0908(1.905 + 0.793)}{0.0862} + \left[0.0552 - \frac{0.0908(0.1036)}{0.0862} \right] \right.$$

$$\left. \cdot \frac{3.81}{\pi} (0.1036^2 - 0.0862^2)^{-1/2} \right\}$$

$$\left\{ \tan^{-1} \left[\frac{0.1036 \tan \dfrac{\pi(1.905)}{2(3.81)} + 0.0862}{(0.1036^2 - 0.0862^2)^{1/2}} \right] \right.$$

Table 13-2 Hydraulic parameters for a BWR

Parameter	$\dot{m}/4$	$\dot{m}/2$	$\dot{m}$	$3/2\,\dot{m}$	$2\,\dot{m}$	$3\,\dot{m}$	$4\,\dot{m}$
				Flow rate (kg/s)			
G_m (kg/m^2 s)	0.503×10^3	1.005×10^3	2.011×10^3	3.017×10^3	4.022×10^3	6.033×10^3	8.044×10^3
Re	0.776×10^5	1.552×10^5	3.105×10^5	4.657×10^5	6.209×10^5	9.314×10^5	1.242×10^6
f	0.0193	0.0168	0.0147	0.0135	0.0128	0.0118	0.0111

$$- \tan^{-1} \left[\frac{0.1036 \tan \dfrac{-\pi(0.794)}{2(3.81)} + 0.0862}{(0.1036^2 - 0.0862^2)^{1/2}} \right] \Bigg\}$$

$$\therefore \Delta p_{\text{gravity}} = 27.527 \times 10^3 - 6.855 \times 10^3 [2.8420 - 1.1362\,(1.277 - 0.7260)]$$

$$\therefore \Delta p_{\text{gravity}} = 12.3 \text{ kPa}$$

Following the same procedure for different values of $\dot{m}$, at a fixed inlet enthalpy, we get the different values of G_m, Re, and f given in Table 13-2.

For a constant total thermal output = 3579 MWt and inlet enthalpy of 1224.1 kJ/kg, we can similarly obtain h_{out}, Z_B, x_{out}, and α_{out}:

$$q = \dot{m}\,(h_{\text{out}} - h_{\text{in}})$$

$$\therefore h_{\text{out}} = h_{\text{in}} + \frac{q \text{ kJ/s}}{\dot{m} \text{ kg/s}}$$

Also:

$$x_{\text{out}} = \frac{h_{\text{out}} - h_f}{h_{fg}} = \frac{h_{\text{out}} - 1277.1}{1492.2}$$

$$\alpha_{\text{out}} = \frac{1}{1 + \dfrac{(1 - x_{\text{out}})}{x_{\text{out}}} \dfrac{\rho_g}{\rho_f}} = \frac{1}{1 + 0.0512 \dfrac{(1 - x_{\text{out}})}{x_{\text{out}}}}$$

Table 13-3 Effect of flow rate on the pressure drop in a BWR

Parameter	$\dot{m}/4$	$\dot{m}/2$	$\dot{m}$	$3/2\,\dot{m}$	$2\,\dot{m}$	$3\,\dot{m}$	$4\,\dot{m}$
				Flow rate (kg/s)			
h_{out} (kJ/kg)	2308.6	1766.7	1495.7	1405.4	1360.3	1315.1	1292.5
x_{out}	0.69	0.33	0.146	0.086	0.056	0.025	0.01
α_{out}	0.98	0.91	0.77	0.65	0.54	0.34	0.17
Z_B(m)	-1.364	-1.133	-0.793	-0.515	-0.264	0.214	0.732
Δp_{fric} (kPa)	6.03	11.14	22.05	34.45	48.75	83.88	128.9
Δp_{acc} (kPa)	4.40	8.36	15.0	19.84	22.94	23.99	17.84
$\Delta p_{\text{gravity}}$ (kPa)	6.47	9.07	12.58	15.14	17.20	20.51	23.34
Δp_{total} (kPa)	16.90	28.61	49.61	69.43	88.89	128.3	170.1

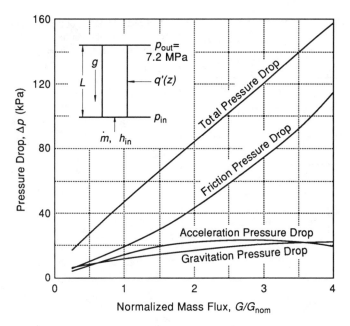

Figure 13-11 Pressure drop components in the BWR channel of Example 13-4. The mass flux at the nominal condition is $G_{nom} = 2000$ kg/m²s.

and

$$Z_B = \frac{L}{\pi} \sin^{-1} \left[-1 + 2 \frac{(h_f - h_{in})}{(h_{out} - h_{in})} \right]$$

$$= \frac{3.81}{\pi} \sin^{-1} \left[-1 + \frac{2(1277.1 - 1224.1)}{(h_{out} - 1224.1)} \right]$$

$$\therefore Z_B = 1.2126 \sin^{-1} \left[-1 + \frac{105.94}{h_{out} - 1224.1)} \right]$$

The values resulting for various flow rates are summarized in Table 13-3. The pressure drop components are plotted in Figure 13-11. The corresponding values of x_{ave}, x_{rise}, x', and x'' are shown in Figure 13-12.

VI TWO-PHASE PRESSURE DROP ALLOWING FOR NONEQUILIBRIUM

The calculations of the pressure drop in a heated channel allowing for nonequilibrium conditions (both nonequal velocity and nonsaturated phases) are discussed in this section. Essentially four flow regions may exist over the entire length of the subchannel

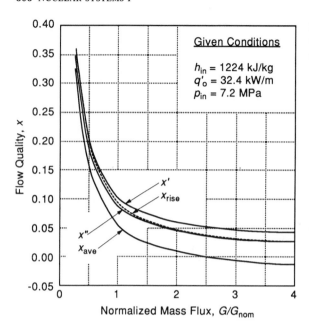

Figure 13-12 Effect of the mass flux on the flow quality for fixed heat flux and inlet enthalpy conditions.

(Fig. 13-13). The existence of these flow regions depends on the heat flux and the inlet conditions. The flow enters as single-phase liquid and may undergo subcooled boiling, bulk boiling, and possibly dryout before exiting the channel. In a BWR fuel assembly, the dryout condition is not encountered, whereas it is encountered in the tube of a once-through steam generator.

The total pressure drop can be obtained as the summation of the pressure drops over each axial region:

$$\Delta p_{total} = \Delta p_{1\phi_\ell} + \Delta p_{SCB} + \Delta p_{BB} + \Delta p_{1\phi_v} \tag{13-57}$$

where:

$\Delta p_{1\phi_\ell}$ = pressure drop in the single-phase liquid region
Δp_{SCB} = pressure drop in the subcooled boiling region
Δp_{BB} = pressure drop in the bulk boiling region
$\Delta p_{1\phi_v}$ = pressure drop in the single-phase vapor region

The presence of each individual flow region is ascertained by evaluating the locations of the transition points Z_{SC}, Z_B, and Z_V, as shown on Figure 13-13. Obviously, if any of these points lies outside the subchannel length, the corresponding flow region and those that follow are absent.

A Position of Flow Regime Transition

Let us now evaluate the locations of the transition points. To simplify the evaluation, assume that the subchannel can be represented by a tube having a diameter equal to

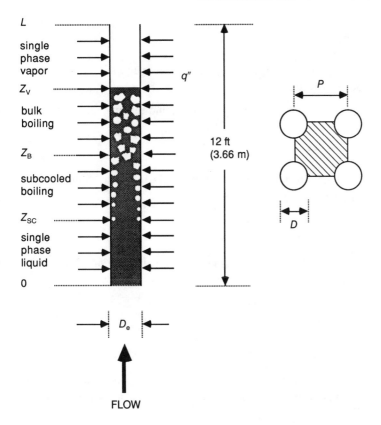

Figure 13-13 Subchannel flow region (left) and cross section (right).

D_e, the subchannel hydraulic diameter. For a tube, the heated diameter (D_h) is equivalent to the hydraulic diameter (D_e), but these diameters are differentiated in the analysis below for generality. Also, the axial heat flux is taken as uniform.

1 Point of transition from single-phase liquid to subcooled flow boiling (Z_{SC}).
This point is the point of bubble detachment according to Saha and Zuber [5] (see Chapter 12). Let:

$$(\Delta T_{sub})_{Z_{SC}} = T_{sat} - T_m(Z_{SC}) \tag{13-58}$$

then:

$$(\Delta T_{sub})_{Z_{SC}} = 0.0022 \left(\frac{q'' D_e}{k_\ell}\right) \qquad \text{for Pe} \leq 70{,}000 \tag{12-21a}$$

$$(\Delta T_{sub})_{Z_{SC}} = 153.8 \left(\frac{q''}{G_m c_p}\right) \qquad \text{for Pe} > 70{,}000 \tag{12-21b}$$

where:

Pe $\equiv$ Peclet number $= G_m D_e c_p / k_\ell$
c_p and k_ℓ = heat capacity and thermal conductivity, respectively, of water
T_m = mean temperature of water

From an energy balance written over the single-phase liquid region:

$$Z_{SC} = \frac{G_m D_e^2 c_p [T_m(Z_{SC}) - T_{in}]}{4q''D_h} \tag{13-59}$$

2 Point of transition from subcooled to bulk boiling (Z_B). At this point the fluid is saturated. An energy balance for this region yields:

$$Z_B = \frac{\dot{m}(h_f - h_{in})}{\pi D_h q''} \tag{13-60}$$

3 Point of transition from bulk boiling to a single-phase vapor region (Z_V). Again, from an energy balance:

$$Z_V = \frac{\dot{m}(h_g - h_{in})}{\pi D_h q''} \tag{13-61}$$

B Flow Quality at Transition Points

To calculate the pressure drop over each region, it is required to evaluate steam flow quality at each transition point. For this purpose we may use the empirical correlation suggested by Levy [2]:

$$x(z) = x_e(z) - x_e(Z_{SC}) \exp \left[\frac{x_e(z)}{x_e(Z_{SC})} - 1 \right] \tag{13-62}$$

where $x_e(z)$ = steam thermodynamic equilibrium quality. The (negative) steam flow quality, $x_e(Z_{SC})$, at the bubble detachment point is given by:

$$x_e(Z_{SC}) = - \left(\frac{c_{pf}(\Delta T_{sub})_{Z_{SC}}}{h_{fg}} \right) \tag{13-63}$$

and $(\Delta T_{sub})_{Z_{SC}}$ is given by Eq. 12-21a or 12-21b.
    The flow quality at Z_{SC} is zero:

$$x(Z_{SC}) = 0 \tag{13-64}$$

The steam flow quality at Z_B is obtained from Eq. 13-62 at $x_e = 0$:

$$x(Z_B) = 0 - x_e(Z_{SC}) \exp (0 - 1) = - x_e(Z_{SC}) \exp (-1) \tag{13-65}$$

    The flow quality at Z_V can be predicted only by applying the critical heat flux relations discussed in Chapter 12.
    We now proceed to calculate each component of pressure drop, i.e., elevation, acceleration, and friction, over each region.

C Gravity Pressure Drop

By definition, the pressure drop due to gravity for the vertical channel is:

$$\Delta p_{grav} = \int \rho_m g dz \qquad (13\text{-}66)$$

1. Δp_{grav} in the single-phase liquid region. The density ρ_m is evaluated at the local temperature (i.e., $\rho_m = \rho_\ell$).
2. Δp_{grav} in the subcooled region. In this region, ρ_m may be assumed to be constant and calculated at:

$$\tfrac{1}{2} (T_{Z_{sc}} + T_{Z_B})$$

3. Δp_{grav} in the bulk boiling region. The density in this region is given by:

$$\rho_m = (1 - \alpha)\rho_f + \alpha\rho_g \qquad (5\text{-}50\text{b})$$

The void fraction α can be calculated from the modified Martinelli void fraction (Eq. 11-95c).

$$\alpha = 1 - X_{tt}/(X_{tt}^2 + 20X_{tt} + 1)^{1/2} \qquad (13\text{-}67\text{a})$$

where from Eq. 11-94:

$$X_{tt} = \left(\frac{\mu_f}{\mu_g}\right)^{0.1} \left(\frac{1 - x}{x}\right)^{0.9} \left(\frac{\rho_g}{\rho_f}\right)^{0.5} \qquad (13\text{-}67\text{b})$$

and x = steam flow quality.

4. Δp_{grav} in the single-phase vapor region. The density (ρ) is evaluated at the local temperature.

D Acceleration Pressure Drop

The acceleration pressure drop is given by (see Eq. 11-101 for the two-phase region):

$$\Delta p_{acc} = \frac{G_m^2}{\rho_f} r_2 \qquad (13\text{-}68)$$

where r_2 is given below for each region.

1. Δp_{acc} in the single-phase liquid region. The pressure drop multiplier in this region is:

$$r_2 = \left(\frac{1}{\rho_{z_{sc}}} - \frac{1}{\rho_{z=0}}\right) \rho_f \qquad (13\text{-}69)$$

where $\rho_{z_{sc}}$ is evaluated at $T_{z_{sc}}$.

2. Δp_{acc} in the subcooled boiling region. A similar acceleration multiplier can be used in this region:

$$r_2 = \left(\frac{1}{\rho_{z_b}} - \frac{1}{\rho_{z_{sc}}} \right) \rho_f \tag{13-70}$$

3. Δp_{acc} in the bulk boiling region. The acceleration multiplier for bulk boiling is more involved, as two-phase flow exists throughout this region:

$$r_2 = \left\{ \left[\frac{(1-x)}{(1-\alpha)\rho_f} + \frac{x}{\alpha\rho_g} \right]_{Z_v} - \left[\frac{(1-x)}{(1-\alpha)\rho_f} + \frac{x}{\alpha\rho_g} \right]_{Z_B} \right\} \rho_f \tag{13-71}$$

If the channel ends prior to dryout, the term Z_v is evaluated at $z = L$.

4. Δp_{acc} in the single-phase vapor region. Equation 13-68 is again applicable with the acceleration multiplier developed at points Z_v and L.

$$r_2 = \left(\frac{1}{\rho_{z=L}} - \frac{1}{\rho_{Z_v}} \right) \rho_f \tag{13-72}$$

E Friction Pressure Drop

The friction pressure drop may be calculated by the relation:

$$\Delta p_{fric} = \int \frac{fG_m^2}{2\rho_m D_e} dz \tag{13-73}$$

1. Δp_{fric} in the single-phase liquid region. In this region f may be obtained from the Darcy–Weisbach correlation:

$$f = 0.316 \, \text{Re}^{-0.25} \tag{9-80}$$

and $\rho_m = \rho_\ell$.

2. Δp_{fric} in the subcooled boiling region. The friction factor calculation in the subcooled and bulk boiling regions should account for the two-phase friction pressure drop. We then proceed to correlate the unknown friction factor in Eq. 13-73 to a series of empirical friction-factor ratios. To do it an isothermal friction factor (f_{iso}) may first be introduced in Eq. 13-73.

$$\Delta p_{fric} = \frac{G_m^2}{2 D_e} \int_{Z_{sc}}^{Z_B} f_{iso} \left(\frac{f}{f_{iso}} \right) \frac{1}{\rho_m} dz \tag{13-74}$$

If $T_m > T_{LB}^*$:

$$\frac{f}{f_{iso}} = 1 + \frac{T_m - T_{LB}^*}{T_{sat} - T_{LB}^*} \left[\left(\frac{f}{f_{iso}} \right)_{sat} - 1 \right] \tag{13-75a}$$

The value of T_{LB}^* in Eq. 13-75a is defined in British units by:

$$T_{LB}^* = T_{sat} + \frac{60(q''/10^6)}{e^{p/900}} - \frac{0.766q''}{h_o} \tag{13-76}$$

If $T_m < T_{LB}^*$:

$$\frac{f}{f_{iso}} = 1 - 0.001 \; \Delta T_m = 1 - 0.001 \; q''/h_o \qquad (13\text{-}75b)$$

when using Eq. 10-94a we get:

$$h_o = 0.023 \left(\frac{k_f}{D_e}\right) Re^{0.8} Pr^{0.4}$$

where q'' is in BTU/hr ft^2, p is in psia, and T is in °F.

The value of $\left(\dfrac{f}{f_{iso}}\right)_{sat}$ in Eq. 13-75a is in fact the value of $\phi_{\ell o}^2$ at 4.2% quality, which is the location at which the flow quality is expected to equal the thermodynamic quality. $\phi_{\ell o}^2$ is given by the modified Martinelli friction multiplier:

$$\phi_{\ell o}^2 = \Omega(p, G_m) \left\{ 1.2 \left[\left(\frac{\rho_f}{\rho_g}\right) - 1 \right] x^{0.824} \right\} + 1.0 \qquad (11\text{-}104)$$

Ω, which is the Jones' correction factor, depends on the pressure and mass flux and is given by the following.

For $G_m \leq 7.0 \times 10^5$ lb/hr-ft^2:

$$\Omega = 1.36 + 0.0005p + 0.1 \left(\frac{G_m}{10^6}\right) - 0.000714p \left(\frac{G_m}{10^6}\right) \qquad (11\text{-}105a)$$

For $G_m > 7.0 \times 10^5$ lb/hr-ft^2:

$$\Omega = 1.26 - 0.0004p + 0.119 \left(\frac{10^6}{G_m}\right) + 0.00028p \left(\frac{10^6}{G_m}\right) \qquad (11\text{-}105b)$$

where p is in psia and G is in lbm/hr-ft^2.

3. Δp_{fric} in the bulk boiling region. Equation 13-73 may be written as:

$$\Delta p_{fric} = \frac{G_m^2 f_{\ell o}}{2\rho_f D_e} \int_{Z_B}^{Z_v} (\phi_{\ell o}^2) \; dz \qquad (13\text{-}77)$$

The friction multiplying factor of Martinelli and Nelson as modified by Jones can be used to define $\phi_{\ell o}^2$, where $f_{\ell o}$ is the appropriate friction factor correlation for the single-phase Re and geometry condition in Chapter 9. Other terms are already defined. For a given p and G_m, when ρ_f and ρ_g are assumed constant, we get:

$$\Delta p_{fric} = \left\{ \Omega \frac{1.2}{1.824} \left[\left(\frac{\rho_f}{\rho_g}\right) - 1 \right] x_{Z_v}^{0.824} + 1 \right\} \frac{G_m^2 f_{\ell o}(Z_v - Z_B)}{2 \; \rho_f D_e} \qquad (13\text{-}78)$$

4. Δp_{fric} in the single-phase vapor region is similar to that of single-phase liquid region.

F Total Pressure Drop

The total pressure drop (without any form pressure losses) is given by:

$$\Delta p_{tot} = \sum_{i=0}^{3} \left[\int_{z_i}^{z_{i+1}} \rho g dz + \frac{G_m^2}{\rho_f} (r_2)_{z_i}^{z_{i+1}} + \int_{z_i}^{z_{i+1}} \frac{\phi_{\ell o}^2 f_{\ell o} G_m^2}{2\rho_f D_e} dz \right] \quad (13\text{-}79)$$

where each numerical index corresponds to each transition point, including inlet and outlet.

Example 13-5 Effect of pressure on pressure drop characteristics

PROBLEM For the interior subchannel of a 7×7 BWR fuel bundle, calculate the pressure drop components of the single-phase, subcooled boiling, and saturated boiling for two cases;

Case A: At operating pressure of 6.89 MPa
Case B: At a low operating pressure of 2.0 MPa

Investigate the effect of varying G_m on Δp, given:

Geometry
Heated length = 3.66 m
Rod diameter = 14.3 mm
Pitch = 18.8 mm
Subchannel flow rate = 0.34 kg/s
Wall heat flux = 1.0 MW/m²
Heat flux axial distribution = uniform
Water inlet temperature
  Case A = 271.3°C
  Case B = 107.1°C

SOLUTION

Case A

    Using the equations defined in section VI, the various components of Δp can be calculated and plotted as shown in Figure 13-14. It is seen that at nominal flow rate the pressure drop is mostly due to saturated boiling, as the saturation condition occurs 0.5 m above the inlet. In this case the inlet conditions already allow subcooled boiling (Fig. 13-15). At a much higher flow rate of about 2.7 kg/s no boiling occurs in the channel. The total pressure drop does not exhibit the S shape, as already demonstrated by the homogeneous flow calculation of Example 13-4. However, as can be seen in Figure 13-16, the friction pressure drop is somewhat higher in this case at nominal flow conditions than in Example 13-4, probably because of the higher flow rate.

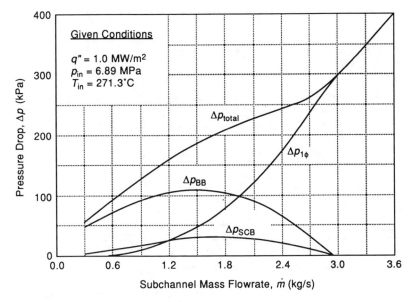

Figure 13-14 The regime components of pressure drop in a 7 × 7 BWR channel as a function of flow (Example 13-5, case A).

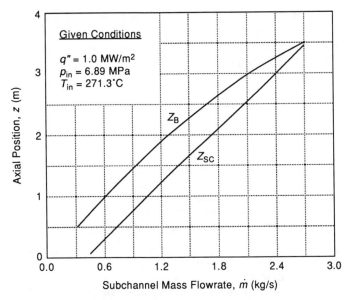

Figure 13-15 Boiling inception and saturation lengths in a 7 × 7 BWR channel as a function of flow (Example 13-5, case A).

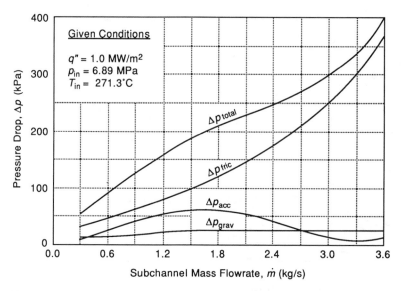

Figure 13-16 Components of total pressure drop in a 7 × 7 BWR channel as a function of flow (Example 13-5, case A).

Case B

The components of total pressure drop for case B (at low pressure of 2.0 MPa) are shown in Figure 13-17. The total pressure drop exhibits the S shape in this case, with maximum pressure drop occurring at 0.16 kg/s, or one-half the nominal flow rate. In this channel the flow is probably unstable between $\dot{m} = 0.16$ and 0.37 kg/s.

REFERENCES

1. Chen, K-F., and Olson, C. A. Analytic solution to verify code predictions of two-phase flow in a boiling water reactor core channel. *Nucl. Technol.* 62:361, 1983.
2. Levy, S. Forced convection subcooled boiling prediction of vapor volumetric fraction. *Int. J. Heat Mass Transfer* 10:247, 1967.
3. Maulbetsch, J. S., and Griffith, P. *A Study of System Induced Instabilities in Forced Convection Flows with Subcooled Boiling.* MIT, Eng. Prof. Lab. Report 5382-35, 1965.
4. Rameau, B., Seiller, J. M., and Lee, K. W. Low heat flux flows sodium boiling experimental results and analysis. Presented at the ASME Winter Annual Meeting, New Orleans, 1984.
5. Saha, P., and Zuber, N. Point of net vapor generation and vapor void fraction in subcooled boiling. Paper B4.7. In: *Proceedings of the 5th International Heat Transfer Conference,* Tokyo, 1974.

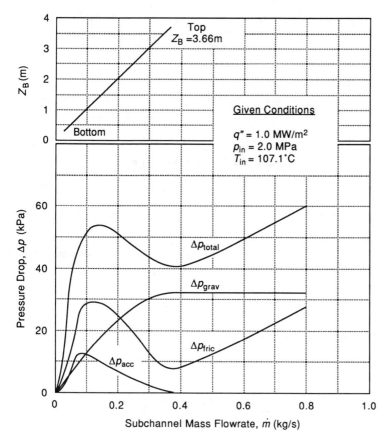

Figure 13-17 Components of total pressure drop of a 7 × 7 BWR channel at reduced pressure and inlet temperature (Example 13-5, case B).

PROBLEMS

Problem 13-1. Heated channel power limits

How much power can be extracted from a PWR with the geometry and operating conditions of Example 13-2 if:

1. The coolant exit temperature is to remain subcooled.
2. The maximum clad temperature is to remain below saturation conditions.
3. The fuel maximum temperature is to remain below the melting temperature of 2400°C (ignore sintering effects).

Answers:

1. $\dot{Q}$ = 6170 MW
2. $\dot{Q}$ = 4070 MW
3. $\dot{Q}$ = 5390 MW

Problem 13-2 Specification of power profile for a given clad temperature (section IV)

Consider a nuclear fuel rod whose cladding outer radius is a. Heat is transferred from the fuel rod to coolant with constant heat-transfer coefficient h. The coolant mass flow rate along the rod is $\dot{m}$. Coolant specific heat c is independent of temperature.

It is desired that the temperature of the outer surface of the fuel rod t (at radius a) be constant, independent of distance z from the coolant inlet end of the fuel rod.

Derive a formula showing how the linear power of the fuel rod q' should vary with z if the temperature at the outer surface of the fuel rod is to be constant.

Answer:

$$\exp\left[-2\pi haz/\dot{m}c\right] = q'(z)/q'_o$$

Problem 13-3 Pressure drop-flow rate characteristic for a fuel channel (section V)

A designer is interested in determining the effect of a 50% channel blockage on the downstream-clad temperatures in a BWR core. The engineering department proposes to assess this effect experimentally by inserting a prototypical BWR channel containing the requisite blockage in the test loop sketched in Figure 13-18 and running the centrifugal pump to deliver prototypic BWR pressure and single-channel flow conditions.

Is the test plan acceptable to you? If not, what changes would you propose and why?

Answer: It is not possible to specify both flow and pressure conditions to be identical to BWR conditions.

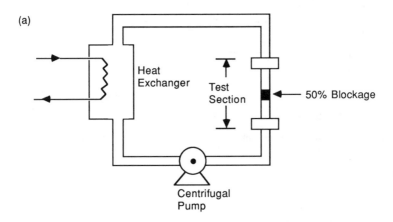

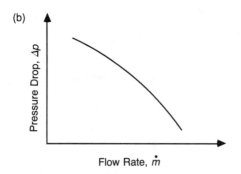

Figure 13-18 Test loop (**a**) and pump characteristic curve (**b**).

Problem 13-4 Pressure drop in a two-phase flow channel (section V)

For the BWR conditions given below, using the HEM model determine the acceleration, frictional, gravitational, and total pressure drop (i.e., ignore the spacer pressure drop).

Compare the frictional pressure drop obtained by the HEM model to that obtained from the Martinelli–Nelson approach, where the friction multiplier is:

$$\phi_{\ell o}^2 = 1 + 1.2 \left(\frac{\rho_f}{\rho_g} - 1 \right) x^{0.824}$$

Geometry: Consider a vertical tube of 17 mm I.D. and 3.8 m length.
Operating conditions: Steady-state conditions at:

Operating pressure = 7550 kPa
Inlet temperature = 275°C
Mass velocity = 1700 kg/m²s
Heat flux (axially constant) = 670 kW/m²

Answers: The pressure drops (HEM model) are as follows:

Δp_{acc} = 12.7 kPa
Δp_{fric} = 13.9 kPa
Δp_{grav} = 16.1 kPa
Δp_{total} = 42.7 kPa

Friction pressure drop (Martinelli–Nelson): Δp_{fric} = 18.0 kPa

Problem 13-5 Critical heat flux for PWR channel (section IV)

Using the conditions of Examples 13-1 and 13-2, determine the minimum critical heat flux ratio (MCHFR). Use the W-3 correlation for PWRs.

Problem 13-6

Consider an 8 × 8 BWR assembly of the characteristics given below.

1. What is the friction pressure drop across the assembly when calculated using the Martinelli–Nelson method?
2. It is desired to operate the assembly such that the maximum void fraction does not exceed 60%. What is the maximum power that can be removed from the assembly?

Assume uniform power horizontally in the fuel pins. Use the Bankoff correlation for estimating the slip ratio. Assume axially linear power profile in the bottom half of the assembly, and a constant power profile in the upper half.

Operating conditions

Fuel pin diameter (D)	12.27 mm
Fuel pin length (L)	4.1 m
Saturation temperature (T_{sat})	287°C
Inlet subcooling ($T_{sat} - T_{in}$)	9.8°C
Pressure (p)	6.89 MPa
Maximum linear heat-generation rate (q_o')	19.0 kW/m
Mass flow rate per assembly ($\dot{m}$)	17.5 kg/s
Assembly outer dimension	129.6 mm

Answers:

1. Δp_{fric} = 26.4 kPa
2. Q = 4.2 MW

Appendixes

NOMENCLATURE

Coordinates

$\vec{i}, \vec{j}, \vec{k}$	Unit vectors along coordinate axes
$\vec{n}$	Unit normal vector directed outward from control surface
$\vec{r}$	Unit position vector
x, y, z	Cartesian
r, θ, z	Cylindrical
r, θ, ϕ	Spherical
u, v, w	Velocity in Cartesian coordinates
v_x, v_y, v_z	
v_r, v_θ, v_z	Velocity in cylindrical coordinates

Extensive and specific properties

		Dimensions in two unit systems*	
Symbol	Description	ML θT	FE plus ML θT
C, c	General property		Varies
E, e	Total energy (internal + kinetic + potential)		E, EM^{-1}
H, h	Enthalpy		E, EM^{-1}
$H°, h°$	Stagnation enthalpy		E, EM^{-1}
M, m	Total mass	M	
S, s	Entropy		E, EM^{-1}
U, u	Internal energy		E, EM^{-1}
$U°, u°$	Stagnation internal energy		E, EM^{-1}
V, ν	Volume	L^3, L^{3}M^{-1}	

*M = mass; θ = temperature; F = force; E = energy; L = length; T = time; − = dimensionless.

619

Dimensionless numbers

Symbol	Description	Location where symbol first appears[†]
Br	Brinkmann number	E 10-6
Ec	Eckert number	P 10-413
Fr	Froude number	E 9-36
Gr	Grashof number	E 10-11
Ku	Kutateladze number	E 11-5
L	Lorentz number	E 10-10
M	Mach number	P 10-444
M_{ij}	Mixing Stanton number	Vol. II
Nu	Nusselt number	E 10-11
Pe	Peclet number	P 10-419
Pr	Prandtl number	E 10-6
Re	Reynolds number	E 9-36
St	Stanton number	E 10-76

*M = mass; θ = temperature; E = energy; − = dimensionless; L = length; T = time; F = force.

[†]A ≡ appendix; T ≡ table; E ≡ equation; P ≡ page; F ≡ figure. For example, E10-6 = equation 10-6.

General notation

Symbol	Description	Dimension in two unit systems*		Location where symbol first appears[†]
		$ML\,\theta T$	FE plus $ML\,\theta T$	
	General english notation			
A, A_f	General area or flow area	L^2		E 2-8
A	Availability function		E	E 4-27
	Atomic mass number	−		E 3-48
A_s	Projected frontal area of spacer	L^2		E 9-89
A_v	Avogadro's number	M^{-1}		E 3-4
	Unobstructed flow area in channels	L^2		E 9-89
A_{fb}	Sum of the area of the total fluid–solid interface and the area of the fluid	L^2		Vol. II
A_{fs}	Area of the total fluid–solid interface within the volume	L^2		Vol. II
a	Atomic fraction	L^{-3}		E 3-14
	Half thickness of plate fuel	L		F 8-9
B	Buildup factor	−		E 3-54
C	Tracer concentration	−		Vol. II
	Constant for friction factor			
C_D	Nozzle coefficient	−		E 9-17b
C_o	Correction factor in drift flux model for nonuniform void distribution	−		E 11-41

*M = mass; θ = temperature; E = energy; − = dimensionless; L = length; T = time; F = force.

[†]A ≡ appendix; T ≡ table; E ≡ equation; P ≡ page; F ≡ figure. For example, E10-6 = equation 10-6.

General notation (*continued*)

Symbol	Description	ML θT	FE plus ML θT	Location where symbol first appears[†]
		Dimension in two unit systems*		
C_S	Spacer drag coefficient	—		E 9-89
C_v	Modified drag coefficient	—		E 9-90
c	Isentropic speed of sound in the fluid	LT^{-1}		E 11-120
c_d	Drag coefficient	—		E 11-24
c_p	Specific heat, constant pressure		$EM^{-1}\theta^{-1}$	E 4-115
c_R	Hydraulic resistance coefficient	—		Vol. II
c_v	Specific heat, constant volume		$EM^{-1}\theta^{-1}$	P 4-121
D	Diameter, rod diameter	L		F 1-12
	Tube diameter	L		E 9-24
D_e	Hydraulic or wetted diameter	L		E 9-55
D_H	Heated diameter	L		E 10-12
D_s	Wire spacer diameter	L		F 1-14
D_V	Volumetric hydraulic diameter	L		P 9-391
D_{ft}	Wall-to-wall distance of hexagonal bundle	L		F 1-14
d	Tube diameter	L		F 9-4
	Diameter of liquid drop	L		E 11-24
E	Neutron kinetic energy		E	E 3-1
F	Modification factor of Chen's correlation	—		E 12-29
	Looseness of bundle packing	—		Vol. II
	Hot spot factor	—		Vol. II
$\vec{F}, \vec{f}, F$	Force, force per unit mass		F, FM^{-1}	E 4-18
F_Q	Heat flux hot channel factor	—		Vol. II
F_v	Void factor	—		E 8-68
F_{ix}	Total drag force in the control volume i in the x direction		F	Vol. II
F_{iz}	Subchannel circumferentially averaged force for vertical flow over the solid surface in the control volume i		F	
$F_{\Delta h}$	Coolant enthalpy rise hot channel factor	—		Vol. II
$F_{\Delta T}$	Coolant temperature rise hot channel factor	—		Vol. II
f	Moody friction factor	—		E 9-50
	Mass fraction of heavy atom in the fuel	—		E 2-13
f'	Fanning friction factor	—		E 9-52
$f_{c.t.}$	Friction factor in a circular tube	—		E 9-85
f_j	Subfactor of x_j	—		Vol. II
$f_{j,y}$	Subfactor relative to parameter x_j affecting the property y	—		Vol. II

(continued)

General notation (*continued*)

Symbol	Description	Dimension in two unit systems*		Location where symbol first appears[†]
		ML θT	*FE* plus *ML θT*	
f_{TP}	Two-phase friction factor	—		E 11-77
f_{tr}	Friction factor for the transverse flow	—		Vol. II
G	Mass flux	$ML^{-2}T^{-1}$		E 5-38a
g	Distance from rod surface to array flow boundary	L		F 1-12
$\vec{g}$ or g	Acceleration due to gravity	LT^{-2}		E 4-23
H	Axial lead of wire wrap	L		E 9-93a
	Head (pump)	L		Vol. II
h	Wall heat-transfer coefficient		$EL^{-2}\theta^{-1}T^{-1}$	E 2-6
	Dimensionless pump head	—		Vol. II
h_g	Gap conductance		$EL^{-2}\theta^{-1}T^{-1}$	E 8-106
h_m^+	Flow mixing-cup enthalpy		EM^{-1}	E 5-52
I	Geometric inertia of the fluid	L^{-1}		E 13-38
	Energy flux		$ET^{-1}L^{-2}$	E 3-50
$\dot{I}$	Irreversibility or lost work		ET^{-1}	E 4-47
$I(\vec{r})$	Indicator function	—		Vol. II
J	Total number of neighboring subchannels	—		Vol. II
$\vec{J}$	Generalized surface source or sink for mass, momentum, and energy			E 4-50
$\vec{j}$ or j	Volumetric flux (superficial velocity)	LT^{-1}		E 5-42
j_{AX}	Flux of species A diffusing through a binary mixture of A and B due to the concentration gradient of A	$ML^{-2}T^{-1}$		Vol. II
K	Total form loss coefficient	—		E 9-23
K_G	Form loss coefficient in transverse direction	—		Vol. II
k	Thermal conductivity		$EL^{-1}\theta^{-1}T^{-1}$	E 4-114
L	Length	L		E 2-5
L_B	Boiling length	L		F 13-1
L_{NB}	Nonboiling length	L		Vol. II
ℓ	Transverse length	L		Vol. II
	Axial dimension	L		Vol. II
ℓ_M, ℓ_H	Mixing length	L		E 10-66
M	Molecular mass	M		P 2-34
$\dot{m}$	Mass flow rate	MT^{-1}		E 4-30a
N	Number of subchannels	—		T 1-4
	Atomic density	L^{-3}		E 2-14
N_p	Total number of rods	—		T 1-4

*M = mass; θ = temperature; E = energy; — = dimensionless; L = length; T = time; F = force.

[†]A ≡ appendix; T ≡ table; E ≡ equation; P ≡ page; F ≡ figure. For example, E10-6 = equation 10-6.

General notation (*continued*)

Symbol	Description	ML θT	FE plus ML θT	Location where symbol first appears[†]
		Dimension in two unit systems*		
N_H, N_w, N_p	Transport coefficient of lumped subchannel	—		Vol. II
N_{rings}	Number of rings in a rod bundle	—		T 1-5
N_{rows}	Number of rows of rods	—		T 1-4
P	Porosity	—		E 8-17
	Decay power		ET^{-1}	E 3-68a,b
	Pitch	L		F 1-12
	Perimeter	L		E 5-64
P_H	Heated perimeter	L		P 10-444
P_R	Power peaking factor	—		Vol. II
P_w	Wetted perimeter	L		E 9-56
ΔP	Clearance on a per-pin basis	L		Vol. II
p	Pressure		FL^{-2}	E 2-8
$\hat{p}$	$\hat{p} = p + \rho g z$		FL^{-2}	Vol. II
Δp^+	$\dfrac{\Delta p}{\rho^*} gL - 1$	—		Vol. II
Q	Heat	ML^2T^{-2}	E	E 4-19
	Volumetric flow rate	L^3T^{-1}		E 5-45
Q'''	Power density		$EL^3 T^{-1}$	P 2-22
$\dot{Q}$	Core power		ET^{-1}	P 2-22
	Heat flow		ET^{-1}	E 7-2a
	Heat-generation rate		ET^{-1}	E 5-124
$\dot{q}$	Rate of energy generated in a pin		ET^{-1}	P 2-22
q'	Linear heat-generation rate		$EL^{-1} T^{-1}$	P 2-22
q'''	Volumetric heat-generation rate		$EL^{-3}T^{-1}$	P 2-22
$\vec{q}'', q''$	Heat flux, surface heat flux		$EL^{-2}T^{-1}$	P 2-22
q''_{cr}	Critical heat flux		$EL^{-2}T^{-1}$	F 2-3
q'_o	Peak linear heat-generation rate		$EL^{-1}T^{-1}$	E 13-14
q'_{rb}	Linear heat generation of equivalent, dispersed heat source		$EL^{-1}T^{-1}$	Vol. II
q'''_{rb}	Equivalent dispersed heat source		$EL^{-3}T^{-1}$	Vol. II
q''_w	Heat flux at the wall		$EL^{-2}T^{-1}$	E 5-152
R	Proportionality constant for hydraulic resistance	$L^{-4}M^n \theta^n$		Vol. II
	Gas constant		$EM^{-1}\theta^{-1}$	T 6-3
	Radius	L		E 2-5
$\bar{R}$	Universal gas constant		$EM^{-1}\theta^{-1}$	AP B634
RR	Reaction rate in a unit volume	$T^{-1}L^{-3}$		P 3-43
$\vec{R}$	Distributed resistance		F	Vol. II
r	Enrichment	—		E 2-15
	Radius	L		T 4-7
r^*	Vapor bubble radius for nucleation	L		E 12-1
r_p	Pressure ratio	—		E 6-97

(*continued*)

General notation (*continued*)

Symbol	Description	Dimension in two unit systems*		Location where symbol first appears[†]
		ML θT	*FE* plus *ML θT*	
S	Surface area	L^2		E 2-1
	Suppression factor of Chen's correlation	—		E 12-32
	Slip ratio	—		E 5-48
S_{ij}	Open flow area in the transverse direction	L^2		Vol. II
$\dot{S}_{gen}$	Rate of entropy generation		$E\,\theta^{-1}$	E 4-25b
S_T	Pitch	L		F 9-28
s_{ij}	Gap within the transverse direction	L		Vol. II
T	Temperature	θ		F 2-1
	Magnitude of as-fabricated clearance or tolerance in an assembly	L		Vol. II
T_o	Reservoir temperature	θ		E 4-27
T_s	Surroundings temperature	θ		E 4-25a
t	Time	T		P 2-25
	Spacer thickness	L		F 1-12
t^*	Peripheral spacer thickness	L		AT J1
t_s	Time after shutdown	T		F 3-8
V	Mean velocity	LT^{-1}		E 2-8
V_m	Mean velocity	LT^{-1}		E 9-44
V_v	Average bundle fluid velocity	LT^{-1}		E 9-90
V_∞	Bubble rise velocity	LT^{-1}		E 11-50
V_{vj}, v_{vj}	Local drift velocity of vapor	LT^{-1}		E 11-32
v	Velocity of a point	LT^{-1}		T 4-2
	Velocity in the lateral direction	LT^{-1}		Vol. II
$\vec{v}$	Velocity vector	LT^{-1}		T 4-2
$v_r, \vec{v}_r$	Relative velocity of the fluid with respect to the control surface	LT^{-1}		E 4-15
$\vec{v}_s$	Velocity of the control volume surface	LT^{-1}		E 4-11
W	Work	ML^2T^{-2}	FL	E 4-19
	Flow rate	MT^{-1}		P 7-273
W_u	Useful work	ML^2T^{-2}	FL	E 4-29
W_{ij}	Crossflow rate per unit length of channel	$MT^{-1}L^{-1}$		Vol. II
$W_{ij}'^{D}$, $W_{ij}'^{M}$, $W_{ij}'^{H}$	Transverse mass flow rate per unit length associated with turbulent mass, momentum, and energy exchange	$MT^{-1}L^{-1}$		Vol. II
W_{ij}^{*M} W_{ij}^{*H}	Transverse mass flow rate associated with both molecular and turbulent momentum and energy exchange	$MT^{-1}L^{-1}$		Vol. II

*M = mass; θ = temperature; E = energy; — = dimensionless; L = length; T = time; F = force.

[†]A ≡ appendix; T ≡ table; E ≡ equation; P ≡ page; F ≡ figure. For example, E10-6 = equation 10-6.

General notation (*continued*)

Symbol	Description	Dimension in two unit systems*		Location where symbol first appears[†]
		ML θT	*FE* plus *ML θT*	
X	Dimensionless radius	—		Vol. II
X^2	Lockart-Martinelli parameter	—		E 11-92
x	Flow quality	—		E 5-35
x_{cr}	Critical quality at CHF	—		Vol. II
x_{st}	Static quality	—		E 5-22
$\Delta x'$	Transverse length	L		Vol. II
Y_i	Preference for downflow in a channel			Vol. II
Z, z	Height	L		P 2-22
Z_B	Boiling boundary	L		E 13-30a
z	Axial position	L		E 2-2
z_c	Position of maximum temperature in clad	L		E 13-25a
z_f	Position of maximum temperature in fuel	L		E 13-28
z_{ij}^T	Turbulent mixing length in COBRA	L		Vol. II
z_{ij}^L	Effective mixing length	L		Vol. II

General Greek symbols

Symbol	Description			Location
α	Confidence level	—		Vol. II
	Local void fraction	—		E 5-11a
	Linear thermal expansion coefficient	θ^{-1}		P 8-309
	Dimensionless angular speed (pump)	—		Vol. II
	Thermal diffusivity	$L^2 T^{-1}$		E 10-43
α_k	Phase density function	—		E 5-1
β	Delayed neutron fraction	—		E 3-65
	Mixing parameter	—		Vol. II
	Volumetric fraction of vapor or gas	—		E 5-51
	Thermal volume expansion coefficient	θ^{-1}		E 4-117
	Ratio of flow area	—		T 9-1
	Dimensionless torque (pump)	—		Vol. II
	Direction angle	—		Vol. II
	Pressure loss parameter	—		P 6-224
Γ	Volumetric vaporization rate	$MT^{-1}L^{-3}$		T 5-1
γ	Fraction of total power generated in the fuel	—		E 3-28
	Specific heat ratio (c_p/c_v)	—		T 6-3
	Porosity	—		Vol. II
δ	Ratio of axial length of grid spacers to axial length of a fuel bundle	—		AP J681

(*continued*)

General notation (*continued*)

Symbol	Description	ML θT	FE plus ML θT	Location where symbol first appears[†]
δ	Gap between fuel and clad	L		E 8-23b
	Thickness of heat-exchanger tube	L		Vol. II
	Thickness of liquid film in annular flow	L		E 11-1
δ^*	Dimensionless film thickness	—		E 11-1
δ_c	Clad thickness	L		F 8-27
δ_g	Gap between fuel and clad	L		E 2-11
δ_T	Thickness of temperature boundary layer	L		F 10-4
ϵ	$\ell^2 dw/dy$ in COBRA code	L^2T^{-1}		Vol. II
	Surface emissivity		$EL^{-2}\theta^{-4}T^{-1}$	E 8-107a
ϵ_H	Eddy diffusivity of energy	L^2T^{-1}		E 10-46
ϵ_M	Momentum diffusivity	L^2T^{-1}		E 9-63
ζ	Thermodynamic efficiency (or effectiveness)	—		E 6-31
η	Pump efficiency	—		Vol. II
η_s	Isentropic efficiency	—		E 6-37
η_{th}	Thermal efficiency	—		E 6-38
θ	Position angle	—		T 4-7
	Two-phase multiplier for mixing	—		Vol. II
	Film temperature drop	θ		Vol. II
θ^*	Influence coefficient	—		Vol. II
λ	Length along channel until fluid reaches saturation	L		Vol. II
λ_c	Taylor instability wavelength	L		E 12-51
μ	Attenuation coefficient	L^{-1}		E 3-53
	Dynamic viscosity	$ML^{-1}T^{-1}$	FTL^{-2}	E 4-84
μ'	Bulk viscosity	$ML^{-1}T^{-1}$	FTL^{-2}	E 4-84
$\hat{\mu}$	Estimated mean of distribution	—		Vol. II
μ_a	Absorption coefficient	L^{-1}		E 3-51
ν	Kinematic viscosity (μ/ρ)	L^2T^{-1}		P 9-361
	Time it takes fluid packet to lose its subcooling	T		Vol. II
	Dimensionless volumetric flow rate (pump)	—		Vol. II
π	Torque		FL	Vol. II
ρ	Density	ML^{-3}		E 2-11
ρ_m^+	Two-phase momentum density	ML^{-3}		E 5-66
Σ	Macroscopic cross section	L^{-1}		E 3-3
σ	Microscopic cross section	L^2		E 3-3
	Normal stress component		FL^{-2}	F 4-8
	Standard deviation	—		T 3-9
	Surface tension		FL^{-1}	E 11-5

*M = mass; θ = temperature; E = energy; − = dimensionless; L = length; T = time; F = force.

[†]A ≡ appendix; T ≡ table; E ≡ equation; P ≡ page; F ≡ figure. For example, E10-6 = equation 10-6.

General notation (*continued*)

Symbol	Description	Dimension in two unit systems*		Location where symbol first appears[†]
		ML θT	*FE* plus *ML θT*	
$\hat{\sigma}$	Estimated standard deviation of distribution	—		Vol. II
τ	Shear stress component		FL^{-2}	F 4-8
	Time after start-up	T		E 3-68a
	Time constant	T		Vol. II
τ_s	Operational time	T		E 3-69a
Φ	Dissipation function		$FL^{-2}T^{-1}$	E 4-107
	Neutron flux	$L^{-2}T^{-1}$		P 3-43
ϕ	Generalized volumetric source or sink for mass, momentum, or energy			E 4-50
	Relative humidity	—		E 7-19
	Azimuthal angle	—		T 4-7
$\phi_{\ell o}^2$	Two-phase friction multiplier based on all-liquid-flow pressure gradient	—		E 11-66
ϕ_ℓ^2	Two-phase frictional multiplier based on pressure gradient of liquid flow	—		E 11-86
ϕ_g^2	Two-phase frictional multiplier based on pressure gradient of gas or vapor flow	—		E 11-86
ξ	Logarithmic energy decrement	—		E 3-41
χ	Energy per fission deposited in the fuel		E	P 3-44
ψ	Force field per unit mass of fluid		FLM^{-1}	E 4-21
	Ratio of eddy diffusivity of heat to momentum	—		E 10-71
ω	Angular speed	T^{-1}		E 9-2b

Subscripts

A	Area			Vol. II
	Annulus			Vol. II
AF	Atmospheric flow			E 6-68
a	Air			E 7-1
	Absorption			P 3-44
acc	Acceleration			E 9-22
avg	Average			F 2-1
B,b	Boiling			E 13-30a
B	Buoyancy			Vol. II
b	Bulk			E 2-6
C	Cold			F 2-1
	Condensate			P 7-273
CD	Condensor			E 6-87
CP	Compressor			E 6-102

(continued)

General notation (*continued*)

Symbol	Description	Dimension in two unit systems*		Location where symbol first appears[†]
		ML θT	*FE* plus *ML* θT	
$\mathcal{Q}_L$	Center line			F 8-9
c	Capture			P 3-43
	Cladding			E 8-38
	Conduction			E 4-113
	Containment			E 7-1
	Contraction			F 9-35
	Coolant			E 6-2
	Core			Vol. II
	Critical point (thermodynamic)			E 9-31
ci	Clad inside			P 8-309
co	Clad outside			E 2-5
cr	Critical flow			E 11-113
crit	Critical			T 6-1
c.m.	Control mass			E 4-17c
c.v.	Control volume			E 4-30
c.t.	Circular tube			E 10-99
d	Downflow			Vol. II
EA	Equivalent annulus			Vol. II
EQUIL	Equilibrium void distribution			Vol. II
e	Expansion			F 9-35
	Extrapolated			E 3-35
	Equilibrium			E 5-53
	Electrical			E 10-10
eff,e	Effective			E 5-129
ex	External			Vol. II
exit, ex, e	Indicating the position of flowing exit			Vol. II
eℓ	Elastic			E 3-39
FL	Flashing			T 7-4
FS	Fuel pellet surface			Vol. II
f	Friction			E 9-91a
	Fluid			Vol. II
	Flow			P 10-444
	Saturated liquid			E 5-53
	Fuel			E 3-19
	Fission			P 3-43
fg	Difference between fluid and gas			T 6-3
fi	Flow without spacers			E 9-82
fo	Fuel outside surface			E 2-5
fric	Friction			T 9-5
fi$_s$	flow including spacers			AT J3
g	Gap between fuel and cladding			E 8-106
	Saturated vapor			E 5-53

*M = mass; θ = temperature; E = energy; $-$ = dimensionless; L = length; T = time; F = force.

[†]A $\equiv$ appendix; T $\equiv$ table; E $\equiv$ equation; P $\equiv$ page; F $\equiv$ figure. For example, E10-6 = equation 10-6.

General notation (*continued*)

Symbol	Description	Dimension in two unit systems*		Location where symbol first appears[†]
		ML θT	*FE* plus *ML θT*	
H, h	Heated			F 2-1
HX	Heat exchanger			E 6-104
h, (heater)	Pressurizer heater			F 7-10
i	Denoting the selected subchannel control volume			Vol. II
	Index of direction			E 5-48
	Inner surface			Vol. II
	Index of streams into or out of the control volume			E 4-30a
in,IN	Indicating the position of flowing in, inlet			E 6-38
iso	Isothermal			E 13-74
iℓ	Inelastic			E 3-59
j	Adjacent subchannel control volume of the subchannel control volume i			Vol. II
	Index of properties			Vol. II
	Index of isotopes			P 3-43
k	Index of phase			E 5-1
L	Lower			F 7-14
	Laminar			E 9-82
ℓ	Liquid phase in a two-phase flow			T 5-1
ℓo	Liquid only			E 11-66
M	Temperature of a quantity of interest			Vol. II
m	Temperature difference of a quantity of interest			Vol. II
	Mixture			E 5-38a
NOM	Nominal			Vol. II
n	Nuclear core			E 7-2d
o	Indicating initial value			E 3-56
	Operating			E 3-68
	Outer surface			Vol. II
	Reservoir conditions			E 4-27
out,OUT	Outlet			P 6-189
P	Pump			E 6-48
p	Primary			E 6-49
	Pore			P 8-301
R	Rated			Vol. II
	Reactor			E 6-53
RO	Condensation as rainout			E 7-128
r	Radiation			E 4-113
	Enrichment			Vol. II

(*continued*)

General notation (*continued*)

Symbol	Description	Dimension in two unit systems*		Location where symbol first appears†
		ML θT	*FE* plus *ML θT*	
rb	Equivalent or dispersed			Vol. II
	Rod bundle			Vol. II
ref	Reference			E 9-23
SC	Condensation at spray drops			E 7-128
SCB	Subcooled boiling			E 13-57
SG, S.G.	Steam generator			E 6-54
SP	Single phase			Vol. II
	Spray			E 7-135
s	Scattering			E 3-39
	Secondary			E 6-49
	Sintered			E 8-98
	Slug flow			Vol. II
	Solid			P 8-301
	Spacer, wire spacer			F 1-14
	Surface, interface			E 4-11
s	Isentropic			E 6-10b
sat	Saturated			T 6-1
st	Structures			E 7-2a
	Static			E 5-22
s′	Interface between continuous vapor and falling liquid droplets			F 7-14
s″	Interface between the continuous vapor phase of the upper volume and the continuous liquid phase of the lower volume			F 7-14
s‴	Interface that separates the discontinuous phase in either the upper or lower volume from the continuous same phase in the other volume			F 7-14
T,t	Turbine			T 6-5
T	Total			P7-240
TD	Theoretical density			E 8-17
TP	Two phase			E 11-67
tb	Transition boiling			E 12-43
th	Thermal			E 6-38
tr	Transverse flow			Vol. II
U	Upper			F 7-14
u	Useful			E 4-28
	Upflow			Vol. II
V	Volume			P 9-386
v	Cavity (void)			E 8-55
	Vapor or gas phase in a two-phase flow			T 5-1

*M = mass; θ = temperature; E = energy; − = dimensionless; L = length; T = time; F = force.

†A ≡ appendix; T ≡ table; E ≡ equation; P ≡ page; F ≡ figure. For example, E10-6 = equation 10-6.

General notation (*continued*)

Symbol	Description	Dimension in two unit systems*		Location where symbol first appears†
		ML θT	*FE* plus *ML* θT	
vj	Local vapor drift			E 11-32
vo	Vapor only			E 11-66
WC	Condensation at the wall			E 7-128
w	Wall			T 5-1
	Water			P 7-240
ϕ	Neutron flux			Vol. II
1ϕ	Single phase			F 13-1
2ϕ	Two phase			F 12-11
ρ	Fuel density			Vol. II
∞	Free stream			F 10-4

Superscripts

D	Hot spot factor resulting from direct contributors			Vol. II
j	Index of isotopes			P 3-43
S	Hot spot factor resulting from statistical contributors			Vol. II
t	Turbulent effect			E 4-132
$\rightarrow$	Vector			P 2-22
$\overline{}$	Spatial average			E 3-41
', ", '''	Per unit length, surface area, volume, respectively			P 2-22
'	Denoting perturbation			E 4-125
TP	Two phase			E 11-77
*	Reference			Vol. II
	Denoting the velocity or enthalpy transported by the diversion crossflow			Vol. II
i	Intrinsic			Vol. II
=	Tensor			E 4-8
$-$, ~	Averaging (time)			E 4-124, E 5-6
o	Stagnation			T 4-2
	Denoting nominal			Vol. II

Special symbols

Δ	Change in, denoting increment			E 2-8
∇	Gradient			E 4-2
δ	Change in, denoting increment			T 3-3
< >	Volumetric averaging			E 2-5
{ }	Area averaging			E 3-20
$\equiv$	Defined as			P 2-22
$\simeq$, $\approx$, $\sim$	Approximately equal to			E 3-38

PHYSICAL AND MATHEMATICAL CONSTANTS

Avogadro's number (A_v)	0.602252×10^{24} molecules/gm mole
	2.731769×10^{26} molecules/lbm mole
Barn	10^{-24}cm^2, 1.0765×10^{-27} ft^2
Boltzmann's constant ($k = \overline{R}/A_v$)	1.38054×10^{-16}erg/°K
	8.61747×10^{-5}eV/°K
Curie	3.70×10^{10} dis/s
Electron charge	4.80298×10^{-10}esu, 1.60210×10^{-20}emu
Faraday's constant	9.648×10^4 coulombs/mole
g_c Conversion factor	1.0 gm cm^2/erg s^2, 32.17 lbm ft/lb$_f$ s^2, 4.17×10^8 lbm ft/lb$_f$ hr^2, 0.9648×10^{18}amu cm^2/MeV s^2
Gravitational acceleration (standard)	32.1739 ft/s^2, 980.665 cm/s^2
Joule's equivalent	778.16 ft-lb$_f$/Btu
Mass–energy conversion	1 amu = 931.478 MeV = 1.41492×10^{-13} Btu = 4.1471×10^{-17} kwhr
	1 gm = 5.60984×10^{26} MeV = 2.49760×10^7 kwhr = 1.04067 Mwd
	1 lbm = 2.54458×10^{32} MeV = 3.86524×10^{16} Btu
Mathematical constants	$e \equiv 2.71828$
	$\pi \equiv 3.14159$
	$ln\ 10 \equiv 2.30259$
Molecular volume	22413.6 cm^3/gm mole, 359.0371 ft^3/lbm mole, at 1 atm and 0°C
Neutron energy	0.0252977 eV at 2200 m/s, 1/40 ev at 2187.017 m/s
Planck's constant	6.6256×10^{-27}erg s, 4.13576×10^{-15}eV s
Rest masses	
Electron	5.48597×10^{-4} amu, 9.10909×10^{-28} gm, 2.00819×10^{-30} lbm
Neutron	1.0086654 amu, $1.6748228 \times 10^{-24}$ gm, 3.692314×10^{-27} lbm
Proton	1.0072766 amu, 1.672499×10^{-24} gm, 3.687192×10^{-27} lbm

Stephan-Boltzmann constant	5.67×10^{-12} W/cm^{2}K^4
Universal gas constant $(\bar{R})$	1545.08 ft-lb$_f$/lbm mole °R
	1.98545 cal/gm mole °K
	1.98545 Btu/lbm mole °R
	8.31434×10^7 erg/gm mole °K
Velocity of light	2.997925×10^{10} cm/s, 9.83619×10^8 ft/s

Source: Adapted from El-Wakil, M. M. *Nuclear Heat Transport* Scranton, PA: International Textbook Co., 1971.

UNIT SYSTEMS

Table	Unit	Reference
C-1	Length	El-Wakil (1971)
C-2	Area	El-Wakil (1971)
C-3	Volume	El-Wakil (1971)
C-4	Mass	El-Wakil (1971)
C-5	Force	Bird et al. (1960)
C-6	Density	El-Wakil (1971)
C-7	Time	El-Wakil (1971)
C-8	Flow	El-Wakil (1971)
C-9	Pressure, Momentum Flux	Bird et al. (1960)
C-10	Work, Energy, Torque	Bird et al. (1960)
C-11	Power	El-Wakil (1971)
C-12	Power Density	El-Wakil (1971)
C-13	Heat Flux	El-Wakil (1971)
C-14	Viscosity, Density × Diffusivity, Concentration × Diffusivity	Bird et al. (1960)
C-15	Thermal Conductivity	Bird et al. (1960)
C-16	Heat-Transfer Coefficient	Bird et al. (1960)
C-17	Momentum, Thermal or Molecular Diffusivity	Bird et al. (1960)
C-18	Surface Tension	—

Bird, R. B., Stewart, W. E., and Lightfoot, E. N. *Transport Phenomena*. New York: Wiley, 1960.

El-Wakil, M. M. *Nuclear Heat Transport*. Scranton, PA: International Textbook Company, 1971.

Relevant SI units for conversion tables

Quantity	Name	Symbol
SI base units		
Length	meter	m
Time	second	s
Mass	kilogram	kg
Temperature	kelvin	°K
Amount of matter	mole	mol
Electric current	ampere	A
SI derived units		
Force	newton	$N = kg\ m/s^2$
Energy, work, heat	joule	$J = N\ m = kg\ m^2/s^2$
Power	watt	$W = J/s$
Frequency	hertz	$Hz = s^{-1}$
Electric charge	coulomb	$C = A\ s$
Electric potential	volt	$V = J/C$
Allowed units (to be used with SI units)		
Time	minute	min
	hour	h
	day	d
Plane angle	degree	°
	minute	′
	second	″
Volume	liter	l
Mass	tone	$t = 1000\ kg$
	atomic mass unit	$u \approx 1.660\ 53 \times 10^{-27}\ kg$
Fluid pressure	bar	$bar = 10^5\ Pa$
Temperature	degree Celsius	°C
Energy	electron volt	$eV \approx 1.602\ 19 \times 10^{-19}\ J$

UTILIZATION OF UNIT CONVERSION TABLES

The column and row units correspond to each other, as illustrated below. Given a quantity in units of a row, multiply by the table value to obtain the quantity in units of the corresponding column.

Example How many meters is 10 cm?

Answer: We desire the quantity in units of meters (the column entry) and have been given the quantity in units of centimeters (the row entry) i.e., 10 cm. Hence meters = 0.01 cm = 0.01 (10) = 0.1

	Columns			
Rows	a Centimeters	b Meters	c	d
a = centimeters		0.01		
b = meters				
c				
d				

Table C-1 Length

	Centimeters (cm)	Meters* (m)	Inches (in.)	Feet+ (ft)	Miles	Microns (μ)	Ångstroms (Å)
cm	1	0.01	0.3937	0.03281	6.214×10^{-4}	10^4	10^8
m	100	1	39.37	3.281	6.214×10^{-6}	10^6	10^{10}
in	2.540	0.0254	1	0.08333	1.578×10^{-5}	2.54×10^4	2.54×10^8
ft	30.48	0.3048	12	1	1.894×10^{-4}	0.3048×10^6	0.3048×10^{10}
miles	1.6093×10^5	1.6093×10^3	6.336×10^4	5.280×10^3	1	1.6093×10^9	1.6093×10^{13}
microns	10^{-4}	10^{-6}	3.937×10^{-5}	3.281×10^{-6}	6.2139×10^{-10}	1	10^4
Ångstroms	10^{-8}	10^{-10}	3.937×10^{-9}	3.281×10^{-10}	6.2139×10^{-14}	10^{-4}	1

*SI units.
+English units.

Table C-2 Area

	cm²	m²*	in.²	ft²+	mile²	acre	barn
cm²	1	10^{-4}	0.155	1.0764×10^{-3}	3.861×10^{-11}	2.4711×10^{-3}	10^{24}
m²	10^4	1	1.550×10^3	10.764	3.861×10^{-7}	2.4711×10^{-4}	10^{28}
in.²	6.4516	6.4516×10^{-4}	1	6.944×10^{-3}	2.491×10^{-10}	1.5944×10^{-7}	6.4516×10^{24}
ft²	929	0.0929	144	1	3.587×10^{-8}	2.2957×10^{-5}	9.29×10^{26}
mile²	2.59×10^{10}	2.59×10^6	4.0144×10^{11}	2.7878×10^7	1	640	2.59×10^{34}
acre	4.0469×10^7	4.0469×10^3	6.2726×10^6	4.356×10^4	1.5625×10^{-3}	1	4.0469×10^{31}
barn	10^{-24}	10^{-28}	1.55×10^{-25}	1.0764×10^{-27}	3.861×10^{-35}	2.4711×10^{-32}	1

*SI units.
+English units.

Table C-3 Volume

	cm³	liters	m³*	in.³	ft³†	cubic yards	U.S. (liq.) gallons	Imperial gallons
cm³	1	10^{-3}	10^{-6}	0.06102	3.532×10^{-5}	1.308×10^{-6}	2.642×10^{-4}	2.20×10^{-4}
liters	10^{3}	1	10^{-3}	61.02	0.03532	1.308×10^{-3}	0.2642	0.220
m³	10^{6}	10^{3}	1	6.102×10^{4}	35.31	1.308	264.2	220.0
in.³	16.39	0.01639	1.639×10^{-5}	1	5.787×10^{-4}	2.143×10^{-5}	4.329×10^{-3}	3.605×10^{-3}
ft³	2.832×10^{4}	28.32	0.02832	1728	1	0.03704	7.481	6.229
cubic yds	7.646×10^{5}	764.6	0.7646	4.666×10^{4}	27.0	1	202.0	168.2
U.S. gals	3.785×10^{3}	3.785	3.785×10^{-3}	231.0	0.1337	4.951×10^{-3}	1	0.8327
Imp. gals	4.546×10^{3}	4.546	4.546×10^{-3}	277.4	0.1605	5.946×10^{-3}	1.201	1

*SI units.
†English units.

Table C-4 Mass

	grams (gm)	kilograms (kg)*	pounds (lbm)†	tons (short)	tons (long)	tons (metric)	atomic mass units (amu)
gm	1	0.001	2.2046×10^{-3}	1.102×10^{-6}	9.842×10^{-7}	10^{-6}	6.0225×10^{23}
kg	1×10^{3}	1	2.2046	1.102×10^{-3}	9.842×10^{-4}	10^{-3}	6.0225×10^{26}
lbm	453.6	0.4536	1	5.0×10^{-4}	4.464×10^{-4}	4.536×10^{-4}	2.7318×10^{26}
tons(s)	9.072×10^{5}	907.2	2.0×10^{3}	1	0.8929	0.9072	5.4636×10^{29}
tons(l)	1.016×10^{6}	1.016×10^{3}	2.240×10^{3}	1.12	1	1.016	6.1192×10^{29}
tons(metric)	10^{6}	1.000	2.2047×10^{3}	1.1023	0.9843	1	6.0225×10^{29}
amu	1.6604×10^{-24}	1.6604×10^{-27}	3.6606×10^{-27}	1.8303×10^{-30}	1.6343×10^{-30}	1.6604×10^{-30}	1

*SI units.
†English units.

Table C-5 Force

	g cm s^{-2} (dynes)	kg m s^{-2}* (newtons)	lbm ft s^{-2} (poundals)	lbf^{+}
dynes	1	10^{-5}	7.2330×10^{-5}	2.2481×10^{-6}
newtons	10^5	1	7.2330	2.2481×10^{-1}
poundals	1.3826×10^4	1.3826×10^{-1}	1	3.1081×10^{-2}
lbf	4.4482×10^5	4.4482	32.1740	1

*SI units.
$^{+}$English units.

Table C-6 Density

	gm/cm^3	kgm/m^3*	lbm/in.^{3}	lbm/ft^{3+}	lbm/U.S. gal	lbm/Imp. gal
gm/cm^3	1	10^3	0.03613	62.43	8.345	10.02
kgm/m^3	10^{-3}	1	3.613×10^{-5}	0.06243	8.345×10^{-3}	0.01002
lbm/in^3	27.68	2.768×10^4	1	1.728×10^3	231	277.4
lbm/ft^3	0.01602	16.02	5.787×10^{-4}	1	0.1337	0.1605
lbm/U.S. gals	0.1198	119.8	4.329×10^{-3}	7.481	1	1.201
lbm/Imp gals	0.09978	99.78	4.605×10^{-3}	6.229	0.8327	1

*SI units.
$^{+}$English units.

Table C-7 Time

	microseconds (μs)	seconds (s)	minutes (min)	hours (hr)	days (d)	years (yr)
μs	1	10^{-6}	1.667×10^{-8}	2.778×10^{-10}	1.157×10^{-11}	3.169×10^{-14}
s	10^{6}	1	1.667×10^{-2}	2.778×10^{-4}	1.157×10^{-5}	3.169×10^{-8}
min	6×10^{7}	60	1	1.667×10^{-2}	6.944×10^{-4}	1.901×10^{-6}
hr	3.6×10^{9}	3.6×10^{3}	60	1	0.04167	1.141×10^{-4}
d	8.64×10^{10}	8.64×10^{4}	1,440	24	1	2.737×10^{-3}
yr	3.1557×10^{13}	3.1557×10^{7}	5.259×10^{5}	8.766×10^{3}	365.24	1

Table C-8 Flow

	cm^3/s	ft^3/min	U.S. gal/min	Imperial gal/min
cm^3/s	1	0.002119	0.01585	0.01320
ft^3/min	472.0	1	7.481	6.229
U.S. gal/min	63.09	0.1337	1	0.8327
Imp gal/min	75.77	0.1605	1.201	1

Table C-9 Pressure, momentum flux

	g cm⁻¹s⁻² (dyne cm⁻²)	Pascal kg m⁻¹s⁻³* (newtons m⁻²)	lbm ft⁻¹s⁻² (poundals ft⁻²)	lbf ft⁻²	lbf in.⁻²† (psia)	Atmospheres (atm)	mm Hg	in. Hg
dyne cm⁻²	1	10^{-1}	6.7197×10^{-2}	2.0886×10^{-3}	1.4504×10^{-5}	9.8692×10^{-7}	7.5006×10^{-4}	2.9530×10^{-5}
newtons m⁻²	10	1	6.7197×10^{-1}	2.0886×10^{-2}	1.4504×10^{-4}	9.8692×10^{-6}	7.5006×10^{-3}	2.9530×10^{-4}
poundals ft⁻²	1.4882×10^{1}	1.4882	1	3.1081×10^{-2}	2.1584×10^{-4}	1.4687×10^{-5}	1.1162×10^{-2}	4.3945×10^{-4}
lbf ft⁻²	4.7880×10^{2}	4.7880×10^{1}	32.1740	1	6.9444×10^{-3}	4.7254×10^{-4}	3.5913×10^{-1}	1.4139×10^{-2}
psia	6.8947×10^{4}	6.8947×10^{3}	4.6330×10^{3}	144	1	6.8046×10^{-2}	5.1715×10^{1}	2.0360
atm	1.0133×10^{6}	1.0133×10^{5}	6.8087×10^{4}	2.1162×10^{3}	14.696	1	760	29.921
mm Hg	1.3332×10^{3}	1.3332×10^{2}	8.9588×10^{1}	2.7845	1.9337×10^{-2}	1.3158×10^{-3}	1	3.9370×10^{-2}
in. Hg	3.3864×10^{4}	3.3864×10^{3}	2.2756×10^{3}	7.0727×10^{1}	4.9116×10^{-1}	3.3421×10^{-2}	25.400	1

*SI units.
†English units.

Table C-10 Work, energy, torque

	g cm² s⁻² (ergs)	kg m² s⁻²* (absolute joules)	lbm ft² s⁻² (ft-poundals)	ft lbf	cal	Btu†	hp-hr	kw-hr
g cm² s⁻²	1	10^{-7}	2.3730×10^{-6}	7.3756×10^{-8}	2.3901×10^{-8}	9.4783×10^{-11}	3.7251×10^{-14}	2.7778×10^{-14}
kg m² s⁻²	10^{7}	1	2.3730×10^{1}	7.3756×10^{-1}	2.3901×10^{-1}	9.4783×10^{-4}	3.7251×10^{-7}	2.7778×10^{-7}
lbm ft² s⁻²	4.2140×10^{5}	4.2140×10^{-2}	1	3.1081×10^{-2}	1.0072×10^{-2}	3.9942×10^{-5}	1.5698×10^{-8}	1.1706×10^{-8}
ft lbf	1.3558×10^{7}	1.3558	32.1740	1	3.2405×10^{-1}	1.2851×10^{-3}	5.0505×10^{-7}	3.7662×10^{-7}
Thermochemical calories‡	4.1840×10^{7}	4.1840	9.9287×10^{1}	3.0860	1	3.9657×10^{-3}	1.5586×10^{-6}	1.1622×10^{-6}
British thermal units	1.0550×10^{10}	1.0550×10^{3}	2.5036×10^{4}	778.16	2.5216×10^{2}	1	3.9301×10^{-4}	2.9307×10^{-4}
Horsepower-hours	2.6845×10^{13}	2.6845×10^{6}	6.3705×10^{7}	1.9500×10^{6}	6.4162×10^{5}	2.5445×10^{3}	1	7.4570×10^{-1}
Absolute kilowatt-hours	3.6000×10^{13}	3.6000×10^{6}	8.5429×10^{7}	2.6552×10^{6}	8.6042×10^{5}	3.4122×10^{3}	1.3410	1

*SI units.
†English units.
‡This unit, abbreviated cal. is used in chemical thermodynamic tables. To convert quantities expressed in International Steam Table calories (abbreviated I.T. cal) to this unit, multiply by 1.000654.

Table C-11 Power

	Ergs/s	Joule/s watt*	kw	Btu/hr†	hp	eV/s
Ergs/s	1	10^{-7}	10^{-10}	3.412×10^{-7}	1.341×10^{-10}	6.2421×10^{11}
Joule/s	10^7	1	10^{-3}	3.412	1.341×10^{-3}	6.2421×10^{18}
kw	10^{10}	10^3	1	3412	1.341	6.2421×10^{21}
Btu/hr	2.931×10^6	0.2931	2.931×10^{-4}	1	3.93×10^{-4}	1.8294×10^{18}
hp	7.457×10^9	745.7	0.7457	2.545×10^3	1	4.6548×10^{21}
eV/s	1.6021×10^{-12}	1.6021×10^{-19}	1.6021×10^{-22}	5.4664×10^{-19}	2.1483×10^{-22}	1

*SI units.
†English units.

Table C-12 Power density

	watt/cm³, kw/lit*	cal/sec cm³	Btu/hr in.³	Btu/hr ft³†	MeV/s cm³
watt/cm³, kw/lit	1	0.2388	55.91	9.662×10^4	6.2420×10^{12}
cal/s cm³	4.187	1	234.1	4.045×10^5	2.613×10^{13}
Btu/hr in³	0.01788	4.272×10^{-3}	1	1728	1.1164×10^{11}
Btu/hr ft³	1.035×10^{-5}	2.472×10^{-6}	5.787×10^{-4}	1	6.4610×10^7
MeV/s cm³	1.602×10^{-13}	3.826×10^{-14}	8.9568×10^{-12}	1.5477×10^{-8}	1

*SI units.
†English units.

Table C-13 Heat flux

	watt/cm²*	cal/s cm²	Btu/hr ft²†	Mev/s cm²
watt/cm²	1	0.2388	3170.2	6.2420×10^{12}
cal/s cm²	4.187	1	1.3272×10^4	2.6134×10^{13}
Btu/hr ft²	3.155×10^{-4}	7.535×10^{-5}	1	1.9691×10^9
Mev/s cm²	1.602×10^{-13}	3.826×10^{-14}	5.0785×10^{-10}	1

*SI units.
†English units.

Table C-14 Viscosity density × diffusivity, concentration × diffusivity

	g cm^{-1} s^{-1} (poises)	kg m^{-1} s^{-1}*	lbm ft^{-1} s^{-1}†	lbf s ft^{-2}	Centipoises	lbm ft^{-1} hr^{-1}
g cm^{-1} s^{-1}	1	10^{-1}	6.7197 × 10^{-2}	2.0886 × 10^{-3}	10^2	2.4191 × 10^2
kg m^{-1} s^{-1}	10	1	6.7197 × 10^{-1}	2.0886 × 10^{-2}	10^3	2.4191 × 10^3
lbm ft^{-1} s^{-1}	1.4882 × 10^1	1.4882	1	3.1081 × 10^{-2}	1.4882 × 10^3	3.60 × 10^3
lbf s ft^{-2}	4.7880 × 10^2	4.7880 × 10^1	32.1740	1	4.7880 × 10^4	1.1583 × 10^3
Centipoises	10^{-2}	10^{-3}	6.7197 × 10^{-4}	2.0886 × 10^{-5}	1	2.4191
lbm ft^{-1} hr^{-1}	4.1338 × 10^{-3}	4.1338 × 10^{-4}	2.7778 × 10^{-4}	8.6336 × 10^{-6}	4.1338 × 10^{-1}	1

*SI units.
†English units.
‡When moles appear in the given and desired units, the conversion factor is the same as for the corresponding mass units.

Table C-15 Thermal conductivity

	g cm s^{-3} °K^{-1} (ergs s^{-1} cm^{-1} °K^{-1})	kg m s^{-3} °K^{-1}* (watts m^{-1} °K^{-1})	lbm ft s^{-3} °F^{-1}†	lbf s^{-1} °F^{-1}	cal s^{-1} cm^{-1} °K^{-1}	Btu hr^{-1} ft^{-1} °F^{-1}†
g cm s^{-3} °K^{-1}	1	10^{-5}	4.0183 × 10^{-5}	1.2489 × 10^{-6}	2.3901 × 10^{-8}	5.7780 × 10^{-6}
kg m s^{-3} °K^{-1}	10^5	1	4.0183	1.2489 × 10^{-1}	2.3901 × 10^{-3}	5.7780 × 10^{-1}
lbm ft s^{-3} °F^{-1}	2.4886 × 10^4	2.4886 × 10^{-1}	1	3.1081 × 10^{-2}	5.9479 × 10^{-4}	1.4379 × 10^{-1}
lbf s^{-1} °F^{-1}	8.0068 × 10^5	8.0068	3.2174 × 10^1	1	1.9137 × 10^{-2}	4.6263
cal s^{-1} cm^{-1} °K^{-1}	4.1840 × 10^7	4.1840 × 10^2	1.6813 × 10^3	5.2256 × 10^1	1	2.4175 × 10^2
Btu hr^{-1} ft^{-1} °F^{-1}	1.7307 × 10^5	1.7307	6.9546	2.1616 × 10^{-1}	4.1365 × 10^{-3}	1

*SI units.
†English units.

Table C-16 Heat-transfer coefficient

	$g\ s^{-3}\ °K^{-1}$	$kg\ s^{-3}\ °K^{-1}$* $(watts\ m^{-2}\ °K^{-1})$	$lbm\ s^{-3}\ °F^{-1}$	$lbf\ ft^{-1}\ s^{-1}\ °F^{-1}$	$cal\ cm^{-2}\ s^{-1}\ °K^{-1}$	$Watts\ cm^{-2}\ °K^{-1}$	$Btu\ ft^{-2}\ hr^{-1}\ °F^{-1}$†
$g\ s^{-3}\ °K^{-1}$	1	10^{-3}	1.2248×10^{-3}	3.8068×10^{-5}	2.3901×10^{-8}	10^{-7}	1.7611×10^{-4}
$kg\ s^{-3}\ °K^{-1}$	10^{3}	1	1.2248	3.8068×10^{-2}	2.3901×10^{-5}	10^{-4}	1.7611×10^{-1}
$lbm\ s^{-3}\ °F^{-1}$	8.1647×10^{2}	8.1647×10^{-1}	1	3.1081×10^{-2}	1.9514×10^{-5}	8.1647×10^{-5}	1.4379×10^{-1}
$lbf\ ft^{-3}\ s^{-3}\ °F^{-1}$	2.6269×10^{4}	2.6269×10^{1}	32.1740	1	6.2784×10^{-4}	2.6269×10^{-3}	4.6263
$cal\ cm^{-2}\ s^{-1}\ °K^{-1}$	4.1840×10^{7}	4.1840×10^{4}	5.1245×10^{4}	1.5928×10^{3}	1	4.1840	7.3686×10^{3}
$Watts\ cm^{-2}\ °K^{-1}$	10^{7}	10^{4}	1.2248×10^{4}	3.8068×10^{2}	2.3901×10^{-1}	1	1.7611×10^{3}
$Btu\ ft^{-2}\ hr^{-1}\ °F^{-1}$	5.6782×10^{3}	5.6782	6.9546	2.1616×10^{-1}	1.3571×10^{-4}	5.6782×10^{-4}	1

*SI units.
†English units.

Table C-17 Momentum, thermal or molecular diffusivity

	$cm^2\ s^{-1}$	$m^2\ s^{-1}$*	$ft^2\ hr^{-1}$†	Centistokes
$cm^2\ s^{-1}$	1	10^{-4}	3.8750	10^2
$m^2\ s^{-1}$	10^4	1	3.8750×10^4	10^6
$ft^2\ hr^{-1}$	2.5807×10^{-1}	2.5807×10^{-5}	1	2.5807×10^1
Centistokes	10^{-2}	10^{-6}	3.8750×10^{-2}	1

*SI units.
†English units.

Table C-18 Surface tension

	N/m*	dyne/cm	lbf/ft†
N/m	1	10^3	6.841×10^{-2}
dyne/cm	0.001	1	6.841×10^{-5}
lbf/ft	14.618	1.4618×10^4	1

*SI units.
†English units.

MATHEMATICAL TABLES

BESSEL FUNCTION*

Some useful derivatives and integrals of Bessel functions are given in Tables D-1 and D-2.

*El-Wakil, M. M. *Nuclear Heat Transport*. Scranton, PA: International Textbook Co., 1971.

Table D-1 Derivatives of Bessel functions

$$\frac{dJ_0(x)}{dx} = -J_1(x) \qquad\qquad \frac{dY_0(x)}{dx} = -Y_1(x)$$

$$\frac{dI_0(x)}{dx} = I_1(x) \qquad\qquad \frac{dK_0(x)}{dx} = -K_1(x)$$

$$\frac{dJ_v(x)}{dx} = J_{v-1}(x) - \frac{v}{x}J_v(x) \qquad\qquad \frac{dY_v(x)}{dx} = Y_{v-1}(x) - \frac{v}{x}Y_v(x)$$

$$= -J_{v+1}(x) + \frac{v}{x}J_v(x) \qquad\qquad = -Y_{v+1}(x) + \frac{v}{x}Y_v(x)$$

$$= \frac{1}{2}[J_{v-1}(x) - J_{v+1}(x)] \qquad\qquad = \frac{1}{2}[Y_{v-1}(x) - Y_{v+1}(x)]$$

$$\frac{dI_v(x)}{dx} = I_{v-1}(x) - \frac{v}{x}I_v(x) \qquad\qquad \frac{dK_v(x)}{dx} = -K_{v-1}(x) - \frac{v}{x}K_v(x)$$

$$= I_{v+1}(x) + \frac{v}{x}I_v(x) \qquad\qquad = -K_{v+1}(x) + \frac{v}{x}K_v(x)$$

$$= \frac{1}{2}[I_{v-1}(x) + I_{v+1}(x)] \qquad\qquad = -\frac{1}{2}[K_{v-1}(x) + K_{v+1}(x)]$$

$$\frac{dx^v J_v(x)}{dx} = x^v J_{v-1}(x) \qquad\qquad \frac{dx^{-v} J_v(x)}{dx} = -x^{-v} J_{v+1}(x)$$

$$\frac{dx^v Y_v(x)}{dx} = x^v Y_{v-1}(x) \qquad\qquad \frac{dx^{-v} Y_v(x)}{dx} = -x^{-v} Y_{v+1}(x)$$

$$\frac{dx^v I_v(x)}{dx} = x^v I_{v-1}(x) \qquad\qquad \frac{dx^{-v} I_v(x)}{dx} = x^{-v} I_{v+1}(x)$$

$$\frac{dx^v K_v(x)}{dx} = -x^v K_{v-1}(x) \qquad\qquad \frac{dx^{-v} K_v(x)}{dx} = -x^{-v} K_{v+1}(x)$$

647

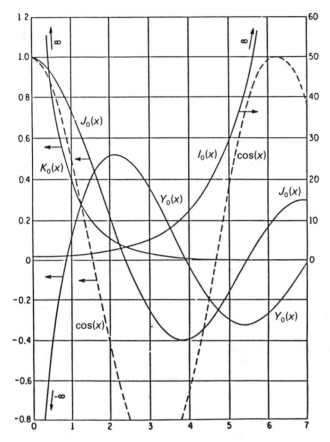

Figure D-1 The four Bessel functions of zero order.

Table D-2 Integrals of Bessel functions

$$\int J_1(x)\, dx = -J_0(x) + C \qquad\qquad \int Y_1(x)\, dx = -Y_0(x) + C$$

$$\int I_1(x)\, dx = I_0(x) + C \qquad\qquad \int K_1(x)\, dx = -K_0(x) + C$$

$$\int x^v J_{v-1}(x)\, dx = x^v J_v(x) + C \qquad \int x^{-v} J_{v+1}(x)\, dx = -x^{-v}J_v(x) + C$$

The Bessel and modified Bessel functions of zero and first order are tabulated in Table D-3 for positive values of x up to $x = 4.0$. Some roots are given below:

Roots of $J_0(x)$: $x = 2.4048,\ 5.5201,\ 8.6537,\ 11.7915,\ \ldots$
Roots of $J_1(x)$: $x = 3.8317,\ 7.0156,\ 10.1735,\ 13.3237,\ \ldots$
Roots of $Y_0(x)$: $x = 0.8936,\ 3.9577,\ 7.0861,\ 10.2223,\ \ldots$
Roots of $Y_1(x)$: $x = 2.1971,\ 5.4297,\ 8.5960,\ 11.7492,\ \ldots$

Table D-3 Some Bessel Functions

x	$J_0(x)$	$J_1(x)$	$Y_0(x)$	$Y_1(x)$	$I_0(x)$	$I_1(x)$	$K_0(x)$	$K_1(x)$
0	1.0000	0.0000	$-\infty$	$-\infty$	1.000	0.0000	∞	∞
0.05	0.9994	0.0250	-1.979	-12.79	1.001	0.0250	3.114	19.91
0.10	0.9975	0.0499	-1.534	-6.459	1.003	0.0501	2.427	9.854
0.15	0.9944	0.0748	-1.271	-4.364	1.006	0.0752	2.030	6.477
0.20	0.9900	0.0995	-1.081	-3.324	1.010	0.1005	1.753	4.776
0.25	0.9844	0.1240	-0.9316	-2.704	1.016	0.1260	1.542	3.747
0.30	0.9776	0.1483	-0.8073	-2.293	1.023	0.1517	1.372	3.056
0.35	0.9696	0.1723	-0.7003	-2.000	1.031	0.1777	1.233	2.559
0.40	0.9604	0.1960	-0.6060	-1.781	1.040	0.2040	1.115	2.184
0.45	0.9500	0.2194	-0.5214	-1.610	1.051	0.2307	1.013	1.892
0.50	0.9385	0.2423	-0.4445	-1.471	1.063	0.2579	0.9244	1.656
0.55	0.9258	0.2647	-0.3739	-1.357	1.077	0.2855	0.8466	1.464
0.60	0.9120	0.2867	-0.3085	-1.260	1.092	0.3137	0.7775	1.303
0.65	0.8971	0.3081	-0.2476	-1.177	1.108	0.3425	0.7159	1.167
0.70	0.8812	0.3290	-0.1907	-1.103	1.126	0.3719	0.6605	1.050
0.75	0.8642	0.3492	-0.1372	-1.038	1.146	0.4020	0.6106	0.9496
0.80	0.8463	0.3688	-0.0868	-0.9781	1.167	0.4329	0.5653	0.8618
0.85	0.8274	0.3878	-0.0393	-0.9236	1.189	0.4646	0.5242	0.7847
0.90	0.8075	0.4059	-0.0056	-0.8731	1.213	0.4971	0.4867	0.7165
0.95	0.7868	0.4234	0.0481	-0.8258	1.239	0.5306	0.4524	0.6560
1.0	0.7652	0.4401	0.0883	-0.7812	1.266	0.5652	0.4210	0.6019
1.1	0.6957	0.4850	0.1622	-0.6981	1.326	0.6375	0.3656	0.5098
1.2	0.6711	0.4983	0.2281	-0.6211	1.394	0.7147	0.3185	0.4346
1.3	0.5937	0.5325	0.2865	-0.5485	1.469	0.7973	0.2782	0.3725
1.4	0.5669	0.5419	0.3379	-0.4791	1.553	0.8861	0.2437	0.3208
1.5	0.4838	0.5644	0.3824	-0.4123	1.647	0.9817	0.2138	0.2774
1.6	0.4554	0.5699	0.4204	-0.3476	1.750	1.085	0.1880	0.2406
1.7	0.3690	0.5802	0.4520	-0.2847	1.864	1.196	0.1655	0.2094
1.8	0.3400	0.5815	0.4774	-0.2237	1.990	1.317	0.1459	0.1826
1.9	0.2528	0.5794	0.4968	-0.1644	2.128	1.448	0.1288	0.1597
2.0	0.2239	0.5767	0.5104	-0.1070	2.280	1.591	0.1139	0.1399
2.1	0.1383	0.5626	0.5183	-0.0517	2.446	1.745	0.1008	0.1227
2.2	0.1104	0.5560	0.5208	-0.0015	2.629	1.914	0.0893	0.1079
2.3	0.0288	0.5305	0.5181	0.0523	2.830	2.098	0.0791	0.0950
2.4	0.0025	0.5202	0.5104	0.1005	3.049	2.298	0.0702	0.0837
2.5	0.0729	0.4843	0.4981	0.1459	3.290	2.517	0.0623	0.0739
2.6	-0.0968	0.4708	0.4813	0.1884	3.553	2.755	0.0554	0.0653
2.7	-0.1641	0.4260	0.4605	0.2276	3.842	3.016	0.0493	0.0577
2.8	-0.1850	0.4097	0.4359	0.2635	4.157	3.301	0.0438	0.0511
2.9	-0.2426	0.3575	0.4079	0.2959	4.503	3.613	0.0390	0.0453
3.0	-0.2601	0.3391	0.3769	0.3247	4.881	3.953	0.0347	0.0402
3.2	-0.3202	0.2613	0.3071	0.3707	5.747	4.734	0.0276	0.0316
3.4	-0.3643	0.1792	0.2296	0.4010	6.785	5.670	0.0220	0.0250
3.6	-0.3918	0.0955	0.1477	0.4154	8.028	6.793	0.6175	0.0198
3.8	-0.4026	0.0128	0.0645	0.4141	9.517	8.140	0.0140	0.0157
4.0	-0.3971	-0.0660	-0.0169	0.3979	11.302	9.759	0.0112	0.0125

DIFFERENTIAL OPERATORS*

* Bird, R. B., Stewart, W. E., and Lightfoot, E. N. *Transport Phenomena*. New York: Wiley, 1960.

Table D-4 Summary of differential operations involving the ∇-operator in rectangular coordinates* (x, y, z)

$$(\nabla \cdot \vec{v}) = \frac{\partial v_x}{\partial x} + \frac{\partial v_y}{\partial y} + \frac{\partial v_z}{\partial z} \tag{A}$$

$$(\nabla^2 s) = \frac{\partial^2 s}{\partial x^2} + \frac{\partial^2 s}{\partial y^2} + \frac{\partial^2 s}{\partial z^2} \tag{B}$$

$$(\overline{\overline{\tau}} : \nabla \vec{v}) = \tau_{xx}\left(\frac{\partial v_x}{\partial x}\right) + \tau_{yy}\left(\frac{\partial v_y}{\partial y}\right) + \tau_{zz}\left(\frac{\partial v_z}{\partial z}\right) + \tau_{xy}\left(\frac{\partial v_x}{\partial y} + \frac{\partial v_y}{\partial x}\right)$$

$$+ \tau_{yz}\left(\frac{\partial v_y}{\partial z} + \frac{\partial v_z}{\partial y}\right) + \tau_{zx}\left(\frac{\partial v_z}{\partial x} + \frac{\partial v_x}{\partial z}\right) \tag{C}$$

$$[\nabla s]_x = \frac{\partial s}{\partial x} \tag{D}$$

$$[\nabla s]_y = \frac{\partial s}{\partial y} \tag{E}$$

$$[\nabla s]_z = \frac{\partial s}{\partial z} \tag{F}$$

$$[\nabla \times \vec{v}]_x = \frac{\partial v_z}{\partial y} - \frac{\partial v_y}{\partial z} \tag{G}$$

$$[\nabla \times \vec{v}]_y = \frac{\partial v_x}{\partial z} - \frac{\partial v_z}{\partial x} \tag{H}$$

$$[\nabla \times \vec{v}]_z = \frac{\partial v_y}{\partial x} - \frac{\partial v_x}{\partial y} \tag{I}$$

$$[\nabla \cdot \overline{\overline{\tau}}]_x = \frac{\partial \tau_{xx}}{\partial x} + \frac{\partial \tau_{xy}}{\partial y} + \frac{\partial \tau_{xz}}{\partial z} \tag{J}$$

$$[\nabla \cdot \overline{\overline{\tau}}]_y = \frac{\partial \tau_{xy}}{\partial x} + \frac{\partial \tau_{yy}}{\partial y} + \frac{\partial \tau_{yz}}{\partial z} \tag{K}$$

$$[\nabla \cdot \overline{\overline{\tau}}]_z = \frac{\partial \tau_{xz}}{\partial x} + \frac{\partial \tau_{yz}}{\partial y} + \frac{\partial \tau_{zz}}{\partial z} \tag{L}$$

$$[\nabla^2 \vec{v}]_x = \frac{\partial^2 v_x}{\partial x^2} + \frac{\partial^2 v_x}{\partial y^2} + \frac{\partial^2 v_x}{\partial z^2} \tag{M}$$

$$[\nabla^2 \vec{v}]_y = \frac{\partial^2 v_y}{\partial x^2} + \frac{\partial^2 v_y}{\partial y^2} + \frac{\partial^2 v_y}{\partial z^2} \tag{N}$$

$$[\nabla^2 \vec{v}]_z = \frac{\partial^2 v_z}{\partial x^2} + \frac{\partial^2 v_z}{\partial y^2} + \frac{\partial^2 v_z}{\partial z^2} \tag{O}$$

$$[\vec{v} \cdot \nabla \vec{v}]_x = v_x \frac{\partial v_x}{\partial x} + v_y \frac{\partial v_x}{\partial y} + v_z \frac{\partial v_x}{\partial z} \tag{P}$$

$$[\vec{v} \cdot \nabla \vec{v}]_y = v_x \frac{\partial v_y}{\partial x} + v_y \frac{\partial v_y}{\partial y} + v_z \frac{\partial v_y}{\partial z} \tag{Q}$$

$$[\vec{v} \cdot \nabla \vec{v}]_z = v_x \frac{\partial v_z}{\partial x} + v_y \frac{\partial v_z}{\partial y} + v_z \frac{\partial v_z}{\partial z} \tag{R}$$

*Operations involving the tensor τ are given for symmetrical τ only.

Table D-5 Summary of differential operations involving the ∇-operator in cylindrical coordinates* (r, θ, z)

$$(\nabla \cdot \vec{v}) = \frac{1}{r}\frac{\partial}{\partial r}(rv_r) + \frac{1}{r}\frac{\partial v_\theta}{\partial \theta} + \frac{\partial v_z}{\partial z} \tag{A}$$

$$(\nabla^2 s) = \frac{1}{r}\frac{\partial}{\partial r}\left(r\frac{\partial s}{\partial r}\right) + \frac{1}{r^2}\frac{\partial^2 s}{\partial \theta^2} + \frac{\partial^2 s}{\partial z^2} \tag{B}$$

$$(\bar{\bar{\tau}} : \nabla\vec{v}) = \tau_{rr}\left(\frac{\partial v_r}{\partial r}\right) + \tau_{\theta\theta}\left(\frac{1}{r}\frac{\partial v_\theta}{\partial \theta} + \frac{v_r}{r}\right) + \tau_{zz}\left(\frac{\partial v_z}{\partial z}\right)$$

$$+ \tau_{r\theta}\left[r\frac{\partial}{\partial r}\left(\frac{v_\theta}{r}\right) + \frac{1}{r}\frac{\partial v_r}{\partial \theta}\right] + \tau_{\theta z}\left(\frac{1}{r}\frac{\partial v_z}{\partial \theta} + \frac{\partial v_\theta}{\partial z}\right) + \tau_{rz}\left(\frac{\partial v_z}{\partial r} + \frac{\partial v_r}{\partial z}\right) \tag{C}$$

$$[\nabla s]_r = \frac{\partial s}{\partial r} \tag{D}$$

$$[\nabla s]_\theta = \frac{1}{r}\frac{\partial s}{\partial \theta} \tag{E}$$

$$[\nabla s]_z = \frac{\partial s}{\partial z} \tag{F}$$

$$[\nabla \times \vec{v}]_r = \frac{1}{r}\frac{\partial v_z}{\partial \theta} - \frac{\partial v_\theta}{\partial z} \tag{G}$$

$$[\nabla \times \vec{v}]_\theta = \frac{\partial v_r}{\partial z} - \frac{\partial v_z}{\partial r} \tag{H}$$

$$[\nabla \times \vec{v}]_z = \frac{1}{r}\frac{\partial}{\partial r}(rv_\theta) - \frac{1}{r}\frac{\partial v_r}{\partial \theta} \tag{I}$$

$$[\nabla \cdot \bar{\bar{\tau}}]_r = \frac{1}{r}\frac{\partial}{\partial r}(r\tau_{rr}) + \frac{1}{r}\frac{\partial}{\partial \theta}\tau_{r\theta} - \frac{1}{r}\tau_{\theta\theta} + \frac{\partial \tau_{rz}}{\partial z} \tag{J}$$

$$[\nabla \cdot \bar{\bar{\tau}}]_\theta = \frac{1}{r}\frac{\partial \tau_{\theta\theta}}{\partial \theta} + \frac{\partial \tau_{r\theta}}{\partial r} + \frac{2}{r}\tau_{r\theta} + \frac{\partial \tau_{\theta z}}{\partial z} \tag{K}$$

$$[\nabla \cdot \bar{\bar{\tau}}]_z = \frac{1}{r}\frac{\partial}{\partial r}(r\tau_{rz}) + \frac{1}{r}\frac{\partial \tau_{\theta z}}{\partial \theta} + \frac{\partial \tau_{zz}}{\partial z} \tag{L}$$

$$[\nabla^2\vec{v}]_r = \frac{\partial}{\partial r}\left(\frac{1}{r}\frac{\partial}{\partial r}(rv_r)\right) + \frac{1}{r^2}\frac{\partial^2 v_r}{\partial \theta^2} - \frac{2}{r^2}\frac{\partial v_\theta}{\partial \theta} + \frac{\partial^2 v_r}{\partial z^2} \tag{M}$$

$$[\nabla^2\vec{v}]_\theta = \frac{\partial}{\partial r}\left(\frac{1}{r}\frac{\partial}{\partial r}(rv_\theta)\right) + \frac{1}{r^2}\frac{\partial^2 v_\theta}{\partial \theta^2} + \frac{2}{r^2}\frac{\partial v_r}{\partial \theta} + \frac{\partial^2 v_\theta}{\partial z^2} \tag{N}$$

$$[\nabla^2\vec{v}]_z = \frac{1}{r}\frac{\partial}{\partial r}\left(r\frac{\partial v_z}{\partial r}\right) + \frac{1}{r^2}\frac{\partial^2 v_z}{\partial \theta^2} + \frac{\partial^2 v_z}{\partial z^2} \tag{O}$$

$$[\vec{v} \cdot \nabla\vec{v}]_r = v_r\frac{\partial v_r}{\partial r} + \frac{v_\theta}{r}\frac{\partial v_r}{\partial \theta} - \frac{v_\theta^2}{r} + v_z\frac{\partial v_r}{\partial z} \tag{P}$$

$$[\vec{v} \cdot \nabla\vec{v}]_\theta = v_r\frac{\partial v_\theta}{\partial r} + \frac{v_\theta}{r}\frac{\partial v_\theta}{\partial \theta} + \frac{v_r v_\theta}{r} + v_z\frac{\partial v_\theta}{\partial z} \tag{Q}$$

$$[\vec{v} \cdot \nabla\vec{v}]_z = v_r\frac{\partial v_z}{\partial r} + \frac{v_\theta}{r}\frac{\partial v_z}{\partial \theta} + v_z\frac{\partial v_z}{\partial z} \tag{R}$$

*Operations involving the tensor τ are given for symmetrical τ only.

Table D-6 Summary of differential operations involving the ∇-operator in spherical coordinates* (r, θ, ϕ)

$$(\nabla \cdot \vec{v}) = \frac{1}{r^2} \frac{\partial}{\partial r}(r^2 v_r) + \frac{1}{r \sin \theta} \frac{\partial}{\partial \theta}(v_\theta \sin \theta) + \frac{1}{r \sin \theta} \frac{\partial v_\phi}{\partial \phi} \tag{A}$$

$$(\nabla^2 s) = \frac{1}{r^2} \frac{\partial}{\partial r}\left(r^2 \frac{\partial s}{\partial r}\right) + \frac{1}{r^2 \sin \theta} \frac{\partial}{\partial \theta}\left(\sin \theta \frac{\partial s}{\partial \theta}\right) + \frac{1}{r^2 \sin^2 \theta} \frac{\partial^2 s}{\partial \phi^2} \tag{B}$$

$$(\overline{\overline{\tau}} : \nabla \vec{v}) = \tau_{rr}\left(\frac{\partial v_r}{\partial r}\right) + \tau_{\theta\theta}\left(\frac{1}{r} \frac{\partial v_\theta}{\partial \theta} + \frac{v_r}{r}\right)$$

$$+ \tau_{\phi\phi}\left(\frac{1}{r \sin \theta} \frac{\partial v_\phi}{\partial \phi} + \frac{v_r}{r} + \frac{v_\theta \cot \theta}{r}\right)$$

$$+ \tau_{r\theta}\left(\frac{\partial v_\theta}{\partial r} + \frac{1}{r} \frac{\partial v_r}{\partial \theta} - \frac{v_\theta}{r}\right) + \tau_{r\phi}\left(\frac{\partial v_\phi}{\partial r} + \frac{1}{r \sin \theta} \frac{\partial v_r}{\partial \phi} - \frac{v_\phi}{r}\right)$$

$$+ \tau_{\theta\phi}\left(\frac{1}{r} \frac{\partial v_\phi}{\partial \theta} + \frac{1}{r \sin \theta} \frac{\partial v_\theta}{\partial \phi} - \frac{\cot \theta}{r} v_\phi\right) \tag{C}$$

$$[\nabla s]_r = \frac{\partial s}{\partial r} \tag{D}$$

$$[\nabla s]_\theta = \frac{1}{r} \frac{\partial s}{\partial \theta} \tag{E}$$

$$[\nabla s]_\phi = \frac{1}{r \sin \theta} \frac{\partial s}{\partial \phi} \tag{F}$$

$$[\nabla \times \vec{v}]_r = \frac{1}{r \sin \theta} \frac{\partial}{\partial \theta}(v_\phi \sin \theta) - \frac{1}{r \sin \theta} \frac{\partial v_\theta}{\partial \phi} \tag{G}$$

$$[\nabla \times \vec{v}]_\theta = \frac{1}{r \sin \theta} \frac{\partial v_r}{\partial \phi} - \frac{1}{r} \frac{\partial}{\partial r}(r v_\phi) \tag{H}$$

$$[\nabla \times \vec{v}]_\phi = \frac{1}{r} \frac{\partial}{\partial r}(r v_\theta) - \frac{1}{r} \frac{\partial v_r}{\partial \theta} \tag{I}$$

$$[\nabla \cdot \overline{\overline{\tau}}]_r = \frac{1}{r^2} \frac{\partial}{\partial r}(r^2 \tau_{rr}) + \frac{1}{r \sin \theta} \frac{\partial}{\partial \theta}(\tau_{r\theta} \sin \theta) + \frac{1}{r \sin \theta} \frac{\partial \tau_{r\phi}}{\partial \phi} - \frac{\tau_{\theta\theta} + \tau_{\phi\phi}}{r} \tag{J}$$

$$[\nabla \cdot \overline{\overline{\tau}}]_\theta = \frac{1}{r^2} \frac{\partial}{\partial r}(r^2 \tau_{r\theta}) + \frac{1}{r \sin \theta} \frac{\partial}{\partial \theta}(\tau_{\theta\theta} \sin \theta) + \frac{1}{r \sin \theta} \frac{\partial \tau_{\theta\phi}}{\partial \phi}$$

$$+ \frac{\tau_{r\theta}}{r} - \frac{\cot \theta}{r} \tau_{\phi\phi} \tag{K}$$

$$[\nabla \cdot \overline{\overline{\tau}}]_\phi = \frac{1}{r^2} \frac{\partial}{\partial r}(r^2 \tau_{r\phi}) + \frac{1}{r} \frac{\partial \tau_{\theta\phi}}{\partial \theta} + \frac{1}{r \sin \theta} \frac{\partial \tau_{\phi\phi}}{\partial \phi} + \frac{\tau_{r\phi}}{r} + \frac{2 \cot \theta}{r} \tau_{\theta\phi} \tag{L}$$

$$[\nabla^2 \vec{v}]_r = \nabla^2 v_r - \frac{2 v_r}{r^2} - \frac{2}{r^2} \frac{\partial v_\theta}{\partial \theta} - \frac{2 v_\theta \cot \theta}{r^2} - \frac{2}{r^2 \sin \theta} \frac{\partial v_\phi}{\partial \phi} \tag{M}$$

$$[\nabla^2 \vec{v}]_\theta = \nabla^2 v_\theta + \frac{2}{r^2} \frac{\partial v_r}{\partial \theta} - \frac{r_\theta}{r^2 \sin^2 \theta} - \frac{2 \cos \theta}{r^2 \sin^2 \theta} \frac{\partial v_\phi}{\partial \phi} \tag{N}$$

$$[\nabla^2 \vec{v}]_\phi = \nabla^2 v_\phi - \frac{v_\phi}{r^2 \sin^2 \theta} + \frac{2}{r^2 \sin \theta} \frac{\partial v_r}{\partial \phi} + \frac{2 \cos \theta}{r^2 \sin^2 \theta} \frac{\partial v_\theta}{\partial \phi} \tag{O}$$

$$[\vec{v} \cdot \nabla \vec{v}]_r = v_r \frac{\partial v_r}{\partial r} + \frac{v_\theta}{r} \frac{\partial v_r}{\partial \theta} + \frac{v_\phi}{r \sin \theta} \frac{\partial v_r}{\partial \phi} - \frac{v_\theta^2 + v_\phi^2}{r} \tag{P}$$

$$[\vec{v} \cdot \nabla \vec{v}]_\theta = v_r \frac{\partial v_\theta}{\partial r} + \frac{v_\theta}{r} \frac{\partial v_\theta}{\partial \theta} + \frac{v_\phi}{r \sin \theta} \frac{\partial v_\theta}{\partial \phi} + \frac{v_r v_\theta}{r} - \frac{v_\phi^2 \cot \theta}{r} \tag{Q}$$

$$[\vec{v} \cdot \nabla \vec{v}]_\phi = v_r \frac{\partial v_\phi}{\partial r} + \frac{v_\theta}{r} \frac{\partial v_\phi}{\partial \theta} + \frac{v_\phi}{r \sin \theta} \frac{\partial v_\phi}{\partial \phi} + \frac{v_\phi v_r}{r} + \frac{v_\theta v_\phi \cot \theta}{r} \tag{R}$$

*Operations involving the tensor τ are given for symmetrical τ only.

THERMODYNAMIC PROPERTIES

Tables

Table E-1 Saturation state properties of steam and water

Temp. (°C)	Pressure bar	Specific volume (m³/kg)		Specific enthalpy (kJ/kg)		p_s (bar)
		Water	Steam	Water	Steam	
0·01	0·006 112	$1·000\ 2 \times 10^{-3}$	206·146	0·000 611	2501	0·00611
10	0·012 271	$1·000\ 4 \times 10^{-3}$	106·422	41·99	2519	0·01227
20	0·023 368	$1·001\ 8 \times 10^{-3}$	57·836	83·86	2538	0·02337
30	0·042 418	$1·004\ 4 \times 10^{-3}$	32·929	125·66	2556	0·04241
40	0·073 750	$1·007\ 9 \times 10^{-3}$	19·546	167·47	2574	0·07375
50	0·123 35	$1·012·1 \times 10^{-3}$	12·045	209·3	2592	0·12335
60	0·199 19	$1·017\ 1 \times 10^{-3}$	7·677 6	251·1	2609	0·19920
70	0·311 61	$1·022\ 8 \times 10^{-3}$	5·045 3	293·0	2626	0·31162
80	0·473 58	$1·029\ 0 \times 10^{-3}$	3·408 3	334·9	2643	0·47360
90	0·701 09	$1·035\ 9 \times 10^{-3}$	2·360 9	376·9	2660	0·70109
100	1·013 25 .	$1·043\ 5 \times 10^{-3}$	1·673 0	419·1	2676	1·01330
110	1·432 7	$1·051\ 5 \times 10^{-3}$	1·210 1	461·3	2691	1·4327
120	1·985 4	$1·060\ 3 \times 10^{-3}$	0·891 71	503·7	2706	1·9854
130	2·701 1	$1·069\ 7 \times 10^{-3}$	0·668 32	546·3	2720	2·7013
140	3·613 6	$1·079\ 8 \times 10^{-3}$	0·508 66	589·1	2734	3·6138
150	4·759 7	$1·090\ 6 \times 10^{-3}$	0·392 57	632·2	2747	4·7600
160	6·180 4	$1·102\ 1 \times 10^{-3}$	0·306 85	675·5	2758	6·1806
170	7·920 2	$1·114\ 4 \times 10^{-3}$	0·242 62	719·1	2769	7·9202
180	10·027	$1·127\ 5 \times 10^{-3}$	0·193 85	763·1	2778	10·027
190	12·553	$1·141\ 5 \times 10^{-3}$	0·156 35	807·5	2786	12·551
200	15·550	$1·156\ 5 \times 10^{-3}$	0·127 19	852·4	2793	15·549
210	19·080	$1·172\ 6 \times 10^{-3}$	0·104 265	897·7	2798	19·077
220	23·202	$1·190\ 0 \times 10^{-3}$	0·086 062	943·7	2802	23·198
230	27·979	$1·208\ 7 \times 10^{-3}$	0·071 472	990·3	2803	27·976
240	33·480	$1·229\ 1 \times 10^{-3}$	0·059 674	1037·6	2803	33·478
250	39·776	$1·251\ 2 \times 10^{-3}$	0·050 056	1085·8	2801	39·776
260	46·941	$1·275\ 5 \times 10^{-3}$	0·042 149	1135·0	2796	46·943
270	55·052	$1·302\ 3 \times 10^{-3}$	0·035 599	1185·2	2790	55·058
280	64·191	$1·332\ 1 \times 10^{-3}$	0·030 133	1236·8	2780	64·202
290	74·449	$1·365\ 5 \times 10^{-3}$	0·025 537	1290	2766	74·861
300	85·917	$1·403\ 6 \times 10^{-3}$	0·021 643	1345	2749	85·927
310	98·694	$1·447\ 5 \times 10^{-3}$	0·018 316	1402	2727	98·700
320	112·89	$1·499\ 2 \times 10^{-3}$	0·015 451	1462	2700	112·89
330	128·64	$1·562 \quad \times 10^{-3}$	0·012 967	1526	2666	128·63
340	146·08	$1·639 \quad \times 10^{-3}$	0·010 779	1596	2623	146·05
350	165·37	$1·741 \quad \times 10^{-3}$	0·008 805	1672	2565	165·35
360	186·74	$1·894 \quad \times 10^{-3}$	0·006 943	1762	2481	186·75
370	210·53	$2·22 \quad \times 10^{-3}$	0·004 93	1892	2331	210·54
374·15	221·2	$3·17 \quad \times 10^{-3}$	0·003 17	2095	2095	221·2

Source: From *U.K. Steam Tables in S.I. Units.* London: Edward Arnold, 1970.
1 bar = 10^5 N/m².

	Water						Steam				Temp.
c_{pf} (kJ/kg°K)	$\sigma \times 10^3$ (N/m)	$\mu f \times 10^6$ (N s/m²)	$\nu_f \times 10^6$ (m²/s)	k_f (W/m°K)	$(Pr)_f$	c_{pg} (kJ/kg°K)	$\mu_g \times 10^6$ (N s/m²)	$\nu_g \times 10^6$ (m²/s)	$k_g \times 10^3$ (W/m°K)	$(Pr)_g$	(°C)
4·218	75·60	1786	1·786	0·569	13·2	1·863	8·105	1672	17·6	0·858	0·01
4·194	74·24	1304	1·305	0·587	9·32	1·870	8·504	905	18·2	0·873	10
4·182	72·78	1002	1·004	0·603	6·95	1·880	8·903	515	18·8	0·888	20
4·179	71·23	798·3	0·802	0·618	5·40	1·890	9·305	306	19·5	0·901	30
4·179	69·61	653·9	0·659	0·631	4·33	1·900	9·701	190	20·2	0·912	40
4·181	67·93	547·8	0·554	0·643	3·56	1·912	10·10	121	20·9	0·924	50
4·185	66·19	467·3	0·473	0·653	2·99	1·924	10·50	80·6	21·6	0·934	60
4·191	64·40	404·8	0·414	0·662	2·56	1·946	10·89	54·9	22·4	0·946	70
4·198	62·57	355·4	0·366	0·670	2·23	1·970	11·29	38·5	23·2	0·959	80
4·207	60·69	315·6	0·327	0·676	1·96	1·999	11·67	27·6	24·0	0·973	90
4·218	58·78	283·1	0·295	0·681	1·75	2·034	12·06	20·2	24·9	0·987	100
4·230	56·83	254·8	0·268	0·684	1·58	2·076	12·45	15·1	25·8	1·00	110
4·244	54·85	231·0	0·245	0·687	1·43	2·125	12·83	11·4	26·7	1·02	120
4·262	52·83	210·9	0·226	0·688	1·31	2·180	13·20	8·82	27·8	1·03	130
4·282	50·79	194·1	0·210	0·688	1·21	2·245	13·57	6·90	28·9	1·05	140
4·306	48·70	179·8	0·196	0·687	1·13	2·320	13·94	5·47	30·0	1·08	150
4·334	46·59	167·7	0·185	0·684	1·06	2·406	14·30	4·39	31·3	1·10	160
4·366	44·44	157·4	0·175	0·681	1·01	2·504	14·66	3·55	32·6	1·13	170
4·403	42·26	148·5	0·167	0·677	0·967	2·615	15·02	2·91	34·1	1·15	180
4·446	40·05	140·7	0·161	0·671	0·932	2·741	15·37	2·40	35·7	1·18	190
4·494	37·81	133·9	0·155	0·664	0·906	2·883	15·72	2·00	37·4	1·21	200
4·550	35·53	127·9	0·150	0·657	0·886	3·043	16·07	1·68	39·4	1·24	210
4·613	33·23	122·4	0·146	0·648	0·871	3·223	16·42	1·41	41·5	1·28	220
4·685	30·90	117·5	0·142	0·639	0·861	3·426	16·78	1·20	43·9	1·31	230
4·769	28·56	112·9	0·139	0·628	0·850	3·656	17·14	1·02	46·5	1·35	240
4·866	26·19	108·7	0·136	0·616	0·859	3·918	17·51	0·876	49·5	1·39	250
4·985	23·82	104·8	0·134	0·603	0·866	4·221	17·90	0·755	52·8	1·43	260
5·134	21·44	101·1	0·132	0·589	0·882	4·575	18·31	0·652	56·6	1·48	270
5·307	19·07	97·5	0·130	0·574	0·902	4·996	18·74	0·565	60·9	1·54	280
5·520	16·71	94·1	0·128	0·558	0·932	5·509	19·21	0·491	66·0	1·61	290
5·794	14·39	90·7	0·127	0·541	0·970	6·148	19·73	0·427	71·9	1·69	300
6·143	12·11	87·2	0·126	0·523	1·024	6·968	20·30	0·372	79·1	1·79	310
6·604	9·89	83·5	0·125	0·503	1·11	8·060	20·95	0·324	87·8	1·92	320
7·241	7·75	79·5	0·124	0·482	1·20	9·580	21·70	0·281	99·0	2·10	330
8·225	5·71	75·4	0·123	0·460	1·35	11·87	22·70	0·245	114	2·36	340
10·07	3·79	69·4	0·121	0·434	1·61	15·8	24·15	0·213	134	2·84	350
15·0	2·03	62·1	0·118	0·397	2·34	27·0	26·45	0·184	162	4·40	360
55	0·47	51·8	0·116	0·340	8·37	107	30·6	0·150	199	16·4	370
∞	0	41·4	0·131	0·240		∞	41·4	0·131	240		374·15

Table E-2 Thermodynamic properties of dry saturated steam pressure

Abs. press. (psia)	Temp. (°F)	Specific volume		Enthalpy			Entropy		
		Sat. liquid (ft³/lbm)	Sat. vapor (ft³/lbm)	Sat. liquid (BTU/lbm)	Evap. (BTU/lbm)	Sat. vapor (BTU/lbm)	Sat. liquid (BTU/lbm °R)	Evap. (BTU/lbm °R)	Sat. vapor (BTU/lbm °R)
1.0	101.74	0.01614	333.6	69.70	1036.3	1106.0	0.1326	1.8456	1.9782
2.0	126.08	0.01623	173.73	93.99	1022.2	1116.2	0.1749	1.7451	1.9200
3.0	141.48	0.01630	118.71	109.37	1013.2	1122.6	0.2008	1.6855	1.8863
4.0	152.97	0.01636	90.63	120.86	1006.4	1127.3	0.2198	1.6427	1.8625
5.0	162.24	0.01640	73.52	130.13	1001.0	1131.1	0.2347	1.6094	1.8441
6.0	170.06	0.01645	61.98	137.96	996.2	1134.2	0.2472	1.5820	1.8292
7.0	176.85	0.01649	53.64	144.76	992.1	1136.9	0.2581	1.5586	1.8167
8.0	182.86	0.01653	47.34	150.79	988.5	1139.3	0.2674	1.5383	1.8057
9.0	188.28	0.01656	42.40	156.22	985.2	1141.4	0.2759	1.5203	1.7962
10	193.21	0.01659	38.42	161.17	982.1	1143.3	0.2835	1.5041	1.7876
14.696	212.00	0.01672	26.80	180.07	970.3	1150.4	0.3120	1.4446	1.7566
15	213.03	0.01672	26.29	181.11	969.7	1150.8	0.3135	1.4415	1.7549
20	227.96	0.01683	20.089	196.16	960.1	1156.3	0.3356	1.3962	1.7319
25	240.07	0.01692	16.303	208.42	952.1	1160.6	0.3533	1.3606	1.7139
30	250.33	0.01701	13.746	218.82	945.3	1164.1	0.3680	1.3313	1.6993
35	259.28	0.01708	11.898	227.91	939.2	1167.1	0.3807	1.3063	1.6870
40	267.25	0.01715	10.498	236.03	933.7	1169.7	0.3919	1.2844	1.6763
45	274.44	0.01721	9.401	243.36	928.6	1172.0	0.4019	1.2650	1.6669
50	281.01	0.01727	8.515	250.09	924.0	1174.1	0.4110	1.2474	1.6585
55	287.07	0.01732	7.787	256.30	919.6	1175.9	0.4193	1.2316	1.6509
60	292.71	0.01738	7.175	262.09	915.5	1177.6	0.4270	1.2168	1.6438
65	297.97	0.01743	6.655	267.50	911.6	1179.1	0.4342	1.2032	1.6374
70	302.92	0.01748	6.206	272.61	907.9	1180.6	0.4409	1.1906	1.6315
75	307.60	0.01753	5.816	277.43	904.5	1181.9	0.4472	1.1787	1.6259
80	312.03	0.01757	5.472	282.02	901.1	1183.1	0.4531	1.1676	1.6207
85	316.25	0.01761	5.168	286.39	897.8	1184.2	0.4587	1.1571	1.6158
90	320.27	0.01766	4.896	290.56	894.7	1185.3	0.4641	1.1471	1.6112
95	324.12	0.01770	4.652	294.56	891.7	1186.2	0.4692	1.1376	1.6068
100	327.81	0.01774	4.432	298.40	888.8	1187.2	0.4740	1.1286	1.6026
110	334.77	0.01782	4.049	305.66	883.2	1188.9	0.4832	1.1117	1.5948

P (psia)	T	v_f	v_g	h_f	h_{fg}	h_g	s_f	s_{fg}	s_g
120	341.25	0.01789	3.728	312.44	877.9	1190.4	0.4916	1.0962	1.5878
130	347.32	0.01796	3.455	318.81	872.9	1191.7	0.4995	1.0817	1.5812
140	353.02	0.01802	3.220	324.82	868.2	1193.0	0.5069	1.0682	1.5751
150	358.42	0.01809	3.015	330.51	863.6	1194.1	0.5138	1.0556	1.5694
160	363.53	0.01815	2.834	335.93	859.2	1195.1	0.5204	1.0436	1.5640
170	368.41	0.01822	2.675	341.09	854.9	1196.0	0.5266	1.0324	1.5590
180	373.06	0.01827	2.532	346.03	850.8	1196.9	0.5325	1.0217	1.5542
190	377.51	0.01833	2.404	350.79	846.8	1197.6	0.5381	1.0116	1.5497
200	381.79	0.01839	2.288	355.36	843.0	1198.4	0.5435	1.0018	1.5453
250	400.95	0.01865	1.8438	376.00	825.1	1201.1	0.5675	0.9588	1.5263
300	417.33	0.01890	1.5433	393.84	809.0	1202.8	0.5879	0.9225	1.5104
350	431.72	0.01913	1.3260	409.69	794.2	1203.9	0.6056	0.8910	1.4966
400	444.59	0.0193	1.1613	424.0	780.5	1204.5	0.6214	0.8630	1.4844
450	456.28	0.0195	1.0320	437.2	767.4	1204.6	0.6356	0.8378	1.4734
500	467.01	0.0197	0.9278	449.4	755.0	1204.4	0.6487	0.8147	1.4634
550	476.94	0.0199	0.8424	460.8	743.1	1203.9	0.6608	0.7934	1.4542
600	486.21	0.0201	0.7698	471.6	731.6	1203.2	0.6720	0.7734	1.4454
650	494.90	0.0203	0.7083	481.8	720.5	1202.3	0.6826	0.7548	1.4374
700	503.10	0.0205	0.6554	491.5	709.7	1201.2	0.6925	0.7371	1.4296
750	510.86	0.0207	0.6092	500.8	699.2	1200.0	0.7019	0.7204	1.4223
800	518.23	0.0209	0.5687	509.7	688.9	1198.6	0.7108	0.7045	1.4153
850	525.26	0.0210	0.5327	518.3	678.8	1197.1	0.7194	0.6891	1.4085
900	531.98	0.0212	0.5006	526.6	668.8	1195.4	0.7275	0.6744	1.4020
950	538.43	0.0214	0.4717	534.6	659.1	1193.7	0.7355	0.6602	1.3957
1000	544.61	0.0216	0.4456	542.4	649.4	1191.8	0.7430	0.6467	1.3897
1100	556.31	0.0220	0.4001	557.4	630.4	1187.7	0.7575	0.6205	1.3780
1200	567.22	0.0223	0.3619	571.7	611.7	1183.4	0.7711	0.5956	1.3667
1300	577.46	0.0227	0.3293	585.4	593.2	1178.6	0.7840	0.5719	1.3559
1400	587.10	0.0231	0.3012	598.7	574.7	1173.4	0.7963	0.5491	1.3454
1500	596.23	0.0235	0.2765	611.6	556.3	1167.9	0.8082	0.5269	1.3351
2000	635.82	0.0257	0.1878	671.7	463.4	1135.1	0.8619	0.4230	1.2849
2500	668.13	0.0287	0.1307	730.6	360.5	1091.1	0.9126	0.3197	1.2322
3000	695.36	0.0346	0.0858	802.5	217.8	1020.3	0.9731	0.1885	1.1615
3206.2	705.40	0.0503	0.0503	902.7	0	902.7	1.0580	0	1.0580

Source: From M. El-Wakil, 1971. Abridged from Keenan, J. H., and Keyes, F. G. *Thermodynamic Properties of Steam*, New York: Wiley, 1937.

Table E-3 Thermodynamic properties of dry saturated steam temperature

Temp. (°F)	Abs. pressure (psia)	Specific volume		Enthalpy			Entropy		
		Sat. liquid (ft³/lbm)	Sat. vapor (ft³/lbm)	Sat. liquid (BTU/lbm)	Evap. (BTU/lbm)	Sat. vapor (BTU/lbm)	Sat. liquid (BTU/lbm °R)	Evap. (BTU/lbm °R)	Sat. vapor (BTU/lbm °R)
32	0.08854	0.01602	3306	0.00	1075.8	1075.8	0.0000	2.1877	2.1877
35	0.09995	0.01602	2947	3.02	1074.1	1077.1	0.0061	2.1709	2.1770
40	0.12170	0.01602	2444	8.05	1071.3	1079.3	0.0162	2.1435	2.1597
45	0.14752	0.01602	2036.4	13.06	1068.4	1081.5	0.0262	2.1167	2.1429
50	0.17811	0.01603	1703.2	18.07	1065.6	1083.7	0.0361	2.0903	2.1264
60	0.2563	0.01604	1206.7	28.06	1059.9	1088.0	0.0555	2.0393	2.0948
70	0.3631	0.01606	867.9	38.04	1054.3	1092.3	0.0745	1.9902	2.0647
80	0.5069	0.01608	633.1	48.02	1048.6	1096.6	0.0932	1.9428	2.0360
90	0.6982	0.01610	468.0	57.99	1042.9	1100.9	0.1115	1.8972	2.0087
100	0.9492	0.01613	350.4	67.97	1037.2	1105.2	0.1295	1.8531	1.9826
110	1.2748	0.01617	265.4	77.94	1031.6	1109.5	0.1417	1.8106	1.9577
120	1.6924	0.01620	203.27	87.92	1025.8	1113.7	0.1645	1.7694	1.9339
130	2.2225	0.01625	157.34	97.90	1020.0	1117.9	0.1816	1.7296	1.9112
140	2.8886	0.01629	123.01	107.89	1014.1	1122.0	0.1984	1.6910	1.8894
150	3.718	0.01634	97.07	117.89	1008.2	1126.1	0.2149	1.6537	1.8685
160	4.741	0.01639	77.29	127.89	1002.3	1130.2	0.2311	1.6174	1.8485
170	5.992	0.01645	62.06	137.90	996.3	1134.2	0.2472	1.5822	1.8293
180	7.510	0.01651	50.23	147.92	990.2	1138.1	0.2630	1.5480	1.8109
190	9.339	0.01657	40.96	157.95	984.1	1142.0	0.2785	1.5147	1.7932
200	11.526	0.01663	33.64	167.99	977.9	1145.9	0.2938	1.4824	1.7762
210	14.123	0.01670	27.82	178.05	971.6	1149.7	0.3090	1.4508	1.7598
212	14.696	0.01672	26.80	180.07	970.3	1150.4	0.3120	1.4446	1.7566
220	17.186	0.01677	23.15	188.13	965.2	1153.4	0.3239	1.4201	1.7440
230	20.780	0.01684	19.382	198.23	958.8	1157.0	0.3387	1.3901	1.7288
240	24.969	0.01692	16.323	208.34	952.2	1160.5	0.3531	1.3609	1.7140
250	29.825	0.01700	13.821	216.48	945.5	1164.0	0.3675	1.3323	1.6998
260	35.429	0.01709	11.763	228.64	938.7	1167.3	0.3817	1.3043	1.6860
270	41.858	0.01717	10.061	238.84	931.8	1170.6	0.3958	1.2769	1.6727
280	49.203	0.01726	8.645	249.06	924.7	1173.8	0.4096	1.2501	1.6597
290	57.556	0.01735	7.461	259.31	917.5	1176.8	0.4234	1.2238	1.6472

300	67.013	0.01745	6.466	269.59	910.1	1179.7	0.4369	1.1980	1.6350
310	77.68	0.01755	5.626	279.92	902.6	1182.5	0.4504	1.1727	1.6231
320	89.66	0.01765	4.914	290.28	894.9	1185.2	0.4637	1.1478	1.6115
330	103.06	0.01776	4.307	300.68	887.0	1187.7	0.4769	1.1233	1.6002
340	118.01	0.01787	3.788	311.13	879.0	1190.1	0.4900	1.0992	1.5891
350	134.63	0.01799	3.342	321.63	870.7	1192.3	0.5029	1.0754	1.5783
360	153.04	0.01811	2.957	332.18	852.2	1194.4	0.5158	1.0519	1.5677
370	173.37	0.01823	2.625	342.79	853.5	1196.3	0.5286	1.0287	1.5573
380	195.77	0.01836	2.335	353.45	844.6	1198.1	0.5413	1.0059	1.5471
390	220.37	0.01850	2.0836	364.17	835.4	1199.6	0.5539	0.9832	1.5371
400	247.31	0.01864	1.8633	374.97	826.0	1201.0	0.5664	0.9608	1.5272
410	276.75	0.01878	1.6700	385.83	816.3	1202.1	0.5788	0.9386	1.5174
420	308.83	0.01894	1.5000	396.77	806.3	1203.1	0.5912	0.9166	1.5078
430	343.72	0.01910	1.3499	407.79	796.0	1203.8	0.6035	0.8947	1.4982
440	381.59	0.01926	1.2171	418.90	785.4	1204.3	0.6158	0.8730	1.4887
450	422.6	0.0194	1.0993	430.1	774.5	1204.6	0.6280	0.8513	1.4793
460	466.9	0.0196	0.9944	441.4	763.2	1204.6	0.6402	0.8298	1.4700
470	514.7	0.0198	0.9009	452.8	751.5	1204.3	0.6523	0.8083	1.4606
480	566.1	0.0200	0.8172	464.4	739.4	1203.7	0.6645	0.7868	1.4513
490	621.4	0.0202	0.7423	476.0	726.8	1202.8	0.6766	0.7653	1.4419
500	680.8	0.0204	0.6749	487.8	713.9	1201.7	0.6887	0.7438	1.4325
520	812.4	0.0209	0.5594	511.9	686.4	1198.2	0.7130	0.7006	1.4136
540	962.5	0.0215	0.4649	536.6	656.6	1193.2	0.7374	0.6568	1.3942
560	1133.1	0.0221	0.3868	562.2	624.2	1186.4	0.7621	0.6121	1.3742
580	1325.8	0.0228	0.3217	588.9	588.4	1177.3	0.7872	0.5659	1.3532
600	1542.9	0.0236	0.2668	610.0	548.5	1165.5	0.8131	0.5176	1.3307
620	1786.6	0.0247	0.2201	646.7	503.6	1150.3	0.8398	0.4664	1.3062
640	2059.7	0.0260	0.1798	678.6	452.0	1130.5	0.8679	0.4110	1.2789
660	2365.4	0.0278	0.1442	714.2	390.2	1104.4	0.8987	0.3485	1.2472
680	2708.1	0.0305	0.1115	757.3	309.9	1067.2	0.9351	0.2719	1.2071
700	3093.7	0.0369	0.0761	823.3	172.1	995.4	0.9905	0.1484	1.1389
705.4	3206.2	0.0503	0.0503	902.7	0	902.7	1.0580	0	1.0580

Source: From El-Wakil, 1971; abridged from Keenan, J. H., and Keyes, F. G. *Thermodynamic Properties of Steam*. New York: Wiley, 1937.

Table E-4 Thermodynamic properties of sodium

Temperature (°R) (Sat. press., psia)		Sat. liquid	Sat. vapor	Temperature of superheated vapor (°R) 800	900	1000	1100	1200	1400	1600	1800	2000	2200	2400	2600	2700
700 (8.7472 × 10⁻⁹)	v	1.7232×10^{-2}	$>10^{10}$													
	h	219.7	2180.5	2203.1	2224.7	2246.4	2268.0	2289.5	2332.7	2375.9	2419.1	2462.3	2505.4	2548.6	2591.8	2613.4
	s	0.6854	3.4866	3.5169	3.5424	3.5652	3.5857	3.6043	3.6381	3.6665	3.6924	3.7148	3.7354	3.7545	3.7713	3.7796
800 (5.0100 × 10⁻⁷)	v	1.7548×10^{-2}	7.4375×10^{8}		8.3835×10^{8}	9.3168×10^{8}	1.0249×10^{9}	1.1180×10^{9}	1.3044×10^{9}	1.4907×10^{9}	1.6771×10^{9}	1.8634×10^{9}	2.0498×10^{9}	2.2361×10^{9}	2.4224×10^{9}	2.5156×10^{9}
	h	252.3	2200.1		2224.4	2246.3	2267.9	2289.7	2332.7	2375.9	2419.1	2462.3	2505.4	2548.6	2591.8	2613.4
	s	0.7290	3.1637		3.1925	3.2155	3.2360	3.2546	3.2884	3.3169	3.3428	3.3652	3.3858	3.4048	3.5217	3.4299
900 (1.1480 × 10⁻⁵)	v	1.7864×10^{-2}	3.6411×10^{7}			4.0267×10^{7}	4.4718×10^{7}	4.8789×10^{7}	5.6924×10^{7}	6.5056×10^{7}	7.3188×10^{7}	8.1320×10^{7}	8.9452×10^{7}	9.7584×10^{7}	1.0572×10^{8}	1.0978×10^{8}
	h	284.3	2217.6			2245.2	2267.7	2289.4	2332.7	2375.9	2419.1	2462.3	2505.4	2548.6	2591.8	2613.4
	s	0.7667	2.9148			2.9440	2.9653	2.9841	3.0179	3.0464	3.0723	3.0947	3.1153	3.1343	3.1511	3.1594
1000 (1.3909 × 10⁻⁴)	v	1.8180×10^{-2}	3.323×10^{6}				3.6834×10^{6}	4.0254×10^{6}	4.6978×10^{6}	5.3693×10^{6}	6.0406×10^{6}	6.7118×10^{6}	7.383×10^{6}	8.0541×10^{6}	8.7253×10^{6}	9.0609×10^{6}
	h	325.9	2232.7				2264.8	2288.6	2332.6	2375.9	2419.1	2462.3	2505.4	2548.6	2591.8	2613.4
	s	0.7999	2.7168				2.7474	2.7680	2.8024	2.8309	2.8568	2.8792	2.8998	2.9188	2.9357	2.9439
1100 (1.0616 × 10⁻³)	v	1.8496×10^{-2}	4.7592×10^{5}					5.2512×10^{5}	6.1515×10^{5}	7.0339×10^{5}	7.9139×10^{5}	8.7935×10^{5}	9.6729×10^{5}	1.0552×10^{6}	1.1432×10^{6}	1.1871×10^{6}
	h	347.0	2245.1					2282.7	2331.7	2375.7	2419.0	2462.3	2505.4	2548.6	2591.8	2613.4
	s	0.8296	2.5551					2.5878	2.6263	2.6552	2.6812	2.7036	2.7243	2.7433	2.7601	2.7684
1200 (5.7398 × 10⁻³)	v	1.8812×10^{-2}	9.5235×10^{4}						1.1345×10^{5}	1.3001×10^{5}	1.4635×10^{5}	1.6263×10^{5}	1.7891×10^{5}	1.9517×10^{5}	2.1144×10^{5}	2.1957×10^{5}
	h	377.7	2254.9						2327.6	2374.7	2418.7	2462.2	2505.4	2548.6	2591.8	2613.4
	s	0.8563	2.4207						2.4778	2.5089	2.5353	2.5578	2.5785	2.5975	2.6143	2.6226
1300 (2.3916 × 10⁻²)	v	1.9128×10^{-2}	2.4520×10^{4}						2.6936×10^{4}	3.1124×10^{4}	3.5095×10^{4}	3.9019×10^{4}	4.2931×10^{4}	4.6838×10^{4}	5.0743×10^{4}	5.2695×10^{4}
	h	408.2	2262.8						2312.2	2371.1	2417.6	2461.7	2505.1	2548.5	2591.8	2613.4
	s	0.8807	2.3073						2.3445	2.3836	2.4115	2.4343	2.4551	2.4742	2.4911	2.4993
1400 (8.1347 × 10⁻²)	v	1.9444×10^{-2}	7.6798×10^{3}							9.0793×10^{3}	1.0292×10^{4}	1.1460×10^{4}	1.2616×10^{4}	1.3767×10^{4}	1.4916×10^{4}	1.5491×10^{4}
	h	438.4	2269.3							2359.9	2414.0	2460.3	2504.5	2548.1	2591.6	2613.2
	s	0.9031	2.2109							2.2715	2.3040	2.3280	2.3491	2.3683	2.3853	2.3935

T										
1500 (2.3351 × 10⁻¹)	v	1.9760×10^{-2}	2.8334×10^{3}	3.1025×10^{3}	3.5625×10^{3}	3.9820×10^{3}	4.3896×10^{3}	4.7929×10^{3}	5.1944×10^{3}	5.3948×10^{3}
	h	468.5	2274.9	2332.6	2404.7	2456.5	2502.7	2547.2	2591.0	2612.8
	s	0.9239	2.1282	2.1651	2.2083	2.2352	2.2573	2.2769	2.2940	2.3023
1600 (5.8425 × 10⁻¹)	v	2.0076×10^{-2}	1.1935×10^{3}		1.4040×10^{3}	1.5823×10^{3}	1.7496×10^{3}	1.9128×10^{3}	2.0743×10^{3}	2.1548×10^{3}
	h	498.5	2280.0		2384.7	2448.0	2498.7	2545.1	2589.7	2611.8
	s	0.9433	2.0567		2.1192	2.1523	2.1765	2.1969	2.2144	2.2228
1700 (1.3170)	v	2.0392×10^{-2}	5.5585×10^{2}		6.0659×10^{2}	6.9378×10^{2}	7.7180×10^{2}	8.4601×10^{2}	9.1858×10^{2}	9.5458×10^{2}
	h	528.5	2285.3		2347.7	2431.3	2490.5	2540.7	2587.1	2609.8
	s	0.9615	1.9948		2.0309	2.0747	2.1031	2.1252	2.1433	2.1519
1800 (2.7164)	v	2.0708×10^{-2}	2.8200×10^{2}			3.2952×10^{2}	3.7033×10^{2}	4.0785×10^{2}	4.4385×10^{2}	4.6158×10^{2}
	h	558.6	2291.1			2402.3	2475.6	2532.5	2582.2	2605.9
	s	0.9786	2.9411			1.9996	2.0347	2.0597	2.0792	2.0882
1900 (5.1529)	v	2.1024×10^{-2}	1.5512×10^{2}			1.6838×10^{2}	1.9197×10^{2}	2.1298×10^{2}	2.3265×10^{2}	2.4224×10^{2}
	h	588.8	2297.2			2359.5	2451.8	2518.8	2573.9	2599.3
	s	0.9949	1.8941			1.9259	1.9701	1.9996	2.0212	2.0209
2000 (9.1533)	v	2.1340×10^{-2}	90.914				1.0543×10^{2}	1.1816×10^{2}	1.2980×10^{2}	1.3539×10^{2}
	h	619.1	2304.1				2417.4	2498.0	2560.9	2588.9
	s	1.0105	1.8530				1.9072	1.9426	1.9673	2.9780
2100 (15.392)	v	2.1656×10^{-2}	56.185				60.665	68.825	76.167	79.656
	h	649.7	2312.1				2372.9	2469.0	2451.9	2573.5
	s	1.0255	1.8171				1.8455	1.8876	1.9164	1.9284
2200 (24.692)	v	2.1972×10^{-2}	36.338					41.754	46.622	48.920
	h	680.7	2321.0					2431.8	2516.2	2552.3
	s	1.0399	1.7855					1.8340	1.8674	1.8811
2300 (38.013)	v	2.2288×10^{-2}	24.454					26.244	29.585	31.163
	h	712.0	2330.7					2388.2	2484.0	2525.0
	s	1.0538	1.7576					1.7820	1.8201	1.8356

Table E-4 Thermodynamic properties of sodium (*continued*)

Temperature (°R) (Sat. press., psia)	Sat. liquid	Sat. vapor	Temperature of superheated vapor (°R)												
			800	900	1000	1100	1200	1400	1600	1800	2000	2200	2400	2600	2700
2400 (56.212)															
v	2.2604×10^{-2}	17.109												19.460	20.580
h	743.8	2341.2												2446.8	2492.6
s	1.0673	1.7329												1.7748	1.7922
2500 (80.236)															
v	2.2920×10^{-2}	12.388												13.219	14.032
h	776.2	2352.6												2406.5	2456.5
s	1.0805	1.7111												1.7321	1.7501
2600 (1.1116×10^2)															
v	2.3236×10^{-2}	9.2328													9.8326
h	809.1	2365.1													2418.1
s	1.0934	1.6919													1.7120
2700 (1.052×10^2)															
v	2.3552×10^{-2}	7.0380													
h	842.7	2378.8													
s	1.1061	1.6751													

Units: v in ft³/lbm, h in BTU/lbm, and s in BTU/lbm °R.

Source: From Meisl, C. J., and A. Shapiro. *Thermodynamic Properties of Alkali Metal Vapors and Mercury* (2nd ed.), General Electric Flight Propulsion Laboratory Report R60FPD358-A, 1960, wherein reprinted from M. El-Wakil: *PowerPlant Technology.* New York: McGraw-Hill, 1984.

Table E-5 Thermodynamic properties of helium

Pressure (psia)	Temperature (°F)					
	100	200	300	400	500	600
14.696						
v	102.23	120.487	138.743	157.00	175.258	193.515
ρ	0.0097820	0.0082997	0.0072076	0.0063694	0.0057059	0.0051676
h	707.73	827.56	952.38	1077.20	1202.02	1326.83
s	6.8421	7.0472	7.2233	7.3776	7.5149	7.6386
50						
v	30.085	35.451	40.817	46.183	51.549	56.915
ρ	0.033239	0.028208	0.024500	0.021653	0.019399	0.017570
h	703.08	827.90	952.72	1077.54	1202.36	1327.18
s	6.2342	6.4393	6.6153	6.7697	6.9070	7.0307
150						
v	10.063	11.8522	13.6407	15.4293	17.2183	19.008
ρ	0.099372	0.084372	0.073310	0.064812	0.058078	0.052610
h	704.08	828.91	953.73	1078.55	1203.37	1328.19
s	5.6886	5.8937	6.0698	6.2241	6.3614	6.4852
400						
v	3.8062	4.4775	5.1487	5.8197	6.4905	7.1616
ρ	0.26273	0.22334	0.194225	0.171831	0.154072	0.139633
h	706.58	831.42	956.24	1081.06	1205.88	1330.70
s	5.2013	5.4065	5.5827	5.7371	5.8744	5.9981
600						
v	2.5546	3.0023	3.44995	3.8973	4.3449	4.7923
ρ	0.39146	0.33308	0.28986	0.25658	0.23016	0.20867
h	708.49	833.33	958.15	1082.97	1207.79	1332.61
s	4.9998	5.2050	5.3813	5.5357	5.6730	5.7968
900						
v	1.7200	2.0187	2.3173	2.6157	2.91399	3.2124
ρ	0.58139	0.49537	0.43154	0.38230	0.34317	0.31129
h	710.29	835.38	960.40	1085.42	1210.42	1335.36
s	4.7981	5.0035	5.1797	5.3342	5.4715	5.5953
1500						
v	1.05192	1.2314	1.4108	1.58994	1.7690	1.9483
ρ	0.95064	0.81207	0.70880	0.62897	0.56528	0.51328
h	715.54	840.77	965.88	1090.92	1215.93	1340.97
s	4.5437	4.7475	4.9257	5.0801	5.2176	5.3414
2500						
v	0.65044	0.75847	0.86635	0.97410	1.08176	1.18947
ρ	1.53741	1.31845	1.15427	1.02659	0.92442	0.84071
h	724.37	849.73	974.95	1100.10	1225.22	1350.29
s	4.2887	4.4928	4.6712	4.8258	4.9634	5.0873
4000						
v	0.42377	0.49161	0.55932	0.62694	0.69444	0.76191
ρ	2.3598	2.0341	1.78789	1.59503	1.44000	1.31248
h	736.48	862.24	987.70	1113.12	1238.46	1363.73
s	4.0531	4.2576	4.4363	4.5912	4.7287	4.8530

Units: v in ft³/lbm, h in BTU/lbm, and s in BTU/lbm °R

From Fabric Filter Systems Study. In: *Handbook of Fabric Filter Technology* (Vol. 1).PB200-648, APTD-0690, National Technical Information Service, December 1970; wherein reprinted from El-Wakil, M. *PowerPlant Technology*. New York: McGraw-Hill, 1984.

Table E-6 Thermodynamic properties of CO_2

p		-75°F	-50°F	0°F	50°F	100°F	150°F	200°F	300°F	400°F	600°F	800°F	1000°F	1200°F	1400°F	1600°F	1800°F
1.00	v	93.90	100.0	112.2	124.4	136.6	148.8	161.0	185.4	209.7	258.5	307.2	356.0	404.8	453.6	502.3	551.0
	h	283.2	288.0	297.8	307.7	318.0	328.4	339.1	361.4	384.7	434.4	487.1	542.4	599.6	658.6	718.8	780.0
	s	1.4772	1.4892	1.5112	1.5316	1.5506	1.5684	1.5852	1.6165	1.6451	1.6969	1.7423	1.7829	1.8197	1.8533	1.8838	1.9123
10.0	v	9.280	9.902	11.15	12.38	13.61	14.84	16.06	18.51	20.96	25.85	30.73	35.61	40.49	45.36	50.24	55.11
	h	282.6	287.5	297.3	307.3	317.7	328.2	339.0	361.3	384.6	434.4	487.1	542.4	599.6	658.6	718.8	780.0
	s	1.3733	1.3853	1.4073	1.4277	1.4467	1.4645	1.4813	1.5126	1.5412	1.5930	1.6384	1.6790	1.7158	1.7494	1.7799	1.8084
20.0	v	4.586	4.904	5.542	6.119	6.778	7.407	8.016	9.247	10.47	12.92	15.36	17.80	20.24	22.68	25.11	27.55
	h	281.9	287.0	296.8	306.8	317.3	327.9	338.8	361.1	384.5	434.3	487.1	542.4	599.6	658.6	718.8	780.0
	s	1.3417	1.3538	1.3759	1.3964	1.4154	1.4332	1.4500	1.4813	1.5099	1.5617	1.6071	1.6477	1.6845	1.7181	1.7486	1.7771
40.0	v	2.239	2.404	2.738	3.053	3.363	3.688	3.993	4.615	5.230	6.458	7.688	8.901	10.12	11.37	12.56	13.78
	h	280.6	285.9	295.8	305.9	316.5	327.4	338.4	360.9	384.3	434.2	487.0	542.4	599.6	658.6	718.8	780.0
	s	1.3088	1.3211	1.3435	1.3642	1.3834	1.4014	1.4184	1.449	1.4787	1.5305	1.5759	1.6165	1.6533	1.6869	1.7174	1.7459
80.0	v		1.154	1.335	1.498	1.657	1.828	1.982	2.298	2.608	3.226	3.839	4.448	5.060	5.670	6.281	6.887
	h		283.8	293.8	304.1	315.1	326.4	337.7	360.2	383.9	434.0	486.9	542.3	599.5	658.6	718.8	780.0
	s		1.2778	1.3044	1.3284	1.3490	1.3679	1.3855	1.4177	1.4468	1.4991	1.5446	1.5852	1.6220	1.6556	1.6861	1.7146
120	v			0.8665	0.9799	1.088	1.208	1.311	1.525	1.734	2.148	2.559	2.966	3.373	3.781	4.188	4.592
	h			291.7	302.2	313.6	325.4	337.0	359.7	383.5	433.8	486.8	542.3	599.5	658.6	718.8	780.0
	s			1.2833	1.3086	1.3297	1.3488	1.3666	1.3993	1.4285	1.4808	1.5263	1.569	1.6037	1.6373	1.6678	1.6963

		A	B	C	D	E	F	G	H	I	J	K	L	M	N
160	v	0.6305	0.7207	0.8033	0.8986	0.9760	1.139	1.297	1.610	1.918	2.224	2.530	2.836	3.141	3.445
	h	289.7	300.4	312.1	324.4	336.3	359.1	383.1	433.6	486.6	542.2	599.5	658.6	718.8	780.0
	s	1.2666	1.2928	1.3154	1.3350	1.3529	1.3857	1.4151	1.4675	1.5133	1.5539	1.5907	1.6243	1.6548	1.6833
200	v	0.4891	0.5652	0.6376	0.7125	0.7748	0.9075	1.035	1.287	1.534	1.779	2.024	2.269	2.513	2.757
	h	287.7	298.6	310.6	323.4	335.6	358.5	382.7	433.4	486.5	542.2	599.5	658.5	718.8	780.0
	s	1.2519	1.2805	1.3038	1.3239	1.3421	1.3753	1.4049	1.4574	1.5033	1.5439	1.5807	1.6143	1.6448	1.6733
240	v	0.3948	0.4614	0.5237	0.5886	0.6407	0.7532	0.8604	1.071	1.273	1.482	1.687	1.891	2.095	2.297
	h	285.6	296.7	309.1	322.4	334.9	358.0	382.3	433.1	486.4	542.1	599.5	658.5	718.8	780.0
	s	1.2395	1.2694	1.2940	1.3145	1.3330	1.3671	1.3963	1.4490	1.4948	1.5356	1.5724	1.6060	1.6365	1.6650
300	v		0.3563	0.4100	0.4636	0.5065	0.5985	0.6868	0.8556	1.021	1.186	1.349	1.513	1.676	1.838
	h		294.0	306.9	320.9	333.9	357.1	381.6	432.8	486.2	542.0	599.4	658.5	718.7	780.0
	s		1.2562	1.2813	1.3029	1.3219	1.3560	1.3862	1.4389	1.4848	1.5256	1.5624	1.5960	1.6265	1.6550
360	v		0.2858	0.3341	0.3780	0.4171	0.4958	0.5693	0.7212	0.8502	0.9874	1.125	1.261	1.397	1.533
	h		291.2	304.6	319.4	332.8	356.3	381.0	432.5	486.0	541.9	599.4	658.5	718.7	779.9
	s		1.2436	1.2699	1.2925	1.3124	1.3475	1.3779	1.4307	1.4766	1.5174	1.5542	1.5878	1.6183	1.6468
440	v		0.2216	0.2652	0.3040	0.3358	0.4022	0.4633	0.5817	0.6950	0.8079	0.9201	1.032	1.142	1.255
	h		287.6	301.6	317.4	331.4	355.1	380.2	432.1	485.8	541.7	599.3	658.4	718.6	779.9
	s		1.2282	1.2559	1.2797	1.3006	1.3370	1.3681	1.4215	1.4675	1.5083	1.5451	1.5787	1.6092	1.6377
520	v		0.1772	0.2174	0.2513	0.2795	0.3374	0.3901	0.4912	0.5881	0.6832	0.7785	0.8733	0.9672	1.062
	h		283.9	298.7	315.4	330.0	354.0	379.9	431.7	485.5	541.5	599.2	658.3	718.6	779.9
	s		1.2148	1.2438	1.2687	1.2905	1.3281	1.3599	1.4138	1.4599	1.5007	1.5375	1.5711	1.6010	1.6301

Table E-6 Thermodynamic properties of CO_2 (continued)

p		−75°F	−50°F	0°F	50°F	100°F	150°F	200°F	300°F	400°F	600°F	800°F	1000°F	1200°F	1400°F	1600°F	1800°F
600	v				0.1452	0.1823	0.2123	0.2383	0.2898	0.3363	0.4250	0.5093	0.5921	0.6747	0.7571	0.8385	0.9202
	h				280.3	295.7	313.4	328.6	352.8	378.6	431.1	485.3	541.4	599.0	658.2	718.6	779.8
	s				1.2020	1.2323	1.2583	1.2809	1.3198	1.3525	1.4071	1.4534	1.4942	1.5310	1.5646	1.5951	1.6236
800	v					0.1196	0.1483	0.1712	0.2126	0.2489	0.3173	0.3812	0.4436	0.5060	0.5680	0.6292	0.6906
	h					288.2	308.4	325.1	350.0	376.5	430.1	484.7	541.0	598.8	658.0	718.4	779.7
	s					1.2111	1.2391	1.2631	1.3041	1.3380	1.3935	1.4404	1.4812	1.5180	1.5516	1.5821	1.6106
1000	v						0.1101	0.1310	0.1663	0.1966	0.2526	0.3048	0.3547	0.4049	0.4545	0.5037	0.5526
	h						303.4	321.6	347.1	374.5	429.1	484.0	540.6	598.5	657.8	718.3	779.6
	s						1.2218	1.2472	1.2903	1.3258	1.3828	1.4302	1.4712	1.5080	1.5416	1.5721	1.6006
1200	v							0.1042	0.1356	0.1621	0.2096	0.2531	0.2953	0.3374	0.3789	0.4199	0.4609
	h							318.4	344.2	372.5	428.1	483.5	540.2	598.2	657.7	718.2	779.5
	s							1.2343	1.2791	1.3158	1.3740	1.4216	1.4628	1.4996	1.5332	1.5637	1.5922
1400	v								0.1136	0.1375	0.1788	0.2160	0.2529	0.2892	0.3249	0.3601	0.3551
	h								341.4	370.4	427.0	482.9	539.8	598.0	657.5	718.0	779.5
	s								1.2703	1.3078	1.3668	1.4145	1.4558	1.4927	1.5263	1.5568	1.5853
1600	v									0.1191	0.1557	0.1898	0.2211	0.2530	0.2843	0.3153	0.3461
	h									367.6	426.0	482.3	539.5	597.7	657.2	717.9	779.4
	s									1.3002	1.3602	1.4083	1.4497	1.4867	1.5193	1.5508	1.5793

p									
1800	v	0.1047	0.1377	0.1675	0.1964	0.2249	0.2528	0.2804	0.3079
	h	364.0	424.9	481.7	539.1	597.4	657.0	717.8	779.3
	s	1.2930	1.3539	1.4023	1.4440	1.4812	1.5148	1.5453	1.5738
2200	v		0.1120	0.1368	0.1605	0.1840	0.2070	0.2296	0.2522
	h		421.7	480.5	538.4	596.9	656.6	717.5	779.2
	s		1.3426	1.3925	1.4344	1.4720	1.5057	1.5362	1.5647
2600	v			0.1156	0.1357	0.1557	0.1752	0.1945	0.2137
	h			479.3	537.6	596.4	656.3	717.2	779.1
	s			1.3834	1.4260	1.4640	1.4981	1.5286	1.5571
3000	v			0.1001	0.1176	0.1349	0.1519	0.1687	0.1856
	h			478.1	536.8	595.8	656.0	717.0	778.8
	s			1.3752	1.4183	1.4569	1.4915	1.5220	1.5505

Source: From Sweigert, Weber, and Allen. *Ind. Eng. Chem.*, 38:185, 1946; wherein reprinted from Perry, J. H. *Chemical Engineer's Handbook* (3rd ed.). New York: McGraw-Hill, 1950.

[†] $s_f = 1.0$ at 32°F. $h_f = 180$ at 32°F. Therefore, according to the bases of Table 208, the entropies of these two CO_2 tables are consistent, but (180 − 36.7) or 143.3 B.t.u./lb. must be added to the enthalpies of saturated CO_2 to make them consistent with those of superheated CO_2.

v = volume (cu ft/lb); h = enthalpy (Btu/lb); s = entropy [Btu/(lb)(°R)]; p = absolute pressure (lb/sq in).

THERMOPHYSICAL PROPERTIES OF
SOME SUBSTANCES

The figure outlining the thermophysical properties of some fluids has been adapted from Poppendiek, H. F., and Sabin, C. M. *Some Heat Transfer Performance Criteria for High Temperature Fluid Systems* (American Society of Mechanical Engineers, 75-WA/HT-103, 1975) except for sodium, which is from *Liquid Metals Handbook* (NAV EXOS P-733 Rev., June 1952, AEC).

Caution: The property values are presented in a comparative manner, which may not yield the accuracy needed for detailed assessments. In such cases, recent tabulated property listings should be consulted.

The table listing physical properties of some solids is from Collier, J. G. *Convective Boiling and Condensation* (2nd ed.). New York: McGraw-Hill, 1981.

Note: For dynamic and kinematic fluid viscosities, see Figures 9-10 and 9-11, respectively. For thermal conductivity of engineering materials, see Figure 10-1.

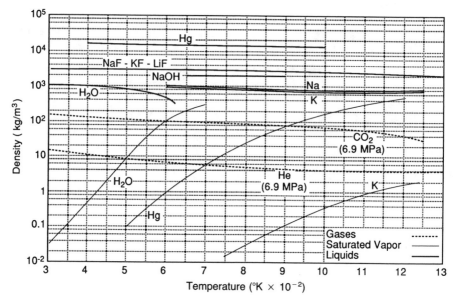

Figure F-1 Density versus temperature.

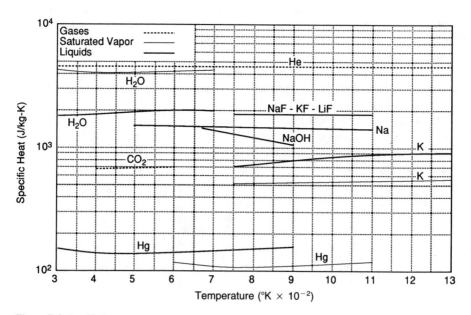

Figure F-2 Specific heat versus temperature.

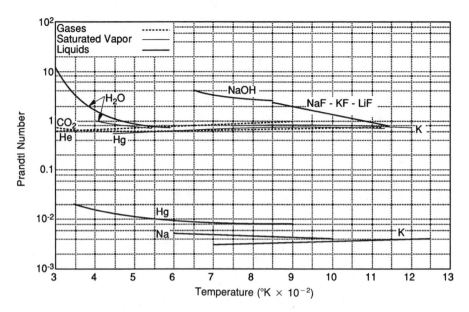

Figure F-3 Prandtl number versus temperature.

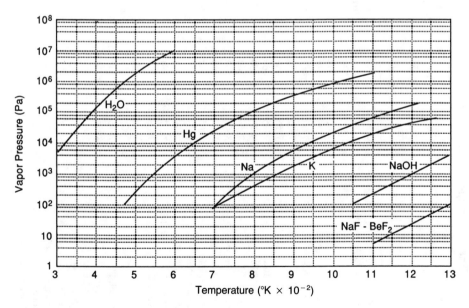

Figure F-4 Vapor pressure versus temperature.

Table F-1 Thermal properties of solids: metals

Metal	Properties at 20°C				k (W/m°C)				
	$\rho \left(\dfrac{kg}{m^3}\right)$	$c_p \left(\dfrac{J}{kg\,°C}\right)$	$k \left(\dfrac{W}{m\,°C}\right)$	$\alpha \left(\dfrac{m^2}{s}\right)$	100°C	200°C	300°C	400°C	600°C
Aluminum, pure	2707	896	204	$8\cdot42 \times 10^{-5}$	206	215	229	249	
Duralumin, 94-96 Al, 3-5 Cu	2787	883	164	$6\cdot68 \times 10^{-5}$	182	194			
Lead	11,370	130	34·6	$2\cdot34 \times 10^{-5}$	33·4	31·5	29·8		
Iron, pure	7897	452	72·7	$2\cdot03 \times 10^{-5}$	67·5	62·3	55·4	48·5	39·8
Iron, wrought, C <0·5%	7849	460	58·9	$1\cdot63 \times 10^{-5}$	57·1	51·9	48·5	45·0	36·4
Iron, cast, C ≈ 4%	7272	419	51·9	$1\cdot70 \times 10^{-5}$					
Carbon steel, C ≈ 0·5%	7833	465	53·7	$1\cdot47 \times 10^{-5}$	51·9	48·5	45·0	41·5	34·6
Carbon steel, C = 1·5%	7753	486	36·4	$0\cdot97 \times 10^{-5}$	36·3	36·3	34·6	32·9	31·2
Nickel steel, 10%	7945	460	26·0	$0\cdot72 \times 10^{-5}$					
Nickel steel, 30%	8073	460	12·1	$0\cdot33 \times 10^{-5}$					
Nickel steel, 50%	8266	460	13·8	$0\cdot36 \times 10^{-5}$					
Nickel steel, 70%	8506	460	26·0	$0\cdot67 \times 10^{-5}$					
Nickel steel, 90%	8762	460	46·7	$1\cdot16 \times 10^{-5}$					
Chrome steel, 1%	7865	460	60·6	$1\cdot67 \times 10^{-5}$	55·4	51·9	46·7	41·5	36·4
Chrome steel, 5%	7833	460	39·8	$1\cdot11 \times 10^{-5}$	38·1	36·4	36·4	32·9	29·4
Chrome steel, 10%	7785	460	31·2	$0\cdot87 \times 10^{-5}$	31·2	31·2	29·4	29·4	31·2
Cr-Ni steel, 18% Cr, 8% Ni	7817	460	16·3	$0\cdot44 \times 10^{-5}$	17·3	17·3	19·0	19·0	22·5
Ni-Cr steel, 20% Ni, 15% Cr	7865	460	14·0	$0\cdot39 \times 10^{-5}$	15·1	15·1	16·3	17·3	19·0
Manganese steel, 2%	7865	460	38·1	$1\cdot05 \times 10^{-5}$	36·4	36·4	36·4	34·6	32·9
Tungsten steel, 2%	7961	444	62·3	$1\cdot76 \times 10^{-5}$	58·9	53·7	48·5	45·0	36·4
Silicon steel, 2%	7673	460	31·2	$0\cdot89 \times 10^{-5}$					
Copper, pure	8954	383	386	$11\cdot2 \times 10^{-5}$	379	374	369	364	353
Bronze, 75 Cu, 25 Sn	8660	343	26·0	$0\cdot86 \times 10^{-5}$					
Brass, 70 Cu, 30 Zn	8522	385	111	$3\cdot41 \times 10^{-5}$	128	144	147	147	
German silver, 62 Cu 15 Ni, 22 Zn	8618	394	24·9	$0\cdot73 \times 10^{-5}$	31·2	39·8	45·0	48·5	
Constantan, 60 Cu, 40 Ni	8922	410	22·7	$0\cdot61 \times 10^{-5}$	22·2	26·0			
Magnesium, pure	1746	1013	171	$9\cdot71 \times 10^{-5}$	168	163	158		
Molybdenum	10,220	251	123	$4\cdot79 \times 10^{-5}$	118	114	111	109	106

Nickel, 99·9% pure	8906	446	90·0	$2·27 \times 10^{-5}$	83·1	72·7	64·0	58·9	113
Silver, 99·9% pure	10,520	234	407	$16·6 \times 10^{-5}$	415	374	362	360	
Tungsten	19,350	134	163	$6·27 \times 10^{-5}$	151	142	133	126	
Zinc, pure	7144	384	112	$4·11 \times 10^{-5}$	109	106	100	93·5	
Tin, pure	7304	227	64·0	$3·88 \times 10^{-5}$	58·9	57·1			

Adapted from Eckert, E. R. G., Drake, Jr., R. M. *Heat and Mass Transfer*. New York: McGraw-Hill, 1959.

DIMENSIONLESS GROUPS OF FLUID MECHANICS
AND HEAT TRANSFER

Name	Notation	Formula	Interpretation in terms of ratio
Biot number	Bo	$\dfrac{hL}{k_s}$	Surface conductance ÷ internal conduction of solid
Cauchy number	Ca	$\dfrac{V^2}{B_s/\rho} = \dfrac{V^2}{a^2}$	Inertia force ÷ compressive force = (Mach number)^2
Eckert number	Ec	$\dfrac{V^2}{c_p\,\Delta T}$	Temperature rise due to energy conversion ÷ temperature difference
Euler number	Eu	$\dfrac{\Delta p}{\rho V^2}$	Pressure force ÷ inertia force
Fourier number	Fo	$\dfrac{kt}{\rho c_p L^2} = \dfrac{\alpha t}{L^2}$	Rate of conduction of heat ÷ rate of storage of energy
Froude number	Fr	$\dfrac{V^2}{gL}$	Inertia force ÷ gravity force
Graetz number	Gz	$\dfrac{D}{L} \cdot \dfrac{V\rho c_p D}{k}$	Re Pr ÷ (L/D); heat transfer by convection in entrance region ÷ heat transfer by conduction
Grashof number	Gr	$\dfrac{g\beta\,\Delta T L^3}{\nu^2}$	Buoyancy force ÷ viscous force
Knudsen number	Kn	$\dfrac{\lambda}{L}$	Mean free path of molecules ÷ characteristic length of an object
Lewis number	Le	$\dfrac{\alpha}{D_c}$	Thermal diffusivity ÷ molecular diffusivity
Mach number	M	$\dfrac{V}{a}$	Macroscopic velocity ÷ speed of sound
Nusselt number	Nu	$\dfrac{hL}{k}$	Temperature gradient at wall ÷ overall temperature difference

(Continued)

Name	Notation	Formula	Interpretation in terms of ratio
Péclet number	Pé	$\dfrac{V\rho c_p D}{k}$	(Re Pr); heat transfer by convection ÷ heat transfer by conduction
Prandtl number	Pr	$\dfrac{\mu c_p}{k} = \dfrac{\nu}{\alpha}$	Diffusion of momentum ÷ diffusion of heat
Reynolds number	Re	$\dfrac{\rho VL}{\mu} = \dfrac{VL}{\nu}$	Inertia force ÷ viscous force
Schmidt number	Sc	$\dfrac{\mu}{\rho D_c} = \dfrac{\nu}{D_c}$	Diffusion of momentum ÷ diffusion of mass
Sherwood number	Sh	$\dfrac{h_D L}{D_c}$	Mass diffusivity ÷ molecular diffusivity
Stanton number	St	$\dfrac{h}{V\rho c_p} = \dfrac{h}{c_p G}$	Heat transfer at wall ÷ energy transported by stream
Stokes number	Sk	$\dfrac{\Delta p L}{\mu V}$	Pressure force ÷ viscous force
Strouhal number	Sl	$\dfrac{L}{tV}$	Frequency of vibration ÷ characteristic frequency
Weber number	We	$\dfrac{\rho V^2 L}{\sigma}$	Inertia force ÷ surface tension force

MULTIPLYING PREFIXES

tera	T	10^{12}
giga	G	10^{9}
mega	M	10^{6}
kilo	k	10^{3}
hecto	h	10^{2}
deca (deka)	da	10^{1}
deci	d	10^{-1}
centi	c	10^{-2}
milli	m	10^{-3}
micro	μ	10^{-6}
nano	n	10^{-9}
pico	p	10^{-12}
femto	f	10^{-15}
atto	a	10^{-18}

LIST OF ELEMENTS

Atomic number	Symbol	Name	Atomic weight*	Atomic number	Symbol	Name	Atomic weight*
1	H	Hydrogen	1.00794 (7)	30	Zn	Zinc	65.39 (2)
2	He	Helium	4.002602 (2)	31	Ga	Gallium	69.723 (4)
3	Li	Lithium	6.941 (2)	32	Ge	Germanium	72.59 (3)
4	Be	Beryllium	9.01218	33	As	Arsenic	74.9216
5	B	Boron	10.811 (5)	34	Se	Selenium	78.96 (3)
6	C	Carbon	12.011	35	Br	Bromine	79.904
7	N	Nitrogen	14.0067	36	Kr	Krypton	83.80
8	O	Oxygen	15.9994 (3)	37	Rb	Rubidium	85.4678 (3)
9	F	Fluorine	18.998403	38	Sr	Strontium	87.62
10	Ne	Neon	20.179	39	Y	Yttrium	88.9059
11	Na	Sodium	22.98977	40	Zr	Zirconium	91.224 (2)
12	Mg	Magnesium	24.305	41	Nb	Niobium	92.9064
13	Al	Aluminum	26.98154	42	Mo	Molybdenum	95.94
14	Si	Silicon	28.0855 (3)	43	Tc	Technetium	[98]
15	P	Phosphorus	30.97376	44	Ru	Ruthenium	101.07 (2)
16	S	Sulfur	32.066 (6)	45	Rh	Rhodium	102.9055
17	Cl	Chlorine	35.453	46	Pd	Palladium	106.42
18	Ar	Argon	39.948	47	Ag	Silver	107.8682 (3)
19	K	Potassium	39.0983	48	Cd	Cadmium	112.41
20	Ca	Calcium	40.078 (4)	49	In	Indium	114.82
21	Sc	Scandium	44.95591	50	Sn	Tin	118.710 (7)
22	Ti	Titanium	47.88 (3)	51	Sb	Antimony	121.75 (3)
23	V	Vanadium	50.9415	52	Te	Tellurium	127.60 (3)
24	Cr	Chromium	51.9961 (6)	53	I	Iodine	126.9045
25	Mn	Manganese	54.9380	54	Xe	Xenon	131.29 (3)
26	Fe	Iron	55.847 (3)	55	Cs	Cesium	132.9054
27	Co	Cobalt	58.9332	56	Ba	Barium	137.33
28	Ni	Nickel	58.69	57	La	Lanthanum	138.9055 (3)
29	Cu	Copper	63.546 (3)	58	Ce	Cerium	140.12

Atomic number	Symbol	Name	Atomic weight*	Atomic number	Symbol	Name	Atomic weight*
59	Pr	Praseodymium	140.9077	83	Bi	Bismuth	208.9804
60	Nd	Neodymium	144.24 (3)	84	Po	Polonium	[209]
61	Pm	Promethium	[145]	85	Al	Astatine	[210]
62	Sm	Samarium	150.36 (3)	86	Rn	Radon	[222]
63	Eu	Europium	151.96	87	Fr	Francium	[223]
64	Gd	Gadolinium	157.25 (3)	88	Ra	Radium	226.0254
65	Tb	Terbium	158.9254	89	Ac	Actinium	227.0278
66	Dy	Dysprosium	162.50 (3)	90	Th	Thorium	232.0381
67	Ho	Holmium	164.9304	91	Pa	Protactinium	231.0359
68	Er	Erbium	167.26 (3)	92	U	Uranium	238.0289
69	Tm	Thulium	168.9342	93	Np	Neptunium	237.0482
70	Yb	Ytterbium	173.04 (3)	94	Pu	Plutonium	[244]
71	Lu	Lutetium	174.967	95	Am	Americium	[243]
72	Hf	Hafnium	178.49 (3)	96	Cm	Curium	[247]
73	Ta	Tantalum	180.9479	97	Bk	Berkelium	[247]
74	W	Tungsten	183.85 (3)	98	Cl	Californium	[251]
75	Re	Rhenium	186.207	99	Es	Einsteinium	[252]
76	Os	Osmium	190.2	100	Fm	Fermium	[257]
77	Ir	Iridium	192.22 (3)	101	Md	Mendelevium	[258]
78	Pl	Platinum	195.08 (3)	102	No	Nobelium	[259]
79	Au	Gold	196.9665	103	Lr	Lawrencium	[260]
80	Hg	Mercury	200.59 (3)	104	Rf	Rutherfordium[†]	[261]
81	Tl	Thallium	204.383	105	Ha	Hahnium[†]	[262]
82	Pb	Lead	207.2	106	Unnamed	Unnamed	[263]

Source: From Walker, F. W., Miller, D. G., and Feiner, F. *Chart of the Nuclides* (13th ed.). Revised 1983, General Electric Co.

*Values in parentheses are the uncertainty in the last digit of the stated atomic weights. Values without a quoted error in parentheses are considered to be reliable to ±1 in the last digit except for Ra, Ac, Pa, and Np, which were not given by Walker et al. (1983). Brackets indicate the most stable or best known isotope.

[†]The names of these elements have not been accepted because of conflicting claims of discovery.

SQUARE AND HEXAGONAL ROD ARRAY DIMENSIONS

I LWR FUEL BUNDLES: SQUARE ARRAYS

Tables J-1 and J-2 present formulas for determining axially averaged unit subchannel and overall bundle dimensions, respectively.

The presentation of these formulas as axially averaged values is arbitrary and reflects the fact that grid-type spacers occupy a small fraction of the axial length of a fuel bundle. Therefore this fraction (δ) has been defined as:

$$\delta = \frac{\text{total axial length of grid spacers}}{\text{axial length of the fuel bundle}}$$

Formulas applicable at the axial grid locations or between grids are easily obtained from Tables J-1 and J-2 by taking $\delta = 1$ or $\delta = 0$, respectively.

Determination of precise dimensions for an LWR assembly would require knowledge of the specific grid configuration used by the manufacturer. Typically, the grid strap thickness at the periphery is slightly enhanced. Here it is taken as thickness t^*. It should also be carefully noted that these formulas are based on the assumptions of a rectangular grid, and no support tabs or fingers. The dimension (g) is the spacing from rod surface to the flow boundary of the assembly. In the grid plane, the segment along g open for flow is $g - t/2$.

II LMFBR FUEL BUNDLES: HEXAGONAL ARRAYS

Tables J-3 and J-4 summarize the formulas for determining axially averaged unit subchannel and overall bundle dimensions, respectively. Fuel pin spacing is illustrated as being performed by a wire wrap. In practice, both grids and wires are used as spacers in LMFBR fuel and blanket bundles. The axially averaged dimensions in these tables are based on averaging the wires over one lead length.

Table J-1 Square arrays: axially averaged unit subchannel dimensions for ductless assembly

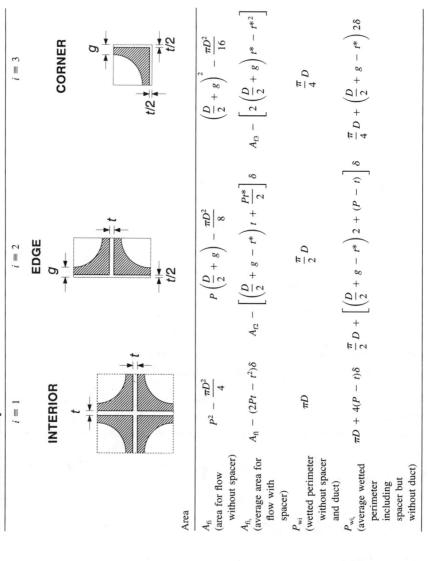

	$i = 1$ INTERIOR	$i = 2$ EDGE	$i = 3$ CORNER
Area			
A_{fi} (area for flow without spacer)	$P^2 - \dfrac{\pi D^2}{4}$	$P\left(\dfrac{D}{2} + g\right) - \dfrac{\pi D^2}{8}$	$\left(\dfrac{D}{2} + g\right)^2 - \dfrac{\pi D^2}{16}$
A_{fi_i} (average area for flow with spacer)	$A_{f1} - (2Pt - t^2)\delta$	$A_{f2} - \left[\left(\dfrac{D}{2} + g - t^*\right)t + \dfrac{Pt^*}{2}\right]\delta$	$A_{f3} - \left[2\left(\dfrac{D}{2} + g\right)t^* - t^{*2}\right]$
P_{wi} (wetted perimeter without spacer and duct)	πD	$\dfrac{\pi D}{2}$	$\dfrac{\pi D}{4}$
P_{wi_i} (average wetted perimeter including spacer but without duct)	$\pi D + 4(P - t)\delta$	$\dfrac{\pi D}{2} + \left[\left(\dfrac{D}{2} + g - t^*\right)2 + (P - t)\right]\delta$	$\dfrac{\pi D}{4} + \left(\dfrac{D}{2} + g - t^*\right)2\delta$

Table J-2 Square arrays: axially averaged overall dimensions for ductless assembly

Assuming grid spacer around N_p rods of thickness t and that $g = \dfrac{P - D}{2}$

1. Total area inside square (A_T):

$$A_T = D_\ell^2$$

where D_ℓ = length of one side of the square.

2. Total average cross-sectional area for flow (A_{fT}):

$$A_{fT} = D_\ell^2 - N_p \frac{\pi}{4} D^2 - N_p \frac{t}{2} P 4\delta + N_p \left(\frac{t}{2}\right)^2 4\delta$$

$$A_{fT} = D_\ell^2 - \left[N_p \frac{\pi}{4} (D^2) + 2\delta(\sqrt{N_p})(D_\ell t) - N_p t^2 \delta \right]$$

because $D_\ell = \sqrt{N_p}\, P$ and $t^* = t$

where:

D = rod diameter
N_p = number of rods
t = interior spacer thickness
t^* = peripheral spacer thickness

3. Total average wetted perimeter (P_{wT})

$$P_{wT} = N_p \pi D + 4\sqrt{N_p}\, D_\ell \delta - 4 N_p t \delta$$

4. Equivalent hydraulic diameter for overall square (D_{eT})

$$D_{eT} = \frac{4 A_{fT}}{P_{wT}} = \frac{4 D_\ell^2 - [N_p \pi D^2 + 8\sqrt{N_p}\, D_\ell t \delta - 4 N_p t^2 \delta]}{N_p \pi D + 4\sqrt{N_p}\, D_\ell \delta - 4 N_p t \delta}$$

Table J-3 Hexagonal array with wire wrap spacer: axially averaged unit subchannel dimensions

	$i \equiv 1$	$i \equiv 2$	$i \equiv 3$
	INTERIOR	**EDGE**	**CORNER**

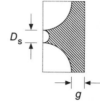

Area

A_{f_i} (area for flow without wire wrap spacers)	$\dfrac{1}{2} P \left(\dfrac{\sqrt{3}}{2} P \right) - \dfrac{\pi D^2}{8} = \dfrac{\sqrt{3}}{4} P^2 - \dfrac{\pi D^2}{8}$	$P \left(\dfrac{D}{2} + g \right) - \dfrac{\pi D^2}{8}$	$\left[\dfrac{1}{\sqrt{3}} \left(\dfrac{D}{2} + g \right)^2 - \dfrac{\pi D^2}{24} \right]$

Note: $g \equiv$ spacer diameter plus wall-pin clearance, if any

A_{f_i} (area for flow including wire wrap spacers)	$A_{f1} - \left(\dfrac{3}{6} \right) \dfrac{\pi}{4} D_s^2 = A_{f_i} - \dfrac{\pi D_s^2}{8}$ (three spacers traverse cell per unit lead; each traverse is 60° of 360°)	$A_{f2} - \left(\dfrac{2}{4} \right) \dfrac{\pi D_s^2}{4}$ (two spacers traverse cell per unit lead; each traverse is 90° of 360°)	$A_{f3} - \left(\dfrac{1}{6} \right) \dfrac{\pi}{4} D_s^2$ (one spacer traverses cell per unit lead; each traverse is 60° of 360°)
P_{w_i} (wetted perimeter including wire wrap spacers)	$\dfrac{\pi D}{2} + \dfrac{\pi D_s}{2}$ (three spacers traverse cell per unit lead; each traverse is 60° of 360°)	$\dfrac{\pi D}{2} + P + \dfrac{\pi D_s}{2}$ (two spacers traverse cell per unit lead; each traverse is 90° of 360°)	$\dfrac{\pi}{6} (D + D_s) + \dfrac{2}{\sqrt{3}} \left(\dfrac{D}{2} + g \right)$ (one spacer traverses cell per unit lead; each traverse is 60° of 360°)

Table J-4 Hexagonal array: axially averaged overall dimensions assuming wire wrap spacers around each rod

1. Total area inside hexagon (A_{hT})

$$A_{hT} = D_{ft}D_\ell + 2\left(\frac{1}{2}\right)D_{ft}D_\ell \sin 30° = \frac{\sqrt{3}}{2}D_{ft}^2$$

Because:

$$D_\ell = \frac{D_{ft}}{2}\frac{1}{\cos 30°} = \frac{\sqrt{3}}{3}D_{ft}$$

where:

D_ℓ = length of one side of the hexagon
D_{ft} = distance across flats of the hexagon; for a bundle considering
      clearances or tolerances between rods and duct

Now:

$$D_\ell = (N_{ps} - 1)(D + D_s) + \frac{2\sqrt{3}}{3}\left(\frac{D}{2} + g\right)$$

$$D_{ft} = 2[(\sqrt{3}/2)N_{rings}(D + D_s) + D/2 + g]$$

D = rod diameter
D_s = wire wrap diameter
g = rod to duct spacing
N_p = number of rods
N_{ps} = number of rods along a side
N_{RINGS} = number of rings

2. Total cross-sectional area for flow (A_{fT})

$$A_{fT} = A_{hT} - N_p \frac{\pi}{4}(D^2 + D_s^2)$$

3. Total wetted perimeter (P_{wT})

$$P_{wT} = 6D_\ell + N_p \pi D + N_p \pi D_s = 2\sqrt{3} D_{ft} + N_p \pi(D + D_s)$$

4. Equivalent diameter for overall hexagonal array (D_{eT})

$$D_{eT} = \frac{4A_{fT}}{P_{wt}} = \frac{2\sqrt{3} D_{ft}^2 - N_p \pi (D^2 + D_s^2)}{2\sqrt{3}D_{ft} + N_p \pi (D + D_s)}$$

INDEX

Page numbers in *italics* indicate figures.
Page numbers followed by t indicate tables.